Lecture Notes in Computer Science 9562

Commenced Publication in 1973
Founding and Former Series Editors:
Gerhard Goos, Juris Hartmanis, and Jan van Leeuwen

More information about this series at http://www.springer.com/series/7410

Eyal Kushilevitz · Tal Malkin (Eds.)

Theory
of Cryptography

13th International Conference, TCC 2016-A
Tel Aviv, Israel, January 10–13, 2016
Proceedings, Part I

Springer

Editors

Eyal Kushilevitz
Department of Computer Science
Technion
Haifa
Israel

Tal Malkin
Department of Computer Science
Columbia University
New York, NY
USA

ISSN 0302-9743 ISSN 1611-3349 (electronic)
Lecture Notes in Computer Science
ISBN 978-3-662-49095-2 ISBN 978-3-662-49096-9 (eBook)
DOI 10.1007/978-3-662-49096-9

Library of Congress Control Number: 2015957796

LNCS Sublibrary: SL4 – Security and Cryptology

Printed on acid-free paper

This Springer imprint is published by SpringerNature
The registered company is Springer-Verlag GmbH Berlin Heidelberg

Preface

The 13th Theory of Cryptography Conference (TCC 2016-A) was held during January 10–13, 2016, at the Suzanne Dellal Center in Tel Aviv, Israel. It was sponsored by the International Association for Cryptographic Research (IACR). The general chairs of the conference were Ran Canetti and Iftach Haitner. We would like to thank them for their hard work in organizing the conference.

The conference received 112 submissions, of which the Program Committee (PC) selected 45 for presentation (with three pairs of papers sharing a single presentation slot per pair). Each submission was reviewed by at least three PC members, often more. The 24 PC members, all top researchers in our field, were helped by 112 external reviewers, who were consulted when appropriate. These proceedings consist of the revised version of the 45 accepted papers. The revisions were not reviewed, and the authors bear full responsibility for the content of their papers.

As in previous years, we used Shai Halevi's excellent web-review software, and are extremely grateful to him for writing it, and for providing fast and reliable technical support whenever we had any questions. Based on the experience from last year, we again made use of the interaction feature supported by the review software, where PC members may directly and anonymously interact with authors. This was used to ask specific technical questions that arise, such as suspected bugs. We felt this was efficient and successful, and are thankful to last year's chairs, Yevgeniy Dodis and Jesper Buus Nielsen, for suggesting this feature, and to Shai Halevi for implementing it.

This was the second year where TCC presented the Test of Time Award to an outstanding paper that was published at TCC at least eight years ago, making a significant contribution to the theory of cryptography, preferably with influence also in other areas of cryptography, theory, and beyond. This year the Test of Time Award Committee selected the following paper, published ten years ago at TCC 2006:

"Calibrating Noise to Sensitivity in Private Data Analysis," by Cynthia Dwork, Frank McSherry, Kobbi Nissim, and Adam Smith.

This paper was selected *for introducing the definition of differential privacy, providing a solid mathematical foundation for a vast body of subsequent work on private data analysis.* The authors were also invited to deliver a talk at TCC 2016-A. The conference also featured two other invited events. First, an invited talk by Yael Kalai and Shafi Goldwasser (delivered by Yael) followed by panel on "cryptographic assumptions." Second, an invited talk by Yevgeniy Dodis. Finally, in addition to regular papers and invited events, the conference also featured a rump session.

We are greatly indebted to many people who were involved in making TCC 2016-A a success. First of all, a big thanks to the most important contributors: all the authors who submitted papers to the conference. Next, we would like to thank the PC members for their hard work, dedication, and diligence in reviewing the papers, verifying the correctness, and in-depth discussion. We are also thankful to the external reviewers for their volunteered hard work and investment in reviewing papers and answering questions, often under time pressure. For running the conference itself, we are very grateful to the general chairs, Ran Canetti and Iftach Haitner, as well as Galit Herzberg and the rest of the local Organizing Committee. Finally, we are thankful to the TCC Steering Committee as well as the entire thriving and vibrant TCC community.

January 2016
Eyal Kushilevitz
Tal Malkin

TCC 2016-A

The 13th Theory of Cryptography Conference

Suzanne Dellal Center, Tel Aviv, Israel
January 10–13, 2016

Sponsored by the *International Association for Cryptographic Research*

General Chairs

Ran Canetti   Tel Aviv University, Israel
     Boston University, USA
Iftach Haitner   Tel Aviv University, Israel

Program Chairs

Eyal Kushilevitz   Technion, Israel
Tal Malkin   Columbia University, USA

Program Commitee

Masayuki Abe   NTT, Japan
Amos Beimel   Ben-Gurion University, Israel
Nir Bitansky   MIT, USA
Andrej Bogdanov   Chinese University of Hong Kong, SAR China
Zvika Brakerski   Weizmann Institute of Science, Israel
Christina Brzuska   Hamburg University of Technology, Germany
Nishanth Chandran   MSR India
Melissa Chase   MSR Redmond, USA
Dana Dachman-Soled   University of Maryland, USA
Yuval Ishai   Technion, Israel
Jonathan Katz   University of Maryland, USA
Hugo Krawczyk   IBM Research, USA
Huijia Lin   UC Santa Barbara, USA
Claudio Orlandi   Aarhus University, Denmark
Omkant Pandey   Drexel University, USA
Valerio Pastro   Columbia University, USA
Leonid Reyzin   Boston University, USA
Guy Rothblum   Samsung Research America, USA
Gil Segev   Hebrew University, Israel
Adam Smith   Pennsylvania State University, USA
Vinod Vaikuntanathan   MIT, USA
Ivan Visconti   University of Salerno, Italy
Brent Waters   UT Austin, USA
Vassilis Zikas   ETH, Switzerland

External Reviewers

Divesh Aggarwal
Prabhanjan Ananth
Daniel Apon
Benny Applebaum
Gilad Asharov
Nuttapong Attrapadung
Pablo Azar
Saikrishna
  Badrinarayanan
Allison Bishop
Elette Boyle
Ignacio Cascudo
David Cash
Binyi Chen
Yilei Chen
Mahdi Cheragchi
Kai-Min Chung
Michele Ciampi
Aloni Cohen
Sandro Coretti
Akshay Degwekar
Gregory Demay
Itai Dinur
Yevgeniy Dodis
Nico Döttling
Antonio Faonio
Sebastian Faust
Victoria Fehr
Dario Fiore
Nils Fleischhacker
Eiichiro Fujisaki
Juan Garay
Ran Gelles
Craig Gentry
Niv Gilboa
Alexander Golovnev
Sergey Gorbunov
Rishab Goyal
Jens Groth

Siyao Guo
Shai Halevi
Prahladh Harsha
Carmit Hazay
Brett Hemenway
Ryo Hiromasa
Justin Holmgren
Ai Ishida
Zahra Jafargholi
Abhishek Jain
Stanislaw Jarecki
Daniel Jost
Tomasz Kazana
Carmen Kempka
Dakshita Khurana
Susumu Kiyoshima
Saleet Klein
Ilan Komargodski
Venkata Koppula
Lucas Kowalczyk
Ranjit Kumaresan
Tancrède Lepoint
Feng-Hao Liu
Tianren Liu
Satya Lokam
Steve Lu
Anna Lysyanskaya
Vadim Lyubashevsky
Mohammad Mahmoody
Hemanta K. Maji
Christian Matt
Eric Miles
Arno Mittelbach
Pratyay Mukherjee
Moni Naor
Jesper Buus Nielsen
Ryo Nishimaki
Adam O'Neill
Miyako Ohkubo

Olya Ohrimenko
Omer Paneth
Sunoo Park
Anat Paskin-Cherniavsky
Giuseppe Persiano
Oxana Poburinnaya
Antigoni Polychroniadou
Tal Rabin
Silas Richelson
Mike Rosulek
Ron Rothblum
Yannis Rouselakis
Alessandra Scafuro
Karn Seth
Luisa Siniscalchi
John Steinberger
Stefano Tessaro
Aishwarya
  Thiruvengadam
Mehdi Tibouchi
Daniel Tschudi
Jalaj Upadhyay
Prashant Vasudevan
Muthu
  Venkitasubramaniam
Daniele Venturi
Dhinakaran
  Vinayagamurthy
Thomas Watson
Hoeteck Wee
Mor Weiss
Daniel Wichs
Keita Xagawa
Eylon Yogev
Ching-Hua Yu
Yu Yu
Mark Zhandry
Hong-Sheng Zhou

Contents – Part I

Multiparty Computation

Contents – Part II

Codes and Interactive Proofs

Limitations of Obfuscation and Obfuscation-Avoiding Constructions

Obfuscation: Impossibility Results and Constructions

Impossibility of VBB Obfuscation with Ideal Constant-Degree Graded Encodings

Rafael Pass[1]([⊠]) and Abhi Shelat[2]

[1] Cornell University, Ithaca, USA
rafael@cs.cornell.edu
[2] University of Virginia, Charlottesville, USA
abhi@virginia.edu

Abstract. A celebrated result by Barak *et al.* (Crypto'01) shows the impossibility of general-purpose *virtual black-box (VBB)* obfuscation in the plain model. A recent work by Canetti, Kalai, and Paneth (TCC'15) extends this impossibility result to the random oracle model (assuming trapdoor permutations).

In contrast, Brakerski-Rothblum (TCC'14) and Barak *et al.* (Euro-Crypt'14) show that in *idealized* graded encoding models, general-purpose VBB obfuscation indeed is possible; these constructions require graded encoding schemes that enable evaluating *high-degree* (polynomial in the size of the circuit to be obfuscated) polynomials on encodings.

We show a complementary impossibility of general-purpose VBB obfuscation in idealized graded encoding models that enable only evaluation of *constant-degree* polynomials (assuming trapdoor permutations).

1 Introduction

The goal of *program obfuscation* is to "scramble" a computer program in order to hide its implementation details (making it hard to "reverse-engineer") while preserving its functionality (i.e., input/output behavior). The most desirable notion of security—*virtual black-box security* (VBB) [BGI+01]—requires that any bit of information an attacker can learn from the obfuscated code can be simulated using only black-box access to the functionality.[1] The celebrated result of Barak *et al.* [BGI+01], however, demonstrates a strong impossibility result

---

R. Pass—Work supported in part by a Microsoft Faculty Fellowship, Google Faculty Award, NSF Award CNS-1217821, NSF Award CCF-1214844, AFOSR Award FA9550-15-1-0262 and DARPA and AFRL under contract FA8750-11-2-0211.

A. Shelat—Work performed while visiting Cornell Tech, and supported by NSF CAREER Award 0845811, NSF TC Award 1111781, NSF TC Award 0939718, DARPA and AFRL under contract FA8750-11-C-0080, Microsoft New Faculty Fellowship, SAIC Scholars Research Award, and Google Faculty Award.

[1] A similar simulation-based, but even stronger, notion of security was previously defined by Hada [Had00]. Even earlier, Canetti [Can97] considered a similar notion of security (without explicitly referring to obfuscation) for the special case of what is now referred to as *point-function obfuscation*.

E. Kushilevitz and T. Malkin (Eds.): TCC 2016-A, Part I, LNCS 9562, pp. 3–17, 2016.
DOI: 10.1007/978-3-662-49096-9_1

regarding VBB obfuscation: they show the existence of families of functions $\{f_s\}$ for which black-box access to f_s (for a randomly chosen s) does not leak any advantage in guessing even a single bit of s, but the code of any program that computes f_s allows recovery of the entire secret s. The idea behind their impossibility result is to consider a function f_s that satisfies two properties (1) the function is not learnable (thus given black-box access to it, it is hard to find a concise representation of it), but (2) on input a program Π that computes the function f_s, $f_s(\Pi)$ reveals some secret. The code of the obfuscated program is thus an input on which the function releases the secret, yet the secret cannot be recovered using just black-box access to the function.

This impossibility result, however, only applies in the *plain* model in which the obfuscated code is a standard circuit that does not make oracle calls to external functionalities (or else, we cannot feed this code as an input to the function). In contrast, Canetti and Vaikuntanathan [CV13] show an obfuscator for NC^1 circuits in an idealized composite-order group with special pseudo-free properties. More recently, Brakerski and Rothblum [BR14] and Barak, Garg, Kalai, Paneth and Sahai [BGK+14], following the breakthrough obfuscation construction of Garg, Gentry, Halevi, Raykova, Sahai, and Waters [GGH+13b][2], demonstrate VBB obfuscation for all polynomial-size circuits in the idealized *graded encoding* [GGH13a] (a.k.a. "approximate" multilinear map [BS03, Rot13]) model.

In the *idealized graded encoding model* [BR14, BGK+14], players have black-box access to a field $\mathbb{F}_p$ (where p is a prime), but they can only perform certain restricted operations on field elements and determine whether an expression evaluates to zero. For instance, the simplest form of graded encodings of [GGH13a] enables computing all polynomials of some (a-priori) bounded polynomial degree, and determine whether the polynomial evaluates to zero; this is referred to as a "zero-test query"[3]. Note that a *generic group* [Sho97] model for $\mathbb{Z}_p^*$ where p is a prime can be viewed as a special-case of an idealized graded encoding model in which operations are restricted to be linear (i.e., degree 1 polynomials). Degree two graded encodings capture idealized groups with bilinear maps.

A natural question is whether weaker idealized models such as the generic group model or idealized groups with bilinear maps suffice for obtaining VBB obfuscation for polynomial-size circuits. This question was first addressed by Lynn, Prabhakaran and Sahai [LPS04] who showed positive obfuscation results for specific functions in the Random Oracle model [BR93] where both the obfuscator and the evaluator have oracle access to a truly random function; they left open the question of whether general-purpose obfuscation in the Random Oracle model is possible. This open question was recently answered in an elegant work by Canetti, Kalai and Paneth [CKP15] who show that the impossibility result of [BGI+01] also extends to the Random Oracle Model. [CKP15] in turn

---

[2] The construction of [GGH+13b] was proved to satisfy the weaker notion of *indistinguishability obfuscation* in an idealized "matrix-multiplication" model.

[3] The constructions in [BR14, BGK+14] require certain additional "set-based" restrictions on polynomials; we return to this in Sect. 2.2.

left open the questions of whether general-purpose VBB obfuscation in more sophisticated idealized models (such as the generic group model) is possible.

Our Results. In this work, we show impossibility of VBB obfuscation in idealized graded encoding models that restrict zero-tests to degree-d polynomials, where d is a *constant*.

Theorem 1 (Informally stated). *Assuming the existence of trapdoor permutations, there exists a family of functions F for which there do not exist VBB obfuscators for F in idealized degree-d graded encoding models, where d is a constant.*

Our theorem stands in contrast with the results of [BR14] and [BGK+14] which indeed show feasibility of general-purpose VBB obfuscation in an idealized graded encoding model that allows for *high-degree* (polynomial in the size of the circuit being obfuscated) zero-test queries.

The obfuscator construction of [BGK+14] also satisfies *subexponential* VBB security (that is security holds also with respect to subexponential-size attackers). Our main theorem extends to rule out general-purpose VBB obfuscation with subexponential security in idealized graded encoding models that allow for n^α-degree zero-test queries (where $\alpha < 1$ and n is the description length of the function being obfuscated).

Follow-up Work. We note that our proof directly generalizes to any graded encoding scheme that operates on elements in a ring (as opposed to $\mathbb{F}_p$) as long as (a) there exists an efficient method for determining the row-rank of a matrix of this ring, and (b) the row-rank of a matrix is polynomially bounded by the column-rank. In follow-up work, Mahmoody, Mohammed, and Nematihaji [MMN15] have extended our techniques to apply to more general rings.

2 Definitions and Preliminaries

2.1 Virtual Black-Box Obfuscation

We recall the definition of approximate VBB obfuscation from Barak *et al.* [BGI+01], and Canetti, Kalai, and Paneth [CKP15], and generalize it for any family of oracles M that are indexed by a security parameter.

Definition 1 (ϵ-Approximate VBB Obfuscation in an Oracle Model [CKP15,BGK+14]). *For a function $\epsilon : \mathbb{N} \to \{0, 1\}$, an obfuscator $\mathcal{O}$ is a secure ϵ-approximate virtual black-box (VBB) obfuscation for the family F in the M-oracle model if it satisfies the following properties:*

– *Approximate Functionality: for all $n \in \mathbb{N}$, $k \in \{0, 1\}^n$:*

$$\Pr\left[\mathcal{O}^{M_{|k|}}(k)(x) \neq F_k(x)\right] \leq 1 - \epsilon(n)$$

where the probability is over the choice of x and the coins of M and $\mathcal{O}$.

– *Virtual Black-Box (VBB): for every poly-size adversary A, there exists a poly-size simulator S and a negligible function μ such that for every* $k \in \{0,1\}^*$:

$$\left| \Pr\left[A^{M_{|k|}}(\mathcal{O}^{M_{|k|}}(k)) = 1 \right] - \Pr\left[S^{F_k}(1^{|k|}) = 1 \right] \right| \le \mu(|k|)$$

where the probability it taken over the coins of M, $\mathcal{O}$, adversary A and the simulator S.

We simply say that $\mathcal{O}$ is a secure VBB obfuscator if $\epsilon = 1$. We further say that $\mathcal{O}$ is a secure (ϵ-approximate) obfuscation in the plain model for the family F if it is a secure (ϵ-approximate) obfuscation for the family F in the $\perp$-oracle model where the $\perp$-oracle returns $\perp$ on every query.

We finally say that $\mathcal{O}$ is subexponentially-secure if the VBB condition holds with respect to any subexponential-size[4] A and a subexponential-size S.

Our definition of subexponentially-secure VBB obfuscation is incomparable to the definition of VBB obfusaction: it is stronger in that we require simulation of subexponential-size attackers, but it is weaker in that we allow the simulator to be subexponential size (even if the attacker is polynomial in size).

We use the following theorem by Bitansky and Paneth [BP13] and its extension which follows by relying on stronger trapdoor permutations. We choose specific constants for simplicity of notation; the theorem holds for any constants.

Theorem 2 ([BP13]). *Assuming the existence of trapdoor permutations, there exists a family of polynomial-time computable functions F such that a polynomial-size 0.8-approximate VBB obfuscator for F does not exist.*

Theorem 3 (scaled version of [BP13]). *Assuming the existence of sub-exponentially secure[5] trapdoor permutations, there exists a family of polynomial-time computable functions F such that a subexponential-size 0.8-approximate subexponentially-secure VBB obfuscator for F does not exist.*

2.2 Idealized Graded Encodings

We now define the ideal level-d graded encoding oracle. For simplicity of notation, we consider an oracle that has the size of the field hard-coded. Our model, inspired by the formalism from [PST14,BR14,BGK+14,Sho97], considers a simple idealized graded encoding oracle which enables players to (a) encode an element v under a "label" l, and receive a random "handle" h in return, and (b) to make "legal" *zero-test queries* on these encodings: a zero-test query is a formal polynomial p on variables $\boldsymbol{h}$, which evaluates to **true** if and only if $p(\boldsymbol{v}) = 0$, where for every i, v_i is the value encoded under handle h_i. The legality of a query is determined by a *legality-predicate* g: $g(p, \boldsymbol{l})$ outputs 1 if the query is deemed

---

[4] That is, whose circuit size is bounded by $T(n) = poly(2^{n^\alpha})$ for any $0 < \alpha < 1$.

[5] That is, security holds against all circuits whose size is bounded by $T(n) = poly(2^{n^\alpha})$ for any $0 < \alpha < 1$.

legal, where l are the labels corresponding to the handles h. In this work we consider a natural class of "well-formed" legality predicates, which, as we shall discuss shortly, generalize all previously used notions of legality.

Definition 2 (Well-formed legality predicate). *Given a set of multi-sets (legal label sets) S define the predicate $g_S(p, l) = 1$ if and only if for every monomial $x_{j_1} \cdots x_{j_d}$ of p, it holds that the multi-set $\{l_{j_1}, \ldots, l_{j_d}\} \in S$. We say that a legality predicate g is well-formed if there exists a set S such that $g = g_S$.*

For instance, to capture:

- idealized groups [Sho97] (where we do not allow any multiplications), consider the predicate g_S corresponding to the set $S = \{\{1\}\}$ (and requiring that all encodings are made under the label 1).
- "simple" d-level graded encodings of [GGH13a], consider the predicate g_S corresponding to the set S where $\{l_{j_1}, \ldots, l_{j_m}\} \in S$ if and only if $\sum_{i \in [m]} l_{j_i} = d$ (and requiring that all encodings are made under a label $l \in [d]$ that represents the element's "level").
- "set-based" d-level graded encodings [GGH13a, BR14, BGK+14], consider the predicate g_S corresponding to the set S where $\{l_{j_1}, \ldots, l_{j_m}\} \in S$ if and only if the disjoint union of labels l_{j_i} where $i \in [m]$ is the set $\{1, 2, \ldots, d\}$, i.e. $\sqcup_{i \in [d]} l_{j_i} = [d]$ (and requiring that all encodings are made under a label l that is a subset of $[d]$).

Additionally, to capture secret-key encodings in which only the obfuscator can create new encodings, we follow [BGK+14] and require that encodings can only occur once upon initialization; after initialization no more encodings can be performed. (In contrast to [BGK+14], however, these encodings can be performed adaptively.)

Definition 3 (Ideal graded encoding oracle). *The oracle $M_q^g = (\mathsf{enc}, \mathsf{zero})$ is a stateful oracle, parameterized by integer q and a legality predicate g, that responds to queries in the following manner:*

1. *Upon initialization and only then, the activator may adaptively make any number of queries of the form $\mathsf{enc}(v, l)$; for each such query, M_q^g picks a uniformly random "handle" $h \in \{0, 1\}^{3|q|}$, stores the tuple (v, l, h) in a list $\mathcal{L}_O$ and returns h.[6] This initialization phase ends if any algorithm other than the activating algorithm makes any query to M^q, or if the activator makes a non enc query. Any subsequent $\mathsf{enc}(\cdot, \cdot)$ queries will be answered with $\perp$.*
2. *On input query $\mathsf{zero}(p)$ where p is a formal polynomial over variables $h_1, \ldots, h_m$, each of which is represented as a string of length $3|q|$ (corresponding to some handle), M_q^g does the following:*

---

[6] In particular, even if the same value v is encoded twice (under the same label), independently random handles are returned for the two encodings. This model thus considers *randomized* graded encodings. Our results also apply to deterministic randomized encodings where the oracle keeps state also during the encoding phase and always returns the *same* handle for an encoding of the value v under the label l.

(a) *For each* $i \in [m]$, *retrieve a tuple* (v_i, l_i, h_i) *from the state* $\mathcal{L}_O$; *if no such tuple exists, it returns* false.

(b) *(Illegal query) If all tuples are retrieved, return* false *if* $g(p, l) \neq 1$

(c) *(Zero test) Finally, return* true *iff* $p(v_1, \ldots, v_n) = 0 \mod q$, *and* false *otherwise.*

3. *(All other queries are answered with $\bot$).*

We say that M is an ideal graded encoding oracle if $M = \{M_{q_1}^{g_1}, M_{q_2}^{g_2}, \ldots\}$, and for every $n \in \mathbb{N}$, q_n is a prime, $|q_n| > n$ and g_n is a well-formed legality predicate. Finally, we say that M is a degree-$d(\cdot)$ ideal graded encoding oracle if for all $n \in \mathbb{N}$, $g_n(p, l)$ returns false when $\deg(p) > d(n)$.

A Remark on the Model. Following [PST14], for simplicity of notation, we do not directly allow players to create new encodings by adding and multiplying old ones as in the definitions of [BR14, BGK+14]. This restriction is without loss of generality since (a) an obfuscator "knows" all values it has previously encoded (since it needs to explicitly provide them to the encoding oracle) so instead of operating on old encodings, it can simply operate on the actual values and simply create a new encoding of the resulting value[7], and (b) when evaluating the obfuscated code, operations on encodings can be simulated by "bogus" independently random handles[8], and emulating zero-test queries by appropriately modifying the zero-test polynomial p to take into account the previously performed operations.

Feasibility of VBB Obfuscation in Idealized Graded Encoding Models. The results of [BR14, BGK+14] demonstrate feasibility of VBB obfuscation in idealized "set-based" graded encoding models that allow zero-test queries with *super-constant* degree.

Theorem 4 ([BR14, BGK+14]). *Under the LWE assumption[9], for every polynomial $p(\cdot)$, there exists a (polynomial-time computable) sequence of well-formed legality predicates $g_1, g_2, \ldots$, such that for any ideal graded encoding oracle $M = \{M_{q_1}^{g_1}, M_{q_2}^{g_2}, \ldots\}$, there exists a polynomial-size obfuscator O[10] such that O is a VBB obfuscator for the class of $p(\cdot)$-sized circuits in the M model.*

---

[7] To make this argument it is important that we allow *adaptive* encodings during the initialization phase, as opposed to a single non-adaptive encoding query as in the definition of [BGK+14].

[8] For this emulation with "bogus" random handles to work, it is important that we consider a model of *randomized* graded encodings (where multiple encodings of the same value are given fresh random handles). In case the encoding is deterministic (and thus encodings of the same value need to be given the same handle) the simulation fails: if the result of the operation yields a value that was previously encoded we should output that handle instead. Nevertheless, as we point out at the end of Sect. 3, our results extend to deal also with deterministic encodings where players can perform operations on the encodings.

[9] [BGK+14] present unconditionally secure obfuscators for $NC1$; the LWE assumption is needed to bootstrap up to polynomial-size circuits.

[10] The only non-uniform advice needed is the prime q_n.

Their construction also satisfies subexponential VBB security assuming an appropriate subexponential strengthening of LWE.

3   Impossibility of VBB Obfuscation

Theorem 5. *Assuming the existence of trapdoor permutations, there exists a family of functions F such that for every constant d and every degree-d ideal graded encoding oracle M, a polynomial-size 0.9-approximate VBB obfuscator for F does not exist in the M oracle model.*

We briefly review the approach of [CKP15] as we will follow the same high-level structure. Their first step is to show that any VBB obfuscator in the Random Oracle model can be transformed into an approximate VBB obfuscator in the plain. They next rely on Theorem 2 to conclude their impossibility result. The first step is achieved by running the original VBB obfuscator in the Random Oracle model by simulating all random oracle queries (with truly random answers). Additionally, to ensure consistency between answers to queries in the obfuscation phase and answers in the execution of the obfuscated code, the obfuscator performs a learning phase in which most *heavy* oracle queries (i.e. oracle queries that are made with high probability when running the obfuscated code on random inputs) are discovered; the answers to the heavy queries are hard-coded into the obfuscated code. This ensures that when the obfuscated code is run on a random input, except with inverse polynomial probability (proportional to the number of random inputs used in the learning phase), the obfuscated code will not make any random oracles queries that were not made during the learning phase (i.e. that are not hard-coded), and as a consequence, the obfuscated code correctly computes the function with high probability. Furthermore, the only difference between the new (plain-model) obfuscator and the original (random-oracle-model) obfuscator is that the former leaks the set of heavy queries; since this leak is something that can be learned by running the obfuscated code of the random-oracle-model obfuscator, VBB security ensures that the same heavy set can be simulated using only black-box access to the function.

As mentioned, we follow the same high-level approach. Our main result (Lemma 6 below) shows how to transform any VBB obfuscator in the constant-degree graded encoding model into an approximate VBB obfuscator in the plain model. The proof of Theorem 5 is then concluded by applying Theorem 2. Just as [CKP15], we run the original (graded-encoding-model) obfuscator and simulate its oracle queries. But it no longer suffices to simply learn all the heavy queries: the obfuscated code may only ask "light" queries (i.e., each query has negligible probability) yet the answer to those queries are correlated (in fact, even determined by) the queries made during the obfuscation phase. For instance, assume that the obfuscator encodes two elements v_1 and v_2, and later the evaluator makes a zero-test query of the form $p(v_1, v_2) = av_1 + bv_2$ where a and b are chosen from some distribution with high min-entropy.

Rather, we show that by running the obfuscated code on sufficiently many random inputs and honestly emulating answers to oracle queries, we can recover a set of linearly independent polynomials in the values $v_1, \ldots, v_\ell$ encoded during the obfuscation phase such that, except with inverse polynomial probability, when the obfuscated code is run on a random input, every zero-test query can be correctly emulated by simply determining whether the zero-test polynomial is a linear combination of polynomials in the stored set. Since the oracle is restricted to answering constant-degree d polynomials, there can be at most $(\ell+1)^d$ monomials in the values $v_1, \ldots, v_\ell$, and thus at most $(\ell+1)^d$ linearly independent polynomials in those values. If we record all zero-test polynomials that evaluate to zero, then after sufficiently many samples, we have either recovered the full basis (which allows one to correctly answer all remaining zero-test queries), or it is unlikely that a new sample will add another linearly independent polynomial, which in turn means that when the obfuscated code is run on a random input, our emulation only fails with small probability. We finally observe that, just as in [CKP15], leaking the set of linearly independent polynomials does not challenge VBB security because this set (just as the set of heavy random oracle queries in the case of [CKP15]) can be learned from just observing the obfuscated code and can thus be simulated.

We now turn to state and formally prove our main lemma, which combined with Theorem 2 directly concludes our main result (i.e., Theorem 5).

Lemma 6 (Main). *For every constant d and every degree-d ideal graded encoding oracle M, if a family of functions F indexed by k has a polynomial-size $\epsilon(|k|)$-approximate VBB obfuscator in the M oracle model, then there exists a polynomial-size $(\epsilon(|k|)+1/|k|)$-approximate[11] VBB obfuscator for F in the plain model.*

Proof. Let $M = \{M_{q_1}^{g_1}, M_{q_2}^{g_2}, \ldots\}$ be a degree-d ideal graded encoding oracle for some constant d. Let $\mathcal{O}$ be an ϵ-approximate obfuscator for family F in the M oracle model that requests encodings of at most $\ell(|k|)$ elements where k is the index for family F; we assume without loss of generality that $\ell(n) \geq 1$. We construct a (non-uniform[12]) polynomial-size $(\epsilon(n)+1/n)$-approximate VBB obfuscator $\mathcal{O}'$ for F in the plain model below.

New obfuscator $\mathcal{O}'(k)$:

1. On input k, run $\mathcal{O}(k)$ and simulate the queries to $M_{q_{|k|}}^{g_k}$ (i.e., answer the initial enc queries by creating a list $\mathcal{L}_O$ of encoded elements as in the definition of $M_{q_{|k|}}^{g_k}$, and answer zero(p) queries by evaluating the polynomial p on the "decoded" elements) to compute the obfuscated program C_k.

---

[11] $1/|k|$ can be replaced by any inverse polynomial by appropriately adjusting the parameters in our proof.

[12] The non-uniformity in our construction is to encode the sequence of primes $q_1, q_2, \ldots$ that is implicit in the oracle M^g. If we model the oracle M with a uniform algorithm that picks the field for each security parameter, then our construction below can also be uniform.

2. If $\mathcal{O}(k)$ did not make any initial encoding queries, simply modify the code of C_k to honestly emulate the M oracle with some hard-coded uniformly chosen randomness (to generate handles), output this modified code, and halt.
3. Otherwise, set $\mathcal{L}_c$ to empty.
4. *Repeat* until there have been $L = (\ell(|k|) + 1)^d |k|$ iterations without any new additions to $\mathcal{L}_c$:
   (a) Sample random input x^j.
   (b) Run $C_k(x^j)$ while simulating zero-test queries to M using the list of encoded elements $\mathcal{L}_O$ from Step 1.
   (c) Additionally, whenever a zero-test query $\mathsf{zero}(p)$ evaluates to true, record the formal polynomial p if it is linearly independent with all previously stored polynomials in $\mathcal{L}_c$.Testing whether p is a linear combination of polynomials in $\mathcal{L}_c$ can be performed efficiently through Gaussian elimination (by viewing each monomial as a separate variable).
5. Output a new circuit C_k' that does the following:
   (a) On input y, run $C_k(y)$.
   (b) If $C_k(y)$ makes a $\mathsf{zero}(p)$ query to M, answer true if p is a linear combination of the polynomials in $\mathcal{L}_c$ and otherwise answer false.

Claim. Obfuscator $\mathcal{O}'$ runs in (non-uniform) polynomial time.

Proof. Recall that $\ell(|k|)$ is an upper bound on the number of encodings. As a consequence, there are at most $(\ell(|k|)+1)^d$ degree-d monomials in the encodings; thus, there can be at most $(\ell(|k|) + 1)^d$ linearly independent zero-test polynomials. Since $\mathcal{O}$ continues iterating until there have been L consecutive iterations with no new additions to $\mathcal{L}_c$, it follows that there can be at most $L \cdot (\ell(|k|)+1)^d$ iterations, each of which can be implemented in polynomial time.

Proposition 1. *The obfuscator $\mathcal{O}'$ is $(\epsilon(n) + 1/n)$-approximately correct.*

Proof. Consider a hybrid obfuscator $\tilde{\mathcal{O}}'$ that proceeds just as $\mathcal{O}'$ except that it *always* outputs a program $\tilde{C}'_k$ that *honestly* simulates the M^g oracle using the state $\mathcal{L}_O$ from Stage 1.

Let Exp_k denote the experiment that consists of running $C_k \leftarrow \mathcal{O}(k)$, picking a uniformly random input $x^* \leftarrow \{0,1\}^{|k|}$, and outputting 1 iff $C_k(x^*) = F_k(x^*)$ (and 0 otherwise). Define Exp'_k and $\widetilde{\mathsf{Exp}}'_k$ in exactly the same way but using $\mathcal{O}'$ and $\tilde{\mathcal{O}}'$ respectively.

Since $\mathcal{O}$ is $\epsilon(n)$-approximately correct, for every $k \in \{0,1\}^n$, we have

$$\Pr[\mathsf{Exp}_k = 0] \le \epsilon(n)$$

We also observe that by construction, for every $k \in \{0,1\}^n$,

$$\Pr[\mathsf{Exp}_k = 0] = \Pr[\widetilde{\mathsf{Exp}}'_k = 0]$$

This directly follows from the observation that the only difference between these experiments is that in $\widetilde{\mathsf{Exp}}$, the obfuscator hard-codes the randomness of M^g

(needed to generate handles) in the obfuscated code in the event that $\mathcal{O}$ did not make any initial encoding queries. But since in the experiment we only evaluate the obfuscated code on a single input, the outputs of the experiments are identically distributed.

Our goal is now to prove that for $k \in \{0,1\}^n$,

$$\Pr[\mathsf{Exp}'_k = 0] \le \Pr[\widetilde{\mathsf{Exp}}'_k = 0] + 1/n$$

which concludes the proof of the proposition.

Note that there is only one difference between the program $\tilde{C}'_k$ produced by $\tilde{\mathcal{O}}'$ and the program C'_k produced by $\mathcal{O}'$ when run on the input x^* in the above experiments:

- $C'_k(x^*)$ may make a zero-test query $\mathsf{zero}(p, \boldsymbol{h})$ that should evaluate to true, but p is not in the span of $\mathcal{L}_c$ (and thus C'_k emulates the answer as false, whereas $\tilde{C}_k$ honestly emulates the answer as true.) Let bad^i denote the event that this happens for the first time when $|\mathcal{L}_c| = i$.

Let us note that $C'_k(x^*)$ can never err in the other direction; that is, it never answers a zero-test query as true when the answer in fact should be false. This follows from the fact that if p is in the span of $\mathcal{L}_c$, then a) all input handles to p correspond to some encoding, and p necessarily evaluates to zero given the encoded value corresponding to those handles, and b) by the wellformedness condition of g, $g(p, \boldsymbol{l})$ necessarily evaluates to true (as p cannot use any monomials not already in use by the polynomials in $\mathcal{L}_c$).

It follows by construction that conditioned on bad^i not happening for any i, experiments Exp'_k and $\widetilde{\mathsf{Exp}}'_k$ proceed identically.

The proof is concluded by the following two claims which show that the probability of any bad event is small. In the following we focus on experiment Exp' but the same arguments straighforwardly hold for $\widetilde{\mathsf{Exp}}'$.

Claim. For every i, $\Pr[\mathsf{bad}^i] \le 1/L$.

Proof. For every *bad* random tape for the experiment that induces event bad^i, we identify at least L unique *good* random tapes obtained by swapping the final run on input x^* with one of the (at least L) sampled iterations (using x_i); furthermore, we show that any two distinct bad executions lead to disjoint sets of good executions. We conclude the claim based on the fact that the fraction of bad tapes is at most $1/L$ and each random tape is equally likely.

Let us now formally specify the mapping Φ from bad tapes to good tapes, and specify an *inverse mapping* Φ^{-1} that given a good tape in the range of Φ recovers the bad tape it was generated from. The existence of such an inverse map shows that any two distinct bad tapes lead to distinct sets of good tapes, as desired.

Recall that by the proof of Claim 3, $m = L \cdot (\ell(n) + 1)^d$ is a bound on the number of iterations in step 4. We define a random tape for the experiment Exp_k as $(\rho, x_1, \ldots, x_m, x^*)$ where $(x_1, \ldots, x_m)$ are the inputs sampled to be used in step (4a) of $\mathcal{O}'$ (note that not all of those samples may be used), x^* is the final input chosen in the experiment, and ρ is the remaining randomness (i.e., the randomness of underlying $\mathcal{O}$ and randomness of $\mathcal{O}'$ in the event that $\mathcal{O}$ did not make the initial encoding queries). Let $q(R)$ denote the number of samples made in step 3 given the random tape R; by construction $L \leq q(R) \leq m$.

We say that a random tape $R = (\rho, x_1, \ldots, x_m, x^*)$ is *bad* if $\mathsf{Exp}'_k(R)$ induces event bad^i; that is, (a) in the evaluations of $C'_k(x_j)$ for $j \in [q(R) - L, \ldots, q(R)]$, there are no linearly independent zero-test polynomials that evaluate to 0, (b) the evaluation of $C'_k(x^*)$ leads to such a linearly independent polynomial that evaluates to 0, and (c) the size of $\mathcal{L}_c$ is i.

Define the mapping $\Phi(R)$ as the set of L random tapes $\Phi(R) = \{R_j\}_{j \in [L]}$ where R_j is constructed by swapping the tth random sample x_t, where $t = (q(R) - j + 1)$, with the last sample x^* as follows:

$$R_j = (\rho, x_1, \ldots, x_{t-1}, x^*, x_{t+1}, \ldots, x_m, x_j)$$

Note that $\mathsf{Exp}'_k(R_j)$ does not induce bad^i since the experiment finds at least $i+1$ linearly independent polynomials that evaluate to zero (and thus $\mathcal{L}_c > i$).

Finally, let $\Phi^{-1}(\cdot)$ be an inverse map that on input a tape R, swaps the last sample in the tape with the first sample x_t that leads to $i+1$ linearly independent polynomials in the set $\mathcal{L}_c$ (and if no such x_t exists simply outputs R). It follows directly by construction that for every bad R, $\Phi^{-1}(\Phi(R)) = R$. (Note that in our definition of the inverse map, we make use of the fact that the event bad^i is parameterized by i.)

By a union bound, it follows from Claim 3 that,

$$\Pr\left[\exists i \text{ s.t. } \mathsf{bad}^i\right] = \Pr\left[\mathsf{bad}^1 \vee \cdots \vee \mathsf{bad}^{\ell'(n)^d}\right] \leq \frac{\ell'(n)^d}{L} = \frac{\ell'(n)^d}{\ell'(n)^d n} = 1/n$$

where $\ell'(n) = \ell(n) + 1$ since as noted in the proof of Claim 3, the maximum size of $\mathcal{L}_c$ is $\ell'(n)^d = (\ell(n) + 1)^d$. This concludes that $\mathcal{O}'$ is $\epsilon(n) + 1/n$ approximately correct.

Proposition 2. *Obfuscator $\mathcal{O}'$ satisfies the virtual-black box property.*

Proof. This proof is essentially identical to the one given in [CKP15] for a similar statement. We include it here to be self-contained. Fix an index k. Given an adversary A' for the new obfuscator $\mathcal{O}'$, we construct a new adversary A^M for the $\mathcal{O}^M$ obfuscator as follows. The new adversary $A^M(C_k)$, on input a circuit C_k produced by the obfuscation $\mathcal{O}^M$ algorithm, simulates steps 2, 3, and 4 of the $\mathcal{O}'$ algorithm by answering all queries using its oracle M (whose answers will be consistent with the oracle used by $\mathcal{O}^M$ to produce C_k).[13] At the end of this simulation, A thus produces a circuit C'_k with exactly the same distribution as the output of $\mathcal{O}'$. Adversary A^M then runs $A'(C'_k)$ (which does not make any oracle queries) and returns the same output. It therefore follows by construction that

$$\Pr\left[A^{M_{|k|}}(\mathcal{O}^{M_{|k|}}(k)) = 1\right] = \Pr\left[A'(\mathcal{O}'(k)) = 1\right]$$

---

[13] Step 2 (i.e., checking whether the initial encoding queries have been made) can be simulated by making a "dummy" $enc(0,0)$ query and checking whether M returns $\perp$.

By the approximate VBB security property of $\mathcal{O}$ for family F, it follows that there exists a simulator S and a negligible function μ such that

$$\Pr\left[A^{M_{|k|}}(\mathcal{O}^{M_{|k|}}(k)) = 1\right] - \Pr\left[S^{F_k}(|k|) = 1\right] \leq \mu(|k|)$$

which immediately implies that

$$\Pr\left[A'(\mathcal{O}'(k)) = 1\right] - \Pr\left[S^{F_k}(|k|) = 1\right] \leq \mu(|k|)$$

and concludes the proposition since S is also a good simulator for A'.

We conclude that $\mathcal{O}'$ is a secure $\epsilon(n) + 1/n$-approximate VBB obfuscator for F (in the plain model). This finishes the proof of Lemma 6.

Remark (extension to "sparse" high-degree zero-test polynomials). This proof uses the constant-degree restriction on the zero-test queries to argue that the number of monomials in encoded values is polynomial. The theorem thus extends to high-degree polynomials as long as the legality predicate restricts these polynomials to be "sparse" in the sense that the *total* number of monomials over which any legal zero-test query is formed must be (a-priori) polynomially bounded. Note that it does not suffice to require that each zero-test query has a small number of monomials. Rather, we require that there exists a small set of monomials that suffices to represent *all* legal zero-test queries.

Remark (extension to "multi-slot" graded encodings). Our result directly extend to "multi-slot" graded encodings (as in [AB15]), which are a model of composite-order graded encodings. In this model, an encoding is a vector of elements; operations on elements are performed component-wise and finally a zero-test can be performed which determines whether the whole vector is 0. Our proof directly extends also to this setting (by simply viewing each component as a separate variable).

Remark (extension to deterministic encodings). Our graded encoding oracle models an idealized *randomized* graded encodings scheme: even if the same value v is encoded twice (under the same label), we get independently random handles for the two encodings. Our proof, however, works in exactly the same way also for *deterministic* randomized encodings, where the oracle keeps state also during the encoding phase and always returns the *same* handle for an encoding of the value v under the label l. This trivially follows since our oracle does not allow players to perform any operations on encodings but simply zero-test queries. As previously mentioned, for the case of randomized graded encodings, it is without loss of generality since operations on encodings can be simulated by "bogus" independently random handles. For the case of deterministic encodings, however, this simulation no longer works: if the result of the operation yields a value that was previously encoded we should output that handle instead. But for the purpose of our proof, we can make the simulation work: Modify the learning phase to keep track of also all handles h "seen", adding them to $\mathcal{L}_c$; additionally, for every operation, make a zero-test query to check whether the value to be encoded after the operation equals the value encoded under any previously stored handle. Next, during the evaluation of the obfuscated code, emulate operations on encodings by first checking (using a zero-test query, which is emulated as before) whether the value to be encoded after the operation equals the value encoded under any previously stored handle, and, if so, outputting this handle, and otherwise outputting a random handle. It follows using the same argument as above that this emulation only fails with inverse polynomial probability.

Remark (extension to rings). We note that our proof directly generalizes to any graded encoding scheme that operates on elements in a ring (as opposed to $\mathbb{F}_p$) as long as (a) there exists an efficient method for determining the row-rank of a matrix of this ring, and (b) the row-rank of a matrix is polynomially bounded by the column-rank. Property (a) is needed to test whether we get a linearly independent polynomial (we used Gaussian elimination for the case of $\mathbb{F}_p$), and property (b) is needed to ensure that the maximum number of linearly independent polynomials is polynomially bounded by the number of monomials (for the case of $\mathbb{F}_p$ row-rank equals column-rank, and thus the number of linearly independent polynomials is bounded by the number of monomials).

4   Impossibility of Subexponential VBB Security

We now consider sub-exponential VBB security and rule out constructions that use n^α-degree zero-test queries for any $0 < \alpha < 1$.

Theorem 7. *Assuming the existence of exponentially-secure trapdoor permutations, there exists a family of polynomial-time computable functions F such that for every $0 < \alpha < 1$, every degree-n^α ideal graded encoding oracle M, a polynomial-size 0.9-approximate VBB obfuscator for F does not exist in the M oracle model.*

Recall that, in contrast, Barak *et al.* [BGK+14] show that for every family F of polynomial-time functions, subexponentially-secure VBB obfuscation is possible using $p(n)$-degree ideal graded encodings where p is a polynomial (under appropriate cryptographic hardness assumptions).

We follow the proof of Theorem 5 and prove the following lemma which combined with Theorem 3 proves the theorem.

Lemma 8. *For every $\alpha < 1$ and degree-n^α ideal graded encoding oracle M, if a family of functions F indexed by k has a polynomial-size $\epsilon(n)$-approximate subexponentially-secure VBB obfuscator in the M oracle model, then there exists a subexponential-size $(\epsilon(n) + 1/n)$-approximate subexponentially-secure VBB obfuscator for F in the plain model.*

Proof (Sketch). The construction is identical to the one in the proof of Lemma 6, except that we set $d = n^\alpha$ (instead of it being a constant), where $n = |k|$.

By the same proof, the size of the new (plain-model) obfuscator is polynomial in $\ell(n)^{n^\alpha} = 2^{n^\alpha \log \ell(n)}$, where $\ell(n)$ is a bound on the number of encoding queries made by the original obfuscator. It follows that the size of the obfuscator is subexponential.

Approximate correctness follows as per the proof of Lemma 6. Finally, subexponential VBB simulation follows in exactly the same way as Lemma 6 by appealing to subexponential VBB security of the original VBB obfuscator.

References

[AB15] Applebaum, B., Brakerski, Z.: Obfuscating circuits via composite-order graded encoding. In: Dodis, Y., Nielsen, J.B. (eds.) TCC 2015, Part II. LNCS, vol. 9015, pp. 528–556. Springer, Heidelberg (2015)

[BGI+01] Barak, B., Goldreich, O., Impagliazzo, R., Rudich, S., Sahai, A., Vadhan, S.P., Yang, K.: On the (im)possibility of obfuscating programs. In: Kilian, J. (ed.) CRYPTO 2001. LNCS, vol. 2139, pp. 1–18. Springer, Heidelberg (2001)

[BGK+14] Barak, B., Garg, S., Kalai, Y.T., Paneth, O., Sahai, A.: Protecting obfuscation against algebraic attacks. In: Nguyen, P.Q., Oswald, E. (eds.) EUROCRYPT 2014. LNCS, vol. 8441, pp. 221–238. Springer, Heidelberg (2014)

[BP13] Bitansky, N., Paneth, O.: On the impossibility of approximate obfuscation and applications to resettable cryptography. In: STOC 2013 (2013)

[BR93] Bellare, M., Rogaway, P.: Random oracles are practical: a paradigm for designing efficient protocols. In: Proceedings of the 1st ACM Conference on Computer and Communications Security, CCS 1993, 3–5 November 1993, Fairfax, Virginia, USA, pp. 62–73 (1993)

[BR14] Brakerski, Z., Rothblum, G.N.: Virtual black-box obfuscation for all circuits via generic graded encoding. In: Lindell, Y. (ed.) TCC 2014. LNCS, vol. 8349, pp. 1–25. Springer, Heidelberg (2014)

[BS03] Boneh, D., Silverberg, A.: Applications of multilinear forms to cryptography. Contemp. Math. **324**(1), 71–90 (2003)

[Can97] Canetti, R.: Towards realizing random oracles: hash functions that hide all partial information. In: Kaliski Jr, B.S. (ed.) CRYPTO 1997. LNCS, vol. 1294, pp. 455–469. Springer, Heidelberg (1997)

[CKP15] Canetti, R., Kalai, Y.T., Paneth, O.: On obfuscation with random oracles. In: Dodis, Y., Nielsen, J.B. (eds.) TCC 2015, Part II. LNCS, vol. 9015, pp. 456–467. Springer, Heidelberg (2015)

[CV13] Canetti, R., Vaikuntanathan, V.: Obfuscating branching programs using black-box pseudo-free groups. Cryptology ePrint Archive, Report 2013/500 (2013). https://eprint.iacr.org/2013/500

[GGH13a] Garg, S., Gentry, C., Halevi, S.: Candidate multilinear maps from ideal lattices. In: Johansson, T., Nguyen, P.Q. (eds.) EUROCRYPT 2013. LNCS, vol. 7881, pp. 1–17. Springer, Heidelberg (2013)

[GGH+13b] Garg, S., Gentry, C., Halevi, S., Raykova, M., Sahai, A., Waters, B.: Candidate indistinguishability obfuscation and functional encryption for all circuits. In: Proceedings of FOCS 2013 (2013)

[Had00] Hada, S.: Zero-knowledge and code obfuscation. In: Okamoto, T. (ed.) ASIACRYPT 2000. LNCS, vol. 1976, pp. 443–457. Springer, Heidelberg (2000)

[LPS04] Lynn, B.Y.S., Prabhakaran, M., Sahai, A.: Positive results and techniques for obfuscation. In: Cachin, C., Camenisch, J.L. (eds.) EUROCRYPT 2004. LNCS, vol. 3027, pp. 20–39. Springer, Heidelberg (2004)

[MMN15] Mahmoody, M., Mohammed, A., Nematihaji, S.: More on impossibility of virtual black-box obfuscation in idealized models. In: TCC 2016 (2015, to appear)

[PST14] Pass, R., Seth, K., Telang, S.: Indistinguishability obfuscation from semantically-secure multilinear encodings. In: Garay, J.A., Gennaro, R. (eds.) CRYPTO 2014, Part I. LNCS, vol. 8616, pp. 500–517. Springer, Heidelberg (2014)

[Rot13] Rothblum, R.D.: On the circular security of bit-encryption. In: Sahai, A. (ed.) TCC 2013. LNCS, vol. 7785, pp. 579–598. Springer, Heidelberg (2013)

[Sho97] Shoup, V.: Lower bounds for discrete logarithms and related problems. In: Fumy, W. (ed.) EUROCRYPT 1997. LNCS, vol. 1233, pp. 256–266. Springer, Heidelberg (1997)

On the Impossibility of Virtual Black-Box Obfuscation in Idealized Models

Mohammad Mahmoody[(⊠)], Ameer Mohammed, and Soheil Nematihaji

University of Virginia, Charlottesville, USA
{mohammad, am8zv, sn8fbt}@virginia.edu

Abstract. The celebrated work of Barak et al. (Crypto'01) ruled out the possibility of *virtual black-box* (VBB) obfuscation for general circuits. The recent work of Canetti, Kalai, and Paneth (TCC'15) extended this impossibility to the random oracle model as well assuming the existence of trapdoor permutations (TDPs). On the other hand, the works of Barak et al. (Crypto'14) and Brakerski-Rothblum (TCC'14) showed that general VBB obfuscation is indeed possible in *idealized graded encoding* models. The recent work of Pass and Shelat (Cryptology ePrint 2015/383) complemented this result by ruling out general VBB obfuscation in idealized graded encoding models that enable evaluation of constant-degree polynomials in finite fields.

In this work, we extend the above two impossibility results for general VBB obfuscation in idealized models. In particular we prove the following two results both assuming the existence of trapdoor permutations:

- There is no general VBB obfuscation in the generic group model of Shoup (Eurocrypt'97) for *any abelian* group. By applying our techniques to the setting of Pass and Shelat we extend their result to any (even non-commutative) finite *ring*.
- There is no general VBB obfuscation in the *random trapdoor permutation oracle* model. Note that as opposed to the random oracle which is an idealized primitive for symmetric primitives, random trapdoor permutation is an idealized public-key primitive.

Keywords: Virtual black-box obfuscation · Idealized models · Graded encoding · Generic group model

1 Introduction

Obfuscating programs to make them "unintelligible" while preserving their functionality is one of the most sought after holy grails in cryptography due to its

M. Mahmoody—Supported by NSF CAREER award CCF-1350939. The work was done in part while the author was visiting the Simons Institute for the Theory of Computing, supported by the Simons Foundation and by the DIMACS-Simons Collaboration in Computing through NSF grant CNS-1523467.

A. Mohammed—Supported by University of Kuwait.

S. Nematihaji—Supported by NSF award CCF-1350939.

E. Kushilevitz and T. Malkin (Eds.): TCC 2016-A, Part I, LNCS 9562, pp. 18–48, 2016.
DOI: 10.1007/978-3-662-49096-9_2

numerous applications. The celebrated work of Barak et al. [BGI+01] was the first to launch a formal study of this notion in its various forms. *Virtual Black-Box* (VBB) obfuscation is a strong form of obfuscation in which the obfuscated code does not reveal any secret bit about the obfuscated program unless that information could already be obtained through a black-box access to the program. It was shown in [BGI+01] that VBB obfuscation is not possible *in general* as there is a family of functions $\mathcal{F}$ that could not be VBB obfuscated. Roughly speaking, $\mathcal{F}$ would consist of circuits C such that given any obfuscation $B = O(C)$ of C, by running B over B itself as input one can obtain a secret s about C that could not be obtained through mere black-box interaction with C. This strong impossibility result, however, did not stop the researchers from exploring the possibility of VBB obfuscation for *special* classes of functions, and positive results for special cases were presented (e.g., [Can97, Wee05]) based on believable computational assumptions.

The work of Lynn, Prabhakaran and Sahai [LPS04] showed the possibility of VBB obfuscation for certain class of functions in the *random oracle model* (ROM). The work of [LPS04] left open whether general purpose obfuscator for all circuits could be obtained in the ROM or not. Note that when we allow the random oracle to be used during the obfuscation phase (and also let the obfuscated code to call the random oracle) the impossibility result of [BGI+01] no longer applies, because the proof of [BGI+01] requires the obfuscated algorithm to be a circuit in the *plain* model where no oracle is accessed. In fact, despite the impossibility of general VBB obfuscation in the plain model, a construction for VBB obfuscation in the ROM could be used as a practical heuristic obfuscation mechanism once instantiated with a strong hash function such as SHA3. This would be in analogy with the way ROM based constructions of other primitives are widely used in practice *despite* the impossibility results of [CGH04].

On a different route, the breakthrough of Garg et al. [GGH+13b] proposed a candidate *indistinguishability obfuscation* (iO), a weaker form of obfuscation compared to VBB for which no impossibility results were (are) known, relying on the so called "approximate multi-linear maps" (MLM) assumption [GGH13a]. Shortly after, it was proved by Barak et al. [BGK+14] and Brakerski and Rothblum [BR14] that the construction of [GGH+13b] could be used to get even VBB secure obfuscation (rather than the weaker variant of iO) in an idealized form of MLMs, called the *graded encoding model*. The VBB obfuscation schemes of [BGK+14, BR14] in idealized models raised new motivations for studying the possibility of VBB obfuscation in such models including the ROM.

Canetti, Kalai, and Paneth [CKP15] proved the first impossibility result for VBB obfuscation in a natural idealized model by ruling out the existence of general purpose VBB obfuscators in the random oracle model, assuming the existence of trapdoor permutations. Their work resolved the open question of [LPS04] negatively. At a technical level, [CKP15] showed how to compile any VBB obfuscator in the ROM into an *approximate* VBB obfuscator in the plain model which preserves the circuit's functionality only for "most" of the inputs. This would rule out VBB obfuscation in plain model (assuming TDPs) since Bitansky and Paneth [BP13] had shown that no approximate VBB obfuscator for general circuits exist if trapdoor permutations exist.

Pass and shelat [Pas15] further studied the possibility of VBB obfuscation in idealized algebraic models in which the positive results of [BGK+14, BR14] were proved. [Pas15] showed that the existence of VBB obfuscation schemes in the graded encoding model highly depends on the degree of polynomials (allowed to be zero tested) in this model. In particular they showed that VBB obfuscation of general circuits is impossible in the graded encoding model of constant-degree polynomials. Their work nicely complemented the positive results of [BGK+14, BR14] that were proved in a similar (graded encoding) model but using super-constant (in fact polynomial in security parameter) polynomials.

We shall emphasize that proving *limitations* of VBB obfuscation or any other primitive in generic models of computation such as the generic group model of Shoup [Sho97] are strong lower-bounds (a la black-box separations [RTV04, IR89]) since such results show that for certain crypto tasks, as long as one uses certain algebraic structures (e.g., an abelian group) in a black-box way as the source of computational hardness, there will always be a *generic* attack that (also treats the underlying algebraic structure in a black-box way and) breaks the constructed scheme. The fact that the proposed attack is generic makes the lower-bound only stronger.

1.1   Our Results

In this work we extend the previous works [CKP15, Pas15] on the impossibility of VBB obfuscation in idealized models of computation and generalize their results to more powerful idealized primitives. We first focus on the generic group model of Shoup [Sho97] (see Definitions 10 and 11) and rule out the existence of general VBB obfuscation in this model for *any finite abelian group*.

Theorem 1 (Informal). *Assuming trapdoor permutations exist, there is no virtual black-box obfuscation for general circuits in the generic group model for any finite abelian group.*

The work of [Pas15] implies a similar lower bound for the case of abelian groups of prime order. We build upon the techniques of [CKP15, Pas15] and extend the result of [Pas15] to arbitrary (even noncyclic) finite abelian groups. See the next section for a detailed description of our techniques for proving this theorem and the next theorems described below.

We then apply our techniques designed to prove Theorem 1 to the setting of graded-encoding model studied in [Pas15] and extend their results to arbitrary finite *rings* (rather than fields) which remained open after their work. Our proof even handles *noncommutative* rings.

Theorem 2 (Informal). *Assuming trapdoor permutations exist, there is no virtual black-box obfuscation for general circuits in ideal degree-$O(1)$ graded encoding model for any finite ring.*

Finally, we generalize the work of [CKP15] beyond random oracles by ruling out general VBB obfuscation in random *trapdoor permutations* (TDP) oracle model. Our result extends to an arbitrary number of levels of hierarchy of trap-doors, capturing idealized version of primitives such as hierarchical identity based encryption [HL02].

Theorem 3 (Informal). *Assuming trapdoor permutations exist, there is no virtual black-box obfuscation for general circuits in the random trapdoor permutation model, even if the oracle provides an unbounded hierarchy of trapdoors.*

Note that the difference between the power of random oracles and random TDPs in cryptography is usually huge, as random oracle is an idealized primitive giving rise to (very efficient) *symmetric* key cryptography primitives, while TDPs could be used to construct many *public-key* objects. Our result indicates that when it comes to VBB obfuscation random TDPs are no more powerful than just random oracles.

Connection to Black-Box Complexity of IO. In a very recent follow-up work by the authors, Rafael Pass, and abhi shelat [MMN+15] it is shown that the results of this work and those of [Pas15] could be used to derive lower-bounds on the assumptions that can be used in a black-box way to construct *indistinguishability* obfuscation. In particular, let $\mathcal{P}$ be a primitive implied by (i.e. exist relative to) random trapdoor permutations, generic abelian group model, or the degree-$O(1)$ graded encoding model; this includes powerful primitives such as exponentially secure TDPs or exponentially secure DDH-type assumptions. [MMN+15] shows that basing iO on $\mathcal{P}$ in a black-box way is either impossible, or it is at least as hard as basing public-key cryptography on one-way functions (in a non-black-box way). Whether or not public-key encryption can be based on one-way functions has remained as one of the most fundamental open questions in cryptography.

1.2   Technical Overview

The high level structure of the proofs of our results follows the high level structure of [CKP15], so we start by recalling this approach. The idea is to convert the VBB obfuscator $O^{\mathcal{I}}$ in the idealized model to an *approximate* VBB obfuscation $\widehat{O}$ in the plain model which gives the correct answer $C(x)$ with high (say, 9/10) probability over the choice of random input x and randomness of obfuscator. The final impossibility follows by applying the result of [BP13] which rules out approximate VBB in the plain model. Following [CKP15] our approximate VBB obfuscator $\widehat{O}$ in the plain model has the following high level structure.

1. **Obfuscation emulation.** Given a circuit C emulate the execution of the obfuscator $O^{\mathcal{I}}$ in the idealized model over input C to get circuit B (running in the idealized model).[1]

---

[1] The emulation here and in next steps would require the idealized model $\mathcal{I}$ to have an efficient "lazy evaluation" procedure. For example lazy evaluation for random oracles chooses a random answer (different from previous ones) given any new query.

2. **Learning phase.** Emulate the execution of B over m random inputs for sufficiently large m. Output B and the view z of the m executions above as the obfuscated code $\widehat{B} = (B, z)$.

3. **Final execution.** Execute the obfuscated code $\widehat{B} = (B, z)$ on new random points using some form of "lazy evaluation" of the oracle while only using the transcript z of the learning phase (and not the transcript of obfuscator O which is kept private) as the partially fixed part of the oracle. The exact solution here depends on the idealized model $\mathcal{I}$, but they all have the following form: if the answer to a new query could be "derived" from z then use this answer, otherwise generate the answer from some simple distribution.

VBB Property. As argued in [CKP15], the VBB property of $\widehat{O}$ follows from the VBB property of O and that the sequence of views z could indeed be sampled by any PPT holding B in the idealized model (by running B on m random inputs), and so it is simulatable (see Lemma 5).

Approximate Correctness. The main challenge is to show the approximate correctness of the new obfuscation procedure in the plain model. The goal here is to show that if the learning phase is run for sufficiently large number of rounds, then its transcript z has enough information for emulating the next (actual) execution *consistently* with but *without* knowing the view of O. In the case that $\mathcal{I}$ is a random oracle [CKP15] showed that it is sufficient to bound the probability of the "bad" event E that the final execution of $\widehat{B} = (B, z)$ on a random input x asks any of the "private" queries of the obfuscator O which is not discovered during the learning phase. The work of [Pas15] studies graded encoding oracle model where the obfuscated code can perform arbitrary zero-test queries for low degree polynomials $p(\cdot)$ defined over the generated labels $s_1, \ldots, s_k$. The oracle will return true if $p(s_1, \ldots, s_k) = 0$ (in which case $p(\cdot)$ is called a zero polynomial) and returns false otherwise. Due to the algebraic structure of the field here, it is no longer enough to learn the heavy queries of the obfuscated code who might now ask its oracle query $p(\cdot)$ from some "flat" distribution while its answer is correlated with previous answers.

Generic Group Model: Proving Theorem 1. To describe the high level ideas of the proof of our Theorem 1 it is instructive to start with the proof of [Pas15] restricted to zero testing degree-1 polynomials and adapt it to the very close model of GGM for $\mathbb{Z}_p$ when p is a prime, since as noted in [Pas15] when it comes to zero-testing linear functions these two models are indeed very close.[2]

Case of $\mathbb{Z}_p$ For Prime p [Pas15]. When we go to the generic group model we can ask addition and labeling queries as well. It can be shown that we do not need to generate any labels during obfuscation and they can be emulated using addition

---

[2] More formally, using the rank argument of [Pas15] it can be shown that for the purpose of obfuscation, the two models are equivalent up to arbitrary small $1/\operatorname{poly}(n)$ completeness error.

queries. Then, by induction, all the returned labels $t_1, \ldots, t_\ell$ for addition queries are linear combinations of $s_1, \ldots, s_k$ with integer coefficients[3] and that is how we represent queries. Suppose we get an addition query $\mathbf{a} + \mathbf{b}$ and want to know the label of the group element determined by (the coefficients) $\mathbf{a} + \mathbf{b} = \mathbf{x}$. Suppose for a moment that we know s is the label for a vector of integers $\mathbf{c}$, and suppose we also know that the difference $\mathbf{x} - \mathbf{c}$ evaluates to zero. In this case, similarly to [CKP15], we can confidently return the label s as the answer to $\mathbf{a} + \mathbf{b}$. To use this idea, at any moment, let W be the space of all (zero) vectors $\boldsymbol{\alpha} - \boldsymbol{\beta}$ such that we have previously discovered same labels for $\boldsymbol{\alpha}$ and $\boldsymbol{\beta}$. Now to answer $\mathbf{a} + \mathbf{b} = \mathbf{x}$ we can go over all previously discovered labels ($\mathbf{c} \mapsto s$) and return s if $\mathbf{x} - \mathbf{c} \in \mathsf{span}(W)$, and return a random label otherwise. The approximate correctness follows from the following two points.

- **The rank argument.** First note that if $\mathbf{x} - \mathbf{c} \in \mathsf{span}(W)$ then the label for the vector $\mathbf{a} + \mathbf{b} = \mathbf{x}$ is indeed s. So we only need worry about cases where $\mathbf{x} - \mathbf{c} \notin \mathsf{span}(W)$ but $\mathbf{x} - \mathbf{c}$ is still zero. The rank argument of [Pas15] shows that this does not happen too often if we repeat the learning phase enough times. The main idea is that if this "bad" event happens, it increases the rank of W, but this rank can increase only k times.
- **Gaussian elimination.** Finally note that the test $\mathbf{x} - \mathbf{c} \notin \mathsf{span}(W)$ can be implemented efficiently using Gaussian elimination when we work in $\mathbb{Z}_p$.

Case of General Cyclic Abelian Groups. We first describe how to extend the above to any cyclic abelian group $\mathbb{Z}_m$ (for possibly non-prime m) as follows.

- **Alternative notion for rank of W.** Unfortunately, when we move to the ring of $\mathbb{Z}_m$ for non-prime m it is no longer a field and we cannot simply talk about the rank of W (or equivalently the dimension of $\mathsf{span}(W)$) anymore.[4] More specifically, similarly to [Pas15], we need such (polynomial) bound to argue that during most of the learning phases the set $\mathsf{span}(W)$ does not grow. To resolve this issue we introduce an alternative notion which here we call $\widetilde{\mathsf{rank}}(W)$ that has the following three properties even when vectors $\mathbf{w} \in W$ are in $\mathbb{Z}_m^k$ (1) If $\mathbf{a} \in \mathsf{span}(W)$ then $\widetilde{\mathsf{rank}}(W) = \widetilde{\mathsf{rank}}(W \cup \{\mathbf{a}\})$, and (2) if $\mathbf{a} \notin \mathsf{span}(W)$ then $\widetilde{\mathsf{rank}}(W) + 1 \leq \widetilde{\mathsf{rank}}(W \cup \{\mathbf{a}\})$, and (3) $1 \leq \widetilde{\mathsf{rank}}(W) \leq k \cdot \log |\mathbb{Z}_m| = k \cdot \log m$. In particular in Lemma 21 we show that the quantity $\widetilde{\mathsf{rank}}(W) := \log |\mathsf{span}(W)|$ has these three properties. These properties together show that $\mathsf{span}(W)$ can only "jump up" $k \cdot \log m$ (or fewer) times during the learning phase, and that property is enough to be able to apply the argument of [Pas15] to show that sufficiently large number of learning phases will bound the error probability by arbitrary $1/\mathsf{poly}(n)$.

---

[3] Even though the summation is technically defined over the group elements, for simplicity we use the addition operation over the labels as well.

[4] Note that this is even the case for $\mathbb{Z}_q$ when q is a prime power, although finite fields have prime power sizes.

- **Solving system of linear equations over** $\mathbb{Z}_m$. Even though m is not necessarily prime, this can still be done using a generalized method for cyclic abelian groups [McC90].

Beyond Cyclic Groups. Here we show how to extend the above argument beyond cyclic groups to arbitrary abelian groups. First note that to solve the Gaussian elimination algorithm for $\mathbb{Z}_m$, we first convert the integer vectors of W into some form of finite module by trivially interpreting the integer vectors of W as vectors in $\mathbb{Z}_m^k$. This "mapping" was also crucially used for bounding $\widetilde{\mathrm{rank}}(W)$.

- **Mapping integers to general abelian** G. When we move to a general abelian group G we again need to have a similar mapping to map W into a "finite" module. Note that we do not know how to solve these questions using integer vectors in W efficiently. In Lemma 9 we show that a generalized mapping $\rho_G(\cdot)\colon \mathbb{Z} \mapsto G$ (generalizing the mapping $\rho_{\mathbb{Z}_m}(x) = (x \bmod m)$ for $\mathbb{Z}_m$) exists for general abelian groups that has the same effect; namely, without loss of generality we can first convert integer vectors in W to vectors in G^k and then work with the new W.
- **The alternative rank argument.** After applying the transformation above over W (to map it into a subset of G^k) we can again define and use the three rank-like properties of $\widetilde{\mathrm{rank}}(\cdot)$ (instead of $\mathrm{rank}(W)$) described above, but here for any finite abelian group G. In particular we use $\widetilde{\mathrm{rank}}(W) := \log |\mathrm{span}_{\mathbb{Z}}(W)|$ where $\mathrm{span}_{\mathbb{Z}}(\cdot)$ is the module spanned by W using *integer* coefficients. Note that even though G is not a ring, multiplying integers with $x \in G$ is naturally defined (see Definition 8).
- **System of linear equations over finite abelian groups.** After the conversion step above, now we need to solve a system of linear equation $\mathbf{x}W = \mathbf{a}$ where elements of $W, \mathbf{a}$ are from G but we are still looking for *integer* vector solutions $\mathbf{x}$. After all, there is no multiplication defined over elements from G. See the full version of this paper in which we give a reduction from this problem (for general finite abelian groups) to the case of $G = \mathbb{Z}_m$ which is solvable in polynomial time [McC90].

Low-Degree Graded Encoding Model: Proving Theorem 2. To prove Theorem 3 for general finite rings, we show how to use the ideas developed for the case of general abelian generic groups discussed above and apply them to the framework of [Pas15] for low-degree graded encoding model as specified in Theorem 2. Recall that here the goal is to detect the zero polynomials by checking their membership in the module $\mathrm{span}(W)$. Since here we deal with polynomials over a ring (or field) R the multiplication is indeed defined. Therefore, if we already know a set of zero polynomials W and want to judge whether $\mathbf{a}$ is also (the vector corresponding to) a zero polynomial, the more natural approach is to solve a system of linear equations $\mathbf{x}W = \mathbf{a}$ over ring R.

Searching for Integer Solutions Again. Unfortunately we are not aware of a polynomial time algorithm to solve $\mathbf{x} \cdot W = \mathbf{a}$ in general finite rings and as we mentioned above even special cases like $R = \mathbb{Z}_m$ are nontrivial [McC90]. Our idea is to try to reduce the problem back to the abelian *groups* by somehow eliminating the ring multiplication. Along this line, when we try to solve $\mathbf{x} \cdot W = \mathbf{a}$, we again restrict ourselves *only* to integer solutions. In other words, we do not multiply inside R anymore, yet we take advantage of the fact that the existence of integer solution to $\mathbf{x} \cdot W = \mathbf{a}$ is *still sufficient* to conclude $\mathbf{a}$ is a zero vector. As we mentioned above, we indeed know how to find integer solutions to such system of linear equations in polynomial time (see [McC90] and the full version of the paper).

Finally note that, we can again use our alternative rank notion of $\widetilde{\mathrm{rank}}(W)$ to show that if we run the learning phase of the obfuscation (in plain model) m times the number of executions in which $\mathrm{span}_{\mathbb{Z}}(W)$ grows is at most $\mathrm{poly}(n)$ (in case of degree-$O(1)$ polynomials). This means that we can still apply the high level structure of the arguments of [Pas15] for the case of finite rings *without* doing Gaussian elimination over rings.

Random Trapdoor Permutation Model: Proving Theorem 3. Here we give the high-level intuition behind our result for the random TDP oracle model.

Recalling the Case of Random Oracles [CKP15]. Recall the high level structure of the proof of [CKP15] for the case of random oracles described above. As we mentioned, [CKP15] showed that to prove approximate correctness it is sufficient to bound the probability of the event E that the final execution of $\widehat{B} = (B, z)$ on a random input x asks any of the queries that is asked by emulated obfuscation O of B (let Q_O denote this set) which is *not* discovered during the learning phase. So if we let Q_E, Q_B, Q_O denote the set of queries asked, respectively, in the final execution, learning, and obfuscation phases, the bad event would be $Q_E \cap (Q_O \setminus Q_B) \neq \varnothing$. This probability could be bound by arbitrary small $1/\mathrm{poly}$ by running the learning phase sufficiently many times. The intuition is that all the "heavy" queries which have a $1/\mathrm{poly}$-chance of being asked by $\widehat{B} = (B, z)$ (i.e., being in Q_E) on a random input x would be learned, and thus the remaining unlearned private queries (i.e., $Q_O \setminus Q_B$) would have a sufficiently small chance of being hit by the execution of $\widehat{B} = (B, z)$ on a random input x.

Warm-up: Random Permutation Oracle. We start by first describing the proof for the simpler case of random permutation oracle. The transformation technique for the random oracle model can be easily adapted to work in the random permutation model as follows. For starters, assume that the random permutation is only provided on one input length k; namely $R \colon \{0,1\}^k \mapsto \{0,1\}^k$. If $k = O(\log n)$ where n is the security parameter, then it means that the whole oracle can be learned during the obfuscation and hardcoded in the obfuscated code, and so R cannot provide any advantage over the plain model. On the other hand if $k = \omega(\log n)$ it means that the range of R is of super-polynomial size. As a result, the same exact procedure proposed in [CKP15] (that assumes R is a random oracle and shows how to securely compile out R from the construction) would also work if R is a random permutation oracle. The reason is that the whole transformation process asks poly(n) number of queries to R and, if the result of the transformation does not work when R is a random permutation, then the whole transformation gives a poly(n) = q query attack to distinguish between whether R is a random permutation or a random function. Such an attack cannot "succeed" with more than negligible probability when the domain of R has super-polynomial size $q^{\omega(1)}$ in the number of queries q.[5]

Random TDP Model. Suppose $T = (G, F, F^{-1})$ is a random trapdoor permutation oracle in which G is a random permutation for generating the public index, F is the family of permutations evaluated using public index, and F^{-1} is the inverse permutation computed using the secret key (see Definition 28 for formal definition and notation used). When the idealized oracle is $T = (G, F, F^{-1})$, we show that it is sufficient to apply the same learning procedure used in the random oracle case over the *normalized* version of the obfuscated algorithm B to get a plain-model execution $\widehat{B}(x)$ that is statistically close to an execution $B^T(x)$ that uses oracle T. This, however, requires careful analysis to prove that inconsistent queries specific to the TDP case occur with small probability.

Indeed, since the three algorithms (emulation, learning, and final execution) are correlated, there is a possibility that the execution of B on the new random input might ask a new query that is *not* in Q_O, and yet still be inconsistent with some query in $Q_O \backslash Q_B$. For example, assume we have a query q of the form $G(sk) = pk$ that was asked during the obfuscation emulation phase (and thus is in Q_O) but was missed in the learning phase (and thus is not in Q_B) and assume that a query of the form $F[pk](x) = y$ was asked during the learning phase (so it is in Q_B). Then, it is possible that during the evaluation of the circuit B, it may ask a query q' of the form $F^{-1}[sk](y)$ and since this is a new query undetermined by previously learned queries, the plain-model circuit $\widehat{B}$ will answer with some random answer y'. Note that in this case, y' would be different from y with very high probability, and thus even though $q \neq q'$, they are together inconsistent with respect to oracle T.

As we show in our case-by-case analysis of the learning heavy queries procedure for the case of trapdoor permutation (in Sect. 4.2), the only bad events that we need to consider (besides hitting unlearned Q_O queries, which was already shown to be unlikely) will be those whose probability of occurring are negligible (we use the lemmas from [GKLM12] as leverage). Due to our normalization procedure, the rest of the cases will be reduced to the case of not learning heavy queries, and this event is already bounded.

---

[5] In general, when the random permutation R is available in *all* input lengths, we can use a mixture of the above arguments by generating all the oracle queries of length $c \log(n)$ (for a sufficiently large constant c) during the obfuscation (in the plain model) and representing this randomness in the obfuscated circuit. This issue also exists in the trapdoor permutation and the generic group models and can be handled exactly the same way.

2   Virtual Black-Box Obfuscation

Below we give a direct formal definition for approximately correct virtual black-box (VBB) obfuscation in idealized models. The (standard) definition of VBB is equivalent to 0-approximate VBB in the plain model where no oracle is accessed.

Definition 4 (Approximate VBB in Idealized Models [BGK+13, CKP15]**).** *For a function $\epsilon(n) \in [0,1]$, a PPT algorithm O is called an ϵ-approximate general purpose VBB obfuscator in the $\mathcal{I}$-ideal model if the following is satisfied:*

- *Approximate Functionality: For any circuit C of size n and input size m*

$$\Pr_{x \leftarrow \{0,1\}^m}[O^{\mathcal{I}}(C)(x) \neq C(x)] \leq \epsilon(n)$$

  where the probability is over the choice of input x, the oracle $\mathcal{I}$, and the internal randomness of O.
- *Virtual Black-Box: For every PPT adversary A, there exists a PPT simulator S and a negligible $\sigma(n)$ such that for all $n \in \mathbb{N}$ and circuits $C \in \{0,1\}^n$:*

$$\left|\Pr[A^{\mathcal{I}}(O^{\mathcal{I}}(C)) = 1] - \Pr[S^C(1^n) = 1]\right| \leq \sigma(n)$$

  where the probability is over $\mathcal{I}$ and the randomness of A, S, and O.

The following lemma is used in [CKP15], and here we state it in an abstract form considering only the VBB security and ignoring the completeness.

Lemma 5 (Preservation of VBB Security). *Let O be a PPT algorithm in the $\mathcal{I}$-ideal model that satisfies VBB security, and let U be a PPT algorithm (in the $\mathcal{I}$-ideal model) that, given input $B = O^{\mathcal{I}}(C)$ for some circuit $C \in \{0,1\}$ of size n, outputs (B, z) where z is some string. If there exists a plain-model PPT algorithm $\widehat{O}$ that, on input C, outputs (B', z') with distribution statistically close to (B, z) conditioned on C then $\widehat{O}$ also satisfies VBB security.*

Proof. To prove that $\widehat{O}$ satisfies the security of VBB obfuscation (regardless of its completeness) we show a reduction that turns any plain-model adversary $\widehat{A}$ that breaks the VBB security of $\widehat{O}$ into an ideal-model adversary $A^{\mathcal{I}}$ against O.

For any fixed circuit C, $A^{\mathcal{I}}$ accepts as input $B = O^{\mathcal{I}}(C)$ then executes $U^{\mathcal{I}}(B)$ to get (B, z). $A^{\mathcal{I}}$ will then run $\widehat{A}$ with input (B, z) then output whatever $\widehat{A}$ outputs. Given the behaviour of $A^{\mathcal{I}}$, we have that:

$$\Pr[A^{\mathcal{I}}(B) = 1] = \Pr[\widehat{A}(B, z) = 1] \tag{1}$$

Furthermore, if we let the statistical distance between (B, z) and $\widehat{O}(C) = (B', z')$ be at most $\epsilon(n)$, then we also have:

$$\left|\Pr[\widehat{A}(B', z') = 1] - \Pr[\widehat{A}(B, z) = 1]\right| \leq \epsilon(n) \tag{2}$$

Since O satisfies the security of VBB in the ideal model, we have that, there is a simulator S for the adversary $A^{\mathcal{I}}$ in the ideal model such that:

$$\left| \Pr[A^{\mathcal{I}}(B) = 1] - \Pr[S^C(1^n) = 1] \right| \leq \mathrm{negl}(n) \tag{3}$$

Now let $\widehat{S}$ be a VBB simulator for $\widehat{A}$. Combining this with Eqs. 1 and 2, and given that $\epsilon(n)$ is a negligible function, we find that $\widehat{O}$ is also VBB-secure using the same simulator S.

3   Impossibility of VBB in Generic Algebraic Models

In this section we will formally state and prove our Theorems 1 and 2 for the generic group and graded encoding models.

3.1   Preliminaries

We start by stating some basic group theoretic notation, facts, and definitions. By $\mathbb{Z}$ we refer to the set of integers. By $\mathbb{Z}_n$ we refer to the additive (or maybe the ring) of integers modulo n. When G is an abelian group, we use $+$ to denote the operation in G. A semigroup $(G, \odot)$ consists of any set G and an associative binary operation $\odot$ over G.

Definition 6. *For semi-groups $(G_1, \odot_1), \ldots, (G_k, \odot_1)$, by the direct product semi-group $(G, \odot) = (G_1 \times \cdots \times G_k, \odot_1 \times \cdots \times \odot_k)$ we refer to the group in which for $g = (g_1, \ldots, g_k) \in G, h = (h_1, \ldots, h_k) \in G$ we define $g \odot h = (g_1 \odot_1 h_1, \ldots, g_k \odot_k h_1)$. If G_i's are groups, their direct product is also a group.*

The following is the fundamental theorem of finitely generated abelian groups restricted to case of finite abelian groups.

Theorem 7 (Characterization of Finite Abelian Groups). *Any finite abelian group G is isomorphic to some group $\mathbb{Z}_{p_1^{\alpha_1}} \times \cdots \times \mathbb{Z}_{p_d^{\alpha_d}}$ in which p_i's are (not necessarily distinct) primes and $\mathbb{Z}_{p_i^{\alpha_i}}$ is the additive group mod $p_i^{\alpha_i}$.*

Definition 8 (Integer vs in-group multiplication for abelian groups). *For integer $a \in \mathbb{Z}$ and $g \in G$ where G is any finite abelian group by $a \cdot g$ we mean adding g by itself $|a|$ times and negating it if $a < 0$. Now let $g, h \in G$ both be from abelian group G and let $G = \mathbb{Z}_{p_1^{\alpha_1}} \times \cdots \times \mathbb{Z}_{p_d^{\alpha_d}}$ where p_i's are primes. If not specified otherwise, by $g \cdot h$ we mean the multiplication of g, h in G interpreted as the multiplicative semigroup that is the direct product of the multiplicative semigroups of $\mathbb{Z}_{p_i^{\alpha_i}}$'s for $i \in [d]$ (where the multiplications in $\mathbb{Z}_{p_i^{\alpha_i}}$ are mod $p_i^{\alpha_i}$).*

Lemma 9 (Mapping integers to abelian groups). *Let $G = \mathbb{Z}_{p_1^{\alpha_1}} \times \cdots \times \mathbb{Z}_{p_d^{\alpha_d}}$. Define $\rho_G \colon \mathbb{Z} \mapsto G$ as $\rho_G(a) = (a_1, \ldots, a_d) \in G$ where $a_i = a \mod p_i^{\alpha_i} \in \mathbb{Z}_{p_i^{\alpha_i}}$. Also for $\mathbf{a} = (a_1, \ldots, a_k) \in \mathbb{Z}^k$ define $\rho_G(\mathbf{a}) = (\rho_G(a_1), \ldots, \rho_G(a_k))$. Then for any $a \in \mathbb{Z}$ and $g \in G = \mathbb{Z}_{p_1^{\alpha_1}} \times \cdots \times \mathbb{Z}_{p_d^{\alpha_d}}$ it still holds that $a \cdot g = \rho_G(a) \cdot g$ where the first multiplication is done according to Definition 8, and the second multiplication is done in G. More generally, if $\mathbf{a} = (a_1, \ldots, a_k) \in \mathbb{Z}^k$, and $\mathbf{g} = (g_1, \ldots, g_k) \in G$, then $\sum_{i \in [k]} a_i g_i = \langle \mathbf{a}, \mathbf{g} \rangle = \langle \rho_G(\mathbf{a}), \mathbf{g} \rangle$.*

3.2   Generic Group Model

We start by formally defining the generic group model.

Definition 10 (Generic Group Model [Sho97]). *Let $(G, \odot)$ be any group of size N and let S be any set of size at least N. The generic group oracle $\mathcal{I}[G \mapsto S]$ (or simply $\mathcal{I}$) is as follows. At first an injective random function $\sigma \colon G \mapsto S$ is chosen, and two type of queries are answered as follows.*

– **Type 1: Labeling Queries.** *Given $g \in G$ oracle returns $\sigma(g)$.*
– **Type 2: Addition Queries.** *Given y_1, y_2, if there exists x_1, x_2 such that $\sigma(x_1) = y_1$ and $\sigma(x_2) = y_2$, it returns $\sigma(x_1 \odot x_2)$. Otherwise it returns $\perp$.*

Definition 11. *[Generic Algorithms in Generic Group Model] Let $A^{\mathcal{I}}$ be an algorithm (or a set of interactive algorithms $A = \{A_1, A_2, \dots\}$) accessing the generic group oracle $\mathcal{I}[G \mapsto S]$. We call $A^{\mathcal{I}}$ generic if it never asks any query (of the second type) that is answered as $\perp$. Namely, only queries are asked for which the labels are previously obtained.*

Remark 12 (Family of Groups). A more general definition allows generic oracle access to a *family* of groups $\{G_1, G_2, \dots\}$ in which the oracle access to each group is provided separately when the index i of G_i is also specified as part of the query and the size of the group G_i is known to the parties. Our negative result of Sect. 3 directly extends to this model as well. We use the above "single-group" definition for sake of simplicity.

Remark 13 (Stateful vs Stateless Oracles and the Multi-Party Setting). Note that in the above definition we used a *stateless* oracle to define the generic group oracle, and we separated the generic nature of the oracle itself from *how* it is used by an algorithm $A^{\mathcal{I}}$. In previous work (e.g., Shoup's original definition [Sho97]) *a stateful* oracle is used such that: it will answer addition queries *only* if the two labels are already *obtained* before.[6]

Note that for "one party" settings in which $A^{\mathcal{I}}$ is a single algorithm, $A^{\mathcal{I}}$ "knows" the labels that it has already obtained from the oracle $\mathcal{I}$, and so w.l.o.g. $A^{\mathcal{I}}$ would never ask any addition queries unless it has previously obtained the labels itself. However, in the multi-party setting, a party might not know the set of labels obtained by other parties. A stateful oracle in this case might reveal some information about other parties' oracle queries if the oracle does *not* answer a query (y_1, y_1) (by returning $\perp$) just because the labels for y_1, y_2 are not obtained so far.

Remark 14 (Equivalence of Two Models for Sparse Encodings). If the encoding of G is sparse in the sense that $|S|/|G| = n^{\omega(1)}$ where n is the security parameter, then the probability that any party could query a correct label before it being returned by oracle through a labeling (type 1) query is indeed negligible. So in this case any algorithm (or set of interactive algorithms) $A^{\mathcal{I}}$ would have a

---

[6] So the oracle might return $\perp$ even if the two labels are in the range of $\sigma(G)$.

behavior that is statistically close to a generic algorithm that would never ask a label in an addition query unless that label is previously obtained from the oracle. Therefore, if $|S|/|G| = n^{\omega(1)}$, we can consider $A^{\mathcal{I}}$ to be an *arbitrary* algorithm (or set of interactive algorithms) in the generic group model $\mathcal{I}$. The execution of A would be statistically close to a "generic execution" in which $A^{\mathcal{I}}$ never asks any label before obtaining it.

In light of Remarks 13 and 14, for simplicity of the exposition we will always assume that the encoding is sparse $|S|/|G| = n^{\omega(1)}$ and so all the generic group model are automatically (statistically close to being) generic.

Theorem 15 (Theorem 1 Formalized). *Let G be any abelian group of size at most $2^{\mathrm{poly}(n)}$. Let O be an obfuscator in the generic group model $\mathcal{I}[G \mapsto S]$ where the obfuscation of any circuit followed by execution of the obfuscated code (jointly) form a generic algorithm. If O is an ϵ-approximate VBB obfuscator in the generic group model $\mathcal{I}[G \mapsto S]$ for poly-size circuits, then for any $\delta = 1/\mathrm{poly}(n)$ there exists an $(\epsilon + \delta)$-approximate VBB obfuscator $\hat{O}$ for poly-size circuits in the plain model.*

Remark 16 (Size of G). Note that if a $\mathrm{poly}(n)$-time algorithm accesses (the labels of the elements of) some group G, it implicitly means that G is at most of $\exp(n)$ size so that its elements could be names with $\mathrm{poly}(n)$ bit strings. We chose, however, to explicitly mention this size requirement $|G| \le 2^{\mathrm{poly}(n)}$ since this upper bound plays a crucial role in our proof for general abelian groups compared to the special case of finite fields.

Remark 17 (Sparse Encodings). If we assume a sparse encoding i.e., $|S|/|G| = n^{\omega(1)}$ (as e.g., is the case in [Pas15] and almost all prior work in generic group model) in Theorem 15 we no longer need to explicitly assume that the obfuscation followed by execution of obfuscated code are in generic form; see Remark 14.

Since [BP13] showed that (assuming TDPs) there is no $(1/2 - 1/\mathrm{poly})$-approximate VBB obfuscator in the plain-model for general circuits, the following corollary is obtained by taking $\delta = \epsilon/2$.

Corollary 18. *If TDPs exist, then there exists no $(1/2 - \epsilon)$-approximate VBB obfuscator O for general circuits in the generic group model for any $\epsilon = 1/\mathrm{poly}(n)$, any finite abelian group G and any label set S of sufficiently large size $|S|/|G| = n^{\omega(1)}$. The result would hold for labeling sets S of arbitrary size if the execution of the obfuscator O followed by the execution of the obfuscated circuit $O(C)$ form a generic algorithm.*

Now we formally prove Theorem 15. We will first describe the algorithm of the obfuscator in the plain model, and then will analyze its properties.

Notation and w.l.o.g. Assumptions. Using Theorem 7 w.l.o.g. we assume that our abelian group G is isomorphic to the additive direct product group $\mathbb{Z}_{p_1^{\alpha_1}} \times \cdots \times \mathbb{Z}_{p_d^{\alpha_d}}$ where p_i's are prime. Let $e_i \in G$ be the vector that is 1 in the i'th coordinate and zero elsewhere. Note that $\{e_1, \ldots, e_k\}$ generates G. We can always assume that the first d labels obtained by O are the labels of $e_1, \ldots, e_d$ and these labels are explicitly passed to the obfuscated circuit $B = O(C)$. Let $k = \text{poly}(n)$ be an upper bound on the running time of the obfuscator O for input C which in turn upper bounds the number of labels obtained during the obfuscation (including the the the d labels for $e_1, \ldots, e_d$). We also assume w.l.o.g. that the obfuscated code never asks any type one (i.e., labeling) oracle queries since it can use the label for $e_1, \ldots, e_d$ to obtain labels of any arbitrary $g = a_1 e_1 + \cdots + a_d e_d$ using a polynomial number of addition (i.e., type two) oracle queries. For $\sigma(g) = s$, $a \in \mathbb{Z}$, and $t = \sigma(a \cdot g)$ we abuse the notation and denote $a \cdot s = t$.

The Construction. Even though the output of the obfuscator is always an actual circuit, we find it easier to first describe how the obfuscator $\widehat{O}$ generates some "data" $\widehat{B}$, and then we will describe how to use $\widehat{B}$ to execute the new obfuscated circuit in the plain model. For simplicity we use $\widehat{B}$ to denote the obfuscated circuit.

How to Obfuscate. *The new obfuscator $\widehat{O}$.* The new obfuscator $\widehat{O}$ uses lazy evaluation to simulate the labeling $\sigma(\cdot)$ oracle. For this goal, it holds a set Q_σ of the generated labels. For any new labeling query $g \in G$ if $\sigma(g) = s$ is already generated it returns s. Otherwise it chooses an unused label s from S uniformly at random and adds the mapping $(g \rightarrow s)$ to Q_σ and returns s. For an addition query (s_1, s_2) it first finds g_1, g_2 such that $\sigma(g_1) = s_1$ an $\sigma(g_2) = s_2$ (which exist since the algorithm that calls the oracle is in generic form) and gets $g = g_1 + g_2$. Now $\widehat{O}$ proceeds as if g is asked as a labeling query and returns the answer. The exact steps of $\widehat{O}$ are as follows.

1. *Emulating obfuscation.* $\widehat{O}$ emulates $O^{\mathcal{I}}(C)$ to get circuit B.
2. *Learning phase 1 (heavy queries):* Set $Q_B = \varnothing$. For $i \in [d]$ let $t_i = \sigma(e_i)$ be the label of $e_i \in G$ which is explicitly passed to B by the obfuscator $O(C)$ and $T = (t_1, \ldots, t_d)$ at the beginning. The length of the sequence T would increase during the steps below but will never exceed k. Choose m at random from $\ell = \lceil \lceil 3 \cdot k \cdot \log(|G|)/\delta \rceil \rceil$. For $i = 1, \ldots, m$ do the following:
   – Choose x_i as a random input for B. Emulate the execution of B on x_i using the (growing) set Q_σ of partial labeling for the lazy evaluation of labels. Note that as we said above, w.l.o.g. B only asks addition (i.e., type two) oracle queries. Suppose B (executed on x_i) needs to answer an addition query (s_1, s_2). If either of the labels $u = s_1$ or $u = s_2$ is *not* already obtained during the learning phase 1 (which means it was obtained during the initial obfuscation phase) append u to the sequence T of discovered labels by $T := (T, u)$. Using induction, it can be shown that for any addition query asked during learning phase 1, at the time of being asked, we would know that the answer to this query will be of the form $\sum_{i \in [k]} a_i \cdot t_i$ for integers a_i. Before seeing why this is the case, let $\mathbf{a}_i = (a_{i,1}, \ldots, a_{i,k})$ be the vector of integer coefficients (of the labels $t_1, t_2, \ldots$) for the answer s that is returned to the i'th query of learning phase 1. We add $(\mathbf{a}_i \rightarrow s)$

to Q_B for the returned label. To see why such vectors exist, let (s_1, s_2) be an addition query asked during this phase, and let $s \in \{s_1, s_2\}$. If the label s is obtained previously during learning phase 1, then the linear form $s = \sum_{i \in [k]} a_i \cdot t_i$ is already stored in Q_B. On the other hand, if s is a new label discovered during an addition (i.e., type two) oracle query which just made $T = (t_1, \ldots, t_{j-1}, t_j = s)$ have length j, then $s = a_i \cdot t_i$ for $a_j = 1$. Finally, if the linear forms for both of (s_1, s_2) in an addition oracle query are known, the linear form for the answer s to this query would be the summation of these vectors.[7]

3. *Learning phase 2 (zero vectors)*: This step does not involve executing B anymore and only generates a set $W = W(Q_B) \subseteq G^k$ of polynomial size. At the beginning of this learning phase let $W = \varnothing$. Then for all $(\mathbf{a}_1 \to s_1) \in Q_B$ and $(\mathbf{a}_2 \to s_2) \in Q_B$, if $s_1 = s_2$, let $\mathbf{a} = \mathbf{a}_1 - \mathbf{a}_2$, and add $\rho_G(\mathbf{a})$ to W where $\rho_G(\mathbf{a})$ is defined in Lemma 9.

4. The output of the obfuscation algorithm will be $\widehat{B} = (B, Q_B, W, T, r)$ where T is the current sequence of discovered labels $(t_1, t_2, \ldots)$ as described in Lemma 9, and r is a sufficiently large sequence of random bits that will be used as needed when we run the obfuscated code $\widehat{B} = (B, Q_B, W, T, r)$ in the plain model.[8]

How to Execute. In this section we describe how to execute $\widehat{B}$ on an input x using (B, Q_B, W, T, r).[9] Before describing how to execute the obfuscated code, we need to define the following algebraic problem.

Definition 19. *[Integer Solutions to Linear Equations over Abelian Groups (iLEAG)] Let G be a finite abelian group. Suppose we are given G (e.g., by describing its decomposition factors according to Theorem 7) an $n \times k$ matrix A with components from G and a vector $\mathbf{b} \in G^k$. We want to find an integer vector $\mathbf{x} \in \mathbb{Z}^n$ such that $\mathbf{x}A = \mathbf{b}$.*

Remark 20 (Integer vs. Ring Solutions). Suppose instead of searching for an integer vector solution $\mathbf{x} \in \mathbb{Z}^n$ we would ask to find $\mathbf{x} \in G^n$ and define multiplication in G according to Definition 8 and call this problem G-LEAG. Then any solution to iLEAG can be directly turned into a solution for G-LEAG by

---

[7] Note that although the sequence T grows as we proceed in learning phase 1, we already now that this sequence will not have length more than d since all of these labels that are discovered while executing the obfuscated code has to be generated by the obfuscator, due to the assumption that the sequential execution of the obfuscator followed by the obfuscated code is in generic form. Therefore we can always consider $\mathbf{a}_i$ to be of dimension k.

[8] Note that even though $W(Q_B)$ could always be derived from Q_B, and even T could be derived from an *ordered* variant of Q_B (in which the order in which Q_B has grown is preserved) we still choose to explicitly represent these elements in the obfuscated $\widehat{B}$ to ease the description of the execution of $\widehat{B}$.

[9] Note that we do not have access to the set Q_σ that was used for consistent lazy evaluation of $\sigma(\cdot)$.

mapping any integer coordinate x_i of $\mathbf{x}$ into G by mapping $\rho_G(x_i)$ of Lemma 9. The converse is true also for $G = \mathbb{Z}_n$, since any $g \in \mathbb{Z}_n$ is also in $\mathbb{Z}$ and it holds that $\rho_G(g) = g \in G$. However, the converse is not true in general for general abelian groups, since there could be members of G that are not in the range of $\rho_G(\mathbb{Z})$. For example let $G = \mathbb{Z}_{p^2} \times \mathbb{Z}_p$ for prime $p > 2$ and let $g = (2,1)$. Note that there is no integer a such that $a \mod p^2 = 2$ but $a \mod p = 1$.

Executing $\widehat{B}$. The execution of $\widehat{B} = (B, Q_B, W, T, r)$ on x will be done *identically* to to the "next" execution during the learning phase 1 of the obfuscation (as if x is the $(m+1)$'st execution of this learning phase) and even the sets $Q_B, W = W(Q_B)$ will grow as the execution proceeds, with the *only* difference described as follows.[10] Suppose we want to answer an addition (i.e., type two) oracle query (s_1, s_2) where for $b = \{1,2\}$ we inductively know that $s_b = \sum_{i \in [k]} a_{b,i} \cdot t_i$. For $b = \{1,2\}$ let $\mathbf{a}_b = (a_{b,1}, \ldots, a_{b,k})$ and let $\mathbf{a} = \mathbf{a}_1 + \mathbf{a}_2$.

- Do the following for all $(\mathbf{b} \to s) \in Q_B$. Let $\mathbf{c} = \mathbf{a} - \mathbf{b}$ and let $\overline{\mathbf{c}} = \rho_G(\mathbf{c}) \in G^k$ as defined in Lemma 9. Let A be a matrix whose rows consists of all vectors in W. Run the polynomial time algorithm presented in the full version of this paper (based on an algorithm of [McC90] for $G = \mathbb{Z}_n$) to see if there is any integer solution $\mathbf{v}$ for $\mathbf{v}A = \overline{\mathbf{c}}$ as an instance of the iLEAG problem defined in Definition 19. If an integer solution $\mathbf{v}$ exists, then return s as the result (recall $(\mathbf{b} \to s) \in Q_B$), break the loop, and continue the execution of $\widehat{B}$. If the loop ended and no such $(\mathbf{b} \to s) \in Q_B$ was found, choose a random label s not in Q_B as the answer, add $(\mathbf{a} \to s)$ to Q_B and continue.

Completeness and the Soundness. In this section we prove the completeness and soundness of the construction of Sect. 3.2.

Size of S. In the analysis below, we will assume w.l.o.g. that the set of labels S has superpolynomial size $|S| = n^{\omega(1)}$. This would immediately hold if the labeing of G is sparse, since it would mean even $|S|/|G| \geq n^{\omega(1)}$. Even if the labeling is not sparse, we will show that w.l.o.g. we can assume that G itself has super-polynomial size (which means that S will be so too). That is because otherwise all the labels in G can be obtained by the obfuscator, the obfuscated code, and the adversary and we will be back to the plain model. More formally, for this case Theorem 15 could be proved through a trivial construction in which the new obfuscator simply generates all the labels of G and plants all of them in the obfuscated code and they will be used by the obfuscated algorithm. More precisely, when the size of G (as a function of security parameter n) is neither of polynomial size $|G| = n^{O(1)}$ nor super-polynomial size $|G| = n^{\omega(1)}$ we can still choose a sufficiently large polynomial $\gamma(n)$ and generate all labels of G when $|G| < \gamma(n)$, and otherwise use the obfuscation of Sect. 3.2.

---

[10] We even allow new labels t_i to be discovered during this execution to be appended to T, even though that would indirectly lead to an abort!

Completeness: Approximate Functionality. Here we prove the the following claim.

Claim. Let $\widehat{B} = (B, Q_B, W, T, r)$ be the output of the obfuscator $\widehat{O}$ given input circuit C with input length α. If we run $\widehat{B}$ over a *random* input according to the algorithm described in Sect. 3.2, then it holds that

$$\Pr_{x \leftarrow \{0,1\}^\alpha, \widehat{B} \leftarrow \widehat{O}(C)} \left[\widehat{B}(x) \neq C(x) \right] \leq \Pr_{x \leftarrow \{0,1\}^\alpha, B \leftarrow O^{\mathcal{I}[G \mapsto S]}(C)} \left[B^{\mathcal{I}[G \mapsto S]}(x) \neq C(x) \right] + \delta$$

over the randomness of $\mathcal{I}[G \mapsto S]$, random choice of x and the randomness of the obfuscators.

Proof. As a mental experiment, suppose we let the learning phase 1 always runs for exactly $\ell + 1 = 1 + \lceil \lceil 3 \cdot k \cdot \log(|G|)/\delta \rceil \rceil$ rounds but only derive the components $(Q_B, W(Q_B), T)$ based on the first m executions. Now, let x_i be the random input used in the i'th execution and y_i be the output of the i'th emulation execution the learning phase 1. Since all the executions of the learning phase 1 are perfect simulations, for every $i \in [\ell]$, and in particular $i = m$, it holds that

$$\Pr[B^{\mathcal{I}[G \mapsto S]}(x) \neq C(x)] = \Pr[y_i \neq C(x)]$$

where probability is over the choice of inputs x, x_i as well as all other randomness in the system. Thus, to prove claim 3.2 it will suffice to prove that

$$|\Pr[y_i \neq C(x)] - \Pr[\widehat{B}(x_i) \neq C(x)]| < \delta.$$

We will indeed do so by bounding the statistical distance between the execution of $\widehat{B}$ vs the $m + 1$'st execution of the learning phase 1 over the same input x_i. Here we will rely on the fact that m is chosen at random from $[\ell]$.

Claim. For random $[\ell]$ the statistical distance between the $m + 1$'st execution of the learning phase 1 (which we call B') and the execution of $\widehat{B}$ over the *same* input x_i is $\leq 2\delta/3 + \mathrm{negl}(n)$.

To prove the above claim we will define three type of bad events over a joint execution of $B' = B_{m+1}$ and $\widehat{B}$ when they are done concurrently and using the same random tapes (and even the input x_i). We will then show that (1) as long as these bad events do not happen the two executions proceed identically, and (2) the total probability of these bad events is at most $2\delta/3 + \mathrm{negl}(n)$. In the following we suppose that the executions of B' and $\widehat{B}$ (over the same random input) has proceeded identically so far. Suppose we want to answer an addition (i.e., type two) oracle query (s_1, s_2) where for $b = \{1, 2\}$ we inductively know that $s_b = \sum_{i \in [k]} a_{b,i} \cdot t_i$. Several things could happen:

- If the execution of $\widehat{B}$ finds $(\mathbf{b} \to s) \in Q_B$ such that when we take $\mathbf{c} = \mathbf{a} - \mathbf{b}$ and let $\overline{\mathbf{c}} = \rho_G(\mathbf{c}) \in G^k$ and let A be a matrix whose rows are vectors in (the current) W, there is an integer solution $\mathbf{v}$ to the iLEAG instance $\mathbf{v}A = \overline{\mathbf{c}}$.

If this happens the execution of $\widehat{B}$ will use $\mathbf{b}$ as the answer. We claim that this is the "correct" answer as B' would also use the same answer. This is because by the definition of W and Lemma 9 for all $\mathbf{w} \in W$ it holds that $\mathbf{w} = (w_1, \ldots, w_k)$ is a "zero vector in G^k" in the sense that summing the (currently discovered labels in) T with coefficients $w_1, \ldots, w_k$ (and multiplication defined according to Definition 8) will be zero. As a result, $\mathbf{v}A = \overline{\mathbf{c}}$ which is a linear combination of vectors in W with integer coefficients will also be a zero vector. Finally, by another application of Lemma 9 it holds that $(c_1, \ldots, c_k) = \mathbf{c} = \mathbf{a} - \mathbf{b}$ is a "zero vector in $\mathbb{Z}^k$" in the sense that summing the (currently discovered labels in) T with *integer* coefficients $c_1, \ldots, c_k$ (and multiplication defined according to Definition 8) will also be zero. Therefore the answer to the query defined by vector $\mathbf{a}$ is equal to the answer defined by vector $\mathbf{b}$ which is s.

– If the above does not happen (and no such $(\mathbf{b} \rightarrow s) \in Q_B$ is found) then either of the following happens. Suppose the answer returned for (s_1, s_2) in execution of B' is s':

- **Bad event E_1:** s' is equal to one of the labels in Q_B. Note that in this case the executions will diverge because $\widehat{B}$ will choose a random label.
- **Bad event E_2:** s' is equal to one of the labels discovered in the emulation of $O^{\mathcal{I}}(C)$ (but not present in the current Q_B).
- **Bad event E_3:** s' is a new label, but the label chosen by $\widehat{B}$ is one of the labels used in the emulation of $O^{\mathcal{I}}(C)$. (Note that in this case the execution of $\widehat{B}$ will not use any previously used labels in Q_B.

It is easy to see that as long as none of the events E_1, E_2, E_3 happen, the execution of B' and $\widehat{B}$ proceeds statistically the same. Therefore, to prove Claim 3.2 and so Claim 3.2 it is sufficient to bound the probability of the events E_1, E_2, E_3 as we do below.

Claim. $\Pr[E_3] < \mathrm{negl}(n)$.

Proof. This is because (as we described at the beginning of this subsection above) the size of S is $n^{\omega(1)}$ but the number of labels discovered in the obfuscation phase is at most $k = \mathrm{poly}(n)$. Therefore the probability that a random label from S after excluding labels in Q_B (which is also of polynomial size) hits one of at most k possible labels is $\leq k/(|S| - |Q_B|) = \mathrm{negl}(n)$. Therefore, the probability that E_3 happens for any of the oracle quries in the execution of $\widehat{B}$ is also $\mathrm{negl}(n)$.

Claim. $\Pr[E_2] < \delta/(3 \log |G|) < \delta/3$.

Proof. We will prove this claim using the randomness of $m \in [\ell]$. Note that every time that a label u is discovered in learning phase 1, this label u cannot be discovered "again", since it will be in Q_B from now on. Therefore, the number of possible indexes of $i \in [\ell]$ such that during the i'th execution of the learning phase 1 we discover a label out of Q_B is at most k. Therefore, over the randomness of $m \leftarrow [\ell]$ the probability that the $m + 1$'st execution discovers any new labels (generated in the obfuscation phase) is at most $k/\ell \leq \delta/(3 \log |G|)$.

Claim. $\Pr[E_1] < \delta/3$.

Proof. Call $i \in [\ell]$ a bad index, if event E_3 happens conditioned on $m = i$ during the execution of B' (which is the $(m+1)$'s execution of learning phase 1). Whenever E_3 happens at any moment, it means that the vector $\overline{c}$ is not currently in $W(Q_B)$, *but it will* be added W just after this query is made. We will show (Lemma 21 below) that the size of $\mathsf{span}_{\mathbb{Z}}(W)$ will at least double after this oracle query for some set $\mathsf{span}_{\mathbb{Z}}(W)$ that depends on W and that $\mathsf{span}_{\mathbb{Z}}(W) \subseteq G^k$, and so $|\mathsf{span}_{\mathbb{Z}}(W)| \leq |G|^k$. As a result the number of bad indexes i will be at most $\log |G|^k = k \log |G|$. Therefore, over the randomness of $m \in [\ell]$ the probability that $m + 1$ is a bad index is at most $k \log |G|/\ell \leq \delta/3$.

Lemma 21. *Let* $W \subseteq G^k$ *for some abelian group* G. *Let* $\mathsf{span}_{\mathbb{Z}}(W) = \{\sum_{\mathbf{w} \in W} a_{\mathbf{w}} \mathbf{w} \mid a_{\mathbf{w}} \in \mathbb{Z}\}$ *be the module spanned by* W *using integer coefficients. If* $\overline{c} \notin \mathsf{span}_{\mathbb{Z}}(W)$, *then it holds that*

$$|\mathsf{span}_{\mathbb{Z}}(W \cup \{\overline{c}\})| \geq 2 \cdot |\mathsf{span}_{\mathbb{Z}}(W)|.$$

Proof. Let $A = \mathsf{span}_{\mathbb{Z}}(W)$ and let $B = \{\overline{c} + \mathbf{w} \mid \mathbf{w} \in \mathsf{span}_{\mathbb{Z}}(W)\}$ be A shifted by $\overline{c}$. It holds that $|A| = |B|$ and $A \cup B \subset \mathsf{span}_{\mathbb{Z}}(W \cup \{\overline{c}\})$. It also holds that $A \cap B = \varnothing$, because otherwise then we would have: $\exists i, j : \mathbf{w} + \overline{c} = \mathbf{w}'$ for $\mathbf{w}, \mathbf{w}' \in \mathsf{span}_{\mathbb{Z}}(W)$ which would mean $\overline{c} = \mathbf{w} - \mathbf{w}' \in \mathsf{span}_{\mathbb{Z}}(W)$ which is a contradiction. Therefore $|\mathsf{span}_{\mathbb{Z}}(W \cup \{\overline{c}\})| \geq |A| + |B| = 2 \cdot |\mathsf{span}_{\mathbb{Z}}(W)|$.

Soundness: VBB Simulatability. To derive the soundness we apply Lemma 5 as follows. O will be the obfuscator in the ideal model and $\widehat{O}$ will be our obfuscator in the plain model where $z' = Q_B, W, T, r$ is the extra information output by $\widehat{O}$. The algorithm U will be a similar algorithm to $\widehat{O}$ but only during its learning phase 1 and 2 starting from an already obfuscated B. However, U will continue generating z' using the actual oracle $\mathcal{I}[G \mapsto S]$ instead of inventing the answers through lazy evaluation. Since the emulation of the oracle during the learning phases, and that all of Q_B, W, T, R could be obtained by only having B (and no secret information about the obfuscation phase are not needed) the algorithm U also has the properties needed for Lemma 5.

Remark 22 (General abelian vs $\mathbb{Z}_n$). Note that when $G = \mathbb{Z}_n$ is cyclic, the mapping $\rho_G \colon \mathbb{Z} \mapsto G$ of Lemma 9 will be equivalent to simply mapping every $a \in \mathbb{Z}$ to $(a \mod n) \in G$. Therefore, Definition 8 generalizes the notion of $\mathbb{Z}_n$ as a ring to general abelian groups, since the multiplication $x \cdot y \mod n$ in $\mathbb{Z}_n$ is the same as a multiplication in which x is interpreted from $\mathbb{Z}$ (as in Definition 8) which is equivalent to doing the multiplication inside G according to by Lemma 9.

3.3   Degree-$O(1)$ Graded Encoding Model

We adapt the following definition from [Pas15] restricted to the degree-d polynomials. For simplicity, as in [Pas15] we also restrict ourselves to the setting in which only the obfuscator generates labels and the obfuscated code only does zero tests, but the proof directly extends to the more general setting of [BGK+14, BR14]. We also use only one finite ring R in the oracle (whose size could in fact depend on the security parameter) but our impossibility result extends to any sequence of finite rings as well.

Definition 23 (Degree-d Ideal Graded Encoding Model). *The oracle* $\mathcal{M}_R^d = (\text{enc}, \text{zero})$ *is stateful and is parameterized by a ring R and a degree d and works in two phases. For each $l \in [d]$, the oracle $\text{enc}(\cdot, l)$ is a random injective function from the ring R to the set of labels $S = \{0, 1\}^{3 \cdot |R|}$.*

1. *Initialization phase: In this phase the oracle answers $\text{enc}(v, l)$ queries and for each query it stores (v, l, h) in a list $\mathcal{L}_O$.*
2. *Zero testing phase: Suppose $p(\cdot)$ is a polynomial whose coefficients are explicitly represented in R and its monomials are represented with labels $h_1, \ldots, h_m$ obtained through $\text{enc}(\cdot, \cdot)$ oracle in phase 1. Given any such query $p(\cdot)$ the oracle answers as follows:*
   (a) *If any h_i is not in $\mathcal{L}_O$ (i.e., it is not obtained in phase 1) return* false.
   (b) *If the degree of $p(\cdot)$ is more than d then return* false.
   (c) *Let $(v_i, l_i, h_i) \in \mathcal{L}_O$. If $p(v_1, \ldots, v_m) = 0$ return* true; *otherwise* false.

Remark 24. Remarks 13 and 14 regarding the stateful vs stateless oracles and the sparsity of the encoding in the context of generic group model apply to the graded encoding model as well. Therefore, as long as the encoding is sparse (which is the case in the definition above whenever $|R|$ is of size $n^{\omega(1)}$) the probability of obtaining any valid label $h = \text{enc}(v, l)$ through any polynomial time algorithm without it being obtained from the oracle previously (by the same party or another party) becomes negligible, and so the model remains essentially equivalent (up to negligible error) even if the oracle does not keep track of which labels are obtained previously through $\mathcal{L}_O$.

We prove the following theorem generalizing a similar result by Pass and shelat [Pas15] who proved this for any finite field; here we prove the theorem for any finite ring.

Theorem 25. *Let R be any ring of size at most $2^{\text{poly}(n)}$. Let O be any ϵ-approximate VBB obfuscator for general circuits in the ideal degree-d graded encoding model $\mathcal{M}_R^d$ for $d = O(1)$ where the initialization phase of $\mathcal{M}_R^d$ happens during the obfuscation phase. Then for any $\delta = 1/\text{poly}(n)$ there is an $(\epsilon + \delta)$-approximate obfuscator $\widehat{O}$ for poly-size circuits in the plain model.*

See Sect. 3.3 for the proof of Theorem 25. As in previous sections, the following corollary is obtained from Theorem 25 by taking $\delta = \epsilon/2$.

Corollary 26. *If TDPs exist, then there exists no $(1/2 - \epsilon)$-approximate VBB obfuscator O for general circuits in the ideal degree-d graded encoding model $\mathcal{M}_R^d$ for any finite ring R of at most exponential size $|R| \leq 2^{\text{poly}(n)}$ and any constant degree d, assuming the initialization phase of $\mathcal{M}_R^d$ happens during the obfuscation phase.*

[Pas15] state their theorem in a more general model where a sequence of fields of growing size are accessed. For simplicity, we state a simplified variant for simplicity of presentation where only one ring is accessed but we let the size of ring R to depend on the security parameter n. Our proof follows the footsteps of [Pas15] but will deviate from their approach when $R \neq \mathbb{Z}_p$ by using some of the ideas employed in Sect. 3.

Proving Theorem 25. Here we sketch the proof assuming the reader is familiar with the proof of Theorem 15 from previous section. The high level structure of the proof remains the same.

Construction. The new obfuscator $\widehat{O}$ will have these phases:

- *Emulating obfuscation.* $\widehat{O}$ emulates $O^{\mathcal{M}_R^d}(C)$ to get circuit B.
- *Learning heavy subspace of space of zero vectors*: The learning phase here will be rather simpler than those of Sect. 3.2 and will be just one phase. Here we repeat the learning phase m times where m is chosen at random from $\ell = \lceil \lceil k \cdot \log(|G|)/\delta \rceil \rceil$. The variables W and T will be the same as in Sect. 3.2 with the difference that W will consist of the vector of coefficients for all polynomials whose zero test answer is true.
- The returned obfuscated code will be $\widehat{B} = (B, W, T, r)$ where r is again the randomness needed to run the obfuscated code.
- *Executing $\widehat{B}$.* To execute $\widehat{B}$ on input x, we answer zero test queries as follows. For any query vector (of coefficients) $\mathbf{a}$ we test whether $\mathbf{a} \in \mathsf{span}_{\mathbb{Z}}(W)$.[11] If $\mathbf{a} \in \mathsf{span}_{\mathbb{Z}}(W)$ then return true, otherwise return false.

Completeness and Soundness.

- The completeness follows from the same argument given for the soundness of Construction 3.2. Namely, the execution of $\widehat{B}$ is identical to the execution of the $m + 1$'s learning phase (as if it exists) up to a point where we return a wrong false answer to an answer that is indeed a zero polynomial. (Note that the converse never happens). However, when such event is about to happen, the size of $\mathsf{span}_{\mathbb{Z}}(W)$ will double. Since the size of $\mathsf{span}_{\mathbb{Z}}(W)$ is at most $|R|^k$, if we choose m at random from $[\ell]$ the probability of the bad event (of returning a wrong false in $m + 1$'st execution) is at most $k \log |R|/\ell = \delta$.
- The soundness follows from Lemma 5 similarly to the way we proved the soundness of the construction of Sect. 3.2.

Extension to Avoid Initialization. In Theorem 25 we have a restriction which says that the initialization phase must happen during the obfuscation phase only. We can extend the proof of Theorem 25 to the case that we don't have this restriction. This entails allowing the obfuscator O and the obfuscated circuit B to ask any type of query (be it initialization phase queries or zero-testing

---

[11] Note that we do *not* solve a system of equations in R and rather search only integer solutions to $\mathbf{x}W = \mathbf{a}$ as we did in Sect. 3.2.

queries) during their execution. The reason that we can avoid this restriction is that, whenever the obfuscated circuit B asks an initialization phase query $enc(v,l)$, we can treat it as a polynomial containing $v.enc(1,l)$ and using that we can find out whether we should answer this query randomly or using one of the previous labels. This is very similar to the method that we employed in the learning and execution phases of generic group model case.

Claim. Let R be any ring of size at most $2^{\text{poly}(n)}$. Let O be any ϵ-approximate VBB obfuscator for general circuits in the ideal degree-d graded encoding model $\mathcal{M}_R^d$ for $d = O(1)$, Then for any $\delta = 1/\text{poly}(n)$ there is an $(\epsilon + \delta)$-approximate obfuscator $\widehat{O}$ for poly-size circuits in the plain model.

Proof. Suppose that obfuscated circuit is B, and let $\{h_i = enc(v_i, l_i)\}_1^n$ be the obfuscator's queries. We already know that n is less than the running time of obfuscator. We might learn some pair of (h_i, v_i) during the learning phase.

Construction. The new ϵ-approximate obfuscator $\widehat{O}$ will have these phases:

- *Emulating obfuscation.* same as previous case.
- *Learning obfuscator's queries and heavy subspace of space of zero vectors*: We do exactly what we did in previous learning phase. Also if obfuscated circuit asked initialization phase queries, we memorize it.
- The returned obfuscated code will be $\widehat{B} = (B, W, T, r)$ where r is again the randomness needed to run the obfuscated code.
- *Executing $\widehat{B}$.* To execute $\widehat{B}$ on input x, we do as follows. If we saw query $enc(v,l)$: First we check, if we memorized query $enc(v,l)$ before, we answer it using memorized queries list otherwise we answer it randomly. Also we treat $enc(v,l)$ as a polynomial $v.enc(1,l)$. We answer zero test queries as follows. For any query vector (of coefficients) $\mathbf{a}$ we test whether $\mathbf{a} \in \text{span}_{\mathbb{Z}}(W)$.[12] If $\mathbf{a} \in \text{span}_{\mathbb{Z}}(W)$, return true, otherwise return false.

Completeness and Soundness.

- The proof of completeness is same as previous case. The only difference is that here we need to be sure that we answer initialization phase query correctly (call it event E). Let j_i be the index such that we saw the query $enc(v_i, l_i)$ for the first time. E happens if we hit one of the index j_i. Since we chose m at random, we can always bound $pr(E)$ by choosing the right l.
- The soundness is same as previous case.

Remark 27. Note that our proof of Theorem 25 does not assume any property for the multiplication (even the associativity!) other than assuming that it is distributive. Distributivity is needed by the proof since we need to be able to conclude that the summation of the vectors of the coefficients of two zero polynomials is also the vector of the coefficients of a zero polynomial; the latter is implied by distributivity.

---

[12] Note that we do *not* solve a system of equations in R and rather search only integer solutions to $\mathbf{x}W = \mathbf{a}$ as we did in Sect. 3.2.

4   Impossibility of VBB in the Random TDP Model

In this section we formally prove Theorem 3 showing that any obfuscator O with access to a random trapdoor permutation oracle T can be transformed into a new obfuscator $\widehat{O}$ in the plain model (no access to an ideal oracle) with some loss in correctness. We start by defining the random trapdoor permutation model and TDP query tuples followed by the formalization of Theorem 3.

Definition 28 (Random Trapdoor Permutation). *For any security parameter n, a random trapdoor permutation (TDP) oracle T_n consists of three subroutines (G, F, F^{-1}) as follows:*

- *$G(\cdot)$ is a random permutation over $\{0,1\}^n$ mapping trapdoors sk to a public indexes pk.*
- *$F[pk](x)$: For any fixed public index pk, $F[pk](\cdot)$ is a random permutation over $\{0,1\}^n$.*
- *$F^{-1}[sk](y)$: For any fixed trapdoor sk such that $G(sk) = pk$, $F^{-1}[sk](\cdot)$ is the inverse permutation of $F[pk](\cdot)$, namely $F^{-1}[sk](F[pk](x)) = x$.*

Definition 29 (TDP query tuple). *Given a random TDP oracle $T_n = (G, F, F^{-1})$, a TDP query tuple consists of three query-answer pairs $(V_G, V_F, V_{F^{-1}})$ where:*

- *$V_G = (sk, pk)$ represents a query to G on input sk and its corresponding answer pk*
- *$V_F = ((pk, x), y)$ represents a query to $F[pk]$ on input x and its corresponding answer y*
- *$V_{F^{-1}} = ((sk, y), x')$ represents a query to $F^{-1}[sk]$ on y and its corresponding answer x'*

We say that a TDP query tuple $(V_G, V_F, V_{F^{-1}})$ is consistent if $x = x'$.

Definition 30 (Partial TDP query tuple). *A partial TDP query tuple is one where one or more of the elements of the tuple are unknown and we denote the missing elements with a period. For example, we say a query set Q contains a TDP query tuple $(\cdot, V_F, \cdot)$ if it contains the query-answer pair $V_F = ((pk, x), y)$ but is missing the query-answer pairs $V_G = (sk, pk)$ and $V_{F^{-1}} = ((sk, y), x')$.*

Theorem 31 (Theorem 3 formalized). *Let O be an ϵ-approximate obfuscator for poly-size circuits in the random TDP oracle model. Then, for any $\delta = 1/\operatorname{poly}(n)$, there exists an $(\epsilon + \delta)$-approximate obfuscator $\widehat{O}$ in the plain model for poly-size circuits.*

Before proving Theorem 31, we state a corollary of this theorem to rule out approximate VBB obfuscation in the ideal TDP model. Since [BP13] showed that assuming TDPs exist, $(1/2 - 1/\operatorname{poly})$-approximate VBB obfuscator does not exist for general circuits, we obtain the following corollary by taking $\delta = \epsilon/2$.

Corollary 32. *If TDPs exist, then there exists no $(1/2 - \epsilon)$-approximate VBB obfuscator O for general circuits in the ideal random TDP model for any $\epsilon = 1/\operatorname{poly}(n)$.*

The proof of Theorem 31 now follows in the next two sections. We will first describe the algorithm of the obfuscator in the plain model, and then will analyze its completeness and soundness.

4.1 The Construction

We first describe how the new obfuscator $\widehat{O}$ generates some data $\widehat{B}$, and then we will show how to use $\widehat{B}$ to run the new obfuscated circuit in the plain model. We also let $l_O, l_B = \operatorname{poly}(n)$, respectively, be the number of queries asked by the obfuscator O and the obfuscated code B to the random trapdoor permutation oracle T. Note that, for simplicity of exposition, we assume the adversary only asks the oracle for queries of size n (i.e. the domain of the permutations in T are of fixed size n). However, as mentioned in Sect. 1.2, we can easily extend the argument to handle $O(\log(n))$-size or $\omega(\log(n))$-size queries to T.

How to Obfuscate

The New Obfuscator $\widehat{O}$ in Plain Model. Given an ϵ-approximate obfuscator O in the random TDP model, we construct a plain-model obfuscator $\widehat{O}$ such that, given a circuit $C \in \{0,1\}^n$, works as follows:

1. *Emulation phase*: Emulate $O^T(C)$. Let Q_O represent the set of queries asked by O^T and their corresponding answers. We initialize $Q_O = \varnothing$. For every query q asked by $O^T(C)$, we would answer the query uniformly at random conditioned on the answers to previous queries.
2. *Canonicalize B*: Let the obfuscated circuit B be the output of $O(C)$. Modify B so that, before asking any query of the form $F^{-1}[sk](y)$, it would first ask $G(sk)$ to get some answer pk followed by $F^{-1}[sk](y)$ to get some answer x then finally asks $F[pk](x)$ to get the expected answer y.
3. *Learning phase*: Set $Q_B = \varnothing$. Let the number of iterations to run the learning phase be $m = 2l_B l_O / \delta$ where $l_B \le |B|$ represents the number of queries asked by B and $l_O \le |O|$ represents the number of queries asked by O. For $i = \{1, ..., m\}$:
   - Choose $x_i \xleftarrow{\$} D_{|C|}$
   - Run $B(x_i)$. For every query q asked by $B(x_i)$:
     - If $(q, a) \in Q_O \cup Q_B$ for some answer a, answer consistently with a
     - Otherwise, answer q uniformly at random and conditioned on the answers of previous related queries in $Q_O \cup Q_B$
     - Let a be the answer to q. If $(q, a) \notin Q_B$, add the pair (q, a) to Q_B
4. The output of the obfuscation algorithm will be $\widehat{B} = (B, Q_B, R)$ where $R = \{r_1, ..., r_{|B|}\}$ is a set of (unused) oracle answers that are generated uniformly at random.

How to Execute. To execute $\widehat{B}$ on an input x using (B, Q_B, R) we simply emulate $B(x)$. For every query q asked by $B(x)$, if $(q, a) \in Q_B$ for some a then return a. Otherwise, answer randomly with one of the answers a in R and add (q, a) to Q_B.

4.2 Completeness and Soundness

Completeness: Approximate Functionality. Consider two separate experiments (real and ideal) that construct the plain-model obfuscator exactly as described in Sect. 4.1 but differ when executing $\widehat{B}$. Specifically, in the real experiment, $\widehat{B}$ emulates $B(x)$ on a random input x using Q_B and R, whereas in the ideal experiment, we execute $\widehat{B}$ and answer $B(x)$'s queries using the actual oracle T instead. In essence, in the real experiment, we can think of the execution as $B^{\widehat{T}}(x)$ where $\widehat{T}$ is the TDP oracle simulated by $\widehat{B}$ using Q_B and R as the oracle's answers (without knowing Q_O, which is part of oracle T). We will contrast the real experiment with the ideal experiment and show that the statistical distance between these two executions is at most δ. In order to achieve this, we will identify the events that differentiate between the executions $B^T(x)$ and $B^{\widehat{T}}(x)$, and to that end we will make use of the following two lemmas:

Lemma 33 ([GKLM12]). *Let B be a canonical oracle-aided algorithm that asks t queries to a TDP oracle T. Let E_G be the event that B asks a query of the form $V_G = (sk, pk)$ after asking either query $V_F = ((pk, x), y)$ and/or $V_{F^{-1}} = ((sk, y), x)$ from the TDP query tuple $(V_G, V_F, V_{F^{-1}})$. Then $\Pr[E_G] \leq O(t^2/2^n)$.*

Lemma 34 ([GKLM12]). *Let B be an oracle-aided algorithm that asks t queries to a TDP T and let Q be the set of queries that B have issued. Then for any new query x, the answer is either (1) determined completely by Q or (2) is drawn from a distribution with a statistical distance of $O(t/2^n)$ away from the uniform distribution.*

Now let q be a new query that is being asked by $B^{\widehat{T}}(x)$. We present a case-by-case analysis of all possible queries to identify the cases that can cause discrepancies between the real and ideal experiments:

- **Case 1:** If q is determined by the queries in Q_B in the real experiment then it is also determined by Q_B in the ideal experiment.
- **Case 2:** If q is not determined by $Q_B \cup Q_O$ in the ideal experiment then it is also not determined by Q_B in the real experiment. In the ideal experiment the query will be answered randomly and consistently with respect to $Q_B \cup Q_O$ whereas in the real experiment the query will be answered randomly and consistently with respect to Q_B. By Lemma 34, the answers will be from a distribution that is statistically close to uniform.
- **Case 3:** If q is not determined by Q_B in the real experiment then, depending on the queries in Q_O, it may or may not be so the ideal experiment:
  - **Case 3a:** The query q is in Q_O. In that case, in the real experiment, the answer would be random whereas in the ideal experiment it would use the correct answer from Q_O.

- **Case 3b**: The query q is of type $V_G = (sk, pk)$ and the corresponding partial TDP query tuple $(., V_F, V_{F^{-1}})$ is in Q_O
- **Case 3c**: The query q is of type $V_F = ((pk, x), y)$ and the corresponding partial TDP query tuple $(V_G, ., V_{F^{-1}})$ is in Q_O
- **Case 3d**: The query q is of type $V_{F^{-1}} = ((sk, y), x)$ and the corresponding partial TDP query tuple $(V_G, V_F, .)$ is in Q_O
- **Case 3e**: The query q is of type $V_F = ((pk, x), y)$ and $V_G = (sk, pk)$ is in Q_B, but $V_{F^{-1}} = ((sk, y), x)$ is in Q_O
- **Case 3f**: The query q is of type $V_{F^{-1}} = ((sk, y), x)$ and $V_G = (sk, pk)$ is in Q_B, but $V_F = ((pk, x), y)$ is in Q_O
- **Case 3g**: The query q is of type $V_F = ((pk, x), y)$ and $V_{F^{-1}} = ((sk, y), x)$ is in Q_B, but $V_G = (sk, pk)$ is in Q_O
- **Case 3h**: The query q is of type $V_{F^{-1}} = ((sk, y), x)$ and $V_F = ((pk, x), y)$ is in Q_B, but $V_G = (sk, pk)$ is in Q_O

We note that the bad events that can cause any differences between the real and ideal experiments are case 2 and parts of case 3. For case 2, Lemma 34 ensures that this event happens with negligible probability. For case 3a, learning heavy queries would diminish the effect of this event. For cases 3b, 3e, and 3f, Lemma 33 ensures that this event happens with negligible probability since V_G was issued *after* V_F and/or $V_{F^{-1}}$ was asked. For cases 3c and 3d, the remaining query from the tuple would have been defined in Q_O and is thus captured during the learning of heavy queries. For case 3g, if V_G and $V_{F^{-1}}$ were asked during the emulation or learning phases, then V_F would also be defined and thus can be learned. However, if $V_{F^{-1}}$ was asked during the execution phase then, due the canonicalization of B, it would have to ask $V_G \in Q_O$ which reduces to case 3a. Similarly, for case 3h, due the canonicalization of B, we would have to ask $V_G \in Q_O$ and this reduces to case 3a once again.

For any x, define $E_k(x)$ to be the event that case k happens and let event $E(x) = (E_2(x) \vee E_{3a}(x) \vee E_{3b}(x) \vee E_{3e}(x) \vee E_{3f}(x))$. Assuming that event E does not happen, the output distributions of $B^T(x)$ and $B^{\widehat{T}}(x)$ are identical. More formally, the probability of correctness for $\widehat{O}$ is:

$$\Pr_x[B^{\widehat{T}}(x) \neq C(x)] = \Pr_x[B^{\widehat{T}}(x) \neq C(x) \wedge \neg E(x)] + \Pr_x[B^{\widehat{T}}(x) \neq C(x) \wedge E_1(x)]$$

$$\leq \Pr_x[B^{\widehat{T}}(x) \neq C(x) \wedge \neg E(x)] + \Pr_x[E(x)]$$

By the approximate functionality of O, we have that:

$$\Pr_x[O^T(C)(x) \neq C(x)] = \Pr_x[B^T(x) \neq C(x)] \leq \epsilon(n)$$

Therefore,

$$\Pr_x[B^{\widehat{T}}(x) \neq C(x) \wedge \neg E(x)] = \Pr_x[B^T(x) \neq C(x) \wedge \neg E(x)] \leq \epsilon$$

We are thus left to show that $\Pr[E(x)] \leq \delta$. By Lemma 34, $\Pr[E_2(x)] \leq \mathrm{negl}(n)$ and by Lemma 33, $\Pr[E_{3b} \vee E_{3e}(x) \vee E_{3f}(x)] \leq \mathrm{negl}(n)$ via a union bound.

The probability of event E_{3a} was already given in [CKP15], but for the sake of completeness we show our version of the analysis here. As a result, we get that $\Pr[E(x)] \leq \delta/2 + \mathrm{negl}(n) \leq \delta$.

Claim. It holds that $\Pr_x[E_{3a}(x)] \leq \delta/2$.

Proof. Let $(q_1, ..., q_{l_B})$ be the sequence of queries asked by $B^{\widehat{T}}(x)$ where $l_B \leq |B|$, and let $q_{i,j}$ be the jth query that is asked by $B^T(x_i)$ during the ith iteration of the learning phase. We define $E_{3a}^j(x)$ to be the event that the jth query of $B(x)$ is in Q_O but not in Q_B. We also define $p_{q,j}$ to be the probability that $q_j = q$ for any query q and $j \in [l_B]$. We can then write the probability of E_{3a} as follows:

$$\Pr_x[E_{3a}(x)] \leq \Pr_x[E_{3a}^1(x) \vee ... \vee E_{3a}^{l_B}(x)]$$

$$= \sum_{j=1}^{l_B} \Pr_x[\neg E_{3a}^1(x) \wedge ... \wedge \neg E_{3a}^{j-1}(x) \wedge E_{3a}^j(x)]$$

$$\leq \sum_{j=1}^{l_B} \sum_{q \in Q_O} \Pr_x[q_j = q \wedge (q_{1,j} \neq q \wedge ... \wedge q_{m,j} \neq q)]$$

$$= \sum_{j=1}^{l_B} \sum_{q \in Q_O} p_{q,j}(1 - p_{q,j})^m \leq \sum_{j=1}^{l_B} \sum_{q \in Q_O} \frac{1}{m} \leq \sum_{j=1}^{l_B} \frac{l_O}{m} = \frac{l_B l_O}{m}.$$

Thus, given that $m = 2l_B l_O/\delta$, we get $\Pr[E_{3a}(x)] \leq \delta/2$.

Soundness: VBB Simulatability. To show that the security property is satisfied, it suffices to provide a PPT algorithm U^T in the ideal TDP model that takes as input $O^T(C)$ for some circuit C and outputs a distribution that is statistically close to the output distribution of $\widehat{O}$. If that is the case, we can invoke Lemma 5 and conclude that $\widehat{O}$ is also VBB-secure.

The description of U is precisely the same as Steps 2–4 of the procedure detailed in Sect. 4.1 except that queries made by $B = O^T(C)$ are answered using oracle T instead of being randomly simulated. If we let (B, Q_B, R) be the output of $U^T(O^T(C))$ then we can easily see that it is identically distributed to the output distribution of $\widehat{O}$ since, in both cases, Q_B has query-answers with consistent and random TDP query tuples. They differ only by how these query answers are generated (U^T answers them using T, while $\widehat{O}$ simulates them using lazy evaluation with respect to some oracle $\widehat{T}$ distributed the same as T).

4.3 Extension to Hierarchical Random TDP

In this section, we reason that the proof for the ideal TDP case can be extended to hierarchical TDP oracles as well. We start by defining how the oracle for the random hierarchical trapdoor permutation primitive changes from Definition 28.

Definition 35 (Random Hierarchical Injective Trapdoor Functions).
For any security parameter n and $l = \text{poly}(n)$, an l-level random hierarchical injective trapdoor function (HTDF) oracle T_n^l consists of $2l + 3$ subroutines $(\{J_i\}_{i=1}^{l+1}, \{K_i\}_{i=0}^{l+1})$ defined as follows:

- *$K_i[\text{ID}_{i-2}, id_{i-1}](td_i)$: An injective function, indexed by identity vector $\text{ID}_{i-2} = (id_0, ..., id_{i-2})$ and id_{i-1}, that accepts as input an i-level trapdoor $td_i \in \{0,1\}^m$ and outputs a randomly chosen identity $id_i \in \{0,1\}^n$ where $m = 10nl$ if $i \in [1, l]$ and $m = n$ (i.e. it is a permutation) if $i = \{0, l+1\}$.*
- *$J_i[\text{ID}_{i-2}, td_{i-1}](id_i)$: An injective function, indexed by identity vector $\text{ID}_{i-2} = (id_0, ..., id_{i-2})$ and td_{i-1} that, given the identity $id_i \in \{0,1\}^n$, outputs the corresponding trapdoor $td_i \in \{0,1\}^m$ where $m = 10nl$ if $i \in [1, l]$ and $m = n$ (i.e. it is a permutation) if $i = \{0, l+1\}$.*

Note that, for any fixed ID_{i-2}, if $td_i = J_i[\text{ID}_{i-2}, td_{i-1}](id_i)$ and $id_{i-1} = K_{i-1}[\text{ID}_{i-3}, id_{i-2}](td_{i-1})$ then $id_i = K_i[\text{ID}_{i-2}, id_{i-1}](td_i)$. In other words, we can think of K_i as the inverse of J_i only if the indices of the two functions match (that is, the trapdoor td_{i-1} indexing J_i corresponds to the identity id_{i-1} indexing K_i).

Remark 36. It is also crucial to note that we used (sparse) injective functions for generating the intermediate levels of trapdoor. Such a change was made in order to obtain interesting primitives from this oracle, such as fully-secure hierarchical identity-based encryption (HIBE). If permutations were used instead, we would only achieve HIBE with security against adversaries that *do not* choose an identity for the permutation F to attack. Furthermore, removing K_i for $i \in [1, l]$ as a way to prevent this attack's capability hinders our ability to perform the canonicalization procedure for the obfuscated circuit.

Remark 37. For the special case of 1-level HTDF (i.e. TDP), we only have three permutations: $K_0, K_1[id_0]$ and $J_1[td_0]$, which correspond to permutations $G, F[pk]$, and $F^{-1}[sk]$, respectively in the language of TDP that we used in Definition 28. Note that here, we would refer to 0-level identities as master public keys and 0-level trapdoors as master secret keys.

We also present a variant of TDP query tuples that generalizes Definition 29 to work with hierarchical injective trapdoor functions.

Definition 38 (HTDF query tuple). *Given a random l-level HTDF oracle $T_n^l = (\{J_i\}_{i=1}^{l+1}, \{K_i\}_{i=0}^{l+1})$, an i-level HTDF query tuple consists of three (possibly) related query-answer pairs $(V_{K_{i-1}}, V_{K_i}, V_{J_i})$ where, for any fixed $\text{ID}_{i-2} = (id_0, ..., id_{i-2})$:*

- *$V_{K_{i-1}} = (td_{i-1}, id_{i-1})$ represents a query to $K_{i-1}[\text{ID}_{i-3}, id_{i-2}]$ on input td_{i-1} and its corresponding answer id_{i-1}*
- *$V_{K_i} = ((id_{i-1}, td_i), id_i)$ represents a query to $K_i[\text{ID}_{i-2}, id_{i-1}]$ on input td_i and its corresponding answer id_i*

- $V_{J_i} = ((td_{i-1}, id_i), td'_i)$ *represents a query to* $J_i[\text{ID}_{i-2}, td_{i-1}]$ *on input* id_i *and its corresponding answer* td'_i

We say that an i-level HTDF query tuple is consistent if $td_i = td'_i$.

Remark 39. For the purposes of comparison, we note that, for the special case of 1-level HTDF (i.e. TDP), we only have TDP query tuples of the form $(V_{K_0}, V_{K_1}, V_{J_1}) = (V_G, V_F, V_F^{-1})$. Thus, $V_G = (sk, pk)$ represents a query to G on $sk = td_0$ and the answer $pk = id_0$, $V_F = ((pk, x), y)$ represents a query to F_{pk} on $x = td_1$ and the answer $y = id_1$, and $V_{F^{-1}} = ((sk, y), x')$ represents a query to F_{sk}^{-1} on y and the answer x', which should be x if the tuple is consistent.

Extension of the Proof. The extension of the impossibility result to random HTDF is straightforward, so we will outline the main differences between the TDP case and describe how to resolve the issues that are related to this oracle. First, we still perform the normalisation procedure on $\widehat{O}$ and B where the query behaviour of these algorithms are modified such that for any query q of the form $J_i[\text{ID}_{i-2}, td_{i-1}](id_i)$, we first ask $K_{i-1}[\text{ID}_{i-3}, id_{i-2}](td_{i-1})$ to get id_{i-1}. This allows us to discover whether we have a query $K_i[\text{ID}_{i-2}, id_{i-1}](td_i)$ whose answer is id_i, in which case we can answer q with td_i. This procedure ensures that all query tuples that contain J_i queries are consistent.

We now turn to verifying whether the proof of approximate functionality for TDP holds in this case as well and, in particular, focus on the event $E(x)$ that was defined Sect. 4.2. The main issue that we have to consider, which is unique to the HTDF case, is the possibility that different consistent TDP query tuples can be related to each other, and an overlap between these queries may cause an inconsistency in one of the tuples. Specifically, an i-level TDP query tuple of the form $(V_{K_{i-1}}, \cdot, \cdot)$ might overlap with an $(i-1)$-level TDP query tuple $(\cdot, \cdot, V_{J_{i-1}})$ from Q_O, where the answer of $V_{K_{i-1}}$ is inconsistent with that of $V_{J_{i-1}}$. However, our normalisation procedure prevents precisely this issue as any TDP query tuple that contains $V_{J_{i-1}}$ must also have $V_{K_{i-1}}$, which means that the queries should not overlap otherwise event E_1 occurs leading to a contradiction to our initial assumption.

Acknowledgement. We thank Victor Shoup and Hendrik W. Lenstra for pointing us out to the literature on solving linear equations over the ring $\mathbb{Z}_n$.

References

[BGI+01] Barak, B., Goldreich, O., Impagliazzo, R., Rudich, S., Sahai, A., Vadhan, S.P., Yang, K.: On the (im)possibility of obfuscating programs. In: Kilian, J. (ed.) CRYPTO 2001. LNCS, vol. 2139, pp. 1–18. Springer, Heidelberg (2001)

[BGK+13] Barak, B., Garg, S., Kalai, Y.T., Paneth, O., Sahai, A.: Protecting obfuscation against algebraic attacks. IACR Cryptology ePrint Archive, 2013:631 (2013)

[BGK+14] Barak, B., Garg, S., Kalai, Y.T., Paneth, O., Sahai, A.: Protecting obfuscation against algebraic attacks. In: Nguyen, P.Q., Oswald, E. (eds.) EUROCRYPT 2014. LNCS, vol. 8441, pp. 221–238. Springer, Heidelberg (2014)

[BP13] Bitansky, N., Paneth, O.: On the impossibility of approximate obfuscation and applications to resettable cryptography. In: Proceedings of the Forty-fifth Annual ACM Symposium on Theory of Computing, STOC 2013, pp. 241–250. ACM, New York (2013)

[BR14] Brakerski, Z., Rothblum, G.N.: Virtual black-box obfuscation for all circuits via generic graded encoding. In: Lindell, Y. (ed.) TCC 2014. LNCS, vol. 8349, pp. 1–25. Springer, Heidelberg (2014)

[Can97] Canetti, R.: Towards realizing random oracles: hash functions that hide all partial information. In: Kaliski Jr, B.S. (ed.) CRYPTO 1997. LNCS, vol. 1294, pp. 455–469. Springer, Heidelberg (1997)

[CGH04] Canetti, R., Goldreich, O., Halevi, S.: The random oracle methodology, revisited. J. ACM **51**(4), 557–594 (2004)

[CKP15] Canetti, R., Tauman Kalai, Y.T., Paneth, O.: On obfuscation with random oracles. Cryptology ePrint Archive, Report 2015/048 (2015). http://eprint.iacr.org/

[GGH13a] Garg, S., Gentry, C., Halevi, S.: Candidate multilinear maps from ideal lattices. In: Johansson, T., Nguyen, P.Q. (eds.) EUROCRYPT 2013. LNCS, vol. 7881, pp. 1–17. Springer, Heidelberg (2013)

[GGH+13b] Garg, S., Gentry, C., Halevi, S., Raykova, M., Sahai, A., Waters, B.: Candidate indistinguishability obfuscation and functional encryption for all circuits. In: 2013 IEEE 54th Annual Symposium on Foundations of Computer Science (FOCS), pp. 40–49. IEEE (2013)

[GKLM12] Goyal, V., Kumar, V., Lokam, S., Mahmoody, M.: On black-box reductions between predicate encryption schemes. In: Cramer, R. (ed.) TCC 2012. LNCS, vol. 7194, pp. 440–457. Springer, Heidelberg (2012)

[HL02] Horwitz, J., Lynn, B.: Toward hierarchical identity-based encryption. In: Knudsen, L.R. (ed.) EUROCRYPT 2002. LNCS, vol. 2332, pp. 466–481. Springer, Heidelberg (2002)

[IR89] Impagliazzo, R., Rudich, S.: Limits on the provable consequences of one-way permutations. In: ACM Symposium on Theory of Computing (STOC), pp. 44–61. ACM Press (1989)

[LPS04] Lynn, B.Y.S., Prabhakaran, M., Sahai, A.: Positive results and techniques for obfuscation. In: Cachin, C., Camenisch, J.L. (eds.) EUROCRYPT 2004. LNCS, vol. 3027, pp. 20–39. Springer, Heidelberg (2004)

[McC90] McCurley, K.S.: The discrete logarithm problem. In: Proceedings of the AMS Symposia in Applied Mathematics: Computational Number Theory and Cryptography, pp. 49–74. American Mathematical Society (1990)

[MMN+15] Mahmoody, M., Mohammed, A., Nematihaji, S., Pass, R., Shelat, A.: Lower bounds on assumptions behind indistinguishability obfuscation (2015, in Submission)

[Pas15] Pass, R., Shelat, A.: Impossibility of vbb obfuscation with ideal constant-degree graded encodings. Cryptology ePrint Archive, Report 2015/383 (2015). http://eprint.iacr.org/

[RTV04] Reingold, O., Trevisan, L., Vadhan, S.P.: Notions of reducibility between cryptographic primitives. In: Naor, M. (ed.) TCC 2004. LNCS, vol. 2951, pp. 1–20. Springer, Heidelberg (2004)

[Sho97] Shoup, V.: Lower bounds for discrete logarithms and related problems. In: Fumy, W. (ed.) EUROCRYPT 1997. LNCS, vol. 1233, pp. 256–266. Springer, Heidelberg (1997)

[Wee05] Wee, H.: On obfuscating point functions. In: Proceedings of the Thirty-Seventh Annual ACM Symposium on Theory of Computing, pp. 523–532. ACM (2005)

Lower Bounds on Assumptions Behind Indistinguishability Obfuscation

Mohammad Mahmoody[1]([⊠]), Ameer Mohammed[1], Soheil Nematihaji[1], Rafael Pass[2], and Abhi Shelat[1]

[1] University of Virginia, Charlottesville, USA
{mohammad,am8zv,sn8fb,shelat}@virginia.edu
[2] Cornell University, Ithaca, USA
rafael@cs.cornell.edu

Abstract. Since the seminal work of Garg et al. (FOCS'13) in which they proposed the first candidate construction for indistinguishability obfuscation (iO for short), iO has become a central cryptographic primitive with numerous applications. The security of the proposed construction of Garg et al. and its variants are proved based on multi-linear maps (Garg et al. Eurocrypt'13) and their idealized model called the graded encoding model (Brakerski and Rothblum TCC'14 and Barak et al. Eurocrypt'14). Whether or not iO could be based on standard and well-studied hardness assumptions has remain an elusive open question.

In this work we prove *lower bounds* on the assumptions that imply iO in a black-box way, based on computational assumptions. Note that any lower bound for iO needs to somehow rely on computational assumptions, because if **P = NP** then statistically secure iO does exist. Our results are twofold:

1. There is no fully black-box construction of iO from (exponentially secure) collision-resistant hash functions unless the polynomial hierarchy collapses. Our lower bound extends to (separate iO from) any primitive implied by a random oracle in a black-box way.

M. Mahmoody—Supported by NSF CAREER award CCF-1350939. The work was done in part while the author was visiting the Simons Institute for the Theory of Computing, supported by the Simons Foundation and by the DIMACS-Simons Collaboration in Cryptography through NSF grant CNS-1523467.
A. Mohammed—Supported by University of Kuwait.
S. Nematihaji—Supported by NSF award CCF-1350939.
R. Pass—Work supported in part by a Microsoft Faculty Fellowship, Google Faculty Award, NSF Award CNS-1217821, NSF Award CCF-1214844, AFOSR Award FA9550-15-1-0262 and DARPA and AFRL under contract FA8750-11-2-0211. The views and conclusions contained in this document are those of the authors and should not be interpreted as representing the official policies, either expressed or implied, of the Defense Advanced Research Projects Agency or the US Government.
A. Shelat—Work performed while visiting Cornell Tech, and supported by NSF CAREER Award 0845811, NSF TC Award 1111781, NSF TC Award 0939718, DARPA and AFRL under contract FA8750-11-C-0080, Microsoft New Faculty Fellowship, SAIC Scholars Research Award, and Google Research Award.

E. Kushilevitz and T. Malkin (Eds.): TCC 2016-A, Part I, LNCS 9562, pp. 49–66, 2016.
DOI: 10.1007/978-3-662-49096-9_3

2. Let $\mathcal{P}$ be any primitive that exists relative to random trapdoor permutations, the generic group model for any finite abelian group, or degree-$O(1)$ graded encoding model for any finite ring. We show that achieving a black-box construction of iO from $\mathcal{P}$ is *as hard as* basing public-key cryptography on one-way functions.

In particular, for any such primitive $\mathcal{P}$ we present a constructive procedure that takes any black-box construction of iO from $\mathcal{P}$ and turns it into a construction of semantically secure public-key encryption form any one-way functions. Our separations hold even if the construction of iO from $\mathcal{P}$ is *semi*-black-box (Reingold, Trevisan, and Vadhan, TCC'04) and the security reduction could access the adversary in a non-black-box way.

Keywords: Indistinguishability obfuscation · Black-box separations

1   Introduction

The celebrated work of Barak et al. [3] initiated a formal study of the notion of program obfuscation which is the process of making programs unintelligible while preserving their functionalities. The main result of [3] was indeed a negative one showing that a strong form of obfuscation, called virtual black-box obfuscation, is indeed impossible for general circuits. The same work [3] also defined a weaker notion of obfuscation, called *indistinguishability* obfuscation (iO). The security of iO only requires that the obfuscation of two equivalent and same-size circuits C_1, C_2 should be computationally indistinguishable in the eyes of efficient adversaries.

The first candidate construction for iO was presented in the breakthrough work of Gentry et al. [12]. [12] showed how to construct iO for $\mathbf{NC_1}$ circuits based on multi-linear assumptions [11], and also showed how to boost iO for $\mathbf{NC_1}$ to iO for general circuits based on the learning with error (LWE) assumption [33]. The work of [12] led to an active area with a long list of results using iO as a "central hub" [36] for cryptographic tasks/primitives and basing them on iO together with one-way functions or other (relatively weak) standard assumptions. Interestingly, as shown by [23] the one-way function itself could be based on iO and the *worst-case* assumption that $\mathbf{NP} \neq \mathbf{BPP}$, leading to tens of applications solely based on iO and $\mathbf{NP} \neq \mathbf{BPP}$.[1]

Assumptions Behind iO. Since the first candidate construction of iO was presented by [12] a few other variants of this constructions with different assumptions have been presented. Brakerski and Rothblum [8] showed that in an idealized model based on multilinear maps, known as the *graded encoding model* one can achieve iO (or even VBB obfuscation) assuming the bounded speedup hypothesis. Barak et al. [2] improved the result of [8] by making the construction unconditionally secure in the graded encoding model. Miles et al. [27] took another step

---

[1] Note that we cannot hope to get OWFs from iO alone without any hardness for class $\mathbf{NP}$ since iO exist unconditionally if $\mathbf{NP} = \mathbf{P}$.

in this direction by making the construction even more secure by allowing unlimited additions across different encoding "levels". Pass et al. [32] constructed iO for NC_1 circuits based on (subexponentially secure) semantically-secure multilinear encodings. Their work was the first result basing iO on falsifiable assumptions [29], however they relied on super-polynomial assumptions. Gentry et al. [15] construct iO based on subgroup elimination assumptions.

Despite all the efforts mentioned above to base iO on hardness assumptions, the assumptions behind the constructions of iO seem to be qualitatively different compared to other cryptographic primitives, even in comparison with very powerful primitives such as fully homomorphic encryption [14,35] which could be constructed from LWE [9]. The recent beautiful work of [1] proved the first *limitation* on the power of iO by ruling out constructions of collision-resistant hash functions from iO and OWF (even if iO uses OWF in a non-black-box way).

To the best of our knowledge, no lower-bounds on the complexity of the assumptions behind iO are proved yet. The same work of [1] ruled out fully black-box constructions of iO from private-key functional encryption schemes (PFE), but this result requires the iO to also handle circuits with PFE gates. We note, however, that obfuscating circuits in *plain* model with no oracle gates is in fact sufficient for all applications of iO. Moreover, since iO exists if $NP = P$, any lower bound on the complexity of (standard definition of) iO in which we only aim at obfuscating circuits in the plain model *needs* to rely on computational assumptions (unless we first prove $NP \neq P$).

In this work we initiate a formal study on the assumption complexity of iO from a lower-bound perspective.[2] We prove our results in the black-box framework of Impagliazzo and Rudich [21] and its refinements by Reingold et al. [34]. Lower bounds against black-box constructions/reductions are considered fundamental due to the abundance of black-box techniques as well as their (typical) efficiency advantage over their non-black-box counterparts.

Since applications of iO almost always lead to non-black-box constructions, it could be argued that a black-box separation for iO is not meaningful. Note, however, that in this work we are not studying which primitives could be constructed from iO in a black-box way. We are instead looking at the complexity of assumptions behind iO. An instructive analogy is zero knowledge proofs for NP (ZKP). Using ZKPs for general NP statements also makes constructions non-black-box (since some piece of code is used as witness used by the prover) yet, we can indeed construct ZKPs for NP from one-way functions in a *fully* black-box way [17,18,28].[3] Therefore, even though iO leads to non-back-box constructions, the construction of iO itself could very much be black-box, and so separating it from classical primitives is also meaningful.

---

[2] As mentioned above, we do not allow oracle gates for the circuits that are going to be obfuscated.

[3] It is also instructive to note that even though ZKP for NP could be constructed from OWFs in a fully black-box way, it is conceivable that a separation would hold if we require a proof system for satisfiability of circuits with oracle gates.

1.1   Our Results

Our first lower bound holds for any primitive implied by a random oracle (e.g., exponentially secure one-way functions or collision-resistant hash functions) and it is proved for *fully* black-box constructions that treat the primitive and the adversary in a black-box way (see Definition 5).

Theorem 1 (Fully black-box separation from primitives implied by random oracle). *Unless the polynomial hierarchy collapses, there is no fully black-box construction of iO from collision-resistant hash functions or more generally any primitive implied by a random oracle in a black-box way.*

Intuition Behind the Proof. To prove Theorem 1 we will first prove a useful lemma (see Lemma 17) which, roughly speaking, asserts that for any pair of circuits C_1, C_2, either (1) a (computationally unbounded) polynomial-query attacker can guess which one is obfuscated in the random oracle model with a probability close to one, or that (2) there is a way to obfuscate them into the same output circuit B. The latter could be used as a witness that C_1 and C_2 compute the same function, assuming that the obfuscation is an iO. Now consider the set of equivalent and same-size circuit $\mathcal{C} = \{(C_0, C_1) \mid C_0 \equiv C_1 \wedge |C_0| = |C_1|\}$. If Case (1) happens for an infinite subset of $\mathcal{C}$, we get a poly-query attacker against iO in the random oracle model which is sufficient for deriving the black-box separations of Theorem 1. On the other hand, if Case (1) happens only for a finite subset of $\mathcal{C}$, we get an efficient procedure to certify the equivalence of two given circuits, implying **NP** $\neq$ **P**. We prove Lemma 17 by reducing it to a result by Mahmoody and Pass [25] who ruled out the existence of non-interactive commitments from one-way functions. Roughly speaking, we construct a non-interactive commitment scheme based on common input (C_1, C_2) in the random oracle model, and we show that: the cheating receiver strategy of [25] implies our Case (1), and the cheating sender strategy of [25] implies our Case (2). The result of [25] shows that either of these strategies always exist. See Sect. 3 for the formal and detailed proof.

Our second lower bound does not rule out black-box construction of iO based on believable assumptions, but shows that achieving such constructions for a large variety of primitives is as hard has solving another long standing open question in cryptography; namely, basing public-key encryption on one-way functions. It also captures a larger class of security reductions known as semi-black-box reductions [34] that allow the security reduction to access the adversary in a non-black-box way (see Definition 6).

Theorem 2 (Hardness of semi-black-box construction). *Let $\mathcal{P}$ be a primitive that provably exists relative to random trapdoor permutation oracle, the generic group model (for any finite abelian group) or the degree-$O(1)$ graded encoding model (for any finite ring). Any semi-black-box construction of iO from $\mathcal{P}$ (constructively) implies a construction of semantically secure public-key encryption from one-way functions.*

Primitives Captured by Theorem 2. Theorem 2 captures a large set of powerful cryptographic primitives that could be constructed in idealized models. For example trapdoor permutations (and any primitive implied by TDPs in a black-box way) trivially exist relative to the idealized model of random TDPs. Even primitives that we do not know how to construct from TDPs in a black-box way (e.g., CCA secure public key encryption) are known to exist in the random TDP model [5]. The generic group model defined by Shoup [37] (see Definition 12) is an idealized model in which (a black-box form of) the DDH assumption holds unconditionally. Therefore, our separation of Theorem 2 covers any primitive that could be constructed from DDH assumption in a black-box way. The same holds for *bilinear* assumptions in the graded encoding model (see Definition 13) of degree 2. Namely, primitives that could be constructed from bilinear assumptions (in a black-box way) exist in the degree $O(1)$ graded encoding model unconditionally. This includes one-round 3-party key-agreement [22], identity based encryption [7], etc.

Intuition Behind the Proof. Our main tool in proving Theorem 2 is the following theorem which is implicit in the recent work of the authors [24,31]. Even though the focus of [24,31] is on virtual black-box obfuscation, the same construction presented in [24,31] for the case of VBB implies the following theorem for iO.

Theorem 3 (Implicit in [24,31]). *The existence of iO in any of the idealized models of: random trapdoor permutation oracle, generic group model for finite abelian groups, or the degree-$O(1)$ graded encoding model for finite rings, implies $1/p(n)$-approximate iO in the plain model for any polynomial $p(n)$.*

We then show that the existence of $(1/6)$-approximate iO and any one-way functions imply the existence of "approximately correct" and "approximately secure" public-key encryption schemes. In order to prove this we employ a construction of Sahai and Waters [36] using which they showed that iO and OWF imply PKE. Here we show that the very same construction, when instantiated using *approximate* iO, leads to "approximately correct" and "approximately secure" public-key encryption. Finally we use a result of Holenstein [19] who showed how to amplify any approximately-correct and approximately-secure PKE into a full fledged (semantically secure) PKE for sufficiently good approximation! See Sect. 4 for the formal and detailed proof.

Remark 4. Our proof of Theorem 1 relies on perfect completeness of iO. Theorem 2 above holds even if with negligible probability over the obfuscator's randomness the obfuscated circuit does not compute the same function. Extending Theorem 1 to allow negligible error over the randomness of the obfuscator remains an interesting open questions.

Previous work on hardness *of black-box constructions.* Theorem 2 has the same spirit as the result by Impagliazzo and Rudich [21] who showed any semi-black-box construction of key agreement from one-way functions would implies $\mathbf{P} \neq \mathbf{NP}$. Therefore, the fact that we are far from proving $\mathbf{P} \neq \mathbf{NP}$ implies that we

as far from basing key agreements on one-way functions in a black-box way.[4]
Similarly, our Theorem 2 shows that as long as we are not able to base public-
key encryption on one-way functions, we cannot base iO on a variety of strong
primitives in a semi-black-box way. Other results of the same flavor exist in
connection with program checkers [6] for **NP**. Mahmoody and Xiao [26] showed
that any construction of one-way functions based on worst-case hardness of **NP**
implies program checkers for **NP** whose existence is one of the long standing
open questions in complexity theory.

Falsifiability of iO. An intriguing open question regarding assumption complexity of iO is whether iO could be based on any "falsifiable" assumption [29].
A falsifiable assumption is one with an efficient challenger security game. The
question is raised since an adversary attacking an iO scheme starts with proposing two *equivalent* circuits and efficient challenger has no direct way to verify
this. Since our primitives used in the theorems above are falsifiable, a separation
of iO from falsifiable assumptions would imply our results for the case of *polynomially secure* primitives. However, constructions of iO based on exponentially
secure falsifiable assumptions are indeed known [32]. Therefore, our results are
interesting even if one can prove that iO cannot be based on falsifiable assumptions. Moreover, the known lower bounds against falsifiable assumptions [16,30]
are proved only for *black-box* proofs of security in which the adversary is used in
a black-box way. Our Theorem 2 holds even for semi-black-box constructions in
which the security reduction could use the adversary in a non-black-box manner.

2   Preliminaries

For circuits C, D, we denote by $C \equiv D$ that they compute the same function.
By $|C|$ we denote the size of the bit representation of C. By a *partial* oracle f we
denote an oracle that is only defined for a subset of possible queries. For random
variables X, Y, by $X \approx Y$ we denote the fact that X and Y are distributed
identically. We call a function $\epsilon(n)$ negligible if $\epsilon(n) < 1/p(n)$ for all polynomial
$p(\cdot)$ and sufficiently large n. We call $\rho(n)$ overwhelming, if $1 - \rho(n)$ is negligible.

2.1   Black-Box Constructions

Definition 5 (Fully black-box constructions [34]). *A fully black-box construction of a primitive $\mathcal{Q}$ from a primitive $\mathcal{P}$ consists of two PPT algorithms
(Q, S) as follows:*

1. *Implementation: For any oracle P that implements $\mathcal{P}$, Q^P implements $\mathcal{Q}$.*
2. *Security reduction: for any oracle P implementing $\mathcal{P}$ and for any (computationally unbounded) oracle adversary A successfully breaking the security of
   Q^P, it holds that $S^{P,A}$ breaks the security of P.*

---

[4] Formalizing semi-black-box constructions interpreting the result of [21] in this context is due to [34].

Primitives with Stronger Hardness. The above definition is for polynomially secure primitives. When the used primitive $\mathcal{P}$ is s-secure for a more quantitative bound $s(n) \gg \mathrm{poly}(n)$, the security reduction S could potentially run in longer running time as well so long as it holds that: when P, A are polynomial time, the total running time of the composed algorithm $S^{P,A}$ is also small enough to be considered a legal attack against the implementation P of $\mathcal{P}$.

In the following more relaxed form of constructions, the security reduction can depend arbitrarily on the adversary but it still treats the implementation of the used primitive in a black-box way.

Definition 6 (Semi-black-box constructions [34]). *A semi-black-box construction of a primitive $\mathcal{Q}$ from a primitive $\mathcal{P}$ is defined similarly to the fully black-box Definition of 5 with the following difference in the security reduction:*

– *For any oracle P implementing $\mathcal{P}$ and any efficient oracle-aided adversary A^P who breaks the security of Q^P it holds that $S^P(A)$ breaks the security of P. Note that since A's description is efficient it could indeed be given to S in a non-black-box way.*

Remark 7. The work of Reingold et al. [34] also defines a "$\forall\exists$" variant of the semi-black-box constructions in which S^P can arbitrarily depend on A^P (rather than depending on it in a unified way). In this work we work with the basic default variant that we also find more natural.

Efficiency of Adversary. We used the term *efficient* in an unspecified way so that it could be applied to complexity classes beyond polynomial time. For example, using a quasi-polynomially secure primitive $\mathcal{P}$ to construct a polynomially secure primitive $\mathcal{Q}$ would require a security primitive that is more relaxed and could lead to a quasi-polynomial (as opposed to polynomial) time attack against P using any polynomial-time attacker against Q^P.

2.2   Indistinguishability Obfuscation

Definition 8 ([3]). *A PPT algorithm O is an* indistinguishability obfuscator *(iO) if the following two hold:*

– *Correctness: For every circuit C, it holds that $\Pr_r[O_r(C) \equiv C] = 1$.*
– *Soundness: For every PPT adversary A there exists a negligible function $\alpha(\cdot)$ such that for every pair of equivalent circuits $C_1 \equiv C_2$ with the same size $|C_1| = |C_2| = n$ it holds that:*

$$\Pr_{r,s,b}[A_s(C_1, C_2, B) = b\colon b \xleftarrow{\$} \{0,1\}, B = O_r(C_b)] \le 1/2 + \alpha(n)$$

where the probability is over the random seeds of the obfuscator O, adversary A and the random bit b.

Definition 9 (Approximate iO). *A PPT O is called an ϵ-approximate iO if it satisfies the same soundness condition and the following modified correctness condition.*

- *Approximate correctness:* $\Pr_{r,x}[O_r(C)(x) \neq C(x)] = \epsilon(|C|)$ *where the probability is over the randomness of the obfuscator as well as the randomly selected input.*

Definition 10 (Fully and semi-black-box constructions of iO). *A fully black-box construction of iO from primitive $\mathcal{P}$ consists of two oracle algorithms (O, S) such that*

- *Implementation (correctness): For every oracle P implementing $\mathcal{P}$, every circuit C, and every randomness r for O it holds that $B = O_r^P(C)$ is an oracle-aided circuit B such that $B^P \equiv C$.*
- *Soundness: For any oracle P implementing $\mathcal{P}$, any $\epsilon \geq 1/\operatorname{poly}(n)$ and any oracle adversary A who ϵ-breaks O^P, it holds that $S^{P,A}(1^{1/\epsilon(n)})$ breaks the security of P.*
  We say that A ϵ-breaks O^P if for an infinite number of pairs of equivalent circuits $C_0 \equiv C_1$ of equal lengths n it holds that

$$\Pr_{r,s}[A_s(C_1, C_2, B): b \xleftarrow{\$} \{0,1\}, B = O_r^P(C_b)] \geq 1/2 + \epsilon(n)$$

  where the probability is over the random seeds of the obfuscator O, adversary A and the random bit b.

A semi-black-box construction of iO from $\mathcal{P}$ is defined similarly, with its soundness defined along the line of Definition 6. Namely, we require that for any efficient adversary A who ϵ-breaks O^P, there is also an efficient adversary breaking the security of P.

2.3   Generic/Idealized Models

Definition 11 (Random Oracle Model). *In the random oracle model, all parties have access to a randomized oracle f such that for each input x, the answer $f(x)$ is uniformly (and independently of the rest of the oracle) distributed over $\{0,1\}^{|x|}$.*

Definition 12 (Generic Group Model [37]). *Let $(G, \odot)$ be any group of size N and let S be any set of size at least N. The generic group oracle $\mathcal{I}[G \mapsto S]$ (or simply $\mathcal{I}$) is as follows. At first an injective random function $\sigma \colon G \mapsto S$ is chosen, and two type of queries are answered as follows.*

- **Labeling Queries.** *Given $g \in G$ oracle returns $\sigma(g)$.*
- **Addition Queries.** *Given y_1, y_2, if there exists x_1, x_2 such that $\sigma(x_1) = y_1$ and $\sigma(x_2) = y_2$, it returns $\sigma(x_1 \odot x_2)$. Otherwise it returns $\perp$.*

Definition 13 (Degree-d Ideal Graded Encoding Model). *The oracle* $\mathcal{M}_R^d = (\text{enc}, \text{zero})$ *is stateful and is parameterized by a ring R and a degree d and works in two phases. For each l the oracle* $\text{enc}(\cdot, l)$ *is a random injective function from the ring R to the set of labels S.*

1. *Initialization phase: In this phase the oracle answers* $\text{enc}(v, l)$ *queries and for each query it stores* (v, l, h) *in a list* $\mathcal{L}_O$.
2. *Zero testing phase: Suppose $p(\cdot)$ is a polynomial whose coefficients are explicitly represented in R and its monomials are represented with labels $h_1, \ldots, h_m$ obtained through* $\text{enc}(\cdot, \cdot)$ *oracle in phase 1. Given any such query $p(\cdot)$ the oracle answers as follows:*
   (a) *If any h_i is not in $\mathcal{L}_O$ (i.e., it is not obtained in phase 1) return* false.
   (b) *If the degree of $p(\cdot)$ is more than d then return* false.
   (c) *Let* $(v_i, l_i, h_i) \in \mathcal{L}_O$. *If $p(v_1, \ldots, v_m) = 0$ return* true, *otherwise* false.

Generic Algorithms. A generic algorithm in the generic group model (resp. graded encoding model) is an algorithm in which no label s is used in an addition (resp. zero testing) query unless it is previously obtained through the oracle itself. In this work we only use *sparse* encodings in which $|S|/|G| = n^{\omega(1)}$ (resp. $|S|/|R| = n^{\omega(1)}$ in the graded encoding model) where n is the security parameter. Therefore, the execution of poly-time algorithms in this model will be statistically close to being generic.

Definition 14 (Primitives in Idealized Models). *We say a primitive $\mathcal{P}$ exists relative to the* randomized *oracle (or idealized model) $\mathcal{I}$ if there is an oracle-aided algorithm P such that:*

1. **Completeness:** *For every instantiation I of $\mathcal{I}$, it holds that P^I implements $\mathcal{P}$ correctly.*
2. **Security:** *Let A be an oracle-aided adversary $A^{\mathcal{I}}$ where the complexity of A is bounded by the specified complexity of the attacks for primitive $\mathcal{P}$. For example if $\mathcal{P}$ is polynomially secure (resp., quasi-polynomially secure), then A runs in in polynomial time (resp., quasi-polynomial time). For every such oracle aided A, with measure one over the sampling of the idealized oracle $I \xleftarrow{\$} \mathcal{I}$, it holds that A does* not *break the security of P^I.*

We call P a black-box *construction of $\mathcal{P}$ relative to $\mathcal{I}$ if the security property holds also in a "black-box" way defined as follows:*

– *Let A be an oracle-aided adversary $A^{\mathcal{I}}$ where the* query complexity of A *is bounded by the specified complexity of the attacks for primitive $\mathcal{P}$. For example if $\mathcal{P}$ is polynomially secure (resp., quasi-polynomially secure), then A only asks a polynomial (resp., quasi-polynomial) number of queries.*
   For every such oracle aided A, with measure one over the sampling of the idealized oracle $I \xleftarrow{\$} \mathcal{I}$, it holds that A does not *break the security of P^I.*

In the definition above, we only require the scheme to be secure after the adversary is fixed. This is along the line of the way the random oracle model is used in cryptography [5], and lets us easily derive certain primitives in idealized models. For example it is easy to see that a random trapdoor permutation, with measure one, is a secure TDP against any fixed adversary of polynomial query complexity. Therefore, TDPs exist in the idealized model of random TDP in a black-box way. In fact stronger results are proved in the literature for other primitives. Impagliazzo and Rudich [21] and Gennaro and Trevisan [13] showed that one-way functions exist relative to the idealized model of random oracle, even if we sample the oracle first and then go over enumerating possible attacks. Chung et al. [10] proved a similar result for collision resistant hash functions.

The following lemma could be verified by inspection.

Lemma 15. *If there is a semi-black-box construction of Q from P then:*

1. *If P exists relative to idealized model $\mathcal{I}$, Q exists relative to $\mathcal{I}$ as well.*
2. *If in addition the construction of P from $\mathcal{I}$ is black-box, then a black-box construction of Q relative to $\mathcal{I}$ exists as well.*

Proof. Let Q be the semi-black-box construction of Q from P. Let P be the implementation of P relative to $\mathcal{I}$. It is easy to see that Q^P is an implementation of Q relative to $\mathcal{I}$. Now we prove the security properties of Lemma 15.

1. Let A be any successful attacker against the implementation of Q^P in the idealized model $\mathcal{I}$. Then by non-zero measure over the choice of $I \xleftarrow{\$} \mathcal{I}$ it holds that A breaks the security of Q^{P^I}. For any such I, the security reduction $S^I(A)$ also breaks the security of P^I. This means that the attacker $S(A) = B$ breaks the security of P^I with non-zero measure over the sampled oracle $I \xleftarrow{\$} \mathcal{I}$. This contradicts the assumption that P is securely realized in $\mathcal{I}$.
2. A similar proof holds for the black-box constructions in idealized models. The only modification is that now we use $B = S^A$ (rather than $S(A)$).

3 Separating iO from Random Oracle Based Primitives

In this section we prove the following formalization of Theorem 1.

Theorem 16 (Theorem 1 formalized). *If $\mathbf{NP} \neq \mathbf{co\text{-}NP}$ then there is no fully black-box construction of iO from any primitive P that exists relative to a random oracle in a black-box way. This includes exponentially secure one-way functions and collision-resistant hash functions.*

To prove Theorem 16 we will first prove a useful lemma (see Lemma 17) which, roughly speaking, asserts that for any pair of circuits C_1, C_2, either an attacker can guess which one is obfuscated in the random oracle model with a probability close to one, or that there is a way to obfuscate them into the same output circuit B. The latter could be used as a witness that C_1 and C_2 compute the same function, assuming that the obfuscation is an iO.

Lemma 17 (Distinguish or Witness). *Let O be an oracle aided randomized polynomial-time algorithm taking circuits as input such that for every length-preserving oracle f and every randomness r it holds that $O_r^f(C) \equiv C$ (i.e., O always outputs circuits with the same input/output functionality as the input circuit C). Then, at least one of the following holds:*

1. *There is an infinite sequence of circuits $(C_0^1, C_1^1), \ldots, (C_0^i, C_1^i), \ldots$ such that $|C_0^i| = |C_1^i|$ for all i, and there exists a (computationally unbounded) poly(n)-query A such that the following holds for all i:*

$$\Pr_{r,s,f,b}[A_s^f(B) = b: b \xleftarrow{\$} \{0,1\}, B = O_r^f(C_b^i)] \geq 1 - 1/n^2$$

where n is the bit size of the circuits: $|C_0^i| = |C_1^i| = n$.
2. **NP = co-NP**.

We will first prove Theorem 16 using Lemma 17, and then we will prove Lemma 17.

Proof (of Theorem 16). In what follows we will always assume **NP $\neq$ co-NP**. We will describe the proof for one-way functions, but it can be verified that the very same proof holds for any primitive that holds relative to random oracles in a black-box way (see Definition 14).

Suppose (O, S) is a fully black-box construction of iO from one-way functions. We use a random oracle f to implement the one-way function required by O. By Lemma 17 and the assumption that **NP $\neq$ co-NP** we know that there is a computationally unbounded attacker A and an infinite sequence of equivalent and same-size circuits $(C_0^1, C_1^1), \ldots, (C_0^i, C_1^i), \ldots$ such that A breaks the security of iO over the challenge circuits (C_0^i, C_1^i) of length $|C_0^i| = n = |C_1^i|$ by guessing which one of them is being obfuscated with probability $\geq 1 - 1/n^2$. Let $\epsilon = 1/4$. By an averaging argument, with probability at least $1 - O(1/n^2)$ over the choice of oracle f, it holds that the probability that A correctly guesses which one of (C_0^i, C_1^i) is being obfuscated is at least $1/2 + \epsilon$. Since the summation $\sum_i 1/i^2 = O(1)$ converges, by Borel-Cantelli lemma, for measure one of the random oracles f it holds that A ϵ-breaks the implemented iO O^f.

Now that A is a "legal" adversary, by definition of fully black-box iO, the security reduction $S^{f,A}$ shall break the one-way property of f. Algorithms A and S are both poly(n)-query attackers, and so the combination $B = S^A$ also asks only a polynomial number of queries to f and succeeds in breaking the one-wayness of f for nonzero measure of samples for f.

The existence of such B, however, is impossible since a random oracle f, with measure one, is secure against attackers who ask only a polynomial number of queries [13, 21].[5]

---

[5] The works of [13, 21] work with polynomial time Turing machines or circuits, however their goal is to *fix* the random oracle f before enumerating the attackers. However, if the attacker is fixed before the sampling of f, the proofs of [13, 21] imply the one-wayness of f with measure one even if the fixed attacker is computationally unbounded.

Now we prove Lemma 17. To prove Lemma 17 we will use the following lemma from [25].

Lemma 18 ([25]). *Suppose S is an oracle-aided PPT algorithm that calls oracle f and takes private input $b \in \{0,1\}$, randomness r, and common input $z \in \{0,1\}^n$ (where n is the security parameter) and outputs $c = S_r^f(z,b)$. For any $\delta = \delta(n) \leq 1/100$, there is a (computationally unbounded) oracle-aided algorithm R such that for all $z \in \{0,1\}^n$ at least one of the following holds.*

1. *If f is the random oracle, $R^f(z,c)$ asks $\mathrm{poly}(n/\delta)$ queries and correctly guesses the random bit b that $S_r^f(z,b)$ used to generate c with probability $\geq 1 - \delta(n)$. Namely:*

$$\Pr_{f,r,b}[R^f(z,c) = b \colon b \xleftarrow{\$} \{0,1\}, c = S_r^f(z,b)] \geq 1 - \delta(n).$$

2. *There is a partial oracle f' of size $\mathrm{poly}(n)$ and two random seeds r_0, r_1 and a message c such that $S_{r_0}^{f'}(z,0) = c = S_{r_1}^{f'}(z,1)$. In other words, there is a message c that could be opened into both $b = 0$ and $b = 1$ using random seeds r_0, r_1, and the queries asked by S during these two possible executions are all described by the partial function f'.*

Remark 19. Mahmoody and Pass [25] proved a more general lemma ruling out (even "somewhere binding") non-interactive commitment schemes in the random oracle model. Lemma 18 is a special case of their result which is still sufficient for us. In the setting of [25] the security parameter is given to the parties in the form of 1^n, but their proof handles parties who in addition receive some $z \in \{0,1\}$ and the parties' behavior could also depend on the given z. For the sake of completeness, we have provided a self contained sketch of the proof of Lemma 18 in Appendix A.

Proof (of Lemma 17)

Consider the set of circuit pairs that are equivalent and of the same size: $\mathcal{C} = \{(C_0, C_1) \mid C_0 \equiv C_1 \wedge |C_0| = |C_1|\}$. We apply Lemma 18 for $\delta = 1/n^2$ as follows. Use $(C_0, C_1) = z \in \mathcal{C}$ as the common input given to both parties. Let S be a sender strategy that, given input bit b, obfuscates C_b and sends out the obfuscated circuit B.

By Lemma 18 for each $(C_0, C_1) = z \in \mathcal{C}$ either of the following holds:

1. $A^f((C_0, C_1), B)$ can guess the random b in the random oracle model correctly with probability at least $1 - 1/n^2$.
2. There is a partial oracle f' of polynomial size and two random strings r_0, r_1 such that $O_{r_b}^{f'}(C_b) = B$ for both $b \in \{0,1\}$.

Note that if $C_0 \not\equiv C_1$ then Case (2) cannot happen as no such (f', r_0, r_1) can exist by perfect completeness of iO. Therefore, if Case (2) happens, the existence of (f', r_0, r_1) serves as an efficiently verifiable proof that $C_0 \equiv C_1$.

Now let $\mathcal{C}_a$ be the subset of $\mathcal{C}$ for which Case (1) holds. There are two cases:

1. C_a is not finite, in which case we have shown that Case 1 of the lemma holds.
2. If C_a is finite, then for all (except a finite number) of $(C_0, C_1) \in C$ we can efficiently prove that C_0, C_1 are equivalent circuits. This would give a proof system for proving the equivalency of two given circuits, but this problem is **co-NP**-complete. Thus, **co-NP = NP**.

4 Hardness of Semi-Black-Box Constructions of iO

In this section, we prove Theorem 2. We will first show that approximate iO is still powerful enough to base public-key cryptography on private-key cryptography. We will use this result and results of [24,31] to derive Theorem 2.

Theorem 20. *The existence of $(1/6)$-approximate iO and any one-way functions imply the existence of semantically secure public-key encryption schemes.*

We first prove Theorem 2 using Theorems 3 and 20.

Proof (of Theorem 2 using Theorems 3 and 20). Let $\mathcal{P}$ be any such primitive with implementation P relative to the idealized model $\mathcal{I}$, and suppose O is any such semi-black-box construction of iO from $\mathcal{P}$. By Lemma 15, we conclude that $O' = O^P$ is a construction of iO in the idealized model $\mathcal{I}$. This, together with Theorem 3 imply that there is a $(1/6)$-approximate iO in the plain model. Finally, by Theorem 20 and the existence of $(1/6)$-approximate iO implies that we can construct semantically secure public-key encryption from one-way functions.

4.1 Proving Theorem 20

In this section we prove that $(1/6)$-approximate iO and one-way functions imply semantically secure public-key encryption. Therefore, any provably secure construction of $(1/6)$-approximate iO would enable us to take any one-way functions and construct a secure public-key encryption scheme from it. In the terminology of [20] it means that Cryptomania collapses to Minicrypt if $(1/6)$-approximate iO exists.

Intuition. Sahai and Waters [36] showed that iO and OWF imply PKE. Here we show that the very same construction, when instantiated using *approximate* iO, leads to "approximately correct" and "approximately secure" public-key encryption. Then, using a result of [19] we amplify the soundness and correctness to get a full fledged semantically secure public key encryption scheme.

Definition 21 (Approximate correctness and security for PKE). *We call a public-key bit-encryption scheme* (Gen, Enc, Dec) *for message space $\{0,1\}$ $\epsilon(n)$-correct if*

$$\Pr[\mathrm{Dec}_{dk}(\mathrm{Enc}_{ek}(b)) = b \colon (ek, dk) \leftarrow \mathrm{Gen}(1^n), b \xleftarrow{\$} \{0,1\}] \geq 1 - \epsilon(n)$$

where the probability is over the randomness of the key generation, encryption, decryption, and the bit b. We call (Gen, Enc, Dec) $\delta(n)$-*secure if for any PPT adversary A, it holds that*

$$\Pr[A(pk, \mathrm{Enc}_{pk}(b)) = b] \leq 1/2 + \delta(n)$$

where the probability is over the randomness of generation, encryption, the adversary, and bit b.

Holenstein [19] showed how to amplify any ϵ-correct and ϵ-secure PKE into a full fledged (semantically secure) PKE for sufficiently small ϵ.

Theorem 22 (Implied by Corollary 7.8 in [19]). *Suppose* (Gen, Enc, Dec) *is ϵ-correct and δ-secure for constants ϵ, δ such that $(1 - 2\epsilon)^2 > 2\delta$. Then there exists a semantically secure PKE.*

Theorem 23 below asserts that approximate iO and one-way functions imply approximately correct and approximately secure PKE.

Theorem 23 (Approximate iO + OWF $\Rightarrow$ Approximate PKE). *If ϵ-approximate iO and one-way functions exist, then there is an ϵ-correct and $(\epsilon + \mathrm{negl}(n))$-secure public-key bit encryption scheme.*

We first prove Theorem 20 using Theorem 23 and then will prove 23.

Proof (of Theorem 20). Because $(1 - 2 \cdot 1/6)^2 > 2 \cdot 1/6$, Theorem 20 follows immediately from Theorem 22 and the following Theorem 23 using $\epsilon = 1/6$.

In the rest of this section we prove Theorem 23.

Proof (of Theorem 23). We show that the very same construction of PKE from iO and OWF presented by Sahai and Waters [36], when instantiated with an ϵ-approximate iO, has the demanded properties of Theorem 23.

Properties of the Construction of [36]. We first describe the abstract properties of the construction of [36] (for PKE using iO and OWFs) and its security proof that we need to know.

– Construction/correctness:
  1. The key generation process generates a circuit C and publishes $O(C) = B$ as public key where O is an iO scheme.
  2. The encryption simply runs B on (r, b) where r is the encryption randomness and b is the bit to be encrypted.
  3. The scheme has completeness 1.
– Security: [36] proves the security of the construction above by showing that no PPT algorithm can distinguish between the following two random variables X_0, X_1 defined as:
  • $X_b \approx (O_s(C), C(r, b)) \approx (B, C(r, b))$ where s is the randomness for O. When clear from the context we drop the randomness s and simply write $O(C)$ denoting it as a random variable over the randomness of O.

It can be verified by inspection that the proof of [36] for indistinguishability of X_0 and X_1 does *not* rely on completeness of the obfuscation O and only relies on its indistinguishability (when applied to circuits with the same functions). We will rely on this feature of the proof of [36] in our analysis.

Below we analyze the correctness and security of the construction of [36] when O is an ϵ-approximate iO.

Correctness. By the definition of ϵ-approximate iO and ϵ-correct bit encryption, and the fact that the [36] construction has perfect completeness when O is iO, it follows that the completeness of the new scheme (when b is also chosen at random) is at least $1 - \epsilon$. Thus the scheme is ϵ-correct.

Security. First recall that for the basic construction of [36] using (perfect) iO, no PPT attacker A can guess b with probability better than $1/2 + \mathrm{negl}(n)$ when b is chosen at random and A is given a sample from the random variable X_b (for random b). As we mentioned above, the proof of this statement does not rely on the correctness of the used iO and only relies on its indistinguishability.

Now we want to bound the distinguishing advantage of PPT adversaries between the following random variables Y_0, Y_1:

- $Y_b \approx (B, B(r, b)) \approx (O_s(C), O_s(C)(r, b))$ where s is the randomness of the obfuscator O.

The difference between Y_b's and X_b's stems from the fact that the public-key $B = O(C)$ no longer computes the same exact function as the circuit C as the obfuscation only guarantees *approximate* correctness. We reduce the analysis of the new scheme to the original analysis of [36].

By the analysis of [36] we already know that if any PPT A is given a sample from X_b for a random b it has at most $1/2 + \mathrm{negl}(n)$ chance of correctly guessing b. Also note that the distributions X_b and Y_b for a *random* b are ϵ-close due to the ϵ-correctness of the obfuscation. More formally, the distributions X_b and Y_b could be defined over the same sampling space using: random seeds of key generation, obfuscation, encryption, and bit b. This way with probability $\geq 1 - \epsilon$ (and by the ϵ-approximate correctness of the obfuscation) the actual sampled values of X_b and Y_b will be equal, and this implies that they are ϵ-close. As a result, when we switch the distribution of the challenge given to the adversary and give a sample of Y_b (for random b) instead of a sample from X_b, the adversary's chance of guessing b correctly can increase at most by ϵ and reach at most $1/2 + \mathrm{negl} + \epsilon$. Therefore, the new scheme is $(\epsilon + \mathrm{negl})$-secure according to Definition 21.

A   Omitted Proofs

For sake of completeness we sketch the proof of Lemma 18.

Proof (Sketch of Lemma 18). Let ϵ be a parameter to be chosen later. Let R be an attacker who maintains a list of "learned" oracle queries $\mathcal{L}$ and, given c

sent by the sender for $b \xleftarrow{\$} \{0, 1\}$ and common input z, it adaptively asks the lexicographically first oracle query $x \notin \mathcal{L}$ that has at least ϵ chance of being asked by sender S conditioned on the knowledge of $(\mathcal{L}, z)$. After asking such x from f, A adds $(x, f(x))$ to $\mathcal{L}$. As long as such query x exists, R asks them. It was shown in [4] that this learning algorithm asks, on average, at most m/ϵ number of queries where $m = \text{poly}(|z|)$ is the number of queries asked by the sender. So as long as $\epsilon = \text{poly}(n/\delta)$ this learning algorithm is efficient.

Now, let $\mathcal{L}$ be the final learned set by R. If conditioned on $\mathcal{L}$ it holds that both of $b = 0$ and $b = 1$ have at least ρ probability of being used by S, then by conditioning on the distribution of the sender's view on $b = 0$ or $b = 1$ all the unlearned queries remain *at most* $\epsilon/\rho = \sigma$-heavy. Now it is easy to see that if we sample a random view for S conditioned on $\mathcal{L}, b = 0$ and $\mathcal{L}, b = 1$ and call them V_0 and V_1, the probability that queries of V_0 and V_1 collide out of $\mathcal{L}$ is at most $m \cdot \sigma \leq m \cdot \epsilon/\rho$. For $\rho > m \cdot \epsilon$ this probability is less than one, which means that if $\rho > m \cdot \epsilon$, then there exists a consistent pair of views for S that he can use to output c for both cases of $b = 0$ and $b = 1$. This means that Case 2 happens.

Now let us assume that Case 2 does not happen. It means that for all executions of the algorithm A, when A is done with learning the ϵ heavy queries, the probability of either $b = 1$ or $b = 0$ conditioned on $\mathcal{L}$ is at most ϵ. This means that A can guess b correctly with probability $1 - \epsilon$.

If we can choose $\rho = O(m/\epsilon)$ and $\epsilon = \delta$ in the argument above (assuming that Case 2 does not happen) we get an attacker A that asks $O(m \cdot \epsilon/\epsilon) = O(m)$ queries. We can alternatively choose smaller ρ and cut A's execution after it asks $O(m/\delta)$ number of queries and use $\epsilon = \delta/10$. By an application of the Markov inequality A will ask more than $100(m/\delta)$ number of queries with probability at most ϵ, and so A will ask at most $O(m/\delta)$ number of queries and will guess b correctly with probability at least $2\epsilon < \delta$.

References

1. Asharov, G., Segev, G.: Limits on the power of indistinguishability obfuscation and functional encryption. Cryptology ePrint Archive, Report 2015/341 (2015). http://eprint.iacr.org/
2. Barak, B., Garg, S., Kalai, Y.T., Paneth, O., Sahai, A.: Protecting obfuscation against algebraic attacks. In: Nguyen, P.Q., Oswald, E. (eds.) EUROCRYPT 2014. LNCS, vol. 8441, pp. 221–238. Springer, Heidelberg (2014)
3. Barak, B., Goldreich, O., Impagliazzo, R., Rudich, S., Sahai, A., Vadhan, S., Yang, K.: On the (im) possibility of obfuscating programs. J. ACM (JACM) **59**(2), 6 (2012)
4. Barak, B., Mahmoody-Ghidary, M.: Lower bounds on signatures from symmetric primitives. In: IEEE Symposium on Foundations of Computer Science (FOCS) (2007)
5. Bellare, M., Rogaway, P.: Random oracles are practical: a paradigm for designing efficient protocols. In: ACM Conference on Computer and Communications Security, pp. 62–73 (1993)
6. Blum, M., Kannan, S.: Designing programs that check their work. J. ACM (JACM) **42**(1), 269–291 (1995)

7. Boneh, D., Franklin, M.: Identity-based encryption from the Weil pairing. SIAM J. Comput. **32**(3), 586–615 (2003). Preliminary version in CRYPTO 2001

8. Brakerski, Z., Rothblum, G.N.: Virtual black-box obfuscation for all circuits via generic graded encoding. In: Lindell, Y. (ed.) TCC 2014. LNCS, vol. 8349, pp. 1–25. Springer, Heidelberg (2014)

9. Brakerski, Z., Vaikuntanathan, V.: Efficient fully homomorphic encryption from (standard) LWE. SIAM J. Comput. **43**(2), 831–871 (2014)

10. Chung, K.-M., Lin, H., Mahmoody, M., Pass, R.: On the power of nonuniformity in proofs of security. In: Proceedings of the 4th Conference on Innovations in Theoretical Computer Science, pp. 389–400. ACM (2013)

11. Garg, S., Gentry, C., Halevi, S.: Candidate multilinear maps from ideal lattices. In: Johansson, T., Nguyen, P.Q. (eds.) EUROCRYPT 2013. LNCS, vol. 7881, pp. 1–17. Springer, Heidelberg (2013)

12. Garg, S., Gentry, C., Halevi, S., Raykova, M., Sahai, A., Waters, B.: Candidate indistinguishability obfuscation and functional encryption for all circuits. In: IEEE Symposium on Foundations of Computer Science (FOCS), pp. 40–49. IEEE (2013)

13. Gennaro, R., Trevisan, L.: Lower bounds on the efficiency of generic cryptographic constructions. In: IEEE Symposium on Foundations of Computer Science (FOCS), pp. 305–313 (2000)

14. Gentry, C.: A fully homomorphic encryption scheme. Ph.D. thesis, Stanford University (2009). http://www.crypto.stanford.edu/craig

15. Gentry, C., Lewko, A.B., Sahai, A., Waters, B.: Indistinguishability obfuscation from the multilinear subgroup elimination assumption. IACR Cryptology ePrint Archive 2014:309 (2014)

16. Gentry, C., Wichs, D.: Separating succinct non-interactive arguments from all falsifiable assumptions. In: Fortnow, L., Vadhan, S.P. (eds.) ACM Symposium on Theory of Computing (STOC), pp. 99–108 (2011)

17. Goldreich, O., Micali, S., Wigderson, A.: Proofs that yield nothing but their validity and a methodology of cryptographic protocol design (extended abstract). In: IEEE Symposium on Foundations of Computer Science (FOCS), pp. 174–187. IEEE (1986)

18. Håstad, J., Impagliazzo, R., Levin, L.A., Luby, M.: A pseudorandom generator from any one-way function. SIAM J. Comput. **28**(4), 1364–1396 (1999)

19. Holenstein, T.: Strengthening key agreement using hard-core sets. Ph.D. thesis, ETH Zurich (2006)

20. Impagliazzo, R.: A personal view of average-case complexity. In: Structure in Complexity Theory Conference, pp. 134–147 (1995)

21. Impagliazzo, R., Rudich, S.; Limits on the provable consequences of one-way permutations. In: ACM Symposium on Theory of Computing (STOC), pp. 44–61. ACM Press (1989)

22. Joux, A.: A one round protocol for tripartite Diffie–Hellman. In: Bosma, W. (ed.) ANTS 2000. LNCS, vol. 1838, pp. 385–393. Springer, Heidelberg (2000)

23. Komargodski, I., Moran, T., Naor, M., Pass, R., Rosen, A., Yogev, E.: One-way functions and (im) perfect obfuscation. In: IEEE Symposium on Foundations of Computer Science (FOCS), pp. 374–383. IEEE (2014)

24. Mahmoody, M., Mohammed, A., Nematihaji, S.: More on impossibility of virtual black-box obfuscation in idealized models. Cryptology ePrint Archive, Report 2015/632 (2015). http://eprint.iacr.org/

25. Mahmoody, M., Pass, R.: The curious case of non-interactive commitments – on the power of black-box vs. non-black-box use of primitives. In: Safavi-Naini, R., Canetti, R. (eds.) CRYPTO 2012. LNCS, vol. 7417, pp. 701–718. Springer, Heidelberg (2012)
26. Mahmoody, M., Xiao, D.: On the power of randomized reductions and the checkability of SAT. In: IEEE Conference on Computational Complexity. IEEE Computer Society (2010)
27. Miles, E., Sahai, A., Weiss, M.: Protecting obfuscation against arithmetic attacks. Cryptology ePrint Archive, Report 2014/878 (2014). http://eprint.iacr.org/
28. Naor, M.: Bit commitment using pseudorandomness. J. Cryptology 4(2), 151–158 (1991)
29. Naor, M.: On cryptographic assumptions and challenges. In: Boneh, D. (ed.) CRYPTO 2003. LNCS, vol. 2729, pp. 96–109. Springer, Heidelberg (2003)
30. Pass, R.: Limits of provable security from standard assumptions. In: Fortnow, L., Vadhan, S.P. (eds.) ACM Symposium on Theory of Computing (STOC), pp. 109–118. ACM (2011)
31. Pass, R., Shelat, A.: Impossibility of VBB obfuscation with ideal constant-degree graded encodings. Cryptology ePrint Archive, Report 2015/383 (2015). http://www.deprint.iacr.org/
32. Pass, R., Seth, K., Telang, S.: Indistinguishability obfuscation from semantically-secure multilinear encodings. In: Garay, J.A., Gennaro, R. (eds.) CRYPTO 2014, Part I. LNCS, vol. 8616, pp. 500–517. Springer, Heidelberg (2014)
33. Regev, O.: On lattices, learning with errors, random linear codes, and cryptography. In: ACM Symposium on Theory of Computing (STOC) (2005)
34. Reingold, O., Trevisan, L., Vadhan, S.P.: Notions of reducibility between cryptographic primitives. In: Naor, M. (ed.) TCC 2004. LNCS, vol. 2951, pp. 1–20. Springer, Heidelberg (2004)
35. Rivest, R.L., Adleman, L., Dertouzos, M.L.: On data banks and privacy homomorphisms. In: Foundations of Secure Computation (Workshop, Georgia Inst. Tech., Atlanta, GA., 1977), pp. 169–179. Academic, New York (1978)
36. Sahai, A., Waters, B.: How to use indistinguishability obfuscation: deniable encryption, and more. In: Proceedings of the 46th Annual ACM Symposium on Theory of Computing, pp. 475–484. ACM (2014)
37. Shoup, V.: Lower bounds for discrete logarithms and related problems. In: Fumy, W. (ed.) EUROCRYPT 1997. LNCS, vol. 1233, pp. 256–266. Springer, Heidelberg (1997)

Indistinguishability Obfuscation: From Approximate to Exact

Nir Bitansky$^{(\boxtimes)}$ and Vinod Vaikuntanathan

MIT CSAIL, Cambridge, USA
{nirbitan,vinodv}@csail.mit.edu

Abstract. We show general transformations from subexponentially-secure approximate indistinguishability obfuscation (IO) where the obfuscated circuit agrees with the original circuit on a $1/2 + \epsilon$ fraction of inputs on a certain samplable distribution, into exact indistinguishability obfuscation where the obfuscated circuit and the original circuit agree on all inputs. As a step towards our results, which is of independent interest, we also obtain an approximate-to-exact transformation for functional encryption. At the core of our techniques is a method for "fooling" the obfuscator into giving us the correct answer, while preserving the indistinguishability-based security. This is achieved based on various types of secure computation protocols that can be obtained from different standard assumptions.

Put together with the recent results of Canetti, Kalai and Paneth (TCC 2015), Pass and Shelat (TCC 2016), and Mahmoody, Mohammed and Nemathaji (TCC 2016), we show how to convert indistinguishability obfuscation schemes in various ideal models into exact obfuscation schemes in the plain model.

1 Introduction

Program obfuscation, the science of making programs "unintelligible" while preserving functionality, has been a holy grail in cryptography for over a decade. While the most natural and intuitively appealing notion of obfuscation, namely *virtual-black-box* (VBB) obfuscation [7], was shown to have strong limitations [7,10,42], the recent work of Garg, Gentry, Halevi, Raykova, Sahai and Waters [35,57] opened new doors by demonstrating that the weaker notion of *indistinguishability obfuscation* (IO) is both very useful and potentially achievable. Since then, a veritable flood of applications has made indistinguishability obfuscation virtually "crypto-complete".

This work was done in part while the authors were visiting the Simons Institute for the Theory of Computing, supported by the Simons Foundation and by the DIMACS/Simons Collaboration in Cryptography through NSF grant CNS-1523467. First author supported in part by NSF Grants CNS-1350619 and CNS-1414119. Second author supported in part by NSF Grants CNS-1350619 and CNS-1414119, Alfred P. Sloan Research Fellowship, Microsoft Faculty Fellowship, the NEC Corporation, and a Steven and Renee Finn Career Development Chair from MIT.

© International Association for Cryptologic Research 2016
E. Kushilevitz and T. Malkin (Eds.): TCC 2016-A, Part I, LNCS 9562, pp. 67–95, 2016.
DOI: 10.1007/978-3-662-49096-9_4

On the flip side, the tremendous power of IO also begets its reliance on strong and untested computational assumptions. Indeed, it has been a major cryptographic quest to come up with a construction of IO based on well-studied computational assumptions. Garg et al. [35] gave the first *candidate* construction of IO, however the construction came as-is, without a security proof. We have recently seen several works [3,14,39,55] that shed light on how a security proof for IO will look like. Pass, Seth and Telang show security of an IO construction based on a "semantic security" assumption on multi-linear maps [33]; Gentry, Lewko, Sahai and Waters [39] (following [40]) show security based on the "multilinear subgroup elimination assumption" on multi-linear maps; Ananth and Jain [3] and Bitansky and Vaikuntanathan [14] show how to construct IO from any functional encryption scheme.

Unfortunately, the first two of these works are based on the mathematical abstraction of multi-linear maps which has had a troubled history so far (with several constructions [19,28,29,34,36,38] and matching attacks [25,27,34, 45,49]), and the last two rely on functional encryption with succinct encryption for which the only known constructions, yet again, use multi-linear maps.

Yet another line of work focuses on proving the security of obfuscators in so-called *idealized models*. In a typical idealized model, both the construction and the adversary have access to an oracle that implements a certain functionality; in the random oracle model [8], this is a random function; in the generic group model [58], this is the functionality of a group; and the most recent entrant to this club, namely the ideal multilinear map model, is an abstraction of the functionality of multilinear maps. Several works [4,6,21,22,60] along this route prove security of (different) constructions of obfuscation (even in the sense of virtual black-box security) in various ideal multi-linear map models.

An even more recent line of work, initiated by Canetti, Kalai, and Paneth [23], investigates how to transform constructions of obfuscation in idealized models into ones in the plain model, where there are no oracles. Indeed, this may lead to an aesthetically appealing avenue to constructing obfuscation schemes:

1. *Construct an obfuscation scheme in an appropriate idealized model.*
2. *"De-idealize" it: translate the ideal model obfuscation scheme into a scheme in the real world.*

Even if eventual constructions of obfuscation schemes do not initially proceed along these lines, we believe that this two-step process is a conceptually appealing route to eventual, mature, constructions of obfuscation schemes. Indeed, constructions in ideal models, while not immediately deployable, typically give us an abstract, high level, understanding.

In more detail, the work of [23] show that any obfuscator in the random oracle model can be converted to an obfuscator in the plain model with the same security properties. Pass and Shelat [54] and subsequently, Mahmoody, Mohammed and Nematihaji [50] extend this to the generic group and ring models respectively, as well as ideal multilinear maps model with *bounded multi-linearity*.

However, the resulting obfuscators suffer from a major drawback: *they only have approximate correctness*. That is, the plain model obfuscator may err on a polynomially large fraction of inputs (or more generally with some polynomial probability when inputs are taken from a given samplable distribution). Roughly speaking, these results proceed by isolating a list of "heavy oracle queries", that is, queries that arise in the evaluation of the obfuscated circuit on a large fraction of inputs. Once the (polynomially large set of) heavy queries are identified, the result of the oracle queries on this set is published as part of the obfuscated circuit. This approach will inherently miss the queries made by a rare set of inputs, resulting in an incorrect evaluation.

While these transformations already have interesting consequences (regarding the impossibility of VBB in these idealised models), the lack of correctness presents a serious obstacle towards fulfilling the above two-step plan. Indeed, it is far from clear that applications of IO will work when we only have *approximate IO* at our disposal. Certainly, one could go through the applications of IO one-by-one, and attempt to re-derive them from approximate IO, but in the absence of automated theorem provers[1], this seems neither particularly efficient nor aesthetically pleasing. This motivates us to ask:

Can approximate indistinguishability obfuscation be made exact?

In other words, we are asking for "one transformation to rule them all", a generic way to compile an approximate obfuscation scheme into a perfectly correct obfuscation scheme, automatically enabling to recover all the applications of IO even given only approximately correct obfuscation.

In this work, we provide exactly such a transformation, under standard additional assumptions. Let us now describe our results in detail.

1.1   Our Results

We say that an obfuscator $\mathsf{ap}\mathcal{O}$ is $(\mathcal{X}, \alpha)$-correct for a given input sampler $\mathcal{X}$ and $\alpha \in [0, 1]$ (which may depend on the security parameter), if it is correct with probability at least α over inputs sampled by $\mathcal{X}$. Security is defined as in the standard setting of (exact) indistinguishability obfuscation. We shall refer to such an obfuscator as an *approximate indistinguishability obfuscator*.

Our main result is that approximate IO with subexponential security for a certain class of samplers can be converted under standard assumptions into *almost exact IO* where for any circuit, with overwhelming probability over the coins of the obfuscator algorithm the resulting obfuscation is correct on *all* inputs. We present two routes towards this result based on different assumptions and with different parameters.

Theorem 1.1 (informal). *Assuming DDH, there exists an input sampler $\mathcal{X}_1$ and a transformation that for any $\alpha \geq \frac{1}{2} + \lambda^{-O(1)}$, converts any $(\mathcal{X}_1, \alpha)$-correct sub-exponentially secure IO scheme for $\mathbf{P}/\mathbf{poly}$ into an almost exact IO scheme for $\mathbf{P}/\mathbf{poly}$.*

---

[1] Graduate students do not count.

Theorem 1.2 (informal). *Assuming sub-exponentially-secure puncturable PRFs in* $\mathbf{NC}^1$*, there exists an input sampler* $\mathcal{X}_2$*, polynomial* $\mathrm{poly}_2(\cdot)$*, and a transformation that for any* $\alpha \geq 1 - \frac{1}{\mathrm{poly}_2(\lambda)}$*, converts any* $(\mathcal{X}_2, \alpha)$*-correct sub-exponentially-secure IO scheme for* $\mathbf{P}/\mathbf{poly}$ *into an almost exact IO scheme for* $\mathbf{P}/\mathbf{poly}$*.*

Since the works of [23,50,54] apply to any efficient sampler $\mathcal{X}$ and any α that is polynomially bounded away from 1, we obtain the following main corollary

Corollary 1.3 (Main Theorems + [23,50,54]). *Assume that there is an indistinguishability obfuscator in either the random oracle model, the ideal generic group/ring model, or ideal multilinear maps model with bounded multilinearity. Then, there is an (almost) exact obfuscator in the plain model.*

We note that our theorems result in IO that may still output an erroneous obfuscator, but only with some negligible probability over the coins of the obfuscator alone. This is analogous to the setting of correcting decryption errors in plain public key encryption [31], and as far as we know is sufficient in all applications. In subsequent work [15], we show that under a worst-case complexity assumption typically used in the setting of derandomization, we could transform any such obfuscator to one that is *perfectly correct*.

We also show how to transform approximate functional encryption into exact functional encryption, where approximate FE is defined analogously to approximate IO with respect to a distribution on the message space and decryption errors. Besides being of independent interest, this transformation will also serve as a building block to obtain the second theorem above.

Theorem 1.4 (Informal). *Assuming weak PRFs in* $\mathbf{NC}^1$*, there exists a message sampler* $\mathcal{X}$*, constant* η*, and a transformation that for any* $\alpha \geq 1 - \eta$*, converts any* $(\mathcal{X}, \alpha)$*-correct FE scheme for* $\mathbf{P}/\mathbf{poly}$ *into an almost exact scheme FE scheme for* $\mathbf{P}/\mathbf{poly}$*.*

We now proceed to provide an overview of our techniques.

1.2 Overview of Our Techniques

The starting point of our constructions comes from the notion of random self-reducibility [1]. That is, imagine that you have an error-prone algorithm A that computes a (Boolean) function F correctly on a $1/2 + \varepsilon$ fraction of inputs. Suppose that there is an efficient randomizer $r(\cdot)$ that maps an input x into a random input $r = r(x)$ such that given $F(r)$, one can efficiently recover $F(x)$. Then, we can turn A into a $\mathbf{BPP}$ algorithm for computing F; namely, $A'(x) = A(r(x))$. The new algorithm computes F correctly for *any* input with high probability over its own random coins. The probability of error can then be made arbitrarily small using standard amplification (i.e., taking majority of $\approx \varepsilon^{-2}$ invocations).

In our setting, F is an arbitrary function, which is likely *not* random self-reducible. Nevertheless, we show how to make the *essence* of this idea work, using various notions of (two-party and multi-party) non-interactive secure function evaluation (SFE) [9,37,59]. Indeed, certain forms of non-interactive SFE (or homomorphic encryption) have been used in several instances in the literature to obtain (sometimes computational) random self-reducibility [5,11,12,26]. The rough idea is that if we can get the obfuscator to homomorphically evaluate a given function on encryptions for some fixed input distribution, then it must also behave correctly with roughly the same probability on encryptions of any arbitrary input. This, however, should be done with care to ensure that homomorphic evaluation does not harm the security of the obfuscator. We next go into more details on how we carry out this agenda.

Our First Construction. Our first construction uses a two-party non-interactive secure function evaluation protocol with security against malicious senders. For simplicity, let us describe this approach in the language of fully homomorphic encryption (FHE). Let (Enc, Dec, Eval) be a (secret-key) FHE scheme (not necessarily compact). (We assume that the randomness of the key generation algorithm acts as the secret key, and avoid explicitly dealing with key generation.)

To exactly obfuscate a circuit C, we use the approximate obfuscator apO to obfuscate the circuit Eval_C that, given as input an encryption of some x, homomorphically computes an encryption of $C(x)$. Assume that $\mathsf{apO}(\mathsf{Eval}_C)$ is correct on a $1/2 + \varepsilon$ fraction of encryptions of 0^n. The key observation is that semantic security of the encryption scheme means that $\mathsf{apO}(\mathsf{Eval}_C)$ is also correct on a $1/2 + \varepsilon - \lambda^{-\omega(1)}$ fraction of encryptions of any x; that is, it outputs $\mathsf{Eval}_C(\mathsf{Enc}(x)) = \mathsf{Enc}(C(x))$. This gives the required randomizer and can be amplified to give us correctness *for every input x*.

The problem with this idea is the security of the final obfuscator. Indeed, $\mathsf{Eval}_C(\mathsf{Enc}(x))$ may reveal information about the circuit C beyond the output $C(x)$. The problem goes even further: since the evaluator in this setting is untrusted, she can try to run the obfuscated circuits with malformed encryptions, potentially making the problem much worse. The solution is to rely on a *maliciously function-hiding* homomorphic encryption scheme. Such an object can be constructed using Yao's garbled circuits combined with an oblivious transfer (OT) protocol secure against malicious receivers (such as the Naor-Pinkas protocol based on the DDH assumption [52]). The evaluation procedure, however, is randomized, but can be derandomized with a pseudo-random function.

While the above works perfectly assuming ideal VBB obfuscation, this is not necessarily the case for IO. Nevertheless, we observe that we can use apO to obfuscate this (de)randomized circuit using the machinery of probabilistic IO [24]. This allows us to show that indistinguishability obfuscation is maintained, but requires going through an exponential number of hybrids, in turn requiring sub-exponential security from apO (and some of the other involved primitives).

Our Second Construction. Our second construction goes through the notion of functional encryption (FE). In a (public-key) FE scheme, the owner of a functional secret key FSK_F can "decrypt" a ciphertext $\mathsf{FE.Enc}(\mathsf{MPK}, m)$ to learn

$F(m)$, but should learn nothing else about m. In an approximately correct FE scheme, the decryption algorithm could err on encryptions of certain messages m, but should be correct with probability $1 - \varepsilon$ on messages m drawn from a (sampleable) distribution $\mathcal{X}$.

We show how to transform an approximately correct FE scheme into an exact FE scheme. Here the main advantage over the setting of approximate IO is that we are only concerned with honestly generated encrypted messages and are not concerned with function hiding. In particular, we can relax the assumptions required for the SFE and rely on (a non-interactive) information-theoretic version of the Ben-Or-Goldwasser-Wigderson multi-party computation protocol for $\mathbf{NC}^1$ [9].

This construction also provides an alternative route for the IO transformation. Concretely, we show that starting from approximate IO, we can first apply the transformation of Garg et al. [35] to obtain approximate FE. For this to work, we need show how to obtain (almost exact) NIZKs and public-key encryption directly from approximate IO, which are required for the transformation. Then, in the second step, we apply our exact-to-approximate transformation for FE, and finally invoke a transformation from (exact) FE to IO [3,14]. The latter transformation requires that the size of the encryption circuit the FE scheme is relatively succinct. In our case, due to the BGW-based SFE, this size grows exponentially in the depth. Fortunately though, in [14], it is shown that this still suffices to obtain IO, assuming also puncturable PRFs in $\mathbf{NC}^1$.

Overall, this leads to a construction of (almost exact) IO from subexponentially-secure approximate IO and subexponentially-secure puncturable PRFs in $\mathbf{NC}^1$ (which in turn can be obtained from standard assumptions such as LWE [16]).

Organization. In Sect. 2, we define the required tools for our transformations, including the forms of SFE that we rely on. In Sect. 3, we describe our first basic transformation from approximate to exact IO. In Sect. 4, we describe our transformation from approximate to exact FE. In Sect. 5, we describe our second transformation for IO, going through our transformation for FE.

2 Preliminaries

The cryptographic definitions in the paper follow the convention of modeling security against non-uniform adversaries. An efficient adversary $\mathcal{A}$ is modeled as a sequence of circuits $\mathcal{A} = \{\mathcal{A}_\lambda\}_{\lambda \in \mathbb{N}}$, such that each circuit $\mathcal{A}_\lambda$ is of polynomial size $\lambda^{O(1)}$ with $\lambda^{O(1)}$ input and output bits. We often omit the subscript λ when it is clear from the context.

When we refer to a randomized algorithm $\mathcal{A}$, we typically do not explicitly denote its random coins, and use the notation $s \leftarrow \mathcal{A}$ or $s \leftarrow \mathcal{A}(x)$ if $\mathcal{A}$ has an extra input x. When we want to be explicit regarding the coins, we shall denote $s \leftarrow \mathcal{A}(r)$, or $s \leftarrow \mathcal{A}(x; r)$, respectively.

Whenever we refer to a circuit class $\mathcal{C} = \{\mathcal{C}_\lambda\}$, we mean that each set $\mathcal{C}_\lambda$ consists of Boolean circuits of size at most $\text{poly}(\lambda)$ for some polynomial $\text{poly}(\cdot)$, defined on the domain $\{0,1\}^{n(\lambda)}$. When referring to inputs $x \in \{0,1\}^{n(\lambda)}$, we often omit λ from the notation.

2.1   Non-interactive Secure Function Evaluation

We consider two-message secure function evaluation (SFE) protocols. Typically, such a protocol consists of two parties (A, B) and has the following syntax. Party A is given input x, encrypts x and sends the encrypted input to B. B given as additional input a function f, homomorphically evaluates f on the encrypted x, and returns the result to A, who can then decrypt the result $f(x)$. The protocol is required to ensure input-privacy for A and function privacy for B (on top of correctness).

Definition 2.1 (Secure Function Evaluation). *A scheme* SFE $=$ (Enc, Eval, Dec), *where* Enc, Eval *are probabilistic and* Dec *is deterministic, is a two-message secure function evaluation protocol for circuit class* $\mathcal{C} = \{\mathcal{C}_\lambda\}$, *where* $\mathcal{C}_\lambda$ *is defined over* $\{0,1\}^{n(\lambda)}$, *if the following requirements hold:*

- **Correctness:** *for any* $\lambda \in \mathbb{N}$, $C \in \mathcal{C}_\lambda$ *and input* $x \in \{0,1\}^n$ *in the domain of* C *it holds that:*

$$\Pr\left[\mathsf{Dec}(\mathsf{R}, \widehat{\mathsf{CT}}) = C(x) \;\middle|\; \begin{array}{l} (\mathsf{CT}, \mathsf{R}) \leftarrow \mathsf{Enc}(x) \\ \widehat{\mathsf{CT}} \leftarrow \mathsf{Eval}(\mathsf{CT}, C) \end{array}\right] \geq 1 - \nu(\lambda),$$

for some negligible $\nu(\cdot)$, *where the probability is over the coin tosses of* Enc *and* Eval.
- **Input Hiding:** *for any polysize distinguisher* $\mathcal{D}$ *there exists a negligible function* $\mu(\cdot)$, *such that for all* $\lambda \in \mathbb{N}$, *and equal size inputs* $x_0, x_1 \in \{0,1\}^n$:

$$|\Pr[\mathcal{D}(\mathsf{CT}_0) = 1] - \Pr[\mathcal{D}(\mathsf{CT}_1) = 1]| \leq \mu(\lambda),$$

  where $\mathsf{CT}_b \leftarrow \mathsf{Enc}(x_b)$.
- **Malicious Function Hiding:** *there exists a (possibly inefficient) function* Ext, *such that for any polysize distinguisher* $\mathcal{D}$ *there exists a negligible function* $\mu(\cdot)$, *such that for all* $\lambda \in \mathbb{N}$, *maliciously chosen* CT^*, *and equal size circuits* $C_0, C_1 \in \mathcal{C}_\lambda$ *that agree on* $x = \mathsf{Ext}(\mathsf{CT}^*)$:

$$\left|\Pr[\mathcal{D}(\widehat{\mathsf{CT}}_0) = 1] - \Pr[\mathcal{D}(\widehat{\mathsf{CT}}_1) = 1]\right| \leq \mu(\lambda),$$

  where $\widehat{\mathsf{CT}}_b \leftarrow \mathsf{Eval}(\mathsf{CT}^*, C_b)$.
  We say that the scheme is δ-function-hiding, *for some concrete negligible function* $\delta(\cdot)$, *if for all poly-size distinguishers, the above indistinguishability gap* $\mu(\lambda)$ *is smaller than* $\delta(\lambda)^{\Omega(1)}$.

Remark 2.2 (strong function privacy). For our most basic transformation from approximate IO to exact IO, we will require $2^{-\sigma(\lambda)} \cdot \lambda^{-\omega(1)}$-function-hiding, where $\sigma(\lambda)$ is the size of encryptions in the scheme. Below, we discuss an instantiation, based on the DDH assumption, that has perfect function-hiding, and thus satisfies this requirement.

Distributed Secure Function Evaluation. We will also consider a notion of two-message distributed function evaluation (DSFE). Such a protocol consists of $k + 2$ parties $(A, B_1, \ldots, B_k, C)$ and has the following syntax. Party A, given input x, shares x into k shares and sends the shares to $B_1, \ldots, B_k$. The parties $B_1, \ldots, B_k$ given as additional input a function f, homomorphically and non-interactively evaluate f on each share, and send the evaluated shares to C, who can then decrypt and obtain the result $f(x)$.

The protocol is required to ensure that each individual share sent by A in the second message hides all information regarding the input x. We also require that C gains no information on the input, except for the output of the function (formally, we will require an indistinguishability-based guarantee analogous to that of functional encryption.) Furthermore, we will require that correctness holds even if some τ fraction of the parties $B_1, \ldots, B_k$ are faulty.

Definition 2.3 (Distributed Secure Function Evaluation). *A scheme* DSFE = (Enc, Eval, Dec), *where* Enc *is probabilistic and* Eval, Dec *are deterministic, is a* (k, τ)-*secure distributed function evaluation protocol for circuit class* $\mathcal{C} = \{\mathcal{C}_\lambda\}$, *where* $\mathcal{C}_\lambda$ *is defined over* $\{0,1\}^n$ *for* $n = n(\lambda)$, $k = k(\lambda)$, *and* $\tau = \tau(\lambda)$, *if the following requirements hold:*

- **Correctness in the Presence of F aults:** *for any* $\lambda \in \mathbb{N}$, $C \in \mathcal{C}_\lambda$ *and input* $x \in \{0,1\}^n$ *in the domain of* C *and any set* $S \in [k]$ *of size smaller than* τk, *and functions* $\{\mathsf{Err}_i : i \in S\}$ *it holds that:*

$$\Pr\left[\mathsf{Dec}(\mathsf{R}, \widehat{\mathsf{CT}}_1, \ldots, \widehat{\mathsf{CT}}_k) = C(x) \,\middle|\, \begin{array}{l} (\mathsf{CT}_1, \ldots, \mathsf{CT}_k, \mathsf{R}) \leftarrow \mathsf{Enc}(x) \\ \forall i \in [k] \setminus S : \widehat{\mathsf{CT}}_i = \mathsf{Eval}(\mathsf{CT}_i, C) \\ \forall i \in S : \widehat{\mathsf{CT}}_i \leftarrow \mathsf{Err}_i(\mathsf{CT}_i) \end{array}\right] \geq 1 - \nu(\lambda),$$

   for some negligible $\nu(\cdot)$, *where the probability is over the coin-tosses of* Enc.
- **Input Hiding:** *for any polysize distinguisher* $\mathcal{D}$ *there exists a negligible function* $\mu(\cdot)$, *such that for all* $\lambda \in \mathbb{N}$, *and equal size inputs* $x_0, x_1 \in \{0,1\}^n$ *and any* $i \in [k]$:

$$|\Pr[\mathcal{D}(\mathsf{CT}_{0,i}) = 1] - \Pr[\mathcal{D}(\mathsf{CT}_{1,i}) = 1]| \leq \mu(\lambda),$$

   where $\mathsf{CT}_{b,i}$ *denotes the* i-*th ciphertext output by* $\mathsf{Enc}(x_b)$.
- **Residual Input Hiding:** *for any polysize distinguisher* $\mathcal{D}$ *there exists a negligible function* $\mu(\cdot)$, *such that for all* $\lambda \in \mathbb{N}$, *inputs* $x_0, x_1 \in \{0,1\}^n$, *and circuit* $C \in \mathcal{C}_\lambda$ *such that* $C(x_0) = C(x_1)$:

$$\left|\Pr[\mathcal{D}(\mathsf{R}_0, \widehat{\mathsf{CT}}_{0,1}, \ldots, \widehat{\mathsf{CT}}_{0,k}) = 1] - \Pr[\mathcal{D}(\mathsf{R}_1, \widehat{\mathsf{CT}}_{1,1}, \ldots, \widehat{\mathsf{CT}}_{1,k}) = 1]\right| \leq \mu(\lambda),$$

   where for $(b, i) \in \{0,1\} \times [k]$, $\widehat{\mathsf{CT}}_{b,i} = \mathsf{Eval}(\mathsf{CT}_{b,i}, C)$, *and* $(\mathsf{CT}_{b,1}, \ldots, \mathsf{CT}_{b,k}, \mathsf{R}_b) \leftarrow \mathsf{Enc}(x_b)$.

Remark 2.4 (difference from SFE). There are two main differences from SFE. The first is in security, in the above we do not require any type of function-hiding, but require residual input-hiding. The second is the functionality: we allow distributed evaluation (with some resilience to faults). The second difference is not essential, and is considered in order to reduce the underlying computational assumptions. In particular, a (non-distributed) SFE with residual input-hiding implies DSFE with $k = 1, \tau = 0$.

Remark 2.5 (deterministic Eval). Jumping ahead, we remark that we will use distributed SFE in a setting where the encryptor is always honest. Since we are not requiring any privacy against the encryptor, we may assume w.l.o.g that Eval is deterministic. Indeed, we can always sample any required randomness as part of the encryption process and embed it in the shares $CT_1, \ldots, CT_k$.

Instantiations. We now mention known instantiations of SFE and DSFE schemes, which we can rely on.

SFE. As mentioned above, for our application, we will require rather strong function-hiding. To instantiate the scheme we may rely on the SFE protocol obtained by using the oblivious transfer protocol of Naor and Pinkas [52] that is based on DDH and is secure against unbounded receivers in conjunction with an information-theoretic variant of Yao's garbled circuit [59] for $\mathbf{NC}^1$ [46]. The resulting SFE scheme is for classes of circuits in $\mathbf{NC}^1$, which will suffice for our purposes. Alternatively, we can use a strong enough computational variant of Yao based on sub-exponential one-way functions, resulting in a construction for all polynomial-size circuits.

More generally, the Naor-Pinkas OT can be replaced with any OT that has statistical function-hiding. In the CRS model, such two-message protocols exist from other standard assumptions as well [56]. While our main transformation is described using SFE in the plain model, it can be naturally extended to the CRS setting (see Remark 3.6).

DSFE. An information-theoretically secure DSFE scheme for circuit classes in $\mathbf{NC}^1$ can be obtained based on a non-interactive variant of the BGW protocol [9] similar to that used in [44]. In the full version of this paper, we outline this variant.In the resulting DSFE scheme, the complexity of encryption does not grow with the size of the circuits evaluated, but does grow exponentially with their maximal depth. As will be discussed later on, this will still be good enough in our context, to bootstrap functional encryption to indistinguishability obfuscation, as shown in [14].

2.2 Symmetric Encryption

A symmetric encryption scheme Sym consists of a tuple of two PPT algorithms (Sym.Enc, Sym.Dec). The encryption algorithm takes as input a symmetric key $SK \in \{0, 1\}^\lambda$, where λ is the security parameter, and a message $m \in \{0, 1\}^*$ of polynomial size in the security parameter, and outputs a ciphertext SCT. The decryption algorithm takes as input (SK, SCT), and outputs the decrypted message m. For this work, we only require one-time security. The detailed definition is standard and is given in the full version of this paper.

2.3   Puncturable Pseudorandom Functions

We consider a simple case of puncturable pseudo-random functions (PRFs) where any PRF may be punctured at a single point. The definition is formulated as in [57], and is satisfied by the Goldreich-Goldwasser-Micali PRF construction [18, 20, 41, 47].

Definition 2.6 (Puncturable PRFs). *Let n, k be polynomially bounded length functions. An efficiently computable family of functions*

$$\mathcal{PRF} = \left\{ \mathsf{PRF}_\mathsf{K} : \{0,1\}^* \to \{0,1\}^\lambda \;\middle|\; \mathsf{K} \in \{0,1\}^{k(\lambda)}, \lambda \in \mathbb{N} \right\},$$

associated with an efficient (probabilistic) key sampler $\mathsf{Gen}_{\mathcal{PRF}}$, is a puncturable PRF if there exists a poly-time puncturing algorithm Punc that takes as input a key K, and a point x^*, and outputs a punctured key $\mathsf{K}\{x^*\}$, so that the following conditions are satisfied:

1. **Functionality is preserved under puncturing:** *For every $x^* \in \{0,1\}^*$,*

$$\Pr_{\mathsf{K} \leftarrow \mathsf{Gen}_{\mathcal{PRF}}(1^\lambda)} \left[\forall x \neq x^* : \mathsf{PRF}_\mathsf{K}(x) = \mathsf{PRF}_{\mathsf{K}\{x^*\}}(x) \;\middle|\; \mathsf{K}\{x^*\} = \mathsf{Punc}(\mathsf{K}, x^*) \right] = 1.$$

2. **Indistinguishability at punctured points:** *for any polysize distinguisher $\mathcal{D}$ there exists a negligible function $\mu(\cdot)$, such that for all $\lambda \in \mathbb{N}$, and any $x^* \in \{0,1\}^*$,*

$$\left| \Pr[\mathcal{D}(x^*, \mathsf{K}\{x^*\}, \mathsf{PRF}_\mathsf{K}(x^*)) = 1] - \Pr[\mathcal{D}(x^*, \mathsf{K}\{x^*\}, u) = 1] \right| \leq \mu(\lambda) \; ,$$

where $\mathsf{K} \leftarrow \mathsf{Gen}_{\mathcal{PRF}}(1^\lambda), \mathsf{K}\{x^*\} = \mathsf{Punc}(\mathsf{K}, x^*),$ and $u \leftarrow \{0,1\}^\lambda$.
We further say that $\mathcal{PRF}$ is δ-secure, for some concrete negligible function $\delta(\cdot)$, if for all polysize distinguishers the above indistinguishability gap $\mu(\lambda)$ is smaller than $\delta(\lambda)^{\Omega(1)}$.

Remark 2.7 (uniform output). For some of our constructions, it will be convenient to assume that the PRF family is *one-universal*; that is, for any fixed x, $\mathsf{PRF}_\mathsf{K}(x)$ is distributed uniformly at random (when K is sampled at random). It is not hard to see that such a puncturable PRF can be easily obtained from any puncturable PRF by adding a random string U to the key and XORing U to every output.

3   Correcting Errors in Indistinguishability Obfuscation

In this section, we define approximate IO and show how to transform any approximate IO to (almost) perfectly correct IO, based on SFE.

3.1   Approximate and Exact IO

We start by defining indistinguishability obfuscation (IO) with almost perfect correctness. The definition is formulated as in [7].

Definition 3.1 (Indistinguishability Obfuscation). *A PPT algorithm $\mathcal{O}$ is said to be an* indistinguishability obfuscator *for a class of circuits $\mathcal{C} = \{\mathcal{C}_\lambda\}$, if it satisfies:*

1. **Almost Perfect Correctness:** *for any security parameter λ and $C \in \mathcal{C}_\lambda$,*

$$\Pr_{\mathcal{O}}\left[\forall x : \mathcal{O}(C, 1^\lambda)(x) = C(x)\right] \geq 1 - 2^{-\lambda} .$$

2. **Indistinguishability:** *for any polysize distinguisher $\mathcal{D}$ there exists a negligible function $\mu(\cdot)$, such that for any two circuits $C_0, C_1 \in \mathcal{C}$ that compute the same function and are of the same size:*

$$\left|\Pr[\mathcal{D}(\mathcal{O}(C_0, 1^\lambda)) = 1] - \Pr[\mathcal{D}(\mathcal{O}(C_1, 1^\lambda)) = 1]\right| \leq \mu(\lambda),$$

where the probability is over the coins of $\mathcal{D}$ and $\mathcal{O}$.
We further say that $\mathcal{O}$ is δ-secure, for some concrete negligible function $\delta(\cdot)$, if for all polysize distinguishers the above indistinguishability gap $\mu(\lambda)$ is smaller than $\delta(\lambda)^{\Omega(1)}$.

We now define an approximate notion of correctness that allows the obfuscated circuit to err with some probability over inputs taken from some samplable distribution.

Definition 3.2 $((\alpha, \mathcal{X})$-correct IO). *For $\alpha(\lambda) \in [0, 1]$ and an ensemble of input samplers $\mathcal{X} = \{\mathcal{X}_\lambda\}$, we say that $\mathcal{O}$ is $(\alpha, \mathcal{X})$-correct if instead of (almost) perfect correctness, it satisfies the following relaxed requirement:*

1. **Approximate Correctness:** *for any security parameter λ, $C \in \mathcal{C}_\lambda$,*

$$\Pr\left[\mathcal{O}(C, 1^\lambda)(x) = C(x) \;\middle|\; x \leftarrow \mathcal{X}_\lambda\right] \geq \alpha(\lambda),$$

where the probability is also over the coins of $\mathcal{O}$.

3.2   The Transformation

We now describe a transformation from approximately correct IO to (almost) perfectly correct IO and analyze it. The transformation is based on SFE satisfying a strong function-hiding guarantee. We discuss an instantiation based on standard computational assumptions in Sect. 3.3.

In Sect. 5, we discuss an alternative transformation through functional encryption based on weaker computational assumptions.

A Worst-Case Approximate Obfuscator. The main step of the transformation is to obtain random self-reducibility; that is, to convert an approximate obfuscator $\mathsf{ap}\mathcal{O}$, which works reasonably well on average for random inputs taken from an appropriate distribution, into a worst-case approximate obfuscator $\mathsf{wc}\mathcal{O}$ that, for any (worst-case) input, works well on average over the random coins of the obfuscator alone. Then, in the second step, we invoke standard "BPP amplification".

Ingredients. In the following, let λ denote a security parameter, let $\varepsilon < 1$ be some constant, $\eta(\lambda) = \lambda^{-\Omega(1)}$ and let $\mathcal{C} = \{\mathcal{C}_\lambda\}$ denote a circuit class. We rely on the following primitives:

- A secure function evaluation scheme SFE for $\mathcal{C}$ that is $2^{-\omega(\sigma(\lambda)+\log\lambda)}$-function-hiding, where $\sigma(\lambda)$ is the length of fresh ciphertexts generated by the encryption algorithm Enc for security parameter λ (and inputs of size $n = n(\lambda)$ in the domain of $\mathcal{C}_\lambda$).

- A $2^{-\tilde{\lambda}^\varepsilon}$-secure puncturable pseudo-random function family $\mathcal{PRF}$, where the security parameter is $\tilde{\lambda} = \omega(\sigma(\lambda) + \log\lambda)^{1/\varepsilon}$.

- A $(\frac{1}{2} + \eta(\lambda), \mathcal{X})$-correct, $2^{-\tilde{\lambda}^\varepsilon}$-secure indistinguishability obfuscator ap$\mathcal{O}$ for $\overline{\mathcal{C}}$, where the security parameter is $\tilde{\lambda} = \omega(\sigma(\lambda) + \log\lambda)^{1/\varepsilon}$. The sampler class $\mathcal{X}$ depends on SFE and the class $\overline{\mathcal{C}}$ depends on SFE, $\mathcal{PRF}$, and $\mathcal{C}$. Both $\mathcal{X}$ and $\overline{\mathcal{C}}$ are specified below as part of the description of the constructed (exact) obfuscator $\mathcal{O}$.

The Worst-Case Obfuscator wc$\mathcal{O}$:
Given a circuit $C : \{0,1\}^n \to \{0,1\}$ and security parameter λ, the obfuscator wc$\mathcal{O}(C, 1^\lambda)$

1. computes a new security parameter $\tilde{\lambda} = \omega(\sigma(\lambda) + \log\lambda)^{1/\varepsilon}$, where $\sigma(\lambda)$ is the length of ciphertexts as defined above,
2. samples a puncturable PRF seed $\mathsf{K} \leftarrow \mathsf{Gen}_{\mathcal{PRF}}(1^{\tilde{\lambda}})$,
3. computes the augmented C-evaluation circuit C_K defined in Fig. 1,
4. outputs an approximate obfuscation $\widetilde{C} \leftarrow \mathsf{ap}\mathcal{O}(C_\mathsf{K}, 1^{\tilde{\lambda}})$.

C_K

Hardwired: the circuit C and the seed K.
Input: a ciphertext CT.

1. derive randomness $r \leftarrow \mathsf{PRF}_\mathsf{K}(\mathsf{CT})$,
2. compute $\widehat{\mathsf{CT}} \leftarrow \mathsf{Eval}(\mathsf{CT}, C; r)$,
3. output $\widehat{\mathsf{CT}}$.

Padding: the circuit is padded to be of size $\ell(|C|, \mathsf{K})$, for some polynomial ℓ specified in the analysis.

Fig. 1. The augmented C-evaluation circuit.

We next describe the how the obfuscation $\widetilde{C}$ is evaluated on any input x via a randomized procedure.

Randomized Evaluation:
Given an obfuscation $\widetilde{C}$, an input $x \in \{0,1\}^n$, and security parameter λ:

1. compute $(\mathsf{CT}, \mathsf{R}) \leftarrow \mathsf{Enc}(x)$,
2. compute $\widehat{\mathsf{CT}} = \widetilde{C}(\mathsf{CT})$,
3. output $y = \mathsf{Dec}(\mathsf{R}, \widehat{\mathsf{CT}})$.

The ensemble of samplers $\mathcal{X}$ consists of samplers $\mathcal{X}^0$ that sample encryptions from $\mathsf{Enc}(0^n)$ whereas the class $\overline{\mathcal{C}}$ consists of circuits C_K as defined in Fig. 1.

Proposition 3.3. wc$\mathcal{O}$ *satisfies:*

1. **Worst-Case Approximate Correctness:** *for any* λ, $C \in \mathcal{C}_\lambda$, $x \in \{0,1\}^n$,

$$\Pr\left[\mathcal{O}(C, 1^\lambda)(x) = C(x)\right] \geq \frac{1}{2} + \eta(\lambda) - \lambda^{-\omega(1)} \ ,$$

   where the probability is over the coins of ap$\mathcal{O}$.
2. **Indistinguishability:** *as in Definition 3.1.*

The intuition behind the proof is outlined in the introduction. We now turn to the actual proof.

Proof. We first prove that the new obfuscator is worst-case approximately correct, and then prove that it is secure.

Correctness. For any $\lambda, n = n(\lambda)$, input $x \in \{0,1\}^n$, let us denote $\mathcal{X}^x := \mathsf{Enc}(x)$ a sampler for encryptions of x. Then, by the input-hiding guarantee of SFE, and the approximate correctness of ap$\mathcal{O}$, we claim that the approximate obfuscation is correct on encryptions of an arbitrary $x \in \{0,1\}^n$ as on encryptions of 0^n. That is, there exists a negligible $\mu(\lambda)$ such that

$$\Pr\left[\widetilde{C}(\mathsf{CT}) = C_\mathsf{K}(\mathsf{CT}) \mid \mathsf{CT} \leftarrow \mathcal{X}^x\right] \geq$$
$$\Pr\left[\widetilde{C}(\mathsf{CT}) = C_\mathsf{K}(\mathsf{CT}) \mid \mathsf{CT} \leftarrow \mathcal{X}^0\right] - \mu(\lambda) \geq$$
$$\frac{1}{2} + \eta(\lambda) - \mu(\lambda),$$

where in both of the above $\mathsf{K} \leftarrow \mathsf{Gen}_{\mathcal{PRF}}(1^{\tilde{\lambda}})$, $\widetilde{C} \leftarrow \mathsf{ap}\mathcal{O}(C_\mathsf{K}, 1^{\tilde{\lambda}})$.

It now follows that decryption is correct with probability noticeably larger than half. Concretely,

$$\Pr\left[\mathsf{Dec}(\mathsf{R}, \widehat{\mathsf{CT}}) = C(x) \ \middle| \ \begin{array}{l} \mathsf{CT}, \mathsf{R} \leftarrow \mathsf{Enc}(x) \\ \widehat{\mathsf{CT}} = \widetilde{C}(\mathsf{CT}) \end{array}\right] \geq$$

$$\Pr\left[\mathsf{Dec}(\mathsf{R}, \widehat{\mathsf{CT}}) = C(x) \ \middle| \ \begin{array}{l} \mathsf{CT}, \mathsf{R} \leftarrow \mathsf{Enc}(x) \\ \widehat{\mathsf{CT}} = C_\mathsf{K}(\mathsf{CT}) \end{array}\right] \cdot \Pr\left[\widetilde{C}(\mathsf{CT}) = C_\mathsf{K}(\mathsf{CT}) \mid \mathsf{CT} \leftarrow \mathcal{X}^x\right] =$$

$$\Pr\left[\mathsf{Dec}(\mathsf{R}, \widehat{\mathsf{CT}}) = C(x) \ \middle| \ \begin{array}{l} \mathsf{CT}, \mathsf{R} \leftarrow \mathsf{Enc}(x) \\ r = \mathsf{PRF}_\mathsf{K}(\mathsf{CT}) \\ \widehat{\mathsf{CT}} = \mathsf{Eval}(\mathsf{CT}, C; r) \end{array}\right] \cdot \Pr\left[\widetilde{C}(\mathsf{CT}) = C_\mathsf{K}(\mathsf{CT}) \mid \mathsf{CT} \leftarrow \mathcal{X}^x\right] \geq$$

$$(1 - \nu(\lambda)) \cdot \left(\frac{1}{2} + \eta(\lambda) - \mu(\lambda)\right),$$

where in all of the above $\mathsf{K} \leftarrow \mathsf{Gen}_{\mathcal{PRF}}(1^{\tilde{\lambda}})$, $\widetilde{C} \leftarrow \mathsf{ap}\mathcal{O}(C_\mathsf{K}, 1^{\tilde{\lambda}})$, and $\nu(\cdot)$ is some negligible function (corresponding to the negligible decryption error of SFE). In the last step, we relied on the fact that for any fixed CT, $\mathsf{PRF}_\mathsf{K}(\mathsf{CT})$ is distributed uniformly at random (Remark 2.7), and the (almost) perfect correctness of SFE.

This completes the proof of correctness.

Security Analysis. Consistently with the notation above, for $\mathsf{K} \leftarrow$ $\mathsf{Gen}_{\mathcal{PRF}}(1^{\tilde{\lambda}})$, and a circuit $C \in \mathcal{C}_\lambda$, we denote by $\tilde{C} \leftarrow \mathsf{apO}(C_\mathsf{K}, 1^{\tilde{\lambda}})$ the corresponding approximate obfuscation of the (derandomized) evaluation circuit. We show that for any polysize distinguisher there exists a neglgible $\mu(\cdot)$, such that for any $C_0, C_1 \in \mathcal{C}_\lambda$ that compute the same function it holds that

$$\left| \Pr[\mathcal{D}(\tilde{C}_0) = 1] - \Pr[\mathcal{D}(\tilde{C}_1) = 1] \right| \le \mu(\lambda) .$$

Roguhly, the above follows from the fact that the output of the two underlying obfuscated circuits on any point $\mathsf{CT} \in \{0,1\}^{\sigma(\lambda)}$ is indistinguishable even given C_0, C_1. Indeed, because the circuits C_0, C_1 compute the same function, by the function-hiding of SFE, for any ciphertext $\mathsf{CT} \in \{0,1\}^{\sigma(\lambda)}$, the evaluated ciphers $\mathsf{Eval}(\mathsf{CT}, C_0)$ and $\mathsf{Eval}(\mathsf{CT}, C_1)$ are indistinguishable. Canetti, Lin, Tessaro, and Vaikuntanathan [24] show that (sub-exponential) IO in conjunction with (sub-exponential) puncuturable PRFs are sufficient in this setting, which they formalize by *probabilistic IO* notion. For the sake of completeness, we next sketch the argument.

We consider a sequence of $2^\sigma + 1$ hybrids $\{\mathcal{H}_\mathsf{CT}\}_{\mathsf{CT} \in \{0, \ldots, 2^\sigma\}}$, where we naturally identify integers in $[2^\sigma]$ with strings in $\{0,1\}^\sigma$. In $\mathcal{H}_\mathsf{CT}$, we obfuscate a circuit $\mathbb{C}_\mathsf{CT}(\mathsf{CT}')$ that computes $C_{0,\mathsf{K}}$ for all $\mathsf{CT}' > \mathsf{CT}$ and $C_{1,\mathsf{K}}$ for all $\mathsf{CT}' \le \mathsf{CT}$; the circuit is padded to size ℓ (as in Fig. 1).

We first note that $\mathbb{C}_0$ computes the same function as $C_{0,\mathsf{K}}$ and that $\mathbb{C}_{2^\sigma}$ computes the same function as $C_{1,\mathsf{K}}$, and thus by the IO security,

$$\left| \Pr\left[\mathcal{D}(\tilde{C}_0) = 1 \right] - \Pr\left[\mathcal{D}(\mathcal{H}_0) = 1 \right] \right| \le 2^{-\tilde{\lambda}^\varepsilon} ,$$

$$\left| \Pr\left[\mathcal{D}(\mathcal{H}_{2^\sigma}) = 1 \right] - \Pr\left[\mathcal{D}(\tilde{C}_1) = 1 \right] \right| \le 2^{-\tilde{\lambda}^\varepsilon} .$$

We show that for any $\mathsf{CT} \in [2^\sigma]$,

$$\left| \Pr\left[\mathcal{D}(\mathcal{H}_{\mathsf{CT}-1}) = 1 \right] - \Pr\left[\mathcal{D}(\mathcal{H}_\mathsf{CT}) = 1 \right] \right| \le O(2^{-\tilde{\lambda}^\varepsilon}) .$$

This follows a standard puncturing argument with respect to the point CT, consisting of:

- puncturing PRF_K at CT, and hardwiring $C_{0,\mathsf{K}}(\mathsf{CT}) = \mathsf{Eval}(\mathsf{CT}, C_0; \mathsf{PRF}_\mathsf{K}(\mathsf{CT}))$, which relies on IO security,
- replacing $\mathsf{PRF}_\mathsf{K}(\mathsf{CT})$ with true randomness, which relies on pseudorandomness at punctured points,
- replacing $\mathsf{Eval}(\mathsf{CT}, C_0)$ with $\mathsf{Eval}(\mathsf{CT}, C_1)$, which relies on function hiding.
- reversing the above steps.

Each of the steps induces a loss of $2^{-\tilde{\lambda}^\varepsilon} = 2^{-\omega(\sigma(\lambda)+\log \lambda)}$ in the indistinguishability gap.

This completes the security analysis.

The (Almost) Exact Obfuscator $\mathcal{O}$: to obtain an (almost) exact obfuscator $\mathcal{O}$ from the worst-case approximate obfuscator we apply standard "BPP amplification". Such a transformation is given in [48, Appendix B]. For the sake of completeness, we sketch it here.

Obfuscation: Given a circuit $C : \{0,1\}^n \rightarrow \{0,1\}$ and security parameter λ, the obfuscator $\mathcal{O}(C, 1^\lambda)$ outputs $N = \frac{\omega(n+\lambda)}{\eta^2(\lambda)}$ obfuscations $\widetilde{C}_1, \ldots, \widetilde{C}_N$, where $\widetilde{C}_i \leftarrow \mathsf{wc}\mathcal{O}(C, 1^\lambda)$, and N random strings $r_1, \ldots, r_N$, where $r_i \leftarrow \{0,1\}^\lambda$.

Evaluation: Given an obfuscation $\left\{ \widetilde{C}_i, r_i \,\middle|\, i \in [N] \right\}$, input $x \in \{0,1\}^n$, and security parameter λ:

1. For $i \in [N]$, invoke the randomized evaluation procedure for $\widetilde{C}_i$, for input x, using randomness r_i, store the result y_i.
2. Output $y = \mathsf{majority}(y_1, \ldots, y_N)$.

Remark 3.4. (deterministic evaluator). Publishing the random strings r_i is done to match the usual obfuscation syntax where the evaluation is deterministic. We may also let the evaluator sample this randomness.

Proposition 3.5. $\mathcal{O}$ *is an (almost) perfectly correct indistinguishability obfuscator.*

Proof (Proof Sketch). We show correctness and security.

Correctness. By a Chernoff bound, for large enough λ, and any $x \in \{0,1\}^n$, the probability that the majority value y among all decrypted $y_1, \ldots, y_N$ is incorrect is bounded by

$$\Pr[y \neq C(x)] \leq 2^{-\Omega(N \cdot \eta^2(\lambda))} \leq 2^{-n+\lambda}.$$

The required correctness follows by a union bound over all inputs in $\{0,1\}^n$.

Security. The obfuscation consists of N independent copies of worst-case obfuscations $\widetilde{C}_i \leftarrow \mathsf{wc}\mathcal{O}(C)$, where $\mathsf{wc}\mathcal{O}$ satisfies indistinguishability. Security thus follows by a standard hybrid argument.

Remark 3.6 (SFE in the CRS model). The above construction can be naturally extended to rely also on non-interactive SFE schemes in the CRS model (rather than the plain model). Indeed, the CRS can be generated by the (honest) obfuscator.

3.3   Instantiating the Scheme

As discussed in Sect. 2.1, we can instantiate the SFE based on the (polynomial) DDH assumption and sub-exponential one-way functions. Sub-exponential one-way functions are also needed here in order to obtain sub-exponentially-secure puncturable PRFs.

We can thus state the following theorem

Theorem 3.7. *Assuming sub-exponentially secure approximate IO for* **P/poly**, *(polynomial) DDH, and sub-exponentially-secure one-way functions, there exists (almost) perfectly correct IO for P/**poly**.*

Alternative instantiations of the above under more computational assumptions [56] can be obtained when extending the scheme to SFE in the CRS model.

4   Correcting Errors in Functional Encryption

In this section, we define approximate FE and show how to transform any approximate FE to (almost) perfectly correct FE, based on DSFE. For the sake of concreteness, we focus on the public-key setting. We also focus on selective-security, which can be generically boosted to adaptive security [2].

4.1   Approximate and Exact FE

We recall the definition of public-key functional encryption (FE) with selective indistinguishability-based security [17,53], and extend the definition to the case of approximate correctness.

A public-key functional encryption (FE) scheme FE, for a function class $\mathcal{F} = \{\mathcal{F}_\lambda\}$ (represented by boolean circuits) and message space $\mathcal{M} = \{\{0,1\}^{n(\lambda)} : \lambda \in \mathbb{N}\}$, consists of four PPT algorithms (FE.Setup, FE.Gen, FE.Enc, FE.Dec) with the following syntax:

- FE.Setup(1^λ): Takes as input a security parameter λ in unary and outputs a (master) public key and a secret key (MPK, MSK).
- FE.Gen(MSK, f): Takes as input a secret key MSK, a function $f \in \mathcal{F}_\lambda$ and outputs a functional key FSK_f.
- FE.Enc(MPK, M): Takes as input a public key MPK, a message $M \in \{0,1\}^{n(\lambda)}$ and outputs an encryption of M.
- FE.Dec(FSK_f, FCT): Takes as input a functional key FSK_f, a ciphertext FCT and outputs $\widehat{M}$.

We next recall the required security properties as well the common (almost) perfect correctness requirement.

Definition 4.1. (Selectively-Secure Public-key FE). *A tuple of PPT algorithms* FE = (FE.Setup, FE.Gen, FE.Enc, FE.Dec) *is a selectively-secure public-key functional encryption scheme, for function class $\mathcal{F} = \{\mathcal{F}_\lambda\}$, and message space $\mathcal{M} = \{\{0,1\}^{n(\lambda)} : \lambda \in \mathbb{N}\}$, if it satisfies:*

1. **Almost Perfect Correctness:** *for every $\lambda \in \mathbb{N}$, message $M \in \{0,1\}^{n(\lambda)}$, and function $f \in \mathcal{F}_\lambda$,*

$$\Pr\left[f(M) \leftarrow \mathsf{FE.Dec}(\mathsf{FSK}_f, \mathsf{FCT}) \,\middle|\, \begin{array}{l} (\mathsf{MPK}, \mathsf{MSK}) \leftarrow \mathsf{FE.Setup}(1^\lambda) \\ \mathsf{FSK}_f \leftarrow \mathsf{FE.Gen}(\mathsf{MSK}, f) \\ \mathsf{FCT} \leftarrow \mathsf{FE.Enc}(\mathsf{MPK}, M) \end{array}\right] \geq 1 - 2^{-\lambda}.$$

2. **Selective-Security:** *for any polysize adversary $\mathcal{A}$, there exists a negligible function $\mu(\lambda)$ such that for any $\lambda \in \mathbb{N}$, it holds that*

$$\mathsf{Adv}_{\mathcal{A}}^{\mathsf{FE}} = \left| \Pr[\mathsf{Expt}_{\mathcal{A}}^{\mathsf{FE}}(1^\lambda, 0) = 1] - \Pr[\mathsf{Expt}_{\mathcal{A}}^{\mathsf{FE}}(1^\lambda, 1) = 1] \right| \leq \mu(\lambda),$$

where for each $b \in \{0,1\}$ and $\lambda \in \mathbb{N}$ the experiment $\mathsf{Expt}_{\mathcal{A}}^{\mathsf{FE}}(1^\lambda, b)$, modeled as a game between the challenger and the adversary $\mathcal{A}$, is defined as follows:

(a) *The adversary submits the challenge message-pair $M_0, M_1 \in \{0,1\}^{n(\lambda)}$ to the challenger.*

(b) *The challenger executes FE.Setup(1^λ) to obtain (MPK, MSK). It then executes FE.Enc(MPK, M_b) to obtain FCT. The challenger sends (MPK, FCT) to the adversary.*

(c) *The adversary submits function queries to the challenger. For any submitted function query $f \in \mathcal{F}_\lambda$, if $f(M_0) = f(M_1)$, the challenger generates and sends FSK$_f \leftarrow$ FE.Gen(MSK, f). In any other case, the challenger aborts.*

(d) *The output of the experiment is the output of $\mathcal{A}$.*

We further say that FE is δ-secure, for some concrete negligible function $\delta(\cdot)$, if for all polysize adversaries the above indistinguishability gap $\mu(\lambda)$ is smaller than $\delta(\lambda)^{\Omega(1)}$.

We now define an approximate notion of correctness that allows decryption to error with some probability over encryption of messages taken from some given distribution.

Definition 4.2. (($\alpha, \mathcal{X}$)-**correct FE**). *For $\alpha(\lambda) \in [0,1]$ and an ensemble of samplers $\mathcal{X} = \{\mathcal{X}_\lambda\}$ with support $\mathcal{M} = \{\{0,1\}^{n(\lambda)} : \lambda \in \mathbb{N}\}$, we say that FE is ($\alpha, \mathcal{X}$)-correct if instead of (almost) perfect correctness, it satisfies the following relaxed requirement:*

1. **Approximate Correctness:** *for every $\lambda \in \mathbb{N}$, and function $f \in \mathcal{F}_\lambda$,*

$$
\Pr\left[f(M) \leftarrow \mathsf{apFE.Dec}(\mathsf{apFSK}_f, \mathsf{FCT}) \middle| \begin{array}{l} (\mathsf{apMPK}, \mathsf{apMSK}) \leftarrow \mathsf{apFE.Setup}(1^\lambda) \\ \mathsf{apFSK}_f \leftarrow \mathsf{apFE.Gen}(\mathsf{apMSK}, f) \\ M \leftarrow \mathcal{X}_\lambda \\ \mathsf{apFCT} \leftarrow \mathsf{apFE.Enc}(\mathsf{apMPK}, M) \end{array} \right] \geq \alpha(\lambda).
$$

4.2   The Transformation

We now describe the transformation from approximately correct FE to (almost) perfectly correct FE and analyze it. The transformation is based on DSFE. We discuss instantiations in Sect. 4.3.

A Worst-Case Approximate FE. As in the case of obfuscation, the main step of the FE transformation is to obtain random self-reducibility; that is, to convert an approximate FE scheme apFE, which works reasonably well on average for random messages taken from some appropriate distribution, into a worst-case approximate scheme wcFE that, for any (worst-case) message, works well on average over the random coins of the obfuscator alone. Then, in the second step, we invoke standard "BPP amplification".

Ingredients. In the following, let λ denote a security parameter, and let $\mathcal{F} = \{\mathcal{F}_\lambda\}$ denote a function class. Consider functions $k(\lambda) \in \mathbb{N}$, and $\rho(\lambda), \eta(\lambda) \in [0,1]$ such that $\eta = \frac{1}{2} - \sqrt{\rho} \in [\frac{1}{\lambda^{O(1)}}, \frac{1}{2} - \frac{1}{\lambda^{O(1)}}]$. We rely on the following primitives:

- A $(k, \sqrt{\rho})$-secure distributed function evaluation scheme DSFE for $\mathcal{C}$. We shall further assume that when encrypting an input, the shares $\mathsf{CT}_1, \ldots, \mathsf{CT}_k$ all have the same marginal distribution (i.e., $\mathsf{CT}_i \equiv \mathsf{CT}_j$).[2]
- A $(1 - \rho, \mathcal{X})$-correct (single-key, selectively-secure) functional encryption scheme apFE = (apFE.Setup, apFE.Gen, apFE.Enc, apFE.Dec) for $\overline{\mathcal{C}}$. The sampler class $\mathcal{X}$ depends on DSFE and the class $\overline{\mathcal{F}}$ depends on DSFE, and $\mathcal{F}$. Both $\mathcal{X}$ and $\overline{\mathcal{F}}$ are specified below as part of the description of the constructed (almost exact) scheme FE.
- A one-time symmetric key encryption scheme Sym = (Sym.Enc, Sym.Dec).

The Worst-Case Scheme wcFE: The scheme wcFE, for function class $\mathcal{F} = \{\mathcal{F}_\lambda\}$ and message space $\mathcal{M} = \{\{0,1\}^{n(\lambda)} : \lambda \in \mathbb{N}\}$, consists of the algorithms (wcFE.Setup, wcFE.Gen, wcFE.Enc, wcFE.Dec) defined as follows:

- wcFE.Setup(1^λ): generate (apMPK, apMSK) $\leftarrow$ apFE.Setup(1^λ). The public key MPK and secret key MSK are accordingly set to be the apMPK and apMSK.
- wcFE.Gen(wcMSK, f): sample SCT $\leftarrow$ Sym.Enc(SK, $0^{\ell \times k}$), where $\ell = \ell(\lambda)$ is a polynomial specified in the security analysis, and SK $\leftarrow \{0,1\}^\lambda$. Consider the augmented f-evaluation function f_{SCT} as defined in Fig. 2. Generate apFSK$_{\mathsf{SCT}} \leftarrow$ apFE.Gen(apMPK, f_{SCT}). The functional key wcFSK$_f$ will consists of the functional key apFSK$_{\mathsf{SCT}}$.
- wcFE.Enc(wcMPK, M):
  1. Compute $(\mathsf{CT}_1, \ldots, \mathsf{CT}_k, \mathsf{R}) \leftarrow$ DSFE.Enc(M),
  2. For $j \in [k]$
     - let ap$M_j = (\text{norm}, \mathsf{CT}_j, \perp, \perp)$
     - generate apFCT$_j \leftarrow$ apFE.Enc(apMPK, apM_j).
  Output wcFCT $= \{\text{apFCT}_1, \ldots, \text{apFCT}_k, \mathsf{R}\}$.
- wcFE.Dec(wcFSK$_f$, wcFCT):
  1. Parse wcFSK$_f =$ apFSK$_{\mathsf{SCT}}$ and wcFCT $= (\text{apFCT}_1, \ldots, \text{apFCT}_k, \mathsf{R})$.
  2. for $j \in [k]$, compute $\widehat{\mathsf{CT}}_j \leftarrow$ apFE.Dec(apFSK$_{\mathsf{SCT}}$, apFCT$_j$).
  3. output $y =$ DSFE.Dec($\mathsf{R}, \widehat{\mathsf{CT}}_1, \ldots, \widehat{\mathsf{CT}}_k$).

The ensemble of samplers $\mathcal{X}$ consists of samplers $\mathcal{X}^0$ that sample FE plaintexts of the form ap$M = (\text{norm}, \mathsf{CT}, \perp, \perp)$ where CT is the first of k ciphertext components sampled from DSFE.Enc(0^n), i.e. it is a share of a zero-encryption in the underlying DSFE scheme. The class $\overline{\mathcal{F}}$ consists of circuits f_{SCT} as in Fig. 2.

Proposition 4.3. wcFE *satisfies:*

1. **Worst-Case Approximate Correctness:** *for every* $\lambda \in \mathbb{N}$, *function* $f \in \mathcal{F}_\lambda$, *and message* $M \in \{0,1\}^n$,

$$\Pr\left[f(M) \leftarrow \text{wcFE.Dec(wcFSK}_f, \text{wcFCT}) \;\middle|\; \begin{array}{l} (\text{wcMPK}, \text{wcMSK}) \leftarrow \text{wcFE.Setup}(1^\lambda) \\ \text{wcFSK}_f \leftarrow \text{wcFE.Gen(wcMSK}, f) \\ \text{wcFCT} \leftarrow \text{wcFE.Enc(wcMPK}, M) \end{array}\right] \geq$$
$$\frac{1}{2} + \eta - \lambda^{-\omega(1)} \ .$$

---

[2] This is just to simplify the construction and is satisfied the instantiation discussed in Sect. 2. In Remark 4.4, we explain how this assumption can be removed (at the cost of complicating the construction).

$$f_{\mathsf{SCT}}$$

Hardwired: the circuit f and a symmetric key ciphertext SCT.
Input $\mathsf{ap}M = (b, \mathsf{CT}, \mathsf{SK}, j)$:

        – a flag bit b,
        – a DSFE ciphertext CT_j,
        – a symmetric encryption key SK.
        – index $j \in [k]$.

1. If $b = \mathbf{norm}$ (normal mode of operation, ignoring inputs SK, j),
    – compute $\widehat{\mathsf{CT}} = \mathsf{Eval}(\mathsf{CT}, f)$.
2. If $b = \mathbf{trap}$ (trapdoor mode of operation, ignoring input CT),
    – compute $(\widehat{\mathsf{CT}}_1, \dots, \widehat{\mathsf{CT}}_k) = \mathsf{Sym.Dec}(\mathsf{SK}, \mathsf{SCT})$,
    – let $\widehat{\mathsf{CT}} := \widehat{\mathsf{CT}}_j$.
3. Output $\widehat{\mathsf{CT}}$.

Fig. 2. The augmented f-evaluation circuit.

2. **Selective Security:** *as in Definition 4.1.*

We now turn to the proof.

Proof. We first prove that the new obfuscator has worst-case approximate correctness, and then prove that it is selectively secure.

Correctness. For any $\lambda, n = n(\lambda)$, message $M \in \{0,1\}^n$, let us denote $\mathcal{X}^M$ a sampler for FE plaintexts of the form $\mathsf{ap}M = (\mathbf{norm}, \mathsf{CT}, \perp, \perp)$ that is defined just like $\mathcal{X}^0$ except that CT is a share of an encryption of M sampled from $\mathsf{DSFE.Enc}(M)$ in the underlying DSFE scheme, rather than a share of an encryption of 0^n.

Then, by the input-hiding guarantee of SFE, and the approximate correctness of apFE, we claim that, for any function $f \in \mathcal{F}$ and corresponding f_{SCT}, decryption in apFE is correct on encryptions of an arbitrary $M \in \{0,1\}^n$ as on encryptions of 0^n. That is, there exists a negligible $\mu(\lambda)$ such that

$$\Pr\left[\mathsf{apFE.Dec}(\mathsf{apFSK}_{f_{\mathsf{SCT}}}, \mathsf{apFCT}) = f_{\mathsf{SCT}}(\mathsf{ap}M) \;\middle|\; \mathsf{ap}M \leftarrow \mathcal{X}^M\right] \geq$$
$$\Pr\left[\mathsf{apFE.Dec}(\mathsf{apFSK}_{f_{\mathsf{SCT}}}, \mathsf{apFCT}) = f_{\mathsf{SCT}}(\mathsf{ap}M) \;\middle|\; \mathsf{ap}M \leftarrow \mathcal{X}^0\right] - \mu \geq$$
$$1 - \rho - \mu \;,$$

where $(\mathsf{apMPK}, \mathsf{apMSK}) \leftarrow \mathsf{apFE.Setup}(1^\lambda)$, $\mathsf{apFSK}_{f_{\mathsf{SCT}}} \leftarrow \mathsf{apFE.Gen}(\mathsf{apMSK}, f_{\mathsf{SCT}})$, $\mathsf{apFCT} \leftarrow \mathsf{apFE.Enc}(\mathsf{apMPK}, \mathsf{ap}M)$, as defined above in the construction of the exact scheme, and $\mathsf{ap}M = (\mathbf{norm}, \mathsf{CT}, \perp, \perp)$.

We now consider alternative samplers $\{\mathcal{X}_j^M \mid j \in [k]\}$ that sample $\mathsf{ap}M_j$ just as in the canonical $\mathcal{X}^M$, except that CT is sampled as the jth share of a DSFE encryption of M (rather than the first). Note that by our assumption that the shares $CT_1, \ldots, CT_k \leftarrow DSFE.Enc(M)$ have the same marginal distribution, the samplers $\mathcal{X}^M, \mathcal{X}_1^M, \ldots, \mathcal{X}_k^M$ all sample from the same distribution. In particular, they satisfy the above statement regarding the probability of correct decryption, satisfied by $\mathcal{X}^M$.

We shall denote by $\mathcal{X}_j^M | CT_j$ the corresponding sampler conditioned on $CT = CT_j$ for some fixed CT_j. We now consider the joint sampler $(apM_1, \ldots, apM_k) \leftarrow \mathcal{X}_{[k]}^M$ where first shares $(CT_1, \ldots, CT_k)$ are sampled from $DSFE.Enc(M)$, and then each apM_j is sampled from $\mathcal{X}_j | CT_j$. Note that this sampler corresponds to the way that encryption is done in our actual scheme wcFE defined above.

Noting that the marginal distribution of each apM_j sampled accordingly to $\mathcal{X}_{[k]}^M$ is the same as $\mathcal{X}_j^M$, it follows that the expected number of successful decryptions for a sample from $\mathcal{X}_{[k]}^M$ can be lower bounded as follows

$$\mathbb{E}\left[|\{j \mid apFE.Dec(apFSK_{f_{SCT}}, apFCT_j) = f_{SCT}(apM_j)\}| \,\Big|\, (apM_1, \ldots, apM_k) \leftarrow \mathcal{X}_{[k]}^M\right] =$$

$$k \cdot \Pr\left[apFE.Dec(apFSK_{f_{SCT}}, apFCT_j) = f_{SCT}(apM_j) \,\Big|\, apM_j \leftarrow \mathcal{X}^M\right] \geq$$

$$k \cdot (1 - \rho - \mu),$$

where $(apMPK, apMSK) \leftarrow apFE.Setup(1^\lambda)$, $apFSK_{f_{SCT}} \leftarrow apFE.Gen(apMSK, f_{SCT})$, $apFCT_j \leftarrow apFE.Enc(apMPK, apM_j)$.

It follows by averaging that with probability at least $1 - \sqrt{\rho} - \frac{2\mu}{\sqrt{\rho}}$ the number of successful decryptions as defined above is larger than $k \cdot (1 - \sqrt{\rho})$. In particular, (for large enough λ) the fraction of faults is below the threshold $\sqrt{\rho}$ allowing to reconstruct $f_{SCT}(apM)$.

Going to our actual encryption scheme wcFE, we now claim that decryption is correct with probability noticeably larger than half. Concretely,

$$\Pr\left[DSFE.Dec(R, \widehat{CT}_1, \ldots, \widehat{CT}_k) = f(M) \,\middle|\, \begin{array}{l} CT_1, \ldots, CT_k, R \leftarrow DSFE.Enc(M) \\ \widehat{CT}_j = apFE.Dec(apFSK_{f_{SCT}}, apFCT_j) \end{array}\right] \geq$$

$$\Pr\left[DSFE.Dec(R, \widehat{CT}_1, \ldots, \widehat{CT}_k) = f(M) \,\middle|\, \begin{array}{l} CT_1, \ldots, CT_k, R \leftarrow DSFE.Enc(M) \\ |\{\widehat{CT}_j = f_{SCT}(apM_j)\}| \geq \sqrt{\rho} \cdot k \end{array}\right] \cdot$$

$$\Pr\left[|\{apFE.Dec(apFSK_{f_{SCT}}, apFCT_j) = f_{SCT}(apM_j)\}| \geq \sqrt{\rho} \cdot k \,\middle|\, CT_1, \ldots, CT_k, R \leftarrow DSFE.Enc(M)\right] =$$

$$\Pr\left[DSFE.Dec(R, \widehat{CT}_1, \ldots, \widehat{CT}_k) = f(M) \,\middle|\, \begin{array}{l} CT_1, \ldots, CT_k, R \leftarrow DSFE.Enc(M) \\ |\{\widehat{CT}_j = DSFE.Eval(CT_j, f)\}| \geq \sqrt{\rho} \cdot k \end{array}\right] \cdot$$

$$\Pr\left[|\{apFE.Dec(apFSK_{f_{SCT}}, apFCT_j) = f_{SCT}(apM_j)\}| \geq \sqrt{\rho} \cdot k \,\middle|\, CT_1, \ldots, CT_k \leftarrow \mathcal{X}_{[k]}^M\right] \geq$$

$$(1 - \nu) \cdot \left(1 - \sqrt{\rho} - \frac{2\mu}{\sqrt{\rho}}\right) \geq \frac{1}{2} + \eta - \lambda^{-\omega(1)} \,,$$

where in all of the above $(apMPK, apMSK) \leftarrow apFE.Setup(1^\lambda)$, $apFSK_{f_{SCT}} \leftarrow apFE.Gen(apMSK, f_{SCT})$, $apM_j = (norm, CT_j, \bot, \bot)$, $apFCT_j \leftarrow apFE.Enc(apMPK, apM_j)$, and $\nu(\cdot)$ is some negligible function (corresponding to the negligible decryption error of DSFE).

This completes the proof of correctness.

Security Analysis. We prove the selective security of wcFE in a sequence of hybrids, showing that any adversary $\mathcal{A}$ cannot tell the case that the challenge is an encryption of M_0 from the case that the challenge is an encryption of M_1, for the corresponding (M_0, M_1) of his choice.

$\mathcal{H}_1$: this corresponds to the usual security game where the challenge is an encryption of M_0.

$\mathcal{H}_2$: here, when generating a key FSK_f for a function f, and accordingly generating an (approximate) key $\mathsf{apFSK}_{f_{\mathsf{SCT}}}$ for the function f_{SCT}, the symmetric ciphertext SCT is not an encryption of $0^{\ell \times k}$ as in the previous hybrid, but rather an encryption of the DSFE evaluation corresponding to the challenge ciphertext. Concretely, consider the generation of the challenge ciphertext FCT^*:

- $\mathsf{FE.Enc}(\mathsf{MPK}, M_0)$:
  1. Compute $(\mathsf{CT}_1^*, \ldots, \mathsf{CT}_k^*, \mathsf{R}^*) \leftarrow \mathsf{DSFE.Enc}(M_0)$,
  2. For $j \in [k]$
     - let $\mathsf{ap}M_j^* = (\mathtt{norm}, \mathsf{CT}_j^*, \perp, \perp)$
     - generate $\mathsf{apFCT}_j^* \leftarrow \mathsf{apFE.Enc}(\mathsf{apMPK}, \mathsf{ap}M_j^*)$.
     Output $\mathsf{FCT}^* = (\mathsf{apFCT}_1^*, \ldots, \mathsf{apFCT}_k^*, \mathsf{R}^*)$.

Then SCT will now encrypt $\widehat{\mathsf{CT}}_{f,1}^*, \ldots, \widehat{\mathsf{CT}}_{f,k}^*$, where $\widehat{\mathsf{CT}}_{f,j}^* = \mathsf{DSFE.Eval}(\mathsf{CT}_j^*, f)$.

Indistinguishability from the previous hybrid follows by the semantic-security of symmetric encryption. (At this point, a corresponding symmetric secret key SK is not present – in all encryptions the symmetric-key slot is set to $\perp$.)

$\mathcal{H}_3$: here, we change the generation of the challenge ciphertext so to invoke the trapdoor mode rather than the normal mode. Concretely, for each $j \in [k]$, we generate $\mathsf{ap}M_j^* = (\mathtt{trap}, \perp, \perp, \mathsf{SK}, j)$, where SK is the symmetric key corresponding SCT.

Indistinguishability from the previous hybrid follows from the security of the underlying scheme apFE. Indeed, at this point, for every function f for which a key $\mathsf{apFSK}_{f_{\mathsf{SCT}}}$ was generated,

$$f_{\mathsf{SCT}}(\mathtt{trap}, \perp, \perp, \mathsf{SK}, j) = f_{\mathsf{SCT}}(\mathtt{norm}, \mathsf{CT}_j, \perp, \perp) = \widehat{\mathsf{CT}}_{f,j}^* \ .$$

$\mathcal{H}_4$: here, we change how the evaluations $\widehat{\mathsf{CT}}_{f,j}^*$ are generated. Recall that in the previous hybrid $\widehat{\mathsf{CT}}_{f,j}^* = \mathsf{DSFE.Eval}(\mathsf{CT}_j^*, f)$, where CT_j^* was generated as part of $(\mathsf{CT}_1^*, \ldots, \mathsf{CT}_k^*, \mathsf{R}^*) \leftarrow \mathsf{DSFE.Enc}(M_0)$. Now, instead of encrypting M_0 in the latter we encrypt M_1.

Indistinguishability now follows from the residual input privacy of the underlying DSFE, since $f(M_0) = f(M_1)$. (Recall, that this is guaranteed also in the presence of R^*, provided that $\mathsf{CT}_1^*, \ldots, \mathsf{CT}_k^*$ are absent from the adversary's view, which is indeed the case in this hybrid.)

$\mathcal{H}_5$–$\mathcal{H}_8$: symmetrically follow the above hybrids in reverse order, until the usual security game where M_1 is encrypted in the challenge.

This completes the security analysis.

Remark 4.4 (removing the assumption on equally-distributed shares). In the above construction we have assumed that the DSFE shares $\mathsf{CT}_1, \ldots, \mathsf{CT}_k$ have the same marginal distribution (for which we have also exhibited an instantiation in Sect. 2.1). To remove this assumption, we could have an instance of an approximate FE scheme apFE_i for each i with respect to the corresponding distribution on CT_i (whereas in the construction above we relied on one instance of an approximate FE defined with respect to the marginal distribution which was the same for all shares).

The (Almost) Exact Scheme FE: to obtain an (almost) exact scheme from the worst-case approximate scheme we again apply standard "BPP amplification". Namely, we consider N parallel copies of the scheme for a large enough N.

Formally, the scheme FE, for function class $\mathcal{F} = \{\mathcal{F}_\lambda\}$ and message space $\mathcal{M} = \{\{0,1\}^{n(\lambda)} : \lambda \in \mathbb{N}\}$, consists of the algorithms (FE.Setup, FE.Gen, FE.Enc, FE.Dec) defined as follows:

- FE.Setup(1^λ): let $N = \frac{\omega(n+\lambda)}{\eta^2}$. For $i \in [N]$, generate $(\mathsf{wcMPK}_i, \mathsf{wcMSK}_i) \leftarrow \mathsf{wcFE.Setup}(1^\lambda)$. The public key MPK and secret key MSK are accordingly set to be all of the public keys $\{\mathsf{wcMPK}_i\}_{i\in[N]}$ and secret keys $\{\mathsf{wcMSK}_i\}_{i\in[N]}$.
- FE.Gen(MSK, f): For $i \in [N]$, generate $\mathsf{wcFSK}_{f,i} \leftarrow \mathsf{wcFE.Gen}(\mathsf{wcMSK}_i, f)$. The functional key FSK_f will consists of the functional keys $\{\mathsf{wcFSK}_{f,i}\}_{i\in[N]}$.
- FE.Enc(MPK, M): For $i \in [N]$, compute $\mathsf{wcFCT}_i \leftarrow \mathsf{wcFE.Enc}(\mathsf{wcMPK}_i, M)$. The ciphertext FCT consists of the ciphertexts $(\mathsf{wcFCT}_1, \ldots, \mathsf{wcFCT}_N)$.
- FE.Dec(FSK_f, FCT):
  1. Parse $\mathsf{FSK}_f = \{\mathsf{wcFSK}_{f,i}\}_{i\in[N]}$ and $\mathsf{FCT} = \{\mathsf{wcFCT}_i\}_{i\in[N]}$.
  2. For $i \in [N]$, compute $y_i = \mathsf{wcFE.Dec}(\mathsf{wcFSK}_{f,i}, \mathsf{wcFCT}_i)$.
  3. Output $y = \mathsf{majority}(y_1, \ldots, y_N)$.

Proposition 4.5. *FE is an (almost) perfectly correct selectively-secure functional encryption scheme.*

Proof (Proof Sketch). We show correctness and security.

Correctness. By a Chernoff bound, for large enough λ, and message $M \in \{0,1\}^n$, the probability that the majority value y among all decrypted $y_1, \ldots, y_N$ is incorrect is bounded by

$$\Pr[y \neq f(M)] \leq 2^{-\Omega(N \cdot \eta^2(\lambda))} \leq 2^{-n+\lambda} .$$

The required correctness follows by a union bound over all messages in $\{0,1\}^n$.

Security. The scheme consists of N independent copies of the worst-case scheme that is selectively secure. Security thus follows by a standard hybrid argument.

4.3   Instantiating the Scheme

As discussed in Sect. 2.1, we can instantiate the DSFE based an information-theoretic variant of BGW for $\mathbf{NC}^1$, resulting in an FE scheme for $\mathbf{NC}^1$. The scheme can then be generically bootstrapped to $\mathbf{P/poly}$ assuming weak PRFs in $\mathbf{NC}^1$ [2].

Theorem 4.6. *Assuming approximate FE for* $\mathbf{P/poly}$ *and weak PRFs in* $\mathbf{NC}^1$, *there exists (almost) perfectly correct FE for* $\mathbf{P/poly}$.

5   An Alternative Transformation for IO Based on FE

Recall that the transformation from (subexponential) approximate IO to (almost) exact IO described in Sect. 3.2 required SFE with function hiding against malicious receivers, and was instantiated based on (polynomial) DDH and subexponential one-way functions. In this section, we show an alternative transformation based on any subexponential puncturable PRF in $\mathbf{NC}^1$. The transformation is based on a combination of the FE transformation from Sect. 4 and known results from the literature.

The high-level idea consists of three basic steps:

1. Start with a (subexponentially-secure) approximate IO and implement directly (subexponentially-secure) approximate FE with compact ciphertexts by following a construction from the exact IO setting [35].
2. Apply the transformation from approximate FE to obtain exact FE with compact ciphertexts, based on weaker assumptions.
3. Apply a transformation from exact FE to (exact) IO [3,14].

Fulfilling this high-level plan requires some care though. The transformation of Garg et al. [35] from IO to FE naturally extends to the approximate setting, but relies on additional assumptions: (exact) public-key encryption and (exact, or rather complete) NIZKs. While these primitives are known based on exact IO [13,57], they do not work in the approximate setting. Nevertheless, we show how these constructions can be extended to imply the exact versions of the primitives (from approximate IO). A second issue that should be addressed is how the approximate FE to exact FE transformation affects the complexity of FE encryption. Indeed, the transformations of [3,14] require certain succinctness properties. We observe that the transformation, when instantiated with the BGW-based DSFE, satisfies the required compactness, when assuming additionally (sub-exponentially-secure) puncturable PRFs in $\mathbf{NC}^1$.

Overall, we prove

Theorem 5.1. *Assuming approximate IO for* $\mathbf{P/poly}$ *and puncturable PRFs in* $\mathbf{NC}^1$, *both with sub-exponential security there exists (almost) perfectly correct IO* $\mathbf{P/poly}$.

We next provide further details.

5.1 Approximate FE from Approximate IO

The starting point for this step is the Garg et al. [35]. To obtain FE from IO and PKE, and NIZKs, the transformation works as follows. Each encryption has the form (e_0, e_1, π), where e_0, e_1 encrypt a message M under two independent copies of a plain (exact) public-key encryption scheme, and π is a proof that (e_0, e_1) are indeed well-formed using an (exact) NIZK with statistical simulation-soundness.

A functional key for a function f is an obfuscation of a circuit $C_{\mathsf{SK}_0, \mathsf{CRS}}$ that given (e_0, e_1, π):

- verifies the correctness of π with respect to the hardwired common reference string CRS,
- if the proof is accepting, decrypts e_0 using the hardwired secret key SK_0 to obtain M,
- and outputs $f(M)$.

It follows readily that if we replace exact IO in this transformation with approximate IO (say while still using exact PKE and NIZKs) the resulting FE scheme would be approximately-correct. Concretely to get α-correct FE for a message sampler $\mathcal{X}$, we start with IO that is α-correct for an input sampler $\mathcal{X}'$ that samples FE encryptions (e_0, e_1, π) of random messages M taken from $\mathcal{X}$.

In fact, even if we start with α-correct versions of PKE and NIZKs we would get $\Omega(\alpha)$-correct FE, however, the security of the FE scheme might no longer hold; indeed, the exact correctness of the PKE and NIZK play an important role in the security proof in [35]. To fill this gap we will show how to obtain exact NIZK and PKE directly from approximate IO. More accurately, we would obtain almost exactly correct versions where the NIZK and PKE are exactly correct with overwhelming probability over the choice of their public parameters (i.e., the common reference string and public-keys), which is sufficient for the security proof in [35].

(Almost) Exact PKE. To obtain (almost) exact PKE, we start with the PKE of Sahai and Waters [57] based on exact IO and one-way functions. Here the public key consists of an obfuscation $\widetilde{C}$ of a circuit C_{K} that given a PRG seed s outputs $\mathsf{PRF}_{\mathsf{K}}(\mathsf{PRG}(s))$ for an appropriately stretching pseudo-random generator and a puncturable PRF. An encryption of M consists of $\mathsf{PRG}(s), M \oplus \widetilde{C}(s)$. Replacing exact IO with α-correct IO in their transformation results in approximate PKE in two senses: (a) the scheme is correct with probability α over an encryption of any message M; (b) it is weakly semantically secure, the probability of guessing a random encrypted message M can be bounded by $\beta = 2^{-|M|} + \lambda^{-\omega(1)} + (1 - \alpha)$. Schemes such as the latter can be corrected using techniques from the literature [31, Theorem 4] so long that $\beta < O(\alpha^4)$, which holds for constant α that is sufficiently close to 1.

In the resulting scheme, the probability of decryption error is over the choice of public-key and the randomness used in encryption. In the same work [31], Dwork, Naor, and Reingold show how to shift the error probability to the choice of the public-key alone; namely, get a scheme where with overwhelming probability over the choice of keys there are no decryption errors at all. This is done as follows, assume the decryption error is bounded by $2^{-\lambda}$, and encryption uses $r(\lambda) = \lambda^{O(1)}$ bits of randomness. We will now publish together with the public key a random string $R \leftarrow \{0,1\}^r$. Encryption will now be done with randomness $R' = R \oplus \mathsf{PRG}(s)$, where $\mathsf{PRG} : \{0,1\}^{\lambda/2} \rightarrow \{0,1\}^r$ is a pseudo-random generator and $s \leftarrow \{0,1\}^{\lambda/2}$ is a random seed. Due to the sparseness of the PRG with probability $2^{-\Omega(\lambda)}$ over the choice of the keys the are no decryption errors. Semantic-security is maintained due to the security of the PRG.

(Almost) Exact NIZK. Statistical simulation-sound NIZKs can be constructed from any NIZK proof and non-interactive commitment schemes in the common reference string model [35]. The same also holds for the case that the NIZK is almost exact (where the resulting SSS NIZK will also be almost exact). The required commitments can be constructed from one-way functions [51]. We now describe how to obtain the required NIZKs from approximate IO.

Concretely, we examine the NIZK construction of Bitansky and Paneth [13] based on exact IO and one-way functions. In their construction, IO is used to implement *invariant signatures* [43], which are in turn used to implement the *hidden-bit model* [32]. Concretely, a verification key VK in their scheme consists of an obfuscated circuit $C_{\mathsf{CRS},\mathsf{K}}$ that given a message $M \in \{0,1\}^n$, computes $(b,r) \leftarrow \mathsf{PRF}_\mathsf{K}(M)$ using a puncturable PRF, and outputs a Naor commitment $\mathsf{C} = \mathsf{COM}_{\mathsf{CRS}}(b,r)$, with respect to common reference string CRS.

Replacing exact IO with α-correct IO preserves two of the guarantees of the invariant signatures: 1) it is invariant in the sense that for every verification key VK and message M, $\mathsf{C} = \mathsf{VK}(M)$ can be opened to a unique bit b, due to the binding of the commitment; 2) it satisfies pseudorandomness of the unique property b, since the obfuscator is as secure as in the exact case. However, now completeness only holds with probability α over random messages M. The implementation of the hidden bit model indeed invokes the invariant signatures for random messages. This leads to a corresponding NIZK with completeness error $(1 - \alpha) \cdot \mathrm{poly}(\lambda)$, for some poly that depends on the NIZK construction (and soundness error $2^{-\lambda}$). Assuming $\alpha > 1 - \frac{1}{\lambda \cdot \mathrm{poly}(\lambda)}$, we can then take say λ^2 independent copies, requiring that the prover succeeds only on a single instance, resulting in a NIZK with completeness error $2^{-\lambda}$ and soundness error $\lambda^2 \cdot 2^{-\lambda}$.

In the resulting scheme, the completeness error is over the choice of the common-reference string and the randomness used by the prover. As before we can use the technique from [31], to shift the error probability to the choice of the CRS alone by sparsifying the coins used by the prover using a PRG. This transformation still maintains computational zero-knowledge due to the pseudo-randomness of the PRG, and has the same unconditional soundness.

A caveat of the latter transformation is that it can only correct a polynomial fraction $1 - \alpha = \lambda^{-\Theta(1)}$ of errors (and not say a constant, as in the previous construction). We stress that in the de-idealized constructions of obfuscation [23,50,54] the error rate can be made an arbitrary small polynomial. Thus the implication to constructions of IO with an ideal assisting oracle still holds.

5.2   FE to IO

Exact FE vs Almost Exact FE. The transformations of [3,14] from FE to IO are naturally described in terms of perfectly correct FE, nevertheless it is easy to verify that they also work starting from FE that is perfectly-correct with overwhelming probability only over the setup phase generating the keys. The resulting IO will be almost perfectly correct.

To almost exact FE given in Sect. 4 can be turned to one that satisfies the above property using again the randomness sparsification technique of [31] described above.

Succinctness. In the previous subsection, we described how to obtain an approximate FE scheme where the complexity of encryption is independent of the circuit and output size of the corresponding functions, as inherited from the exact scheme of [35]. To fulfill our approach we need to make sure that applying our transformation to exact FE still preserves certain succinctness properties required by the transformations in [3,14]. Concretely, we note that our approximate to exact FE transformation inherits its succinctness from the underlying DSFE scheme. As discussed in 4.3, using the BGW-based DSFE, incurs a $2^{O(d)}$ overhead in the complexity of encryption, where d is the maximal depth of any circuit in the class, but is otherwise as efficient. Fortunately, Bitansky and Vaikuntanathan [14] show that this is still sufficient for a variant of their transformation from FE to IO, under the additional assumption of sub-exponentially-secure puncturable PRFs in $\mathbf{NC}^1$.

Acknowledgements. We thank Ilan Komargodsky for pointing out [48, Appendix B], and the anonymous TCC reviewers for their comments.

References

1. Abadi, M., Feigenbaum, J., Kilian, J.: On hiding information from an oracle. J. Comput. Syst. Sci. **39**(1), 21–50 (1989)
2. Ananth, P., Brakerski, Z., Segev, G., Vaikuntanathan, V.: The trojan method in functional encryption: From selective to adaptive security, generically. IACR Cryptology ePrint Archive 2014, 917 (2014)
3. Ananth, P., Jain, A.: Indistinguishability obfuscation from compact functional encryption. In: Gennaro, R., Robshaw, M. (eds.) CRYPTO 2015. LNCS, vol. 9215, pp. 308–326. Springer, Heidelberg (2015)
4. Applebaum, B., Brakerski, Z.: Obfuscating circuits via composite-order graded encoding. In: Dodis, Y., Nielsen, J.B. (eds.) TCC 2015, Part II. LNCS, vol. 9015, pp. 528–556. Springer, Heidelberg (2015)
5. Applebaum, B., Ishai, Y., Kushilevitz, E.: Computationally private randomizing polynomials and their applications. Comput. Complex. **15**(2), 115–162 (2006)
6. Barak, B., Garg, S., Kalai, Y.T., Paneth, O., Sahai, A.: Protecting obfuscation against algebraic attacks. In: Nguyen, P.Q., Oswald, E. (eds.) EUROCRYPT 2014. LNCS, vol. 8441, pp. 221–238. Springer, Heidelberg (2014)
7. Barak, B., Goldreich, O., Impagliazzo, R., Rudich, S., Sahai, A., Vadhan, S.P., Yang, K.: On the (im)possibility of obfuscating programs. J. ACM **59**(2), 6 (2012)

8. Bellare, M., Rogaway, P.: Random oracles are practical: a paradigm for designing efficient protocols. In: Denning, D.E., Pyle, R., Ganesan, R., Sandhu, R.S., Ashby, V. (eds.) CCS 1993, pp. 62–73. ACM, Fairfax (1993)

9. Ben-Or, M., Goldwasser, S., Wigderson, A.: Completeness theorems for non-cryptographic fault-tolerant distributed computation (extended abstract). In: Proceedings of the 20th Annual ACM Symposium on Theory of Computing, pp. 1–10. Chicago, Illinois, USA, 2–4 May 1988

10. Bitansky, N., Canetti, R., Cohn, H., Goldwasser, S., Kalai, Y.T., Paneth, O., Rosen, A.: The impossibility of obfuscation with auxiliary input or a universal simulator. In: Garay, J.A., Gennaro, R. (eds.) CRYPTO 2014, Part II. LNCS, vol. 8617, pp. 71–89. Springer, Heidelberg (2014)

11. Bitansky, N., Goldwasser, S., Jain, A., Paneth, O., Vaikuntanathan, V., Waters, B.: Time-lock puzzles from randomized encodings. IACR Cryptology ePrint Archive 2015, 514 (2015)

12. Bitansky, N., Paneth, O.: From the impossibility of obfuscation to a new non-black-box simulation technique. In: 53rd Annual IEEE Symposium on Foundations of Computer Science, FOCS 2012, pp. 223–232. New Brunswick, NJ, USA, 20–23 October 2012

13. Bitansky, N., Paneth, O.: ZAPs and non-interactive witness indistinguishability from indistinguishability obfuscation. In: Dodis, Y., Nielsen, J.B. (eds.) TCC 2015, Part II. LNCS, vol. 9015, pp. 401–427. Springer, Heidelberg (2015)

14. Bitansky, N., Vaikuntanathan, V.: Indistinguishability obfuscation from functional encryption. In: FOCS (2015)

15. Bitansky, N., Vaikuntanathan, V.: A note on perfect correctness by derandomization (2015)

16. Boneh, D., Lewi, K., Montgomery, H., Raghunathan, A.: Key homomorphic PRFs and their applications. In: Canetti, R., Garay, J.A. (eds.) CRYPTO 2013, Part I. LNCS, vol. 8042, pp. 410–428. Springer, Heidelberg (2013)

17. Boneh, D., Sahai, A., Waters, B.: Functional encryption: a new vision for public-key cryptography. Commun. ACM 55(11), 56–64 (2012)

18. Boneh, D., Waters, B.: Constrained pseudorandom functions and their applications. In: Sako, K., Sarkar, P. (eds.) ASIACRYPT 2013, Part II. LNCS, vol. 8270, pp. 280–300. Springer, Heidelberg (2013)

19. Boneh, D., Wu, D.J., Zimmerman, J.: Immunizing multilinear maps against zeroizing attacks. IACR Cryptology ePrint Archive 2014, 930 (2014)

20. Boyle, E., Goldwasser, S., Ivan, I.: Functional signatures and pseudorandom functions. In: Krawczyk, H. (ed.) PKC 2014. LNCS, vol. 8383, pp. 501–519. Springer, Heidelberg (2014)

21. Brakerski, Z., Rothblum, G.N.: Black-box obfuscation for d-CNFs. In: Naor, M. (ed.) Innovations in Theoretical Computer Science, ITCS 2014, pp. 235–250. ACM, Princeton, 12–14 January 2014

22. Brakerski, Z., Rothblum, G.N.: Virtual black-box obfuscation for all circuits via generic graded encoding. In: Lindell, Y. (ed.) TCC 2014. LNCS, vol. 8349, pp. 1–25. Springer, Heidelberg (2014)

23. Jager, T.: Verifiable random functions from weaker assumptions. In: Dodis, Y., Nielsen, J.B. (eds.) TCC 2015, Part II. LNCS, vol. 9015, pp. 121–143. Springer, Heidelberg (2015)

24. Canetti, R., Lin, H., Tessaro, S., Vaikuntanathan, V.: Obfuscation of probabilistic circuits and applications. In: Dodis, Y., Nielsen, J.B. (eds.) TCC 2015, Part II. LNCS, vol. 9015, pp. 468–497. Springer, Heidelberg (2015)

25. Cheon, J.H., Han, K., Lee, C., Ryu, H., Stehlé, D.: Cryptanalysis of the multilinear map over the integers. In: Oswald, E., Fischlin, M. (eds.) EUROCRYPT 2015. LNCS, vol. 9056, pp. 3–12. Springer, Heidelberg (2015)
26. Chung, K.-M., Kalai, Y., Vadhan, S.: Improved delegation of computation using fully homomorphic encryption. In: Rabin, T. (ed.) CRYPTO 2010. LNCS, vol. 6223, pp. 483–501. Springer, Heidelberg (2010)
27. Coron, J.-S., Gentry, C., Halevi, S., Lepoint, T., Maji, H.K., Miles, E., Raykova, M., Sahai, A., Tibouchi, M.: Zeroizing without low-level zeroes: new MMAP attacks and their limitations. In: Gennaro, R., Robshaw, M. (eds.) CRYPTO 2015. LNCS, vol. 9215, pp. 247–266. Springer, Heidelberg (2015)
28. Coron, J.-S., Lepoint, T., Tibouchi, M.: Practical multilinear maps over the integers. In: Canetti, R., Garay, J.A. (eds.) CRYPTO 2013, Part I. LNCS, vol. 8042, pp. 476–493. Springer, Heidelberg (2013)
29. Coron, J.-S., Lepoint, T., Tibouchi, M.: New multilinear maps over the integers. In: Gennaro, R., Robshaw, M. (eds.) CRYPTO 2015. LNCS, vol. 9215, pp. 267–286. Springer, Heidelberg (2015)
30. Dodis, Y., Nielsen, J.B. (eds.): TCC 2015. Security and Cryptology, vol. 9015. Springer, Heidelberg (2015)
31. Dwork, C., Naor, M., Reingold, O.: Immunizing encryption schemes from decryption errors. In: Cachin, C., Camenisch, J.L. (eds.) EUROCRYPT 2004. LNCS, vol. 3027, pp. 342–360. Springer, Heidelberg (2004)
32. Feige, U., Lapidot, D., Shamir, A.: Multiple noninteractive zero knowledge proofs under general assumptions. SIAM J. Comput. 29(1), 1–28 (1999)
33. Garg, S., Gentry, C., Halevi, S.: Candidate multilinear maps from ideal lattices and applications. IACR Cryptology ePrint Archive 2012, 610 (2012)
34. Garg, S., Gentry, C., Halevi, S.: Candidate multilinear maps from ideal lattices. In: Johansson, T., Nguyen, P.Q. (eds.) EUROCRYPT 2013. LNCS, vol. 7881, pp. 1–17. Springer, Heidelberg (2013)
35. Garg, S., Gentry, C., Halevi, S., Sahai, A., Raikova, M., Waters, B.: Candidate indistinguishability obfuscation and functional encryption for all circuits. In: FOCS (2013)
36. Garg, S., Gentry, C., Halevi, S., Zhandry, M.: Fully secure functional encryption without obfuscation. IACR Cryptology ePrint Archive, 2014, 666 (2014)
37. Gentry, C.: Fully homomorphic encryption using ideal lattices. In: STOC, pp. 169–178 (2009)
38. Gentry, C., Gorbunov, S., Halevi, S.: Graph-induced multilinear maps from lattices. In: Dodis, Y., Nielsen, J.B. (eds.) TCC 2015, Part II. LNCS, vol. 9015, pp. 498–527. Springer, Heidelberg (2015)
39. Gentry, C., Lewko, A.B., Sahai, A., Waters, B.: Indistinguishability obfuscation from the multilinear subgroup elimination assumption. IACR Cryptology ePrint Archive, 2014, 309 (2014)
40. Gentry, C., Lewko, A., Waters, B.: Witness encryption from instance independent assumptions. In: Garay, J.A., Gennaro, R. (eds.) CRYPTO 2014, Part I. LNCS, vol. 8616, pp. 426–443. Springer, Heidelberg (2014)
41. Goldreich, O., Goldwasser, S., Micali, S.: How to construct random functions. J. ACM 33(4), 792–807 (1986)
42. Goldwasser, S., Kalai, Y.T.: On the impossibility of obfuscation with auxiliary input. In: FOCS, pp. 553–562. IEEE Computer Society (2005)
43. Goldwasser, S., Ostrovsky, R.: Invariant signatures and non-interactive zero-knowledge proofs are equivalent. In: Brickell, E.F. (ed.) CRYPTO 1992. LNCS, vol. 740, pp. 228–245. Springer, Heidelberg (1993)

44. Gorbunov, S., Vaikuntanathan, V., Wee, H.: Functional encryption with bounded collusions via multi-party computation. In: Safavi-Naini, R., Canetti, R. (eds.) CRYPTO 2012. LNCS, vol. 7417, pp. 162–179. Springer, Heidelberg (2012)

45. Yupu, H., Jia, H.: Cryptanalysis of GGH map. IACR Cryptology ePrint Archive, 2015, 301 (2015)

46. Ishai, Y., Kushilevitz, E.: Perfect constant-round secure computation via perfect randomizing polynomials. In: Widmayer, P., Triguero, F., Morales, R., Hennessy, M., Eidenbenz, S., Conejo, R. (eds.) ICALP 2002. LNCS, vol. 2380, pp. 244–256. Springer, Heidelberg (2002)

47. Kiayias, A., Papadopoulos, S., Triandopoulos, N., Zacharias, T.: Delegatable pseudorandom functions and applications. In: Sadeghi, A.-R., Gligor, V.D., Yung, M. (eds.) CCS 2013, pp. 669–684. ACM, New York (2013)

48. Komargodski, I., Moran, T., Naor, M., Pass, R., Rosen, A., Yogev, E.: One-way functions and (im)perfect obfuscation. In: 55th IEEE Annual Symposium on Foundations of Computer Science, FOCS 2014, pp. 374–383. IEEE Computer Society, Philadelphia, PA, USA, 18–21 October 2014

49. Lee, H.T., Seo, J.H.: Security analysis of multilinear maps over the integers. In: Garay, J.A., Gennaro, R. (eds.) CRYPTO 2014, Part I. LNCS, vol. 8616, pp. 224–240. Springer, Heidelberg (2014)

50. Mahmoody, M., Mohammed, A., Nematihaji, S.: More on impossibility of virtual black-box obfuscation in idealized models. In: TCC (2016). http://eprint.iacr.org/

51. Naor, M.: Bit commitment using pseudorandomness. J. Cryptology 4(2), 151–158 (1991)

52. Naor, M., Pinkas, B.: Efficient oblivious transfer protocols. In: SODA, pp. 448–457 (2001)

53. O'Neill, A.: Definitional issues in functional encryption. Cryptology ePrint Archive, Report 2010/556 (2010)

54. Pass, R., Shelat, A.: Impossibility of VBB obfuscation with ideal constant-degree graded encodings. In: TCC (2016)

55. Pass, R., Seth, K., Telang, S.: Indistinguishability obfuscation from semantically-secure multilinear encodings. In: Garay, J.A., Gennaro, R. (eds.) CRYPTO 2014, Part I. LNCS, vol. 8616, pp. 500–517. Springer, Heidelberg (2014)

56. Peikert, C., Vaikuntanathan, V., Waters, B.: A framework for efficient and composable oblivious transfer. In: Wagner, D. (ed.) CRYPTO 2008. LNCS, vol. 5157, pp. 554–571. Springer, Heidelberg (2008)

57. Sahai, A., Waters, B.: How to use indistinguishability obfuscation: deniable encryption, and more. In: Shmoys, D.B. (ed.) STOC, pp. 475–484. ACM, New York (2014)

58. Shoup, V.: Lower bounds for discrete logarithms and related problems. In: Fumy, W. (ed.) EUROCRYPT 1997. LNCS, vol. 1233, pp. 256–266. Springer, Heidelberg (1997)

59. Yao, A.C.-C.: How to generate and exchange secrets (extended abstract). In: 27th Annual Symposium on Foundations of Computer Science, pp. 162–167. IEEE Computer Society, Toronto, Canada, 27–29 October 1986

60. Zimmerman, J.: How to obfuscate programs directly. In: Oswald, E., Fischlin, M. (eds.) EUROCRYPT 2015. LNCS, vol. 9057, pp. 439–467. Springer, Heidelberg (2015)

Output-Compressing Randomized Encodings and Applications

Huijia Lin[1](✉), Rafael Pass[2], Karn Seth[2], and Sidharth Telang[2]

[1] University of California at Santa Barbara, Santa Barbara, USA
rachel.lin@cs.ucsb.edu
[2] Cornell University, Ithaca, USA
{rafael,karn,sidtelang}@cs.cornell.edu

Abstract. We consider *randomized encodings (RE)* that enable encoding a Turing machine Π and input x into its "randomized encoding" $\hat{\Pi}(x)$ in sublinear, or even polylogarithmic, time in the running-time of $\Pi(x)$, *independent of its output length*. We refer to the former as *sublinear RE* and the latter as *compact RE*. For such efficient RE, the standard simulation-based notion of security is impossible, and we thus consider a weaker (distributional) indistinguishability-based notion of security: Roughly speaking, we require indistinguishability of $\hat{\Pi}_0(x_0)$ and $\hat{\Pi}_0(x_1)$ as long as Π_0, x_0 and Π_1, x_1 are sampled from some distributions such that $\Pi_0(x_0), \mathsf{Time}(\Pi_0(x_0))$ and $\Pi_1(x_1), \mathsf{Time}(\Pi_1(x_1))$ are indistinguishable.

We show the following:
- **Impossibility in the Plain Model:** Assuming the existence of subexponentially secure one-way functions, subexponentially-secure sublinear RE does not exists. (If additionally assuming subexponentially-secure iO for circuits we can also rule out polynomially-secure sublinear RE.) As a consequence, we rule out also puncturable iO for Turing machines (even those without inputs).
- **Feasibility in the CRS model and Applications to iO for circuits:** Subexponentially-secure sublinear RE in the CRS model and one-way functions imply iO for circuits through a simple construction generalizing GGM's PRF construction. Additionally, any compact (even with sublinear compactness) functional encryption essentially directly yields a sublinear RE in the CRS model, and as such we get an alternative, modular, and simpler proof of the results of [AJ15,BV15] showing that subexponentially-secure sublinearly compact FE implies iO. We further show other ways of instantiating sublinear RE in the CRS model (and thus also iO): under the subexponential LWE

H. Lin—Work supported in part by a NSF award CNS-1514526.

R. Pass—Work supported in part by a Alfred P. Sloan Fellowship, Microsoft New Faculty Fellowship, NSF Award CNS-1217821, NSF CAREER Award CCF-0746990, NSF Award CCF-1214844, AFOSR YIP Award FA9550-10-1-0093, and DARPA and AFRL under contract FA8750-11-2-0211. The views and conclusions contained in this document are those of the authors and should not be interpreted as representing the official policies, either expressed or implied, of the Defense Advanced Research Projects Agency or the US Government.

© International Association for Cryptologic Research 2016
E. Kushilevitz and T. Malkin (Eds.): TCC 2016-A, Part I, LNCS 9562, pp. 96–124, 2016.
DOI: 10.1007/978-3-662-49096-9_5

assumption, it suffices to have a subexponentially secure FE schemes with just *sublinear ciphertext* (as opposed to having sublinear encryption time).

- **Applications to iO for Unbounded-input Turing machines:** Subexponentially-secure compact RE for natural *restricted* classes of distributions over programs and inputs (which are not ruled out by our impossibility result, and for which we can give candidate constructions) imply **iO** for *unbounded-input* Turing machines. This yields the first construction of **iO** for unbounded-input Turing machines that does not rely on (public-coin) differing-input obfuscation.

1 Introduction

The beautiful notion of a *randomized encoding (RE)*, introduced by Ishai and Kushilevitz [IK00], aims to trade the computation of a "complex" (deterministic) function Π on a given input x for the computation of a "simpler" randomized function—the "encoding algorithm"—whose output distribution $\hat{\Pi}(x)$ encodes $\Pi(x)$ (from which $\Pi(x)$ can be efficiently decoded, or "evaluated"). Furthermore, the encoding $\hat{\Pi}(x)$ should not reveal anything beyond $\Pi(x)$; this is referred to as the *privacy*, or *security*, property of randomized encodings and is typically defined through the simulation paradigm [GMR89].

Most previous work have focused on randomized encodings where encodings can be computed in lower parallel-time complexity than what is required for computing the original function Π. For instance, all log-space computations have *perfectly-secure* randomized encodings in $\mathbf{NC}^0$ [IK00, IK02a, AIK04], and assuming low-depth pseudo-random generators, this extends to all polynomial-time computations (with computational security) [AIK06, Yao82]. Such randomized encodings have been shown to have various applications to parallel cryptography, secure computation, verifiable delegation, etc. (see [App11] for a survey).

Bitansky, Garg, Lin, Pass and Telang [BGL+15] recently initiated a study of *succinct randomized encodings* where we require that the *time* required to compute $\hat{\Pi}(x)$ is smaller than the time required to compute $\Pi(x)$; their study focused on functions Π that have *single-bit* outputs. [BGL+15, CHJV14, KLW14] show that subexponentially-secure indistinguishability obfuscators (**iO**) [BGI+01, GGH+13] and one-way functions[1] imply the existence of such succinct randomized encodings for all polynomial-time Turing machines that output just a single bit.

We here further the study of such objects, focusing on functions Π with *long* outputs. Given a description of a Turing machine Π and an input x, we consider two notions of efficiency for randomized encodings $\hat{\Pi}(x)$ of $\Pi(x)$ with running time T.

- *compact RE*: Encoding time (and thus also size of the encodings) is $\text{poly}(|\Pi|, |x|, \log T)$

---

[1] The one-way function assumption can be weakened to assume just that NP $\not\subseteq$ *io*BPP [KMN+14].

– *sublinear RE*: Encoding time (and thus also size) is bounded by $\text{poly}(|\Pi|, |x|) * T^{1-\epsilon}$, for some $\epsilon > 0$.

We assume without loss of generality that the randomized encoding $\hat{\Pi}(x)$ of Π, x itself is a program, and that the decoding/evaluation algorithm simply executes $\hat{\Pi}(x)$.

It is easy to see that for such notions of efficiency, the standard simulation-based notion of security is impossible to achieve—roughly speaking, the simulator given just $\Pi(x)$ needs to output a "compressed" version of it, which is impossible if $\Pi(x)$ has high pseudo-Kolmogorov complexity (e.g., if Π is a PRG); we formalize this argument in Theorem 14 in Sect. 6. Consequently, we consider weaker indistinguishability-based notions of privacy. One natural indistinguishability based notion of privacy simply requires that encoding $\hat{\Pi}_0(x_0)$ and $\hat{\Pi}_1(x_1)$ are indistinguishable as long as $\Pi_0(x_0) = \Pi_1(x_1)$ and $\text{Time}(\Pi_0(x_0)) = \text{Time}(\Pi_1(x_1))$, where $\text{Time}(\Pi(x))$ is the running-time of $\Pi(x)$; such a notion was recently considered by Ananth and Jain [AJ15]. In this work, we consider a stronger notion which requires indistinguishability of $\hat{\Pi}_0(x_0)$ and $\hat{\Pi}_0(x_1)$ as long as Π_0, x_0 and Π_1, x_1 are sampled from some distributions such that $\Pi_0(x_0), \text{Time}(\Pi_0(x_0))$ and $\Pi_1(x_1), \text{Time}(\Pi_1(x_1))$ are indistinguishable. We refer to this notion as *distributional indistinguishability security*, and note that it easily follows that the standard simulation-based security implies distributional indistinguishability security.

The goal of this paper is to investigate compact and sublinear RE satisfying the above-mentioned distributional indistinguishability notion. For the remainder of the introduction, we refer to randomized encodings satisfying distributional indistinguishability security as simply *RE*. For comparison, we refer to randomized encodings with the weaker (non-distributional) indistinguishability security as *weak RE*.

COMPACT RE V.S. OBFUSCATION. Before turning to describe our results, let us point out that RE can be viewed as (a degenerate form) of obfuscation for special classes of programs.

Recall that an indistinguishability obfuscator (iO) [BGI+01, GGH+13] is a method $\mathcal{O}$ for "scrambling" a program Π into $\mathcal{O}(\Pi)$ such that for any two functionally equivalent programs Π_0, Π_1 (that is, their outputs and run-time are the same on all inputs,) $\mathcal{O}(\Pi_0)$ is indistinguishable from $\mathcal{O}(\Pi_1)$. iO for Turing machines [BGI+01, BCP14, ABG+13] additionally requires that the size of the obfuscated code does not grow (more than polylogarithmically) with the running-time of the Turing machine.

We may also consider a useful strengthening of this notion—which we call "puncturable iO"—which, roughly speaking, requires indistinguishability of $\mathcal{O}(\Pi_0)$ and $\mathcal{O}(\Pi_1)$ as long as Π_0 and Π_1 differ *on at most one input* x^* and their outputs on input x^* are indistinguishable. More precisely, we say that a distribution D is *admissible* if there exists some x^* such that a) for every triple (Π_0, Π_1, Π) in the support of D, and every $x \neq x^*$, it holds that $\Pi_0(x) = \Pi_1(x) = \Pi(x)$, and b) $(\Pi, \Pi_0(x^*))$ and $(\Pi, \Pi_1(x^*))$ are computationally indistinguishable when (Π_0, Π_1, Π) are sampled randomly from D.

Puncturable **iO** requires indistinguishability of $\mathcal{O}(\Pi_0)$ and $iO(\Pi_1)$ for Π_0, Π_1 sampled from any admissible distribution. Interestingly, for the case of *circuits*, puncturable **iO** is equivalent to (standard) **iO**.[2] Indeed, such a notion is implicitly used in the beautiful and powerful punctured-program paradigm by Sahai and Waters [SW14], and all its applications. (In this context, think of Π as the "punctured" version of the programs Π_0, Π_1.)

In the case of Turing machines, when restricting to the degenerate case of Turing machines with no inputs (or more precisely, we only consider the execution of $\Pi()$ on the "empty" input), the notion of **iO** for Turing machines is equivalent to the notion of a compact *weak* RE. Compact RE, on the other hand, is equivalent to puncturable **iO** for Turing machines (without inputs). (Jumping ahead, as we shall see, for the case of Turing machines it is unlikely that puncturable **iO** is equivalent to standard **iO**.)

1.1 Our Results

IO FROM SUBLINEAR RE. We start by showing that sublinear RE is an extremely useful primitive: Subexponentially-secure sublinear RE implies indistinguishability obfuscators for all polynomial-size circuits.

Theorem 1. *The existence of subexponentially-secure sublinear RE and one-way functions implies the existence of subexponentially-secure iO for circuits.*

Before continuing, let us mention that Theorem 1 is related to a recent beautiful result by Ananth and Jain [AJ15] which shows that *under the LWE assumption*, subexponentially-secure compact RE (satisfying only the weak indistinguishability security) implies **iO** for circuits. Their construction goes from RE to *functional encryption* (FE) [BSW11], and then from FE to **iO**; (the first step relies on previous constructions of FE [GKP+13a, GVW13], while the second step relies on a sequence of complex transformations and analysis). In contrast, the proof of Theorem 1 directly constructs **iO** from RE in a surprisingly simple way: We essentially use the GGM construction [GGM86] that builds a PRF from a PRG using a tree, but replace the PRG with a RE. Let us explain in more details below.

Consider a program Π taking n-bit inputs. We consider a binary tree where the leaves are randomized encodings of the function applied to all possible inputs, and each node in the tree is a randomized encoding that generates its two children. More precisely, given a sequence of bits $x_1, \cdots, x_i$, let $\tilde{\Pi}_{R, x_1, \cdots, x_i}$ denote an (input-less) program that

– if $i = n$ simply outputs a RE of the program Π and input $(x_1, \cdots, x_n)$ using R as randomness, and

---

[2] To see this, consider a hybrid program $\Pi^y(x)$ that runs $\Pi(x)$ if $x \neq x^*$ and otherwise (i.e., if $x = x^*$ outputs y). By the **iO** property we have that for every Π, Π_0, Π_1 in the support of D, $\mathcal{O}(\Pi^{\Pi_b(x^*)})$ is indistinguishable from $\mathcal{O}(\Pi_b)$. Thus, if $\mathcal{O}(\Pi_0)$, $\mathcal{O}(\Pi_1)$ are distinguishable, so are $\mathcal{O}(\Pi^{\Pi_0(x^*)})$, $\mathcal{O}(\Pi^{\Pi_1(x^*)})$, which contradicts indistinguishability of $(\Pi, \Pi_0(x^*))$ and $(\Pi, \Pi_1(x^*))$.

– otherwise, after expanding R_0, R_1, R_2, R_3 from R using a PRG, outputs randomized encodings of (input-less) programs $\tilde{\Pi}_{R_0,x_1,\cdots,x_i,0}$ and $\tilde{\Pi}_{R_1,x_1,\cdots,x_i,1}$ using respectively R_2, R_3 as randomness.

We associate each node in the binary tree that has index $x_1, \cdots, x_i$ with a randomized encoding of the program $\tilde{\Pi}_{R,x_1,\cdots,x_i}$, denoted as $\hat{\Pi}_{R,x_1,\cdots,x_i}$. In particular, the root of the tree is associated with a randomized encoding $\hat{\Pi}$ of the (initial) program $\tilde{\Pi}_R$ hardwired with a randomly chosen R.

The obfuscation of Π is now a program with the "root" $\hat{\Pi}$ hardcoded, and given an input x, computes the path from the root to the leaf x – by recursively evaluating the randomized encodings associated with nodes on the path – and finally outputs the evaluation of the leaf. More precisely, on input x, evaluate $\hat{\Pi}$ to obtain $\hat{\Pi}_0, \hat{\Pi}_1$, next evaluate $\hat{\Pi}_{x_1}$ to obtain $\hat{\Pi}_{x_1,0}, \hat{\Pi}_{x_1,1}$, so on and so forth until $\hat{\Pi}_{x_1,\cdots,x_n}$ is evaluated, yielding the output $\Pi(x_1, \cdots, x_n)$.

Note that for any two functionally equivalent programs, the randomized encodings associated with individual leaf node are computationally indistinguishable by the indistinguishability security property (the non-distributional version suffices here). Then, by the distributional indistinguishability security, the randomized encodings associated with tree nodes one layer above are also indistinguishable. Thus, by induction, it follows that the roots are indistinguishable, which implies that obfuscations of functionally equivalent programs are indistinguishable. Let us note that the reason that subexponential security is needed is that each time we go up one level in the tree (in the inductive argument), we lose at least a factor 2 in the indistinguishability gap (as each node generates two randomized encodings, its children). Hence, we need to ensure that encodings are at least poly(2^n)-indistinguishable, which can be done by scaling up the security parameter.

ON THE EXISTENCE OF COMPACT AND SUBLINEAR RE. We next turn to investigating the existence of compact and sublinear RE. We show—assuming just the existence of subexponentially-secure one-way functions—*impossibility* of subexponentially-secure sublinear (and thus also compact) RE.[3]

Theorem 2. *Assume the existence of subexponentially secure one-way functions. Then, there do not exists subexponentially-secure sublinear RE.*

As observed above, compact RE can be interpreted as a stronger notion of iO (which we referred to as *puncturable iO*) for "degenerate" input-less Turing machines, and as such Theorem 2 rules out (assuming just one-way functions) such a natural strengthening of iO for (input-less) Turing machines. We note that this impossibility stands in contrast with the case of *circuits* where puncturable iO is equivalent to iO.

We remark that although it may seem like Theorem 2 makes Theorem 1 pointless, it turns out that Theorem 1 plays a crucial role in the proof of Theorem 2:

---

[3] This result was established after hearing that Bitansky and Paneth had ruled out compact RE assuming public-coin differing-input obfuscation for Turing Machines and collision-resistant hashfunctions. We are very grateful to them for informing us of their result.

Theorem 2 is proven by first ruling out sublinear (even just polynomially-secure) RE *assuming* iO and one-way functions. Next, by using Theorem 1, the iO assumption comes for free if considering subexponentially-secure RE. That is, assuming one-way functions, we have the following paradigm:

$$\text{sub-exp secure sublinear RE} \overset{\text{Theorem 1}}{\implies} \text{iO} \implies \text{impossibility of (poly secure) sublinear RE}$$

Let us now briefly sketch how to rule out sublinear RE assuming iO and one-way functions (as mentioned, Theorem 2 is then deduced by relying on Theorem 1). The idea is somewhat similar to the non-black-box zero-knowledge protocol of Barak [Bar01].

Let $\Pi_{s,u}^b$ be a program that takes no input and outputs a sufficiently long pseudo-random string $y = \mathsf{PRG}(s)$ and an indistinguishability obfuscation $\tilde{R}_y^b$ (generated using pseudo-random coins $\mathsf{PRG}(u)$) of the program R_y^b. The program R_y^b takes input Σ of length $|y|/2$, and outputs b iff Σ, when interpreted as an input-less Turing machine, generates y; in all other cases, it outputs $\perp$.[4] We note that the size of the program $\Pi_{s,u}^b$ is linear in the security parameter λ, whereas the pseudo-random string y it generates could have length $|y| = \lambda^\alpha$ for any sufficiently large constant α.

Consider the pair of distributions $\Pi_{U_\lambda, U_\lambda}^0$ and $\Pi_{U_\lambda, U_\lambda}^1$ that samples respectively programs $\Pi_{s,u}^0$ and $\Pi_{s,u}^1$ as described above with random s and u. We first argue that their outputs are computationally indistinguishable. Recall that the output of $\Pi_{s,u}^b$ is a pair $(y, \tilde{R}_y^b)$. By the pseudorandomness of PRG, this output distribution is indistinguishable from $(X, \tilde{R}_X^b)$ where X a uniformly distributed random variable over λ^α bit strings. With overwhelming probability X has high Kolmogorov complexity, and when this happens R_X^b is functionally equivalent to the program $R_\perp$ that always outputs $\perp$. Therefore, by the security of the iO, the output of programs sampled from $\Pi_{U_\lambda, U_\lambda}^b$ is computationally indistinguishable to $(X, \tilde{R}_\perp)$, and hence outputs of $\Pi_{U_\lambda, U_\lambda}^0$ and $\Pi_{U_\lambda, U_\lambda}^1$ are indistinguishable.

Let us now turn to showing that randomized encodings of $\Pi_{U_\lambda, U_\lambda}^0$ and $\Pi_{U_\lambda, U_\lambda}^1$ can be distinguished. Recall that a randomized encoding $\hat{\Pi}^b$ of $\Pi_{U_\lambda, U_\lambda}^b$ itself can be viewed as a (input-less) program that outputs $(y, \tilde{R}_y^b)$. Given $\hat{\Pi}^b$, the distinguisher can thus first evaluate $\hat{\Pi}^b$ to obtain $(y, \tilde{R}_y^b)$ and next evaluate $\tilde{R}_y^b(\hat{\Pi}^b)$ to attempt to recover b. Note that $\hat{\Pi}^b$ clearly is a program that generates y (as its first input); furthermore, if the RE scheme is compact, the length of the program $|\hat{\Pi}^b|$ is bounded by $\mathrm{poly}(\lambda, \log \lambda^\alpha)$, which is far smaller than $|y|/2 = \lambda^\alpha/2$ when α is sufficiently large. Therefore, $\Sigma = \hat{\Pi}^b$ is indeed an input that makes $\tilde{R}_y^b$ output b, enabling the distinguisher to distinguish $\hat{\Pi}^0$ and $\hat{\Pi}^1$ with probability close to 1!

Finally, if the RE is only sublinear, the length of the encoding $|\hat{\Pi}^b|$ is only sublinear in the output length, in particular, bounded by $\mathrm{poly}(\lambda)(\lambda^\alpha)^{1-\epsilon}$ for

---

[4] To enable this, we require iO for bounded-input Turing machines, whereas Theorem 1 only gives us iO for circuits. However, by the results of [BGL+15, CHJV14, KLW14] we can go from iO for circuits to iO for bounded-inputs Turing machines.

some constant $\epsilon > 0$. If $\alpha > 1/(1 - \epsilon)$ (which clearly happens if ϵ is sufficiently small), then we do not get enough "compression" for the above proof to go through. We circumvent this problem by composing a sublinear RE with itself a sufficient (constant) number of times—to compose once, consider creating randomized encoding of the randomized encoding of a function, instead of the function itself; each time of composition reduces the size of the encoding to be w.r.t. a smaller exponent $1 - \epsilon'$. Therefore, it is without loss of generality to assume that ϵ is any sufficiently *big* constant satisfying $\alpha << 1/(1 - \epsilon)$; so the desired compression occurs.

SUBLINEAR RE IN THE CRS MODEL FROM SUBLINEAR FE. Despite Theorem 2, not all is lost. We remark that any sublinear functional encryption scheme (FE) [AJ15, BV15] almost directly yields a sublinear RE in the *Common Reference String (CRS)* model; roughly speaking, an FE scheme is called sublinear if the encryption time is sublinear in the size of the circuit that can be evaluated on the encrypted message.

Theorem 3. *Assume the existence of subexponentially-secure sublinear (resp. compact) FE. Then there exists a subexponentially-secures sublinear (resp. compact) RE in the CRS model.*

Furthermore, Theorem 1 straightforwardly extends also to RE in the CRS model. Taken together, these result provide an alternative, modular, simpler proof of the recent results of Ananth and Jain [AJ15] and Bitansky and Vaikuntanathan [BV15] showing that subexponentially-secure sublinear FE implies subexponentially-secure **iO**. (All these approaches, including a related work by Brakerski, Komargodski and Segev [BKS15] have one thing in common though: they all proceed by processing inputs one bit at a time, and hard-coding parts of input to the program.)

Theorem 4 (informal, alternative proof of [BV15, AJ15]**).** *Assume the existence of subexponentially-secure sublinear FE. Then there exists a subexponentially-secure **iO** for circuits.*

But there are also other ways to instantiate sublinear RE in the CRS model. We show that under the *subexponential LWE assumption* (relying on [GKP+13a, ABSV14, GVW13]) sublinear RE in the CRS model can be based on a significantly weaker notion of sublinear FE—namely FE schemes where the encryption time may be fully polynomial (in the size of the circuit to be evaluated) but only the *size of the ciphertext* is sublinear in the circuit size—we refer to this notion as a *FE with sublinear ciphertexts*. Roughly speaking, we show this by (1) transforming the "succinct" FE (i.e. compact FE for 1-bit outputs) of [GKP+13a, ABSV14] into an RE which depends linearly on the output length but only polylogarithmically on the running time, (2) transforming an FE with sublinear ciphertext into an RE with "large" running-time but short output, and (3) finally composing the two randomized encodings (i.e., computing the step 1 RE of the step 2 RE).

Combining this result with (the CRS-extended version of) Theorem 1, we get:

Theorem 5 (informal). *Assume the existence of subexponentially-secure FE with sublinear ciphertexts and the subexponential LWE assumption. Then there exists a subexponentially-secure **iO** for circuits.*

TOWARD TURING MACHINE OBFUSCATION WITH UNBOUNDED INPUTS.   We finally address the question of constructing indistinguishability obfuscators for Turing machines with *unbounded* inputs. (For the case of Turing machine obfuscation with unbounded-length inputs, the same obfuscated code needs to work for every input-length, and in particular, the size of the obfuscated code cannot grow with it.) Although it is known that subexponentially secure **iO** for circuits implies **iO** for Turing machines with *bounded inputs lengths* [BGL+15, CHJV14, KLW14], the only known construction of **iO** for Turing machines with unbounded inputs relies on (public-coin) differing-input obfuscation for circuits and (public-coin) SNARKs [BCP14, ABG+13, IPS15]—these are strong "extractability" assumptions (and variants of them are known to be implausible [BCPR13, GGHW13, BP15]).

We note that the construction from Theorem 1 easily extends to show that subexponentially-secure *compact* RE implies **iO** for Turing machines with unbounded input: instead of having a binary tree, we have a ternary tree where the "third" child of a node is always a leaf; that is, for a tree node corresponding to $x_1, \cdots, x_i$, its third child is associated with a randomized encoding of program Π, and input $(x_1, \cdots, x_i)$, which can be evaluated to obtain output $\Pi(x_1, \cdots x_i)$. Then, by using a tree of *super-polynomial* depth, we can handle any polynomial-length input. Note that since obfuscating a program only involves computing the root RE (as before), the obfuscation is still efficient. Moreover, for any input, we still compute the output of the program in time polynomial in the length of the input by evaluating the "third" child of the node when all input bits have been processed.[5]

But as shown in Theorem 2, compact RE cannot exist (assuming one-way functions)! However, just as for the case of differing-inputs obfuscation and SNARKs, we may assume the existence of compact RE for *restricted* types of "nice" distributions (over programs and inputs), for which impossibility does not hold, yet the construction in Theorem 1 still works. We formalize one natural class of such distributions, and may assume that the **iO** for bounded-input Turing machines construction of [KLW14] (based on **iO** for circuits) yields such a compact RE (for the restricted class of distributions). This yields a new candidate construction of unbounded input Turing machines (based on a very different type of assumption than known constructions).

---

[5] Proving security becomes slightly more problematic since there is no longer a polynomial bound on the depth of the tree (recall that we required poly(2^n)-indistinguishable RE to deal with inputs of length n). Thus issue, however, can be dealt with by using larger and larger security parameters for RE that are deeper down in the tree.

2   Preliminaries

Let $\mathcal{N}$ denote the set of positive integers, and $[n]$ denote the set $\{1, 2, \ldots, n\}$. We denote by PPT probabilistic polynomial time Turing machines. The term **negligible** is used for denoting functions that are (asymptotically) smaller than one over any polynomial. More precisely, a function $\nu(\cdot)$ from non-negative integers to reals is called **negligible** if for every constant $c > 0$ and all sufficiently large n, it holds that $\nu(n) < n^{-c}$.

TURING MACHINE NOTATION. For any Turing machine Π, input $x \in \{0, 1\}^*$ and time bound $T \in \mathbb{N}$, we denote by $\Pi^T(x)$ the output of Π on x when run for T steps. We refer to $\{\mathcal{M}_\lambda\}_{\lambda \in \mathbb{N}}$ as a class of Turing machines. One particular class we will consider is the class of Turing machines that have 1-bit output. We call such a machine a Boolean Turing machine. Throughout this paper, by *Turing machine* we refer to a machine with *multi-bit* output unless we explicitly mention it to be a Boolean Turing machine.

2.1   Concrete Security

Definition 1 $((\lambda_0, S(\cdot))$-indistinguishability). *A pair of distributions X, Y are S-indistinguishable for some $S \in \mathbb{N}$ if every S-size distinguisher D it holds that*

$$| \Pr[x \xleftarrow{\$} X : D(x) = 1] - \Pr[y \xleftarrow{\$} Y : D(y) = 1]| \leq \frac{1}{S}$$

A pair of ensembles $\{X_\lambda\}$, $\{Y_\lambda\}$ are $(\lambda_0, S(\cdot))$-indistinguishable for some $\lambda_0 \in \mathbb{N}$ and $S : \mathbb{N} \to \mathbb{N}$ if for every security parameter $\lambda > \lambda_0$, the distributions X_λ and Y_λ are $S(\lambda)$ indistinguishable.

DISCUSSION ON $(\lambda_0, S(\cdot))$-INDISTINGUISHABILITY: We remark that the above definition requires that there is a universal λ_0 that works for all distinguisher D. A seemingly weaker variant could switch the order of quantifiers and only require that for every distinguisher D there is a λ_0. We show that the above definition is w.l.o.g, since it is implied by the following standard definition with auxiliary inputs in the weaker fashion.

Let U be a universal TM that on an input x and a circuit C computes $C(x)$. Let $S' : N \to N$ denote the run time $S'(S)$ of U on input a size S circuit.

Definition 2. *A pair of ensembles $\{X_\lambda\}$, $\{Y_\lambda\}$ are $S(\cdot)$-indistinguishable if for every $S' \circ S(\cdot)$-time uniform TM distinguisher D, there exists a $\lambda_0 \in N$, such that, for every security parameter $\lambda > \lambda_0$, and every auxiliary input $z = z_\lambda \in \{0, 1\}^*$,*

$$| \Pr[x \xleftarrow{\$} X_\lambda : D(1^\lambda, x, z) = 1] - \Pr[y \xleftarrow{\$} Y_\lambda : D(1^\lambda, y, z) = 1]| \leq \frac{1}{S(\lambda)}$$

This definition implies $(\lambda_0, S(\cdot))$-indistinguishability. Consider a distinguisher D that on input $(1^\lambda, x, z)$ runs the universal TM $U(x, z)$, and let λ_U be the constant associated with it. For any $\lambda > \lambda_U$, and every $S(\lambda)$-size circuit C, by setting the auxiliary input $z = C$, the above definition implies that the distinguishing gap by C is at most $1/S(\lambda)$. Therefore, λ_U is effectively the universal constant that works for all (circuit) distinguisher.

Below, we state definitions of cryptographic primitives using $(\lambda_0, S(\cdot))$ indistinguishability. Traditional polynomial or sub-exponential security can be directly derived from such more concrete definitions as follows:

Definition 3 (Polynomial Indistinguishability). *A pair of ensembles* $\{X_\lambda\}$, $\{Y_\lambda\}$ *are polynomially indistinguishable if for every polynomial* $p(\cdot)$, *there is a constant* $\lambda_p \in N$, *such that, the two ensembles are* $(\lambda_p, p(\cdot))$-*indistinguishable.*

Definition 4 (Sub-exponential Indistinguishability). *A pair of ensembles* $\{X_\lambda\}$, $\{Y_\lambda\}$ *are sub-exponentially indistinguishable, if there is a sub-exponential function* $S(\lambda) = 2^{\lambda^\varepsilon}$ *with* $\varepsilon \in (0,1)$ *and a constant* $\lambda_0 \in N$, *such that, the two ensembles are* $(\lambda_0, S(\cdot))$-*indistinguishable.*

2.2 Standard Cryptographic Primitives

Definition 5 (Pseudorandom Generator). *A deterministic PT uniform machine* PRG *is a pseudorandom generator if the following conditions are satisfied:*

Syntax. *For every* $\lambda, \lambda' \in \mathbb{N}$ *and every* $r \in \{0,1\}^\lambda$, PRG(r, λ') *outputs* $r' \in \{0,1\}^{\lambda'}$

$(\lambda_0, S(\cdot))$**-Security.** *For every function* $p(\cdot)$, *such that,* $p(\lambda) \leq S(\lambda)$ *for all* λ, *the following ensembles are* $(\lambda_0, S(\cdot))$ *indistinguishable*

$$\left\{ r \xleftarrow{\$} \{0,1\}^\lambda : \mathsf{PRG}(r, p(\lambda)) \right\} \quad \left\{ r' \xleftarrow{\$} \{0,1\}^{p(\lambda)} \right\}$$

2.3 Indistinguishability Obfuscation

In this section, we recall the definition of indistinguishability obfuscation for Turing machines from [BGI+01, BCP14, ABG+13]. Following [BCP14], we considers two notions of obfuscation for Turing machines. The first definition, called *bounded-input* indistinguishability obfuscation, only requires the obfuscated program to work for inputs of *bounded length* and furthermore the size of the obfuscated program may depend polynomially on this input length bound. (This is the notion achieved in [BGL+15, CHJV14, KLW14] assuming subexponentially-secure iO for circuits and one-way functions.)

The second notion considered in [BCP14] is stronger and requires the obfuscated program to work on any arbitrary polynomial length input (and the size of the obfuscated machine thus only depends on the program size and security parameter). We refer to this notion as *unbounded-input* indistinguishability obfuscation. (This stronger notion of unbounded-input indistinguishability obfuscator for Turing machines is only known to be achievable based on strong "extractability assumptions"—namely, (public-coin) differing-input obfuscation for circuits and (public-coin) SNARKs [BCP14,ABG+13,IPS15], variants of which are known to be implausible [BCPR13,GGHW13,BP15]).

Definition 6 (Indistinguishability Obfuscator ($i\mathcal{O}$) for a class of Turing machines). *An indistinguishability obfuscator for a class of Turing machines $\{\mathcal{M}_\lambda\}_{\lambda \in \mathbb{N}}$ is a uniform machine that behaves as follows:*

$\hat{\Pi} \leftarrow i\mathcal{O}(1^\lambda, \Pi, T)$: $i\mathcal{O}$ *takes as input a security parameter 1^λ, the Turing machine to obfuscate $\Pi \in \mathcal{M}_\lambda$ and a time bound T for Π. It outputs a Turing machine $\hat{\Pi}$.*

We require the following conditions to hold.

Correctness: *For every $\lambda \in N$, $\Pi_\lambda \in \mathcal{M}_\lambda$, input x_λ and time bound T_λ,*

$$\Pr[(\tilde{\Pi} \xleftarrow{\$} i\mathcal{O}(1^\lambda, \Pi_\lambda, T_\lambda) : \tilde{\Pi}(x_\lambda) = \Pi^T(x_\lambda)] = 1.$$

Efficiency: *The running times of $i\mathcal{O}$ and $\hat{\Pi}$ are bounded as follows: There exists polynomial p such that for every security parameter λ, Turing machine $\Pi \in \mathcal{M}_\lambda$, time bound T and every obfuscated machine $\hat{\Pi} \leftarrow i\mathcal{O}(1^\lambda, \Pi, T)$ and input x, we have that*

$$\mathsf{Time}_{i\mathcal{O}}(1^\lambda, \Pi, T) \leq p(\lambda, |\Pi|, \log T)$$
$$\mathsf{Time}_{\hat{\Pi}}(x) \leq p(\lambda, |\Pi|, |x|, T)$$

$(\lambda_0, S(\cdot))$**-Security:** *For every ensemble of pairs of Turing machines and time bounds $\{\Pi_{0,\lambda}, \Pi_{1,\lambda}, T_\lambda\}$ where for every $\lambda \in \mathbb{N}$, $\Pi_0 = \Pi_{0,\lambda}$, $\Pi_1 = \Pi_{1,\lambda}$, $T = T_\lambda$, satisfying the following*

$$\Pi_0, \Pi_1 \in \mathcal{M}_\lambda \quad |\Pi_0| = |\Pi_1| \leq \mathrm{poly}(\lambda) \quad T \leq \mathrm{poly}(\lambda)$$
$$\forall x, \Pi_0^T(x) = \Pi_1^T(x),$$

the following ensembles are $(\lambda_0, S(\cdot))$-indistinguishable

$$\{i\mathcal{O}(1^\lambda, \Pi_{0,\lambda}, T_\lambda)\} \; \{i\mathcal{O}(1^\lambda, \Pi_{1,\lambda}, T_\lambda)\} .$$

Definition 7 (Unbounded-input indistinguishability obfuscator for Turing machines). *An unbounded-input indistinguishability obfuscator for Turing machines $i\mathcal{O}(\cdot, \cdot, \cdot)$ is simply an indistinguishability obfuscator for the class of all Boolean Turing machines.*

Remark 1 (Obfuscation for Boolean Turing machines is without loss of generality). *The above definition is equivalent to one that considers the class of all Turing machines. Any Turing machine with output length m can be represented as a Boolean Turing machine that takes in an additional input $i \in [m]$ and returns the ith bit of the m-bit long output.*

Definition 8 (Bounded-input indistinguishability obfuscator for Turing machines). *A bounded-input indistinguishability obfuscator for Turing machines* $i\mathcal{O}(\cdot, \cdot, \cdot, \cdot)$ *is a uniform machine such that for every polynomial* p, $i\mathcal{O}(p, \cdot, \cdot, \cdot)$ *is an indistinguishability obfuscator for the class of Turing machines* $\{\mathcal{M}_\lambda\}$ *where* $\mathcal{M}_\lambda$ *are machines that accept only inputs of length* $p(\lambda)$. *Additionally,* $i\mathcal{O}(p, 1^\lambda, \Pi, T)$ *is allowed to run in time* $\mathrm{poly}(p(\lambda) + \lambda + |\Pi| + \log T)$.

2.4 Functional Encryption

Definition 9 (Selectively-secure Single-Query Public-key Functional Encryption). *A tuple of PPT algorithms* $(\mathsf{FE.Setup}, \mathsf{FE.Enc}, \mathsf{FE.Dec})$ *is a selectively-secure functional encryption scheme for a class of circuits* $\{\mathcal{C}_\lambda\}$ *if it satisfies the following properties.*

Completeness. *For every* $\lambda \in \mathbb{N}$, $C \in \mathcal{C}_\lambda$ *and message* $m \in \{0,1\}^*$,

$$\Pr\left[\begin{array}{l} (mpk, msk) \leftarrow \mathsf{FE.Setup}(1^\lambda) \\ \quad c \leftarrow \mathsf{FE.Enc}(1^\lambda, m) \qquad : C(m) \leftarrow \mathsf{FE.Dec}(sk_C, c) \\ sk_C \leftarrow \mathsf{FE.KeyGen}(msk, C) \end{array}\right] = 1$$

$(\lambda_0, S(\cdot))$**-Selective-security.** *For every ensemble of circuits and pair of messages* $\{C_\lambda, m_{0,\lambda}, m_{1,\lambda}\}$ *where* $C_\lambda \in \mathcal{C}_\lambda$, $|C_\lambda|, |m_{0,\lambda}|, |m_{1,\lambda}| \le \mathrm{poly}(\lambda)$, *and* $C_\lambda(m_{0,\lambda}) = C_\lambda(m_{1,\lambda})$, *the following ensembles of distributions* $\{D_{0,\lambda}\}$ *and* $\{D_{1,\lambda}\}$ *are* $(\lambda_0, S(\cdot))$*-indistinguishable.*

$$D_{b,\lambda} = \left(\begin{array}{c} (mpk, msk) \leftarrow \mathsf{FE.Setup}(1^\lambda) \\ c \leftarrow \mathsf{FE.Enc}(1^\lambda, m_{b,\lambda}) \qquad : mpk, c, sk_C \\ sk_C \leftarrow \mathsf{FE.KeyGen}(msk, C_\lambda) \end{array}\right)$$

We note that in this work, we only need the security of the functional encryption scheme to hold with respect to statically chosen challenge messages and functions.

Definition 10 (Compact Functional Encryption). *We say a functional encryption scheme is* compact *if it additionally satisfies the following requirement:*

Compactness. *The running time of* $\mathsf{FE.Enc}$ *is bounded as follows.*
There exists a polynomial p *such that for every security parameter* $\lambda \in \mathbb{N}$ *and message* $m \in \{0,1\}^*$, $\mathsf{Time}_{\mathsf{FE.Enc}}(1^\lambda, m) \le p(\lambda, |m|, polylog(s))$, *where* $s = \max_{C \in \mathcal{C}_\lambda} |C|$.
Furthermore, we say the functional encryption scheme has sub-linear compactness *if there exists a polynomial* p *and constant* $\epsilon > 0$ *such that for every security parameter* $\lambda \in \mathbb{N}$ *and message* $m \in \{0,1\}^*$, $\mathsf{Time}_{\mathsf{FE.Enc}}(1^\lambda, m) \le p(\lambda, |m|)s^{1-\epsilon}$.

We also define a notion of succinctness, as follows:

Definition 11 (Succinct Functional Encryption). *A compact functional encryption scheme for a class of circuits that output only a single bit is called a* succinct *functional encryption scheme.*

Theorem 6 [GKP+13a]. *Assuming (sub-exponentially secure) LWE, there exists a (sub-exponentially secure) succinct functional encryption scheme for* NC^1.

We note that [GKP+13a] do not explicitly consider sub-exponentially secure succinct functional encryption, but their construction satisfies it (assuming sub-exponentially secure LWE).

Theorem 7 [GKP+13a, ABSV14]. *Assuming the existence of symmetric-key encryption with decryption in* NC^1 *(resp. sub-exponentially secure) and succinct FE for* NC^1 *(resp. sub-exponentially secure), there exists succinct FE for* P/poly *(resp. sub-exponentially secure).*

We also consider an even weaker notion of sublinear-compactness, where only the ciphertext size is sublinear in the size bound s of the function being evaluation, but the encryption time can depend polynomially on s.

Definition 12 (Weakly Sublinear Compact Functional Encryption). *We say a functional encryption scheme for a class of circuits* $\{\mathcal{C}_\lambda\}$ *is weakly sublinear compact if there exists* $\epsilon > 0$ *such that for every* $\lambda \in \mathbb{N}$, $pk \leftarrow \mathsf{FE.Setup}(1^\lambda)$ *and* $m \in \{0,1\}^*$ *we have that*

$$\mathsf{Time}_{\mathsf{FE.Enc}}(pk, m) = \mathrm{poly}(\lambda, |m|, s)$$

$$\mathsf{outlen}_{\mathsf{FE.Enc}}(pk, m) = s^{1-\epsilon} \cdot \mathrm{poly}(\lambda, |m|)$$

where $s = \max_{C \in \mathcal{C}_\lambda} |C|$.

3 Randomized Encoding Schemes

Roughly speaking, randomized encoding schemes encodes a computation of a program Π on an input x, into an encoded computation $(\hat{\Pi}, \hat{x})$, with the following two properties: First, the encoded computation evaluates to the same output $\Pi(x)$, while leaking no other information about Π and x. Second, the encoding is "simpler" to compute than the original computation. In the literature, different measures of simplicity have been considered. For instance, in the original works by [IK02a, AIK06], the depth of computation is used and it was shown that any computation in P can be encoded in NC_1 using Yao's garbled circuits [Yao82]. A recent line of works [BGL+15, CHJV14, KLW14] uses the time-complexity as the measure and show that any *Boolean* Turing machine computation can be encoded in time poly-logarithmic in the run-time of the computation.

Traditionally, the security of randomized encoding schemes are capture via simulation. In this work, we consider a new *distributional* indistinguishability-based security notion, and show that it is implied by the transitional simulation security. Additionally, we further explore how compact the encoded computation can be: Similar to the recent works [BGL+15, CHJV14, KLW14], we consider encoding whose size depends poly-logarithmically on the run-time of the encoded computation; but differently, we directly consider Turing machines with arbitrary length outputs, and require the size of the encoding to be independent of the output length. Such scheme is called a compact randomized encoding scheme.

3.1 Distributional Indistinguishability Security

In this paper, we study randomized encoding for all Turing machine computation, whose encoding size is independent of the output length of the computation—we say such randomized encoding schemes are **compact**. Towards this, we must consider weaker security notions than simulation security, and indistinguishability-based security notions are natural candidates. One weaker notion that has been considered in the literature requires encoding of two computation, (Π_1, x_1) and (Π_2, x_2) with the same output $\Pi_1(x_1) = \Pi_2(x_2)$, to be indistinguishable. In this work, we generalize this notion to, what called *distributional* indistinguishability security—this notion requires encoding of computations sampled from two distributions, $(\Pi_1, x_1) \xleftarrow{\$} D_1$ and $(\Pi_2, x_2) \xleftarrow{\$} D_2$, to be indistinguishable, provided that their outputs are indistinguishable.

Definition 13 (Randomized Encoding Scheme for a Class of Turing Machines). *A Randomized Encoding scheme* RE *for a class of Turing machines* $\{\mathcal{M}_\lambda\}$ *consists of two algorithms,*

- $(\hat{\Pi}, \hat{x}) \xleftarrow{\$} \mathsf{Enc}(1^\lambda, \Pi, x, T)$: *On input a security parameter* 1^λ, *Turing machine* $\Pi \in \mathcal{M}_\lambda$, *input* x *and time bound* T, Enc *generates an encoded machine* $\hat{\Pi}$ *and encoded input* $\hat{x}$.
- $y = \mathsf{Eval}(\hat{\Pi}, \hat{x})$: *On input* $(\hat{\Pi}, \hat{x})$ *produced by* $\mathsf{Enc}, \mathsf{Eval}$ *outputs* y.

Correctness: *The two algorithms* Enc *and* Eval *satisfy the following correctness condition: For all security parameters* $\lambda \in \mathbb{N}$, *Turing machines* $\Pi \in \mathcal{M}_\lambda$, *inputs* x *and time bounds* T, *it holds that,*

$$\Pr[(\hat{\Pi}, \hat{x}) \xleftarrow{\$} \mathsf{Enc}(1^\lambda, \Pi, x, T) : \ \mathsf{Eval}(\hat{\Pi}, \hat{x}) = \Pi^T(x))] = 1$$

Definition 14 (Distributional $(\lambda_0, S(\cdot))$-Indistinguishability Security). *A randomized encoding scheme* RE *for a class of Turing machines* $\{\mathcal{M}_\lambda\}$ *satisfies distributional* $(\lambda_0, S(\cdot))$-*indistinguishability security, (or* $(\lambda_0, S(\cdot))$-*ind-security for short) if the following is true w.r.t. some constant* $c > 0$:

For every ensembles of distributions $\{D_{0,\lambda}\}$ *and* $\{D_{1,\lambda}\}$ *with the following property:*

1. *there exists a polynomial* B, *such that, for every* $b \in \{0, 1\}$, $D_{b,\lambda}$ *is a distribution over tuples of the form* (Π_b, x_b, T_b), *where* Π_b *is a Turing machine,* x_b *is an input and* T_b *is a time bound, and* $\lambda, |\Pi_b|, |x_b|, T_b \leq B(\lambda)$.

2. *there exist an integer $\lambda_0' \geq \lambda_0$, and a function S' with $S'(\lambda) \leq S(\lambda)$ for all λ, such that, the following ensembles of output distributions are $(\lambda_0', S'(\cdot))$-indistinguishable,*

$$\left\{ (\Pi_0, x_0, T_0) \xleftarrow{\$} \mathcal{D}_{0,\lambda} : \Pi_0^{T_0}(x_0), T_0, |\Pi_0|, |x_0| \right\}$$

$$\left\{ (\Pi_1, x_1, T_1) \xleftarrow{\$} \mathcal{D}_{1,\lambda} : \Pi_1^{T_1}(x_1), T_1, |\Pi_1|, |x_1| \right\}$$

the following ensembles of encoding is $(\lambda_0', S''(\cdot))$-indistinguishable, where $S''(\lambda) = \frac{S'(\lambda)}{\lambda^c} - B(\lambda)^c$.

$$\left\{ (\Pi_0, x_0, T_0) \xleftarrow{\$} \mathcal{D}_{0,\lambda} : \mathsf{Enc}(1^\lambda, \Pi_0, x_0, T_0) \right\}$$

$$\left\{ (\Pi_1, x_1, T_1) \xleftarrow{\$} \mathcal{D}_{1,\lambda} : \mathsf{Enc}(1^\lambda, \Pi_1, x_1, T_1) \right\}$$

For convenience, in the rest of the paper, we directly refer to distributional indistinguishability security as indistinguishability security. The above concrete security directly gives the standard polynomial and sub-exponential security.

Definition 15 (Polynomial and Sub-exponential Indistinguishability Security). *A randomized encoding scheme* RE *for a class of Turing machines $\{\mathcal{M}_\lambda\}$ satisfies polynomial ind-security, if it satisfies $(\lambda_p, p(\cdot))$-indistinguishability security for every polynomial p and some $\lambda_p \in N$. Furthermore, it satisfies sub-exponential ind-security if it satisfies $(\lambda_0, S(\cdot))$-indistinguishability security for $S(\lambda) = 2^{\lambda^\varepsilon}$ with some $\varepsilon \in (0, 1)$.*

We note that, by definition, it holds that any randomized encoding scheme that is $(\lambda_0, S(\cdot))$-ind-secure, is also $(\lambda_0', S'(\cdot))$-ind-secure for any $\lambda_0' \geq \lambda_0$ and S' s.t. $S'(\lambda) \leq S(\lambda)$ for every λ. Therefore, naturally, sub-exponential ind-security is stronger than polynomial ind-security.

In the full version, we show that RE schemes with ind-security are composable just as RE schemes with simulation security are.

3.2   Compactness and Sublinear Compactness

With indistinguishability-security, we now define compact randomized encoding schemes for all Turing machines, whose time-complexity of encoding is independent of the output length.

Definition 16 (Compact Randomized Encoding for Turing machines). *A $(\lambda_0, S(\cdot))$-ind-secure compact randomized encoding scheme for Turing machines, is a randomized encoding scheme with $(\lambda_0, S(\cdot))$-indistinguishability security for the class of all Turing machines, with the following efficiency:*

- *For every security parameter λ, Turing machine Π, input x, time bound T and every encoded pair $(\hat{\Pi}, \hat{x}) \leftarrow \mathsf{Enc}(1^\lambda, \Pi, x, T)$, it holds*

$$\mathsf{Time}_{\mathsf{Enc}}(1^\lambda, \Pi, x, T) = \mathrm{poly}(\lambda, |\Pi|, |x|, \log T)$$

$$\mathsf{Time}_{\mathsf{Eval}}(\hat{\Pi}, \hat{x}) = \mathrm{poly}(\lambda, |\Pi|, |x|, T)$$

In this work, we also consider a weaker variant of the above compactness requirement, where the encoding time is sub-linear (instead of poly-logarithmic) in the computation time. For our results a compact randomized encoding scheme with sub-linear efficiency will suffice.

Definition 17 (Sub-linear Compactness of Randomized Encoding schemes). *We say a randomized encoding scheme* $\mathsf{RE} = (\mathsf{Enc}, \mathsf{Eval})$ *for a class of Turing machines* $\{\mathcal{M}_\lambda\}$ *has sub-linear compactness if the efficiency requirement on* Enc *in Definition 16 is relaxed to: For some constant* $\varepsilon \in (0, 1)$,

$$\mathsf{Time}_{\mathsf{Enc}}(1^\lambda, \Pi, x, T) \leq \mathrm{poly}(\lambda, |\Pi|, |x|) \cdot T^{1-\epsilon}$$

4 Unbounded-Input IO from Compact RE

In this section, we define our succinct indistinguishability obfuscator for Turing machines. Let $\mathsf{RE} = (\mathsf{Enc}, \mathsf{Eval})$ be a compact randomized encoding scheme for Turing machines with sub-exponential indistinguishability security. Let c be the constant for the security loss associated with the indistinguishability security of RE. We assume without loss of generality that $\mathsf{Enc}(1^\lambda, \cdot, \cdot)$ requires a random tape of length λ. Let PRG be a sub-exponentially secure pseudorandom generator and let ϵ be the constant associated with the sub-exponential security of PRG.

For every $\lambda \in \mathbb{N}$, $D \leq 2^\lambda$, define

$$l(\lambda, -1) = \lambda$$

$$l(\lambda, D) = l(\lambda, D - 1) + (2d\lambda)^{1/\epsilon}$$

where $d > 0$ is any constant strictly greater than c.

Construction 1. *Consider a Turing machine* Π, *security parameter* $\lambda \in \mathbb{N}$, *and time bound* T *of* Π. *For every partial input* $s \in \{0, 1\}^*$ *with* $|s| \leq 2^\lambda$ *and* $R \in \{0, 1\}^{2l(\lambda, |s|)}$, *we recursively define a Turing machine* $\widetilde{\Pi}_{s,R}$ *to be as follows:*

When $|s| < 2^\lambda$:
On the empty input, $\widetilde{\Pi}_{s,R}$ *outputs:*

$$\mathsf{Enc}(1^{l(\lambda, |s|+1)}, \widetilde{\Pi}_{s0, R_0}, T'(\lambda, |s| + 1, |\Pi|, \log(T)); R_1)$$
$$\mathsf{Enc}(1^{l(\lambda, |s|+2)}, \widetilde{\Pi}_{s1, R_2}, T'(\lambda, |s| + 1, |\Pi|, \log(T)); R_3)$$
$$\mathsf{Enc}(1^{l(\lambda, |s|+1)}, \Pi, s, T; R_4)$$

where $(R_0, R_1, R_2, R_3, R_4) \leftarrow \mathsf{PRG}(R, 5 \cdot 2l(\lambda, |s| + 1))$ *and* T' *is some fixed polynomial in* $\lambda, |s| + 1, |\Pi|$ *and* $\log(T)$. *In the special case when* $|s| = 2^\lambda - 1$, *the time bound used in the first two encodings is set to* T.
On all other inputs, $\widetilde{\Pi}_{s,R}$ *outputs* $\bot$.
When $|s| = 2^\lambda$:
On the empty input, $\widetilde{\Pi}_{s,R}$ *outputs* $\mathsf{Enc}(1^{l(\lambda, |s|+1)}, \Pi, s, T; R)$. *On all other inputs,* $\widetilde{\Pi}_{s,R}$ *outputs* $\bot$.

We define $T'(\cdot, \cdot, \cdot, \cdot)$ (corresponding to the bound placed on the running time of $\widetilde{\Pi}_{s,R}$) to be the smallest polynomial such that for all λ, $s \in \{0,1\}^{\leq 2^\lambda}$, $R \in \{0,1\}^{2l(\lambda,|s|)}$, Π and T,

$$T'(\lambda, |s|, |\Pi|, \log(T)) \geq p(\lambda_{|s|+1}, |\widetilde{\Pi}_{s0,R}|, 0, \log(T'_{|s|+1}))$$
$$+ p(\lambda_{|s|+1}, |\widetilde{\Pi}_{s1,R}|, 0, \log(T'_{|s|+1}))$$
$$+ p(\lambda_{|s|+1}, |\Pi|, |s|, \log(T))$$
$$+ \mathsf{Time}_{\mathsf{PRG}}(R, 5 \cdot 2l(\lambda, |s| + 1))$$

where $\lambda_{|s|+1} = l(\lambda, |s| + 1)$, $T'_{|s|+1} = T'(\lambda, |s| + 1, |\Pi|, \log(T))$ (corresponding to the security parameter and time bound used for each of $\widetilde{\Pi}_{s0,R_0}$ and $\widetilde{\Pi}_{s1,R_1}$), $\mathsf{Time}_{\mathsf{PRG}}$ is the bound on the running time of the PRG, and $p(\cdot, \cdot, \cdot, \cdot)$ is the bound on $\mathsf{Time}_{\mathsf{Enc}}$ from the compactness of RE. We note that the polynomial T' exists because p is a polynomial, each of $\lambda_{|s|+1}$ and $|\widetilde{\Pi}_{s,R}|$ are of size polynomial in $\lambda, |s|$ and $|\Pi|$, and the self-dependence of $T'(\lambda, |s|, |\Pi|, \log(T))$ on $T'_{|s|+1}$ is only poly-logarithmic.

<u>REMARK:</u> *We note that $|\widetilde{\Pi}_{s,R}|$ is always $poly(\lambda, |\Pi|, |s|, \log(T))$. This is because $\widetilde{\Pi}_{s,R}$ is fully described by λ, Π, s, R and T, and the size of each of these is bounded by $poly(\lambda, |\Pi|, |s|, \log(T))$.*

Given this definition of $\widetilde{\Pi}_{s,R}$, we define our indistinguishability obfuscator as follows:

Construction 2 (Indistinguishability Obfuscator). *On input $\lambda \in \mathbb{N}$, Turing machine Π and time bound T, define $\widetilde{\Pi}$, the indistinguishability obfuscation of Π, to be*

$$\widetilde{\Pi} = \boldsymbol{iO}(1^\lambda, \Pi, T) = \mathsf{Enc}(1^{l(\lambda,0)}, \widetilde{\Pi}_{\epsilon,R}, T'(\lambda, 0, |\Pi|, \log(T)))$$

where ϵ is the empty string, and $R \xleftarrow{\$} \{0,1\}^{2l(\lambda,0)}$ and T' a fixed polynomial in $\lambda, |\Pi|$ and $\log(T)$, as described above.

<u>EVALUATION:</u> The algorithm to evaluate $\widetilde{\Pi}$ on input $x \in \{0,1\}^d, d < 2^\lambda$ proceeds as follows:

1. For every $0 \leq i \leq d$, compute encodings of $\widetilde{\Pi}_{x_{\leq i},R}$ successively, starting with $\widetilde{\Pi}$, an encoding of $\widetilde{\Pi}_{\epsilon,R}$, and subsequently, for every $0 < i \leq d$, computing the encoding of $\widetilde{\Pi}_{x_{\leq i},R}$ by evaluating the encoding of $\widetilde{\Pi}_{x_{<i},R}$, and selecting the encoding of $\widetilde{\Pi}_{x_{\leq i},R}$ from its output.
2. Evaluate the encoding of $\widetilde{\Pi}_{x,R} = \widetilde{\Pi}_{x_{\leq d},R}$ and obtain from its output $(\hat{\Pi}, \hat{x}) = \mathsf{Enc}(1^{l(\lambda,|x|+1)}, \Pi, x, T; R_4)$.
3. Run $\mathsf{Eval}(\hat{\Pi}, \hat{x})$ to obtain $\Pi(x)$.

We defer analysis of the correctness, running time, and compactness of our iO construction to the full version of our paper [LPST15].

4.1 Security Proof

Theorem 8. *Let* $(\mathsf{Enc}, \mathsf{Eval})$ *be a sub-exponentially-indistinguishability-secure, compact randomized encoding scheme and let* PRG *be a sub-exponentially-secure pseudorandom generator. Then the indistinguishability obfuscator defined in Construction 2 is subexponentially-secure.*

Proof. Consider any pair of ensembles of Turing machines and time bounds $\{\Pi^0_\lambda, \Pi^1_\lambda, T_\lambda\}$ where for every $\lambda \in \mathbb{N}$, $\Pi^0 = \Pi^0_\lambda$, $\Pi^1 = \Pi^1_\lambda$, $T = T_\lambda$,

$$|\Pi^0| = |\Pi^1| \le \mathrm{poly}(\lambda) \quad |T| \le \mathrm{poly}(\lambda)$$
$$\forall x, \Pi^{0,T}(x) = \Pi^{1,T}(x)$$

We first introduce some notation to describe the distributions of randomized encodings generated by $i\mathcal{O}(1^\lambda, \Pi^0_\lambda, T_\lambda)$ and $i\mathcal{O}(1^\lambda, \Pi^1_\lambda, T_\lambda)$. For $\lambda \in \mathbb{N}$, $s \in \{0,1\}^*, |s| \le 2^\lambda$, we define the following distributions

$$D_{\lambda,0,s} = \mathsf{Enc}(1^{l(\lambda,|s|)}, \widetilde{\Pi}^0_{s,R}, T')$$
$$D_{\lambda,1,s} = \mathsf{Enc}(1^{l(\lambda,|s|)}, \widetilde{\Pi}^1_{s,R}, T')$$

where R is uniformly random, T' is as described in Construction 1 and $\widetilde{\Pi}^b_{s,R}$ is defined for the Turing machine Π^b_λ, security parameter λ and time bound T_λ. We will show something stronger than the theorem statement. In particular, we have the following claim.

Claim. There exists $\lambda_0, \epsilon \in \mathbb{N}$ such that for every $\lambda > \lambda_0$, for every $s \in \{0,1\}^*, |s| \le 2^\lambda$ we have that the distributions $D_{\lambda,0,s}$ and $D_{\lambda,1,s}$ are $S(\lambda)$ indistinguishable where $S(\lambda) \ge 10 \cdot 2^{l(\lambda,|s|-1)^\epsilon}$.

Using the above claim with s as the empty string and recalling $l(\lambda, 0) = \lambda$, the theorem statement follows. Therefore, in the remainder of the proof, we prove the above claim.

PROOF OF CLAIM. Let ϵ be the larger of the constants associated with the sub-exponential security of the pseudorandom generator PRG and the indistinguishability security of the encoding scheme $(\mathsf{Enc}, \mathsf{Eval})$ (these constants are also named ϵ in their respective security definitions). Similarly, We consider λ_0 to be large enough so that the security of the encoding scheme $(\mathsf{Enc}, \mathsf{Eval})$ and the pseudorandom generator PRG is applicable. We will actually require a larger λ_0 so that certain asymptotic conditions (depending only on the polynomial size bounds of Π^0_λ, Π^1_λ and T_λ) hold, which we make explicit in the remainder of the proof. For every $\lambda > \lambda_0$, we prove the claim by induction on $|s|$. Our base case will be when $|s| = 2^\lambda$ and in the inductive step we show the claim holds for all s of a particular length d, if it holds for all s of length $d + 1$.

INDUCTION STATEMENT, FOR A FIXED $\lambda > \lambda_0$: For every $s \in \{0,1\}^{\le 2^\lambda}$, the distributions $D_{\lambda,0,s}$ and $D_{\lambda,1,s}$ are $10 \cdot 2^{l(\lambda,|s|-1)^\epsilon}$ indistinguishable.

BASE CASE: $|s| = 2^\lambda$. In this case, recall that the output of $\widetilde{\Pi}^b_{s,R}$ is simply $(\hat{\Pi}^{b,T}_\lambda, \hat{s})$. We first claim that, for all s, $(\hat{\Pi}^{0,T}_\lambda \hat{s})$ and $(\hat{\Pi}^{1,T}_\lambda, \hat{s})$ are $2^{\lambda'^\epsilon}$ indistinguishable where $\lambda' = l(\lambda, |s|)$, as follows.

Recall that the output of evaluating $\hat{\Pi}^{b,T}_\lambda, \hat{s}$ is simply $\Pi^{b,T}_\lambda(s)$. Since we have that $\Pi^{0,T}_\lambda(s) = \Pi^{1,T}_\lambda(s)$ for all s, we can apply the security of the randomized encoding scheme. More concretely, since the output (point) distributions are identical, they are $10 \cdot 2^{\lambda'^\epsilon}$-indistinguishable where $\lambda' = l(\lambda, |s| + 1)$. Let $B(\cdot)$ be a polynomial such that $B(\lambda')$ bounds from above $|\Pi^b|, |s|$ and T. By the security of the encoding scheme, the encodings $(\hat{\Pi}^{0,T}_\lambda \hat{s})$ and $(\hat{\Pi}^{1,T}_\lambda \hat{s})$ are S' indistinguishable where

$$S' \geq \frac{10 \cdot 2^{l(\lambda, |s|+1)^\epsilon}}{l(\lambda, |s| + 1)^c} - B(l(\lambda, |s| + 1))^c \geq \frac{10 \cdot 2^{l(\lambda, |s|+1)^\epsilon}}{l(\lambda, |s| + 1)^d} \geq 10 \cdot 2^{l(\lambda, |s|)^\epsilon}$$

where the first inequality holds for sufficiently large λ and in the second inequality, we use the fact that $l(\lambda, |s|+1) = l(\lambda, |s|) + \lambda^{d/\epsilon}$. Thus $(\hat{\Pi}^{0,T}_\lambda, \hat{s})$ and $(\hat{\Pi}^{1,T}_\lambda, \hat{s})$ are $10 \cdot 2^{l(\lambda, |s_\lambda|)^\epsilon}$-indistinguishable.

Now, recall that the output of $\widetilde{\Pi}^b_{s,R}$ is simply $(\hat{\Pi}^{b,T}_\lambda, \hat{s})$. By the above argument, we have that, for all s, $(\hat{\Pi}^{0,T}_\lambda \hat{s})$ and $(\hat{\Pi}^{1,T}_\lambda, \hat{s})$ are $2^{\lambda'^\epsilon}$-indistinguishable where $\lambda' = l(\lambda, |s|)$. Let B' be the polynomial such that $B'(l(\lambda, |s|))$ bounds $|\widetilde{\Pi}^b_{s,R}|$ and the running time of $\widetilde{\Pi}^b_{s,R}$. The encodings $D_{\lambda,0,s}$ and $D_{\lambda,1,s}$ are S' indistinguishable where

$$S' \geq \frac{10 \cdot 2^{l(\lambda, |s|)^\epsilon}}{l(\lambda, |s|)^c} - B'(l(\lambda, |s|))^c \geq \frac{10 \cdot 2^{l(\lambda, |s|+1)^\epsilon}}{l(\lambda, |s|)^d} \geq 10 \cdot 2^{l(\lambda, |s|-1)^\epsilon}$$

where, as before, the first inequality holds for sufficiently large λ and in the second inequality, we use the fact that $l(\lambda, |s| + 1) = l(\lambda, |s|) + \lambda^{d/\epsilon}$. Hence the claim holds for $|s| = 2^\lambda$.

INDUCTIVE STEP: $|s| < 2^\lambda$. By the induction hypothesis, we assume the claim holds for all s' such that $|s'| = |s| + 1$. Recall that the output of $\widetilde{\Pi}^b_{s,R}$ (where $R \xleftarrow{\$} \{0,1\}^{2l(\lambda, |s|)}$) is

$$\mathsf{Enc}(1^{l(\lambda, |s|+1)}, \widetilde{\Pi}^b_{s0,R_0}, T'; R_1)$$
$$\mathsf{Enc}(1^{l(\lambda, |s|+1)}, \widetilde{\Pi}^b_{s1,R_2}, T'; R_3)$$
$$\mathsf{Enc}(1^{l(\lambda, |s|+1)}, \Pi^b_\lambda, s, T; R_4)$$

where $(R_0, R_1, R_2, R_3, R_4) \leftarrow \mathsf{PRG}(R, 5 \cdot 2l(\lambda, |s| + 1))$. Let H^b denote the above output distribution. We will show H^0 and H^1 are indistinguishable by a hybrid argument as follows.

- Let G_1 be a hybrid distribution exactly as H^0 except that $(R_0, R_1, R_2, R_3, R_4) \xleftarrow{\$} \{0,1\}^{5 \cdot 2l(\lambda, |s|+1)}$. We claim that for both the distributions H^0 and G_1 are $5 \cdot 2^{\lambda'^\epsilon}$ indistinguishable where $\lambda' = l(\lambda, |s|)$.

This follows from the PRG security as follows: any size $5 \cdot 2^{\lambda'^\epsilon}$ adversary A that distinguishes H^0 and G_1 can be turned into an adversary A' that can break the PRG security with seed length $2\lambda'$ with the same advantage. A' has Π_λ^0, Π_λ^1, T_λ and s hardcoded in it. Hence, the size of A' is

$$5 \cdot 2^{\lambda'^\epsilon} + \text{poly}(\lambda) + \text{poly}(|s|) \leq 5 \cdot 2^{\lambda'^\epsilon} + \text{poly}(\lambda') \leq 2^{(2\lambda')^\epsilon}$$

where the last inequality holds when λ is sufficiently large. Hence, A' breaks the $2^{(2\lambda')^\epsilon}$-security of PRG and we have a contradiction.

Writing out the components of G_1, we have that it is identical to

$$G_1 \equiv D_{\lambda,0,s0}, D_{\lambda,0,s1}, \text{Enc}(1^{l(\lambda,|s|+1)}, \Pi_\lambda^0, s, T_\lambda; R)$$

– Let G_2 be a hybrid distribution obtained by modifying the first component of G_1 as follows.

$$G_2 \equiv D_{\lambda,1,s0}, D_{\lambda,0,s1}, \text{Enc}(1^{l(\lambda,|s|+1)}, \Pi_\lambda^0, s, T_\lambda; R)$$

We show that G_1 and G_2 are $5 \cdot 2^{\lambda'^\epsilon}$ indistinguishable. This follows from the induction hypothesis as follows: any size $5 \cdot 2^{\lambda'^\epsilon}$ adversary A that distinguishes G_1 and G_2 with advantage better than $1/(5 \cdot 2^{\lambda'^\epsilon})$ can be turned into an adversary A' that can distinguish $D_{\lambda,0,s0}$ and $D_{\lambda,1,s0}$ with the same advantage. As before, A' has Π_λ^0, Π_λ^1, T_λ and s hardcoded in it, and therefore the size of A' is at most $5 \cdot 2^{\lambda'^\epsilon} + \text{poly}(\lambda') \leq 10 \cdot 2^{\lambda'^\epsilon}$. Hence, A' breaks the induction hypothesis that says $D_{\lambda,0,s0}$ and $D_{\lambda,1,s0}$ are $10 \cdot 2^{\lambda'^\epsilon}$-indistinguishable.

– Similarly, let G_3 be a hybrid distribution obtained by modifying the second component of G_2 as follows.

$$G_3 \equiv D_{\lambda,1,s0}, D_{\lambda,1,s1}, \text{Enc}(1^{l(\lambda,|s|+1)}, \Pi_\lambda^0, s, T_\lambda; R)$$

Similarly as above, we have that G_2 and G_3 are $5 \cdot 2^{\lambda'^\epsilon}$-indistinguishable.

– Let G_4 be a hybrid distribution obtained by modifying the third component of G_3 as follows.

$$G_4 \equiv D_{\lambda,1,s0}, D_{\lambda,1,s1}, \text{Enc}(1^{l(\lambda,|s|+1)}, \Pi_\lambda^1, s, T_\lambda; R)$$

We show G_3 and G_4 are $5 \cdot 2^{\lambda'^\epsilon}$-indistinguishable. First, since $\Pi_\lambda^{0,T}(s) = \Pi_\lambda^{1,T}(s)$, by the security of the encoding scheme, we have that the encodings that form the third component of G_3 and G_4 are S' indistinguishable where, similar to the base case, $B(l(\lambda,|s|))$ bounds from above $|\Pi_\lambda^b|$, $|s|$ and T

$$S' \geq \frac{10 \cdot 2^{l(\lambda,|s|)^\epsilon}}{l(\lambda,|s|)^c} - B(l(\lambda,|s|))^c \geq \frac{10 \cdot 2^{l(\lambda,|s|)^\epsilon}}{l(\lambda,|s|)^d} \geq 10 \cdot 2^{l(\lambda,|s|-1)^\epsilon}$$

Hence by a similar argument as before, the hybrid distributions are $5 \cdot 2^{\lambda'^\epsilon}$-indistinguishable.

- Finally we observe that G_4 and H^1 are $5 \cdot 2^{\lambda'^\epsilon}$-indistinguishable just as G_1 and H^0 were. By a simple hybrid argument, we have that H^0 and H^1 are $2^{\lambda'^\epsilon}$-indistinguishable.

Recall that H^0 and H^1 are the distributions of outputs of $\widetilde{\Pi}^0_{s,R}$ and $\widetilde{\Pi}^1_{s,R}$ respectively. By the security of the randomized encoding scheme, the encodings of these machines, i.e. $D_{\lambda,0,s}$ and $D_{\lambda,1,s}$ are $S'(\lambda)$-indistinguishable where

$$S'(\lambda) \geq \frac{2^{l(\lambda,|s|)^\epsilon}}{l(\lambda,|s|)^c} - B'(l(\lambda,|s|)^c \geq \frac{2^{l(\lambda,|s|)^\epsilon}}{l(\lambda,|s|)^d} \geq \frac{2^{l(\lambda,|s|-1)^\epsilon} \cdot 2^{(2d\lambda)}}{2^{d\lambda} \cdot (2d\lambda)^{d/\epsilon}} \geq 10 \cdot 2^{l(\lambda,|s|-1)^\epsilon}$$

where $B'(l(\lambda,|s|))$ bounds from above $|\Pi^b_{s,R}|$ and T'. The second inequality holds for sufficiently large λ. In the third inequality, we use the fact that $l(\lambda,|s|) \leq |s|(2d\lambda)^{1/\epsilon} \leq 2^\lambda (2d\lambda)^{1/\epsilon}$ and the last inequality holds for sufficiently large λ.

4.2 Nice Distributions

Later in Sect. 6, we show that compact RE does not exist for general distributions in the plain model. However, here we observe that the above construction of unbounded input IO relies only on compact RE for certain "special purpose" distributions that is not ruled out by the impossibility result in Sect. 6. We now abstract out the structure of these special purpose distributions. Let RE = (Enc, Dec) be a randomized encoding scheme; we define "nice" distributions w.r.t. RE.

0-nice distributions: We say that a pair of distribution ensembles $\{\mathcal{D}_{0,\lambda}\}$ and $\{D_{1,\lambda}\}$ are *0-nice* if $D_{0,\lambda}$ always outputs a fixed tuple (Π_0, x, T) while $D_{1,\lambda}$ always outputs a fixed tuple (Π_1, x, T), satisfying that $\Pi_0^T(x) = \Pi_1^T(x)$.

k-nice distributions: We say that a pair of distribution ensembles $\{\mathcal{D}_{0,\lambda}\}$ and $\{\mathcal{D}_{1,\lambda}\}$ are *k-nice* if there exist some $\ell = \mathrm{poly}(\lambda)$ pairs of distributions $(\{\mathcal{E}^i_{0,\lambda}\}, \{\mathcal{E}^i_{1,\lambda}\})_{i \in [\ell]}$, where the ith pair is k^i-nice with $k^i \leq k - 1$, such that, $\mathcal{D}_{b,\lambda}$ samples tuple (Π_b, x_b, T_b) satisfying the following:

- For each $i \in [\ell]$, sample $(\Lambda^i_b, z^i_b, T^i_b) \xleftarrow{\$} \mathcal{E}^i_{b,\lambda}$.
- The output of $\Pi_b(x_b)$ consists of ℓ randomized encodings, where the ith encoding is in the support of $\mathsf{Enc}(1^{\lambda'}, \Lambda^i_b, z^i_b, T^i_b)$, for some $\lambda' = \mathrm{poly}(\lambda)$.

Finally, we say that a pair of distribution ensembles $\{\mathcal{D}_{0,\lambda}\}$ and $\{\mathcal{D}_{1,\lambda}\}$ are *nice* w.r.t. RE if they are *k-nice* w.r.t. RE for some integer k.

Our construction of unbounded input IO and its analysis in previous sections relies only on compact RE for nice distribution ensembles. Hence we can refine Theorem 8 to the following:

Proposition 1. *Assume the existence of a compact randomized encoding scheme* RE *which is sub-exponentially-indistinguishability-secure for every pair of distribution ensemble that are nice w.r.t.* RE; *assume further the existence of sub-exponentially secure one-way functions. Then, there is an unbounded-input indistinguishability obfuscator for Turing machines.*

We stress again that compact RE for nice distributions is not ruled out by the impossibility result in Sect. 6. Hence, we obtain unbounded input IO from a new assumption different from the extactability assumptions used in previous work [BCP14, ABG+13, IPS15].

CANDIDATE CONSTRUCTION: Finally, we describe a candidate construction of compact RE for nice distributions using the KLW indistinguishability obfuscator for bounded-input Boolean Turing machines: Given input $(1^\lambda, M, x, T)$, the encoding is an obfuscation, using the KLW scheme, of the program $\Pi_{M,x}$ that on input $i \in [T]$ outputs the ith bit of the output $M^T(x)$. Since $\Pi_{M,x}$ is Boolean, the KLW obfuscator can be applied, and the encoding time is $\text{poly}(\lambda, |M|, |x|, \log T)$ (hence compact). By the security of indistinguishability obfuscation, for any M_1, x_1 and M_2, x_2 with identical outputs, their encodings are indistinguishable, and thus this construction is a weak compact RE. We here consider it also a candidate construction for compact RE with distributional indistinguishability.

BOUNDED-INPUT IO FROM SUBLINEAR RE: We note that relying on a very similar construction as above, a randomized encoding scheme with only sublinear compactness (as opposed to full compactness) can be used to construct a bounded-input indistinguishability obfuscator for Turing machines. We refer the reader to the full version of this paper [LPST15] for more details.

5   Bounded-Input IO from Compact RE in the CRS Model

In this section we consider compact RE schemes for Turing machines in the *common reference string* (CRS) model. We show that (1) such encoding schemes can be constructed from compact functional encryption for circuits, and that (2) such encoding schemes suffice to get IO for circuits, which then by [KLW14] suffices to get bounded-input IO for Turing machines.

5.1   Randomized Encoding Schemes in the CRS Model

We first formally define a randomized encoding scheme for a class of Turing machines in the CRS model. In this model, a one-time setup is performed which takes (in addition to the security parameter) a bound on machine size, input length, running time and output length. Only computations that respect these bounds can be encoded using this setup. The setup outputs a *long* CRS (the length is polynomial in the aforementioned bounds) and a *short* public encoding key (which depends only on the security parameter). The public encoding key is used by the encoding algorithm, which produces encodings that are *compact* as before. The CRS is used by the evaluation algorithm.

Definition 18 (Randomized Encoding Schemes in the CRS Model). *A Randomized Encoding scheme RE for a class of Turing machines $\{\mathcal{M}_\lambda\}$ in the CRS model consists of the following algorithms:*

- $(\text{crs}, pk) \xleftarrow{\$} \text{Setup}(1^\lambda, 1^m, 1^n, 1^T, 1^l)$: Setup *gets as input (in unary) the security parameter* λ, *a machine size bound* m, *input length bound* n, *time bound* T *and output length bound* l.
- $\hat{\Pi}_x \xleftarrow{\$} \text{Enc}(pk, \Pi, x)$: Enc *is probabilistic and gets as input a public key* pk *generated by* Setup, *Turing machine* $\Pi \in \mathcal{M}_\lambda$ *and input* x. *It outputs an encoding* $\hat{\Pi}_x{}^6$.
- $y \leftarrow \text{Eval}(\hat{\Pi}_x, \text{crs})$: *On input* $\hat{\Pi}_x$ *produced by* Enc *and* crs *produced by* Setup, Eval *outputs* y.

Correctness: *For every security parameters* $\lambda \in \mathbb{N}$, $m, n, T, l \in \mathbb{N}$, *Turing machine* $\Pi \in \mathcal{M}_\lambda$ *and input* x, *such that,* $|\Pi| \leq m$, $|x| \leq n$, *and* $|\Pi^T(x)| \leq l$, *we have that*

$$\Pr\left[\begin{array}{l} (\text{crs}, pk) \xleftarrow{\$} \text{Setup}(1^\lambda, 1^m, 1^n, 1^T, 1^l) \\ \hat{\Pi}_x \xleftarrow{\$} \text{Enc}(pk, \Pi, x) \end{array} : \text{Eval}(\hat{\Pi}_x, \text{crs}) = \Pi^T(x) \right] = 1$$

In the CRS model, it is possible to have a compact RE for all Turing machines with simulation security.

Definition 19. *A randomized encoding scheme* RE *for a class of Turing machines* $\{\mathcal{M}_\lambda\}$ *in the CRS model satisfies* $(\lambda_0, S(\cdot))$-**simulation security,** *if there exists a PPT algorithm* Sim *and a constant* c, *such that, for every ensemble* $\{\Pi_\lambda, x_\lambda, m_\lambda, n_\lambda, l_\lambda, T_\lambda\}$ *where* $\Pi_\lambda \in \mathcal{M}_\lambda$ *and* $|\Pi_\lambda|, |x_\lambda|, m_\lambda, n_\lambda, l_\lambda, T_\lambda \leq B(\lambda)$ *for some polynomial* B, *the following ensembles are* $(\lambda_0, S'(\lambda))$ *indistinguishable, with* $S'(\lambda) = S(\lambda) - B(\lambda)^c$ *for all* $\lambda \in N$.

$$\left\{ (\text{crs}, \text{pk}) \xleftarrow{\$} \text{Setup}(1^\lambda, 1^m, 1^n, 1^T, 1^l), \ \hat{\Pi}_x \xleftarrow{\$} \text{Enc}(pk, \Pi, x) \ : \ (\text{crs}, pk, \hat{\Pi}_x) \right\}$$

$$\left\{ (\text{crs}, pk, \hat{\Pi}_x) \xleftarrow{\$} \text{Sim}(1^\lambda, \Pi^T(x), 1^{|\Pi|}, 1^{|x|}, 1^m, 1^n, 1^T, 1^l) \ : \ (\text{crs}, pk, \hat{\Pi}_x) \right\}$$

where subscripts of security parameter are suppressed.

Definition 20 (Compactness and Sublinear Compactness in the CRS model). *A randomized encoding scheme* RE = (Setup, Enc, Eval) *for Turing machines in the CRS model is* compact *(or sublinear compact) if* Setup *is PPT, and* Enc *and* Eval *have the same efficiency as their counterparts in a compact (or sublinear compact) randomized encoding scheme for Turing machines in the plain model.*

Remark 2. *We note that a distributional-indistinguishability notion of security (analogous to Definition 14) can be defined for randomized encoding schemes in the CRS model. In the full version of this paper [LPST15], we provide this definition and show* (λ_0, S)-*simulation security implies* (λ_0, S)-*indistinguishability security both in the plain model and the CRS model.*

---

[6] Encoding $\hat{\Pi}_x$ can be viewed as the combination of the program encoding $\hat{\Pi}$ and the input encoding $\hat{x}$ of Definition 13.

5.2   Succinctness and Weak-Compactness

We also consider a different weakening of compactness, called succinctness [BGL+15], where encoding time can depend linearly on the length of the output (but only polylogarithmically on the time bound T).

Definition 21 (Succinct Randomized Encoding for Turing machines [BGL+15]). *A succinct randomized encoding scheme for Turing machines in the CRS model is succinct if it has the following efficiency:*

- *For every security parameters $\lambda \in \mathbb{N}$, $m, n, T, l \in \mathbb{N}$, Turing machine $\Pi \in \mathcal{M}_\lambda$ and input x, such that, $|\Pi| \leq m$, $|x| \leq n$, and $|\Pi^T(x)| \leq l$, every public key $pk \leftarrow \mathsf{Setup}(1^\lambda, 1^m, 1^n, 1^T, 1^l)$ and every encoding $\hat{\Pi}_x \leftarrow \mathsf{Enc}(1^\lambda, \Pi, x, T)$, it holds*

$$\mathsf{Time}_{\mathsf{Setup}}(1^\lambda, 1^m, 1^n, 1^T, 1^l) = \mathrm{poly}(\lambda, m, n, T, l)$$

$$\mathsf{Time}_{\mathsf{Enc}}(pk, \Pi, x) = \ell \cdot \mathrm{poly}(\lambda, |\Pi|, |x|, \log T)$$

$$\mathsf{Time}_{\mathsf{Eval}}(\hat{\Pi}, \hat{x}) = \mathrm{poly}(\lambda, m, n, T)$$

We finally consider a notion of RE that is weaker than sublinear-compactness, where we allow the encoding time to be polynomially dependent on the time bound T, but still require the encoding size be sub-linear in T. We call such RE schemes *weakly sublinear compact*.

Definition 22 (Weakly Sublinear Compact Randomized Encoding scheme). *We say a randomized encoding scheme $\mathsf{RE} = (\mathsf{Setup}, \mathsf{Enc}, \mathsf{Eval})$ in the CRS model for a class of Turing machines $\{\mathcal{M}_\lambda\}$ is weakly sublinear compact if the efficiency requirement on Enc in Definition 21 is changed to: For some constant $\varepsilon \in (0, 1)$,*

$$\mathsf{Time}_{\mathsf{Enc}}(pk, \Pi, x) = \mathrm{poly}(\lambda, |\Pi|, |x|, T)$$

$$\mathsf{outlen}_{\mathsf{Enc}}(pk, \Pi, x) = T^{1-\epsilon} \cdot \mathrm{poly}(\lambda, |\Pi|, x|)$$

Next, we observe that RE schemes satisfying the notions defined above (*i.e.* succinctness and weak sublinear compactness) can be composed to get a RE scheme satisfying sub-linear compactness. In particular, by composing a succinct RE scheme with a weakly compact RE scheme, one can obtain a sub-linearly compact RE scheme. We defer the proof to the full version of the paper.

Theorem 9. *Assume the existence of pseudorandom generators. If there is a succinct RE scheme and a weakly sublinear compact RE scheme for Turing machines, then there is a sub-linearly compact randomized encoding scheme for Turing machines.*

5.3   Randomized Encodings with CRS from Compact Functional Encryption

In this section we construct RE schemes in the CRS model from Compact Functional encryption schemes and pseudorandom generators.

Let (FE.Setup, FE.Enc, FE.Dec) be a public key, compact functional encryption scheme for $\mathbf{P}/poly$, and let PRG be a pseudorandom generator. We define a randomized encoding scheme in the CRS model (Setup, Enc, Eval) as follows.

The setup algorithm Setup($1^\lambda, 1^m, 1^n, 1^T, 1^l$):

- Setup first generates keys for the functional encryption scheme $(mpk, msk) \leftarrow$ FE.Setup(1^λ) and samples a uniformly random string $s \leftarrow \{0,1\}^\lambda$.
- Next, it generates the string $c \leftarrow 0^l \oplus$ PRG(s, l). That is, it encrypts 0^l using a one-time pad with the key coming from PRG(s, l)
- Let U be the universal circuit that on input (Π, x) where $|\Pi| \leq m$ and $|x| \leq n$ runs machine Π on x for at most T steps and outputs the first l bits of the tape as output. We define a circuit $C_{U,c}$, that has the string c and circuit U hardcoded in it, as follows.
  1. $C_{U,c}$ takes as input (Π, x, s', b) where (Π, x) satisfies the size constraints as described above, $s' \in \{0,1\}^\lambda$ and $b \in \{0,1\}$.
  2. If $b = 0$ then $C_{U,c}$ outputs $U(\Pi, x)$.
  3. Otherwise $C_{U,c}$ outputs $c \oplus$ PRG(s').
- Setup runs $sk_C \leftarrow$ FE.KeyGen($msk, C_{U,c}$) and outputs sk_C as the common reference string crs and mpk as the public encoding key pk.

The encoding algorithm Enc(pk, Π, x): Enc parses pk as the functional public key mpk and runs $ct \leftarrow$ FE.Enc($mpk, (\Pi, x, 0^\lambda, 0)$). Enc outputs the functional ciphertext ct as the encoding $\hat{\Pi}_x$.

The evaluation algorithm Eval($\hat{\Pi}_x$, crs): Eval parses $\hat{\Pi}_x$ as a functional ciphertext ct and crs as the functional secret key $sk_{C_{U,c}}$. Eval runs $y \leftarrow$ FE.Dec($sk_{C_{U,c}}, ct$) and outputs y.

The correctness of the above encoding scheme follows directly from that of the underlying functional encryption scheme. When a randomized encoding of (Π, x) is evaluated, it outputs the result of running the universal circuit U on (Π, x) that is $\Pi^T(x)$. Also the efficiency properties of the above scheme follow directly from the compactness properties of the functional encryption scheme. For example, if the functional encryption scheme we start from has sub-linear compactness (the ciphertext size is sub-linear in the circuit size of the function for which the functional secret keys are generated) then we get an encoding scheme with sub-linear compactness.

We have the following theorem. We refer the reader to the full version for the proof.

Theorem 10. *Let* (FE.Setup, FE.Enc, FE.Dec) *be a public key functional encryption scheme for* $\mathbf{P}/poly$ *with* $(\lambda_0, S(\cdot))$ *selective security, and let* PRG *be a pseudorandom generator with* $(\lambda_0, S(\cdot))$ *security. The randomized encoding scheme defined above is* $(\lambda_0, \frac{S(\cdot)}{4})$-*simulation secure.*

Corollary 1. *If there exists a public key, compact (resp. succinct, weakly sub-linear compact) functional encryption for* P/poly *scheme with selective security,* *and a secure PRG, then there exists a compact (resp. succinct[7], weakly sublinear compact) randomized encoding scheme for Turing machines in the CRS model that is simulation secure.*

The above theorem and corollary also work in the regime of sub-exponential security. That is, starting with a functional encryption scheme and pseudorandom generator that are sub-exponentially secure we obtain a RE scheme with sub-exponential security.

The following corollary is obtained by combining Corollary 1 with Theorem 6 and Theorem 7. While we use this corollary in our results, we believe it is of independent interest too. Succinct RE schemes for Turing machines were shown by [BGL+15] to have a variety of applications. However the only known construction of it ([KLW14]) relies on **iO** for circuits. We observe that in the CRS model, succinct RE schemes can be based simply on LWE.

Corollary 2. *Assuming LWE (resp. with sub-exponential hardness), there exists a succinct RE scheme for Turing machines in the CRS model with (resp. sub-exponential) simulation security.*

Finally, the following corollary shows that, assuming LWE, weakly sublinear compact FE is sufficient to construct sublinearly-compact RE in the CRS model. This corollary follows by combining Corollary 1, which shows that weakly sublinear compact FE implies weakly sublinear compact RE in the CRS model, Corollary 2, which constructs succinct RE in the CRS model from LWE, and finally Theorem 9, which shows that weakly sublinear compact RE and succinct RE can be combined to produce sublinearly-compact RE in the CRS model.

Corollary 3. *Assuming LWE (resp. with sub-exponential hardness), if there exists a weakly sublinear compact FE scheme for P/poly (resp. with sub-exponential security), then there exists a sublinearly-compact RE scheme for Turing machines in the CRS model with (resp. sub-exponential) simulation security.*

5.4   IO for Circuits from RE in the CRS Model

In this section we show that compact RE schemes for Turing machines in the CRS model implies **iO** for circuits; combining with the result of [KLW14] that **iO** for circuits implies **iO** for (bounded-input) Turing machines, we obtain the following theorem:

Theorem 11. *Assume the existence of sub-exponentially secure one-way functions. If there exists a sublinearly compact randomized encoding scheme in the CRS model with sub-exponential simulation security, then there exists an bounded-input indistinguishability obfuscator for Turning machines.*

---

[7] We note that for succinct RE, we first apply the transformation from succinct FE to get succinct RE with 1-bit output, and to encode Turing Machines with multi-bit outputs, we generate one such RE for each output bit.

We note that the theorem also holds w.r.t. sublinearly compact randomized encoding scheme in the CRS model, satisfying, weaker, distributional indistinguishability security, *with auxiliary inputs* (i.e., Definition 14 w.r.t. distributions $\{D_{b,\lambda}\}$ that additionally samples an auxiliary input z_b, and the security requirement is that if the output distributions together with the auxiliary inputs are indistinguishable, then the encodings together with the auxiliary inputs are also indistinguishable, with appropriate security loss). Since the distributional indistinguishability security is implied by simulation security, and in the CRS model, we can construct sublinearly compact RE with simulation security from sublinearly compact FE schemes, for simplicity, we directly state and prove the theorem w.r.t. simulation security.

The construction and proof is very similar to that of unbounded-input **iO** from compact RE schemes in the plain model presented in Sect. 4. We refer the reader to the full version [LPST15] for more details.

5.5   Summary of Results Using RE in the CRS Model

We observe that by combining Theorem 11 with Corollary 1, we reprove the results of [AJ15, BV15]

Theorem 12. *Assuming the existence of compact functional encryption with subexponential security, there exists a bounded-input indistinguishability obfuscator for Turing Machines.*

Further, we get the following new result, as a consequence of Corollary 3 and Theorem 11:

Theorem 13. *Assuming the existence of weakly sublinear compact functional encryption with subexponential security and LWE with subexponential security, there exists a bounded-input indistinguishability obfuscator for Turing Machines.*

6   Impossibility of Compact RE

In this section, we mention several impossibility results related to sublinear (and hence compact) RE with different security. We refer the reader to the full version [LPST15] for the proofs.

Theorem 14. *The following impossibility results hold in the* plain *model:*

1. *Sublinear randomized encoding schemes with (polynomial) simulation security do not exist, assuming one-way functions.*
2. *Sublinear randomized encoding schemes with sub-exponential indistinguishability security do not exist, assuming sub-exponentially secure one-way functions.*
3. *Sublinear randomized encoding schemes with (polynomial) indistinguishability security do not exist, assuming bounded-input iO for Turing machines and one-way functions.*

Acknowledgment. We are extremely grateful to Nir Bitansky and Omer Paneth for informing us of their impossibility result for compact RE assuming differing-input obfuscation, SNARKs and collision-resistant hash functions; this results was the inspiration behind our main impossibility result. We are also very grateful to them for many delightful and insightful discussions.

References

[ABG+13] Ananth, P., Boneh, D., Garg, S., Sahai, A., Zhandry, M.: Differing-inputs obfuscation and applications. IACR Cryptology ePrint Archive 2013:689 (2013)

[ABSV14] Ananth, P., Brakerski, Z., Segev, G., Vaikuntanathan,V.: The trojan method in functional encryption: From selective to adaptive security. Technical report, generically. Cryptology ePrint Archive, Report 2014/917 (2014)

[AIK04] Applebaum, B., Ishai, Y., Kushilevitz, E.: Cryptography in nc^0. In: FOCS, pp. 166–175 (2004)

[AIK06] Applebaum, B., Ishai, Y., Kushilevitz, E.: Computationally private randomizing polynomials and their applications. Comput. Complex. **15**(2), 115–162 (2006)

[AJ15] Ananth, P., Jain, A.: Indistinguishability obfuscation from compact functional encryption. IACR Cryptology ePrint Archive 2015:173 (2015)

[App11] Applebaum, B.: Randomly encoding functions: a new cryptographic paradigm. In: Fehr, S. (ed.) ICITS 2011. LNCS, vol. 6673, pp. 25–31. Springer, Heidelberg (2011)

[Bar01] Barak, B.: How to go beyond the black-box simulation barrier. In: FOCS, pp. 106–115 (2001)

[BCP14] Boyle, E., Chung, K.-M., Pass, R.: On extractability obfuscation. In: Lindell, Y. (ed.) TCC 2014. LNCS, vol. 8349, pp. 52–73. Springer, Heidelberg (2014)

[BCPR13] Bitansky, N., Canetti, R., Paneth, O., Rosen, A.: Indistinguishability obfuscation vs. auxiliary-input extractable functions: One must fall. IACR Cryptology ePrint Archive, 2013:641 (2013)

[BGI+01] Barak, B., Goldreich, O., Impagliazzo, R., Rudich, S., Sahai, A., Vadhan, S.P., Yang, K.: On the (Im)possibility of obfuscating programs. In: Kilian, J. (ed.) CRYPTO 2001. LNCS, vol. 2139, pp. 1–18. Springer, Heidelberg (2001)

[BGL+15] Bitansky, N., Garg, S., Lin, H., Pass, R., Telang, S.: Succinct randomized encodings and their applications. IACR Cryptology ePrint Archive, 2015:356 (2015)

[BKS15] Brakerski, Z., Komargodski, I., Segev, G.: From single-input to multi-input functional encryption in the private-key setting. IACR Cryptology ePrint Archive, 2015:158 (2015)

[BP15] Bitansky, N., Paneth, O.: ZAPs and non-interactive witness indistinguishability from indistinguishability obfuscation. In: Dodis, Y., Nielsen, J.B. (eds.) TCC 2015, Part II. LNCS, vol. 9015, pp. 401–427. Springer, Heidelberg (2015)

[BSW11] Boneh, D., Sahai, A., Waters, B.: Functional encryption: definitions and challenges. In: Ishai, Y. (ed.) TCC 2011. LNCS, vol. 6597, pp. 253–273. Springer, Heidelberg (2011)

[BV15] Bitansky, N., Vaikuntanathan, V.: Indistinguishability obfuscation from functional encryption. IACR Cryptology ePrint Archive, 2015:163 (2015)

[CHJV14] Canetti, R., Holmgren, J., Jain, A., Vaikuntanathan, V.: Indistinguishability obfuscation of iterated circuits and RAM programs. IACR Cryptology ePrint Archive, 2014:769 (2014)

[GGH+13] Garg, S., Gentry, C., Halevi, S., Raykova, M., Sahai, A., Waters, B.: Candidate indistinguishability obfuscation and functional encryption for all circuits. In: Proceedings of FOCS 2013 (2013)

[GGHW13] Garg, S., Gentry, C., Halevi, S., Wichs, D.: On the implausibility of differing-inputs obfuscation and extractable witness encryption with auxiliary input. In: Garay, J.A., Gennaro, R. (eds.) CRYPTO 2014, Part I. LNCS, vol. 8616, pp. 518–535. Springer, Heidelberg (2014)

[GGM86] Goldreich, O., Goldwasser, S., Micali, S.: How to construct random functions. J. ACM **33**(4), 792–807 (1986)

[GKP+13a] Goldwasser, S., Kalai, Y.T., Popa, R.A., Vaikuntanathan, V., Zeldovich, N.: Reusable garbled circuits and succinct functional encryption. In: Symposium on Theory of Computing Conference, STOC 2013, Palo Alto, CA, USA, June 1–4, 2013, pp. 555–564 (2013)

[GMR89] Goldwasser, S., Micali, S., Rackoff, C.: The knowledge complexity of interactive proof systems. SIAM J. Comput. **18**(1), 186–208 (1989)

[GVW13] Gorbunov, S., Vaikuntanathan, V., Wee, H.: Attribute-based encryption for circuits. In: Symposium on Theory of Computing Conference, STOC 2013, Palo Alto, CA, USA, June 1–4, 2013, pp. 545–554 (2013)

[IK00] Ishai, Y., Kushilevitz, E.: Randomizing polynomials: A new representation with applications to round-efficient secure computation. In: 41st Annual Symposium on Foundations of Computer Science, FOCS 2000, 12–14 November 2000, Redondo Beach, California, USA, pp. 294–304 (2000)

[IK02a] Ishai, Y., Kushilevitz, E.: Perfect constant-round secure computation via perfect randomizing polynomials. In: Widmayer, P., Triguero, F., Morales, R., Hennessy, M., Eidenbenz, S., Conejo, R. (eds.) ICALP 2002. LNCS, vol. 2380, pp. 244–256. Springer, Heidelberg (2002)

[IPS15] Ishai, Y., Pandey, O., Sahai, A.: Public-coin differing-inputs obfuscation and its applications. In: Dodis, Y., Nielsen, J.B. (eds.) TCC 2015, Part II. LNCS, vol. 9015, pp. 668–697. Springer, Heidelberg (2015)

[KLW14] Koppula, V., Lewko, A.B., Waters, B.: Indistinguishability obfuscation for turing machines with unbounded memory. Technical report, Cryptology ePrint Archive, Report 2014/925 (2014). http://eprint.iacr.org

[KMN+14] Komargodski, I., Moran, T., Naor, M., Pass, R., Rosen, A., Yogev, E.: One-way functions and (im)perfect obfuscation (2014)

[LPST15] Lin, H., Pass, R., Seth, K., Telang, S.: Output-compressing randomized encodings and applications. Cryptology ePrint Archive, Report 2015/720 (2015). http://eprint.iacr.org/

[SW14] Sahai, A., Waters, B.: How to use indistinguishability obfuscation:Deniable encryption, and more. In: Proceedings of STOC 2014 (2014)

[Yao82] Yao, A.C.-C.: Protocols for secure computations (extended abstract). In: 23rd Annual Symposium on Foundations of Computer Science, Chicago, Illinois, USA, 3–5 November 1982, pp. 160–164 (1982)

Functional Encryption for Turing Machines

Prabhanjan Ananth$^{(\boxtimes)}$ and Amit Sahai

Department of Computer Science and Center for Encrypted Functionalities,
University of California, Los Angeles, USA
prabhanjan.va@gmail.com, sahai@cs.ucla.edu

Abstract. In this work, we construct an adaptively secure functional encryption for Turing machines scheme, based on indistinguishability obfuscation for circuits. Our work places no restrictions on the types of Turing machines that can be associated with each secret key, in the sense that the Turing machines can accept inputs of unbounded length, and there is no limit to the description size or the space complexity of the Turing machines.

Prior to our work, only special cases of this result were known, or stronger assumptions were required. More specifically, previous work (implicitly) achieved selectively secure FE for Turing machines with a-priori bounded input based on indistinguishability obfuscation (STOC 2015), or achieved FE for general Turing machines only based on knowledge-type assumptions such as public-coin differing-inputs obfuscation (TCC 2015).

A consequence of our result is the first constructions of *succinct* adaptively secure garbling schemes (even for circuits) in the standard model. Prior succinct garbling schemes (even for circuits) were only known to be adaptively secure in the random oracle model.

1 Introduction

Contemporary cloud-based computing systems demand encryption schemes that go far beyond the traditional goal of merely securing a communication channel. The notion of functional encryption, first conceived under the name of Attribute-Based Encryption in [SW05] and formalized later in the works of [BSW11,O'N10], has emerged as a powerful form of encryption well-suited to many contemporary applications (see [BSW11,BSW12] for further discussion

P. Ananth—This work was partially supported by grant #360584 from the Simons Foundation.

A. Sahai—Research supported in part from a DARPA/ARL SAFEWARE award, NSF Frontier Award 1413955, NSF grants 1228984, 1136174, 1118096, and 1065276, a Xerox Faculty Research Award, a Google Faculty Research Award, an equipment grant from Intel, and an Okawa Foundation Research Grant. This material is based upon work supported by the Defense Advanced Research Projects Agency through the ARL under Contract W911NF-15-C-0205. The views expressed are those of the author and do not reflect the official policy or position of the Department of Defense, the National Science Foundation, or the U.S. Government.

E. Kushilevitz and T. Malkin (Eds.): TCC 2016-A, Part I, LNCS 9562, pp. 125–153, 2016.
DOI: 10.1007/978-3-662-49096-9_6

of application scenarios for functional encryption). A functional encryption (FE) scheme allows a user possessing a key associated with a function f to recover the output $f(x)$, given an encryption of x. The intuitive security guarantee of a FE scheme dictates that the only information about x revealed to the user is $f(x)$. Furthermore, if the user obtains keys for many functions $f_1, \ldots f_k$, then the user should only learn $f_1(x), \ldots, f_k(x)$ and nothing more. It turns out that formalizing security using a simulation-based definition leads to impossibility results [BSW11, AGVW13]; however, there are sound adaptive indistinguishability-based formulations [BSW11] that also imply simulation-based security in restricted settings [CIJ+13]. Following most recent work on FE [GGH+13, Wat15, GGHZ14, ABSV15], we will focus on achieving this strong indistinguishability-based notion of security here.

In this work, we address the following basic question:

"Is FE possible for functions described by arbitrary Turing machines?"

Previous Work and Its Limitations. There have been many works on functional encryption over the past few years but a satisfying answer to this question has remained elusive.

The first constructions of FE considered only limited functions, such as inner product [KSW08]. The first constructions of FE that allowed for more general functions considered the setting where the adversary can just request a single (or a bounded number of) key queries [SS10, GVW12], but only for functions represented by circuits. A major advance occurred in the work of [GGH+13], which constructed an FE scheme allowing for functions specified by arbitrary circuits, with no bound on key queries, based on indistinguishability obfuscation (iO) for circuits. Since this work, the assumption of iO for circuits has become the staple assumption in this area.

However, [GGH+13] and other FE results deal with functionalities represented by *circuits* – and representing functions as circuits gives rise to two major drawbacks. The first drawback is that a circuit representation takes the worst case running time on every input. Research to deal with this issue was initiated by Goldwasser et al. [GKP+13], and there have been several recent works [BGL+15, CHJV15, KLW15, CCC+15], that (implicitly or explicitly) give rise to FE schemes with input-specific runtimes based on iO for circuits.

The second drawback is that the input length of the function is a-priori bounded. In many scenarios, especially involving large datasets, having an a-priori bound is clearly unreasonable. For example, if functional encryption is used for allowing a researcher to perform some data analysis on hospital records, then having a bound on input length would require that there be an a-priori bound, at the time of setting up the encryption scheme, on the length of encrypted hospital records, which seems quite unreasonable. In general, we would like to represent the function being computed as a Turing Machine, that can accept inputs of arbitrary length. The problem of constructing FE schemes which can handle messages of unbounded length has remained largely open: the recent works of [BGL+15, CHJV15, KLW15] construct iO for Turing Machines only with bounded input length, where the bound must be specified at the time of obfuscating the Turing Machine. If this iO method is combined, for example, with the FE construction recipe of [GGH+13], then this would only yield FE for functions with a bound on input length specified at the time of setting up the FE scheme.

There have been works [BCP14, IPS15] on overcoming the issue of a priori bounded input lengths but these are based on strong knowledge-type assumptions called differing inputs obfuscation [BGI+12, BCP14, ABG+13] or more recently public-coin differing inputs obfuscation [IPS15]. Our main contribution is developing new technical approaches that allow us to remove the need for such assumptions, and use only iO for circuits[1].

Results and Technical Overview. We prove the following informal theorem.

Theorem 1 (Informal). *There exists a public-key FE scheme, assuming the existence of indistinguishability obfuscation and one-way functions, that satisfies the following properties:*

1. *There is no a priori bound on the number of functional keys issued.*
2. *The secret keys correspond to Turing machines.*
3. *It achieves adaptive security.*
4. *There is no a priori bound on length of the plaintext and the size of the Turing machine.*
5. *The running time of encryption is independent of the Turing machine size. The running time of the key generation is independent of the plaintext size.*

A corollary of the above theorem is the first construction of succinct adaptively secure garbling schemes for TMs (with indistinguishability-based security) in the standard model. By succinctness, we mean that the size of the input encoding is independent of the function (circuit or TM) size. Prior solutions were either shown in the random oracle model [BHR12, AIKW15] or under restricted settings [BGG+14].

We now give a roadmap for the overall approach and the techniques we use to achieve our result. To gather some ideas towards achieving our goal of adaptive FE for TMs, we first focus on the simplest possible scenario of FE for Turing machines: adversary can make only a single ciphertext query and a function query, and furthermore we work in the secret-key setting. We call a FE scheme satisfying this security notion to be 1-CT 1-Key Private-key FE.

Initial Goal: Adaptive 1-CT 1-Key Private-key FE for TMs. To build an adaptive 1-CT 1-key private-key FE for TMs scheme, we first take inspiration from the corresponding FE for *circuits* constructions known in the literature to see what tools might be helpful here. Sahai and Seyalioglu [SS10] and Gorbunov et al. [GVW12] give constructions using the tool of randomized encodings (RE) of computation. A randomized encoding is a representation of a function along with an input that is simpler to compute than the function itself. Further this representation reveals only the output of the function and nothing else. In other words, given functions f_1, f_2 and inputs x_1, x_2 such that $f_1(x_1) = f_2(x_2)$,

---

[1] We stress that despite recent cryptanalytic progress, iO candidates such as [BGK+14] remain beyond the reach of any known cryptanalytic technique.

it should be the case that the encoding of (f_1, x_1) should be computationally indistinguishable from an encoding of (f_2, x_2). Such randomized encodings for TMs were recently constructed in [BGL+15, CHJV15, KLW15], based on iO for circuits.

The essential difference between a randomized encoding and what we need for a 1-CT 1-key FE scheme concerns two additional features that we would need from the randomized encoding:

- First, we need the randomized encoding to be computable *separately* for the function and the input. That is, given only f, it should be possible to compute an encoding $\hat{f}$; and given only x, it should be possible to compute an encoding $\hat{x}$; such that $(\hat{f}, \hat{x})$ constitute a randomized encoding of (f, x). We need this because the ciphertext will be akin to the encoding of the input, whereas the private key will be akin to the encoding of the function. This is essentially the notion of a decomposable randomized encoding [AIK06].
- Then, more crucially, we also need to strengthen our notion of security: In a standard randomized encoding scheme, the adversary needs to declare f_1, f_2, x_1, x_2 all at the beginning, and then we have the guarantee that $(\hat{f_1}, \hat{x_1})$ is computationally indistinguishable to $(\hat{f_2}, \hat{x_2})$. However, for an FE scheme, even with just "selective" security, the adversary is given the power to adaptively specify at least f_1, f_2 after it has seen the encodings $\hat{x_1}$ and $\hat{x_2}$. More generally, we would like to have security where the adversary can choose whether it would like to specify f_1, f_2 first or x_1, x_2 first.

It turns out that achieving these two properties is relatively straightforward when dealing with randomized encodings of circuits using Yao's garbled circuits [Yao86]. It is not so straightforward for us in the context of TMs and adaptive security, as we explain below.

To see why our situation is nontrivial and to get intuition about the obstacles we must overcome, let us first consider a *failed attempt* to achieve these properties by trying to apply the generic transformation, which was formalized in the work of Bellare et al. [BHR12], to achieve adaptive security: in this attempt, the new input encoding and new function encoding will now be $(\hat{x} \oplus R, S)$ and $(R, \hat{f} \oplus S)$, respectively, where R and S are random strings. The idea behind this transformation is as follows: no matter what the adversary queries for (input or function) in the beginning, it is just given two random strings (R, S). When the adversary makes the other query, the simulator would know at this point both the input and the function. Hence, it would obtain the corresponding encodings $\hat{f}$ and $\hat{x}$ from the ordinary security of the randomized encoding scheme. Now, the simulator would respond to the adversary by giving $(\hat{x} \oplus R, \hat{f} \oplus S)$ thus successfully simulating the game. The problem with this solution for us lies in the *sizes* of the encodings. If we look at the strings R and S, they are as long as the length of $\hat{x}$ and $\hat{f}$ respectively. This would mean that the size of the new input encoding (resp., new function encoding) depends on the function length (resp., input length) – which violates our main goal of achieving FE without restrictions on input length!

Revisiting the KLW Randomized Encoding. In order to achieve our goal, we will need to look at the specifics of the decomposable RE for TMs construction in [KLW15]. We then develop new ideas specific to the construction that help us achieve adaptive security. Before we do that, we revisit the KLW randomized encoding at a high level, sufficient for us to explain the new ideas in our work. The encoding procedure of a Turing machine M and input x consists of the following two main steps:

1. The storage tape of the TM is initialized with the encryption of x. It then builds an accumulator storage tree on the ciphertext. The accumulator storage tree resembles a Merkle hash tree with the additional property that this tree is unconditionally sound for a select portion of the storage. The root of the tree is then authenticated.
2. A program that computes the next step function of the Turing machine M is then designed. This program enables computation of M one step at a time. This program has secrets that enable decrypting encrypted tape symbols and also to perform some checks on the input encrypted symbol. To hide the secrets, this program is obfuscated.

The decoding just involves running the next message function repeatedly on the computation obtained so far until the Turing Machine terminates. At this point, the decode algorithm will output whatever the Turing Machine outputs.

First Step Towards Adaptivity: 3-Stage KLW. The main issue with trying to use the random masking technique was that we were trying to use randomness to mask the entire input encoding or the function encoding, which could be of unbounded length. So our main goal will be to find a way to achieve adaptivity where randomness need only be used to mask *bounded* portions of the encoding.

As a first step towards achieving this, we want to symmetrize how we treat the input x and the function f. We do this by treating both x and f as being inputs to a Universal Turing Machine U, where U is both of bounded size and is entirely known a-priori, such that $U(f, x) = f(x)$.

That is, we have three algorithms[2]: InpEnc outputs an encoding of input x, FnEnc outputs an encoding of f, and UTMEnc outputs a TM encoding of UTM.

A natural approach would be to try to use the KLW scheme sketched above to achieve the goal. The only difference is that, unlike the original KLW scheme, in the 3-stage KLW scheme, the input encoding is split into two encodings (InpEnc and FnEnc) and so there must be a way to stitch the input encodings into one. We develop a mechanism, called combiner, to achieve this goal. A combiner is an algorithm that combines two input encodings into one input encoding. Furthermore, the combiner algorithm we develop is succinct; it only takes a portion of the two encodings (of say, x and f) and spits out an element that together with the encodings of x and f represent $x\|f$. Note, however, that the combiner algorithm needs secret information in order to perform its combining

---

[2] The actual algorithms as presented in the technical section is slightly different. We chose to present it this way in the introduction for intuitive clarity.

role correctly. The key to constructing this combiner is the accumulator storage scheme of KLW. Recall that the accumulator storage on $(x\|f)$ was essentially a binary tree on $x\|f$. We modify this accumulator storage such that the storage tree on $(x\|f)$ can be built by first building a storage tree on x, then building a separate independent storage tree on f, and then joining both these two trees by making them children of a root node. Once we have this tool, developing our combiner algorithm is easy: the input encoding of x consists of a storage tree on an encryption of x, encoding of f consists of a storage tree on the encryption of f. The combine algorithm then takes *only* the root nodes of both these two trees and creates a new root node which is the parent of these two root nodes. The combiner then signs on the root node as a means of authenticating the fact that this new root node was created legally.

We are almost ready to now apply the random masking technique to achieve adaptive security by masking our new succinct representations. However, there is a problem: the combiner algorithm. In 3-stage KLW, once we have encodings of x and f, before we can have a randomized encoding, these two encodings need to be combined using secret information. This is not allowed in a randomized encoding, where the decode algorithm must be public.

Getting rid of combiner: 2-ary FE for TMs (1-CT 1-Key Setting). Since we need to eliminate the need for the combiner algorithm, we start by trying to delegate the combine operation to the decoder. We can attempt to do so by including an obfuscated version of the combiner program as part of the encoding itself, where obfuscation is needed since the combiner procedure contains some secret values that have to be hidden. By itself, however, this approach does not work, because the adversary who now possesses the obfuscated combine program can now illegally combine different storages (other than those corresponding to x and f) – we term this type of attack as a *mixed storage attack*.

To prevent mixed storage attacks, we use splittable signatures: the challenger can sign the root of the storage of x as well as the root of the storage of f. The obfuscated program now only outputs the combined value if the signatures can be verified correctly. By using splittable signatures, we can argue that the adversary is prevented from mixed storage attacks relying only on indistinguishability obfuscation for circuits.

Once we have the obfuscated combiner program, the next issue is whether the obfuscated combiner should be included as part of InpEnc or FnEnc. Including it in either of them will cause problems because the simulator needs to simulate the appropriate parameters in the combiner algorithm and it can do that only after looking at both the InpEnc and FnEnc queries. Here we can (finally!) apply the random masking technique since the size of the combiner is independent of the size of the input as well as the function and thus the length of the random mask needed is small. The resulting scheme that we get is a 2-ary FE [GGG+14] for TMs, where the adversary can only make a single message and key query – note that it is essentially the same as 3-stage KLW scheme except that it does not have the combiner algorithm.

Using some additional but similar ideas, we can show that the algorithms FnEnc and UTMEnc can be combined into one encoding. The result is a scheme with an input encoding, function encoding and a decode algorithm with the security guarantee that the input query and the function query can be made adaptively, which is precisely the goal we had started off with.

Boosting Mechanism: 1-Key 1-CT (Private-Key) FE to Many-Key (Public-Key) FE. Now that we have achieved the goal of single-ciphertext single-key private key FE for TMs, the next direction is to explore whether there is any way to combine this with other known tools to obtain a public-key FE with unbounded number of function queries. We give a mechanism of combining the 1-Key 1-CT FE scheme with other FE schemes that are defined for *circuits* to obtain a public-key FE scheme for Turing machines. Further, our resulting FE scheme is such that it is adaptively secure assuming only that the 1-Key 1-CT FE scheme is adaptively secure. The high level approach is that the ciphertexts and the functional keys are designed such that every ciphertext-functional key pair gives rise to a unique instantiation of single-ciphertext single-key private FE. This is reminiscent of the approach of Waters [Wat15], later revisited by [ABSV15], in the context of constructing adaptively secure FE for circuits.

Our boosting mechanism, however, diverges in several ways from the previous works of [Wat15, ABSV15]. First, we note that just syntactically, our boosting mechanism is the first such mechanism that uses only 1-Key 1-CT FE as a building block; in contrast, for example, [ABSV15] needed *many*-Key 1-CT FE as a building block.

Zooming in on the main new idea we develop for our boosting mechanism, we find that it is used exactly to deal with the fact that unbounded inputs that must be embedded in ciphertexts. Note that all previous FE schemes placed an a-priori bound on the inputs to be encrypted in ciphertexts. Therefore, to build our encryption mechanism, we cannot use previous FE encryption to encode inputs. We also cannot directly use the 1-Key 1-CT FE, since this scheme can only support a single key and a single ciphertext. To resolve this dilemma, we note that even though previous FE schemes could not handle inputs of unbounded length, previous FE schemes can handle *keys* corresponding to arbitrary-length circuits. Therefore, crucially in our boosting procedure, when encrypting an input x, we actually prepare a circuit H_x that has x built into it, and then use an existing FE scheme to prepare a key corresponding to H_x. Here we make use of the Brakerski-Segev [BS14] transformation to guarantee that the key for H_x does not leak x. We utilize a new layer of indirection, where this circuit H_x expects to receive as input the master secret key of a 1-Key 1-CT FE scheme, and then uses this master secret key to create a 1-Key 1-CT encryption of x. In this way, the final FE scheme that we construct inherits the security of the 1-Key 1-CT encryption scheme, but a fresh and independent instance of the 1-Key 1-CT scheme is created for each pair of (input, function) that is ever considered within our final FE scheme.

Subsequent Work. Recently, Nimishaki, Wichs and, Zhandry [NWZ15] construct a traitor tracing scheme which allows for embedding user information in the issued keys. One of the main tools used to construct this primitive is an adaptively secure FE scheme. As a first step, they show how to achieve a traitor tracing scheme from a private linear broadcast encryption (PLBE) scheme defined for a large identity space. In the next step, they show how to design a PLBE scheme from adaptive FE.

2 Preliminaries

We denote λ to be the security parameter. We say that a function $\mu(\lambda)$ is negligible if for any polynomial $p(\lambda)$ it holds that $\mu(\lambda) < 1/p(\lambda)$ for all sufficiently large $\lambda \in \mathbb{N}$. We use the notation negl to denote a negligible function.

We assume that the reader is familiar with the notion of Turing machines, standard cryptographic notions of pseudorandom functions and symmetric encryption schemes. We use the convention that a Turing machine also outputs the time it takes to execute. As a consequence, if we have $M_0(x) = M_1(x)$ then it means that not only are the outputs same but even the running times are the same.

2.1 Functional Encryption for Turing Machines

We now define the notion of functional encryption (FE) for Turing machines. This notion differs from the traditional notion of FE for circuits (to be defined later) in that the functional keys are associated to Turing machines as against circuits. Further, the functional keys can be used to decrypt ciphertexts of messages of arbitrary length and the decryption time depends only the running time of the Turing machine on the message.

A public-key functional encryption scheme, defined for a message space $\mathcal{M}$ and a class of Turing machines $\mathcal{F}$, consists of four PPT algorithms FE = (Setup, KeyGen, Enc, Dec) described as follows.

- Setup(1^λ): The setup algorithm takes as input the security parameter λ in unary and outputs a public key-secret key pair (PK, MSK).
- KeyGen(MSK, $f \in \mathcal{F}$): The key generation algorithm takes as input the master secret key MSK, a Turing machine $f \in \mathcal{F}^3$, and outputs a functional key sk_f.
- Enc(PK, $m \in \mathcal{M}$): The encryption algorithm takes as input the public key PK, a message $m \in \mathcal{M}$ and outputs a ciphertext CT.
- Dec(sk_f, CT): The decryption algorithm takes as input the functional key sk_f, a ciphertext CT and outputs $\hat{m}$.

The FE scheme defined above, in addition to correctness and security, needs to satisfy the efficiency property. All these properties are defined below.

Correctness. The correctness notion of a FE scheme dictates that there exists a negligible function $\mathsf{negl}(\lambda)$ such that for all sufficiently large $\lambda \in \mathbb{N}$,

---

[3] We use the same notation to denote the function as well as the Turing machine representing the function f.

for every message $m \in \mathcal{M}$, and for every Turing machine $f \in \mathcal{F}$ it holds that $\Pr[f(m) \leftarrow \mathsf{Dec}(\mathsf{KeyGen}(\mathsf{MSK}, f), \mathsf{Enc}(\mathsf{PK}, m))] \geq 1 - \mathsf{negl}(\lambda)$, where $(\mathsf{PK}, \mathsf{MSK}) \leftarrow \mathsf{Setup}(1^\lambda)$, and the probability is taken over the random choices of all algorithms.

Efficiency. The efficiency property of a public-key FE scheme says that the algorithm Setup on input 1^λ should run in time polynomial in λ, KeyGen on input the Turing machine f (along with master secret key) should run in time polynomial in $(\lambda, |f|)$, Enc on input a message m (along with the public key) should run in time polynomial in $(\lambda, |m|)$. Finally, Dec on input a functional key of f and an encryption of m should run in time polynomial in $(\lambda, |f|, |m|, \mathsf{timeTM}(f, m)))$.

Security. The security notion we define is identical to the indistinguishability-based security notion defined for circuits.

Definition 1. *A public-key functional encryption scheme* $\mathsf{FE} = (\mathsf{Setup}, \mathsf{KeyGen}, \mathsf{Enc}, \mathsf{Dec})$ *over a class of Turing machines* $\mathcal{F}$ *and a message space* $\mathcal{M}$ *is **adaptively secure** if for any PPT adversary* $\mathcal{A}$ *there exists a negligible function* $\mu(\lambda)$ *such that for all sufficiently large* $\lambda \in \mathbb{N}$, *the advantage of* $\mathcal{A}$ *is defined to be*

$$\mathsf{Adv}_{\mathcal{A}}^{\mathsf{FE}} = \left| \mathsf{Prob}[\mathsf{Expt}_{\mathcal{A}}^{\mathsf{FE}}(1^\lambda, 0) = 1] - \mathsf{Prob}[\mathsf{Expt}_{\mathcal{A}}^{\mathsf{FE}}(1^\lambda, 1) = 1] \right| \leq \mu(\lambda),$$

where for each $b \in \{0, 1\}$ *and* $\lambda \in \mathbb{N}$ *the experiment* $\mathsf{Expt}_{\mathcal{A}}^{\mathsf{FE}}(1^\lambda, b)$, *modeled as a game between the challenger and the adversary* $\mathcal{A}$, *is defined as follows:*

1. *The challenger first executes* $\mathsf{Setup}(1^\lambda)$ *to obtain* $(\mathsf{PK}, \mathsf{MSK})$. *It then sends* PK *to the adversary.*
2. ***Query Phase I:*** *The adversary submits a Turing machine query* f *to the challenger. The challenger sends back* sk_f *to the adversary, where* sk_f *is the output of* $\mathsf{KeyGen}(\mathsf{MSK}, f)$.
3. ***Challenge Phase:*** *The adversary submits a message-pair* (m_0, m_1) *to the challenger. The challenger checks whether* $f(m_0) = f(m_1)$ *for all Turing machine queries* f *made so far. If this is not the case, the challenger aborts. Otherwise, the challenger sends back* $\mathsf{CT} = \mathsf{Enc}(\mathsf{MSK}, m_b)$.
4. ***Query Phase II:*** *The adversary submits a Turing machine query* f *to the challenger. The challenger generates* sk_f, *where* sk_f *is the output of* $\mathsf{KeyGen}(\mathsf{MSK}, f)$. *It sends* sk_f *to the adversary only if* $f(m_0) = f(m_1)$, *otherwise it aborts.*
5. *The output of the experiment is* b', *where* b' *is the output of* $\mathcal{A}$.

We can also consider a weaker notion, termed as *selective security*, where the adversary has to submit the challenge message pair at the beginning of the game itself even before it receives the public parameters and such a FE scheme is said to be *selectively secure*.

Private Key Setting. We can analogously define the notion of FE for TMs in the private-key setting. The difference between the public-key setting and the private-key setting is that in the private-key setting, the encryptor needs to know the master secret key to encrypt the messages. We provide the formal definition of private-key FE for TMs in the full version [AS15].

2.2 (Compact) FE for Circuits

Public-Key FE. One of the building blocks in our construction of FE for TMs is a public-key FE for circuits (i.e., the functions are represented as circuits). We now recall its definition from [BSW11, O'N10].

A public-key functional encryption (FE) scheme PubFE, defined for a class of functions $\mathcal{F} = \{\mathcal{F}_\lambda\}_{\lambda \in \mathbb{N}}$ and message space $\mathcal{M} = \{\mathcal{M}_\lambda\}_{\lambda \in \mathbb{N}}$, is represented by four PPT algorithms, namely (Setup, KeyGen, Enc, Dec). The input length of any $f \in \mathcal{F}_\lambda$ is the same as the length of any $m \in \mathcal{M}_\lambda$. The description of these four algorithms is given below.

- Setup(1^λ): It takes as input a security parameter λ in unary and outputs a public key-secret key pair (PK, MSK).
- KeyGen(MSK, $f \in \mathcal{F}_\lambda$): It takes as input a secret key MSK, a function $f \in \mathcal{F}_\lambda$ and outputs a functional key sk_f.
- Enc(PK, $m \in \mathcal{M}_\lambda$): It takes as input a public key PK, a message $m \in \mathcal{M}_\lambda$ and outputs an encryption of m.
- Dec(sk_f, CT): It takes as input a functional key sk_f, a ciphertext CT and outputs $\hat{m}$.

We require the FE scheme to satisfy the efficiency property in addition to the traditional properties of correctness and security.

Correctness. The correctness property says that there exists a negligible function $\mathsf{negl}(\lambda)$ such that for all sufficiently large $\lambda \in \mathbb{N}$, for every message $m \in \mathcal{M}_\lambda$, and for every function $f \in \mathcal{F}_\lambda$ it holds that $\Pr[f(m) \leftarrow \mathsf{Dec}(\mathsf{KeyGen}(\mathsf{MSK}, f),$ $\mathsf{Enc}(\mathsf{PK}, m))] \geq 1 - \mathsf{negl}(\lambda)$, where (PK, MSK) $\leftarrow$ Setup(1^λ), and the probability is taken over the random choices of all algorithms.

Efficiency. At a high level, the efficiency property says that the setup and the encryption algorithm is independent of the size of the circuits for which functional keys are produced. More formally, the running time of the setup algorithm, Setup(1^λ) is a polynomial in just the security parameter λ and the encryption algorithm, Enc(PK, m) is a polynomial in only the security parameter λ and length of the message, $|m|$.

An FE scheme that satisfies the above efficiency property is termed as compact FE. It was shown by [AJ15, BV15] that iO is implied by (sub-exponentially hard) compact FE. However, we don't place any sub exponential hardness requirement on compact FE in our work.

Remark 1. We note that the definitions of FE for circuits commonly used in the literature do not have the above efficiency property.

Security. The security definition is modeled as a game between the challenger and the adversary as before.

Definition 2. *A public-key functional encryption scheme* FE = (Setup, KeyGen, Enc, Dec) *over a function space* $\mathcal{F} = \{\mathcal{F}_\lambda\}_{\lambda \in \mathbb{N}}$ *and a message space* $\mathcal{M} = \{\mathcal{M}_\lambda\}_{\lambda \in \mathbb{N}}$ *is an* **adaptively-secure public-key functional encryption scheme** *if for any PPT adversary* $\mathcal{A}$ *there exists a negligible function* $\mu(\lambda)$ *such that for all sufficiently large* $\lambda \in \mathbb{N}$, *the advantage of* $\mathcal{A}$ *is defined to be*

$$\mathsf{Adv}^{\mathsf{FE}}_{\mathcal{A}} = \left| \mathrm{Prob}[\mathsf{Expt}^{\mathsf{FE}}_{\mathcal{A}}(1^\lambda, 0) = 1] - \mathrm{Prob}[\mathsf{Expt}^{\mathsf{FE}}_{\mathcal{A}}(1^\lambda, 1) = 1] \right| \leq \mu(\lambda),$$

where for each $b \in \{0, 1\}$ *and* $\lambda \in \mathbb{N}$ *the experiment* $\mathsf{Expt}^{\mathsf{FE}}_{\mathcal{A}}(1^\lambda, b)$, *modeled as a game between the challenger and the adversary* $\mathcal{A}$, *is defined as follows:*

1. *The challenger first executes* Setup(1^λ) *to obtain* (PK, MSK). *It then sends* PK *to the adversary.*
2. *Query Phase I: The adversary submits a function query* f *to the challenger. The challenger sends back* sk_f *to the adversary, where* sk_f *is the output of* KeyGen(MSK, f).
3. *Challenge Phase: The adversary submits a message-pair* (m_0, m_1) *to the challenger. The challenger checks whether* $f(m_0) = f(m_1)$ *for all function queries* f *made so far. If this is not the case, the challenger aborts. Otherwise, the challenger sends back* CT = Enc(MSK, m_b).
4. *Query Phase II: The adversary submits a function query* f *to the challenger. The challenger generates* sk_f, *where* sk_f *is the output of* KeyGen(MSK, f). *It sends* sk_f *to the adversary only if* $f(m_0) = f(m_1)$, *otherwise it aborts.*
5. *The output of the experiment is* b', *where* b' *is the output of* $\mathcal{A}$.

We define the FE scheme to be selectively secure if the adversary has to declare the challenge message pair even before it receives the public parameters.

Function-Private Private Key FE. We now give an analogous definition of FE for circuits in the private-key setting. In particular, we focus on the private-key FE that is function-private.

A function-private private-key functional encryption (FE) scheme PrivFE, defined for a class of functions $\mathcal{F} = \{\mathcal{F}_\lambda\}_{\lambda \in \mathbb{N}}$ and message space $\mathcal{M} = \{\mathcal{M}_\lambda\}_{\lambda \in \mathbb{N}}$, is represented by four PPT algorithms, namely (PrivFE.Setup, PrivFE.KeyGen, PrivFE.Enc, PrivFE.Dec). The input length of any $f \in \mathcal{F}_\lambda$ is the same as the length of any $m \in \mathcal{M}_\lambda$.

We give the description of the four algorithms below.

- PrivFE.Setup(1^λ): It takes as input a security parameter λ in unary and outputs a secret key PrivFE.MSK.
- PrivFE.KeyGen(PrivFE.MSK, $f \in \mathcal{F}_\lambda$): It takes as input a secret key PrivFE.MSK, a function $f \in \mathcal{F}_\lambda$ and outputs a functional key PrivFE.sk_f.
- PrivFE.Enc(PrivFE.MSK, $m \in \mathcal{M}_\lambda$): It takes as input a secret key PrivFE.MSK, a message $m \in \mathcal{M}_\lambda$ and outputs an encryption of m.
- PrivFE.Dec(PrivFE.sk_f, CT): It takes as input a functional key PrivFE.sk_f, a ciphertext CT and outputs $\hat{m}$.

We require the above function-private private key FE scheme to satisfy the correctness, efficiency and the function privacy properties of the above FE scheme.

Correctness. The correctness notion of a function-private private-key FE scheme dictates that there exists a negligible function $\mathsf{negl}(\lambda)$ such that for all sufficiently large $\lambda \in \mathbb{N}$, for every message $m \in \mathcal{M}_\lambda$, and for every function $f \in \mathcal{F}_\lambda$ it holds that $\Pr[f(m) \leftarrow \mathsf{PrivFE.Dec}(\mathsf{PrivFE.KeyGen}(\mathsf{PrivFE.MSK}, f), \mathsf{PrivFE.Enc}(\mathsf{PrivFE.MSK}, m))] \geq 1 - \mathsf{negl}(\lambda)$, where $\mathsf{PrivFE.MSK} \leftarrow \mathsf{PrivFE.Setup}(1^\lambda)$, and the probability is taken over the random choices of all algorithms.

Efficiency. At a high level, the efficiency property says that the setup algorithm and the encryption algorithm is independent of the size of the circuits for which functional keys are produced. More formally, the running time of $\mathsf{PrivFE.Setup}(1^\lambda)$ is just a polynomial in the security parameter λ, and $\mathsf{PrivFE.Enc}(\mathsf{PrivFE.MSK}, m)$ is a polynomial in only the security parameter λ and length of the message, $|m|$.

Function Privacy. We now recall the definition of function privacy in private key FE as defined by Brakerski, and Segev [BS14]. Note that the function privacy property below subsumes the usual notion of security (when only one function is submitted).

Definition 3. *A private-key functional encryption scheme* $\mathsf{PrivFE} = (\mathsf{PrivFE.Setup}, \mathsf{PrivFE.KeyGen}, \mathsf{PrivFE.Enc}, \mathsf{PrivFE.Dec})$ *over a function space* $\mathcal{F} = \{\mathcal{F}_\lambda\}_{\lambda \in \mathbb{N}}$ *and a message space* $\mathcal{M} = \{\mathcal{M}_\lambda\}_{\lambda \in \mathbb{N}}$ *is a **function-private adaptively-secure private-key FE scheme** if for any PPT adversary* $\mathcal{A}$ *there exists a negligible function* $\mu(\lambda)$ *such that for all sufficiently large* $\lambda \in \mathbb{N}$, *the advantage of* $\mathcal{A}$ *is defined to be*

$$\mathsf{Adv}_{\mathcal{A}}^{\mathsf{PrivFE}} = \left| \mathsf{Prob}[\mathsf{Expt}_{\mathcal{A}}^{\mathsf{PrivFE}}(1^\lambda, 0) = 1] - \mathsf{Prob}[\mathsf{Expt}_{\mathcal{A}}^{\mathsf{PrivFE}}(1^\lambda, 1) = 1] \right| \leq \mu(\lambda),$$

where for each $b \in \{0, 1\}$ *and* $\lambda \in \mathbb{N}$ *the experiment* $\mathsf{Expt}_{\mathcal{A}}^{\mathsf{PrivFE}}(1^\lambda, b)$, *modeled as a game between the challenger and the adversary* $\mathcal{A}$, *is defined as follows:*

1. *The challenger first executes* $\mathsf{PrivFE.MSK} \leftarrow \mathsf{PrivFE.Setup}(1^\lambda)$. *The adversary then makes the following message queries and function queries in no particular order.*
   - **Message queries:** *The adversary submits a message-pair* (m_0, m_1) *to the challenger. In return, the challenger sends back* $\mathsf{CT} = \mathsf{PrivFE.Enc}(\mathsf{PrivFE.MSK}, m_b)$.
   - **Function queries:** *The adversary then makes functional key queries. For every function-pair query* (f_0, f_1), *the challenger sends* $\mathsf{PrivFE.sk}_{f_b}$ *to the adversary, where* $\mathsf{PrivFE.sk}_{f_b}$ *is the output of* $\mathsf{PrivFE.KeyGen}(\mathsf{PrivFE.MSK}, f_b)$ *only if* $f_0(m_0) = f_1(m_1)$, *for all message-pair queries* (m_0, m_1). *Otherwise, it aborts.*
2. *The output of the experiment is* b', *where* b' *is the output of* $\mathcal{A}$.

We define a function-private private key FE to be selectively secure if the adversary has to declare all the challenge message pairs at the beginning of the security game.

Remark 2. We note that we can define a private-key FE scheme without the function privacy property, analogous to the public-key FE.

Single-Key Setting. A single-key function-private functional encryption scheme (in the private-key setting) is a functional encryption scheme, where the adversary in the security game (either selective or adaptive) is allowed to query for only one function. There are several known constructions [SS10, GVW12, GKP+12] but none of them satisfy the efficiency property of our FE definition – in particular, the size of the ciphertexts in these constructions grow with the circuit size (for which functional keys are computed). We later describe how to obtain a single-key scheme that indeed satisfies the efficiency property.

3   Adaptive 1-Key 1-Ciphertext FE for TMs

One of the main tools in our constructions is a single-key single-ciphertext FE for TMs in the private key setting. In the security game, the adversary only gets to make a single message and function query. Since we are interested in adaptive security, the message and the function query can be made in any order. In the language of randomized encodings (RE), this primitive is nothing but an adaptively secure *succinct* decomposable RE. The formal definition of single-ciphertext single-key FE for TMs is provided in the full version [AS15].

In the adaptive security game of single-ciphertext single-key FE, the adversary can only make a single function query and a single challenge message query. We define this notion for the case when the functions are represented by Turing machines.

As before, we can define a single-ciphertext single-key private-key FE to be *selectively-secure* if the adversary has to declare the challenge message pair even before he submits the function query.

We now proceed to build this tool based on iO and one-way functions. Towards this end, we first consider the notion of private key multi-ary functional encryption (FE) [GGG+14] for TMs. Multi-ary FE is a generalization of FE where the functions can take more than one input. We are interested in the restricted setting when the adversary only makes a single function and message query. Moreover, we restrict ourselves to the 2-ary setting, i.e., the arity of the functions is 2. We refer to this notion as 2-ary FE for TMs. We describe this notion formally in Sect. 3.1. Prior to this work, we knew how to construct this only based on (public coins) differing inputs obfuscation. Later we show how to construct this primitive assuming just iO for circuits and one-way functions.

3.1   Semi-Adaptive 2-Ary FE for TMs: 1-Key 1-Ciphertext Setting

The formal description of the 2-ary FE for TMs is given below. A 2-ary FE for a class of Turing machines $\mathcal{F}$ consists of four PPT algorithms, 2FE = (2FE.Setup, 2FE.Enc, 2FE.KeyGen, 2FE.Dec), as described below.

- 2FE.Setup(1^λ): On input the security parameter λ, the algorithm 2FE.Setup outputs a master secret key 2FE.MSK.
- 2FE.KeyGen(2FE.MSK, M): On input the master secret key 2FE.MSK and Turing machine $M \in \mathcal{F}$, it outputs the key 2FE.sk$_M$.
- 2FE.Enc(2FE.MSK, x, b): On input the master secret key 2FE.MSK, message $x \in \{0,1\}^*$ and position $b \in \{0,1\}$, it outputs 2FE.CT$_x$.
- 2FE.Dec(2FE.sk$_M$, 2FE.CT$_x$, 2FE.CT$_y$): On input the functional key 2FE.sk$_M$ and ciphertexts 2FE.CT$_x$ and 2FE.CT$_y$, it outputs the value z.

Remark 3. The bit b essentially indicates the position with respect to which the message needs to be encrypted. For convenience sake, we refer to the first position as the 0th position and the second position as the 1st position.

For the above notion to be interesting, a 2-ary FE for TMs scheme is required to satisfy the following correctness, efficiency and security properties.

Correctness: This property ensures that the output of 2FE.Dec(2FE.sk$_M$, 2FE.CT$_x$, 2FE.CT$_y$) is always $M(x, y)$ where (i) 2FE.MSK $\leftarrow$ 2FE.Setup(1^λ), (ii) 2FE.sk$_M$ $\leftarrow$ 2FE.KeyGen(2FE.MSK, M), (iii) 2FE.CT$_x$ $\leftarrow$ 2FE.Enc(2FE.MSK, $x, 0$) and (iv) 2FE.CT$_y$ $\leftarrow$ 2FE.Enc(2FE.MSK, $y, 1$).

Efficiency: This property says that the size of the ciphertexts (resp., functional key) depend solely on the size of the message (resp., machine) and the security parameter. That is, the complexity of 2FE.Enc(2FE.MSK, x, b) is a polynomial in $(\lambda, |x|)$ and the complexity of 2FE.KeyGen(2FE.MSK, M) is a polynomial in $(\lambda, |M|)$. Furthermore, we require that the complexity of 2FE.Dec(2FE.sk$_M$, 2FE.CT$_x$, 2FE.CT$_y$) is just a polynomial in $(\lambda, |x|, |y|, |M|, t)$, where t is the time taken by M to execute on the input (x, y).

Semi-Adaptive Security: The security guarantee states that the adversary cannot distinguish joint ciphertexts of (x_0, y_0) from the joint ciphertexts of (x_1, y_1) given the functional key of M, as long as $M(x_0, y_0) = M(x_1, y_1)$. Note that we adopt the convention that the Turing machine also outputs its running time and thus this alone ensures that the execution time of $M(x_0, y_0)$ is the same as the execution time of $M(x_1, y_1)$. Depending on the order of the message and the Turing machine queries the adversary can make, there are many ways to model the security of a 2-ary FE scheme. We adopt the notion where the adversary can make the message queries corresponding to 0th and 1st position in an adaptive manner but the TM query should be made *only after both the message queries*. We term this notion *semi-adaptive* security.

Suppose $\mathcal{A}$ be any PPT adversary. We define an experiment Expt$_{\mathcal{A}}$SemiAd below.

Expt$_{\mathcal{A}}^{\mathsf{SemiAd}}$($1^\lambda$):

1. The challenger first executes 2FE.Setup(1^λ) to obtain 2FE.MSK. It then chooses a bit b at random.
2. The following two bullets are executed in an arbitrary order (depending on the choice of the adversary).

- The adversary submits the message query (x_0, x_1), corresponding to 0th position, to the challenger. The challenger responds with $2\mathsf{FE.CT}_x \leftarrow 2\mathsf{FE.Enc}(2\mathsf{FE.MSK}, x_0, 0)$ if $b = 0$ else it responds with $2\mathsf{FE.CT}_x \leftarrow 2\mathsf{FE.Enc}(2\mathsf{FE.MSK}, x_1, 0)$.
- The adversary submits the message query (y_0, y_1), corresponding to 1st position, to the challenger. The challenger responds with $2\mathsf{FE.CT}_y \leftarrow 2\mathsf{FE.Enc}(2\mathsf{FE.MSK}, y_0, 1)$ if $b = 0$ else it responds with $2\mathsf{FE.CT}_y \leftarrow 2\mathsf{FE.Enc}(2\mathsf{FE.MSK}, y_1, 1)$.

3. After both the message queries, the adversary then submits a Turing machine M to the challenger. The challenger aborts if either (i) $M(x_0, y_0) \neq M(x_1, y_1)$ or (ii) $|x_0| \neq |x_1|$ or (iii) $|y_0| \neq |y_1|$. If it has not aborted, it executes $2\mathsf{FE.sk}_M \leftarrow 2\mathsf{FE.KeyGen}(2\mathsf{FE.MSK}, M)$. It then sends $2\mathsf{FE.sk}_M$ to the adversary.

4. The adversary outputs b'.

The experiment outputs 1 if $b = b'$, otherwise it outputs 0.

We now define the semi-adaptive security notion.

Definition 4. *A 2-ary FE scheme is semi-adaptive secure if for any PPT adversary $\mathcal{A}$, we have that the probability that the output of the experiment $\mathsf{Expt}_{\mathcal{A}}^{\mathsf{SemiAd}}$ is 1 is at most $1/2 + \mathsf{negl}(\lambda)$, for any negligible function negl.*

3.2 Adaptive FE from Semi-adaptive 2-Ary FE for TMs

We now show how to achieve *adaptively secure* single-ciphertext single-key FE starting from a *semi-adaptively secure* 2-ary FE for TMs. Recall that in the semi-adaptive security game of 2-ary FE, the key query can be made only after the message queries but however, the message queries corresponding to the first and the second position can be made in an adaptive manner. This leads to the main idea behind our construction – symmetrization of the input and the TM. That is, the adaptive FE functional key of a machine M is the 2-ary FE encryption of M w.r.t the 1st position and the adaptive FE encryption of a message m is essentially the 2-ary FE encryption of m w.r.t the 0th position. This takes care of the adaptivity issue. To facilitate the execution of M on m, a 2-ary FE key of a universal TM (UTM) is also provided. The question is whether we include the 2-ary FE key of UTM in the ciphertext or the functional key. This is crucial because the UTM key can only be provided by the challenger after seeing the queries corresponding to both the 0th and 1st position. To solve this issue, we additively secret share the UTM key across both the ciphertext and the functional key. This gives the challenger leeway to provide a random string as part of the response to the first query and by providing the appropriate secret share in the second response it can reveal the UTM key – at this point the challenger has seen both m and M. The formal scheme is described next.

Consider a 2-ary FE for TMs, denoted by $2\mathsf{FE} = (2\mathsf{FE.Setup}, 2\mathsf{FE.KeyGen}, 2\mathsf{FE.Enc}, 2\mathsf{FE.Dec})$, for a class of Turing machines $\mathcal{F}$. We construct a single-ciphertext single-key FE, $\mathsf{OneCTKey}$, for the same class $\mathcal{F}$.

Denote by $\mathsf{UTM} = \mathsf{UTM}_\lambda$, the universal Turing machine, that takes as input a Turing machine M, message m and outputs $M(m)$ if it halts within 2^λ steps else it outputs $\perp$. Further, we denote by ℓ_{UTM} to be the length of the output of a 2FE key of UTM.

$\mathsf{OneCTKey.Setup}(1^\lambda)$: On input the security parameter λ, it first executes $\mathsf{2FE.Setup}(1^\lambda)$ to obtain the master secret key $\mathsf{2FE.MSK}$. It also picks a random string R in $\{0,1\}^{\ell_{\mathsf{UTM}}}$. It outputs the secret key $\mathsf{OneCTKey.MSK} = (\mathsf{2FE.MSK}, R)$ as the master secret key.

$\mathsf{OneCTKey.KeyGen}(\mathsf{OneCTKey.MSK}, M \in \mathcal{F})$: On input the master secret key $\mathsf{OneCTKey.MSK} = (\mathsf{2FE.MSK}, R)$, and a Turing machine $M \in \mathcal{F}$, it executes 2-ary FE encryption of M w.r.t 0th position, $\mathsf{2FE.Enc}(\mathsf{2FE.MSK}, M, 0)$, to obtain $\mathsf{2FE.CT}_M$. It then computes a 2-ary FE key of UTM by generating $\mathsf{2FE.sk}_{\mathsf{UTM}} \leftarrow \mathsf{2FE.KeyGen}(\mathsf{2FE.MSK}, \mathsf{UTM}_\lambda)$. Finally, it outputs the functional key $\mathsf{OneCTKey}.sk_M = (\mathsf{2FE.CT}_M, \mathsf{2FE.sk}_{\mathsf{UTM}} \oplus R)$.

$\mathsf{OneCTKey.Enc}(\mathsf{OneCTKey.MSK}, m)$: On input the master secret key $\mathsf{OneCTKey.MSK} = (\mathsf{2FE.MSK}, R)$, and message m, it generates a 2-ary FE encryption of m by executing $\mathsf{2FE.CT}_m \leftarrow \mathsf{2FE.Enc}(\mathsf{2FE.MSK}, m, 1)$. It outputs the ciphertext $\mathsf{OneCTKey.CT} = (\mathsf{2FE.CT}_m, R)$.

$\mathsf{OneCTKey.Dec}(\mathsf{OneCTKey}.sk_M, \mathsf{OneCTKey.CT})$: On input the functional key $\mathsf{OneCTKey}.sk_M = (\mathsf{2FE.CT}_M, S)$ and ciphertext $\mathsf{OneCTKey.CT} = (\mathsf{2FE.CT}_m, R)$. It computes $S \oplus R$ to obtain $\mathsf{2FE.sk}_{\mathsf{UTM}}$. It then executes $\mathsf{2FE.Dec}(\mathsf{2FE.sk}_{\mathsf{UTM}}, \mathsf{2FE.CT}_M, \mathsf{2FE.CT}_m)$ to obtain z. Finally, it outputs z.

We prove the following theorem. The proof of the theorem is available in the full version [AS15].

Theorem 2. *The scheme* $\mathsf{OneCTKey}$ *satisfies correctness, efficiency and adaptive security properties.*

3.3 Constructing Semi-adaptive 2-Ary FE for TMs: Overview

Lets begin with the following simple idea: the 2-ary FE encryption of x w.r.t 0th position will just be a standard public key encryption of x_0. Since this encryption should not be malleable, we provide an authentication of the ciphertext. Similarly, the 2-ary FE encryption of y w.r.t 1st position is also a public key encryption of y along with its authentication. The functional key of M is an obfuscated program that takes as input an encrypted tape symbol; decrypts it; executes the next message function and then outputs an encryption of the new symbol. The evaluation is performed by executing next message function one step at a time while updating the storage tape which is initialized to the encryptions of x and y along with their respective authentications.

This however suffers from consistency issues. An adversary could re-use encrypted storage tape values of the current tape in the future steps. It would seem that using signatures to bind the time step to the tape symbol should solve this problem. In fact, if we had virtual black box obfuscation this idea would work. However, we are stuck with indistinguishability obfuscation and it is not clear how to make this work – signatures in general aren't compatible with iO because signatures guarantee computational soundness whereas iO demands information theoretic soundness. Looking back at the literature, we notice that Koppula-Lewko-Waters had to deal with similar issues in their recent work on randomized encodings (RE)[4] for TMs [KLW15]. The template of their construction comprises of two components as described below. The actual construction of KLW has more intricate details involved from what is presented below but to keep the discussion at an intuitive level, we choose to describe it this way.

Let M and x be the input to the encoding procedure.

- **Storage tree**: Encrypt x using a public key encryption scheme. Initialize the work tape with this ciphertext. Compute a storage tree on this ciphertext. The root of the storage tree along with the current time step (which is initially 0) is then signed using a signature scheme. This signature serves as an authentication of the work tape and the current time step.
- **Obfuscated next message program**: The obfuscated program takes as input an encrypted tape symbol (leaf node), its path to the root of the storage tree and the signature on the root. It performs few checks to test whether the encrypted tape symbol is valid. It then decrypts the encrypted tape symbol, computes the next message function of the TM M and then re-encrypts the output tape symbol. Finally, it computes the new root of the storage tree (this can be done by just having the appropriate path from the new tape symbol leading up to the root) and signs it.

There are two main hurdles in using the above template for our construction of 2-ary FE for TMs: (i) the TM only takes a single input in the above template whereas in our setting the TM takes two inputs. Moreover, we require that the TM and the inputs are encoded separately and, (ii) the security notion considered by KLW is *weak-selective* – the adversary is required to declare both the TM and the input at the beginning of the game. On the other hand the security notion we consider is stronger. Because of these two main reasons, we employ new techniques to achieve our construction.

Ciphertext Combiner Mechanism. As remarked earlier, we require that the TM and the inputs are encoded separately. We exploit the fact that inherently KLW has two components – storage tree and obfuscated next message program – that depend upon the input and the TM separately. But note that we have two inputs and so we need to further split the storage tree component. The tree structure automatically allows for such a decomposition. We compute a storage tree on the (encrypted) 0th position input and another tree on the (encrypted) 1st position input. We can then combine the roots of both the trees, during the decryption phase, to obtain a new root. But the root of the new tree needs to

---

[4] A randomized encoding of a machine M and input x is an encoding of $M(x)$ that takes much less time to compute than $M(x)$. Furthermore, the encoding should only reveal $M(x)$ and nothing more.

be authenticated and this operation needs to be public. We could provide the decryptor the signing key but then we end up sacrificing security!

To overcome this problem, we provide a combiner program, as part of one of the ciphertexts, that takes as input two nodes in the tree and outputs a new node along with a signature. This signature is signed using a signing key which is part of the combiner program. Of course the combiner program needs to be obfuscated to hide the signing key. As we will see later in the actual construction, we require "iO-compatible" signatures a.k.a splittable signatures scheme of KLW to make this idea work.

While using combiner seems to solve the problem, the next question is in which ciphertext do we include the combiner? We will see next that this becomes crucial for our proof of security.

Ensuring Semi-adaptivity. Suppose we decide to include the combiner as part of the 0th ciphertext. In line with the techniques used in proving the security using iO, we require that in the proof of security we hardwire the resulting (combined) root node in the combiner. But this is not possible if the 0th position challenge message is requested before the 1st position challenge message. The same problem occurs if we include the combiner as part of the 1st position ciphertext – the adversary can now query for the 1st position challenge ciphertext first and then query the 0th position challenge message.

This conundrum can be tackled by using deniable encryption. We can compute a deniable encryption of combiner in one ciphertext and in the other ciphertext we open the deniable ciphertext. This gives us the flexibility to open the ciphertext to whatever message we want depending on the adversary's queries. While this solves the problem, we can replace deniable encryption with a much simpler tool – one-time pad! We compute a one-time pad of the combiner with randomness R in one ciphertext and the other ciphertext contains just R. This solves our problem just like the case of deniable encryption.

We present a high level and a simplified description of the 2-ary FE scheme below. The formal description is more involved and is presented in full version [AS15] where we present the construction in a modular fashion by first describing an intermediate primitive that we call 3-stage KLW.

1. **Setup**: Generate a master signing key-verification key pair (SK, VK). Also generate two auxiliary signature key-verification key pairs (SK_x, VK_x) and (SK_y, VK_y). Generate the public parameters PP of the storage tree. Compute a random string R of appropriate length. The public key is PP while the master secret key is $(SK_x, SK_y, VK_x, VK_y, SK, VK, R)$.

2. **Key generation of M**: Generate an obfuscated next message program of M whose functionality is as in the above high level description. The pair (SK, VK) is hardwired inside the obfuscated program.

3. **Encryption of x w.r.t 0th position**: Compute a storage tree on x. Sign the root of the tree rt_x using SK_x to obtain σ_x. Compute the obfuscated combiner program $S = \mathsf{Comb} \oplus R$ whose description is as given above. Output $(\mathsf{rt}_x, \sigma_x, S)$.

4. **Encryption of** y **w.r.t** 1st **position**: Compute a storage tree on y. Sign the root of the tree rt_y using SK_y to obtain σ_y. Output $(\mathsf{rt}_y, \sigma_y, R)$.

5. **Decryption**: First, compute $S \oplus R$ to recover Comb. Then execute Comb on inputs $((\mathsf{rt}_x, \sigma_x), (\mathsf{rt}_y, \sigma_y))$ to obtain the joint root rt accompanied by the signature σ computed using SK. Once this is done, using the joint tree and obfuscated next message program of M, execute the decode procedure of KLW to recover the answer.

4   Adaptive FE for TMs

We show how to obtain an adaptively secure public key functional encryption scheme for Turing machines. To achieve this, we use a public key FE scheme for circuits, single-key FE scheme for circuits and single-key single-ciphertext FE for Turing machines.

Tools. We use the following tools to achieve the transformation.

- (Compact) Public key FE scheme for circuits, denoted by PubFE = (PubFE.Setup, PubFE.KeyGen, PubFE.Enc, PubFE.Dec). It suffices for us that PubFE is selectively secure.
- (Compact) Function-private Single-key FE scheme for circuits, denoted by OneKey = (OneKey.Setup, OneKey.KeyGen, OneKey.Enc, OneKey.Dec). It suffices for us that OneKey is selectively secure.
- Single-key single-ciphertext FE scheme for Turing machines, denoted by OneCTKey = (OneCTKey.Setup, OneCTKey.KeyGen, OneCTKey.Enc, OneCTKey.Dec). We require that OneCTKey is adaptively secure.
- Psuedorandom function family, F.
- Symmetric encryption scheme with pseudorandom ciphertexts, denoted by Sym = (Sym.Setup, Sym.Enc, Sym.Dec).

Instantiations of the Tools. We gave an construction of single-key single-ciphertext FE for Turing machines satisfying adaptive security in Sect. 3. We can instantiate the compact public-key FE scheme using the construction of [GGH+13, Wat15] (here, we refer to the post-challenge FE construction of [Wat15]). This construction can be based on indistinguishability obfuscation and other standard assumptions. Lastly, we can instantiate a function-private single key FE by, first, applying the function-privacy transformation by Brakerski-Segev [BS14] on the public-key FE[5]. The resulting FE is a private-key FE which is also function-private. And, a function-private single-key FE in the private key setting is a special case of function-private private key FE. Note that this instantiation can be based off the same assumptions as public-key FE (this is because, [BS14] does not add any additional assumptions).

Intuition. We view our construction as a transformation from adaptively secure 1-CT 1-key FE scheme into one that can handle unbounded collusions.

---

[5] The function-privacy transformation was defined for private key FE but a public key FE can be transformed into a private key FE in a straightforward way.

Even though in general we don't know any way of achieving this, we show that by leveraging additional tools we can attain this goal. These additional tools, as mentioned above, are multi-key FE schemes that are only selective secure.

The key idea is as follows: we give a mechanism to generate a *unique key* corresponding to a pair of ciphertext (of m) and functional key (of f) in the resulting adaptive multi-key FE scheme. This unique key would correspond to the master secret key of the adaptive 1-CT 1-Key FE scheme. At this point, we can generate functional keys of f and ciphertext of m w.r.t this unique key. Implementing this mechanism using iO, in the context of FE for circuits, was introduced by Waters [Wat15]. We show how to implement the same, in the more general context of FE for TMs, but using just a multi-key FE. We highlight that in general we don't know how to replace the use of iO with multi-key FE since FE does not offer function hiding.

At the high level, the construction proceeds as follows. A ciphertext of m "communicates" a PRF key K to a functional key of f. This communication is enabled by a multi-key FE scheme. The functional key using K and hard-wired values, derives the master secret key OneCTKey.MSK of a 1-CT 1-Key FE scheme. If then computes a functional key of f w.r.t OneCTKey.MSK. But the ciphertext of m does not contain an encryption w.r.t OneCTKey.MSK! And so this key has to be "communicated" from functional key back to the ciphertext. To do this, we will use another instantiation of selectively secure FE scheme. Here, we note that it suffices to consider just a single-key scheme and that too in the private key setting. Once we have this instantiation, the functional key can now generate a single-key FE encryption of OneCTKey.MSK. The single-key FE functional key, which will now be part of the ciphertext, will take as input encryption of OneCTKey.MSK and outputs an encryption of m w.r.t OneCTKey.MSK. Finally, we can just run the decryption algorithm of OneCTKey to obtain the answer. We illustrate a simple example, when a single ciphertext and functional key is released, in Fig. 1.

Our construction has more details that we present below.

Construction. We now describe the construction. We denote the FE for TMs scheme, that we construct, to be FE = (Setup, KeyGen, Enc, Dec).

Setup(1^λ): Execute PubFE.Setup(1^λ) to obtain (PubFE.MSK, PubFE.PK). Output the secret key-public key pair (MSK = PubFE.MSK, PK = PubFE.PK).

KeyGen(MSK = PubFE.MSK, f): Draw C_E at random[6]. Denote τ to be $\overline{(\tau_0||\tau_1||\tau_2||\tau_3)}$, where τ_i for $i \in \{0,1,2,3\}$ is picked at random. Execute

---

[6] The length of C_E is determined as follows. Denote by $|f|$, the size of the Turing machine representing f. Denote by $\ell_{OneCTKey}$, the length of the ciphertext obtained by encrypting a message of length $|f|$, using OneCTKey.Enc. Denote by ℓ_{OneKey}, the length of the ciphertext obtained by encrypting a message of length $\lambda + 2$, using OneKey.Enc. Further, denote by ℓ_{Sym} to be the length of the ciphertext obtained by encrypting a message of length $\ell_{OneCTKey} + \ell_{OneKey}$, using Sym.Enc. We set the length of C_E to be ℓ_{Sym}.

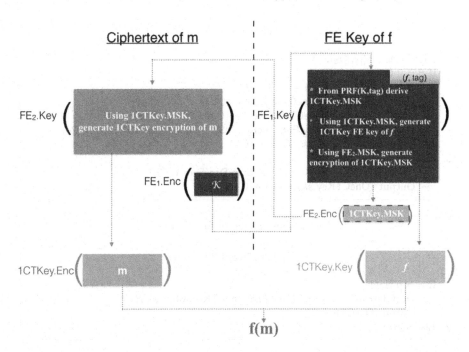

Fig. 1. The ciphertext of m has two components – the first component is a single-key FE (denoted by $\mathsf{FE_2}$) functional key and the second component is a multi-key FE (denoted by $\mathsf{FE_1}$) encryption of a PRF key K. The function key of f is just a $\mathsf{FE_1}$ functional key of the program described in the figure. The arrows indicate the flow of execution of decryption of the ciphertext of m using the functional key of f.

$\mathsf{PubFE.KeyGen}(\mathsf{PubFE.MSK}, G[f, C_E, \tau])$, where $G[f, C_E, \tau]$ is described in Fig. 2, to obtain $\mathsf{PubFE}.sk_G$. Output $sk_f = \mathsf{PubFE}.sk_G$.

$\underline{\mathsf{Enc}(\mathsf{PK} = \mathsf{PubFE.PK}, m)}$:

- Draw a PRF key K at random from $\{0,1\}^\lambda$.
- Execute $\mathsf{OneKey.Setup}(1^\lambda)$ to obtain $\mathsf{OneKey.MSK}$.
- Execute $\mathsf{OneKey.KeyGen}(\mathsf{OneKey.MSK}, H[m])$ to obtain $\mathsf{OneKey}.sk_H$, where $H[m]$ is defined in Fig. 3.
- Execute $\mathsf{PubFE.Enc}(\mathsf{PubFE.PK}, (\mathsf{OneKey.MSK}, \mathsf{K}, \bot, 0))$ to obtain $\mathsf{PubFE.CT}$.

Finally, output $\mathsf{CT} = (\mathsf{OneKey}.sk_H, \mathsf{PubFE.CT})$.

$\underline{\mathsf{Dec}(sk_f = sk_G, \mathsf{CT} = (\mathsf{OneKey}.sk_H, \mathsf{PubFE.CT}))}$:

- Execute $\mathsf{PubFE.Dec}(\mathsf{PubFE}.sk_G, \mathsf{PubFE.CT})$ to obtain $(\mathsf{OneCTKey}.sk_f, \mathsf{OneKey.CT})$.
- Execute $\mathsf{OneKey.Dec}(\mathsf{OneKey}.sk_H, \mathsf{OneKey.CT})$ to obtain $\mathsf{OneCTKey.CT}$.
- Execute $\mathsf{OneCTKey.Dec}(\mathsf{OneCTKey}.sk_f, \mathsf{OneCTKey.CT})$ to obtain $\hat{m}$.

Output $\hat{m}$.

$$G[f, C_E, \tau](\mathsf{OneKey.MSK}, \mathsf{K}, \mathsf{Sym}.k, \beta)$$

1. Parse τ as $(\tau_0 \| \tau_1 \| \tau_2 \| \tau_3)$
2. If $\beta = 0$ then
    - $R_i \leftarrow \mathsf{PRF}(\mathsf{K}, \tau_i)$, for $i \in \{0, 1, 2, 3\}$
    - $\mathsf{OneCTKey.MSK} \leftarrow \mathsf{OneCTKey.Setup}(1^\lambda; R_0)$
    - $\mathsf{OneCTKey}.sk_f \leftarrow \mathsf{OneCTKey.KeyGen}(\mathsf{OneCTKey.MSK}, f; R_1)$
    - $\mathsf{OneKey.CT} \leftarrow \mathsf{OneKey.Enc}(\mathsf{OneKey.MSK}, (\mathsf{OneCTKey.MSK}, R_2, 0); R_3)$
    - Output $(\mathsf{OneCTKey}.sk_f, \mathsf{OneKey.CT})$
3. Else,
    - $(\mathsf{OneCTKey}.sk_f, \mathsf{OneKey.CT}) \leftarrow \mathsf{Sym.Dec}(\mathsf{Sym}.k, C_E)$
    - Output $(\mathsf{OneCTKey}.sk_f, \mathsf{OneKey.CT})$

Fig. 2. Description of function G.

$$H[m](\mathsf{OneCTKey.MSK}, R, \alpha)$$

1. If $\alpha = 0$ then
    - $\mathsf{OneCTKey.CT} \leftarrow \mathsf{OneCTKey.Enc}(\mathsf{OneCTKey.MSK}, m; R)$
    - Output $\mathsf{OneCTKey.CT}$
2. Else, output $\perp$.

Fig. 3. Description of function H.

We prove the following theorem that establishes the proof of security of the above scheme.

Theorem 3. *Assuming the selective security of* $\mathsf{PubFE}, \mathsf{OneKey}$, *adaptive security of* $\mathsf{OneCTKey}$, *security of* F, Sym, *we have that the scheme* FE *is adaptively secure.*

Since the proof is involved, we choose to first present the proof of selective security of FE. We then point out the (minor) changes that need to be made to prove the adaptive security of FE. We give a sketch of the proof of the above scheme in Sect. 5 and the formal proof is provided in the full version [AS15]. We also present the proof of correctness and efficiency in the full version.

5 Proof of Theorem 3: Overview

To explain the proof intuition, we restrict ourselves to the setting when the adversary makes only a single message and key query.

In the first hybrid, the challenger uses a bit b picked at random, to generate the challenge ciphertext as in the (selective) security notion. By using the security of many primitives (listed in the theorem statement), we then move to a hybrid where the bit b is information-theoretically hidden from the adversary. At this point, the probability that the adversary guesses the bit b is $1/2$.

And thus the probability that the adversary guesses b correctly in the first hybrid is at most $1/2 + \mathsf{negl}(\lambda)$.

Hybrid$_0$: This corresponds to the real experiment when the challenger uses the bth message in the challenge message pair query to compute the challenge ciphertext, where the bit b is picked at random. The output of the hybrid is the same as the output of the adversary.

Hybrid$_1$: In this hybrid, the values corresponding to the challenge ciphertext are hardwired in the "C_E" component of all the functional keys.

That is, the challenger upon receiving a function query f, first samples a symmetric key $\mathsf{Sym}.k^*$. It generates an encryption of the message $(\mathsf{OneCTKey.MSK}, R_2, 0)$ with respect to the single-key FE scheme. Call this ciphertext, $\mathsf{OneKey.CT}$. It then samples a functional key of f with respect to the single-key single-ciphertext FE scheme. Call this functional key, $\mathsf{OneCTKey}.sk_f$. It is important to note here that, the (pseudo)randomness used in the generation of $\mathsf{OneKey.CT}$ and $\mathsf{OneCTKey}_f$ is as described in the scheme. Finally, it computes a symmetric encryption of $(\mathsf{OneKey.CT}, \mathsf{OneCTKey}.sk_f)$ using the key $\mathsf{Sym}.k$. The resulting ciphertext will be assigned to C_E and then the challenger proceeds as in the previous hybrid.

The indistinguishability of Hybrid$_0$ and Hybrid$_1$ follows from the security of symmetric encryption scheme.

Hybrid$_2$: In this hybrid, the mode is switched from $\beta = 0$ to $\beta = 1$.

Upon receiving a challenge message query (m_0, m_1), the challenger computes the challenge ciphertext as follows. Recall that there are two components in the ciphertext – namely, the single-key FE functional key and the public-key FE ciphertext. The challenger computes the single-key FE functional key as in the previous hybrid. However, it generates the public-key FE ciphertext to be an encryption of $(\bot, \bot, \mathsf{Sym}.k^*, 1)$ instead of $(\mathsf{OneKey.MSK}^*, \mathsf{K}^*, \bot, 0)$, as in Hybrid$_1$. The rest of the hybrid is the same as the previous hybrid.

The indistinguishability of Hybrid$_1$ and Hybrid$_2$ follows from the security of public-key FE scheme. This is because the output of G (Fig. 2) on input $(\bot, \bot, \mathsf{Sym}.k^*, 1)$ is nothing but the decryption of C_E. And by our choice of C_E, this is the same as the output of G on input $(\mathsf{OneKey.MSK}^*, \mathsf{K}^*, \bot, 0)$.

Hybrid$_3$: The hardwired values in the "C_E" components of all the functional keys are now computed using randomness drawn from a uniform distribution. Recall that in the previous hybrid, the single-key ciphertext and the single-key single-ciphertext FE encrypted in C_E were computed using pseudorandom values.

The indistinguishability of Hybrid$_2$ and Hybrid$_3$ follows from the security of pseudorandom function family.

Hybrid$_4$: A branch encrypting message m_0 (the 0th message in the challenge message query) is introduced in the function H.

The challenger upon receiving the challenge message query (m_0, m_1), first computes a single-key FE functional key of the function $H^*[m_0, m_b, v]$, as

---

$$H^*[m, m', v](\mathsf{OneCTKey.MSK}, R, \alpha)$$

1. If $\alpha = 0$ then
   - $\mathsf{OneCTKey.CT} \leftarrow \mathsf{OneCTKey.Enc}(\mathsf{OneCTKey.MSK}, m; R)$
   - Output $\mathsf{OneCTKey.CT}$
2. If $\alpha = 1$ then
   - $\mathsf{OneCTKey.CT} \leftarrow \mathsf{OneCTKey.Enc}(\mathsf{OneCTKey.MSK}, m'; R)$
   - Output $\mathsf{OneCTKey.CT}$
3. Else, output v.

Fig. 4. Description of hybrid function H^*.

described in Fig. 4. Here, b is the challenge bit, picked at random by the challenger. The program H^* is the same as H except that it contains an additional branch. The rest of the hybrid is the same as Hybrid_3.

The indistinguishability of Hybrid_3 and Hybrid_4 follows from the function-privacy property of single-key FE scheme. To see why, let us look at the messages that are encrypted under the single-key FE scheme (note that each encryption is part of the "C_E" component of some functional key). We observe that each message is of the form $(\mathsf{OneCTKey.MSK}, R, 0)$. From the descriptions of H and H^*, it follows that the output of H on $(\mathsf{OneCTKey.MSK}, R, 0)$ is the same as the output of H^* on $(\mathsf{OneCTKey.MSK}, R, 0)$.

Hybrid_5: We switch the mode of α from 0 to 1 in the OneKey ciphertexts output by all the functional keys.

The challenger, upon receiving a functional query f, first generates a single-key FE ciphertext to be an encryption of $(\mathsf{OneCTKey.MSK}, R, 1)$, where $\mathsf{OneCTKey.MSK}$ is as generated in the previous hybrids. The resulting ciphertext along with the single-key single-ciphertext FE functional key is then encrypted, using the symmetric key encryption, to obtain C_E. The rest of the functional key is then generated as previously.

The indistinguishability of Hybrid_4 and Hybrid_5 is more complex and involves more intermediate hybrids and thus we defer the explanation.

Hybrid_6: We change the $\alpha = 0$ branch in the function H to encrypt the message m_0 instead of m_b.

The challenger upon receiving a message query (m_0, m_1), first generates a single-key FE functional key of $H^*[m_0, m_0, v]$. It then generates the public key FE encryption as in previous hybrids. The rest of the hybrid is as in Hybrid_5.

The indistinguishability of Hybrid_5 and Hybrid_6 follows from the function privacy property of single-key FE scheme. To see why, we look at the messages encrypted in the single-key FE ciphertexts. We first note all these ciphertexts are part of "C_E" component of some functional key. Further, each message is of the form $(\mathsf{OneCTKey.MSK}, R, 1)$. Thus, the output of $H^*[m_b, m_0, v]$ is the same as the output of $H^*[m_0, m_0, v]$.

Observe that the challenge bit b is no longer used. This combined with the indistinguishability of consecutive hybrids proves that the probability that $\mathcal{A}$ wins in Hybrid_1 is at most $1/2 + \mathsf{negl}(\lambda)$. This proves the security of FE.

6   Future Directions

The works of [AJ15, BV15, AJS15] show the equivalence of (sub-exponentially secure) FE and iO for the case of circuits. It would be interesting to explore the possibility of the equivalence of FE for Turing machines and iO for Turing machines (with no restriction on the input length). One direct consequence of a feasibility result in this direction is establishing the existence of iO for Turing machines based on iO for circuits. The current feasibility results on iO for Turing machines are based on knowledge assumptions.

Acknowledgements. We thank Brent Waters for collaboration at early stages of this project and several discussions. We also thank the anonymous reviewers of TCC 2016-A for their useful suggestions. This work was done in part while the authors were visiting the Simons Institute for the Theory of Computing, supported by the Simons Foundation and by the DIMACS/Simons Collaboration in Cryptography through NSF grant CNS-1523467.

A   Tools Used in [KLW15]

We recall the key tools, namely, positional accumulators, iterators and splittable signatures, used in the work of Koppula et al. [KLW15].

We now describe the syntax of the tools below. We refer the reader to [KLW15] for the correctness and the security definitions.

A.1   Positional Accumulators

The notion of *positional accumulators* is defined below. A positional accumulator for message space Msg_λ consists of the following algorithms.

$\mathsf{SetupAcc}(1^\lambda, T) \to \mathsf{PP}_{\mathsf{Acc}}, w_0, store_0$ The setup algorithm takes as input a security parameter λ in unary and an integer T in binary representing the maximum number of values that can stored. It outputs public parameters $\mathsf{PP}_{\mathsf{Acc}}$, an initial accumulator value w_0, and an initial storage value $store_0$.

$\mathsf{EnforceRead}(1^\lambda, T, (m_1, \mathrm{INDEX}_1), \ldots, (m_k, \mathrm{INDEX}_k), \mathrm{INDEX}^*) \to (\mathsf{PP}_{\mathsf{Acc}}, w_0, store_0)$. The setup enforce read algorithm takes as input a security parameter λ in unary, an integer T in binary representing the maximum number of values that can be stored, and a sequence of symbol, index pairs, where each index is between 0 and $T-1$, and an additional INDEX^* also between 0 and $T-1$. It outputs public parameters $\mathsf{PP}_{\mathsf{Acc}}$, an initial accumulator value w_0, and an initial storage value $store_0$.

$\mathsf{EnforceWrite}(1^\lambda, T, (m_1, \mathrm{INDEX}_1), \ldots, (m_k, \mathrm{INDEX}_k)) \to \mathsf{PP}_{\mathsf{Acc}}, w_0, store_0$ The setup enforce write algorithm takes as input a security parameter λ in unary, an integer T in binary representing the maximum number of values that can be stored, and a sequence of symbol, index pairs, where each index is between 0 and $T-1$. It outputs public parameters $\mathsf{PP}_{\mathsf{Acc}}$, an initial accumulator value w_0, and an initial storage value $store_0$.

PrepRead($\mathsf{PP_{Acc}}, store_{in}, \mathrm{INDEX}$) $\rightarrow$ m, π The prep-read algorithm takes as input the public parameters $\mathsf{PP_{Acc}}$, a storage value $store_{In}$, and an index between 0 and $T - 1$. It outputs a symbol m (that can be ϵ) and a value π.

PrepWrite($\mathsf{PP_{Acc}}, store_{in}, \mathrm{INDEX}$) $\rightarrow$ aux The prep-write algorithm takes as input the public parameters $\mathsf{PP_{Acc}}$, a storage value $store_{In}$, and an index between 0 and $T - 1$. It outputs an auxiliary value aux.

VerifyRead($\mathsf{PP_{Acc}}, w_{in}, m_{read}, \mathrm{INDEX}, \pi$) $\rightarrow$ $\{True, False\}$ The verify-read algorithm takes as input the public parameters $\mathsf{PP_{Acc}}$, an accumulator value w_{in}, a symbol, m_{read}, an index between 0 and $T - 1$, and a value π. It outputs $True$ or $False$.

WriteStore($\mathsf{PP_{Acc}}, store_{in}, \mathrm{INDEX}, m$) $\rightarrow$ $store_{Out}$ The write-store algorithm takes in the public parameters, a storage value $store_{in}$, an index between 0 and $T - 1$, and a symbol m. It outputs a storage value $store_{out}$.

Update($\mathsf{PP_{Acc}}, w_{in}, m_{write}, \mathrm{INDEX}, aux$) $\rightarrow$ w_{Out} or $Reject$ The update algorithm takes in the public parameters $\mathsf{PP_{Acc}}$, an accumulator value w_{in}, a symbol m_{write}, and index between 0 and $T - 1$, and an auxiliary value aux. It outputs an accumulator value w_{out} or $Reject$.

A.2 Iterators

In this subsection, we now describe the notion of cryptographic iterators. As remarked earlier, iterators essentially consist of states that are updated on the basis of the messages received. We describe its syntax below.

Syntax. Let ℓ be any polynomial. An iterator $\mathsf{PP_{Itr}}$ with message space $\mathsf{Msg}_\lambda = \{0, 1\}^{\ell(\lambda)}$ and state space $\mathsf{SplScheme}_\lambda$ consists of three algorithms - SetupItr, ItrEnforce and Iterate defined below.

SetupItr($1^\lambda, T$) The setup algorithm takes as input the security parameter λ (in unary), and an integer bound T (in binary) on the number of iterations. It outputs public parameters $\mathsf{PP_{Itr}}$ and an initial state $v_0 \in \mathsf{SplScheme}_\lambda$.

ItrEnforce($1^\lambda, T, \boldsymbol{m} = (m_1, \ldots, m_k)$) The enforced setup algorithm takes as input the security parameter λ (in unary), an integer bound T (in binary) and k messages $(m_1, \ldots, m_k)$, where each $m_i \in \{0, 1\}^{\ell(\lambda)}$ and k is some polynomial in λ. It outputs public parameters $\mathsf{PP_{Itr}}$ and a state $v_0 \in \mathsf{SplScheme}$.

Iterate($\mathsf{PP_{Itr}}, v_{in}, m$) The iterate algorithm takes as input the public parameters $\mathsf{PP_{Itr}}$, a state v_{in}, and a message $m \in \{0, 1\}^{\ell(\lambda)}$. It outputs a state $v_{out} \in \mathsf{SplScheme}_\lambda$.

For simplicity of notation, the dependence of ℓ on λ will not be explicitly mentioned. Also, for any integer $k \leq T$, we will use the notation $\mathsf{Iterate}^k(\mathsf{PP_{Itr}}, v_0, (m_1, \ldots, m_k))$ to denote $\mathsf{Iterate}(\mathsf{PP_{Itr}}, v_{k-1}, m_k)$, where $v_j = \mathsf{Iterate}(\mathsf{PP_{Itr}}, v_{j-1}, m_j)$ for all $1 \leq j \leq k - 1$.

A.3 Splittable Signatures

We describe the syntax of the splittable signatures scheme below.

Syntax. A splittable signature scheme SplScheme for message space Msg consists of the following algorithms:

SetupSpl(1^λ) The setup algorithm is a randomized algorithm that takes as input the security parameter λ and outputs a signing key SK, a verification key VK and *reject-verification key* VK$_{\text{rej}}$.

SignSpl(SK, m) The signing algorithm is a deterministic algorithm that takes as input a signing key SK and a message $m \in$ Msg. It outputs a signature σ.

VerSpl(VK, m, σ) The verification algorithm is a deterministic algorithm that takes as input a verification key VK, signature σ and a message m. It outputs either 0 or 1.

SplitSpl(SK, m^*) The splitting algorithm is randomized. It takes as input a secret key SK and a message $m^* \in$ Msg. It outputs a signature $\sigma_{\text{one}} =$ SignSpl(SK, m^*), a one-message verification key VK$_{\text{one}}$, an all-but-one signing key SK$_{\text{abo}}$ and an all-but-one verification key VK$_{\text{abo}}$.

SignSplAbo(SK$_{\text{abo}}$, m) The all-but-one signing algorithm is deterministic. It takes as input an all-but-one signing key SK$_{\text{abo}}$ and a message m, and outputs a signature σ.

KLW described various security notions corresponding to the above splittable signatures scheme. We describe only one of the properties that will be useful for this work. This security notion is termed as VK$_{\text{one}}$ indistinguishability and states that given a signature on a message m, an adversary should not be able to distinguish the verification key VK from the split verification key VK$_{\text{one}}$, that is computed as a result of applying SplitSpl on the signing key and message m.

References

[ABG+13] Ananth, P., Boneh, D., Garg, S., Sahai, A., Zhandry, M.: Differing-inputs obfuscation and applications. IACR Cryptology ePrint Arch. 2013, 689 (2013)

[ABSV15] Ananth, P., Brakerski, Z., Segev, G., Vaikuntanathan, V.: From selective to adaptive security in functional encryption. In: Gennaro, R., Robshaw, M. (eds.) CRYPTO 2015. LNCS, vol. 9216, pp. 657–677. Springer, Heidelberg (2015)

[AGVW13] Agrawal, S., Gorbunov, S., Vaikuntanathan, V., Wee, H.: Functional encryption: new perspectives and lower bounds. In: Canetti, R., Garay, J.A. (eds.) [CG13], pp. 500–518

[AIK06] Applebaum, B., Ishai, Y., Kushilevitz, E.: Computationally private randomizing polynomials and their applications. Comput. Complex. **15**(2), 115–162 (2006)

[AIKW15] Applebaum, B., Ishai, Y., Kushilevitz, E., Waters, B.: Encoding functions with constant online rate, or how to compress garbled circuit keys. SIAM J. Comput. **44**(2), 433–466 (2015)

[AJ15] Ananth, P., Jain, A.: Indistinguishability obfuscation from compact functional encryption. In: Gennaro, R., Robshaw, M. (eds.) CRYPTO 2015. LNCS, vol. 9215, pp. 308–326. Springer, Heidelberg (2015)

[AJS15] Ananth, P., Jain, A., Sahai, A.: Indistinguishability obfuscation from functional encryption for simple functions. Cryptology ePrint Archive, Report 2015/730 (2015). http://eprint.iacr.org/

[AS15] Ananth, P., Sahai, A.: Functional encryption for turing machines. Cryptology ePrint Archive, Report 2015/776 (2015). http://eprint.iacr.org/

[BCP14] Boyle, E., Chung, K.-M., Pass, R.: On extractability obfuscation. In: Lindell, Y. (ed.) TCC 2014. LNCS, vol. 8349, pp. 52–73. Springer, Heidelberg (2014)

[BGG+14] Boneh, D., Gentry, C., Gorbunov, S., Halevi, S., Nikolaenko, V., Segev, G., Vaikuntanathan, V., Vinayagamurthy, D.: Fully key-homomorphic encryption, arithmetic circuit ABE and compact garbled circuits. In: Nguyen, P.Q., Oswald, E. (eds.) EUROCRYPT 2014. LNCS, vol. 8441, pp. 533–556. Springer, Heidelberg (2014)

[BGI+12] Barak, B., Goldreich, O., Impagliazzo, R., Rudich, S., Sahai, A., Vadhan, S.P., Yang, K.: On the (im)possibility of obfuscating programs. J. ACM 59(2), 6 (2012)

[BGK+14] Barak, B., Garg, S., Kalai, Y.T., Paneth, O., Sahai, A.: Protecting obfuscation against algebraic attacks. In: Nguyen, P.Q., Oswald, E. (eds.) EUROCRYPT 2014. LNCS, vol. 8441, pp. 221–238. Springer, Heidelberg (2014)

[BGL+15] Bitansky, N., Garg, S., Lin, H., Pass, R., Telang, S.: Succinct randomized encodings and their applications. In: STOC (2015)

[BHR12] Bellare, M., Hoang, V.T., Rogaway, P.: Adaptively secure garbling with applications to one-time programs and secure outsourcing. In: Wang, X., Sako, K. (eds.) ASIACRYPT 2012. LNCS, vol. 7658, pp. 134–153. Springer, Heidelberg (2012)

[BS14] Brakerski, Z., Segev, G.: Function-private functional encryption in the private-key setting. Cryptology ePrint Archive, Report 2014/550 (2014)

[BSW11] Boneh, D., Sahai, A., Waters, B.: Functional encryption: definitions and challenges. In: Ishai, Y. (ed.) TCC 2011. LNCS, vol. 6597, pp. 253–273. Springer, Heidelberg (2011)

[BSW12] Boneh, D., Sahai, A., Waters, B.: Functional encryption: a new vision for public-key cryptography. Commun. ACM 55(11), 56–64 (2012)

[BV15] Bitansky, N., Vaikuntanathan, V.: Indistinguishability obfuscation from functional encryption. In: FOCS (2015)

[CCC+15] Chen, Y.-C., Chow, S.S.M., Chung, K.-M., Lai, R.W.F., Lin, W.-K., Zhou, H.-S.: Computation-trace indistinguishability obfuscation and its applications. IACR Cryptology ePrint Arch. 2015, 406 (2015)

[CG13] Canetti, R., Garay, J.A. (eds.): CRYPTO 2013, Part II. LNCS, vol. 8043. Springer, Heidelberg (2013)

[CHJV15] Canetti, R., Holmgren, J., Jain, A., Vaikuntanathan, V.: Indistinguishability obfuscation of iterated circuits and ram programs. In: STOC (2015)

[CIJ+13] De Caro, A., Iovino, V., Jain, A., O'Neill, A., Paneth, O., Persiano, G.: On the achievability of simulation-based security for functional encryption. In: Canetti, R., Garay, J.A. (eds.) [CG13], pp. 519–535

[GGG+14] Goldwasser, S., Gordon, S.D., Goyal, V., Jain, A., Katz, J., Liu, F.-H., Sahai, A., Shi, E., Zhou, H.-S.: Multi-input functional encryption. In: Nguyen, P.Q., Oswald, E. (eds.) EUROCRYPT 2014. LNCS, vol. 8441, pp. 578–602. Springer, Heidelberg (2014)

[GGH+13] Garg, S., Gentry, C., Halevi, S., Raykova, M., Sahai, A., Waters, B.: Candidate indistinguishability obfuscation and functional encryption for all circuits. In: FOCS (2013)

[GGHZ14] Garg, S., Gentry, C., Halevi, S., Zhandry, M.: Fully secure functional encryption without obfuscation. IACR Cryptology ePrint Arch. 2014, 666 (2014)

[GKP+12] Goldwasser, S., Kalai, Y.T., Popa, R.A., Vaikuntanathan, V., Zeldovich, N.: Succinct functional encryption and applications: reusable garbled circuits and beyond. IACR Cryptology ePrint Arch. 2012, 733 (2012)

[GKP+13] Goldwasser, S., Kalai, Y.T., Popa, R.A., Vaikuntanathan, V., Zeldovich, N.: Reusable garbled circuits and succinct functional encryption. In: STOC (2013)

[GVW12] Gorbunov, S., Vaikuntanathan, V., Wee, H.: Functional encryption with bounded collusions via multi-party computation. In: Safavi-Naini, R., Canetti, R. (eds.) CRYPTO 2012. LNCS, vol. 7417, pp. 162–179. Springer, Heidelberg (2012)

[IPS15] Ishai, Y., Pandey, O., Sahai, A.: Public-coin differing-inputs obfuscation and its applications. In: Dodis, Y., Nielsen, J.B. (eds.) TCC 2015, Part II. LNCS, vol. 9015, pp. 668–697. Springer, Heidelberg (2015)

[KLW15] Koppula, V., Lewko, A.B., Waters, B.: Indistinguishability obfuscation for turing machines with unbounded memory. In: STOC (2015)

[KSW08] Katz, J., Sahai, A., Waters, B.: Predicate encryption supporting disjunctions, polynomial equations, and inner products. In: Smart, N.P. (ed.) EUROCRYPT 2008. LNCS, vol. 4965, pp. 146–162. Springer, Heidelberg (2008)

[NWZ15] Nishimaki, R., Wichs, D., Zhandry, M.: Anonymous traitor tracing: how to embed arbitrary information in a key. Cryptology ePrint Archive, Report 2015/750 (2015). http://eprint.iacr.org/

[O'N10] O'Neill, A.: Definitional issues in functional encryption. IACR Cryptology ePrint Arch. 2010, 556 (2010)

[SS10] Sahai, A., Seyalioglu, H.: Worry-free encryption: functional encryption with public keys. In: Proceedings of the 17th ACM Conference on Computer and Communications Security, pp. 463–472. ACM (2010)

[SW05] Sahai, A., Waters, B.: Fuzzy identity-based encryption. In: Cramer, R. (ed.) EUROCRYPT 2005. LNCS, vol. 3494, pp. 457–473. Springer, Heidelberg (2005)

[Wat15] Waters, B.: A punctured programming approach to adaptively secure functional encryption. In: Gennaro, R., Robshaw, M. (eds.) CRYPTO 2015. LNCS, vol. 9216, pp. 678–697. Springer, Heidelberg (2015)

[Yao86] Yao, A.C.-C.: How to generate and exchange secrets (extended abstract). In: 27th Annual Symposium on Foundations of Computer Science, Toronto, Canada, 27–29 October 1986, pp. 162–167. IEEE Computer Society (1986)

Differential Privacy

The Complexity of Computing the Optimal Composition of Differential Privacy

Jack Murtagh$^{(\boxtimes)}$ and Salil Vadhan

Center for Research on Computation and Society,
John A. Paulson School of Engineering and Applied Sciences,
Harvard University, Cambridge, MA, USA
{jmurtagh,salil}@seas.harvard.edu

Abstract. In the study of differential privacy, composition theorems (starting with the original paper of Dwork, McSherry, Nissim, and Smith (TCC'06)) bound the degradation of privacy when composing several differentially private algorithms. Kairouz, Oh, and Viswanath (ICML'15) showed how to compute the optimal bound for composing k arbitrary (ϵ, δ)-differentially private algorithms. We characterize the optimal composition for the more general case of k arbitrary $(\epsilon_1, \delta_1), \ldots, (\epsilon_k, \delta_k)$-differentially private algorithms where the privacy parameters may differ for each algorithm in the composition. We show that computing the optimal composition in general is #P-complete. Since computing optimal composition exactly is infeasible (unless FP=#P), we give an approximation algorithm that computes the composition to arbitrary accuracy in polynomial time. The algorithm is a modification of Dyer's dynamic programming approach to approximately counting solutions to knapsack problems (STOC'03).

Keywords: Differential privacy · Composition · Computational complexity · Approximation algorithms

1 Introduction

Differential privacy is a framework that allows statistical analysis of private databases while minimizing the risks to individuals in the databases. The idea is that an individual should be relatively unaffected whether he or she decides to join or opt out of a research dataset. More specifically, the probability distribution of outputs of a statistical analysis of a database should be nearly identical to the distribution of outputs on the same database with a single person's data removed. Here the probability space is over the coin flips of the randomized differentially private algorithm that handles the queries. To formalize this, we call

A full version of this paper is available on arXiv [10].

J. Murtagh–Supported by NSF grant CNS-1237235 and a grant from the Sloan Foundation.

S. Vadhan– Supported by NSF grant CNS-1237235, a grant from the Sloan Foundation, and a Simons Investigator Award.

E. Kushilevitz and T. Malkin (Eds.): TCC 2016-A, Part I, LNCS 9562, pp. 157–175, 2016.
DOI: 10.1007/978-3-662-49096-9_7

two databases D_0, D_1 with n rows each *neighboring* if they are identical on at least $n-1$ rows, and define differential privacy as follows:

Definition 1.1 (Differential Privacy [2,3]). *A randomized algorithm M is (ϵ, δ)-differentially private if for all pairs of neighboring databases D_0 and D_1 and all output sets $S \subseteq \mathrm{Range}(M)$*

$$\Pr[M(D_0) \in S] \leq e^\epsilon \Pr[M(D_1) \in S] + \delta$$

where the probabilities are over the coin flips of the algorithm M.

In the practice of differential privacy, we generally think of ϵ as a small, non-negligible, constant (e.g. $\epsilon = .1$). We view δ as a "security parameter" that is cryptographically small (e.g. $\delta = 2^{-30}$). One of the important properties of differential privacy is that if we run multiple distinct differentially private algorithms on the same database, the resulting composed algorithm is also differentially private, albeit with some degradation in the privacy parameters (ϵ, δ). In this paper, we are interested in quantifying the degradation of privacy under composition. We will denote the composition of k differentially private algorithms $M_1, M_2, \ldots, M_k$ as $(M_1, M_2, \ldots, M_k)$ where

$$(M_1, M_2, \ldots, M_k)(x) = (M_1(x), M_2(x), \ldots, M_k(x)).$$

A handful of composition theorems already exist in the literature. The first basic result says:

Theorem 1.2 (Basic Composition [2]). *For every $\epsilon \geq 0$, $\delta \in [0,1]$, and (ϵ, δ)-differentially private algorithms $M_1, M_2, \ldots, M_k$, the composition $(M_1, M_2, \ldots, M_k)$ satisfies $(k\epsilon, k\delta)$-differential privacy.*

This tells us that under composition, the privacy parameters of the individual algorithms "sum up," so to speak. We care about understanding composition because in practice we rarely want to release only a single statistic about a dataset. Releasing many statistics may require running multiple differentially private algorithms on the same database. Composition is also a very useful tool in algorithm design. Often, new differentially private algorithms are created by combining several simpler algorithms. Composition theorems help us analyze the privacy properties of algorithms designed in this way.

Theorem 1.2 shows a linear degradation in global privacy as the number of algorithms in the composition (k) grows and it is of interest to improve on this bound. If we can prove that privacy degrades more slowly under composition, we can get more utility out of our algorithms under the same global privacy guarantees. Dwork, Rothblum, and Vadhan gave the following improvement on the basic summing composition above [5].

Theorem 1.3 (Advanced Composition [5]). *For every $\epsilon > 0, \delta, \delta' > 0$, $k \in \mathbb{N}$, and (ϵ, δ)-differentially private algorithms $M_1, M_2, \ldots, M_k$, the composition $(M_1, M_2, \ldots, M_k)$ satisfies $(\epsilon_g, k\delta + \delta')$-differential privacy for*

$$\epsilon_g = \sqrt{2k \ln(1/\delta')} \cdot \epsilon + k \cdot \epsilon \cdot (e^\epsilon - 1).$$

Theorem 1.3 shows that privacy under composition degrades by a function of $O(\sqrt{k \ln(1/\delta')})$ which is an improvement if $\delta' = 2^{-O(k)}$. It can be shown that a degradation function of $\Omega(\sqrt{k \ln(1/\delta)})$ is necessary even for the simplest differentially private algorithms, such as randomized response [11].

Despite giving an asymptotically correct upper bound for the global privacy parameter, ϵ_g, Theorem 1.3 is not exact. We want an exact characterization because, beyond being theoretically interesting, constant factors in composition theorems can make a substantial difference in the practice of differential privacy. Furthermore, Theorem 1.3 only applies to "homogeneous" composition where each individual algorithm has the same pair of privacy parameters, (ϵ, δ). In practice we often want to analyze the more general case where some individual algorithms in the composition may offer more or less privacy than others. That is, given algorithms $M_1, M_2, \ldots, M_k$, we want to compute the best achievable privacy parameters for $(M_1, M_2, \ldots, M_k)$. Formally, we want to compute the function:

$$\text{OptComp}(M_1, M_2, \ldots, M_k, \delta_g) = \inf\{\epsilon_g : (M_1, M_2, \ldots, M_k) \text{ is } (\epsilon_g, \delta_g)\text{-DP}\}.$$

It is convenient for us to view δ_g as given and then compute the best ϵ_g, but the dual formulation, viewing ϵ_g as given, is equivalent (by binary search). Actually, we want a function that depends only on the privacy parameters of the individual algorithms:

$$\text{OptComp}((\epsilon_1, \delta_1), (\epsilon_2, \delta_2), \ldots, (\epsilon_k, \delta_k), \delta_g) =$$
$$\sup\{\text{OptComp}(M_1, M_2, \ldots, M_k, \delta_g) : M_i \text{ is } (\epsilon_i, \delta_i)\text{-DP} \ \forall i \in [k]\}.$$

In other words we want OptComp to give us the minimum possible ϵ_g that maintains privacy for every sequence of algorithms with the given privacy parameters (ϵ_i, δ_i). This definition refers to the case where the sequence of algorithms $(M_1, \ldots, M_k)$ and the pair of neighboring databases (D_0, D_1) on which they are applied are fixed, but we show that the same optimal bound holds even if the algorithms and databases are chosen adaptively, i.e. M_i and databases (D_0, D_1) are chosen adaptively based on the outputs of $M_1, \ldots, M_{i-1}$. (See Sect. 2 for a formal definition.)

A result from Kairouz, Oh, and Viswanath [9] characterizes OptComp for the homogeneous case.

Theorem 1.4 (Optimal Homogeneous Composition [9]). *For every $\epsilon \geq 0$ and $\delta \in [0, 1)$, $\text{OptComp}((\epsilon, \delta)_1, (\epsilon, \delta)_2, \ldots, (\epsilon, \delta)_k, \delta_g) = (k - 2i)\epsilon$, where i is the largest integer in $\{0, 1, \ldots, \lfloor k/2 \rfloor\}$ such that*

$$\frac{\sum_{l=0}^{i-1} \binom{k}{l} \left(e^{(k-l)\epsilon} - e^{(k-2i+l)\epsilon}\right)}{(1 + e^\epsilon)^k} \leq 1 - \frac{1 - \delta_g}{(1 - \delta)^k} .$$

With this theorem the authors exactly characterize the composition behavior of differentially private algorithms with a polynomial-time computable solution.

The problem remains to find the optimal composition behavior for the more general heterogeneous case. Kairouz, Oh, and Viswanath also provide an upper bound for heterogeneous composition that generalizes the $O(\sqrt{k \ln(1/\delta')})$ degradation found in Theorem 1.3 for homogeneous composition but do not comment on how close it is to optimal.

1.1   Our Results

We begin by extending the results of Kairouz, Oh, and Viswanath [9] to the general heterogeneous case.

Theorem 1.5 (Optimal Heterogeneous Composition). *For all $\epsilon_1, \ldots, \epsilon_k \geq$ 0 and $\delta_1, \ldots, \delta_k, \delta_g \in [0,1), \mathrm{OptComp}((\epsilon_1, \delta_1), (\epsilon_2, \delta_2), \ldots, (\epsilon_k, \delta_k), \delta_g)$ equals the least value of ϵ_g such that*

$$\frac{1}{\prod_{i=1}^{k}(1 + e^{\epsilon_i})} \sum_{S \subseteq \{1,\ldots,k\}} \max\left\{ e^{\sum_{i \in S} \epsilon_i} - e^{\epsilon_g} \cdot e^{\sum_{i \notin S} \epsilon_i}, 0 \right\} \leq 1 - \frac{1 - \delta_g}{\prod_{i=1}^{k}(1 - \delta_i)}. \quad (1)$$

Theorem 1.5 exactly characterizes the optimal composition behavior for any arbitrary set of differentially private algorithms. It also shows that optimal composition can be computed in time exponential in k by computing the sum over $S \subseteq \{1, \ldots, k\}$ by brute force. Of course in practice an exponential-time algorithm is not satisfactory for large k. Our next result shows that this exponential complexity is necessary:

Theorem 1.6. *Computing* $\mathrm{OptComp}$ *is $\#P$-complete, even on instances where $\delta_1 = \delta_2 = \ldots = \delta_k = 0$ and $\sum_{i \in [k]} \epsilon_i \leq \epsilon$ for any desired constant $\epsilon > 0$.*

Recall that $\#P$ is the class of counting problems associated with decision problems in NP. So being $\#P$-complete means that there is no polynomial-time algorithm for OptComp unless there is a polynomial-time algorithm for counting the number of satisfying assignments of boolean formulas (or equivalently for counting the number of solutions of all NP problems). So there is almost certainly no efficient algorithm for OptComp and therefore no analytic solution. Despite the intractability of exact computation, we show that OptComp can be approximated efficiently.

Theorem 1.7. *There is a polynomial-time algorithm that given $\epsilon_1, \ldots, \epsilon_k \geq$ $0, \delta_1, \ldots \delta_k, \delta_g \in [0,1)$, and $\eta > 0$, outputs ϵ^* where*

$$\mathrm{OptComp}((\epsilon_1, \delta_1), \ldots, (\epsilon_k, \delta_k), \delta_g) \leq \epsilon^* \leq \mathrm{OptComp}((\epsilon_1, \delta_1), \ldots, (\epsilon_k, \delta_k), e^{-\eta/2} \cdot \delta_g) + \eta \,.$$

The algorithm runs in $O\left(\log\left(\frac{k}{\eta} \sum_{i=1}^{k} \epsilon_i\right) \frac{k^2}{\eta} \sum_{i=1}^{k} \epsilon_i\right)$ time assuming constant-time arithmetic operations.

Note that we incur a relative error of η in approximating δ_g *and* an additive error of η in approximating ϵ_g. Since we always take ϵ_g to be non-negligible or even constant, we get a very good approximation when η is polynomially small or even a constant. Thus, it is acceptable that the running time is polynomial in $1/\eta$.

In addition to the results listed above, our proof of Theorem 1.5 also provides a somewhat simpler proof of the Kairouz-Oh-Viswanath homogeneous composition theorem (Theorem 1.4 [9]). The proof in [9] introduces a view of differential privacy through the lens of hypothesis testing and uses geometric arguments. Our proof relies only on elementary techniques commonly found in the differential privacy literature.

Practical Application. The theoretical results presented here were motivated by our work on an applied project called "Privacy Tools for Sharing Research Data"[1]. We are building a system that will allow researchers with sensitive datasets to make differentially private statistics about their data available through data repositories using the Dataverse[2] platform [1,8]. Part of this system is a tool that helps both data depositors and data analysts distribute a global privacy budget across many statistics. Users select which statistics they would like to compute and are given estimates of how accurately each statistic can be computed. They can also redistribute their privacy budget according to which statistics they think are most valuable in their dataset. We implemented the approximation algorithm from Theorem 1.7 and integrated it with this tool to ensure that users get the most utility out of their privacy budget.

2   Technical Preliminaries

A useful notation for thinking about differential privacy is defined below.

Definition 2.1. *For two discrete random variables Y and Z taking values in the same output space S, the δ-approximate max-divergence of Y and Z is defined as:*

$$D_\infty^\delta(Y\|Z) \equiv \max_S \left[\ln \frac{\Pr[Y \in S] - \delta}{\Pr[Z \in S]}\right].$$

Notice that an algorithm M is (ϵ, δ) differentially private if and only if for all pairs of neighboring databases, D_0, D_1, we have $D_\infty^\delta(M(D_0)\|M(D_1)) \leq \epsilon$. The standard fact that differential privacy is closed under "post processing" [3,4] now can be formulated as:

Fact 2.2. *If $f\colon S \to R$ is any randomized function, then*

$$D_\infty^\delta(f(Y)\|f(Z)) \leq D_\infty^\delta(Y\|Z).$$

---

[1] privacytools.seas.harvard.edu.
[2] dataverse.org.

Adaptive Composition. The composition results in our paper actually hold for a more general model of composition than the one described above. The model is called k-fold adaptive composition and was formalized in [5]. We generalize their formulation to the heterogeneous setting where privacy parameters may differ across different algorithms in the composition.

The idea is that instead of running k differentially private algorithms chosen all at once on a single database, we can imagine an adversary adaptively engaging in a "composition game." The game takes as input a bit $b \in \{0,1\}$ and privacy parameters $(\epsilon_1, \delta_1), \ldots, (\epsilon_k, \delta_k)$. A randomized adversary A, tries to learn b through k rounds of interaction as follows: on the ith round of the game, A chooses an (ϵ_i, δ_i)-differentially private algorithm M_i and two neighboring databases $D_{(i,0)}, D_{(i,1)}$. A then receives an output $y_i \leftarrow M_i(D_{(i,b)})$ where the internal randomness of M_i is independent of the internal randomness of $M_1, \ldots, M_{i-1}$. The choices of $M_i, D_{(i,0)}$, and $D_{(i,1)}$ may depend on $y_0, \ldots, y_{i-1}$ as well as the adversary's own randomness.

The outcome of this game is called the *view of the adversary*, V^b which is defined to be $(y_1, \ldots, y_k)$ along with A's coin tosses. The algorithms M_i and databases $D_{(i,0)}, D_{(i,1)}$ from each round can be reconstructed from V^b. Now we can formally define privacy guarantees under k-fold adaptive composition.

Definition 2.3. *We say that the sequences of privacy parameters $\epsilon_1, \ldots, \epsilon_k \geq 0, \delta_1, \ldots, \delta_k \in [0,1)$ satisfy (ϵ_g, δ_g)-differential privacy under adaptive composition if for every adversary A we have $D_\infty^{\delta_g}(V^0 \| V^1) \leq \epsilon_g$, where V^b represents the view of A in composition game b with privacy parameter inputs $(\epsilon_1, \delta_1), \ldots, (\epsilon_k, \delta_k)$.*

Computing Real-Valued Functions. Many of the computations we discuss involve irrational numbers and we need to be explicit about how we model such computations on finite, discrete machines. Namely when we talk about computing a function $f : \{0,1\}^* \to \mathbb{R}$, what we really mean is computing f to any desired number q bits of precision. More precisely, given x, q, the task is to compute a number $y \in \mathbb{Q}$ such that $|f(x) - y| \leq \frac{1}{2^q}$. We measure the complexity of algorithms for this task as a function of $|x| + q$.

3   Characterization of OptComp

Following [9], we show that to analyze the composition of arbitrary (ϵ_i, δ_i)-DP algorithms, it suffices to analyze the composition of the following simple variant of randomized response [11].

Definition 3.1 ([9]). *Define a randomized algorithm $\tilde{M}_{(\epsilon,\delta)} : \{0,1\} \to \{0,1,2,3\}$ as follows, setting $\alpha = 1 - \delta$:*

$$
\begin{array}{ll}
\Pr[\tilde{M}_{(\epsilon,\delta)}(0) = 0] = \delta & \Pr[\tilde{M}_{(\epsilon,\delta)}(1) = 0] = 0 \\
\Pr[\tilde{M}_{(\epsilon,\delta)}(0) = 1] = \alpha \cdot \frac{e^\epsilon}{1+e^\epsilon} & \Pr[\tilde{M}_{(\epsilon,\delta)}(1) = 1] = \alpha \cdot \frac{1}{1+e^\epsilon} \\
\Pr[\tilde{M}_{(\epsilon,\delta)}(0) = 2] = \alpha \cdot \frac{1}{1+e^\epsilon} & \Pr[\tilde{M}_{(\epsilon,\delta)}(1) = 2] = \alpha \cdot \frac{e^\epsilon}{1+e^\epsilon} \\
\Pr[\tilde{M}_{(\epsilon,\delta)}(0) = 3] = 0 & \Pr[\tilde{M}_{(\epsilon,\delta)}(1) = 3] = \delta
\end{array}
$$

Note that $\tilde{M}_{(\epsilon,\delta)}$ is in fact (ϵ,δ)-DP. Kairouz, Oh, and Viswanath showed that $\tilde{M}_{(\epsilon,\delta)}$ can be used to simulate the output of every (ϵ,δ)-DP algorithm on adjacent databases.

Lemma 3.2 ([9]). *For every (ϵ,δ)-DP algorithm M and neighboring databases D_0, D_1, there exists a randomized algorithm T such that $T(\tilde{M}_{(\epsilon,\delta)}(b))$ is identically distributed to $M(D_b)$ for $b = 0$ and $b = 1$.*

Proof. We provide a new proof of this lemma in the full version of the paper [10].

Since $\tilde{M}_{(\epsilon,\delta)}$ can simulate any (ϵ,δ) differentially private algorithm and it is known that post-processing preserves differential privacy (Fact 2.2), it follows that to analyze the composition of arbitrary differentially private algorithms, it suffices to analyze the composition of $\tilde{M}_{(\epsilon_i,\delta_i)}$'s:

Lemma 3.3. *For all $\epsilon_1,\ldots,\epsilon_k \geq 0, \delta_1,\ldots,\delta_k, \delta_g \in [0,1)$,*

$$\mathrm{OptComp}((\epsilon_1,\delta_1),\ldots,(\epsilon_k,\delta_k),\delta_g) = \mathrm{OptComp}(\tilde{M}_{(\epsilon_1,\delta_1)},\ldots,\tilde{M}_{(\epsilon_k,\delta_k)},\delta_g).$$

Proof. Since $\tilde{M}_{(\epsilon_1,\delta_1)},\ldots,\tilde{M}_{(\epsilon_k,\delta_k)}$ are $(\epsilon_1,\delta_1),\ldots,(\epsilon_k,\delta_k)$-differentially private, we have:

$$
\begin{aligned}
&\mathrm{OptComp}((\epsilon_1,\delta_1),\ldots,(\epsilon_k,\delta_k),\delta_g) \\
=\ &\sup\{\mathrm{OptComp}(M_1,\ldots,M_k,\delta_g)\colon M_i \text{ is } (\epsilon_i,\delta_i)\text{-DP } \forall i \in [k]\} \\
\geq\ &\mathrm{OptComp}(\tilde{M}_{(\epsilon_1,\delta_1)},\ldots,\tilde{M}_{(\epsilon_k,\delta_k)},\delta_g)\ .
\end{aligned}
$$

For the other direction, it suffices to show that for every $M_1,\ldots,M_k$ that are $(\epsilon_1,\delta_1),\ldots,(\epsilon_k,\delta_k)$-differentially private, we have

$$\mathrm{OptComp}(M_1,\ldots,M_k,\delta_g) \leq \mathrm{OptComp}(\tilde{M}_{(\epsilon_1,\delta_1)},\ldots,\tilde{M}_{(\epsilon_k,\delta_k)})\ .$$

That is,

$$\inf\{\epsilon_g\colon (M_1,\ldots,M_k) \text{ is } (\epsilon_g,\delta_g)\text{-DP}\} \leq \inf\{\epsilon_g\colon (\tilde{M}_{(\epsilon_1,\delta_1)},\ldots,\tilde{M}_{(\epsilon_k,\delta_k)}) \text{ is } (\epsilon_g,\delta_g)\text{-DP}\}.$$

So suppose $(\tilde{M}_{(\epsilon_1,\delta_1)},\ldots,\tilde{M}_{(\epsilon_k,\delta_k)})$ is (ϵ_g,δ_g)-DP. We will show that $(M_1,\ldots,M_k)$ is also (ϵ_g,δ_g)-DP. Taking the infimum over ϵ_g then completes the proof.

We know from Lemma 3.2 that for every pair of neighboring databases D_0, D_1, there must exist randomized algorithms $T_1,\ldots,T_k$ such that $T_i(\tilde{M}_{(\epsilon_i,\delta_i)}(b))$ is identically distributed to $M_i(D_b)$ for all $i \in \{1,\ldots,k\}$. By hypothesis we have

$$D_\infty^{\delta_g}\left((\tilde{M}_{(\epsilon_1,\delta_1)}(0),\ldots,\tilde{M}_{(\epsilon_k,\delta_k)}(0))\|(\tilde{M}_{(\epsilon_1,\delta_1)}(1),\ldots,\tilde{M}_{(\epsilon_k,\delta_k)}(1))\right) \leq \epsilon_g\ .$$

Thus by Fact 2.2 we have:

$$
\begin{aligned}
&D_\infty^{\delta_g}\left((M_1(D_0),\ldots,M_k(D_0))\|(M_1(D_1),\ldots,M_k(D_1))\right) = \\
&D_\infty^{\delta_g}\left((T_1(\tilde{M}_{(\epsilon_1,\delta_1)}(0)),\ldots,T_k(\tilde{M}_{(\epsilon_k,\delta_k)}(0)))\|(T_1(\tilde{M}_{(\epsilon_1,\delta_1)}(1)),\ldots,T_k(\tilde{M}_{(\epsilon_k,\delta_k)}(1)))\right) \leq \epsilon_g.
\end{aligned}
$$

Now we are ready to characterize OptComp for an arbitrary set of differentially private algorithms.

Proof (Proof of Theorem 1.5). Given $(\epsilon_1, \delta_1), \ldots, (\epsilon_k, \delta_k)$ and δ_g, let $\tilde{M}^k(b)$ denote the composition $(\tilde{M}_{(\epsilon_1, \delta_1)}(b), \ldots, \tilde{M}_{(\epsilon_k, \delta_k)}(b))$ and let $\tilde{P}_b^k(x)$ be the probability mass function of $\tilde{M}^k(b)$, for $b = 0$ and $b = 1$. By Lemma 3.3, $\text{OptComp}((\epsilon_1, \delta_1), \ldots, (\epsilon_k, \delta_k), \delta_g)$ is the smallest value of ϵ_g such that:

$$\delta_g \geq \max_{Q \subseteq \{0,1,2,3\}^k} \{\tilde{P}_0^k(Q) - e^{\epsilon_g} \cdot \tilde{P}_1^k(Q)\}.$$

Given ϵ_g, the set $S \subseteq \{0,1,2,3\}^k$ that maximizes the right-hand side is

$$S = S(\epsilon_g) = \left\{ x \in \{0,1,2,3\}^k \mid \tilde{P}_0^k(x) \geq e^{\epsilon_g} \cdot \tilde{P}_1^k(x) \right\}.$$

We can further split $S(\epsilon_g)$ into $S(\epsilon_g) = S_0(\epsilon_g) \cup S_1(\epsilon_g)$ with

$$S_0(\epsilon_g) = \left\{ x \in \{0,1,2,3\}^k \mid \tilde{P}_1^k(x) = 0 \right\}.$$

$$S_1(\epsilon_g) = \left\{ x \in \{0,1,2,3\}^k \mid \tilde{P}_0^k(x) \geq e^{\epsilon_g} \cdot \tilde{P}_1^k(x), \text{ and } \tilde{P}_1^k(x) > 0 \right\}.$$

Note that $S_0(\epsilon_g) \cap S_1(\epsilon_g) = \emptyset$. We have $\tilde{P}_1^k(S_0(\epsilon_g)) = 0$ and $\tilde{P}_0^k(S_0(\epsilon_g)) = 1 - \Pr[\tilde{M}^k(0) \in \{1,2,3\}^k] = 1 - \prod_{i=1}^k (1 - \delta_i)$. So

$$\tilde{P}_0^k(S(\epsilon_g)) - e^{\epsilon_g}\tilde{P}_1^k(S(\epsilon_g)) = \tilde{P}_0^k(S_0(\epsilon_g)) - e^{\epsilon_g}\tilde{P}_1^k(S_0(\epsilon_g)) + \tilde{P}_0^k(S_1(\epsilon_g)) - e^{\epsilon_g}\tilde{P}_1^k(S_1(\epsilon_g))$$

$$= 1 - \prod_{i=1}^k (1 - \delta_i)^k + \tilde{P}_0^k(S_1(\epsilon_g)) - e^{\epsilon_g}\tilde{P}_1^k(S_1(\epsilon_g)).$$

Now we just need to analyze $\tilde{P}_0^k(S_1(\epsilon_g)) - e^{\epsilon_g}\tilde{P}_1^k(S_1(\epsilon_g))$. Notice that $S_1(\epsilon_g) \subseteq \{1,2\}^k$ because for all $x \in S_1(\epsilon_g)$, we have $\tilde{P}_0(x) > \tilde{P}_1(x) > 0$. So we can write:

$$\tilde{P}_0^k(S_1(\epsilon_g)) - e^{\epsilon_g} \cdot \tilde{P}_1^k(S_1(\epsilon_g))$$

$$= \sum_{y \in \{1,2\}^k} \max \left\{ \prod_{i: y_i = 1} \frac{(1 - \delta_i)e^{\epsilon_i}}{1 + e^{\epsilon_i}} \cdot \prod_{i: y_i = 2} \frac{(1 - \delta_i)}{1 + e^{\epsilon_i}} - e^{\epsilon_g} \prod_{i: y_i = 1} \frac{(1 - \delta_i)}{1 + e^{\epsilon_i}} \cdot \prod_{i: y_i = 2} \frac{(1 - \delta_i)e^{\epsilon_i}}{1 + e^{\epsilon_i}}, 0 \right\}$$

$$= \prod_{i=1}^k \frac{1 - \delta_i}{1 + e^{\epsilon_i}} \sum_{y \in \{0,1\}^k} \max \left\{ \frac{e^{\sum_{i=1}^k \epsilon_i}}{e^{\sum_{i=1}^k y_i \epsilon_i}} - e^{\epsilon_g} \cdot e^{\sum_{i=1}^k y_i \epsilon_i}, 0 \right\}.$$

Putting everything together yields:

$$\delta_g \geq \tilde{P}_0^k(S_0(\epsilon_g)) - e^{\epsilon_g}\tilde{P}_1^k(S_0(\epsilon_g)) + \tilde{P}_0^k(S_1(\epsilon_g)) - e^{\epsilon_g}\tilde{P}_1^k(S_1(\epsilon_g))$$

$$= 1 - \prod_{i=1}^k (1 - \delta_i) + \frac{\prod_{i=1}^k (1 - \delta_i)}{\prod_{i=1}^k (1 + e^{\epsilon_i})} \sum_{S \subseteq \{1,\ldots,k\}} \max \left\{ e^{\sum_{i \in S} \epsilon_i} - e^{\epsilon_g} \cdot e^{\sum_{i \notin S} \epsilon_i}, 0 \right\}.$$

We have characterized the optimal composition for an arbitrary set of differentially private algorithms $(M_1, \ldots, M_k)$ under the assumption that the algorithms are chosen in advance and all run on the same database. Next we show that OptComp under this restrictive model of composition is actually equivalent under the more general k-fold adaptive composition discussed in Sect. 2.

Theorem 3.4. *The privacy parameters $\epsilon_1, \ldots, \epsilon_k \geq 0, \delta_1, \ldots, \delta_k \in [0, 1)$, satisfy (ϵ_g, δ_g)-differential privacy under adaptive composition if and only if* $\mathrm{OptComp}((\epsilon_1, \delta_1), \ldots, (\epsilon_k, \delta_k), \delta_g) \leq \epsilon_g$.

Proof. First suppose the privacy parameters $\epsilon_1, \ldots, \epsilon_k, \delta_1, \ldots, \delta_k$ satisfy (ϵ_g, δ_g)-differential privacy under adaptive composition. Then $\mathrm{OptComp}((\epsilon_1, \delta_1), \ldots, (\epsilon_k, \delta_k), \delta_g) \leq \epsilon_g$ because adaptive composition is more general than the composition defining OptComp.

Conversely, suppose $\mathrm{OptComp}((\epsilon_1, \delta_1), \ldots, (\epsilon_k, \delta_k), \delta_g) \leq \epsilon_g$. In particular, this means $\mathrm{OptComp}(\tilde{M}_{(\epsilon_1, \delta_1)}, \ldots, \tilde{M}_{(\epsilon_k, \delta_k)}, \delta_g) \leq \epsilon_g$. To complete the proof, we must show that the privacy parameters $\epsilon_1, \ldots, \epsilon_k, \delta_1, \ldots, \delta_k$ satisfy (ϵ_g, δ_g)-differential privacy under adaptive composition.

Fix an adversary A. On each round i, A uses its coin tosses r and the previous outputs $y_1, \ldots, y_{i-1}$ to select an (ϵ_i, δ_i)-differentially private algorithm $M_i = M_i^{r, y_1, \ldots, y_{i-1}}$ and neighboring databases $D_0 = D_0^{r, y_1, \ldots, y_{i-1}}, D_1 = D_1^{r, y_1, \ldots, y_{i-1}}$. Let V^b be the view of A with the given privacy parameters under composition game b for $b = 0$ and $b = 1$.

Lemma 3.2 tells us that there exists an algorithm $T_i = T_i^{r, y_1, \ldots, y_{i-1}}$ such that $T_i(\tilde{M}_{(\epsilon_i, \delta_i)}(b))$ is identically distributed to $M_i(D_b)$ for both $b = 0, 1$ for all $i \in [k]$. Define $\hat{T}(z_1, \ldots, z_k)$ for $z_1, \ldots, z_k \in \{0, 1, 2, 3\}$ as follows:

1. Randomly choose coins r for A
2. For $i = 1, \ldots, k$, let $y_i \leftarrow T_i^{r, y_1, \ldots, y_{i-1}}(z_i)$
3. Output $(r, y_1, \ldots, y_k)$

Notice that $\hat{T}(\tilde{M}_{(\epsilon_1, \delta_1)}(b), \ldots, \tilde{M}_{(\epsilon_k, \delta_k)}(b))$ is identically distributed to V^b for both $b = 0, 1$. By hypothesis we have

$$D_\infty^{\delta_g}\left((\tilde{M}_{(\epsilon_1, \delta_1)}(0), \ldots, \tilde{M}_{(\epsilon_k, \delta_k)}(0)) \| (\tilde{M}_{(\epsilon_1, \delta_1)}(1), \ldots, \tilde{M}_{(\epsilon_k, \delta_k)}(1)) \right) \leq \epsilon_g.$$

Thus by Fact 2.2 we have:

$$D_\infty^{\delta_g}(V^0 \| V^1) = D_\infty^{\delta_g}\left(\hat{T}(\tilde{M}_{(\epsilon_1, \delta_1)}(0), \ldots, \tilde{M}_{(\epsilon_k, \delta_k)}(0)) \| \hat{T}(\tilde{M}_{(\epsilon_1, \delta_1)}(1), \ldots, \tilde{M}_{(\epsilon_k, \delta_k)}(1)) \right) \leq \epsilon_g.$$

4 Hardness of OptComp

$\#P$ is the class of all counting problems associated with decision problems in NP. It is a set of functions that count the number of solutions to some NP problem. More formally:

Definition 4.1. *A function* $f\colon \{0,1\}^* \to \mathbb{N}$ *is in the class* #P *if there exists a polynomial* $p\colon \mathbb{N} \to \mathbb{N}$ *and a polynomial time algorithm* M *such that for every* $x \in \{0,1\}^*$:

$$f(x) = \left| \left\{ y \in \{0,1\}^{p(|x|)} \colon M(x,y) = 1 \right\} \right| .$$

Definition 4.2. *A function* g *is called* #P-*hard if every function* $f \in$ #P *can be computed in polynomial time given oracle access to* g. *That is, evaluations of* g *can be done in one time step.*

If a function is #P-hard, then there is no polynomial-time algorithm for computing it unless there is a polynomial-time algorithm for counting the number of solutions of all NP problems.

Definition 4.3. *A function* f *is called* #P-*easy if there is some function* $g \in$ #P *such that* f *can be computed in polynomial time given oracle access to* g.

If a function is both #P-hard and #P-easy, we say it is #P-complete. Proving that computing OptComp is #P-complete can be broken into two steps: showing that it is #P-easy and showing that it is #P-hard.

Lemma 4.4. *Computing OptComp is* #P-*easy.*

Proof. A proof of this statement can be found in the full version of the paper [10]. ▯

Next we show that computing OptComp is also #P-hard through a series of reductions. We start with a multiplicative version of the partition problem that is known to be #P-complete by Ehrgott [7]. The problems in the chain of reductions are defined below.

Definition 4.5. #INT-PARTITION *is the following problem: given a set* $Z = \{z_1, z_2, \ldots, z_k\}$ *of positive integers, count the number of partitions* $P \subseteq [k]$ *such that*

$$\prod_{i \in P} z_i - \prod_{i \notin P} z_i = 0 .$$

All of the remaining problems in our chain of reductions take inputs $\{w_1, \ldots, w_k\}$ where $1 \leq w_i \leq e$ is the Dth root of a positive integer for all $i \in [k]$ and some positive integer D. All of the reductions we present hold for every positive integer D, including $D = 1$ when the inputs are integers. The reason we choose D to be large enough such that our inputs are in the range $[1, e]$ is because in the final reduction to OptComp, ϵ_i values in the proof are set to $\ln(w_i)$. We want to show that our reductions hold for reasonable values of ϵ's in a differential privacy setting so throughout the proofs we use w_i's $\in [1, e]$ to correspond to ϵ_i's $\in [0, 1]$ in the final reduction. It is important to note though that the reductions still hold for any choice of positive integer D and thus any range of ϵ's ≥ 0.

Definition 4.6. #PARTITION *is the following problem: given a number* $D \in \mathbb{N}$ *and a set* $W = \{w_1, w_2, \ldots, w_k\}$ *of real numbers where for all* $i \in [k]$, $1 \leq w_i \leq e$ *is the Dth root of a positive integer, count the number of partitions* $P \subseteq [k]$ *such that*

$$\prod_{i \in P} w_i - \prod_{i \notin P} w_i = 0.$$

Definition 4.7. #T-PARTITION *is the following problem: given a number* $D \in \mathbb{N}$ *and a set* $W = \{w_1, w_2, \ldots, w_k\}$ *of real numbers where for all* $i \in [k]$, $1 \leq w_i \leq e$ *is the Dth root of a positive integer and a* positive *real number* T, *count the number of partitions* $P \subseteq [k]$ *such that*

$$\prod_{i \in P} w_i - \prod_{i \notin P} w_i = T.$$

Definition 4.8. #SUM-PARTITION: *given a number* $D \in \mathbb{N}$ *and a set* $W = \{w_1, w_2, \ldots, w_k\}$ *of real numbers where for all* $i \in [k]$, $1 \leq w_i \leq e$ *is the Dth root of a positive integer and a real number* $r > 1$, *find*

$$\sum_{P \subseteq [k]} \max \left\{ \prod_{i \in P} w_i - r \cdot \prod_{i \notin P} w_i, 0 \right\}.$$

We prove that computing OptComp is #P-hard by the following series of reductions:

$$\text{#INT-PARTITION} \leq \text{#PARTITION} \leq \text{#T-PARTITION} \leq \text{#SUM-PARTITION} \leq \text{OptComp}.$$

Since #INT-PARTITION is known to be #P-complete [7], the chain of reductions will prove that OptComp is #P-hard.

Lemma 4.9. *For every constant* $c > 1$, #PARTITION *is #P-hard, even on instances where* $\prod_i w_i \leq c$.

Proof. Given an instance of #INT-PARTITION, $\{z_1, \ldots, z_k\}$, we show how to find the solution in polynomial time using a #PARTITION oracle. Set $D = \lceil \log_c(\prod_i z_i) \rceil$ and $w_i = \sqrt[D]{z_i} \; \forall i \in [k]$. Note that $\prod_i w_i = (\prod_i z_i)^{1/D} \leq c$. Let $P \subseteq [k]$:

$$\prod_{i \in P} w_i = \prod_{i \notin P} w_i \iff \left(\prod_{i \in P} w_i \right)^D = \left(\prod_{i \notin P} w_i \right)^D$$

$$\iff \prod_{i \in P} z_i = \prod_{i \notin P} z_i.$$

There is a one-to-one correspondence between solutions to the #PARTITION problem and solutions to the given #INT-PARTITION instance. We can solve #INT-PARTITION in polynomial time with a #PARTITION oracle. Therefore #PARTITION is #P-hard.

Lemma 4.10. *For every constant $c > 1$, #T-PARTITION is #P-hard, even on instances where $\prod_i w_i \le c$.*

Proof. Let $c > 1$ be a constant. We will reduce from #PARTITION, so consider an instance of the #PARTITION problem, $W = \{w_1, w_2, \ldots, w_k\}$. We may assume $\prod_i w_i \le \sqrt{c}$ since $\sqrt{c}$ is also a constant greater than 1.

Set $W' = W \cup \{w_{k+1}\}$, where $w_{k+1} = \prod_{i=1}^{k} w_i$. Notice that $\prod_{i=1}^{k+1} w_i \le (\sqrt{c})^2 = c$. Set $T = \sqrt{w_{k+1}} \, (w_{k+1} - 1)$. Now we can use a #T-PARTITION oracle to count the number of partitions $Q \subseteq \{1, \ldots, k+1\}$ such that

$$\prod_{i \in Q} w_i - \prod_{i \notin Q} w_i = T \ .$$

Let $P = Q \cap \{1, \ldots, k\}$. We will argue that $\prod_{i \in Q} w_i - \prod_{i \notin Q} w_i = T$ if and only if $\prod_{i \in P} w_i = \prod_{i \notin P} w_i$, which completes the proof. There are two cases to consider: $w_{k+1} \in Q$ and $w_{k+1} \notin Q$.

Case 1: $w_{k+1} \in Q$. In this case, we have:

$$w_{k+1} \cdot \left(\prod_{i \in P} w_i \right) - \prod_{i \notin P} w_i = \prod_{i \in Q} w_i - \prod_{i \notin Q} w_i = T = \sqrt{w_{k+1}} \, (w_{k+1} - 1)$$

$$\Longleftrightarrow \left(\prod_{i \in [k]} w_i \right) \left(\prod_{i \in P} w_i \right) - \prod_{i \in [k]} w_i = \sqrt{\prod_{i \in [k]} w_i} \left(\prod_{i \in [k]} w_i - 1 \right) \left(\prod_{i \in P} w_i \right) \quad \text{multiplied both sides by } \prod_{i \in P} w_i$$

$$\Longleftrightarrow \left(\prod_{i \in P} w_i - \sqrt{\prod_{i \in [k]} w_i} \right) \left(\prod_{i \in [k]} w_i \prod_{i \in P} w_i + \sqrt{\prod_{i \in [k]} w_i} \right) = 0 \quad \text{factored quadratic in } \prod_{i \in P} w_i$$

$$\Longleftrightarrow \prod_{i \in P} w_i = \sqrt{\prod_{i \in [k]} w_i}$$

$$\Longleftrightarrow \prod_{i \notin P} w_i = \prod_{i \in P} w_i \ .$$

So there is a one-to-one correspondence between solutions to the #T-PARTITION instance W' where $w_{k+1} \in Q$ and solutions to the original #PARTITION instance W.

Case 2: $w_{k+1} \notin Q$. Solutions now look like:

$$\prod_{i \in P} w_i - \prod_{i \in [k]} w_i \prod_{i \notin P} w_i = \sqrt{\prod_{i \in [k]} w_i} \left(\prod_{i \in [k]} w_i - 1 \right) .$$

One way this can be true is if $w_i = 1$ for all $i \in [k]$. We can check ahead of time if our input set W contains all ones. If it does, then there are $2^k - 2$ partitions that yield equal products (all except $P = [k]$ and $P = \emptyset$) so we can just output $2^k - 2$ as the solution and not even use our oracle. The only other way to satisfy the above expression is for $\prod_{i \in P} w_i > \prod_{i \in [k]} w_i$ which cannot happen because $P \subseteq [k]$. So there are no solutions in the case that $w_{k+1} \notin Q$.

Therefore the output of the #T-PARTITION oracle on W' is the solution to the #PARTITION problem. So #T-PARTITION is #P-hard.

Lemma 4.11. *For every constant $c > 1$, #SUM-PARTITION is #P-hard even on instances where $\prod_i w_i \leq c$.*

Proof. We will use a #SUM-PARTITION oracle to solve #T-PARTITION given a set of Dth roots of positive integers $W = \{w_1, \ldots, w_k\}$ and a positive real number T. Notice that for every $z > 0$:

$$\prod_{i \in P} w_i - \prod_{i \notin P} w_i = z \implies \prod_{i \in P} w_i - \frac{\prod_{i \in [k]} w_i}{\prod_{i \in P} w_i} = z$$

$$\implies \exists\, j \in \mathbb{Z}^+ \text{such that } \sqrt[D]{j} - \frac{\prod_{i \in [k]} w_i}{\sqrt[D]{j}} = z.$$

Above, j must be a positive integer, which tells us that the gap in products from every partition must take a particular form. This means that for a given D and W, #T-PARTITION can only be non-zero on a discrete set of possible values of $T = z$. Given z, we can find a $z' > z$ such that the above has no solutions in the interval (z, z'). Specifically, solve the above quadratic for $\sqrt[D]{j}$ (where j may or may not be an integer), let $j' = \lfloor j + 1 \rfloor > j$, and $z' = \sqrt[D]{j'} - \frac{\prod_i w_i}{\sqrt[D]{j'}}$. We use this property twice in the proof.

Define $P^z \equiv \{P \subseteq [k] \mid \prod_{i \in P} w_i - \prod_{i \notin P} w_i \geq z\}$. As described above we can find the interval (T, T') of values above T with no solutions. Then, for every $c \in (T, T')$:

$$\left| \left\{ P \subseteq [k] \mid \prod_{i \in P} w_i - \prod_{i \notin P} w_i = T \right\} \right| = \left| P^T \setminus P^c \right|$$

$$= \frac{1}{T} \left(\sum_{P \in P^T \setminus P^c} \left(\prod_{i \in P} w_i - \prod_{i \notin P} w_i \right) \right)$$

$$= \frac{1}{T} \left(\sum_{P \in P^T} \left(\prod_{i \in P} w_i - \prod_{i \notin P} w_i \right) - \sum_{P \in P^c} \left(\prod_{i \in P} w_i - \prod_{i \notin P} w_i \right) \right).$$

We now show how to find $\sum_{P \in P^z} \left(\prod_{i \in P} w_i - \prod_{i \notin P} w_i \right)$ for any $z > 0$ using the #SUM-PARTITION oracle. Once we have this procedure, we can run it for $z = T$ and $z = c$ and plug the outputs into the expression above to solve the #T-PARTITION problem. We want to set the input r to the #SUM-PARTITION oracle such that:

$$\prod_{i \in P} w_i - r \cdot \prod_{i \notin P} w_i \geq 0 \iff \prod_{i \in P} w_i - \prod_{i \notin P} w_i \geq z.$$

Solving this expression for r gives:

$$r_z = \frac{4 \prod_{i \in [k]} w_i}{\left(\sqrt{z^2 + 4 \prod_{i \in [k]} w_i} - z \right)^2}.$$

Below we check that this setting satisfies the requirement.

$$\prod_{i\in P} w_i - \frac{4\prod_{i\in[k]} w_i}{\left(\sqrt{z^2 + 4\prod_{i\in[k]} w_i} - z\right)^2}\cdot \prod_{i\notin P} w_i \geq 0 \iff 1 - \frac{4\left(\prod_{i\notin P} w_i\right)^2}{\left(\sqrt{z^2 + 4\prod_{i\in[k]} w_i} - z\right)^2} \geq 0$$

$$\iff \sqrt{z^2 + 4\prod_{i\in[k]} w_i} \geq 2\prod_{i\notin P} w_i + z$$

$$\iff 4\prod_{i\in[k]} w_i \geq 4\left(\prod_{i\notin P} w_i\right)^2 + 4z\prod_{i\notin P} w_i$$

$$\iff \prod_{i\in P} w_i - \prod_{i\notin P} w_i \geq z.$$

So we have $P^z = \left\{P \subseteq [k] \mid \prod_{i\in P} w_i - r_z\cdot \prod_{i\notin P} w_i \geq 0\right\}$ but this does not necessarily mean that

$$\sum_{P\in P^z}\left(\prod_{i\in P} w_i - \prod_{i\notin P} w_i\right) = \sum_{P\in P^z}\left(\prod_{i\in P} w_i - r_z\cdot \prod_{i\notin P} w_i\right).$$

The sum on the left-hand side without the r_z coefficient is what we actually need to compute. To get this we again use the discreteness of potential solutions to find $z'' \neq z$ such that $P^z = P^{z''}$. We just pick z'' from the interval (z, z') of values above z that cannot possibly contain solutions to #T-PARTITION.

Running our #SUM-PARTITION oracle for r_z and $r_{z''}$ will output:

$$\sum_{P\in P^z}\left(\prod_{i\in P} w_i - r_z\cdot \prod_{i\notin P} w_i\right)$$

$$\sum_{P\in P^z}\left(\prod_{i\in P} w_i - r_{z''}\cdot \prod_{i\notin P} w_i\right)$$

This is just a system of two equations with two unknowns and it can be solved for $\sum_{P\in P^z}\prod_{i\in P} w_i$ and $\sum_{P\in P^z}\prod_{i\notin P} w_i$ separately. Then we can reconstruct $\sum_{P\in P^z}\left(\prod_{i\in P} w_i - \prod_{i\notin P} w_i\right)$. Running this procedure for $z = T$ and $z = c$ gives us all of the information we need to count the number of solutions to the #T-PARTITION instance we were given. We can solve #T-PARTITION in polynomial time with four calls to a #SUM-PARTITION oracle. Therefore #SUM-PARTITION is #P-hard.

Now we prove that computing OptComp is #P-complete.

Proof (Proof of Theorem 1.6). We have already shown that computing OptComp is #P-easy. Here we prove that it is also #P-hard, thereby proving #P-completeness.

Given an instance D, $W = \{w_1, \ldots, w_k\}, r$ of #SUM-PARTITION, where $\forall i \in [k]$, w_i is the Dth root of an integer and $\prod_i w_i \leq c$, set $\epsilon_i = \ln(w_i) \ \forall i \in [k]$, $\delta_1 = \delta_2 = \ldots \delta_k = 0$ and $\epsilon_g = \ln(r)$. Note that $\sum_i \epsilon_i = \ln(\prod_i w_i) \leq \ln(c)$. Since we can take c to be an arbitrary constant greater than 1, we can ensure that $\sum_i \epsilon_i \leq \epsilon$ for an arbitrary $\epsilon > 0$.

Again we will use the version of OptComp that takes ϵ_g as input and outputs δ_g. After using an OptComp oracle to find δ_g we know the optimal composition Eq. 1 from Theorem 1.5 is satisfied:

$$\frac{1}{\prod_{i=1}^{k}(1 + e^{\epsilon_i})} \sum_{S \subseteq \{1,\ldots,k\}} \max\left\{ e^{\sum_{i \in S} \epsilon_i} - e^{\epsilon_g} \cdot e^{\sum_{i \notin S} \epsilon_i}, 0 \right\} = 1 - \frac{1 - \delta_g}{\prod_{i=1}^{k}(1 - \delta_i)} = \delta_g.$$

Thus we can compute:

$$\delta_g \cdot \prod_{i=1}^{k}(1 + e^{\epsilon_i}) = \sum_{S \subseteq \{1,\ldots,k\}} \max\left\{ e^{\sum_{i \in S} \epsilon_i} - e^{\epsilon_g} \cdot e^{\sum_{i \notin S} \epsilon_i}, 0 \right\}$$

$$= \sum_{S \subseteq \{1,\ldots,k\}} \max\left\{ \prod_{i \in S} w_i - r \cdot \prod_{i \notin S} w_i, 0 \right\}.$$

This last expression is exactly the solution to the instance of #SUM-PARTITION we were given. We solved #SUM-PARTITION in polynomial time with one call to an OptComp oracle. Therefore computing OptComp is #P-hard.

5   Approximation of OptComp

Although we cannot hope to efficiently compute the optimal composition for a general set of differentially private algorithms (assuming P$\neq$NP or even FP$\neq$ #P), we show in this section that we can approximate OptComp arbitrarily well in polynomial time.

Theorem 1.7 (Restated). *There is a polynomial-time algorithm that given* $\epsilon_1, \ldots, \epsilon_k \geq 0, \delta_1, \ldots \delta_k, \delta_g \in [0, 1)$, *and* $\eta > 0$, *outputs* ϵ^* *where*

$$\text{OptComp}((\epsilon_1, \delta_1), \ldots, (\epsilon_k, \delta_k), \delta_g) \leq \epsilon^* \leq \text{OptComp}((\epsilon_1, \delta_1), \ldots, (\epsilon_k, \delta_k), e^{-\eta/2} \cdot \delta_g) + \eta .$$

The algorithm runs in $O\left(\log\left(\frac{k}{\eta} \sum_{i=1}^{k} \epsilon_i\right) \frac{k^2}{\eta} \sum_{i=1}^{k} \epsilon_i\right)$ *time assuming constant-time arithmetic operations.*

We prove this theorem using the following three lemmas:

Lemma 5.1. *Given non-negative integers $a_1, \ldots, a_k$, B and weights $w_1, \ldots, w_k \in \mathbb{R}$, one can compute*

$$\sum_{\substack{S \subseteq [k] \text{ s.t.} \\ \sum_{i \in S} a_i \leq B}} \prod_{i \in S} w_i$$

in time $O(Bk)$.

Notice that the constraint in Lemma 5.1 is the same one that characterizes knapsack problems. Indeed, the algorithm we give for computing $\sum_{S \subseteq [k]} \prod_{i \in S} w_i$ is a slight modification of the known pseudo-polynomial time algorithm for counting knapsack solutions, which uses dynamic programming. Next we show that we can use this algorithm to approximate OptComp.

Lemma 5.2. *Given $\epsilon_1, \ldots, \epsilon_k, \epsilon^* \geq 0, \delta_1, \ldots \delta_k, \delta_g \in [0, 1)$, if $\epsilon_i = a_i \epsilon_0 \; \forall i \in \{1, \ldots, k\}$ for non-negative integers a_i and some $\epsilon_0 > 0$, then there is an algorithm that determines whether or not $\mathrm{OptComp}((\epsilon_1, \delta_1), \ldots, (\epsilon_k, \delta_k), \delta_g) \leq \epsilon^*$ that runs in time $O\left(\frac{k}{\epsilon_0} \sum_{i=1}^k \epsilon_i\right)$.*

In other words, if the ϵ values we are given are all integer multiples of some ϵ_0, we can determine whether or not the composition of those privacy parameters is (ϵ^*, δ_g)-DP in pseudo-polynomial time for every $\epsilon^* \geq 0$. This means that given any inputs to OptComp, if we discretize and polynomially bound the ϵ_i's, then we can check if the parameters satisfy any global privacy guarantee in polynomial time. Once we have this, we only need to run binary search over values of ϵ^* to find the optimal one. In other words, we can solve OptComp exactly for a slightly different set of ϵ_i's. The next lemma tells us that the output of OptComp on this different set of ϵ_i's can be used as a good approximation to OptComp on the original ϵ_i's.

Lemma 5.3. *For all $\epsilon_1, \ldots, \epsilon_k, c \geq 0$ and $\delta_1, \ldots, \delta_k, \delta_g \in [0, 1)$:*

$$\mathrm{OptComp}((\epsilon_1 + c, \delta_1), \ldots, (\epsilon_k + c, \delta_k), \delta_g) \leq \mathrm{OptComp}((\epsilon_1, \delta_1), \ldots, (\epsilon_k, \delta_k), e^{-kc/2} \cdot \delta_g) + kc \; .$$

Next we prove the three lemmas and then show that Theorem 1.7 follows.

Proof (Proof of Lemma 5.1). We modify Dyer's algorithm for approximately counting solutions to knapsack problems [6]. The algorithm uses dynamic programming. Given non-negative integers $a_1, \ldots, a_k$, B and weights $w_1, \ldots, w_k \in \mathbb{R}$, define

$$F(r, s) = \sum_{\substack{S \subseteq [r] \text{ s.t.} \\ \sum_{i \in S} a_i \leq s}} \prod_{i \in S} w_i \; .$$

We want to compute $F(k, B)$. We can find this by tabulating $F(r, s)$ for $(0 \leq r \leq k, \; 0 \leq s \leq B)$ using the recursion:

$$F(r, s) = \begin{cases} 1 & \text{if } r = 0 \\ F(r - 1, s) + w_r F(r - 1, s - a_r) & \text{if } r > 0 \text{ and } a_r \leq s \\ F(r - 1, s) & \text{if } r > 0 \text{ and } a_r > s. \end{cases}$$

Each cell $F(r, s)$ in the table can be computed in constant time given earlier cells $F(r', s')$ where $r' < r$. Thus filling the entire table takes time $O(Bk)$.

Proof (Proof of Lemma 5.2). Given $\epsilon_1, \ldots, \epsilon_k, \epsilon^* \geq 0$ such that $\epsilon_i = a_i \epsilon_0 \; \forall i \in \{1, \ldots, k\}$ for non-negative integers a_i and some $\epsilon_0 > 0$, and $\delta_1, \ldots \delta_k, \delta_g \in [0, 1)$, Theorem 1.5 tells us that answering whether or not

$$\text{OptComp}((\epsilon_1, \delta_1), \ldots, (\epsilon_k, \delta_k), \delta_g) \leq \epsilon^*$$

is equivalent to answering whether or not the following inequality holds:

$$\frac{1}{\prod_{i=1}^{k}(1 + e^{\epsilon_i})} \sum_{S \subseteq \{1, \ldots, k\}} \max \left\{ e^{\sum_{i \in S} \epsilon_i} - e^{\epsilon^*} \cdot e^{\sum_{i \notin S} \epsilon_i}, 0 \right\} \leq 1 - \frac{1 - \delta_g}{\prod_{i=1}^{k}(1 - \delta_i)} .$$

The right-hand side and the coefficient on the sum are easy to compute given the inputs so in order to check the inequality, we will show how to compute the sum. Define

$$K = \left\{ T \subseteq [k] \mid \sum_{i \notin T} \epsilon_i \geq \epsilon^* + \sum_{i \in T} \epsilon_i \right\}$$

$$= \left\{ T \subseteq [k] \mid \sum_{i \in T} \epsilon_i \leq \left(\sum_{i=1}^{k} \epsilon_i - \epsilon^* \right)/2 \right\}$$

$$= \left\{ T \subseteq [k] \mid \sum_{i \in T} a_i \leq B \right\} \text{ for } B = \left\lfloor \left(\sum_{i=1}^{k} \epsilon_i - \epsilon^* \right)/2\epsilon_0 \right\rfloor$$

and observe that by setting $T = S^c$, we have

$$\sum_{S \subseteq \{1, \ldots, k\}} \max \left\{ e^{\sum_{i \in S} \epsilon_i} - e^{\epsilon^*} \cdot e^{\sum_{i \notin S} \epsilon_i}, 0 \right\} = \sum_{T \in K} \left(\left(e^{\sum_{i=1}^{k} \epsilon_i} \cdot \prod_{i \in T} e^{-\epsilon_i} \right) - \left(e^{\epsilon^*} \cdot \prod_{i \in T} e^{\epsilon_i} \right) \right).$$

We just need to compute this last expression and we can do it for each term separately since K is a set of knapsack solutions. Specifically, setting $w_i = e^{-\epsilon_i} \; \forall i \in [k]$, Lemma 5.1 tells us that we can compute $\sum_{T \subseteq [k]} \prod_{i \in T} w_i$ subject to $\sum_{i \in T} a_i \leq B$, which is equivalent to $\sum_{T \in K} \prod_{i \in T} e^{-\epsilon_i}$.

To compute $\sum_{T \in K} \prod_{i \in T} e^{\epsilon_i}$, we instead set $w_i = e^{\epsilon_i}$ and run the same procedure. Since we used the algorithm from Lemma 5.1, the running time is $O(Bk) = O\left(\frac{k}{\epsilon_0} \sum_{i=1}^{k} \epsilon_i \right)$.

Proof (Proof of Lemma 5.3). Let $\text{OptComp}((\epsilon_1, \delta_1), \ldots, (\epsilon_k, \delta_k), e^{-kc/2} \cdot \delta_g) = \epsilon_g$. From Eq. 1 in Theorem 1.5 we know:

$$\frac{1}{\prod_{i=1}^{k}(1 + e^{\epsilon_i})} \sum_{S \subseteq \{1, \ldots, k\}} \max \left\{ e^{\sum_{i \in S} \epsilon_i} - e^{\epsilon_g} \cdot e^{\sum_{i \notin S} \epsilon_i}, 0 \right\} \leq 1 - \frac{1 - e^{-kc/2} \cdot \delta_g}{\prod_{i=1}^{k}(1 - \delta_i)} .$$

Multiplying both sides by $e^{kc/2}$ gives:

$$\frac{e^{kc/2}}{\prod_{i=1}^{k}(1+e^{\epsilon_i})} \sum_{S\subseteq\{1,\ldots,k\}} \max\left\{e^{\sum_{i\in S}\epsilon_i} - e^{\epsilon_g}\cdot e^{\sum_{i\notin S}\epsilon_i}, 0\right\} \leq e^{kc/2}\cdot\left(1 - \frac{1-e^{-kc/2}\cdot\delta_g}{\prod_{i=1}^{k}(1-\delta_i)}\right)$$

$$\leq 1 - \frac{1-\delta_g}{\prod_{i=1}^{k}(1-\delta_i)}.$$

The above inequality together with Theorem 1.5 means that showing the following will complete the proof:

$$\sum_{S\subseteq\{1,\ldots,k\}} \max\left\{e^{\sum_{i\in S}(\epsilon_i+c)} - e^{\epsilon_g+kc}\cdot e^{\sum_{i\notin S}(\epsilon_i+c)}, 0\right\}$$

$$\leq \frac{e^{kc/2}\cdot\prod_{i=1}^{k}(1+e^{\epsilon_i+c})}{\prod_{i=1}^{k}(1+e^{\epsilon_i})} \sum_{S\subseteq\{1,\ldots,k\}} \max\left\{e^{\sum_{i\in S}\epsilon_i} - e^{\epsilon_g}\cdot e^{\sum_{i\notin S}\epsilon_i}, 0\right\}.$$

Since $(1+e^{\epsilon_i+c})/(1+e^{\epsilon_i}) \geq e^{c/2}$ for every $\epsilon_i > 0$, it suffices to show:

$$\sum_{S\subseteq\{1,\ldots,k\}} \max\left\{e^{\sum_{i\in S}(\epsilon_i+c)} - e^{\epsilon_g+kc}\cdot e^{\sum_{i\notin S}(\epsilon_i+c)}, 0\right\} \leq$$

$$\sum_{S\subseteq\{1,\ldots,k\}} e^{kc}\cdot\max\left\{e^{\sum_{i\in S}\epsilon_i} - e^{\epsilon_g}\cdot e^{\sum_{i\notin S}\epsilon_i}, 0\right\}.$$

This inequality holds term by term. If a right-hand term is zero $\left(\sum_{i\in S}\epsilon_i \leq \epsilon_g + \sum_{i\notin S}\epsilon_i\right)$, then so is the corresponding left-hand term $\left(\sum_{i\in S}(\epsilon_i+c) \leq \epsilon_g + kc + \sum_{i\notin S}(\epsilon_i+c)\right)$. For the nonzero terms, the factor of e^{kc} ensures that the right-hand terms are larger than the left-hand terms.

Proof (Proof of Theorem 1.7). Lemma 5.2 tells us that we can determine whether a set of privacy parameters satisfies some global differential privacy guarantee if the ϵ values are discretized. Notice that then we can solve OptComp exactly for a discretized set of ϵ values by running binary search over values of ϵ^* until we find the minimum ϵ^* that satisfies (ϵ^*, δ_g)-DP.

Given $\epsilon_1, \ldots, \epsilon_k, \epsilon^*$, and an additive error parameter $\eta > 0$, set $a_i = \left\lfloor\frac{k}{\eta}\epsilon_i\right\rfloor, \epsilon'_i = \frac{\eta}{k}\cdot a_i \ \forall i \in [k]$. With these settings, the a_i's are non-negative integers and the ϵ'_i values are all integer multiples of $\epsilon_0 = \eta/k$. Lemma 5.2 tells us that we can determine if the new privacy parameters with ϵ' values satisfy (ϵ^*, δ_g)-DP in time $O\left(\frac{k^2}{\eta}\sum_{i=1}^{k}\epsilon_i\right)$. Running binary search over values of ϵ^* will then compute OptComp$((\epsilon'_1, \delta_1), \ldots, (\epsilon'_k, \delta_k), \delta_g) = \epsilon'_g$ exactly in time $O\left(\log\left(\frac{k}{\eta}\sum_{i=1}^{k}\epsilon_i\right)\frac{k^2}{\eta}\sum_{i=1}^{k}\epsilon_i\right)$.

Notice that $\epsilon_i - \eta/k \leq \epsilon'_i \leq \epsilon_i \ \forall i \in [k]$. Lemma 5.3 says that the outputted ϵ'_g is at most OptComp$((\epsilon_1, \delta_1), \ldots, (\epsilon_k, \delta_k), e^{-\eta/2}\cdot\delta_g) + \eta$ as desired.

References

1. Crosas, M.: The dataverse network®: an open-source application for sharing, discovering and preserving data. D-lib Mag. **17**(1), 2 (2011)
2. Dwork, C., Kenthapadi, K., McSherry, F., Mironov, I., Naor, M.: Our data, ourselves: privacy via distributed noise generation. In: Vaudenay, S. (ed.) EUROCRYPT 2006. LNCS, vol. 4004, pp. 486–503. Springer, Heidelberg (2006)
3. Dwork, C., McSherry, F., Nissim, K., Smith, A.: Calibrating noise to sensitivity in private data analysis. In: Halevi, S., Rabin, T. (eds.) TCC 2006. LNCS, vol. 3876, pp. 265–284. Springer, Heidelberg (2006)
4. Dwork, C., Roth, A.: The algorithmic foundations of differential privacy. Found. Trends Theor. Comput. Sci. **9**(3–4), 211–407 (2013)
5. Dwork, C., Rothblum, G.N., Vadhan, S.: Boosting and differential privacy. In: 51st IEEE Symposium on Foundations of Computer Science, pp. 51–60. IEEE (2010)
6. Dyer, M.: Approximate counting by dynamic programming. In: 35th ACM Symposium on Theory of Computing, pp. 693–699. ACM (2003)
7. Ehrgott, M.: Approximation algorithms for combinatorial multicriteria optimization problems. Int. Trans. Oper. Res. **7**(1), 5–31 (2000)
8. King, G.: An introduction to the dataverse network as an infrastructure for data sharing. Sociol. Methods Res. **36**(2), 173–199 (2007)
9. Kairouz, P., Oh, S., Viswanath. P.: The composition theorem for differential privacy. In: 32nd International Conference on Machine Learning, pp. 1376–1385 (2015)
10. Murtagh, J., Vadhan, S.: The Complexity of Computing the Optimal Composition of Differential Privacy (2015). http://arxiv.org/abs/1507.03113
11. Warner, S.L.: Randomized response: a survey technique for eliminating evasive answer bias. J. Am. Stat. Assoc. **60**(309), 63–69 (1965)

Order-Revealing Encryption and the Hardness of Private Learning

Mark Bun[1][(✉)] and Mark Zhandry[2]

[1] Harvard University, Cambridge, MA, USA
mbun@seas.harvard.edu
[2] Massachusetts Institute of Technology, Cambridge, MA, USA
mzhandry@gmail.com

Abstract. An order-revealing encryption scheme gives a public procedure by which two ciphertexts can be compared to reveal the ordering of their underlying plaintexts. We show how to use order-revealing encryption to separate computationally efficient PAC learning from efficient (ε, δ)-differentially private PAC learning. That is, we construct a concept class that is efficiently PAC learnable, but for which every efficient learner fails to be differentially private. This answers a question of Kasiviswanathan et al. (FOCS '08, SIAM J. Comput. '11).

To prove our result, we give a generic transformation from an order-revealing encryption scheme into one with strongly correct comparison, which enables the consistent comparison of ciphertexts that are not obtained as the valid encryption of any message. We believe this construction may be of independent interest.

Keywords: Differential privacy · Learning theory · Order-revealing encryption

1 Introduction

Many agencies hold sensitive information about individuals, where statistical analysis of this data could yield great societal benefit. The line of work on differential privacy [20] aims to enable such analysis while giving a strong formal guarantee on the privacy afforded to individuals. Noting that the framework of computational learning theory captures many of these statistical tasks, Kasiviswanathan et al. [37] initiated the study of *differentially private learning*. Roughly speaking, a differentially private learner is required to output a classification of labeled examples that is accurate, but does not change significantly based on the presence or absence of any individual example.

The early positive results in private learning established that, ignoring computational complexity, any concept class is privately learnable with a number

---

M. Bun—Supported by an NDSEG fellowship and NSF grant CNS-1237235.

M. Zhandry—Work done while the author was a graduate student at Stanford University. Supported by the DARPA PROCEED program.

E. Kushilevitz and T. Malkin (Eds.): TCC 2016-A, Part I, LNCS 9562, pp. 176–206, 2016.
DOI: 10.1007/978-3-662-49096-9_8

of samples logarithmic in the size of the concept class [37]. Since then, a number of works have improved our understanding of the sample complexity – the minimum number of examples – required by such learners to simultaneously achieve accuracy and privacy. Some of these works showed that privacy incurs an inherent additional cost in sample complexity; that is, some concept classes require more samples to learn privately than they require to learn without privacy [1,2,13,16,17,25]. In this work, we address the complementary question of whether there is also a *computational* price of differential privacy for learning tasks, for which much less is known. The initial work of Kasiviswanathan et al. [37] identified the important question of whether any efficiently PAC learnable concept class is also efficiently privately learnable, but only limited progress has been made on this question since then [1,44].

Our main result gives a strong negative answer to this question. We exhibit a concept class that is efficiently PAC learnable, but under plausible cryptographic assumptions cannot be learned efficiently and privately. To prove this result, we establish a connection between private learning and *order-revealing encryption*. We construct a new order-revealing encryption scheme with strong correctness properties that may be of independent learning-theoretic and cryptographic interest.

1.1 Differential Privacy and Private Learning

We first recall Valiant's (distribution-free) PAC model for learning [54]. Let $\mathcal{C}$ be a *concept class* consisting of concepts $c : X \to \{0,1\}$ for a data universe X. A learner L is given n samples of the form $(x_i, c(x_i))$ where the x_i's are drawn i.i.d. from an unknown distribution, and are labeled according to an unknown concept c. The goal of the learner is to output a *hypothesis* $h : X \to \{0,1\}$ from a hypothesis class $\mathcal{H}$ that approximates c well on the unknown distribution. That is, the probability that h disagrees with c on a fresh example from the unknown distribution should be small – say, less than 0.05. The hypothesis class $\mathcal{H}$ may be different from $\mathcal{C}$, but in the case where $\mathcal{H} \subseteq \mathcal{C}$ we call L a *proper* learner. Moreover, we say a learner is *efficient* if it runs in time polynomial in the description size of c and the size of its examples.

Kasiviswanathan et al. [37] defined a private learner to be a PAC learner that is also differentially private. Two samples $S = \{(x_1, b_1), \ldots, (x_n, b_n)\}$ and $S' = \{(x'_1, b'_1), \ldots, (x'_n, b'_n)\}$ are said to be *neighboring* if they differ on exactly one example, which we think of as corresponding to one individual's information. A randomized learner $L : (X \times \{0,1\})^n \to \mathcal{H}$ is (ε, δ)-*differentially private* if for all neighboring datasets S and S' and all sets $T \subseteq \mathcal{H}$,

$$\Pr[L(S) \in T] \le e^{\varepsilon} \Pr[L(S') \in T] + \delta.$$

The original definition of differential privacy [20] took $\delta = 0$, a case which is called *pure* differential privacy. The definition with positive δ, called *approximate* differential privacy, first appeared in [19] and has since been shown to enable substantial accuracy gains. Throughout this introduction, we will think of ε as a small constant, e.g. $\varepsilon = 0.1$, and $\delta = o(1/n)$.

Kasiviswanathan et al. [37] gave a generic "Private Occam's Razor" algorithm, showing that any concept class $\mathcal{C}$ can be privately (properly) learned using $O(\log |\mathcal{C}|)$ samples. Unfortunately, this algorithm runs in time $\Omega(|\mathcal{C}|)$, which is exponential in the description size of each concept. With an eye toward designing efficient private learners, Blum et al. [5] made the powerful observation that any efficient learning algorithm in the *statistical queries* (SQ) framework of Kearns [39] can be efficiently simulated with differential privacy. Moreover, Kasiviswanathan et al. [37] showed that the efficient learner for the concept class of parity functions based on Gaussian elimination can also be implemented efficiently with differential privacy. These two techniques – SQ learning and Gaussian elimination – are essentially the only methods known for computationally efficient PAC learning. The fact that these can both be implemented privately led Kasiviswanathan et al. [37] to ask whether *all* efficiently learnable concept classes could also be efficiently learned with differential privacy.

Beimel et al. [1] made partial progress toward this question in the special case of pure differential privacy with proper learning, showing that the sample complexity of efficient learners can be much higher than that of inefficient ones. Specifically, they showed that assuming the existence of pseudorandom generators with exponential stretch, there exists for any $\ell(d) = \omega(\log d)$ a concept class over $\{0, 1\}^d$ for which every efficient proper private learner requires $\Omega(d)$ samples, but an inefficient proper private learner only requires $O(\ell(d))$ examples. Nissim [44] strengthened this result substantially for "representation learning," where a proper learner is further restricted to output a canonical representation of its hypothesis. He showed that, assuming the existence of one-way functions, there exists a concept class that is efficiently representation learnable, but not efficiently privately representation learnable (even with approximate differential privacy). With Nissim's kind permission, we give the details of this construction in Sect. 5.

Despite these negative results for proper learning, one might still have hoped that any efficiently learnable concept class could be efficiently *improperly* learned with privacy. Indeed, a number of works have shown that, especially with differential privacy, improper learning can be much more powerful than proper learning. For instance, Beimel et al. [1] showed that under pure differential privacy, the simple class of Point functions (indicators of a single domain element) requires $\Omega(d)$ samples to privately learn properly, but only $O(1)$ samples to privately learn improperly. Moreover, computational separations are known between proper and improper learning even without privacy considerations. Pitt and Valiant [46] showed that unless $\mathbf{NP} = \mathbf{RP}$, k-term DNF are not efficiently properly learnable, but they are efficiently improperly learnable [54].

Under plausible cryptographic assumptions, we resolve the question of Kasiviswanathan et al. [37] in the negative, even for improper learners. The assumption we need is the existence of "strongly correct" order-revealing encryption (ORE) schemes, described in Sect. 1.3.

Theorem 1 (Informal). *Assuming the existence of strongly correct ORE, there exists an efficiently computable concept class* EncThresh *that is efficiently*

PAC learnable, but not efficiently learnable by any (ε, δ)-*differentially private algorithm.*

We stress that this result holds even for improper learners and for the relaxed notion of approximate differential privacy. We remark that cryptography has played a major role in shaping our understanding of the computational complexity of learning in a number of models (e.g. [40,41,49,54]). It has also been used before to show separations between what is efficiently learnable in different models (e.g. [4,50]).

1.2   Our Techniques

We give an informal overview of the construction and analysis of the concept class EncThresh.

We first describe the concept class of thresholds Thresh and its simple PAC learning algorithm. Consider the domain $[N] = \{1, \ldots, N\}$. Given a number $t \in [N]$, a threshold concept c_t is defined by $c_t(x) = 1$ if and only if $x \leq t$. The concept class of thresholds admits a simple and efficient proper PAC learning algorithm L_{Thresh}. Given a sample $\{(x_1, c_t(x_1)), \ldots, (x_n, c_t(x_n))\}$ labeled by an unknown concept c_t, the learner L_{Thresh} identifies the largest positive example x_{i^*} and outputs the hypothesis $h = c_{x_{i^*}}$. That is, L_{Thresh} chooses the threshold concept that minimizes the empirical error on its sample. To achieve a small constant error on *any* underlying distribution on examples, it suffices to take $n = O(1)$ samples. Moreover, this learner can be modified to guarantee differential privacy by instead randomly sampling a threshold hypothesis with probability that decays exponentially in the empirical error of the hypothesis [37,42]. The sampling can be performed in polynomial time, and requires only a modest blow-up in the learner's sample complexity.

A simple but important observation about L_{Thresh} – which, crucially, is not true of the differentially private version – is that it is completely oblivious to the actual numeric values of its examples, or even to the fact that the domain is $[N]$. In fact, L_{Thresh} works equally well on any totally-ordered domain on which it can efficiently compare examples. In an extreme case, the learner L_{Thresh} still works when its examples are encrypted under an *order-revealing encryption* (ORE) scheme, which guarantees that L_{Thresh} is able to learn the order of its examples, but nothing else about them. Up to small technical modifications, our concept class EncThresh is exactly the class Thresh where examples are encrypted under an ORE scheme.

For EncThresh to be efficiently PAC learnable, it must be learnable even under distributions that place arbitrary weight on examples corresponding to invalid ciphertexts. To this end, we require a "strong correctness" condition on our ORE scheme. The strong correctness condition ensures that all ciphertexts, even those that are not obtained as encryptions of messages, can be compared in a consistent fashion. This condition is not met by current constructions of ORE, and one of the technical contributions of this work is a generic transformation from weakly correct ORE schemes to strongly correct ones.

While a learner similar to L_{Thresh} is able to efficiently PAC learn the concept class EncThresh, we argue that it cannot do so while preserving differential privacy with respect to its examples. Intuitively, the security of the ORE scheme ensures that essentially the only thing a learner for EncThresh can do is output a hypothesis that compares an example to one it already has. We make this intuition precise by giving an algorithm that traces the hypothesis output by any efficient learner back to one of the examples used to produce it. This formalization builds conceptually on the connection between differential privacy and traitor-tracing schemes (see Sect. 1.4), but requires new ideas to adapt to the PAC learning model.

1.3   Order-Revealing Encryption

Motivated by the task of answering range queries on encrypted databases, an *order-revealing* encryption (ORE) scheme [7,8] is a special type of symmetric key encryption scheme where it is possible to publicly sort *ciphertexts* according to the order of the *plaintexts*. More precisely, the plaintext space of the scheme is the set of integers $[N] = \{1, ..., N\}$,[1] and in addition to the *private* encryption and decryption procedures Enc, Dec, there is a public comparison procedure Comp that takes as input two ciphertexts, and reveals the order of the corresponding plaintexts. The notion of *best-possible semantic security*, defined in Boneh et al. [8], intuitively captures the requirement that, given a collection of ciphertexts, no information about the plaintexts is learned, *except* for the ordering.

Known Constructions of Order-Revealing Encryption. Relatively few constructions of order-revealing encryption are known, and all constructions are currently based on strong assumptions. Order-revealing encryption can be seen as a special case of 2-input *functional encryption*, also known as *property preserving encryption* [45]. In such a scheme, there are several functions $f_1, ..., f_k$, and given two ciphertexts c_0, c_1 encrypting m_0, m_1, it is possible to learn $f_i(m_0, m_1)$ for all $i \in [k]$. General *multi-input* functional encryption schemes can be obtained from indistinguishability obfuscation [30] or multilinear maps [8]. It is also possible to build ORE from *single-input* functional encryption with function privacy, which means that f is kept secret. Such schemes can be built from regular single-input schemes without function privacy [12], and such single-input schemes can also be built from obfuscation [27] or multilinear maps [28].

It is known that the forms of functional encryption discussed above actually imply obfuscation [3], meaning that all the assumptions from which we can currently build order-revealing encryption imply obfuscation. However, we stress that ORE appears to be much, much weaker than obfuscation or functional encryption: only a single very simple functionality is supported, namely comparison. In particular the functionality does not support evaluating cryptographic primitives on the plaintext, a feature required of essentially all of the interesting applications of obfuscation/functional encryption. Therefore, we conjecture

---

[1] More generally, any totally-ordered plaintext space can be considered.

that ORE can actually be based on significantly weaker assumptions. One way or another, it is important to resolve the status of ORE relative to obfuscation and other strong primitives: if ORE can be based on mild assumptions, it would strengthen our impossibility result, and likely lead to more efficient ORE constructions that can actually be used in practice. If ORE actually implies obfuscation or other similarly strong primitives, then ORE could be a path to building more efficient obfuscation with better security. Our work demonstrates that, in addition to having real-world practical motivations, ORE is also an interesting theoretical object.

Unfortunately, the above constructions are not quite sufficient for our purposes. The issue arises from the fact that our learner needs to work for *any* distribution on ciphertexts, even distributions whose support includes malformed ciphertexts. Unfortunately, previous constructions only achieve a weak form of correctness, which guarantees that encrypting two messages and then comparing the ciphertexts using Comp produces the same result (with overwhelming probability) as comparing the plaintexts directly. This requirement only specifies how Comp works on *valid* ciphertexts, namely actual encryptions of messages. Moreover, correctness is only guaranteed for these messages with overwhelming probability, meaning even some valid ciphertexts may cause Comp to misbehave.

For our learner, this weak form of correctness means, for some distributions that place significant weight on bad ciphertexts, the comparison procedure is completely useless, and thus the learner will fail for these distributions.

We therefore need a stronger correctness guarantee. We need that, for any two *ciphertexts*, the comparison procedure is consistent with decrypting the two ciphertexts and comparing the resulting plaintexts. This correctness guarantee is meaningful even for improperly generated ciphertexts.

We note that none of the existing constructions of order-revealing encryption outlined above satisfy this stronger notion. For the obfuscation-based schemes, ciphertexts consist of obfuscated programs. In these schemes, it is easy to describe invalid ciphertexts where the obfuscated program performs incorrectly, causing the comparison procedure to output the wrong result. In the multilinear map-based schemes, the underlying instantiation use current "noisy" multilinear maps, such as [26]. An invalid ciphertext could, for example, have too much noise, which will cause the comparison procedure to behave unpredictably.

Obtaining Strong Correctness. We first argue that, for all existing ORE schemes, the scheme can be modified so that Comp is correct for all *valid* ciphertexts. We then give a generic conversion from any ORE scheme with weakly correct comparison, including the tweaked existing schemes, into a strongly correct scheme. We simply modify the ciphertext by adding a non-interactive zero-knowledge (NIZK) proof that the ciphertext is well-formed, with the common reference string added to the public comparison key. Then the decryption and comparison procedures check the proof(s), and only output the result (either decryption or comparison) if the proof(s) are valid. The (computational) zero-knowledge property of the NIZK implies that the addition of the proof to the ciphertext does not affect security. Meanwhile, NIZK soundness implies that any ciphertext accepted by the decryption and comparison procedures must be valid, and the weak correctness property of the underlying ORE implies that for valid ciphertexts, decryption and comparison are consistent. The result is that comparisons are consistent with decryption *for all* ciphertexts, giving strong correctness.

As we need strong correctness for every ciphertext, even hard-to-generate ones, we need the NIZK proofs to have perfect soundness, as opposed to computational soundness. Such NIZK proofs were built in [32].

We note also that the conversion outlined above is not specific to ORE, and applies more generally to functional encryption schemes.

1.4   Related Work

Hardness of Private Query Release. One of the most basic and well-studied statistical tasks in differential privacy is the problem of releasing answers to *counting queries*. A counting query asks, "what fraction of the records in a dataset D satisfy the predicate q?". Given a collection of k counting queries $q_1, \ldots, q_k$ from a family $\mathcal{Q}$, the goal of a query release algorithm is to release approximate answers to these queries while preserving differential privacy. A remarkable result of Blum et al. [6], with subsequent improvements by [21,23,33–35,48], showed that an arbitrary sequence of counting queries can be answered accurately with differential privacy even when k is exponential in the dataset size n. Unfortunately, all of these algorithms that are capable of answering more than n^2 queries are inefficient, running in time exponential in the dimensionality of the data. Moreover, several works [10,21,52] have gone on to show that this inefficiency is likely inherent.

These computational lower bounds for private query release rely on a connection between the hardness of private query release and *traitor-tracing schemes*, which was first observed by Dwork et al. [21]. Traitor-tracing schemes were introduced by Chor, Fiat, and Naor [18] to help digital content producers identify pirates as they illegally redistribute content. Traitor-tracing schemes are conceptually analogous to the example reidentification scheme we use to obtain our hardness result for private learning. Instantiating this connection with the traitor-tracing scheme of Boneh, Sahai, and Waters [9], which relies on certain assumptions in bilinear groups, Dwork et al. [21] exhibited a family of $2^{\tilde{O}(\sqrt{n})}$ queries for which no efficient algorithm can produce a data structure which could be used to answer all queries in this family. Very recently, Boneh and Zhandry [10] constructed a new traitor-tracing scheme based on indistinguishability obfuscation that yields the same infeasibility result for a family of $n \cdot 2^{O(d)}$ queries on records of size d. Extending this connection, Ullman [52] constructed a specialized traitor-tracing scheme to show that no efficient private algorithm can answer more than $\tilde{O}(n^2)$ arbitrary queries that are given as input to the algorithm.

Dwork et al. [21] also showed strong lower bounds against private algorithms for producing *synthetic data*. Synthetic data generation algorithms produce a new "fake" dataset, whose rows are of the same type as those in the original dataset, with the promise that the answers to some restricted set of queries on the synthetic dataset well-approximate the answers on the original dataset. Assuming the existence of one-way functions, Dwork et al. [21] exhibited an efficiently computable collection of queries for which no efficient private algorithm can produce useful synthetic data. Ullman and Vadhan [53] refined this result to hold even for extremely simple classes of queries.

Nevertheless, the restriction to synthetic data is significant to these results, and they do not rule out the possibility that other privacy-preserving data structures can be used to answer large families of restricted queries. In fact, when the synthetic data restriction is lifted, there are algorithms (e.g. [15,22,36,51]) that answer queries from certain exponentially large families in subexponential time. One can view the problem of synthetic data generation as analogous to proper learning. In both cases, placing natural syntactic restrictions on the output of an algorithm may in fact come at the expense of utility or computational efficiency.

Efficiency of SQ Learning. Feldman and Kanade [24] addressed the question of whether information-theoretically efficient SQ learners – i.e., those making polynomially many queries – could be made computationally efficient. One of their main negative results showed that unless **NP = RP**, there exists a concept class with polynomial query complexity that is not efficiently SQ learnable. Moreover, this concept class is efficiently PAC learnable, which suggests that the restriction to SQ learning can introduce an inherent computational cost.

We show that the concept class EncThresh can be learned (inefficiently) with polynomially many statistical queries. The result of Blum et al. [5] discussed above, showing that SQ learning algorithms can be efficiently simulated by differentially private algorithms, thus shows that EncThresh also separates SQ learners making polynomially many queries from computationally efficient SQ learners.

Corollary 1 (Informal). *Assuming the existence of strongly correct ORE, the concept class EncThresh is efficiently PAC learnable and has polynomial SQ query complexity, but is not efficiently SQ learnable.*

While our proof relies on much stronger hardness assumptions, it reveals ORE as a new barrier to efficient SQ learning. As discussed in more detail in Sect. 3.3, even though their result is about computational hardness, Feldman and Kanade's choice of a concept class relies crucially on the fact that parities are hard to learn in the SQ model even information-theoretically. By contrast, our concept class EncThresh is computationally hard to SQ learn for a reason that appears fundamentally different than the information-theoretic hardness of SQ learning parities.

Learning from Encrypted Data. Several works have developed schemes for training, testing, and classifying machine learning models over encrypted data (e.g. [11,31]). In a model use case, a client holds a sensitive dataset, and uploads an encrypted version of the dataset to a cloud computing service. The cloud service then trains a model over the encrypted data and produces an encrypted classifier it can send back to the client, ideally without learning anything about the examples it received. The notion of privacy afforded to the individuals in the dataset here is complementary to differential privacy. While the cloud service does not learn anything about the individuals in the dataset, its output might still depend heavily on the data of certain individuals.

In fact, our non-differentially private PAC learner for the class EncThresh exactly performs the task of learning over encrypted data, producing a classifier without learning anything about its examples beyond their order (this addresses the difficulty of implementing comparisons from prior work [31]). Thus one can interpret our results as showing that not only are these two notions of privacy for machine learning training complementary, but that they may actually be in conflict. Moreover, the strong correctness guarantee we provide for ORE (which applies more generally to multi-input functional encryption) may help enable the theoretical study of learning from encrypted data in other PAC-style settings.

2   Preliminaries and Definitions

2.1   PAC Learning and Private PAC Learning

For each $k \in \mathbb{N}$, let X_k be an instance space (such as $\{0,1\}^k$), where the parameter k represents the size of the elements in X_k. Let C_k be a set of boolean functions $\{c : X_k \to \{0,1\}\}$. The sequence $(X_1, C_1), (X_2, C_2), \ldots$ represents an infinite sequence of learning problems defined over instance spaces of increasing dimension. We will generally suppress the parameter k, and refer to the problem of learning C as the problem of learning C_k for every k.

A learner L is given examples sampled from an unknown probability distribution $\mathcal{D}$ over X, where the examples are labeled according to an unknown *target concept* $c \in C$. The learner must select a hypothesis h from a hypothesis class $\mathcal{H}$ that approximates the target concept with respect to the distribution $\mathcal{D}$. More precisely,

Definition 1. *The* generalization error *of a hypothesis* $h : X \to \{0,1\}$ *(with respect to a target concept* c *and distribution* $\mathcal{D}$*) is defined by* $\mathrm{error}_{\mathcal{D}}(c,h) = \Pr_{x \sim \mathcal{D}}[h(x) \neq c(x)]$*. If* $\mathrm{error}_{\mathcal{D}}(c,h) \leq \alpha$ *we say that* h *is an* α-good *hypothesis for* c *on* $\mathcal{D}$*.*

Definition 2 (PAC Learning [54]). *Algorithm* $L : (X \times \{0,1\})^n \to \mathcal{H}$ *is an* (α, β)-accurate PAC learner *for the concept class* C *using hypothesis class* $\mathcal{H}$ *with sample complexity* n *if for all target concepts* $c \in C$ *and all distributions* $\mathcal{D}$ *on* X, *given an input of* n *samples* $S = ((x_1, c(x_1)), \ldots, (x_n, c(x_n)))$, *where each* x_i *is drawn i.i.d. from* $\mathcal{D}$, *algorithm* L *outputs a hypothesis* $h \in \mathcal{H}$ *satisfying* $\Pr[\mathrm{error}_{\mathcal{D}}(c,h) \leq \alpha] \geq 1 - \beta$. *The probability here is taken over the random choice of the examples in* S *and the coin tosses of the learner* L.

The learner L *is* efficient *if it runs in time polynomial in the size parameter* k, *the representation size of the target concept* c, *and the accuracy parameters* $1/\alpha$ *and* $1/\beta$. *Note that a necessary (but not sufficient) condition for* L *to be efficient is that its sample complexity* n *is polynomial in the learning parameters.*

If $\mathcal{H} \subseteq \mathcal{C}$ then L is called a proper *learner*. *Otherwise, it is called an* improper *learner.*

Kasiviswanathan et al. [37] defined a *private learner* as a PAC learner that is also differentially private. Recall the definition of differential privacy:

Definition 3. *A learner* $L : (X \times \{0,1\})^n \to \mathcal{H}$ *is* (ε, δ)-differentially private *if for all sets* $T \subseteq \mathcal{H}$*, and neighboring sets of examples* $S \sim S'$*,*

$$\Pr[L(S) \in T] \leq e^\varepsilon \Pr[L(S') \in T] + \delta.$$

The technical object that we will use to show our hardness results for differential privacy is what we call an *example reidentification scheme*. It is analogous to the hard-to-sanitize database distributions [21,53] and re-identifiable database distributions [14] used in prior works to prove hardness results for private query release, but is adapted to the setting of computational learning. In the first step, an algorithm $\mathsf{Gen}_{\mathsf{ex}}$ chooses a concept and a sample S labeled according to that concept. In the second step, a learner L receives either the sample S or the sample S_{-i} where an appropriately chosen example i is replaced by a junk example, and learns a hypothesis h. Finally, an algorithm $\mathsf{Trace}_{\mathsf{ex}}$ attempts to use h to identify one of the rows given to L. If $\mathsf{Trace}_{\mathsf{ex}}$ succeeds at identifying such a row with high probability, then it must be able to distinguish $L(S)$ from $L(S_{-i})$, showing that L cannot be differentially private. We formalize these ideas below.

Definition 4. *An* (α, ξ)-example reidentification scheme *for a concept class* $\mathcal{C}$ *consists of a pair of algorithms,* $(\mathsf{Gen}_{\mathsf{ex}}, \mathsf{Trace}_{\mathsf{ex}})$ *with the following properties.*

$\mathsf{Gen}_{\mathsf{ex}}(k, n)$ *Samples a concept* $c \in \mathcal{C}_k$ *and an associated distribution* $\mathcal{D}$*. Draws i.i.d. examples* $x_1, \ldots, x_n \leftarrow_R \mathcal{D}$*, and a fixed value* x_0*. Let* S *denote the labeled sample* $((x_1, c(x_1)), \ldots, (x_n, c(x_n)))$*, and for any index* $i \in [n]$*, let* S_{-i} *denote the sample with the pair* $(x_i, c(x_i))$ *replaced with* $(x_0, c(x_0))$*.*

$\mathsf{Trace}_{\mathsf{ex}}(h)$ *Takes state shared with* $\mathsf{Gen}_{\mathsf{ex}}$ *as well as a hypothesis* h *and identifies an index in* $[n]$ *(or* $\perp$ *if none is found).*

The scheme obeys the following "completeness" and "soundness" criteria on the ability of $\mathsf{Trace}_{\mathsf{ex}}$ *to identify an example given to a learner* L*.*

Completeness. A good hypothesis can be traced to some example. That is, for every efficient learner L*,*

$$\Pr[\mathrm{error}_{\mathcal{D}}(c, h) \leq \alpha \wedge \mathsf{Trace}_{\mathsf{ex}}(h) = \perp] \leq \xi.$$

Here, the probability is taken over $h \leftarrow_R L(S)$ *and the coins of* $\mathsf{Gen}_{\mathsf{ex}}$ *and* $\mathsf{Trace}_{\mathsf{ex}}$*.*

Soundness. For every efficient learner L*,* $\mathsf{Trace}_{\mathsf{ex}}$ *cannot trace* i *from the sample* S_{-i}*. That is, for all* $i \in [n]$*,*

$$\Pr[\mathsf{Trace}_{\mathsf{ex}}(h) = i] \leq \xi$$

for $h \leftarrow_R L(S_{-i})$*.*

We may sometimes relax the completeness condition to hold only under certain restrictions on L's output (e.g. L is a proper learner or L is a representation learner). In this case, we say the $(\mathsf{Gen_{ex}}, \mathsf{Trace_{ex}})$ is an example reidentification scheme for (properly, representation) learning a class C.

Theorem 2. *Let* $(\mathsf{Gen_{ex}}, \mathsf{Trace_{ex}})$ *be an* (α, ξ)-*example reidentification scheme for a concept class* C. *Then for every* $\beta > 0$ *and polynomial* $n(k)$, *there is no efficient* (ε, δ)-*differentially private* (α, β)-*PAC learner for* C *using* n *samples when*

$$\delta < \left(\frac{1 - \beta - \xi}{n}\right) - e^\varepsilon \xi.$$

In a typical setting of parameters, we will take $\alpha, \beta, \varepsilon = O(1)$ and $\delta, \xi = o(1/n)$, in which case the inequality in Theorem 2 will be satisfied for sufficiently large n.

Proof. Suppose instead there were a computationally efficient (ε, δ)-differentially private (α, β)-PAC learner L for C using n samples. Then there exists an $i \in [n]$ such that $\Pr[\mathsf{Trace_{ex}}(L(S)) = i] \geq (1 - \beta - \xi)/n$. However, since L is differentially private,

$$\Pr[\mathsf{Trace_{ex}}(L(S_{-i})) = i] \geq e^{-\varepsilon} \left(\frac{1 - \beta - \xi}{n} - \delta\right) > \xi(n),$$

which contradicts the soundness of $(\mathsf{Gen_{ex}}, \mathsf{Trace_{ex}})$.

2.2 Order-Revealing Encryption

Definition 5. *An Order-Revealing Encryption (ORE) scheme is a tuple* $(\mathsf{Gen}, \mathsf{Enc}, \mathsf{Dec}, \mathsf{Comp})$ *of algorithms where:*

- $\mathsf{Gen}(1^\lambda, 1^\ell)$ *is a randomized procedure that takes as inputs a security parameter* λ *and plaintext length* ℓ, *and outputs a secret encryption/decryption key* sk *and public parameters* pars.
- $\mathsf{Enc}(\mathsf{sk}, m)$ *is a potentially randomized procedure that takes as input a secret key* sk *and a message* $m \in \{0,1\}^\ell$, *and outputs a ciphertext* c.
- $\mathsf{Dec}(\mathsf{sk}, c)$ *is a deterministic procedure that takes as input a secret key* sk *and a ciphertext* c, *and outputs a plaintext message* $m \in \{0,1\}^\ell$ *or a special symbol* $\perp$.
- $\mathsf{Comp}(\mathsf{pars}, c_0, c_1)$ *is a deterministic procedure that "compares" two ciphertexts, outputting either ">", "<", "=", or* $\perp$.

Correctness. An ORE scheme must satisfy two separate correctness requirements:

- **Correct Decryption:** This is the standard notion of correctness for an encryption scheme, which says that decryption succeeds. We will only consider *strongly* correct decryption, which requires that decryption *always* succeeds. For all security parameters λ and message lengths ℓ,

$$\Pr[\mathsf{Dec}(\mathsf{sk}, \mathsf{Enc}(\mathsf{sk}, m)) = m : (\mathsf{sk}, \mathsf{pars}) \leftarrow \mathsf{Gen}(1^\lambda, 1^\ell)] = 1.$$

- **Correct Comparison:** We require that the comparison function succeeds. We will consider two notions, namely *strong* and *weak*. In order to define these notions, we first define two auxiliary functions:
  - $\mathsf{Comp}_{plain}(m_0, m_1)$ is just the plaintext comparison function. That is, for $m_0 < m_1$, $\mathsf{Comp}_{plain}(m_0, m_1) = $ " $<$ ", $\mathsf{Comp}_{plain}(m_1, m_0) = $ " $>$ ", and $\mathsf{Comp}_{plain}(m_0, m_0) = $ " $=$ ".
  - $\mathsf{Comp}_{ciph}(\mathsf{sk}, c_0, c_1)$ is a ciphertext comparison function which uses the secret key. If first computes $m_b = \mathsf{Dec}(\mathsf{sk}, c_b)$ for $b = 0, 1$. If either $m_0 = \perp$ or $m_1 = \perp$ (in other words, if either decryption failed), then Comp_{ciph} outputs $\perp$. If both $m_0, m_1 \neq \perp$, then the output is $\mathsf{Comp}_{plain}(m_0, m_1)$.

Now we define our comparison correctness notions:

- **Weakly Correct Comparison:** This informally requires that comparison is consistent with encryption. For all security parameters λ, message lengths ℓ, and messages $m_0, m_1 \in \{0, 1\}^\ell$,

$$\Pr\left[\begin{array}{l} \mathsf{Comp}(\mathsf{pars}, c_0, c_1) \\ = \mathsf{Comp}_{plain}(m_0, m_1) \end{array} : \begin{array}{l} (\mathsf{sk}, \mathsf{pars}) \leftarrow \mathsf{Gen}(1^\lambda, 1^\ell) \\ c_b \leftarrow \mathsf{Enc}(\mathsf{sk}, m_b) \end{array} \right] = 1.$$

In particular, for correctly generated ciphertexts, Comp never outputs $\perp$.

- **Strongly Correct Comparison:** This informally requires that comparison is consistent with *decryption*. For all security parameters λ, message lengths ℓ, and ciphertexts c_0, c_1,

$$\Pr\left[\begin{array}{l} \mathsf{Comp}(\mathsf{pars}, c_0, c_1) \\ = \mathsf{Comp}_{ciph}(\mathsf{sk}, c_0, c_1) \end{array} : (\mathsf{sk}, \mathsf{pars}) \leftarrow \mathsf{Gen}(1^\lambda, 1^\ell) \right] = 1.$$

Security. For security, we will consider a relaxation of the "best possible" security notion of Boneh et al. [8]. Namely, we only consider static adversaries that submit all queries at once. "Best possible" security is a modification of the standard notion of CPA security for symmetric key encryption to block trivial attacks. That is, since the comparison function always leaks the order of the plaintexts, the left and right sets of challenge messages must have the same order. In our relaxation where all challenge messages are queried at once, we can therefore assume without loss of generality that the left and right sequences of messages are sorted in ascending order. For simplicity, we will also disallow the adversary from querying on the same message more than once. This gives the following definition:

Definition 6. *An ORE scheme* $(\mathsf{Gen}, \mathsf{Enc}, \mathsf{Dec}, \mathsf{Comp})$ *is statically secure if, for all efficient adversaries* $\mathcal{A}$, $|\Pr[W_0] - \Pr[W_1]|$ *is negligible, where* W_b *is the event that* $\mathcal{A}$ *outputs 1 in the following experiment:*

- $\mathcal{A}$ *produces two message sequences* $m_1^{(L)} < m_2^{(L)} < \cdots < m_q^{(L)}$ *and* $m_1^{(R)} < m_2^{(R)} < \cdots < m_q^{(R)}$

– The challenger runs $(\mathsf{sk}, \mathsf{pars}) \leftarrow \mathsf{Gen}(1^\lambda, 1^\ell)$. It then responds to $\mathcal{A}$ with pars, as well as $c_1, \ldots, c_q$ where

$$c_i = \begin{cases} \mathsf{Enc}(\mathsf{sk}, m_i^{(L)}) & \text{if } b = 0 \\ \mathsf{Enc}(\mathsf{sk}, m_i^{(R)}) & \text{if } b = 1 \end{cases}$$

– $\mathcal{A}$ outputs a guess b' for b.

We also consider a weaker definition, which only allows the sequences $m_i^{(L)}$ and $m_i^{(R)}$ to differ at a single point:

Definition 7. An ORE scheme $(\mathsf{Gen}, \mathsf{Enc}, \mathsf{Dec}, \mathsf{Comp})$ is statically single-challenge secure if, for all efficient adversaries $\mathcal{A}$, $|\Pr[W_0] - \Pr[W_1]|$ is negligible, where W_b is the event that $\mathcal{A}$ outputs 1 in the following experiment:

– $\mathcal{A}$ produces a sequence of messages $m_1 < m_2 < \cdots < m_q$, and challenge messages m_L, m_R such that $m_i < m_L < m_R < m_{i+1}$ for some $i \in [q-1]$.
– The challenger runs $(\mathsf{sk}, \mathsf{pars}) \leftarrow \mathsf{Gen}(1^\lambda, 1^\ell)$. It then responds to $\mathcal{A}$ with pars, as well as $c_1, \ldots, c_q$ where $c_i = \mathsf{Enc}(\mathsf{sk}, m_i)$ and

$$c^* = \begin{cases} \mathsf{Enc}(\mathsf{sk}, m_L) & \text{if } b = 0 \\ \mathsf{Enc}(\mathsf{sk}, m_R) & \text{if } b = 1 \end{cases}$$

– $\mathcal{A}$ outputs a guess b' for b.

We now argue that these two definitions are equivalent up to some polynomial loss in security.

Theorem 3. $(\mathsf{Gen}, \mathsf{Enc}, \mathsf{Dec}, \mathsf{Comp})$ is statically secure if and only if it is statically single-challenge secure.

Proof. We prove that single-challenge security implies many-challenge security through a sequence of hybrids. Each hybrid will only differ in the messages m_i that are encrypted, and each adjacent hybrid will only differ in a single message. The first hybrid will encrypt $m_i^{(L)}$, and the last hybrid will encrypt $m_i^{(R)}$. Thus, by applying the single-challenge security for each hybrid, we conclude that the first and last hybrids are indistinguishable, thus showing many-challenge security.

Hybrid j for $j \leq q$.

$$m_i = \begin{cases} \min(m_i^{(L)}, m_i^{(R)}) & \text{if } i \leq j \\ m_i^{(L)} & \text{if } i > j \end{cases}$$

First, notice that all the m_i are in order since both sequences $m_i^{(L)}$ and $m_i^{(R)}$ are in order. Second, the only difference between **Hybrid** $(j-1)$ and **Hybrid** j is that $m_j = m_j^{(L)}$ in **Hybrid** $(j-1)$ and $m_j = \min(m_j^{(L)}, m_j^{(R)})$ in **Hybrid** j. Thus, single-challenge security implies that each adjacent hybrid is indistinguishable. Moreover, for j where $m_j^{(L)} < m_j^{(R)}$, the two hybrids are actually identical.

Hybrid j for $j > q$.

$$m_i = \begin{cases} \min(m_i^{(L)}, m_i^{(R)}) & \text{if } i \leq 2q - j \\ m_i^{(R)} & \text{if } i > 2q - j \end{cases}$$

Again, notice that all the m_i are in order. Moreover, the only different between **Hybrid** $(2q - j)$ and **Hybrid** $(2q - j + 1)$ is that $m_j = \min(m_j^{(L)}, m_j^{(R)})$ in **Hybrid** $(2q - j)$ and $m_j = m_j^{(R)}$ in **Hybrid** $(2q - j + 1)$. Thus, single-challenge security implies that each adjacent hybrid is indistinguishable. Moreover, for j where $m_j^{(L)} > m_j^{(R)}$, the two hybrids are actually identical.

3 The Concept Class EncThresh and Its Learnability

Let (Gen, Enc, Dec, Comp) be a statically secure ORE scheme with strongly correct comparison. We define a concept class EncThresh, which intuitively captures the class of threshold functions where examples are encrypted under the ORE scheme. Throughout this discussion, we will take $N = 2^\ell$ and regard the plaintext space of the ORE scheme to be $[N] = \{1, \ldots, N\}$. Ideally we would like, for each threshold $t \in [N + 1]$ and each (sk, pars) $\leftarrow$ Gen(1^λ), to define a concept

$$f_{t,\text{sk},\text{pars}}(c) = \begin{cases} 1 & \text{if } \text{Dec}_{\text{sk}}(c) < t \\ 0 & \text{otherwise.} \end{cases}$$

However, we need to make a few technical modifications to ensure that EncThresh is efficiently PAC learnable.

1. In order for the learner to be able to use the comparison function Comp, it must be given the public parameters pars generated by the ORE scheme. We address this in the natural way by attaching a set of public parameters to each example. Moreover, we define EncThresh so that each concept is supported on the single set of public parameters that corresponds to the secret key used for encryption and decryption.
2. Only a subset of binary strings form valid (sk, pars) pairs that are actually produced by Gen in the ORE scheme. To represent concepts, we need a reasonable encoding scheme for these valid pairs. The encoding scheme we choose is the polynomial-length sequence of random coin tosses used by the algorithm Gen to produce (sk, pars).

We now formally describe the concept class EncThresh. Each concept is parameterized by a string r, representing the coin tosses of the algorithm Gen, and a threshold $t \in [N + 1]$ for $N = 2^\ell$. In what follows, let (sk^r, pars^r) be the keys output by Gen($1^\lambda, 1^\ell$) when run on the sequence of coin tosses r. Let

$$f_{t,r}(\text{pars}, c) = \begin{cases} 1 & \text{if } (\text{pars} = \text{pars}^r) \wedge (\text{Dec}(\text{sk}^r, c) \neq \perp) \wedge (\text{Dec}(\text{sk}^r, c) < t) \\ 0 & \text{otherwise.} \end{cases}$$

Notice that given t and r, the concept $f_{t,r}$ can be efficiently evaluated. The description length k of the instance space $X_k = \{0, 1\}^k$ is polynomial in the security parameter λ and plaintext length ℓ.

3.1 An Efficient PAC Learner for EncThresh

We argue that EncThresh is efficiently PAC learnable by formalizing the argument from the introduction. Because we need to include the ORE public parameters in each example, the PAC learner L (Algorithm 1) for EncThresh actually works in two stages. In the first stage, L determines whether there is significant probability mass on examples corresponding to some public parameters pars. Recall that each concept in EncThresh is supported on exactly one such set of parameters. If there is no significant mass on any pars, then the all-zeroes hypothesis is a good hypothesis. On the other hand, if there is a heavy set of parameters, the learner L applies Comp using those parameters to learn a good comparator.

Theorem 4. *Let $\alpha, \beta > 0$. There exists a PAC learning algorithm L for the concept class EncThresh achieving error α and confidence $1 - \beta$. Moreover, L is efficient (running in time polynomial in the parameters $k, 1/\alpha, \log(1/\beta)$).*

Algorithm 1. Learner L for EncThresh

---

1. Request examples $\{(\mathsf{pars}_1, c_1, b_1), \ldots, (\mathsf{pars}_n, c_n, b_n)\}$ for $n = \lceil \log(1/\beta)/\alpha \rceil$.
2. Identify an i for which $b_i = 1$ and set $\mathsf{pars}^* = \mathsf{pars}_i$; if no such i exists, return $h \equiv 0$.
3. Let $G = \{j : \mathsf{pars}_j = \mathsf{pars}^*, b_j = 1\}$. Let $j^* \in G$ be an index with $\mathsf{Comp}(\mathsf{pars}^*, c_j, c_{j^*}) \in \{<, =, \perp\}$ for all $j \in G$.
4. Return h defined by

$$h(\mathsf{pars}, c) = \begin{cases} 1 & \text{if } (\mathsf{pars} = \mathsf{pars}^*) \wedge (\mathsf{Comp}(\mathsf{pars}^*, c, c_{j^*}) \in \{<, =\}) \\ 0 & \text{otherwise.} \end{cases}$$

Proof. Fix a target concept $f_{t,r} \in \mathsf{EncThresh}_k$ and a distribution $\mathcal{D}$ on examples. First observe that the learner L always outputs a hypothesis with one-sided error, i.e. we always have $h \leq f_{t,r}$ pointwise. Also observe that $f_{t',r} \leq f_{t,r}$ pointwise for any $t' < t$. These both follow from the strong correctness of the ORE scheme. Let $(\mathsf{sk}^r, \mathsf{pars}^r)$ denote the keys output by $\mathsf{Gen}(1^\lambda, 1^\ell)$ when run on the sequence of coin tosses r. Let POS denote the set of examples (pars, c) on which $f_{t,r}(\mathsf{pars}, c) = 1$. We divide the analysis of the learner in to two cases based on the weight $\mathcal{D}$ places on POS.

Case 1: $\mathcal{D}$ places weight at least α on POS. Define $\hat{t} \in [N + 1]$ as the largest $\hat{t} \leq t$ such that $\mathrm{error}_{\mathcal{D}}(f_{\hat{t},r}, f_{t,r}) \geq \alpha$. Such a $\hat{t}$ is guaranteed to exist since $f_{0,r}$ is the all-zeros function, and therefore $\mathrm{error}_{\mathcal{D}}(f_{0,r}, f_{t,r})$ is equal to the weight $\mathcal{D}$ places on POS, which is at least α.

Suppose $f_{\hat{t}+1,r} \leq h$ pointwise. Since h has one-sided error (that is, $h \leq f_{t,r}$ pointwise), we have $\mathrm{error}_{\mathcal{D}}(f_{\hat{t}+1,r}, f_{t,r}) = \mathrm{error}_{\mathcal{D}}(f_{\hat{t}+1,r}, h) + \mathrm{error}_{\mathcal{D}}(h, f_{t,r})$, or

$$\mathrm{error}_{\mathcal{D}}(h, f_{t,r}) = \mathrm{error}_{\mathcal{D}}(f_{\hat{t}+1,r}, f_{t,r}) - \mathrm{error}_{\mathcal{D}}(f_{\hat{t}+1,r}, h) \leq \mathrm{error}_{\mathcal{D}}(f_{\hat{t}+1,r}, f_{t,r}) < \alpha.$$

Therefore, it suffices to show that $f_{\hat{t}+1,r} \leq h$ with probability at least $1 - \beta$. This is guaranteed as long as L receives a sample $(\mathsf{pars}^r, c_i, 1)$ with $\hat{t} \leq \mathsf{Dec}(\mathsf{sk}^r, c_i) < t$. In other words, $f_{t,r}(\mathsf{pars}^r, c_i) = 1$ and $f_{\hat{t},r}(\mathsf{pars}^r, c_i) = 0$. Since $f_{\hat{t},r} \leq f_{t,r}$ pointwise, such samples exactly account for the error between $f_{\hat{t},r}$ and $f_{t,r}$. Thus since $\mathrm{error}_{\mathcal{D}}(f_{\hat{t},r}, f_{t,r}) \geq \alpha$, for each i it must be that $\hat{t} \leq \mathsf{Dec}(\mathsf{sk}^r, c_i) < t$ with probability at least α. The learner L therefore receives *some* sample c_i with $\hat{t} \leq \mathsf{Dec}(\mathsf{sk}^r, c_i) < t$ with probability at least $1 - (1 - \alpha)^n \geq 1 - \beta$ (since we took $n \geq \log(1/\beta)/\alpha$).

Case 2: $\mathcal{D}$ places less than α weight on POS. Then the identically zero hypothesis has error at most α, so the claim holds because $0 \leq h \leq f_{t,r}$.

3.2   Hardness of Privately Learning EncThresh

We now prove the hardness of privately learning EncThresh by constructing an example reidentification scheme for this concept class. Recall that an example reidentification scheme consists of two algorithms, $\mathsf{Gen}_{\mathsf{ex}}$, which selects a distribution, a concept, and examples to give to a learner, and $\mathsf{Trace}_{\mathsf{ex}}$ which attempts to identify one of the examples the learner received.

Our example reidentification scheme yields a hard distribution even for *weak-learning*, where the error parameter α is taken to be inverse-polynomially close to $1/2$.

Theorem 5. *Let $\gamma(n)$ and $\xi(n)$ be noticeable functions. Let $(\mathsf{Gen}, \mathsf{Enc}, \mathsf{Dec}, \mathsf{Comp})$ be a statically single-challenge secure ORE scheme. Then there exists an (efficient) $(\alpha = \frac{1}{2} - \gamma, \xi)$-example reidentification scheme $(\mathsf{Gen}_{\mathsf{ex}}, \mathsf{Trace}_{\mathsf{ex}})$ for the concept class* EncThresh.

We start with an informal description of the scheme $(\mathsf{Gen}_{\mathsf{ex}}, \mathsf{Trace}_{\mathsf{ex}})$. The algorithm $\mathsf{Gen}_{\mathsf{ex}}$ sets up the parameters of the ORE scheme, chooses the "middle" threshold concept corresponding to $t = N/2$, and sets the distribution on examples to be encryptions of uniformly random messages (together with the correct public parameters needed for comparison). Let $m_1 < m_2 < \cdots < m_n$ denote the sorted sequence of messages whose encryptions make up the sample produced by $\mathsf{Gen}_{\mathsf{ex}}$ (with overwhelming probability, they are indeed distinct). We can thus break the plaintext space up into buckets of the form $B_i = [m_i, m_{i+1})$. Suppose L is a (weak) learner that produces a hypothesis h with advantage γ over random guessing. Such a hypothesis h must be able to distinguish encryptions of messages $m \leq t$ from encryptions of messages $m > t$ with advantage γ. Thus, there must be a pair of adjacent buckets B_{i-1}, B_i for which h can distinguish encryptions of messages from B_{i-1} from encryptions from B_i with advantage $\frac{\gamma}{n}$.

This observation leads to a natural definition for $\mathsf{Trace}_{\mathsf{ex}}$: locate a pair of adjacent buckets B_{i-1}, B_i that h distinguishes, and output the identity i of the example separating those buckets. Completeness of the resulting scheme, i.e. the fact that some example is reidentified when L succeeds, follows immediately from the preceding discussion. We argue soundness, i.e. that an example absent from L's sample is not identified, by reducing to the static security of the ORE scheme. The intuition is that if L is not given example i, then it should not be able to distinguish encryptions from bucket B_{i-1} from encryptions from bucket B_i.

To make the security reduction somewhat more precise, suppose for the sake of contradiction that there is an efficient algorithm L that violates the soundness of $(\mathsf{Gen}_{\mathsf{ex}}, \mathsf{Trace}_{\mathsf{ex}})$ with noticeable probability ξ. That is, there is some i such that even without example i, the algorithm L manages to produce (with probability ξ) a hypothesis h that distinguishes B_{i-1} from B_i. A natural first attempt to violate the security of the ORE is to construct an adversary that challenges on the message sequences $m_1 < \cdots < m_{i-1} < m_i^{(L)} < m_{i+1}, <, m_n$ and $m_1 < \cdots < m_{i-1} < m_i^{(R)} < m_{i+1} < \cdots < m_n$, where $m_i^{(L)}$ is randomly chosen from B_{i-1} and $m_i^{(R)}$ is randomly chosen from B_i. Then if h can distinguish B_{i-1} from B_i, the adversary can distinguish the two sequences. Unfortunately, this approach fails for a somewhat subtle reason. The hypothesis h is only guaranteed to distinguish B_{i-1} from B_i *with probability* ξ. If h fails to distinguish the buckets – or distinguishes them in the opposite direction – then the adversary's advantage is lost.

To overcome this issue, we instead rely on the security of the ORE for sequences that differ on *two* messages. For the "left" challenge, our adversary samples two messages from the same randomly chosen bucket, B_{i-1} or B_i (in addition to requesting encryptions of $m_1, \ldots, m_{i-1}, m_i, \ldots, m_n$). For the "right" challenge, it samples one message from each bucket B_{i-1} and B_i. Let c^0 and c^1 be the ciphertexts corresponding to thee challenge messages. If h agrees on c^0 and c^1, then this suggests the messages are from the same bucket, and the adversary should guess "left". On the other hand, if h disagrees on c^0 and c^1, then the adversary should guess "right". If h distinguishes the buckets B_{i-1} and B_i, this adversary does strictly better than random guessing. On the other hand, even if h fails to distinguish the buckets, the adversary does at least as well as random guessing. So overall, it still has a noticeable advantage at the ORE security game.

We now give the formal proof of Theorem 5.

Proof. We construct an example reidentification scheme for EncThresh as follows. The algorithm $\mathsf{Gen}_{\mathsf{ex}}$ fixes the threshold $t = N/2$ and samples $(\mathsf{sk}^r, \mathsf{pars}^r) \leftarrow_{\mathsf{R}} \mathsf{Gen}(1^\lambda, 1^\ell)$, yielding a concept $f_{t,r}$. Let $\mathcal{D}$ be the distribution $(\mathsf{pars}^r, \mathsf{Enc}(\mathsf{sk}^r, m))$ for uniformly random $m \in [N]$. Let $m_1', \ldots, m_n' \leftarrow_{\mathsf{R}} [N]$, and let $m_1 \leq \cdots \leq m_n$ be the result of sorting the m_i'. Let $m_0 = 0$ and $m_{n+1} = N$. Since $n = \mathrm{poly}(k) \ll N$, these random messages will be well-spaced. In particular, with overwhelming probability, $|m_{i+1} - m_i| > 1$ for every i, so we assume this is the case in what follows. $\mathsf{Gen}_{\mathsf{ex}}$ then sets the samples to be $(x_1 = (\mathsf{pars}^r, \mathsf{Enc}(\mathsf{sk}^r, m_1')), \ldots, x_n = (\mathsf{pars}^r, \mathsf{Enc}(\mathsf{sk}^r, m_n')))$. Let $x_0 = (\mathsf{pars}^r, \mathsf{Enc}(\mathsf{sk}^r, m_0))$ be a "junk" example.

The algorithm $\mathsf{Trace}_{\mathsf{ex}}$ creates buckets $B_i = [m_i, m_{i+1}]$. For each i, let

$$p_i = \Pr_{m \in B_i, \text{coins of Enc}} [h(\mathsf{pars}^r, \mathsf{Enc}(\mathsf{sk}, m)) = 1].$$

By sampling random choices of m in each bucket, $\mathsf{Trace}_{\mathsf{ex}}$ can efficiently compute a good estimate $\hat{p}_i \approx p_i$ for each i (Lemma 1). It then accuses the least i for which $\hat{p}_{i-1} - \hat{p}_i \geq \frac{\gamma}{n}$, and $\perp$ if none is found.

Lemma 1. *Let* $K = \frac{8n^2}{\gamma^2} \log(9n/\xi)$. *For each* $i = 0, \dots, n$, *let*

$$\hat{p}_i = \frac{1}{K} \sum_{j=1}^{K} h(x_j)$$

where $x_j = (\mathsf{pars}^r, \mathsf{Enc}(\mathsf{sk}^r, m_j))$ *for i.i.d.* $m_1, \dots, m_K \leftarrow_R B_i$. *Then* $|\hat{p}_i - p_i| \leq \frac{\gamma}{4n}$ *for every* i *with probability at least* $1 - \xi/4$.

Proof. By a Chernoff bound, the probability that any given $\hat{p}_i$ deviates from p_i by more than $\frac{\gamma}{4n}$ is at most $2\exp(-K\gamma^2/8n^2) \leq \frac{\xi}{4(n+1)}$. The lemma follows by a union bound.

We first verify completeness for this scheme. Let L be a learner for EncThresh using n examples. If the hypothesis h produced by L is $(\frac{1}{2} - \gamma)$-good, then there exists $i_0 < i_1$ such that $p_{i_0} - p_{i_1} \geq 2\gamma$. If this is the case, then there must be an i for which $p_{i-1} - p_i \geq \frac{2\gamma}{n}$. Then with probability all but $\xi(n)/2$ over the estimates $\hat{p}_i$, we have $\hat{p}_{i-1} - \hat{p}_i \geq \frac{\gamma}{n}$, so some index is accused.

Now we verify soundness. Fix a PPT L, and let $j^* \in [n]$. Suppose L violates the soundness of the scheme with respect to j^*, i.e.

$$\Pr_{h \leftarrow_R L(S_{-j^*}), \text{coins of } \mathsf{Gen}_{\mathsf{ex}}} [\mathsf{Trace}_{\mathsf{ex}}(h) = j^*] > \xi.$$

We will use L to construct an adversary $\mathcal{A}$ for the ORE scheme that succeeds with noticeable advantage. It suffices to build an adversary for the static (many-challenge) security of ORE, with Theorem 3 showing how to convert it to a single-challenge adversary. This many-challenge adversary is presented as Algorithm 2. (While not explicitly stated, the adversary should halt and output a random guess whenever the messages it samples are not well-spaced.)

Let i^* be such that $m_{i^*} = m'_{j^*}$. With probability at least ξ over the parameters $(\mathsf{sk}^r, \mathsf{pars}^r)$, the choice of messages, the choice of the hypothesis h, and the coins of $\mathsf{Trace}_{\mathsf{ex}}$, there is a gap $\hat{p}_{i^*-1} - \hat{p}_{i^*} \geq \frac{\gamma}{n}$. Hence, by Lemma 1, there is a gap $p_{i^*-1} - p_{i^*} \geq \frac{\gamma}{2n}$ with probability at least $\frac{\xi}{2}$.

We now calculate the advantage of the adversary $\mathcal{A}$. Fix a hypothesis h. For notational simplicity, let $p = p_{i^*-1}$ and let $q = p_{i^*}$. Let $y_0 = h(\mathsf{pars}^r, c_{i^*}^0)$ and $y_1 = h(\mathsf{pars}^r, c_{i^*}^1)$. Then the adversary's success probability is:

$$\Pr[b' = b] = \frac{1}{2}(\Pr[y_0 = y_1 | b = 0] + \Pr[y_0 \neq y_1 | b = 1])$$

$$= \frac{1}{2}\left(\frac{1}{2}(p^2 + (1-p)^2 + q^2 + (1-q)^2) + (1 - pq - (1-p)(1-q))\right)$$

$$= \frac{1}{2} + \frac{1}{2}(p - q)^2.$$

Algorithm 2. ORE adversary $\mathcal{A}$

---

1. Sample $m'_1, \ldots, m'_n \leftarrow_R [N]$, and let $m_1 \leq \cdots \leq m_n$ be the result of sorting the m'_j. Let π be the permutation on $\{1, \ldots, n\}$ such that $m_{\pi(j)} = m'_j$. Let $m_0 = 0$. Let $i^* = \pi(j^*)$ so that $m_{i^*} = m'_{j^*}$.
2. Construct pairs (m_L^0, m_L^1) and (m_R^0, m_R^1) as follows. Let $B_0 = (m_{i^*-1}, m_{i^*})$ and $B_1 = (m_{i^*}, m_{i^*+1})$. Sample $m_L^0 \leq m_L^1$ at random from the same B_j, for a random choice of $j \in \{0, 1\}$. Sample $m_R^0 \leftarrow_R B_0$ and $m_R^1 \leftarrow_R B_1$.
3. Challenge on the pair of sequences $m_0, m_1, \ldots, m_{i^*-1}, m_L^1, m_L^2, m_{i^*}, \ldots, m_n$ and $m_0, m_1, \ldots, m_{i^*-1}, m_R^1, m_R^2, m_{i^*}, \ldots, m_n$, receiving ciphertexts $c_1, \ldots, c_{i^*}^0$, $c_{i^*}^1, \ldots, c_n$. For $j \neq j^*$, let $c'_j = c_{\pi(j)}$ so that c'_j is an encryption of m'_j.
4. Set $t = N/2$ and let

$$
\begin{aligned}
S_{-j^*} = \{ & (\mathsf{pars}^r, c'_1, \chi(m'_1 \leq t)), \ldots, (\mathsf{pars}^r, c'_{j^*-1}, \chi(m'_{j^*-1} \leq t)), \\
& (\mathsf{pars}^r, c_0, 1), (\mathsf{pars}^r, c'_{j^*+1}, \chi(m'_{j^*+1} \leq t)), \ldots, \\
& (\mathsf{pars}^r, c'_n, \chi(m'_n \leq t)) \} \\
= \{ & (\mathsf{pars}^r, c_{\pi(1)}, \chi(m_{\pi(1)} \leq t)), \ldots, (\mathsf{pars}^r, c_{\pi(j^*-1)}, \chi(m_{\pi(j^*-1)} \leq t)), \\
& (\mathsf{pars}^r, c_0, 1), (\mathsf{pars}^r, c_{\pi(j^*+1)}, \chi(m_{\pi(j^*+1)} \leq t)), \ldots, \\
& (\mathsf{pars}^r, c_{\pi(n)}, \chi(m_{\pi(n)} \leq t)) \}
\end{aligned}
$$

Obtain $h \leftarrow_R L(S_{-j^*})$.
5. Guess $b' = 0$ if $h(\mathsf{pars}^r, c_{i^*}^0) = h(\mathsf{pars}^r, c_{i^*}^1)$. Otherwise guess $b' = 1$.

Thus if $p - q \geq \frac{\gamma}{2n}$, then the adversary's advantage is at least $\frac{\gamma^2}{4n^2}$. On the other hand, even for arbitrary values of p, q, the advantage is still nonnegative. Therefore, the advantage of the strategy is at least $\frac{\xi\gamma^2}{8n^2} - \mathrm{negl}(k)$ (the $\mathrm{negl}(k)$ term coming from the assumption that the m'_i sampled where distinct), which is a noticeable function of the parameter k. This contradicts the static security of the ORE scheme.

3.3 The SQ Learnability of EncThresh

The statistical query (SQ) model is a natural restriction of the PAC model by which a learner is able to measure statistical properties of its examples, but cannot see the individual examples themselves. We recall the definition of an SQ learner.

Definition 8 (SQ learning) [39]. *Let $c : X \rightarrow \{0,1\}$ be a target concept and let $\mathcal{D}$ be a distribution over X. In the SQ model, a learner is given access to a statistical query oracle $\mathsf{STAT}(c, \mathcal{D})$. It may make queries to this oracle of the form (ψ, τ), where $\psi : X \times \{0,1\} \rightarrow \{0,1\}$ is a query function and $\tau \in (0,1)$ is an error tolerance. The oracle $\mathsf{STAT}(c, \mathcal{D})$ responds with a value v such that $|v - \Pr_{x \in \mathcal{D}}[\psi(x, c(x)) = 1]| \leq \tau$. The goal of a learner is to produce, with probability at least $1 - \beta$, a hypothesis $h : X \rightarrow \{0,1\}$ such that $\mathrm{error}_{\mathcal{D}}(c, h) \leq \alpha$. The query functions must be efficiently evaluable, and the tolerance τ must be lower bounded by an inverse polynomial in k and $1/\alpha$.*

The query complexity *of a learner is the worst-case number of queries it issues to the statistical query oracle. An SQ learner is efficient if it also runs in time polynomial in* $k, 1/\alpha, 1/\beta$.

Feldman and Kanade [24] investigated the relationship between query complexity and computational complexity for SQ learners. They exhibited a concept class $\mathcal{C}$ which is efficiently PAC learnable and SQ learnable with polynomially many queries, but assuming $\mathbf{NP} \neq \mathbf{RP}$, is not efficiently SQ learnable. Concepts in this concept class take the form

$$g_{\phi,y}(x, x') = \begin{cases} \mathrm{PAR}_y(x') & \text{if } x = \phi \\ 0 & \text{otherwise.} \end{cases}$$

Here, $\mathrm{PAR}_y(x')$ is the inner product of y and x' modulo 2. The concept class $\mathcal{C}$ consists of $g_{\phi,y}$ where ϕ is a satisfiable 3-CNF formula and y is the lexicographically first satisfying assignment to ϕ. The efficient PAC learner for parities based on Gaussian elimination shows that $\mathcal{C}$ is also efficiently PAC learnable. It is also (inefficiently) SQ learnable with polynomially many queries: either the all-zeroes hypothesis is good, or an SQ learner can recover the formula ϕ bit-by-bit and determine the satisfying assignment y by brute force. On the other hand, because parities are information-theoretically hard to SQ learn, the satisfying assignment y remains hidden to an SQ learner unless it is able to solve 3-SAT.

In this section, we show that the concept class EncThresh shares these properties with $\mathcal{C}$. Namely, we know that EncThresh is efficiently PAC learnable and because it is not efficiently privately learnable, it is not efficiently SQ learnable [5]. We can also show that EncThresh has an SQ learner with polynomial query complexity. Making this observation about EncThresh is of interest because the hardness of SQ learning EncThresh does not seem to be related to the (information-theoretic) hardness of SQ learning parities.

Proposition 1. *The concept class* EncThresh *is (inefficiently) SQ learnable with polynomially many queries.*

As with $\mathcal{C}$ there are two cases. In the first case, the target distribution places nearly zero weight on examples with pars = pars^r, and so the all-zeroes hypothesis is good. In the second case, the target distribution places noticeable weight on these examples, and our learner can use statistical queries to recover the comparison parameters pars^r bit-by-bit. Once the public parameters are recovered, our learner can determine a corresponding secret key by brute force. Lemma 2 below shows that any corresponding secret key – even one that is not actually sk^r – suffices. The learner can then use binary search to determine the threshold value t.

Proof. Let $f_{t,r}$ be the target concept, $\mathcal{D}$ be the target distribution, and α be the target error rate. With the statistical query $(x \times b \mapsto b, \alpha/4)$, we can determine whether the all-zeroes hypothesis is accurate. That is, if we receive a value that is less than $\alpha/2$, then $\Pr_{x \in \mathcal{D}}[f_{t,r}(x) = 1] \leq \alpha$. If not, then we know that $\Pr_{x \in \mathcal{D}}[f_{t,r}(x) = 1] \geq \alpha/4$, so $\mathcal{D}$ places significant weight on examples prefixed with pars^r. Suppose now that we are in the latter case.

Let $m = |\text{pars}|$. For $i = 1, \ldots, m$, define $\psi_i(\text{pars}, c, b) = 1$ if $\text{pars}_i = 1$ and $b = 1$, and $\psi_i(\text{pars}, c, b) = 0$ otherwise. Then by asking the queries $(\psi_i, \alpha/16)$, we can determine each bit pars_i^r of pars^r.

Now by brute force search, we determine a secret key sk for which $(\text{sk}, \text{pars}^r) \in \text{Range(Gen)}$. The recovered secret key sk may not necessarily be the same as sk^r. However, the following lemma shows that sk and sk^r are functionally equivalent:

Lemma 2. *Suppose* $(\text{Gen}, \text{Enc}, \text{Dec}, \text{Comp})$ *is a strongly correct ORE scheme. Then for any pair* $(\text{sk}_1, \text{pars}), (\text{sk}_2, \text{pars}) \in \text{Range(Gen)}$, *we have that* $\text{Dec}_{\text{sk}_1}(c) = \text{Dec}_{\text{sk}_2}(c)$ *for all ciphertexts* c.

With the secret key sk in hand, we now conduct a binary search for the threshold t. Recall that we have an estimate v for the weight that $f_{t,r}$ places on positive examples, i.e. $|v - \Pr_{x \in \mathcal{D}}[f_{t,r}(x) = 1]| \leq \alpha/4$. Starting at $t_1 = N/2$, we issue the query $(\varphi_1, \alpha/4)$ where $\varphi_1(\text{pars}, c, b) = 1$ iff $\text{pars} = \text{pars}^r$ and $\text{Dec}(\text{sk}, c) < t$. Let h_{t_1} denote the hypothesis

$$h_{t_1}(\text{pars}, c) = \begin{cases} 1 & \text{if } (\text{pars} = \text{pars}^r) \wedge (\text{Dec}(\text{sk}, c) \neq \perp) \wedge (\text{Dec}(\text{sk}, c) < t_1) \\ 0 & \text{otherwise.} \end{cases}$$

Thus, the query $(\varphi_1, \alpha/4)$ approximates the weight h_{t_1} places on positive examples. Let the answer to this query be v_1. If $|v_1 - v| \leq \alpha/2$, then we can halt and output the good hypothesis h_{t_1}. Otherwise, if $v_1 < v - \alpha/2$, we set the next threshold to $t_2 = 3N/4$, and if $v_1 > v + \alpha/2$, we set the next threshold to $t_2 = N/4$. We recurse up to $\log N = \ell = \text{poly}(k)$ times, yielding a good hypothesis for $f_{t,r}$.

Proof (Proof of Lemma 2). Suppose the lemma is not true. First suppose that there exists a ciphertext c such that $\text{Dec}(\text{sk}_1, c) = p_1 < p_2 = \text{Dec}(\text{sk}_2, c)$. Let $c' \in \text{Enc}(\text{sk}_1, p_2)$. Then by strong correctness applied to the parameters $(\text{sk}_1, \text{pars})$, we must have $\text{Comp}(\text{pars}, c, c') = $ "$<$". Now by strong correctness applied to $(\text{sk}_2, \text{pars})$, we must have $\text{Dec}(\text{sk}_2, c') > p_2$. Thus, $p_1 < \text{Dec}(\text{sk}_1, c') = p_2 < \text{Dec}(\text{sk}_2, c')$. Repeating this argument, we obtain a contradiction because the message space is finite.

Now suppose instead that there is a ciphertext c for which $\text{Dec}(\text{sk}_1, c) = p \in [N]$, but $\text{Dec}(\text{sk}_2, c) = \perp$. Let $c' \in \text{Enc}(\text{sk}_1, p')$ for some $p' > p$. Then $\text{Comp}(\text{pars}, c, c') = $ "$<$" by strong correctness applied to $(\text{pars}, \text{sk}_1)$. But $\text{Comp}(\text{pars}, c, c') = $ "$\perp$" by strong correctness applied to $(\text{pars}, \text{sk}_2)$, again yielding a contradiction.

4   ORE with Strong Correctness

We now explain how to obtain ORE with strongly correct comparison, as all prior ORE schemes only satisfy the weaker notion of correctness. The lack of strong correctness is easiest to see with the scheme of Boneh et al. [8].

The protocol is built from current multilinear map constructions, which are noisy. If the noise terms grow too large, the correctness of the multilinear map is not guaranteed. The comparison function in [8] is computed by performing multilinear operations, and for correctly generated ciphertexts, the operations will give the right answer. However, there exist ciphertexts, namely those with very large noise, for which the comparison function gives an incorrect output. The result is that the comparison operation is not guaranteed to be consistent with decrypting the ciphertexts and comparing the plaintexts.

As described in the introduction, we give a generic conversion from any ORE scheme with weakly correct comparison into a strongly correct scheme. We simply modify the encryption algorithm by adding a non-interactive zero-knowledge (NIZK) proof that the resulting ciphertext is well-formed. Then the decryption and comparison procedures check the proof(s), and only output a non-$\bot$ result (either decryption or comparison) if the proof(s) are valid.

Instantiating our Scheme. In our construction, we need the (weak) correctness of the underlying ORE scheme to hold with probability one. However, the existing protocols only have correctness with overwhelming probability, so some minor adjustments need to be made to the protocols. This is easiest to see in the ORE scheme of Boneh et al. [8]. The Boneh et al. scheme uses noisy multilinear maps [26] which may introduce errors. Therefore, the protocol described in [8] only achieves the (weak) correctness property with overwhelming probability, whereas we will require (weak) correctness with probability 1 for the conversion. However, it is straightforward to generate the parameters for the protocol in such a way as to completely eliminate errors. Essentially, the parameters in the protocol have an error term that is generated by a (discrete) Gaussian distribution, which has unbounded support. Instead, we truncate the Gaussian, resulting in a noise distribution with bounded support. By truncating sufficiently far from the center, the resulting distribution is also statistically close to the full Gaussian, so security of the protocol with truncated noise follows from the security of the protocol with un-truncated noise. By truncating the noise distribution, it is straightforward to set parameters so that no errors can occur.

It is similarly straightforward to modify current obfuscation candidates, which are also built from multilinear maps, to obtain perfect (weak) correctness by truncating the noise distributions. Thus, our scheme has instantiations using multilinear maps or iO.

4.1 Conversion from Weakly Correct ORE

We describe our generic conversion from an order-revaling encryption scheme with weak correctness using NIZKs. We will need the following additional tools:

Perfectly Binding Commitments. A perfectly binding commitment Com is a randomized algorithm with two properties. The first is perfect binding, which states that if $\mathsf{Com}(m; r) = \mathsf{Com}(m'; r')$, then $m = m'$. The second requirement is computational hiding, which states that the distributions $\mathsf{Com}(m)$ and $\mathsf{Com}(m')$ are computationally indistinguishable for any messages m, m'. Such commitments can be built, say, from any injective one-way function.

Perfectly Sound NIZK. A NIZK protocol consists of three algorithms:

- Setup(1^λ) is a randomized algorithm that outputs a common reference string crs.
- Prove(crs, x, w) takes as input a common reference string crs, an NP statement x, and a witness w, and produces a proof π.
- Ver(crs, x, π) takes as input a common reference string crs, statement x, and a proof π, and outputs either accept or reject.

We make three requirements for a NIZK:

- **Perfect Completeness.** For all security parameters λ and any true statement x with witness w,

$$\Pr[\mathsf{Ver}(\mathsf{crs}, x, \pi) = \mathsf{accept} : \mathsf{crs} \leftarrow \mathsf{Setup}(1^\lambda); \pi \leftarrow \mathsf{Prove}(\mathsf{crs}, x, w)] = 1.$$

- **Perfect Soundness.** For all security parameters λ, any *false* statement x and any (invalid) proof π,

$$\Pr[\mathsf{Ver}(\mathsf{crs}, x, \pi) = \mathsf{accept} : \mathsf{crs} \leftarrow \mathsf{Setup}(1^\lambda)] = 0.$$

- **Computational Zero Knowledge.** There exists a simulator $\mathcal{S}_1, \mathcal{S}_2$ such that for any computationally bounded adversary $\mathcal{A}$, the quantity

$$\| \Pr[\mathcal{A}^{\mathsf{Prove}(\mathsf{crs}, \cdot, \cdot)}(\mathsf{crs}) = 1 : \mathsf{crs} \leftarrow \mathsf{Setup}(1^\lambda)]$$
$$- \Pr[\mathcal{A}^{Sim(\mathsf{crs}, \tau, \cdot, \cdot)}(\mathsf{crs}) = 1 : (\mathsf{crs}, \tau) \leftarrow \mathcal{S}_1(1^\lambda)]\|$$

is negligible, where $Sim(\mathsf{crs}, \tau, x, w)$ outputs $\mathcal{S}_2(\mathsf{crs}, \tau, x)$ if w is a valid witness for x, and $Sim(\mathsf{crs}, \tau, x, w) = \bot$ if w is invalid.

NIZKs satisfying these requirements can be built from bilinear maps [32].

The Construction. We now give our conversion. Let (Setup, Prove, Ver) be a perfectly sound NIZK and (Gen', Enc', Dec', Comp') and ORE with *weakly* correct comparison. We will assume that Enc' is deterministic; if not, we can derandomize Enc' using a pseudorandom function. Let Com be a perfectly binding commitment. We construct a new ORE scheme (Gen, Enc, Dec, Comp) with *strongly* correct comparison:

- Gen($1^\lambda, 1^\ell$): run (sk', pars') $\leftarrow$ Gen'($1^\lambda, 1^\ell$). Let $\sigma = \mathsf{Com}(\mathsf{sk}; r)$ for randomness r, and run crs $\leftarrow$ Setup(1^λ). Then the secret key is sk = (sk', r, crs) and the public parameters are pars = (pars', σ, crs).
- Enc(sk, m): Compute $c' = \mathsf{Enc}'(\mathsf{sk}', m)$. Let $x_{c'}$ be the statement $\exists \hat{m}, \hat{\mathsf{sk}}', \hat{r} : \sigma = \mathsf{Com}(\hat{\mathsf{sk}}', \hat{r}) \wedge c' = \mathsf{Enc}'(\hat{\mathsf{sk}}', \hat{m})$. Run $\pi_{c'} = \mathsf{Prove}(\mathsf{crs}, x_{c'}, (m, \mathsf{sk}', r))$. Output the ciphertext $c = (c', \pi_{c'})$.

- $\mathsf{Dec}(\mathsf{sk}, c)$: Write $c = (c', \pi_{c'})$. If $\mathsf{Ver}(\mathsf{crs}, x_{c'}, \pi_{c'}) = \mathsf{reject}$, output $\perp$. Otherwise, output $m = \mathsf{Dec}'(\mathsf{sk}', c')$.
- $\mathsf{Comp}(\mathsf{pars}, c_0, c_1)$; Write $c_b = (c'_b, \pi_{c'_b})$ and $\mathsf{pars} = (\mathsf{pars}', \sigma, \mathsf{crs})$. If $\mathsf{Ver}(\mathsf{crs}, x_{c'_b}, \pi_{c'_b}) = \mathsf{reject}$ for either $b = 0, 1$, then output $\perp$. Otherwise, output $\mathsf{Comp}'(\mathsf{pars}', c'_0, c'_1)$.

Correctness. Notice that, for each plaintext m, the ciphertext component $c' = \mathsf{Enc}'(\mathsf{sk}', m)$ is the *unique* value such that $\mathsf{Dec}(\mathsf{sk}, (c', \pi)) = m$ for some proof π. Moreover, the completeness of the zero knowledge proof implies that $\mathsf{Enc}(\mathsf{sk}, m)$ outputs a valid proof. Decryption correctness follows.

For strong comparison correctness, consider two ciphertexts c_0, c_1 where $c_b = (c'_b, \pi_{c'_b})$. Suppose both proofs $\pi_{c'_b}$ are valid, which means that verification passes when running Comp and so $\mathsf{Comp}(\mathsf{pars}, c_0, c_1) = \mathsf{Comp}'(\mathsf{pars}', c'_0, c'_1)$. Verification also passes when decrypting c_b, and so $\mathsf{Dec}(\mathsf{sk}, c_b) = \mathsf{Dec}'(\mathsf{sk}', c'_b)$.

Since the proofs are valid, $c'_b = \mathsf{Enc}'(\mathsf{sk}', m_b)$ for some m_b for both $b = 0, 1$. The weak correctness of comparison for $(\mathsf{Gen}', \mathsf{Enc}', \mathsf{Dec}', \mathsf{Comp}')$ implies that $\mathsf{Comp}'(\mathsf{pars}', c'_0, c'_1) = \mathsf{Comp}_{plain}(m_0, m_1)$. The decryption correctness of $(\mathsf{Gen}', \mathsf{Enc}', \mathsf{Dec}', \mathsf{Comp}')$ then implies that $\mathsf{Dec}(\mathsf{sk}', c'_b) = m_b$, and therefore $\mathsf{Dec}(\mathsf{sk}, c_b) = m_b$. Thus $\mathsf{Comp}_{ciph}(\mathsf{sk}, c_0, c_1) = \mathsf{Comp}_{plain}(m_0, m_1)$. Putting it all together, $\mathsf{Comp}(\mathsf{pars}, c_0, c_1) = \mathsf{Comp}_{ciph}(\mathsf{sk}, c_0, c_1)$, as desired.

Now suppose one of the proofs $\pi_{c'_b}$ are invalid. Then $\mathsf{Comp}(\mathsf{pars}, c_0, c_1) = \perp$ and $\mathsf{Dec}(\mathsf{sk}, c_b) = \perp$. This means $\mathsf{Comp}_{ciph}(\mathsf{sk}, c_0, c_1) = \perp = \mathsf{Comp}(\mathsf{pars}, c_0, c_1)$, as desired.

Security. To prove security, we first use the zero-knowledge simulator to simulate the proofs π'_c without using a witness (namely, the secret decryption key). Then we use the hiding property of the commitment to replace σ with a commitment to 0. At this point, the entire game can be simulated using an Enc' oracle, and so the security reduces to the security of Enc'.

Theorem 6. *If* $(\mathsf{Gen}', \mathsf{Enc}', \mathsf{Dec}', \mathsf{Comp}')$ *is a (statically) secure ORE,* $(\mathsf{Setup}, \mathsf{Prove}, \mathsf{Ver})$ *is computationally zero knowledge, and* Com *is computationally hiding, then* $(\mathsf{Gen}, \mathsf{Enc}, \mathsf{Dec}, \mathsf{Comp})$ *is a statically secure ORE.*

Proof. We will prove security through a sequence of hybrids. Let $\mathcal{A}$ be an adversary with advantage ϵ in breaking the static security of $(\mathsf{Gen}, \mathsf{Enc}, \mathsf{Dec}, \mathsf{Comp})$.

Hybrid 0. This is the real experiment, where $\sigma \leftarrow \mathsf{Com}(\mathsf{sk})$, $\mathsf{crs} \leftarrow \mathsf{Setup}(1^\lambda)$, and the proofs $\pi_{c'}$ are answered using Prove and valid witnesses. $\mathcal{A}$ has advantage ϵ in distinguishing the left and right ciphertexts.

Hybrid 1. This is the same as **Hybrid 0**, except that crs is generated as $(\mathsf{crs}, \tau) \leftarrow \mathcal{S}_1(1^\lambda)$, and all proofs are generated using $\mathcal{S}_2(\mathsf{crs}, \tau, \cdot)$. The zero knowledge property of $(\mathsf{Setup}, \mathsf{Prove}, \mathsf{Ver})$ shows that this is indistinguishable from **Hybrid 0**.

Hybrid 2. This is the same as **Hybrid 1**, except that $\sigma \leftarrow \mathsf{Com}(0)$. Since the randomness for computing σ is not needed for simulation, this change is undetectable using the hiding of Com.

Thus the advantage of $\mathcal{A}$ in **Hybrid 2** is at least $\epsilon - \mathrm{negl}$ for some negligible function negl. Now consider the following adversary $c\mathcal{B}$ that attempts to break the security of $(\mathsf{Gen}', \mathsf{Enc}', \mathsf{Dec}', \mathsf{Comp}')$. $\mathcal{B}$ simulates $\mathcal{A}$, and forwards the message sequences $m_1^{(L)} < m_2^{(L)} < \cdots < m_q^{(L)}$ and $m_1^{(R)} < m_2^{(R)} < \cdots < m_q^{(R)}$ produced by $\mathcal{A}$ to its own challenger. In response, it receives pars', and ciphertexts c_i', where c_i' encrypts either $m_i^{(L)}$ if $b = 0$ or $m_i^{(R)}$ if $b = 1$, for a random bit b chosen by the challenger.

$\mathcal{B}$ now generates $\sigma \leftarrow \mathsf{Com}(0)$, $(\mathsf{crs}, \tau) \leftarrow \mathcal{S}_1(1^\lambda)$, and lets $\mathsf{pars} = (\mathsf{pars}', \sigma, \mathsf{crs})$. It also computes $\pi_{c_i'} \leftarrow \mathcal{S}_2(\mathsf{crs}, \tau, x_{c_i'})$, and defines $c_i = (c_i', \pi_{c_i'})$, and gives pars and the c_i to $\mathcal{A}$. Finally when $\mathcal{A}$ outputs a guess b' for b, $\mathcal{B}$ outputs the same guess b'.

We see that the view of $\mathcal{A}$ as a subroutine of $\mathcal{B}$ is exactly the same view as in **Hybrid 2**. Thus, $b' = b$ with probability at least $\epsilon - \mathrm{negl}$. The security of $(\mathsf{Gen}', \mathsf{Enc}', \mathsf{Dec}', \mathsf{Comp}')$ implies that this quantity, and hence ϵ, must be negligible. Thus $\mathcal{A}$ must have negligible advantage in breaking the security of $(\mathsf{Gen}, \mathsf{Enc}, \mathsf{Dec}, \mathsf{Comp})$.

5   A Separation for Representation Learning

In this section, we show how to construct a concept class $\mathsf{ValidSig}$ that separates efficient *representation* learning from efficient private representation learning, assuming only the existence of one-way functions. Here by "representation learning" we mean a restricted form of proper learning where a learner must output a particular representation (i.e. encoding) of a hypothesis h in the concept class $\mathcal{C}$. As with proper learning, this is a natural syntactic restriction to place on a learner: for instance, if one wants to learn linear threshold functions (LTF), it makes sense to require a learner to produce the actual coefficients of an LTF, rather than an arbitrary circuit that happens to compute an LTF.

The construction is based on the following elegant idea due to Kobbi Nissim [44]. Suppose $H : D \rightarrow R$ is a cryptographic hash function with the property that given $x_1, \ldots, x_n$ with $y = H(x_1) = \cdots = H(x_n)$, it is infeasible for an efficient adversary to find another x for which $H(x) = y$. Consider the concept class $\mathsf{HashPoint}$ consisting of the concepts

$$f_x(x') = \begin{cases} 1 & \text{if } H(x) = H(x') \\ 0 & \text{otherwise.} \end{cases}$$

for every $x \in R$. The representation of a concept f_x is the point x. The concept class $\mathsf{HashPoint}$ is very easy to learn (by representation) without privacy: a learner can identify any positive example x_i and output the representation x_i. Since $H(x_i) = H(x)$, the concept f_{x_i} is actually equal to the target concept

f_x. On the other hand, a learner that identifies an index x^* for which $f_{x^*} = f_x$ cannot be differentially private, since the security of the hash function means it is infeasible to produce such an x^* that is not present in the sample.

Note that this argument breaks down if one tries to show that HashPoint is not privately properly learnable. While it is infeasible to privately produce a representation x^* for which f_{x^*} is a good hypothesis, the hypothesis $h(x) = \chi(H(x) = h(x_i))$ is equal as a function to every good f_{x^*}. Moreover, this hypothesis can be constructed privately as long as the sample contains sufficiently many positive examples.

We make this discussion formal by constructing a concept class ValidSig based on *super-secure digital signature schemes*, which can be constructed from one-way functions. Our use of signatures to derive hardness results for private proper learning is very analogous to prior hardness results for synthetic data generation [21,53].

Definition 9. *A digital signature scheme is a triple of algorithms* (Gen, Sign, Ver) *where*

- Gen(1^λ) *produces a key pair* (sk, vk).
- Sign(sk, m) *takes the private signing key* sk *and a message* $m \in \{0,1\}^*$ *and produces a signature* σ *for the message* m.
- Ver(vk, m, σ) *takes the public verification key* vk, *a message* m, *and a signature* σ, *and (deterministically) outputs a bit indicating whether* σ *is a valid signature for* m.

The correctness property of a digital signature scheme is that for every (sk, vk) $\leftarrow_R$ Gen(1^λ), *every message* $m \in \{0,1\}^*$, *and every signature* $\sigma \leftarrow_R$ Sign(sk, m), *we have* Ver(vk, m, σ) = 1.

Definition 10. *A digital signature scheme is* super-secure *under adaptive chosen-plaintext attacks if all efficient adversaries* $\mathcal{A}$ *win the following weak forgery game with negligible probability:*

- *The challenger samples* (sk, vk) $\leftarrow_R$ Gen(1^λ).
- *The adversary* $\mathcal{A}$ *is given* vk *and oracle access to* Sign(sk, ·). *It adaptively queries the signing oracle, obtaining a sequence of message-signature pairs* A. *It then outputs a forgery* (m^*, σ^*).
- *The value of the game is 1 iff* Ver(vk, m^*, σ^*) = 1 *and* (m^*, σ^*) $\notin$ A.

It is known that super-secure digital signature schemes can be constructed from one-way functions [29,38,43,47].

We now describe our concept class ValidSig. Let (Gen, Sign, Ver) be a super-secure digital signature scheme. We define a concept class ValidSig as follows. Fix the message length ℓ. For every (vk, m, σ) with $m \in \{0,1\}^\ell$ and Ver(vk, m, σ) = 1, define the concept

$$f_{\mathsf{vk},m,\sigma}(\mathsf{vk}', m', \sigma') = \begin{cases} 1 & \text{if } (\mathsf{vk} = \mathsf{vk}') \wedge (\mathsf{Ver}(\mathsf{vk}, m', \sigma') = 1) \\ 0 & \text{otherwise.} \end{cases}$$

For convenience, we also include the all-zeroes hypothesis in ValidSig, with representation $\perp$.

Theorem 7. *Let $\alpha, \beta > 0$. There exists a proper PAC learning algorithm L for the concept class ValidSig achieving error α and confidence $1 - \beta$. Moreover, L is efficient (running in time polynomial in the parameters $k, 1/\alpha, \log(1/\beta)$).*

Algorithm 3. Learner L for ValidSig

---

1. Request examples $\{((\text{vk}'_1, m'_1, \sigma'_1), b_1), \ldots, ((\text{vk}'_n, m'_n, \sigma'_n), b_n)\}$ for $n = \lceil \log(1/\beta)/\alpha \rceil$.
2. Identify an i for which $b_i = 1$ and return the representation $(\text{vk}'_i, m'_i, \sigma'_i)$. If no such i exists, return $\perp$ representing the all-zeroes hypothesis.

Proof. Fix a target concept $f_{\text{vk},m,\sigma} \in \text{ValidSig}_k$ and a distribution $\mathcal{D}$ on examples. Let POS denote the set of examples $(\text{vk}', m', \sigma')$ on which $f_{\text{vk},m,\sigma}(\text{vk}', m', \sigma') = 1$. We divide the analysis of the learner into three cases based on the weight $\mathcal{D}$ places on the sets POS.

Case 1: $\mathcal{D}$ places at least α weight on POS. Then L receives a positive example with probability at least $1 - (1 - \alpha)^n \geq 1 - \beta$, and is thus able to identify a concept that equals the target concept.

Case 2: $\mathcal{D}$ places less than α weight on POS. If L gets a positive example, then the analysis of Case 1 applies. Otherwise, the all-zeroes hypothesis is α-good.

We now prove the hardness of properly privately learning ValidSig by constructing an example reidentification scheme for properly learning this concept class. Our example reidentification scheme yields a hard distribution even when the error parameter α is taken to be inverse-polynomially close to 1.

Theorem 8. *Let $\gamma(n)$ and $\xi(n)$ be noticeable functions. Let (Gen, Sign, Ver) be a super-secure digital signature scheme. Then there exists an (efficient) $(\alpha = 1 - \gamma, \xi)$-example reidentification scheme $(\text{Gen}_{\text{ex}}, \text{Trace}_{\text{ex}})$ for representation learning the concept class ValidSig.*

Proof. We construct an example reidentification scheme for ValidSig as follows. The algorithm Gen_{ex} samples $(\text{sk}, \text{vk}) \leftarrow_{\text{R}} \text{Gen}(1^\lambda)$, a message $m \in \{0,1\}^\ell$, and a signature $\sigma \leftarrow_{\text{R}} \text{Sign}(\text{sk}, m)$, yielding a concept $f_{\text{vk},m,\sigma}$. Let $\mathcal{D}$ be the distribution of $(\text{vk}, m, \text{Sign}(\text{sk}, m))$ for random $m \leftarrow_{\text{R}} \{0,1\}^\ell$. Gen_{ex} then samples $x_0, x_1, \ldots, x_n$ i.i.d. from $\mathcal{D}$. Given a representation $(\text{vk}^*, m^*, \sigma^*)$, the algorithm Trace_{ex} simply identifies an index i for which $x_i = (\text{vk}^*, m^*, \sigma^*)$, and outputs $\perp$ if none is found.

We first verify completeness. Let L be a learner for ValidSig using n examples. If the representation $(\text{vk}^*, m^*, \sigma^*)$ produced by L represents an $(1 - \gamma)$-good hypothesis, then it must be the case that $\text{vk}^* = \text{vk}$ and $\text{Ver}(\text{vk}, m^*, \sigma^*) = 1$. Thus, if L violates the completeness condition, it can be used to construct the weak forgery adversary $\mathcal{A}$ (Algorithm 4) that succeeds with noticeable probability ξ.

Algorithm 4. Weak forgery adversary $\mathcal{A}$

---

1. Query the signing oracle on random messages $m'_1, \ldots, m'_n \leftarrow_{\mathrm{R}} \{0,1\}^\ell$, obtaining signatures $\sigma'_1, \ldots, \sigma'_n$.
2. Run L on the labeled examples $((\mathsf{vk}, m'_1, \sigma'_1), 1), \ldots, ((\mathsf{vk}, m'_n, \sigma'_n), 1)$, obtaining a representation (m^*, σ^*).
3. Output the forgery (m^*, σ^*).

Now we verify soundness. Observe that for any i, the sample S_{-i} contains no information about message m_i. Therefore, the learner has a $2^{-\ell} = \mathrm{negl}(k)$ probability at producing a representation containing message m_i, proving soundness.

Acknowledgements. We gratefully acknowledge Kobbi Nissim and Salil Vadhan for helpful discussions about this work, and also thank Salil Vadhan for suggestions on its presentation.

References

1. Beimel, A., Kasiviswanathan, S.P., Nissim, K.: Bounds on the sample complexity for private learning and private data release. In: Micciancio, D. (ed.) TCC 2010. LNCS, vol. 5978, pp. 437–454. Springer, Heidelberg (2010)
2. Beimel, A., Nissim, K., Stemmer, U.: Private learning and sanitization: pure vs. approximate differential privacy. In: Raghavendra, P., Raskhodnikova, S., Jansen, K., Rolim, J.D.P. (eds.) RANDOM 2013 and APPROX 2013. LNCS, vol. 8096, pp. 363–378. Springer, Heidelberg (2013)
3. Bitansky, N., Vaikuntanathan, V.: Indistinguishability obfuscation from functional encryption. In: FOCS (2015)
4. Blum, A.: Separating distribution-free and mistake-bound learning models over the Boolean domain. SIAM J. Comput. **23**(5), 990–1000 (1994). http://dx.doi.org/10.1137/S009753979223455X
5. Blum, A., Dwork, C., McSherry, F., Nissim, K.: Practical privacy: the SuLQ framework. In: PODS, pp. 128–138 (2005)
6. Blum, A., Ligett, K., Roth, A.: A learning theory approach to non-interactive database privacy. In: STOC, pp. 609–618 (2008)
7. Boldyreva, A., Chenette, N., O'Neill, A.: Order-preserving encryption revisited: improved security analysis and alternative solutions. In: Rogaway, P. (ed.) CRYPTO 2011. LNCS, vol. 6841, pp. 578–595. Springer, Heidelberg (2011)
8. Boneh, D., Lewi, K., Raykova, M., Sahai, A., Zhandry, M., Zimmerman, J.: Semantically secure order-revealing encryption: multi-input functional encryption without obfuscation. In: Oswald, E., Fischlin, M. (eds.) EUROCRYPT 2015. LNCS, vol. 9057, pp. 563–594. Springer, Heidelberg (2015)
9. Boneh, D., Sahai, A., Waters, B.: Fully collusion resistant traitor tracing with short ciphertexts and private keys. In: Vaudenay, S. (ed.) EUROCRYPT 2006. LNCS, vol. 4004, pp. 573–592. Springer, Heidelberg (2006)
10. Boneh, D., Zhandry, M.: Multiparty key exchange, efficient traitor tracing, and more from indistinguishability obfuscation. In: Garay, J.A., Gennaro, R. (eds.) CRYPTO 2014, Part I. LNCS, vol. 8616, pp. 480–499. Springer, Heidelberg (2014)

11. Bost, R., Popa, R.A., Tu, S., Goldwasser, S.: Machine learning classification over encrypted data. IACR Cryptology ePrint Archive 2014:331 (2014). http://eprint.iacr.org/2014/331

12. Brakerski, Z., Segev, G.: Function-private functional encryption in the private-key setting. In: Dodis, Y., Nielsen, J.B. (eds.) TCC 2015, Part II. LNCS, vol. 9015, pp. 306–324. Springer, Heidelberg (2015)

13. Bun, M., Nissim, K., Stemmer, U., Vadhan, S.P.: Differentially private release and learning of threshold functions. In: Foundations of Computer Science, FOCS 2015, 18–20 October 2015, Berkeley, CA, USA, pp. 634–649 (2015)

14. Bun, M., Ullman, J., Vadhan, S.P.: Fingerprinting codes and the price of approximate differential privacy. In: Symposium on Theory of Computing, STOC 2014, May 31 - June 03, 2014, New York, NY, USA, pp. 1–10 (2014)

15. Chandrasekaran, K., Thaler, J., Ullman, J., Wan, A.: Faster private release of marginals on small databases. In: ITCS 2014 (2014)

16. Chaudhuri, K., Hsu, D.: Sample complexity bounds for differentially private learning. In: Kakade, S.M., von Luxburg, U. (eds.) COLT, JMLR Proceedings, vol. 19, pp. 155–186. JMLR.org (2011)

17. Chaudhuri, K., Hsu, D., Song, S.: The large margin mechanism for differentially private maximization. In: Advances in Neural Information Processing Systems 27: Annual Conference on Neural Information Processing Systems 2014, 8–13 December 2014, Montreal, Quebec, Canada, pp. 1287–1295 (2014)

18. Chor, B., Fiat, A., Naor, M.: Tracing Traitors. In: Desmedt, Y.G. (ed.) CRYPTO 1994. LNCS, vol. 839, pp. 257–270. Springer, Heidelberg (1994)

19. Dwork, C., Kenthapadi, K., McSherry, F., Mironov, I., Naor, M.: Our data, ourselves: privacy via distributed noise generation. In: Vaudenay, S. (ed.) EUROCRYPT 2006. LNCS, vol. 4004, pp. 486–503. Springer, Heidelberg (2006)

20. Dwork, C., McSherry, F., Nissim, K., Smith, A.: Calibrating noise to sensitivity in private data analysis. In: Halevi, S., Rabin, T. (eds.) TCC 2006. LNCS, vol. 3876, pp. 265–284. Springer, Heidelberg (2006)

21. Dwork, C., Naor, M., Reingold, O., Rothblum, G.N., Vadhan, S.P.: On the complexity of differentially private data release: efficient algorithms and hardness results. In: STOC, pp. 381–390 (2009)

22. Dwork, C., Nikolov, A., Talwar, K.: Using convex relaxations for efficiently and privately releasing marginals. In: Proceedings of the Thirtieth Annual Symposium on Computational Geometry, SOCG 2014, pp. 261:261–261:270. ACM, New York (2014). http://doi.acm.org/10.1145/2582112.2582123

23. Dwork, C., Rothblum, G.N., Vadhan, S.P.: Boosting and differential privacy. In: FOCS, pp. 51–60 (2010)

24. Feldman, V., Kanade, V.: Computational bounds on statistical query learning. In: The 25th Annual Conference on Learning Theory, COLT 2012, 25–27 June 2012, Edinburgh, Scotland, pp. 16.1-16.22 (2012). http://www.jmlr.org/proceedings/papers/v23/feldman12a/feldman12a.pdf

25. Feldman, V., Xiao, D.: Sample complexity bounds on differentially private learning via communication complexity. CoRR abs/1402.6278 (2014)

26. Garg, S., Gentry, C., Halevi, S.: Candidate multilinear maps from ideal lattices. In: Johansson, T., Nguyen, P.Q. (eds.) EUROCRYPT 2013. LNCS, vol. 7881, pp. 1–17. Springer, Heidelberg (2013)

27. Garg, S., Gentry, C., Halevi, S., Raykova, M., Sahai, A., Waters, B.: Candidate indistinguishability obfuscation and functional encryption for all circuits. In: Proceedings of FOCS (2013)

28. Garg, S., Gentry, C., Halevi, S., Zhandry, M.: Fully secure functional encryption without obfuscation (2014)
29. Goldreich, O.: Foundations of Cryptography: Basic Applications, vol. 2. Cambridge University Press, Cambridge (2004)
30. Goldwasser, S., Gordon, S.D., Goyal, V., Jain, A., Katz, J., Liu, F.-H., Sahai, A., Shi, E., Zhou, H.-S.: Multi-input functional encryption. In: Nguyen, P.Q., Oswald, E. (eds.) EUROCRYPT 2014. LNCS, vol. 8441, pp. 578–602. Springer, Heidelberg (2014)
31. Graepel, T., Lauter, K., Naehrig, M.: ML confidential: machine learning on encrypted data. In: Kwon, T., Lee, M.-K., Kwon, D. (eds.) ICISC 2012. LNCS, vol. 7839, pp. 1–21. Springer, Heidelberg (2013)
32. Groth, J., Ostrovsky, R., Sahai, A.: New techniques for noninteractive zero-knowledge. J. ACM **59**(3), 11:1–11:35 (2012). http://doi.acm.org/10.1145/2220357.2220358
33. Gupta, A., Roth, A., Ullman, J.: Iterative constructions and private data release. In: Cramer, R. (ed.) TCC 2012. LNCS, vol. 7194, pp. 339–356. Springer, Heidelberg (2012)
34. Hardt, M., Ligett, K., McSherry, F.: A simple and practical algorithm for differentially private data release. In: NIPS, pp. 2348–2356 (2012)
35. Hardt, M., Rothblum, G.N.: A multiplicative weights mechanism for privacy-preserving data analysis. In: FOCS, pp. 61–70 (2010)
36. Hardt, M., Rothblum, G.N., Servedio, R.A.: Private data release via learning thresholds. In: SODA, pp. 168–187 (2012)
37. Kasiviswanathan, S.P., Lee, H.K., Nissim, K., Raskhodnikova, S., Smith, A.: What can we learn privately? SIAM J. Comput. **40**(3), 793–826 (2011)
38. Katz, J., Koo, C.Y.: On constructing universal one-way hash functions from arbitrary one-way functions. IACR Cryptology ePrint Archive 2005, p. 328 (2005)
39. Kearns, M.: Efficient noise-tolerant learning from statistical queries. J. ACM **45**(6), 983–1006 (1998). http://doi.acm.org/10.1145/293347.293351
40. Kearns, M., Valiant, L.: Cryptographic limitations on learning Boolean formulae and finite automata. J. ACM **41**(1), 67–95 (1994). http://doi.acm.org/10.1145/174644.174647
41. Kharitonov, M.: Cryptographic lower bounds for learnability of Boolean functions on the uniform distribution. J. Comput. Syst. Sci. **50**(3), 600–610 (1995). http://dx.doi.org/10.1006/jcss.1995.1046
42. McSherry, F., Talwar, K.: Mechanism design via differential privacy. In: Proceedings of the 48th Annual IEEE Symposium on Foundations of Computer Science, FOCS 2007, pp. 94–103. IEEE Computer Society, Washington, DC (2007). http://dx.doi.org/10.1109/FOCS.2007.41
43. Naor, M., Yung, M.: Universal one-way hash functions and their cryptographic applications. In: Proceedings of the Twenty-First Annual ACM Symposium on Theory of Computing, STOC 1989, pp. 33–43. ACM, New York (1989). http://doi.acm.org/10.1145/73007.73011
44. Nissim, K.: Personal communication, July 2014
45. Pandey, O., Rouselakis, Y.: Property preserving symmetric encryption. In: Pointcheval, D., Johansson, T. (eds.) EUROCRYPT 2012. LNCS, vol. 7237, pp. 375–391. Springer, Heidelberg (2012)
46. Pitt, L., Valiant, L.G.: Computational limitations on learning from examples. J. ACM **35**(4), 965–984 (1988). http://doi.acm.org/10.1145/48014.63140

47. Rompel, J.: One-way functions are necessary and sufficient for secure signatures. In: Proceedings of the Twenty-Second Annual ACM Symposium on Theory of Computing, STOC 1990, pp. 387–394. ACM, New York (1990). http://doi.acm.org/10.1145/100216.100269

48. Roth, A., Roughgarden, T.: Interactive privacy via the median mechanism. In: STOC, pp. 765–774 (2010)

49. Servedio, R.A.: Computational sample complexity and attribute-efficient learning. J. Comput. Syst. Sci. **60**(1), 161–178 (2000). http://dx.doi.org/10.1006/jcss.1999.1666

50. Servedio, R.A., Gortler, S.J.: Equivalences and separations between quantum and classical learnability. SIAM J. Comput. **33**(5), 1067–1092 (2004). http://dx.doi.org/10.1137/S0097539704412910

51. Thaler, J., Ullman, J., Vadhan, S.: Faster algorithms for privately releasing marginals. In: Czumaj, A., Mehlhorn, K., Pitts, A., Wattenhofer, R. (eds.) ICALP 2012, Part I. LNCS, vol. 7391, pp. 810–821. Springer, Heidelberg (2012)

52. Ullman, J.: Answering $n^{2+o(1)}$ counting queries with differential privacy is hard. In: STOC, pp. 361–370 (2013)

53. Ullman, J., Vadhan, S.: PCPs and the hardness of generating private synthetic data. In: Ishai, Y. (ed.) TCC 2011. LNCS, vol. 6597, pp. 400–416. Springer, Heidelberg (2011)

54. Valiant, L.G.: A theory of the learnable. Commun. ACM **27**(11), 1134–1142 (1984). http://doi.acm.org/10.1145/1968.1972

LWR and LPN

LWR and LPN

On the Hardness of Learning with Rounding over Small Modulus

Andrej Bogdanov[1]([⊠]), Siyao Guo[1], Daniel Masny[2], Silas Richelson[3],
and Alon Rosen[4]

[1] Department of Computer Science and Engineering,
Chinese University of Hong Kong, Hong Kong, China
{andrejb,syguo}@cse.cuhk.edu.hk
[2] Horst-Görtz Institute for IT Security and Faculty of Mathematics,
Ruhr-Universität Bochum, Bochum, Germany
daniel.masny@ruhr-uni-bochum.de
[3] Department of Electrical Engineering and Computer Science,
MIT, Cambridge, USA
silas.richelson@gmail.com
[4] Efi Arazi School of Computer Science, IDC Herzliya, Herzliya, Israel
alon.rosen@idc.ac.il

Abstract. We show the following reductions from the learning with errors problem (LWE) to the learning with rounding problem (LWR): (1) Learning the secret and (2) distinguishing samples from random strings is at least as hard for LWR as it is for LWE for efficient algorithms if the number of samples is no larger than $O(q/Bp)$, where q is the LWR modulus, p is the rounding modulus, and the noise is sampled from any distribution supported over the set $\{-B, \ldots, B\}$.

Our second result generalizes a theorem of Alwen, Krenn, Pietrzak, and Wichs (CRYPTO 2013) and provides an alternate proof of it. Unlike Alwen et al., we do not impose any number theoretic restrictions on the modulus q. The first result also extends to variants of LWR and LWE over polynomial rings. The above reductions are sample preserving and run in time $\text{poly}(n, q, m)$.

As additional results we show that (3) distinguishing any number of LWR samples from random strings is of equivalent hardness to LWE whose noise distribution is uniform over the integers in the range $[-q/2p, \ldots, q/2p)$ provided q is a multiple of p and (4) the "noise flooding" technique for converting faulty LWE noise to a discrete Gaussian distribution can be applied whenever $q = \Omega(B\sqrt{m})$.

Part of this work done while authors were visiting IDC Herzliya, supported by the European Research Council under the European Union's Seventh Framework Programme (FP 2007-2013), ERC Grant Agreement n. 307952. The first and second author were supported in part by RGC GRF grants CUHK410112 and CUHK410113. The third author was supported by DFG Research Training Group GRK 1817/1. The fifth author was supported by ISF grant no.1255/12 and by the ERC under the EU's Seventh Framework Programme (FP/2007-2013) ERC Grant Agreement n. 307952. Work in part done while the author was visiting the Simons Institute for the Theory of Computing, supported by the Simons Foundation and by the DIMACS/Simons Collaboration in Cryptography through NSF grant #CNS-1523467.

E. Kushilevitz and T. Malkin (Eds.): TCC 2016-A, Part I, LNCS 9562, pp. 209–224, 2016.
DOI: 10.1007/978-3-662-49096-9_9

1   Introduction

1.1   Learning with Rounding

The learning with rounding (LWR) problem, introduced by Banerjee, Peikert, and Rosen [BPR12], concerns the cryptographic properties of the function $f_\mathbf{s}\colon \mathbb{Z}_q^n \to \mathbb{Z}_p$ given by

$$f_\mathbf{s}(\mathbf{x}) = \lfloor \langle \mathbf{x}, \mathbf{s} \rangle \rceil_p = \lfloor (p/q) \cdot \langle \mathbf{x}, \mathbf{s} \rangle \rceil$$

where $\mathbf{s} \in \mathbb{Z}_q^n$ is a secret key, $\langle \mathbf{x}, \mathbf{s} \rangle$ is the inner product of $\mathbf{x}$ and $\mathbf{s}$ mod q, and $\lfloor \cdot \rceil$ denotes the closest integer. In this work we are interested in the algorithmic hardness of the tasks of learning the secret $\mathbf{s}$ and of distinguishing $f_\mathbf{s}$ from a random function given uniform and independent samples of the form $(\mathbf{x}, f_\mathbf{s}(\mathbf{x}))$.

Learning with rounding was proposed as a deterministic variant of the learning with errors (LWE) problem [Reg05]. In this problem $f_\mathbf{s}$ is replaced by the randomized function $g_\mathbf{s}\colon \mathbb{Z}_q^n \to \mathbb{Z}_q$ given by $g_\mathbf{s}(\mathbf{x}) = \langle \mathbf{x}, \mathbf{s} \rangle + e$, where e is sampled from some error distribution over $\mathbb{Z}_q$ independently for every input $\mathbf{x} \in \mathbb{Z}_q^n$.

In spite of the superficial similarities between the two problems, the cryptographic hardness of LWE is much better understood. Extending works of Regev [Reg05], Peikert [Pei09], and others, Brakerski et al. [BLP+13] gave a polynomial-time reduction from finding an approximate shortest vector in an arbitrary lattice to the task of distinguishing $g_\mathbf{s}$ from a random function given access to uniform and independent samples $(\mathbf{x}, g_\mathbf{s}(\mathbf{x}))$ when e is drawn from the discrete Gaussian distribution of sufficiently large standard deviation. Their reduction is versatile in two important aspects. First, it is meaningful for any modulus q that exceeds the standard deviation of the noise. Second, it does not assume a bound on the number of samples given to the distinguisher.

In contrast, the hardness of the learning with rounding problem has only been established for restricted settings of the parameters. In their work Banerjee, Peikert, and Rosen show that if $f_\mathbf{s}$ can be efficiently distinguished from a random function given m random samples with advantage δ, then so can $g_\mathbf{s}$ with advantage $\delta - O(mBp/q)$, where the noise e is supported on the range of integers $\{-B, \ldots, B\}$ modulo q. From here one can conclude the hardness of distinguishing $f_\mathbf{s}$ from a random function given m random samples assuming the hardness of learning with errors, but only when the modulus q is of an exponential order of magnitude in the security parameter.

Alwen et al. [AKPW13] give a reduction from LWE to the same problem assuming that q_{max} is at least as large as $2nmBp$ and q_{max}^2 does not divide q, where q_{max} is the largest prime divisor of q. This reduction can be meaningful even for values of q that are polynomially related to the security parameter. For example, when q is a prime number then the improvement over the reduction of Banerjee, Peikert, and Rosen is substantial.

However, the result of Alwen et al. does not apply to all (sufficiently large) values of the modulus q. For example it does not cover values of q that are powers of two. In this case the rounding function is particularly natural as it outputs the first $\log p$ significant bits of q in binary representation. Moreover, rounding with a small prime q necessarily introduces noticeable bias, consequently requiring some form of deterministic extraction. Finally, the work of Alwen et al. does not include a treatment of the significantly more efficient ring variant of LWR.

1.2 Our Results

We establish the cryptographic hardness of the function $f_{\mathbf{s}}$ in the following three settings:

One-Wayness: In Theorem 1 in Sect. 2 we show that any algorithm that recovers the secret $\mathbf{s}$ from m independent random samples of the form $(\mathbf{x}, f_{\mathbf{s}}(\mathbf{x}))$ with probability at least ε also recovers the secret $\mathbf{s}$ from m independent random samples of the form $(\mathbf{x}, \lfloor g_{\mathbf{s}}(\mathbf{x}) \rceil_p)$ with probability at least $\varepsilon^2/(1 + 2Bp/q)^m$.

Therefore, if the function $G(\mathbf{x}_1, \ldots, \mathbf{x}_m, \mathbf{s}) = (\mathbf{x}_1, \ldots, \mathbf{x}_m, g_{\mathbf{s}}(\mathbf{x}_1), \ldots, g_{\mathbf{s}}(\mathbf{x}_m))$ is one-way under some B-bounded distribution (*i.e.* if the search version of LWE is hard) then we conclude that

$$F(\mathbf{x}_1, \ldots, \mathbf{x}_m, \mathbf{s}) = (\mathbf{x}_1, \ldots, \mathbf{x}_m, f_{\mathbf{s}}(\mathbf{x}_1), \ldots, f_{\mathbf{s}}(\mathbf{x}_m))$$

is also one-way, as long as $q \geq 2mBp$.

In Theorem 2 in Sect. 2.2 we show that the ring variants of the LWE and LWR problems (defined in that section) are related in an analogous manner.

Pseudorandomness: In Theorem 3 in Sect. 3 we show that if there exists an efficient distinguisher that tells apart m independent random samples $(\mathbf{x}, g_{\mathbf{s}}(\mathbf{x}))$ from m independent random samples of the form $(\mathbf{x}, \lfloor u \rceil_p)$, then LWE secrets can be learned efficiently assuming $q \geq 2mBp$.

In particular, when p divides q, the above function F is a pseudorandom generator assuming the hardness of learning with errors.

Theorem 3 improves upon several aspects of the work of Alwen et al.: First, we do not impose any number-theoretic restrictions on q; second, they require the stronger condition $q \geq 2nmBp$; third, unlike theirs, our reduction is sample preserving; and fourth, we believe our proof is considerably simpler. On the other hand, the complexity of their reduction has a better dependence on the modulus q and the distinguishing probability.

Hardness of learning from samples with uniform noise: In Theorem 5 in Sect. 4 we give an efficient reduction that takes as input independent random samples of the form $(\mathbf{x}, g_{\mathbf{s}}(\mathbf{x}))$ and produces independent random samples of the form $(\mathbf{x}, f_{\mathbf{s}}(\mathbf{x}))$ provided that p divides q and the noise e of $g_{\mathbf{s}}$ is uniformly distributed over the integers in the range $[-q/2p, \ldots, q/2p)$. Therefore if $f_{\mathbf{s}}$ can be distinguished efficiently from a random function for any number of independent random samples, so can $g_{\mathbf{s}}$. By a reduction of Chow [Cho13] in the other direction (Theorem 6), the two problems are in fact equivalent. These reductions do not impose any additional restriction on p, q and the number of LWR samples m.

The learning with errors problem under this noise distribution is not known to be as hard as the learning with errors problem with discrete

Gaussian noise when the number of samples is unbounded in terms of q and n. The existence of a reduction to the case of discrete Gaussian noise is an interesting open problem.

Noise flooding: In addition, our technique allows for an improved analysis of noise flooding. The noise flooding technique is ubiquitous in the LWE cryptographic literature. Roughly speaking, it is used to rerandomize a faulty sample $(\mathbf{x}, \langle \mathbf{x}, \mathbf{s} \rangle + e_{\mathsf{bad}})$ into one of the form $(\mathbf{x}, \langle \mathbf{x}, \mathbf{s} \rangle + e_{\mathsf{good}})$ where e_{good} is distributed according to the error distribution implicit in $g_\mathbf{s}(\cdot)$, while e_{bad} is not. Most of the time, the desired error distribution is a discrete Gaussian over $\mathbb{Z}_q$ whereas e_{bad} is some arbitrary B-bounded element in $\mathbb{Z}_q$. The most common method is to draw a fresh Gaussian error e and set $e_{\mathsf{good}} = e_{\mathsf{bad}} + e$ which results in the distribution of e_{good} being within statistical distance B/σ of the desired Gaussian. However, this requires choosing parameters in order to ensure that $B/\sigma \geq B/q$ is small. In particular, it requires setting q to be larger than any polynomial in the security parameter. Even worse, often the bound B is polynomially related to the standard deviation σ' of another discrete Gaussian used in the construction. This means that q/σ' also grows faster than any polynomial in the security parameter, which is not ideal as the quantity q/σ' corresponds to the strength of assumption one is making on the hardness of the underlying lattice problem. In Sect. 5 we use techniques from Sect. 2 to give a simple proof that noise flooding can be used whenever $q = \Omega(B\sqrt{m})$. In particular, it can be used even when q is polynomial in the security parameter.

Conventions. We write $x \leftarrow X$ for a uniform sample from the set X, $R(\mathbf{x})$ for the function $(R(x_1), \ldots, R(x_n))$, and $\mathbb{Z}_q^{n*}$ for the set of vectors in $\mathbb{Z}_q^n$ which are not zero-divisors. Namely, $\mathbb{Z}_q^{n*} = \{\mathbf{x} \in \mathbb{Z}_q^n : \gcd(x_1, \ldots, x_n, q) = 1\}$. All algorithms are assumed to be randomized.

2   One-Wayness of LWR

In this section we prove the following theorem. We say a distribution over $\mathbb{Z}_q$ is B-*bounded* if it is supported over the interval of integers $\{-B, \ldots, B\}$, where $B \leq (q-1)/2$. We say a B-bounded distribution e is *balanced* if $\Pr[e \leq 0] \geq 1/2$ and $\Pr[e \geq 0] \geq 1/2$.

Theorem 1. *Let p, q, n, m, and B be integers such that $q > 2pB$. For every algorithm* Learn,

$$\Pr_{\mathbf{A}, \mathbf{s}, \mathbf{e}}[\mathsf{Learn}(\mathbf{A}, \lfloor \mathbf{As} + \mathbf{e} \rceil_p) = \mathbf{s}] \geq \frac{\Pr_{\mathbf{A}, \mathbf{s}}[\mathsf{Learn}(\mathbf{A}, \lfloor \mathbf{As} \rceil_p) = \mathbf{s}]^2}{(1 + 2pB/q)^m},$$

where $\mathbf{A} \leftarrow \mathbb{Z}_q^{m \times n}$, the noise $\mathbf{e}$ is independent over all m coordinates, B-bounded and balanced in each coordinate, and $\mathbf{s}$ is chosen from any distribution supported on $\mathbb{Z}_q^{n*}$.

The assumptions made on the secret and error distribution in Theorem 1 are extremely mild. The condition $\mathbf{s} \in \mathbb{Z}_q^{n*}$ is satisfied for at least a $1 - O(1/2^n)$ fraction of secrets $s \leftarrow \mathbb{Z}_q^n$. While a B-bounded error distribution may not be balanced, it can always be converted to a $2B$-bounded and balanced error distribution by a suitable constant shift. The discrete Gaussian distribution of standard deviation σ is $e^{-\Omega(t^2)}$-statistically close to being $t\sigma$-bounded and balanced for every $t \geq 1$.

Theorem 2 in Sect. 2.2 concerns the ring variants of the LWR and LWE problems and will be proved in an analogous manner.

We now outline the proof of Theorem 1. Let $\mathsf{X_s}$ denote the distribution of a single LWR sample $\mathbf{a}, \lfloor \langle \mathbf{a}, \mathbf{s} \rangle \rceil_p$ where $\mathbf{a} \leftarrow \mathbb{Z}_q^n$ and $\mathsf{Y_s}$ denote the distribution of a single rounded LWE sample $\mathbf{a}, \lfloor \langle \mathbf{a}, \mathbf{s} \rangle + e \rceil_p$. To prove Theorem 1 we will fix $\mathbf{s}$ and look at the ratio of probabilities of any possible instance under the product distributions $\mathsf{X_s^m}$ and $\mathsf{Y_s^m}$, respectively. If this ratio was always bounded by a sufficiently small quantity K,[1] then it would follow that the success probability of any search algorithm for LWR does not deteriorate by more than a factor of $1/K$ when it is run on rounded LWE instances instead.

While it happens that there are exceptional instances for which the ratio of probabilities under $\mathsf{X_s^m}$ and $\mathsf{Y_s^m}$ can be large, our proof of Theorem 1 will show that such instances cannot occur too often under the rounded LWE distribution and therefore does not significantly affect the success probability of the inversion algorithm. This can be showed by a standard probabilistic analysis, but we opt instead to work with a measure of distributions that is particularly well suited for bounding ratios of probabilities: the Rényi divergence.

The role of Rényi divergence in our analysis accounts for our quantitative improvement over the result of Banerjee, Peikert, and Rosen, who used the measure of statistical distance in its place. Rényi divergence has been used in a related context: Bai, Langlois, Lepoint, Stehlé and Steinfeld [BLL+15] use it to obtain tighter bounds for several lattice-based primitives.

2.1   Proof of Theorem 1

Given two distributions X and Y over Ω, the power of their Rényi divergence[2] is $\mathrm{RD}_2(\mathsf{X}\|\mathsf{Y}) = \mathbb{E}_{a \leftarrow \mathsf{X}}[\Pr[\mathsf{X} = a]/\Pr[\mathsf{Y} = a]]$.

Lemma 1. *Let $\mathsf{X_s}$ be the distribution of a single LWR sample and let $\mathsf{Y_s}$ be that of a single rounded LWE sample. Assume $B < q/2p$. For every $\mathbf{s} \in \mathbb{Z}_q^{n*}$ and every noise distribution that is B-bounded and balanced, $\mathrm{RD}_2(\mathsf{X_s}\|\mathsf{Y_s}) \leq 1 + 2Bp/q$.*

Proof. By the definition of Rényi divergence,

$$\mathrm{RD}_2(\mathsf{X_s}\|\mathsf{Y_s}) = \mathbb{E}_{\mathbf{a} \leftarrow \mathbb{Z}_q^n} \frac{\Pr\left[\mathsf{X_s} = (\mathbf{a}, \lfloor \langle \mathbf{a}, \mathbf{s} \rangle \rceil_p)\right]}{\Pr\left[\mathsf{Y_s} = (\mathbf{a}, \lfloor \langle \mathbf{a}, \mathbf{s} \rangle \rceil_p)\right]} = \mathbb{E}_{\mathbf{a} \leftarrow \mathbb{Z}_q^n} \frac{1}{\Pr_e\left[\lfloor \langle \mathbf{a}, \mathbf{s} \rangle + e \rceil_p = \lfloor \langle \mathbf{a}, \mathbf{s} \rangle \rceil_p\right]}.$$

---

[1] Levin [Lev86] calls this condition K-domination.

[2] Rényi divergences [vEH14] are a class of measures parametrized by a real number $\alpha > 1$. The definition we give specializes α to 2, which is sufficient for our analysis.

Let $\mathsf{BAD_s}$ be the set $\{\mathbf{a} \in \mathbb{Z}_q^n : |\langle \mathbf{a}, \mathbf{s} \rangle - \frac{q}{p}\lfloor\langle \mathbf{a}, \mathbf{s}\rangle\rceil_p| < B\}$. These are the $\mathbf{a}$ for which $\langle \mathbf{a}, \mathbf{s} \rangle$ is dangerously close to the rounding boundary. When $\mathbf{a} \notin \mathsf{BAD_s}$, $\Pr_e\left[\lfloor\langle \mathbf{a}, \mathbf{s}\rangle + e\rceil_p = \lfloor\langle \mathbf{a}, \mathbf{s}\rangle\rceil_p\right] = 1$. Since $\gcd(s_1, \ldots, s_n, q) = 1$, the inner product $\langle \mathbf{a}, \mathbf{s} \rangle$ is uniformly distributed over $\mathbb{Z}_q$, so $\Pr[\mathbf{a} \in \mathsf{BAD_s}] \leq (2B-1)p/q$. When $\mathbf{a} \in \mathsf{BAD_s}$, the event $\lfloor\langle \mathbf{a}, \mathbf{s}\rangle + e\rceil_p = \lfloor\langle \mathbf{a}, \mathbf{s}\rangle\rceil_p$ still holds at least in one of the two cases $e \leq$ or $e \geq 0$. By our assumptions on the noise distribution, $\Pr_e\left[\lfloor\langle \mathbf{a}, \mathbf{s}\rangle + e\rceil_p = \lfloor\langle \mathbf{a}, \mathbf{s}\rangle\rceil_p\right] \geq 1/2$. Conditioning over the event $\mathbf{a} \in \mathsf{BAD_s}$, we conclude that

$$\mathrm{RD}_2(\mathsf{X_s}\|\mathsf{Y_s}) \leq 1 \cdot \Pr[\mathbf{a} \notin \mathsf{BAD_s}] + 2 \cdot \Pr[\mathbf{a} \in \mathsf{BAD_s}] \leq 1 + \frac{2Bp}{q}. \qquad \square$$

To complete the proof of Theorem 1 we need two elementary properties of Rényi divergence.

Claim. For any two distributions X and Y, (1) $\mathrm{RD}_2(\mathsf{X}^m\|\mathsf{Y}^m) = \mathrm{RD}_2(\mathsf{X}\|\mathsf{Y})^m$ and (2) for any event E, $\Pr[\mathsf{Y} \in E] \geq \Pr[\mathsf{X} \in E]^2/\mathrm{RD}_2(\mathsf{X}\|\mathsf{Y})$.

Proof. Property (1) follows immediately from independence of the m samples. Property (2) is the Cauchy-Schwarz inequality applied to the functions

$$f(a) = \frac{\Pr[\mathsf{X} = a]}{\sqrt{\Pr[\mathsf{Y} = a]}}; \text{ and } g(a) = \sqrt{\Pr[\mathsf{Y} = a]}. \qquad \square$$

Proof (Proof of Theorem 1). Fix $\mathbf{s}$ such that $\gcd(\mathbf{s}, q) = 1$ and the randomness of Learn. By Lemma 1 and part (1) of Claim 2.1, $\mathrm{RD}_2(\mathsf{X_s^m}\|\mathsf{Y_s^m}) \leq (1 + 2Bp/q)^m$. Letting E be the event $\{(\mathbf{A}, \mathbf{y}): \mathsf{Learn}(\mathbf{A}, \mathbf{y}) = \mathbf{s}\}$, by part (2) of Claim 2.1,

$$\Pr_{\mathbf{A},\mathbf{e}}[\mathsf{Learn}(\mathbf{A}, \lfloor\mathbf{As} + \mathbf{e}\rceil_p) = \mathbf{s}] \geq \frac{\Pr_{\mathbf{A}}[\mathsf{Learn}(\mathbf{A}, \lfloor\mathbf{As}\rceil_p) = \mathbf{s}]^2}{(1 + 2pB/q)^m}.$$

To obtain the theorem, we average over $\mathbf{s}$ and the randomness of Learn and apply the Cauchy-Schwarz inequality. $\qquad \square$

2.2 Hardness over Rings

For many applications it is more attractive to use a ring version of LWR (RLWR). Banerjee, Peikert, and Rosen [BPR12] introduced it together with LWR. It brings the advantage of reducing the entropy of $\mathbf{A}$ for same sized $\lfloor\mathbf{As} + \mathbf{e}\rceil_p$. In the following theorem, we give a variant of Theorem 1 for the RLWR based on the hardness of ring LWE. This theorem is not needed for the remaining sections of the paper.

Theorem 2. *Let p, q, n, k, B be integers such that $q > 2pB$. Let R_q be the ring $\mathbb{Z}_q[x]/g(x)$ where g is a polynomial of degree n over $\mathbb{Z}_q$ and f be an arbitrary function over R_q. For every algorithm Learn,*

$$\Pr_{a,s,\mathbf{e}}[\mathsf{Learn}(a, \lfloor as + \mathbf{e}\rceil_p) = f(s)] \geq \frac{\Pr_{a,s}[\mathsf{Learn}(a, \lfloor as\rceil_p) = f(s)]^2}{(1 + 2pB/q)^{nk}},$$

where $\mathbf{a} \leftarrow R_q^k$, the noise $\mathbf{e}$ is independent over all k coordinates, B-bounded and balanced in each coordinate, and s is chosen from any distribution supported on the set of all units in R_q.

An element in $R_q = \mathbb{Z}_q[x]/g(x)$ can be represented as a polynomial (in x) of degree less than n with coefficients in $\mathbb{Z}_q$. Here, for $a \in R_q$, $\lfloor a \rfloor_p$ is an element in $\mathbb{Z}_p[x]/g(x)$ obtained by applying the function $\lfloor \cdot \rfloor_p$ to each of coefficient of a separately. A distribution over ring R_q is B-bounded and balanced if every coefficient is drawn independently from a B-bounded and balanced distribution over $\mathbb{Z}_q$.

The bound in Theorem 2 matches the bound in Theorem 1 since k can be chosen such that nk is on the order of m. Theorem 2 follows from Claim 2.1 and the following variant of Lemma 1.

Lemma 2. *Assume $B < q/2p$. For every unit $s \in R_q$ and noise distribution χ that is B-bounded and balanced over R_q, $\mathrm{RD}_2(\mathsf{X}_s \| \mathsf{Y}_s) \leq (1 + 2pB/q)^n$ where X_s is the random variable $(a, \lfloor a \cdot s \rfloor_p)$ and Y_s is the random variable $(a, \lfloor a \cdot s \rfloor_p + e)$ with $a \leftarrow R_q$ and $e \leftarrow \chi$.*

Since the proof is very similar to the proof of Lemma 1, we defer it to Appendix A.

3   Pseudorandomness of LWR

In this section we prove the following Theorem. We will implicitly assume that algorithms have access to the prime factorization of the modulus q throughout this section.

Theorem 3. *For every $\varepsilon > 0$, n, m, $q > 2pB$, and algorithm Dist such that*

$$\left| \Pr_{\mathbf{A},\mathbf{s}}\left[\mathsf{Dist}\left(\mathbf{A}, \lfloor \mathbf{As} \rfloor_p\right) = 1\right] - \Pr_{\mathbf{A},\mathbf{u}}\left[\mathsf{Dist}\left(\mathbf{A}, \lfloor \mathbf{u} \rfloor_p\right) = 1\right] \right| \geq \varepsilon, \qquad (1)$$

where $\mathbf{A} \leftarrow \mathbb{Z}_q^{m \times n}$, $\mathbf{s} \leftarrow \{0,1\}^n$ and $\mathbf{u} \leftarrow \mathbb{Z}_q^m$ there exists an algorithm Learn that runs in time polynomial in n, m, the number of divisors of q, and the running time of Dist such that

$$\Pr_{\mathbf{A},\mathbf{s}}\left[\mathsf{Learn}\left(\mathbf{A}, \mathbf{As} + \mathbf{e}\right) = \mathbf{s}\right] \geq \left(\frac{\varepsilon}{4qm} - \frac{2^n}{p^m}\right)^2 \cdot \frac{1}{(1 + 2Bp/q)^m} \qquad (2)$$

for any noise distribution $\mathbf{e}$ that is B-bounded and balanced in each coordinate.

One unusual aspect of this theorem is that the secret is a uniformly distributed *binary* string in $\mathbb{Z}_q^n$. This assumption can be made essentially without loss of generality: Brakerski et al. [BLP+13] show that under discrete Gaussian noise, learning a binary secret in $\{0,1\}^n$ from LWE samples is as hard as learning a secret uniformly sampled from $\mathbb{Z}_q^{\Omega(n/\log q)}$. The assumption (1) can also be stated with $\mathbf{s}$ sampled uniformly from $\mathbb{Z}_q^n$: In Sect. 3.4 we show that distinguishing LWR samples from random ones is no easier for uniformly distributed secrets than it is for any other distribution on secrets, including the uniform distribution over binary secrets. (When q is prime, the proof of Theorem 3 can be carried out for s uniformly distributed over $\mathbb{Z}_q^n$ so these additional steps are not needed.)

To prove Theorem 3 we follow a sequence of standard steps originating from Yao [Yao82], Goldreich and Levin [GL89]: In Lemma 3 we convert the distinguisher Dist into a predictor that given a sequence of LWR samples and a label $\mathbf{a}$ guesses the inner product $\langle \mathbf{a}, \mathbf{s} \rangle$ in $\mathbb{Z}_q$ with significant advantage. In Lemma 4 we show how to use this predictor to efficiently learn the entries of the vector $\mathbf{s}$ modulo q' for some divisor $q' > 1$ of q. If the entries of the secret $\mathbf{s}$ are bits, $\mathbf{s}$ is then fully recovered given LWR samples. By Theorem 1 the learner's advantage does not deteriorate significantly when the LWR samples are replaced by LWE samples.

Our proof resembles the work of Micciancio and Mol [MM11] who give, to the best of our knowledge, the only sample preserving search-to-decision reduction for LWE (including its variants). Unlike our theorem, theirs imposes certain number-theoretic restrictions on q. Also, while Micciancio and Mol work with a problem that is "dual" to LWE, we work directly with LWR samples.

3.1   Predicting the Inner Product

Lemma 3. *For all ε (possibly negative), n, m, q, every polynomial-time function R over $\mathbb{Z}_q$, and every algorithm Dist such that*

$$\Pr_{\mathbf{A},\mathbf{s}}\big[\mathsf{Dist}\big(\mathbf{A}, R(\mathbf{As})\big) = 1\big] - \Pr_{\mathbf{A},\mathbf{u}}\big[\mathsf{Dist}\big(\mathbf{A}, R(\mathbf{u})\big) = 1\big] = \varepsilon,$$

there exists an algorithm Pred whose running time is polynomial in its input size and the running time of Dist such that

$$\Pr_{\mathbf{A},\mathbf{s},\mathbf{a}}\big[\mathsf{Pred}\big(\mathbf{A}, R(\mathbf{As}), \mathbf{a}\big) = \langle \mathbf{a}, \mathbf{s} \rangle\big] = \frac{1}{q} + \frac{\varepsilon}{mq}.$$

where the probabilities are taken over $\mathbf{A} \leftarrow \mathbb{Z}_q^{m \times n}$, $\mathbf{u} \leftarrow \mathbb{Z}_q^m$, the random coins of the algorithms, and secret $\mathbf{s}$ sampled from an arbitrary distribution.

Here, $R(\mathbf{y})$ is the vector obtained by applying R to every coordinate of the vector $\mathbf{y}$.

Proof. Consider the following algorithm Pred. On input $(\mathbf{A}, \mathbf{b}) = ((\mathbf{a}_1, b_1), \ldots, (\mathbf{a}_m, b_m))$ $(\mathbf{a}_j \in \mathbb{Z}_q^n, b_j \in \mathbb{Z}_q)$ and $\mathbf{a} \in \mathbb{Z}_q^n$:

1. Sample a random index $i \leftarrow \{1, \ldots, m\}$ and a random $c \leftarrow \mathbb{Z}_q$.
2. Obtain $\mathbf{A}', \mathbf{b}'$ from $\mathbf{A}, \mathbf{b}$ by replacing $\mathbf{a}_i$ with $\mathbf{a}$, b_i with $R(c)$, and every b_j for $j > i$ with an independent element of the form $R(u_j), u_j \leftarrow \mathbb{Z}_q$.
3. If $\mathsf{Dist}(\mathbf{A}', \mathbf{b}') = 1$, output c. Otherwise, output a uniformly random element in $\mathbb{Z}_q$.

Let $\mathbf{h}_i = \left(R(\langle \mathbf{a}_1, \mathbf{s}\rangle), \ldots, R(\langle \mathbf{a}_i, \mathbf{s}\rangle), R(u_{i+1}), \ldots, R(u_m)\right) \in \mathbb{Z}_p^m$, for i ranging from 0 to m. Then $\mathbf{h}_m = R(\mathbf{As})$ and $\mathbf{h}_0 = R(\mathbf{u})$ so by the assumption on Dist it follows that

$$\mathbb{E}_i\left[\Pr_{\mathbf{A},\mathbf{s},\mathbf{u}}\left[\mathsf{Dist}(\mathbf{A}, \mathbf{h}_i) = 1\right] - \Pr_{\mathbf{A},\mathbf{s},\mathbf{u}}\left[\mathsf{Dist}(\mathbf{A}, \mathbf{h}_{i-1}) = 1\right]\right] = \frac{\varepsilon}{m}.$$

Conditioned on the choice of i,

$$\Pr\left[\mathsf{Pred}(\mathbf{A}, \mathbf{b}, \mathbf{a}) = \langle \mathbf{a}, \mathbf{s}\rangle\right]$$

$$= \Pr\left[\mathsf{Dist}(\mathbf{A}', \mathbf{b}') = 1 \text{ and } c = \langle \mathbf{a}, \mathbf{s}\rangle\right] + \frac{1}{q} \cdot \Pr\left[\mathsf{Dist}(\mathbf{A}', \mathbf{b}') \neq 1\right]$$

$$= \frac{1}{q} \cdot \Pr\left[\mathsf{Dist}(\mathbf{A}', \mathbf{b}') = 1 \mid c = \langle \mathbf{a}, \mathbf{s}\rangle\right] + \frac{1}{q} \cdot \Pr\left[\mathsf{Dist}(\mathbf{A}', \mathbf{b}') \neq 1\right]$$

$$= \frac{1}{q} + \frac{1}{q} \cdot \left(\Pr\left[\mathsf{Dist}(\mathbf{A}', \mathbf{b}') = 1 \big| c = \langle \mathbf{a}, \mathbf{s}\rangle\right] - \Pr\left[\mathsf{Dist}(\mathbf{A}', \mathbf{b}') = 1\right]\right)$$

when $\mathbf{b} = R(\mathbf{As})$, the distribution $(\mathbf{A}', \mathbf{b}')$ is the same as $(\mathbf{A}, \mathbf{h}_{i-1})$ while $(\mathbf{A}', \mathbf{b}')$ conditioned on $c = \langle \mathbf{a}, \mathbf{s}\rangle$ is the same as $(\mathbf{A}, \mathbf{h}_i)$. Averaging over i yields the desired advantage of Pred. □

3.2   Learning the Secret

Lemma 4. *There exists an oracle algorithm* List *such that for every algorithm* Pred *satisfying* $|\Pr[\mathsf{Pred}(\mathbf{a}) = \langle \mathbf{a}, \mathbf{s}\rangle] - 1/q| \geq \varepsilon$, $\mathsf{List}^{\mathsf{Pred}}(\varepsilon)$ *outputs a list of entries* $(q', \mathbf{s}')$ *containing at least one such that* $q' > 1$, q' *divides* q, *and* $\mathbf{s}' = \mathbf{s} \bmod q'$ *in time polynomial in* n, $1/\varepsilon$, *and the number of divisors of* q *with probability at least* $\varepsilon/4$. *The probabilities are taken over* $\mathbf{a} \leftarrow \mathbb{Z}_q^n$, *any distribution on* $\mathbf{s}$, *and the randomness of the algorithms.*

When q is a prime number, the conclusion of the theorem implies that the list must contain the secret $\mathbf{s}$. When q is a composite, the assumption does not in general guarantee full recovery of $\mathbf{s}$. For example, the predictor $\mathsf{Pred}(a) = \langle \mathbf{a}, \mathbf{s}\rangle \bmod q'$ has advantage $\varepsilon = (q' - 1)/q$ but does not distinguish between pairs of secrets that are congruent modulo q'. In this case List cannot hope to learn any information on $\mathbf{s}$ beyond the value $\mathbf{s}$ modulo q'.

The proof of Lemma 4 makes use of the following result of Akavia, Goldwasser, and Safra [AGS03] on learning heavy Fourier coefficients, extending work of Kushilevitz, Mansour, and others. Recall that the Fourier coefficients of a function $h: \mathbb{Z}_q^n \to \mathbb{C}$ are the complex numbers $\hat{h}(\mathbf{a}) = \mathbb{E}_{\mathbf{x} \leftarrow \mathbb{Z}_q^n}[h(\mathbf{x})\omega^{-\langle \mathbf{a}, \mathbf{x}\rangle}]$, where $\omega = e^{2\pi i/q}$ is a primitive q-th root of unity. Our functions of interest all map into the unit complex circle $\mathbb{T} = \{c \in \mathbb{C}: |c| = 1\}$, so we specialize the result to this setting.

Theorem 4 (Akavia et al. [AGS03]). *There is an algorithm* AGS *that given query access to a function* $h: \mathbb{Z}_q^n \to \mathbb{T}$ *outputs a list of size at most* $2/\varepsilon^2$ *which contains all* $\mathbf{a} \in \mathbb{Z}_q^n$ *such that* $|\hat{h}(\mathbf{a})| \geq \varepsilon$ *in time polynomial in* n, $\log q$, *and* $1/\varepsilon$ *with probability at least* $1/2$.

We will also need the following property of the Fourier transform of random variables. For completeness the proof is given below.

Claim. For every random variable Z over $\mathbb{Z}_q$ there exists a nonzero r in $\mathbb{Z}_q$ such that $|E[\omega^{rZ}]| \geq |\Pr[Z = 0] - 1/q|$.

Proof (Proof of Lemma 4). We first replace Pred by the following algorithm: Sample a uniformly random unit (invertible element) u from $\mathbb{Z}_q^*$ and output $u^{-1}\mathsf{Pred}(u\mathbf{a})$. This transformation does not affect the advantage of Pred but ensures that for fixed $\mathbf{s}$ and randomness of Pred, the value $E_{\mathbf{a}}[\omega^{r(\mathsf{Pred}(\mathbf{a}) - \langle \mathbf{a}, \mathbf{s} \rangle)}]$ is the same for all r with the same $\gcd(r, q)$.

Algorithm List works as follows: For every divisor $r < q$ of q run AGS with oracle access to the function $h_r(\mathbf{a}) = \omega^{r \cdot \mathsf{Pred}(\mathbf{a})}$ and output $(q' = q/r, \mathbf{s}'/r \mod q')$ for every $\mathbf{s}'$ in the list produced by AGS.

We now assume Pred satisfies the assumption of the lemma and analyze List. By Claim 3.2 there exists a nonzero $r \in \mathbb{Z}_q$ such that $|E[\omega^{r(\mathsf{Pred}(\mathbf{a}) - \langle \mathbf{a}, \mathbf{s} \rangle)}]| \geq \varepsilon$. By Markov's inequality and the convexity of the absolute value, with probability at least $\varepsilon/2$ over the choice of $\mathbf{s}$ and the randomness of Pred $|E_{\mathbf{a}}[\omega^{r(\mathsf{Pred}(\mathbf{a}) - \langle \mathbf{a}, \mathbf{s} \rangle)}]|$ is at least $\varepsilon/2$. We fix $\mathbf{s}$ and the randomness of Pred and assume this is the case. By our discussion on Pred, the expectation of interest is the same for all r with the same $\gcd(r, q)$, so we may and will assume without loss of generality that r is a divisor of q.

Since $E_{\mathbf{a}}[\omega^{r(\mathsf{Pred}(\mathbf{a}) - \langle \mathbf{a}, \mathbf{s} \rangle)}] = \hat{h}_r(r\mathbf{s})$, by Theorem 4, the r-th run of AGS outputs $r\mathbf{s}$ with probability at least $1/2$. Since $(r\mathbf{s})/r \mod q' = \mathbf{s} \mod q'$ it follows that the entry $(q', \mathbf{s} \mod q')$ must appear in the output of List with probability at least $(1/2)(\varepsilon/2) = \varepsilon/4$. Regarding time complexity, List makes a call to AGS for every divisor of q except q, so its running time is polynomial in n and the number of divisors of q. $\qquad\qquad\square$

Proof (Proof of Claim 3.2). Let $\varepsilon = \Pr[Z = 0] - 1/q$ and $h(a) = q(\Pr[Z = a] - \Pr[U = a])$, where $U \leftarrow \mathbb{Z}_q$ is a uniform random variable. By Parseval's identity from Fourier analysis,

$$\sum_{r \in \mathbb{Z}_q} |\hat{h}(r)|^2 = E_{a \leftarrow \mathbb{Z}_q}[h(a)^2] \geq \frac{1}{q}h(0)^2 = q\varepsilon^2.$$

On the left hand side, after normalizing we obtain that $\hat{h}(r) = E[\omega^{-rZ}] - E[\omega^{-rU}]$. Therefore $\hat{h}(0) = 0$, so $|\hat{h}(r)|^2 = |E[\omega^{-rZ}]|^2$ must be at least as large as $q\varepsilon^2/(q-1)$ for at least one nonzero value of r, giving a slightly stronger conclusion than desired. $\qquad\qquad\square$

3.3   Proof of Theorem 3

On input $(\mathbf{A}, \mathbf{b})$, algorithm Learn runs $\mathsf{List}^{\mathsf{Pred}(\mathbf{A}, \lfloor \mathbf{b} \rfloor_p, \cdot)}(\varepsilon/2qm)$ and outputs any $\mathbf{s} \in \{0, 1\}^n$ appearing in the list such that $\lfloor \mathbf{As} \rfloor_p = \lfloor \mathbf{b} \rfloor_p$ (or the message fail if no such $\mathbf{s}$ exists). By Theorem 1,

$$\Pr[\mathsf{Learn}(\mathbf{A}, \lfloor \mathbf{As} + \mathbf{e} \rfloor_p) = \mathbf{s}] \geq \frac{\Pr[\mathsf{Learn}(\mathbf{A}, \lfloor \mathbf{As} \rfloor_p) = \mathbf{s}]^2}{(1 + 2Bp/q)^m}.$$

For $\mathsf{Learn}(\mathbf{A}, \lfloor \mathbf{As} \rceil_p)$ to output $\mathbf{s}$ it is sufficient that $\mathbf{s}$ appears in the output of $\mathsf{List}^{\mathsf{Pred}(\mathbf{A}, \lfloor \mathbf{As} \rceil_p, \cdot)}(\varepsilon/2qm)$ and that no other $\mathbf{s}' \in \{0,1\}^n$ satisfies $\lfloor \mathbf{As}' \rceil_p = \lfloor \mathbf{As} \rceil_p$. By Lemmas 3 and 4, the list contains $\mathbf{s} \bmod q'$ for some q' with probability at least $\varepsilon/4qm$. Since $\mathbf{s}$ is binary, $\mathbf{s} \bmod q' = \mathbf{s}$. By a union bound, the probability that some $\lfloor \mathbf{As}' \rceil_p = \lfloor \mathbf{As} \rceil_p$ for some $\mathbf{s}' \neq \mathbf{s}$ is at most $2^n p^{-m}$ and so

$$\Pr[\mathsf{Learn}(\mathbf{A}, \lfloor \mathbf{As} + \mathbf{e} \rceil_p) = \mathbf{s}] \geq \frac{(\varepsilon/4qm - 2^n p^{-m})^2}{(1 + 2Bp/q)^m}.$$

3.4 Rerandomizing the Secret

Lemma 5. *Let S be any distribution supported on $\mathbb{Z}_q^{n*}$. For every function R on $\mathbb{Z}_q$, there is a polynomial-time transformation that (1) maps the distribution $(\mathbf{A}, R(\mathbf{As}))_{\mathbf{A} \leftarrow \mathbb{Z}_q^{m \times n}, \mathbf{s} \leftarrow S}$ to $(\mathbf{A}, R(\mathbf{As}))_{\mathbf{A} \leftarrow \mathbb{Z}_q^{m \times n}, \mathbf{s} \leftarrow \mathbb{Z}_q^{n*}}$ and (2) maps the distribution $(\mathbf{A}, R(\mathbf{u}))_{\mathbf{A} \leftarrow \mathbb{Z}_q^{m \times n}, \mathbf{u} \leftarrow \mathbb{Z}_q^{m}}$ to itself.*

In particular, it follows that the distinguishing advantage (1) can be preserved when the secret is chosen uniformly from $\mathbb{Z}_q^{n*}$ instead of uniformly from $\{0,1\}^n - \{0^n\}$. The sets $\mathbb{Z}_q^{n*}$ and $\{0,1\}^n - \{0^n\}$ can be replaced by $\mathbb{Z}_q^n$ and $\{0,1\}^n$, respectively, if we allow for failure with probability $O(2^{-n})$.

To prove Lemma 5 we need a basic fact from algebra. We omit the easy proof.

Claim. Multiplication by an $n \times n$ invertible matrix over $\mathbb{Z}_q$ is a transitive action on $\mathbb{Z}_q^{n*}$.

Proof (Proof of Lemma 5). Choose a uniformly random invertible matrix $\mathbf{P} \in \mathbb{Z}_q^{n \times n}$ and apply the map $f(\mathbf{a}, b) = (\mathbf{Pa}, b)$ to every row. Clearly this map satisfies the second condition. For the first condition, we write $f(\mathbf{a}, R(\langle \mathbf{a}, \mathbf{s} \rangle)) = (\mathbf{Pa}, R(\langle \mathbf{a}, \mathbf{s} \rangle))$, which is identically distributed as $(\mathbf{a}, R(\langle \mathbf{a}, \mathbf{P}^{-t}\mathbf{s} \rangle))$. By Claim 3.4, for every $\mathbf{s}$ in the support of S the orbit of $\mathbf{P}^{-t}\mathbf{s}$ is $\mathbb{Z}_q^{n*}$, so by symmetry $\mathbf{P}^{-t}\mathbf{s}$ is uniformly random in $\mathbb{Z}_q^{n*}$. Therefore the first condition also holds. □

4 Equivalence of LWR and LWE with Uniform Errors

When the number of LWR samples is not a priori bounded, we show that the pseudorandomness (resp. one-wayness) of LWR follows from the pseudorandomness (resp. one-wayness) of LWE with a uniform noise distribution over the range of integers $[-\frac{q}{2p}, \ldots, \frac{q}{2p})$. We use a rejection sampling based approach to reject LWE samples which are likely to be rounded to the wrong value in $\mathbb{Z}_p$. This comes at the cost of throwing away samples, and indeed the sample complexity of our reduction grows with q.

Theorem 5. *Let p and q be integers such that p divides q. Then there is a reduction R with query access to independent samples such that for every $\mathbf{s} \in \mathbb{Z}_q^{n*}$:*

- *Given query access to samples* $(\mathbf{a}, \langle \mathbf{a}, \mathbf{s} \rangle + e)$, $\mathbf{a} \leftarrow \mathbb{Z}_q^n$, $e \leftarrow [-\frac{q}{2p}, \ldots, \frac{q}{2p}) \subset \mathbb{Z}_q$, R *outputs samples from the distribution* $(\mathbf{a}, \lfloor \langle \mathbf{a}, \mathbf{s} \rangle \rceil_p)$, $\mathbf{a} \leftarrow \mathbb{Z}_q^n$,
- *Given query access to uniform samples* $(\mathbf{a}, u)$, $\mathbf{a} \leftarrow \mathbb{Z}_q^n$, $u \leftarrow \mathbb{Z}_q$, R *outputs a uniform sample* $(\mathbf{a}, v)$, $\mathbf{a} \leftarrow \mathbb{Z}_q^n$, $v \leftarrow \mathbb{Z}_p$.

In both cases, the expected running time and sample complexity of the reduction is $O(q/p)$.

Proof. We view the set $(q/p)\mathbb{Z}_p$ as a subset of $\mathbb{Z}_q$. The reduction R queries its oracle until it obtains the first sample $(\mathbf{a}, b) \in \mathbb{Z}_q^n \times \mathbb{Z}_q$ such that b is in the set $(q/p)\mathbb{Z}_p$ and outputs $(\mathbf{a}, (p/q)b) \in \mathbb{Z}_q^n \times \mathbb{Z}_p$. In both cases of interest b is uniformly distributed in $\mathbb{Z}_q$, so the expected number of query calls until success is q/p.

When the queried samples are uniformly distributed in $\mathbb{Z}_q^n \times \mathbb{Z}_q$, the output is also uniformly distributed in $\mathbb{Z}_q^n \times \mathbb{Z}_p$. For queried samples of the form $(\mathbf{a}, \langle \mathbf{a}, \mathbf{s} \rangle + e)$, we calculate the probability mass function of the output distribution. For every possible output $(\mathbf{a}', b')$, we have

$$\Pr\left[\text{R outputs } (\mathbf{a}', b')\right] = \Pr\left[\mathbf{a} = \mathbf{a}' \text{ and } \langle \mathbf{a}, \mathbf{s} \rangle + e = b' \mid \langle \mathbf{a}, \mathbf{s} \rangle + e \in (q/p)\mathbb{Z}_p\right]$$

$$= \Pr_{\mathbf{a}}[\mathbf{a} = \mathbf{a}'] \cdot \frac{\Pr_e\left[\langle \mathbf{a}, \mathbf{s} \rangle + e = (q/p)b' \mid \mathbf{a} = \mathbf{a}'\right]}{\Pr_e\left[\langle \mathbf{a}, \mathbf{s} \rangle + e \in (q/p)\mathbb{Z}_p \mid \mathbf{a} = \mathbf{a}'\right]}$$

$$= q^{-n} \cdot \begin{cases} \frac{p/q}{p/q}, & \text{if } (q/p)b' - \langle \mathbf{a}', \mathbf{s} \rangle \in [-\frac{q}{2p}, \ldots, \frac{q}{2p}) \\ 0, & \text{otherwise.} \end{cases}$$

$$= \begin{cases} q^{-n}, & \text{if } b' = \lfloor \langle \mathbf{a}', \mathbf{s} \rangle \rceil_p \\ 0, & \text{otherwise.} \end{cases}$$

This is the probability mass function of the distribution $(\mathbf{a}, \lfloor \langle \mathbf{a}, \mathbf{s} \rangle \rceil_p)$, as desired. $\qquad\square$

The following theorem whose proof appears in the M.Eng. thesis of Chow [Cho13] shows that distinguishing LWR samples from uniform and inverting LWR samples are not substantially harder than they are for LWE samples under the above noise distribution.

Theorem 6. *For all m, n, p, q such that p divides q, and ε (possibly negative), and polynomial-time algorithm* Dist *such that*

$$\Pr_{\mathbf{A}, \mathbf{s}}\left[\mathsf{Dist}(\mathbf{A}, \mathbf{A}\mathbf{s} + \mathbf{e}) = 1\right] - \Pr_{\mathbf{A}, \mathbf{u}}\left[\mathsf{Dist}(\mathbf{A}, \mathbf{u}) = 1\right] = \varepsilon,$$

there exists a polynomial time algorithm Dist' *such that*

$$\Pr_{\mathbf{A}, \mathbf{s}}\left[\mathsf{Dist}'(\mathbf{A}, \lfloor \mathbf{A}\mathbf{s} \rceil_p) = 1\right] - \Pr_{\mathbf{A}, \mathbf{u}}\left[\mathsf{Dist}'(\mathbf{A}, \lfloor \mathbf{u} \rceil_p) = 1\right] = \frac{\varepsilon}{q},$$

where $\mathbf{A} \leftarrow \mathbb{Z}_q^{m \times n}$, *the noise* $\mathbf{e}$ *is independent over all m coordinates and uniform over the set* $[-\frac{q}{2p}, \ldots, \frac{q}{2p}) \subseteq \mathbb{Z}_q$ *in each coordinate, and* $\mathbf{s}$ *is chosen from any distribution supported on* $\mathbb{Z}_q^{n*}$.

Proof. Consider the following algorithm Dist'.

On input $(\mathbf{A}, \mathbf{b}) = ((\mathbf{a}_1, b_1), \ldots, (\mathbf{a}_m, b_m))$ $(\mathbf{a}_j \in \mathbb{Z}_q^n,\ b_j \in \mathbb{Z}_p)$ and $\mathbf{a} \in \mathbb{Z}_q^n$:

1. Sample a random $\mathbf{r} \leftarrow \mathbb{Z}_q^n$ and a random $\mathbf{c} \leftarrow \mathbb{Z}_q^m$.
2. Obtain $\mathbf{A}', \mathbf{b}' \in \mathbb{Z}_q^{m \times n} \times \mathbb{Z}_q^m$ from $\mathbf{A}, \mathbf{b}$ by letting $\mathbf{A}' = \mathbf{A} - \mathbf{c} \bullet \mathbf{r}$, $\mathbf{b}' = \frac{q}{p} \cdot \mathbf{b} - \mathbf{c}$.
3. If $\mathsf{Dist}(\mathbf{A}', \mathbf{b}') = 1$, output 1. Otherwise, output 0.

Here, $\mathbf{c} \bullet \mathbf{r}$ is the outer product of the vectors $\mathbf{c}$ and $\mathbf{r}$.

When $\mathbf{b} = \lfloor \mathbf{u} \rceil_p$, $(\mathbf{A}', \mathbf{b}')$ is distributed as $(\mathbf{A}', \mathbf{u})$. When $\mathbf{b} = \lfloor \mathbf{As} \rceil_p$, we can write

$$
\begin{aligned}
(\mathbf{A}', \mathbf{b}') &= (\mathbf{A} - \mathbf{c} \bullet \mathbf{r}, \tfrac{q}{p} \cdot \lfloor \mathbf{As} \rceil_p - \mathbf{c}) \\
&= (\mathbf{A}', \tfrac{q}{p} \cdot \lfloor \mathbf{A}'\mathbf{s} + \mathbf{c} \cdot \langle \mathbf{r}, \mathbf{s} \rangle \rceil_p - \mathbf{c}) \\
&= (\mathbf{A}', \mathbf{A}'\mathbf{s} + \mathbf{c} \cdot \langle \mathbf{r}, \mathbf{s} \rangle - \{\mathbf{A}'\mathbf{s} + \mathbf{c} \cdot \langle \mathbf{r}, \mathbf{s} \rangle\}_p - \mathbf{c}) \\
&= (\mathbf{A}', \mathbf{A}'\mathbf{s} + \mathbf{c} \cdot (\langle \mathbf{r}, \mathbf{s} \rangle - 1) - \{\mathbf{A}'\mathbf{s} + \mathbf{c} \cdot \langle \mathbf{r}, \mathbf{s} \rangle\}_p)
\end{aligned}
$$

where $\{x\}_p = x - \frac{q}{p} \cdot \lfloor x \rceil_p$. Conditioned on $\langle \mathbf{r}, \mathbf{s} \rangle = 1$, $(\mathbf{A}', \mathbf{b}')$ is distributed as $(\mathbf{A}', \mathbf{A}'\mathbf{s} + \{\mathbf{u}\}_p)$, which is the same as $(\mathbf{A}', \mathbf{A}'\mathbf{s} + \mathbf{e})$ where each coordinate of $\mathbf{e}$ is uniformly drawn from set $[-\frac{q}{2p}, \ldots, \frac{q}{2p}) \subseteq \mathbb{Z}_q$. In this case Dist has distinguishing advantage ε. Conditioned on $\langle \mathbf{r}, \mathbf{s} \rangle \neq 1$, $(\mathbf{A}', \mathbf{b}')$ is distributed uniformly over $\mathbb{Z}_q^{m \times n} \times \mathbb{Z}_q^m$ and Dist has zero distinguishing advantage. Since for any $\mathbf{s} \in \mathbb{Z}_q^{n*}$, the probability that $\langle \mathbf{r}, \mathbf{s} \rangle = 1$ equals $1/q$ over the random choice of $\mathbf{r}$, the overall distinguishing advantage is ε/q. $\qquad \square$

5 Noise Flooding

In this section, let χ_σ denote the discrete Gaussian distribution on $\mathbb{Z}_q$ with standard deviation σ: $\chi_\sigma(x)$ is proportional to $\exp(-\pi(x/\sigma)^2)$. Often in applications of LWE, one is given a sample $(\mathbf{a}, b)$ with $b = \langle \mathbf{a}, \mathbf{s} \rangle + e$ for $e \leftarrow \chi_\sigma$ and by performing various arithmetic operations obtains a new pair $(\mathbf{a}', b')$ with $b' = \langle \mathbf{a}', \mathbf{s} \rangle + e'$. Sometimes, the noise quantity e' obtained is not distributed according to a Gaussian, but is only subject to an overall bound on its absolute value. If the proof of security needs $(\mathbf{a}', b')$ to be an LWE instance, then sometimes the "noise flooding" technique is used where a fresh Gaussian $x \leftarrow \chi_{\sigma'}$ is drawn and b' is set to $\langle \mathbf{a}', \mathbf{s} \rangle + e' + x$. As long as $e' + \chi_{\sigma'} \approx \chi_{\sigma'}$ the resulting $(\mathbf{a}', b')$ is statistically close to a fresh LWE instance. This technique in some form or another appears in many places, for example [AIK11, GKPV10, DGK+10, OPW11]. Unfortunately, $e' + \chi_{\sigma'} \approx_s \chi_{\sigma'}$ requires q to be large and so the applications also carry this requirement. In this section we bound the continuous analogue of Rényi divergence between $e' + \chi_{\sigma'}$ and $\chi_{\sigma'}$ and show that the noise flooding technique can be used even when q is polynomial in the security parameter, as long as the number of samples is also bounded.

We remark that our main result in this section, Corollary 1, follows from general results in prior work which bound the Rényi divergence between Gaussians. For example, Lemma 4.2 of [LSS14] implies Corollary 1 below. However, we are unaware of a theorem in the literature with a simple statement which subsumes Corollary 1. We include a proof for completeness.

Claim. Let Ψ_α be the continuous Gaussian on $\mathbb{R}$ with standard deviation α: $\Psi_\alpha(x) = \alpha^{-1} e^{-\pi(x/\alpha)^2}$. Then for any $\beta \in \mathbb{R}$,

$$\mathrm{RD}_2(\beta + \Psi_\alpha \| \Psi_\alpha) = e^{2\pi(\beta/\alpha)^2}.$$

Proof. We have

$$\mathrm{RD}_2(\beta + \Psi_\alpha \| \Psi_\alpha) = \alpha^{-1} \int_{-\infty}^{\infty} e^{-\left(\pi/\alpha^2\right)\left[2(x-\beta)^2 - x^2\right]} dx$$

$$= \alpha^{-1} \cdot e^{2\pi\left(\beta/\alpha\right)^2} \int_{-\infty}^{\infty} e^{-\left(\pi/\alpha^2\right)\left[(x-2\beta)^2\right]} dx$$

$$= e^{2\pi\left(\beta/\alpha\right)^2}.$$

We have used the substitution $u = x - 2\beta$ and the identity $\int_{\mathbb{R}} e^{-\pi c u^2} du = c^{-1/2}$ for all $c > 0$.   □

Corollary 1. *Fix $m, q, k \in \mathbb{Z}$, a bound B, and a standard deviation σ such that $B < \sigma < q$. Moreover, let $e \in \mathbb{Z}_q$ be such that $|e| \le B$. If $\sigma = \Omega(B\sqrt{m/\log k})$, then*

$$\mathrm{RD}_2\left((e + \chi_\sigma)^m \| \chi_\sigma^m\right) = \mathrm{poly}(k)$$

where X^m denotes m independent samples from X.

Proof. Rényi divergence cannot grow by applying a function to both distributions. Since the discrete Gaussians $e + \chi_\sigma$ and χ_σ are obtained from the continuous Gaussians $\beta + \Psi_\alpha$ and Ψ_α by scaling and rounding, where $\beta = |e|/q$ and $\alpha = \sigma/q$, we see that

$$\mathrm{RD}_2\left(e + \chi_\sigma \| \chi_\sigma\right) \le \mathrm{RD}_2\left(\beta + \Psi_\alpha \| \Psi_\alpha\right) = \exp\left(2\pi(\beta/\alpha)^2\right) \le \exp\left(2\pi(B/\sigma)^2\right).$$

Therefore, $\mathrm{RD}_2\left((e + \chi_\sigma)^m \| \chi_\sigma^m\right) \le \exp\left(2\pi m(B/\sigma)^2\right)$, and the result follows.   □

Acknowledgement. We would like to thank Damien Stehlé for sharing an early version of [BLL+15] and useful suggestions, Daniele Micciancio for insightful comments on an earlier version of the manuscript. We also thank the anonymous TCC 2016A reviewers for useful suggestions.

A   Proof of Lemma 2

Proof. By the definition of Rényi divergence,

$$\mathrm{RD}_2(\mathsf{X}_s \| \mathsf{Y}_s) = \mathbb{E}_{a \leftarrow R_q} \frac{\Pr\left(\mathsf{X}_s = (a, \lfloor a \cdot s \rceil_p)\right)}{\Pr\left(\mathsf{Y}_s = (a, \lfloor a \cdot s \rceil_p)\right)}$$

$$= \mathbb{E}_{a \leftarrow R_q} \frac{1}{\Pr_{e \leftarrow \chi}\left(\lfloor a \cdot s + e \rceil_p = \lfloor a \cdot s \rceil_p\right)}.$$

We define the set $\mathsf{border}_{p,q}(B) = \left\{ x \in \mathbb{Z}_q : \left| x - \frac{q}{p} \lfloor x \rceil_p \right| < B \right\}$. For a ring element $a \in R_q$, we use a_i denote the ith coefficient in the power basis. For $t = 0, \ldots, n$ and for any $t \in \{0, \ldots, n\}$, we define the set $\mathsf{BAD}_{s,t} = \left\{ a \in R_q : |\{i \in [n], (a \cdot s)_i \in \mathsf{border}_{p,q}(B)\}| = t \right\}$. These are the a for which $a \cdot s$ has exactly t coefficients which are dangerously close to the rounding boundary. Fix arbitrary t and $a \in \mathsf{BAD}_{s,t}$. For any $i \in [n]$ such that $(a \cdot s)_i \notin \mathsf{border}_{p,q}(B)$, $\mathrm{Pr}_{e_i}[\lfloor (a \cdot s)_i + e_i \rceil_p = \lfloor (a \cdot s)_i \rceil_p] = 1$. For any $i \in [n]$ such that $(a \cdot s)_i \in \mathsf{border}_{p,q}(B)$, the event $\lfloor (a \cdot s)_i + e_i \rceil_p = \lfloor (a \cdot s)_i \rceil_p$ still holds in one of the two cases $e_i \in [-B, \ldots, 0]$ and $e_i \in [0, \ldots, B]$. By the assumption on the noise distribution $\mathrm{Pr}_{e_i}[\lfloor (a \cdot s)_i + e_i \rceil_p = \lfloor (a \cdot s)_i \rceil_p] \geq 1/2$. Because e is independent over all coefficients and a has exactly t coefficients in $\mathsf{border}_{p,q}(B)$, $\mathrm{Pr}_{e \leftarrow \chi}\left(\lfloor a \cdot s + e \rceil_p = \lfloor a \cdot s \rceil_p \right) \geq \frac{1}{2^t}$. Because s is a unit in R_q so that $a \cdot s$ is uniform over R_q and $\mathrm{Pr}[a \in \mathsf{BAD}_{s,t}] \leq \binom{n}{t} \left(1 - \frac{|\mathsf{border}_{p,q}(B)|}{q} \right)^{n-t} \left(\frac{|\mathsf{border}_{p,q}(B)|}{q} \right)^t$. Conditioning over the event $a \in \mathsf{BAD}_{s,t}$, we conclude

$$\mathrm{RD}_2\left(\mathsf{X}_s \| \mathsf{Y}_s \right) \leq \sum_{t=0}^{n} 2^t \cdot \mathrm{Pr}[a \in \mathsf{BAD}_{s,t}] = \left(1 + \frac{|\mathsf{border}_{p,q}(B)|}{q} \right)^n.$$

The desired conclusion follows from $|\mathsf{border}_{p,q}(B)| \leq 2pB$. □

References

[AGS03] Akavia, A., Goldwasser, S., Safra, S.: Proving hard-core predicates using list decoding. In: 44th Symposium on Foundations of Computer Science (FOCS 2003), 11–14 October 2003, Cambridge, MA, USA, Proceedings, pp. 146–157. IEEE Computer Society (2003)

[AIK11] Applebaum, B., Ishai, Y., Kushilevitz, E.: How to garble arithmetic circuits. In: IEEE 52nd Annual Symposium on Foundations of Computer Science, FOCS 2011, Palm Springs, CA, USA, October 22–25, 2011, pp. 120–129 (2011)

[AKPW13] Alwen, J., Krenn, S., Pietrzak, K., Wichs, D.: Learning with rounding, revisited. In: Canetti, R., Garay, J.A. (eds.) CRYPTO 2013, Part I. LNCS, vol. 8042, pp. 57–74. Springer, Heidelberg (2013)

[BLL+15] Bai, S., Langlois, A., Lepoint, T., Stehlé, D., Steinfeld, R.: Improved security proofs in lattice-based cryptography: using the rényi divergence rather than the statistical distance. IACR Cryptology ePrint Arch. **2015**, 483 (2015)

[BLP+13] Brakerski, Z., Langlois, A., Peikert, C., Regev, O., Stehlé, D.: Classical hardness of learning with errors. In: Boneh, D., Roughgarden, T., Feigenbaum, J. (eds.) STOC, pp. 575–584. ACM (2013)

[BPR12] Banerjee, A., Peikert, C., Rosen, A.: Pseudorandom functions and lattices. In: Pointcheval, D., Johansson, T. (eds.) EUROCRYPT 2012. LNCS, vol. 7237, pp. 719–737. Springer, Heidelberg (2012)

[Cho13] Chow, C.-W.: On algorithmic aspects of the learning with errors problem and its variants. Master's thesis, The Chinese University of Hong Kong, September 2013

[DGK+10] Dodis, Y., Goldwasser, S., Tauman Kalai, Y., Peikert, C., Vaikuntanathan, V.: Public-key encryption schemes with auxiliary inputs. In: Micciancio, D. (ed.) TCC 2010. LNCS, vol. 5978, pp. 361–381. Springer, Heidelberg (2010)

[GKPV10] Goldwasser, S., Kalai, Y.T., Peikert, C., Vaikuntanathan, V.: Robustness of the learning with errors assumption. In: Innovations in Computer Science - ICS 2010, Tsinghua University, Beijing, China, January 5–7, 2010. Proceedings, pp. 230–240 (2010)

[GL89] Goldreich, O., Levin, L.A.: A hard-core predicate for all one-way functions. In: Johnson, D.S. (ed.) Proceedings of the 21st Annual ACM Symposium on Theory of Computing, May 14–17, 1989, Seattle, Washigton, USA, pp. 25–32. ACM (1989)

[Lev86] Levin, L.A.: Average case complete problems. SIAM J. Comput. **15**(1), 285–286 (1986)

[LSS14] Langlois, A., Stehlé, D., Steinfeld, R.: GGHLite: more efficient multilinear maps from ideal lattices. In: Nguyen, P.Q., Oswald, E. (eds.) EUROCRYPT 2014. LNCS, vol. 8441, pp. 239–256. Springer, Heidelberg (2014)

[MM11] Micciancio, D., Mol, P.: Pseudorandom knapsacks and the sample complexity of LWE search-to-decision reductions. In: Rogaway [Rog11], pp. 465–484

[OPW11] O'Neill, A., Peikert, C., Waters, B.: Bi-deniable public-key encryption. In: Rogaway [Rog11], pp. 525–542

[Pei09] Peikert, C.: Public-key cryptosystems from the worst-case shortest vector problem: extended abstract. In: Mitzenmacher, M. (ed.) STOC, pp. 333–342. ACM (2009)

[Reg05] Regev, O.: On lattices, learning with errors, random linear codes, and cryptography. In: Gabow, H.N., Fagin, R. (eds.) STOC, pp. 84–93. ACM (2005)

[Rog11] Rogaway, P. (ed.): CRYPTO 2011. LNCS, vol. 6841, pp. 277–296. Springer, Heidelberg (2011)

[vEH14] van Erven, T., Harremoës, P.: Rényi divergence and Kullback-Leibler divergence. IEEE Trans. Inf. Theor. **60**(7), 3797–3820 (2014)

[Yao82] Yao, A.C.-C.: Theory and applications of trapdoor functions (extended abstract). In: 23rd Annual Symposium on Foundations of Computer Science, Chicago, Illinois, USA, 3–5 November 1982, pp. 80–91 (1982)

Two-Round Man-in-the-Middle Security from LPN

David Cash[1]([✉]), Eike Kiltz[2], and Stefano Tessaro[3]

[1] Rutgers University, New Brunswick, USA
david.cash@cs.rutgers.edu
[2] Ruhr University Bochum, Bochum, Germany
[3] University of California, Santa Barbara, USA

Abstract. Secret-key authentication protocols have recently received a considerable amount of attention, and a long line of research has been devoted to devising efficient protocols with security based on the hardness of the learning-parity with noise (LPN) problem, with the goal of achieving low communication and round complexities, as well as highest possible security guarantees.

In this paper, we construct 2-round authentication protocols that are secure against sequential man-in-the-middle (MIM) attacks with tight reductions to LPN, Field-LPN, or other problems. The best prior protocols had either loose reductions and required 3 rounds (Lyubashevsky and Masny, CRYPTO'13) or had a much larger key (Kiltz et al., EURO-CRYPT'11 and Dodis et al., EUROCRYPT'12). Our constructions follow from a new generic deterministic and round-preserving transformation enhancing actively-secure protocols of a special form to be sequentially MIM-secure while only adding a limited amount of key material and computation.

Keywords: Secret-key authentication · Man-in-the-Middle security · LPN · Field LPN

1 Introduction

This paper constructs efficient provably-secure protocols for *secret-key* authentication, i.e., for the basic cryptographic task where one party, called the *prover*, proves to another – the *verifier* – that they share the same key. Theoretical constructions of such protocols (with strong security, to be defined below) exist from any one-way function. Moreover, practical two-round protocols can be built from any message-authentication code (MAC) by having one party authenticate a random challenge, and can be instantiated efficiently for example assuming AES-128 is unpredictable.

In contrast, this paper contributes to a line of work [10,13,15,16,18–20,22] on building provably-secure authentication protocols with security reductions to the *learning parity with noise (LPN)* and related problems that are as efficient as possible, meaning that key-size, communication, and rounds are minimized.

© International Association for Cryptologic Research 2016
E. Kushilevitz and T. Malkin (Eds.): TCC 2016-A, Part I, LNCS 9562, pp. 225–248, 2016.
DOI: 10.1007/978-3-662-49096-9_10

LPN problem provides confidence in security due to the failure to find polynomial-time algorithms for it and its variants, despite wide interest [5,7,21], and finding constructions of cryptographic primitives based on LPN has given rise to a substantial body of works [2,6,14,17].

The motivation behind LPN-based authentication protocols is their potential to be implemented with different efficiency characteristics from protocols with security reductions to blockcipher security or to problems from number theory and related fields. For instance, the parallel nature of LPN-based protocols seems difficult to achieve with factoring or discrete-log type assumption. The potential efficiency benefits of LPN-based implementations are a subject of ongoing research, which has identified some advantageous scenarios [15] but also invented faster attacks [21]. We thus focus on developing techniques for protocol design and theoretical analysis that beat previous asymptotic runtimes, key sizes, and round complexity of protocols with similar security reductions. We make no specific claims of more efficient protocols in specific deployment scenarios.

Concurrently to the above, the recent interest on secret-key authentication has also motivated attempts to develop a better understanding of its foundations, providing theoretical constructions based on concrete number-theoretic assumptions like the *Decisional Diffie-Hellman* (DDH) assumption, or general assumptions like weak pseudorandom functions [10,22]. We will also contribute to these lines of work with new constructions.

But before we turn to describing our contributions in detail, we first give an overview of different security notions for secret-key authentication, as well as of previous works.

SECURITY NOTIONS. Several security notions for secret-key authentication protocols have been considered, inspired by corresponding notions for the task of public-key authentication [12]. The weakest, *passive security*, says that an attacker should not be able to fool the verifier after observing several sessions between an honest prover and an honest verifier. This seems unreasonably weak for most settings, so the stronger *man-in-the-middle (MIM) security* notion says that no attacker should be able to cause the verifier to accept in any session where a message has been changed. Realizing that MIM security from LPN seems difficult to achieve efficiently, several works instead targeted an intermediate notion called *active security* which says that the attacker cannot fool the verifier after interacting with the prover arbitrarily and observing sessions passively.

THE LPN ASSUMPTION (AND ITS VARIANTS). Recall that for parameters $\ell \in \mathbb{N}$ and $0 \le \gamma \le \frac{1}{2}$, the (decisional) *Learning Parity with Noise (LPN)* problem $\mathsf{LPN}_{\ell,\gamma}$ is the problem of distinguishing a polynomial number of samples of the form $(\mathbf{r}_i, \mathbf{r}_i^T \mathbf{s} + e_i)$, for a common random secret $\mathbf{s} \in \{0,1\}^\ell$, random vector $\mathbf{r}_i \in \{0,1\}^\ell$, and random bit e_i (taking value one with probability γ), from samples of the form $(\mathbf{r}_i, b_i)$, where b_i is a random bit. The corresponding $\mathsf{LPN}_{\ell,\gamma}$ assumption is that no efficient (i.e., polynomial-time) attacker can distinguish between the two distributions, except with negligible advantage. Ignoring the obvious differences in the error distributions, this is the modulo 2 variant of the learning with error problem introduced in [27].

We are also going to consider a variant of the LPN problem, introduced and studied in [15], called *Field LPN*. The Field-LPN$_{\ell,\gamma}$ problem is very similar, however samples have the form $(\mathbf{r}_i, \mathbf{r}_i \circ \mathbf{s} + \mathbf{e}_i)$ or $(\mathbf{r}_i, \mathbf{r}'_i)$, where $\circ$ denotes multiplication of ℓ-bit vectors interpreted as elements of the extension field $\mathbb{F}_{2^\ell}$, $\mathbf{e}_i$ is a random vector where each component is 1 independently with probability γ, and $\mathbf{r}'_i$ is uniform.

PRIOR CONSTRUCTIONS. Let us briefly outline the landscape of earlier works on secret-key authentication. Table 1 summarizes some of these results.

Juels and Weis [18] first pointed out that a very simple two-round secret-key authentication protocol by Hopper and Blum [16], called the *HB protocol*, enjoys very low hardware complexity, and is hence amenable to implementations on RFID tags. Moreover, they proved that it is passively secure under the LPN assumption. Also in [18], they proposed a further three-round protocol, called HB$^+$, which was proven actively secure in its sequential version under the LPN assumption, and later the proof was extended to its parallel version by Katz *et al.* [19]. The round complexity was then reduced to *two* rounds by a new protocol of Kiltz *et al.* [20], and in contrast to HB$^+$, this latter protocol enjoys a tight reduction to the hardness of LPN. Heyse *et al.* [15] then proposed an even more efficient two-round protocol, called *Lapin*, based on the hardness of the *field* LPN problem. We stress that three-round protocols are less attractive than two-round ones since the prover needs to keep a *state* (beyond the secret key), which is problematic on lightweight devices like RFID tags.

In contrast, progress has been significantly harder in the context of MIM security. On the one hand, researchers have attempted to design multiple HB-like protocols with MIM security [8,11,13,25] without or only partial security proofs. Otherwise, provably MIM-secure constructions all in fact provide a full message-authentication code (MAC) secure under LPN or Field LPN [10,15,20]. Unfortunately, these constructions are significantly less efficient than the existing actively secure protocols mentioned above.

While following [3] the notion of MIM security traditionally allows an attacker to interact with arbitrarily many instances of the prover and the verifier concurrently, Lyubashevsky and Masny [22] recently considered the notion of *sequential MIM* (sMIM) security, which slightly weakens MIM to only allow the attacker to interfere with non-overlapping sequential sessions. They argue this notion is sufficient for situations in which keys are managed to never allow parallel session, and the sMIM notion is an interesting technical step towards improving authentication protocol security beyond active security while maintaining efficiency. Moreover, existing MIM attacks against actively secure protocols are often sequential (e.g., [26]). They give new protocols based on LPN and field-LPN that nearly match the complexities of actively-secure ones, but all require three rounds and suffer from a non-tight reduction to the underlying problem.

With respect to other assumptions, we also note that efficient three-round constructions from DDH and weak pseudorandom functions have also been given, achieving both active security [10] and sMIM security [22]. Two-round MIM secure protocols from PRGs have recently been proposed [9].

All of our constructions come with reductions running in polynomial time and succeeding with probability polynomially proportional to that of a given attack. One may consider looser reductions via so-called *complexity leveraging* where the reduction loses an exponential factor, with the view that one can enlarge the security parameter to compensate for the loss. Indeed one can prove (say) the AUTH$_2$ protocol from [20] as a fully-secure MAC with an exponential loss of security. A concrete instantiation of the result (assuming the BKW attack complexity is optimal [7]) will be more efficient than the other approaches we have outlined.

Polynomial reductions, however, are preferred as they are more robust to algorithmic advances against the underlying problems. Achieving them is, in our view, an interesting theoretical challenge that requires new techniques. In an implementation it is not clear to the authors if either approach (leveraging or polynomial reductions) is necessarily more secure given the many factors one must consider.

OUR CONTRIBUTIONS. We provide the to-date most efficient constructions of sMIM-secure authentication protocols based on the hardness of LPN, as well as on other assumptions. Our constructions are *two* rounds and the first message consists of a truly random challenge, and enjoy tight security reduction to the underlying assumption.

We improve upon the round complexity of existing sMIM-secure protocols without increasing key length and communication complexity, and without resorting to complexity leveraging. See Table 1 for a comparison of two of our new protocols to prior work. Note that our protocols are only a small constant factor less efficient than the best known actively (or even passively) secure protocols.

At the high level, our constructions follow from a generic transformation that upgrades a two round protocol of a special form to be sMIM-secure without introducing significant overhead. The required form is not especially contrived, but requires some care in its formalization and we present examples of such protocols to obtain our instantiations. We note that our reduction does not employ rewinding or forking lemmas like [22], and is tighter and (arguably, to our taste) simpler as a result.

Our first construction achieves sMiM security with a tight reduction to LPN, two rounds of communication, and only a modest increase in either key size or communication over [22]. Our second construction, from *Field LPN*, matches the key size and communication of prior work in two rounds instead of three and has a tight reduction. In fact, for an appropriate choice of components, the second construction can be understood as a two-round version of the three-round protocol from [22], though their proof does not cover the two round version.

We also provide a simple construction of a two-round sMIM secure authentication protocol based on the DDH assumption, where the prover response consists of *two* group elements. Interestingly, the same construction was proven MIM secure under the (less standard) Gap-CDH assumption in [10].

Our last construction is based on an arbitrary weak PRF. The complexity of the construction is comparable to the one building a MAC from a weak PRF,

Table 1. Authentication protocols based on LPN-related Assumptions. The security column lists the best possible security reduction from the given assumption, where q is the number of tag and verification queries. (The two MAC_2 protocols are even secure in the full MiM model.) The complexity column lists the key sizes and communication complexity of the protocol (with lower-order terms dropped), where ℓ parameterizes the hardness of the assumption. All LPN-based protocols offer a trade-off between key size and communication, which is listed in the last two columns. $^{(*)}$: Reductions to active security only considered one challenge session, and thus did not have the factor q. We state the bound for q challenge sessions for a fair comparison to MiM security. $^{(**)}$: We remark that the key size of the LPN-based protocol in [22] is ℓ^2 but one may be able to reduce it to $O(\ell)$ by using an *almost* pairwise independent hash function.

Protocol	Rounds	Security			Complexity		Compl. trade-off	
		Assumption	Active$^{(*)}$	sMiM	key size	com.	key size	com.
HB [16]	2	$LPN_{\ell,\gamma}$	–	–	ℓ	ℓ^2	ℓ^2	2ℓ
HB$^+$ [18]	3	$LPN_{\ell,\gamma}$	$q\sqrt{\epsilon}$	–	2ℓ	$2\ell^2$	$2\ell^2$	3ℓ
AUTH$_2$ [20]	2	$LPN_{\ell,\gamma}$	$q\epsilon$	–	2ℓ	ℓ^2	$2\ell^2$	2ℓ
Lapin [15]		Field-LPN$_{\ell,\gamma}$			2ℓ	2ℓ	–	–
MAC$_2$ [20]	2	$LPN_{\ell,\gamma}$	$q\epsilon$	$q\epsilon$	$3\ell^2$	ℓ^2	ℓ^3	4ℓ
Lapin+MAC$_2$ [15]		Field-LPN$_{\ell,\gamma}$			ℓ^2	4ℓ	–	–
LM [22]	3	$LPN_{\ell,\gamma}$	$q\sqrt{\epsilon}$		ℓ^2 (**)	ℓ^2	ℓ^2	3ℓ
		Field-LPN$_{\ell,\gamma}$			4ℓ	3ℓ	–	–
This work	2	$LPN_{\ell,\gamma}$	$q\epsilon$		5ℓ	ℓ^2	$2\ell^2$	3ℓ
		Field-LPN$_{\ell,\gamma}$			4ℓ	3ℓ	–	–

using for example the constructions in [1, 23, 24]. However, our new protocols enjoy much better parallelism when compared to the naive approach, and is hence interesting on its own right. It is also fair to point out that [22] accomplishes in three rounds the harder task of finding a generic construction from a (randomized) *weak* PRF. We observe however that the only known concrete instantiations of weak PRFs are based on LPN/LWE-type assumptions as well as on DDH, and for all these concrete instantiations our constructions are more efficient.

We remark that it is not hard to see that our proofs do not show (full) MIM security, but we are not aware of an explicit MIM attack against the protocols.

ORGANIZATION. Section 2 contains basic definitions used below. In Sect. 3 we describe our transformation from weaker protocols of a special form, and in Sect. 4 we give several instantiations of the transformation.

2   Preliminaries

For a set $\mathcal{X}$, $x \xleftarrow{\$} \mathcal{X}$ denotes sampling x from $\mathcal{X}$ according to the uniform distribution. We use bold lowercase letters for vectors and bold uppercase letters for matrices, e.g., $\mathbf{x} \in \mathbb{F}_2^\ell$, $\mathbf{X} \in \mathbb{F}_2^{\ell \times n}$. For $\mathbf{c} \in \mathbb{F}_2^\ell$, let $\mathbf{M_c}$ denote the matrix of the linear map $l_{\mathbf{c}}$ implementing the finite field multiplication with $\mathbf{c}$ when interpreted as an element in $\mathbb{F}_{2^\ell}$.[1]

SYMMETRIC AUTHENTICATION SYNTAX. We are going to consider secret-key *authentication* protocols, where a prover proves to a verifier that they hold the same secret key over two or more rounds.

More formally, an r-*round symmetric authentication protocol* with associated key space $\mathcal{K}$ is a triple of algorithms $\mathsf{Auth} = (\mathsf{Gen}, \mathsf{P}, \mathsf{V})$ with the following properties:

- Key Generation. The probabilistic key-generation algorithm $K \leftarrow \mathsf{Gen}(1^k)$ takes as input a security parameter $k \in \mathbb{N}$ (in unary) and outputs a secret key $K \in \mathcal{K}$.
- Interactive Execution. The probabilistic interactive algorithms P and V, which we refer to as the *prover* and the *verifier*, take both as input a secret key $K \in \mathcal{K}$, synchronously interact with each other over r rounds, and finally V always receives the last message and outputs a decision $\mathsf{out}(\mathsf{P}_K, \mathsf{V}_K) \in \{\mathsf{accept}, \mathsf{reject}\}$.

We say that Auth has completeness error α if for all $k \in \mathbb{N}$, $\Pr[\mathsf{out}(\mathsf{P}_K, \mathsf{V}_K) = \mathsf{reject}; K \leftarrow \mathsf{Gen}(1^k)] \leq \alpha$. In this paper, we will focus on the simpler case of *two-round* protocols, where additionally the first message is a random challenge $c \in \mathcal{C}$ for some set $\mathcal{C}$. We call such protocols *two-round random-challenge* secret-key authentication protocols. In particular, in such protocols the prover simplifies to a probabilistic algorithm P_K, taking the challenge and the secret key K, and producing the message t to be sent back to the adversary. Moreover, for a challenge $c \in \mathcal{C}$ and response t from the prover, the verifier is fully specified by an algorithm $\mathsf{V}_K(c, t) \in \{\mathsf{accept}, \mathsf{reject}\}$.

SECURITY. Several security notions for symmetric-key authentication protocols have been considered in the literature. The weakest one, *passive security*, says that an attacker should not be able to fool the verifier after observing several sessions between a honest prover and a honest verifier. The stronger notion called *active security* says that the attacker cannot fool the verifier after interacting with the prover arbitrarily and observing sessions passively.

This paper targets the security notion of *(sequential) security against man-in-the-middle attacks* (or s-mim security, for short). Here, the adversary acts as a man-in-the middle in a sequence of independent sessions between the prover and the verifier, all with the same secret key. The adversary wins whenever it manages to let the verifier accept in some session *and* has changed at least one of the messages sent by the prover or the verifier. We are going to formalize this notion for the relevant case of two-round protocols with random challenge.

Concretely, we describe this security notion via the following game S-MIM for an attacker A and a two-round random-challenge authentication protocol $\mathsf{Auth} = (\mathsf{Gen}, \mathsf{P}, \mathsf{V})$ with challenge set $\mathcal{C}$.

---

[1] This representation is unique once the irreducible polynomial f defining $\mathbb{F}_{2^n} = \mathbb{F}_2[x]/(f)$ is fixed.

main S-MIM:	**Procedure** $P(c')$:
$\text{sid} \leftarrow 0$	If $c'[\text{sid}] = \perp$ then
$K \xleftarrow{\$} \text{Gen}(1^k)$	$\quad c'[\text{sid}] \leftarrow c', \ t[\text{sid}] \xleftarrow{\$} P_K(c')$
Run $A^{C(),P(),V()}(1^k)$	Ret $t[\text{sid}]$
Ret $\exists i: (c[i], t[i]) \neq (c'[i], t'[i]) \wedge d[i] = \text{accept}$	
	Procedure $V(t')$:
Procedure $C()$:	$t'[\text{sid}] \leftarrow t', \ c \xleftarrow{\$} C()$
If $c[\text{sid}] = \perp$ then	$d[\text{sid}] \leftarrow V_K(c, t'[\text{sid}])$
$\quad c[\text{sid}] \xleftarrow{\$} C$	$\text{sid} \leftarrow \text{sid} + 1$
Ret $c[\text{sid}]$	Ret $d[\text{sid}]$

In the game, the attacker makes calls to three oracles, $C(\cdot), P(\cdot)$ and $V(\cdot)$. All oracles use a global variable sid to "synchronize" the sessions being simulated. The first oracle returns, for every session, a new random challenge. The oracle $P(c')$ runs the prover on input c' and returns the response t. Oracle $V(t')$ checks that t' is a valid response for the current session challenge $c[\text{sid}]$ (obtained by calling $C()$), and increases the session number. Note that there is a unique value $c[\text{sid}]$ defined in every session, and P only provides (at most) one valid challenge-tag pair (c', t) per session. The s-mim advantage is $\mathbf{Adv}_{\text{Auth}}^{\text{s-mim}}(A) = \Pr\left[\text{S-MIM}_{\text{Auth}}^A \Rightarrow \text{true}\right]$, and we say that Auth is (t, r, ϵ)-s-mim-secure if for all attackers A with time complexity t and running at most r sessions, we have $\mathbf{Adv}_{\text{Auth}}^{\text{s-mim}}(A) \leq \epsilon$.

HASH FUNCTIONS. Our constructions rely on almost pairwise-independent hash functions.

Definition 1 (Almost pairwise-independent hash functions). *For $\delta \geq 1$, a function* $H : \mathcal{K}_H \times \mathcal{X} \to \mathcal{Y}$ *is δ-almost pairwise-independent if*

$$\Pr\left[H_{K_H}(x) = y \wedge H_{K_H}(x') = y'\right] \leq \frac{\delta}{|\mathcal{Y}|^2}$$

for all distinct $x, x' \in \mathcal{X}$ *and all* $y, y' \in \mathcal{Y}$, *and where* $K_H \xleftarrow{\$} \mathcal{K}_H$. *Moreover, by itself,* $H_{K_H}(x)$ *is uniformly distributed over* $\mathcal{Y}$.

The requirement that a single input has uniformly distributed output is not common, but will be useful in applications and satisfied by the construction given below. Moreover, Definition 1 implies *adaptive* security, i.e., when given x, $H_{K_H}(x) = y$, for any x' and y' chosen *adaptively* depending on y, the probability that $H_{K_H}(x') = y'$ is at most $\delta/|\mathcal{Y}|$.

Lemma 2. *If* H *is δ-almost pairwise-independent, then for every (unbounded) adversary A and every $x \in \mathcal{X}$, we have*

$$\Pr[H_{K_H}(x') = y' \wedge x' \neq x : K_H \xleftarrow{\$} \mathcal{K}_H, \ (x', y') \xleftarrow{\$} A(H_{K_H}(x), x)] \leq \frac{\delta}{|\mathcal{Y}|}.$$

Proof. Assume wlog that A is deterministic, and let $x'(x,y)$ and $y'(x,y)$ be the values of x' and y' output by A on inputs y, x, where $x'(x,y) \neq x$ by assumption. Then,

$$\Pr\left[\mathsf{H}_{K_\mathsf{H}}(x') = y' \ : \ K_\mathsf{H} \overset{\$}{\leftarrow} \mathcal{K}_\mathsf{H}, \ (x', y') \overset{\$}{\leftarrow} A(\mathsf{H}_{K_\mathsf{H}}(x), x)\right]$$
$$= \sum_y \Pr\left[\mathsf{H}_{K_\mathsf{H}}(x) = y \wedge \mathsf{H}_{K_\mathsf{H}}(x'(x,y)) = y'(x,y)\right],$$

which is smaller than $|\mathcal{Y}| \cdot \frac{\delta}{|\mathcal{Y}|^2} = \frac{\delta}{|\mathcal{Y}|}$.   □

A CONSTRUCTION. We will make use of the following key-length efficient construction of a δ-almost-pairwise independent function, where $\mathcal{K}_\mathsf{H} = \mathbb{F}^2$, $\mathcal{Y} = \mathbb{F}$ and $\mathcal{X} = \mathbb{F}^\ell$ for some finite field $\mathbb{F}$. The function, given $K_\mathsf{H} = (a, b) \in \mathbb{F}^2$ and input $x = (x_0, \ldots, x_{\ell-1}) \in \mathbb{F}^\ell$, outputs $\mathsf{H}_{a,b}(x) = \sum_{i=0}^{\ell-1} x_i \circ a^i + b$.

Lemma 3. *The function* H *above is* δ*-almost pairwise independent for* $\delta = \ell - 1$.

The folklore proof is given for completeness.

Proof. Fix $x = (x_0, x_1, \ldots, x_{\ell-1})$ and $x' = (x'_0, x'_1, \ldots, x'_{\ell-1})$. Also, we define the polynomial $p_x(a) = \sum_{i=0}^{\ell-1} x_i \circ a^i$, and analogously, define $p_{x'}(a)$. Given two $y, y' \in \mathbb{F}$, we look at the number of keys (a, b) such that $p_x(a) + b = y$ and $p_{x'}(a) + b = y'$. This in particular implies that a needs to satisfy

$$p_x(a) - p_{x'}(a) = \sum_{i=0}^{\ell-1}(x_i - x'_i) \circ a^i = y - y',$$

and since there exists i with $x_i \neq x'_i$, note that by the Schwartz-Zippel lemma there are at most $\ell - 1$ solutions a with the above property, since $p_x(a) - p_{x'}(a)$ is a polynomial of degree at most $\ell - 1$. Each such a defines a unique b, and thus there are overall at most $\ell - 1$ solutions, and each one of them is taken with probability $|\mathbb{F}|^2$.

Finally, note that the distribution of $\mathsf{H}_{a,b}(x)$ is, by itself, uniform, because the term b is uniform, and thus completely blinds the output.   □

3   Generic Construction

This section presents our main result, a generic construction of a two-round sequential MIM-secure authentication protocol Auth. Our construction relies on a simpler two-round symmetric authentication protocol Auth' used as a component and which satisfies a particular form of security, in addition to having a structured tag space, as we discuss next. Later below, we will provide several instantiations of this generic construction in Sect. 4 via constructions of Auth' based on a set of different assumptions.

3.1 Tools

Our construction is going to rely on an authentication protocol $\mathsf{Auth} = (\mathsf{Gen}, \mathsf{P}, \mathsf{V})$ whose responses given by the prover (which we call *tags*, following existing conventions in the literature) $\tau \xleftarrow{\$} \mathsf{P}_K(c)$ are composed of two distinct components $\tau = (\tau_1, \tau_2) \in \mathcal{T}_1 \times \mathcal{T}_2$. We refer to τ_1 and τ_2 as the *left* and *right* tag, respectively. In addition to this, we are going to require that the protocol satisfies two new properties which we now introduce and discuss.

TAG SPARSITY. The first property is a *combinatorial* property on the tag space of Auth. We are going to require that given any challenge c, any secret key K, and any *left* component of the tag τ_1, there are only few right components τ_2 such that $\tau = (\tau_1, \tau_2)$ is a valid tag for challenge c and key K. This is captured formally by the following definition.

Definition 4 (Right tag-sparsity). *For an $\epsilon = \epsilon(k)$, we say that $\mathsf{Auth} = (\mathsf{Gen}, \mathsf{P}, \mathsf{V})$ with tags in $\mathcal{T}_1 \times \mathcal{T}_2$, challenge space $\mathcal{C}$, and key space $\mathcal{K}$ has ϵ-sparse right tags (or alternatively, Auth has ϵ-right tag sparsity) if*

$$\Pr\left[\mathsf{V}_K(c, (\tau_1, \tau_2)) = \mathsf{accept}; \ \tau_2 \xleftarrow{\$} \mathcal{T}_2\right] \leq \epsilon$$

for all $c \in \mathcal{C}$, $K \in \mathcal{K}$, and $\tau_1 \in \mathcal{T}_1$.

Note that one equivalent formulation is that for all K, c, and τ_1, there are at most $\epsilon \cdot |\mathcal{T}_2|$ valid τ_2.

ROR-CMA SECURITY. We also consider a new property called *real-or-random right-tag chosen-message security* (or ror-cma security, for short), which is specific to protocols as above with tag space $\mathcal{T}_1 \times \mathcal{T}_2$. It considers a game where an attacker first receives a challenge c^*, then can obtain prover tags for arbitrary challenges of its choice, and at the end can issue *exactly one* verification query for the challenge c^*. The notion demands that the attacker cannot distinguish this game from another game where queries for challenges $c \neq c^*$ have the right tag τ_2 replaced by a *random* element from the same set. Formally, we introduce the following two games – denoted $\mathsf{ROR\text{-}CMA}(0), \mathsf{ROR\text{-}CMA}(1)$ – involving Auth as well as an adversary A which outputs a decision value in $\{\mathsf{true}, \mathsf{false}\}$ at the end of the game:

main ROR-CMA(b):	**Procedure** T(c):
$K \xleftarrow{\$} \mathsf{Gen}(1^k)$	$(\tau_1, \tau_2^1) \xleftarrow{\$} \mathsf{P}_K(c), \ \tau_2^0 \xleftarrow{\$} \mathcal{T}_2$
$c^* \xleftarrow{\$} \mathcal{C}$	If $c = c^*$ then
$(\tau^*, state) \xleftarrow{\$} A^{\mathrm{T}(\cdot)}(c^*)$	Ret $\tau = (\tau_1, \tau_2^1)$
$d \leftarrow \mathsf{V}_K(c^*, \tau^*)$	Else ret $\tau = (\tau_1, \tau_2^b)$
Ret $A(state, d)$	

Then, for an attacker A and a two-round protocol Auth, we define the ror-cma *advantage* as

$$\mathbf{Adv}_{\mathsf{Auth}}^{\mathsf{ror\text{-}cma}}(A) = \Pr[\mathsf{ROR\text{-}CMA}_{\mathsf{Auth}}^A(0) \Rightarrow \mathsf{true}] - \Pr[\mathsf{ROR\text{-}CMA}_{\mathsf{Auth}}^A(1) \Rightarrow \mathsf{true}] \ .$$

Accordingly, we say that Auth is (t, q, ϵ)-ror-cma-secure if for all t-time attackers A issuing at most q queries to oracle $\mathrm{T}(\cdot)$, we have $\mathbf{Adv}_{\mathsf{Auth}}^{\mathsf{ror\text{-}cma}}(A) \leq \epsilon$.

RELATION TO ACTIVE SECURITY. We stress that ror-cma security and negligible right-tag sparsity, when achieved simultaneously, do not even imply passive security. Indeed, it is easy to modify any protocol with these two properties into one accepting tags of the form $(\tau_1, 0)$ for every K and c (and hence becoming completely insecure) without invalidating these two properties. However, any such protocol can easily be enhanced to be secure against *active* adversaries by blinding τ_2 with a secret field element K, either via addition or multiplication. (Note that negligible right-tag sparsity implies that the set of right tags has overwhelming size.)

Nonetheless, in order to better understand our construction below, it is important to observe why the resulting protocol is *not* necessarily s-mim secure. Consider e.g. the protocol such that $\mathsf{P}_K(c) = (\tau_1, \tau_2 = \mathsf{PRF}_K(\tau_1 \| c))$ for a random τ_1 and pseudorandom function PRF with key K and n-bit output, and for which V_K accepts (τ_1, τ_2) on input c if and only if $\mathsf{PRF}_K(\tau_1 \| c)$ has Hamming distance at most 1 from τ_2. One can verify that this protocol is ror-cma secure and has negligible right-tag sparsity. But when the above tranformation is applied, resulting in tags $(\tau_1, \tau_2 = \mathsf{PRF}_K(\tau_1 \| c) + K')$, an attacker can easily derive a new valid tag for c as $(\tau_1, \tau_2 + \Delta)$ for any weight-one Δ – hence breaking s-mim security. (Similar counterexamples can be built when blinding via multiplication.)

3.2   The Generic Construction

We now turn to describing our generic construction transforming a ror-cma-secure two-round random challenge authentication protocol Auth' with ϵ-right tag sparsity (for a small ϵ) into a sequential MIM secure two-round authentication protocol.

DESCRIPTION. Let $\mathsf{Auth}' = (\mathsf{Gen}', \mathsf{P}', \mathsf{V}')$ be two-round authentication protocol with associated key space $\mathcal{K}$, challenge space $\mathcal{C}$, and split tag space $\mathcal{T} = \mathcal{T}_1 \times \mathcal{T}_2$, where we assume that $\mathcal{T}_2 = \mathbb{F}$ is a finite field.[2] We will use $+$ and $\circ$ to denote addition and multiplication of field elements, respectively. Let $\mathsf{H} : \mathcal{K}_{\mathsf{H}} \times \mathcal{T}_1 \to \mathbb{F}$ be a hash function. We build a 2-round symmetric authentication protocol $\mathsf{Auth} = (\mathsf{Gen}, \mathsf{P}, \mathsf{V})$ as follows. (The protocol Auth inherits the completeness error of Auth'.)

- Key Generation. The key-generation algorithm $\mathsf{Gen}(1^k)$ picks a key $K_{\mathsf{H}} \xleftarrow{\$} \mathcal{K}_{\mathsf{H}}$ for H, an element $K_{\mathbb{F}} \xleftarrow{\$} \mathbb{F} \setminus \{0\}$, and generates a key $K' \xleftarrow{\$} \mathsf{Gen}'(1^k)$ for Auth'. The key is $K = (K', K_{\mathsf{H}}, K_{\mathbb{F}})$.
- Challenge. The challenge is generated by the verifier V as $c \xleftarrow{\$} \mathcal{C}$.

---

[2] This is w.l.o.g., as we can always represent $\mathcal{T}_2$ as a bit-string $\{0,1\}^t$ for some $t \in \mathbb{N}$ which we associate with $\mathbb{F}_{2^t}$.

- Response. The response $\sigma = (\sigma_1, \sigma_2)$ to challenge $c \in \mathcal{C}$ is computed by the prover P by first running $\tau = (\tau_1, \tau_2) \xleftarrow{\$} \mathsf{P}'_{K'}(c)$ and

$$\sigma = (\sigma_1, \sigma_2) = (\tau_1, \tau_2 \circ K_{\mathbb{F}} + \mathsf{H}_{K_{\mathsf{H}}}(\tau_1)) \in \mathcal{T}_1 \times \mathbb{F}.$$

- Verify. Given challenge c and response $\sigma = (\sigma_1, \sigma_2)$, the verifier V reconstructs

$$\tau = (\tau_1, \tau_2) = (\sigma_1, \, (\sigma_2 - \mathsf{H}_{K_{\mathsf{H}}}(\sigma_1)) \circ K_{\mathbb{F}}^{-1})$$

and returns the decision $\{\mathsf{accept}, \mathsf{reject}\} \leftarrow \mathsf{V}'_{K'}(c, \tau)$.

OVERHEAD. We note that our transformation does *not* increase the tag size of the underlying protocol Auth', and thus retains its communication complexity. Moreover, the key length increases by adding $K_{\mathbb{F}}$ and K_{H}. Below, we will show that H can be instantiated with the hash-function construction given in Sect. 2, and thus these two additional keys consist overall of *three* field elements.

3.3   Security

The following theorem establishes the concrete security of our generic construction. In particular, it says that as long as for sufficiently small δ and ϵ, H is δ-almost pairwise independent and Auth' has both ϵ-right-tag sparsity and is ror-cma-secure, then the construction is s-mim-secure.

Theorem 5 (Security of the generic construction). *Assume that H is δ-almost universal and that* Auth' *satisfies ϵ-right tag sparsity and has completeness error α. Then, for all* s-mim-*attackers A invoking at most r sessions, there exists a* ror-cma-*attack B such that*

$$\mathbf{Adv}^{\text{s-mim}}_{\mathsf{Auth}}(A) \leq r \cdot \left(\mathbf{Adv}^{\text{ror-cma}}_{\mathsf{Auth}'}(B) + \frac{r}{|\mathcal{C}|} + \epsilon\delta \frac{|\mathbb{F}|}{|\mathbb{F}| - 1} + r \cdot \alpha \right),$$

where B has running time approximately equal to that of A, and makes at most r queries to its oracle. In other words, if Auth' *is (t, r, ϵ)-ror-cma-secure, then* Auth *is $(t', r, r \cdot (\epsilon + r/|\mathcal{C}| + \epsilon\delta|\mathbb{F}|/(|\mathbb{F}| - 1)))$-s-mim-secure, where $t' \approx t$.*

Proof. Let A be an attacker for game S-MIM which calls its oracles for at most r sessions. In the following, we are going to upper bound $\mathbf{Adv}^{\text{s-mim}}_{\mathsf{Auth}}(A) = \Pr\left[\text{S-MIM}^A_{\mathsf{Auth}} \Rightarrow \mathsf{true} \right]$. The proof proceeds via a sequence of games.

As our first step, we prove that it is sufficient to consider the *first* round where the attacker alters the communication between prover and verifier, and the latter still accepts. Formally, for all $\mathsf{sid}^* \in \{1, \ldots, r\}$, let $\text{WIN}_{\mathsf{sid}^*}$ be the event that in the experiment S-MIM$^A_{\mathsf{Auth}}$ session sid^* is the first session where the attacker makes the verifier non-trivially accept (and thus $d[\mathsf{sid}^*] = \mathsf{accept}$) with $(c'[\mathsf{sid}^*], t'[\mathsf{sid}^*]) \neq (c[\mathsf{sid}^*], t[\mathsf{sid}^*])$. In particular, for all $\mathsf{sid} < \mathsf{sid}^*$ we either have $(c[\mathsf{sid}], t[\mathsf{sid}]) = (c'[\mathsf{sid}], t'[\mathsf{sid}])$ or $d[\mathsf{sid}] = \mathsf{reject}$. Moreover, let

---

main G_{sid^*}:

$\mathsf{sid} \leftarrow 0$

$K_{\mathsf{Auth}'} \xleftarrow{\$} \mathsf{Gen}'(1^k)$

$K_{\mathsf{H}} \xleftarrow{\$} \mathcal{K}_{\mathsf{H}}$

$K_{\mathbb{F}} \xleftarrow{\$} \mathbb{F} \setminus \{0\}$

Run $A^{\mathrm{C}(\cdot),\mathrm{P}(\cdot),\mathrm{V}(\cdot)}(1^k)$

Ret $((c[\mathsf{sid}^*], \sigma[\mathsf{sid}^*])$

$(c'[\mathsf{sid}^*], \sigma'[\mathsf{sid}^*]))$

$\wedge\ (d[\mathsf{sid}^*] = \mathsf{accept})$

Procedure C():

If $c[\mathsf{sid}] = \bot$ then

$\quad c[\mathsf{sid}] \xleftarrow{\$} \mathcal{C}$

Ret $c[\mathsf{sid}]$

Procedure P(c'):

If $c'[\mathsf{sid}] = \bot$ then $c'[\mathsf{sid}] \leftarrow c'$ else ret $\bot$

$(\tau_1, \tau_2) \xleftarrow{\$} \mathrm{P}'_{K_{\mathsf{Auth}'}}(c')$, $\sigma_2 \leftarrow \tau_2 \circ K_{\mathbb{F}} + \mathsf{H}_{K_{\mathsf{H}}}(\tau_1)$

$\sigma[\mathsf{sid}] \leftarrow (\tau_1, \sigma_2)$

Ret $\sigma[\mathsf{sid}]$

$\neq$ **Procedure** V($\sigma' = (\sigma'_1, \sigma'_2)$):

$d[\mathsf{sid}] \leftarrow \mathsf{reject}$, $c[\mathsf{sid}] \xleftarrow{\$} \mathrm{C}()$

If $\mathsf{sid} < \mathsf{sid}^*$ and $(c'[\mathsf{sid}], \sigma') = (c[\mathsf{sid}], \sigma[\mathsf{sid}])$

then

$\quad d[\mathsf{sid}] \leftarrow \mathsf{accept}$

If $\mathsf{sid} = \mathsf{sid}^*$ then

$\quad \sigma'[\mathsf{sid}] \leftarrow (\sigma'_1, \sigma'_2)$

$\quad \tau'_1 \leftarrow \sigma'_1,\ \tau'_2 \leftarrow (\sigma'_2 - \mathsf{H}_{K_{\mathsf{H}}}(\sigma'_1)) \circ K_{\mathbb{F}}^{-1}$

$\quad d[\mathsf{sid}] \leftarrow \mathrm{V}'_{K_{\mathsf{Auth}'}}(c[\mathsf{sid}], (\tau'_1, \tau'_2))$

$\mathsf{sid} \leftarrow \mathsf{sid} + 1$

Ret $d[\mathsf{sid}]$

Fig. 1. Game G_{sid^*} for $\mathsf{sid}^* \in \{1, \ldots, r\}$ in the proof of Theorem 5. All oracles return $\bot$ if $\mathsf{sid} > \mathsf{sid}^*$.

$\mathrm{WIN} = \bigcup_{\mathsf{sid}^*=1}^{r} \mathrm{WIN}_{\mathsf{sid}^*}$ be the event that $\text{S-MIM}_{\mathsf{Auth}}^A$ outputs true in the first place. Clearly, the r events $\mathrm{WIN}_1, \ldots, \mathrm{WIN}_r$ are disjoint, and therefore

$$\Pr[\mathrm{WIN}] = \Pr\left[\bigcup_{\mathsf{sid}^*=1}^{r} \mathrm{WIN}_{\mathsf{sid}}^*\right] = \sum_{\mathsf{sid}^*=1}^{r} \Pr[\mathrm{WIN}_{\mathsf{sid}}^*].$$

As our first step, we introduce r new games $G_1, \ldots, G_r$, where G_{sid^*} only allows the adversary A to execute sid^* sessions, and the verifier returns reject for the first $\mathsf{sid}^* - 1$ sessions unless the adversary A has been simply forwarding honestly generated messages. A formal description of G_{sid^*} is given in Fig. 1. There, we implicitly assume that all oracles return $\bot$ whenever $\mathsf{sid} > \mathsf{sid}^*$. It is easy to see that by construction, $\Pr[G_{\mathsf{sid}^*}^A \Rightarrow \mathsf{true}] \geq \Pr[\mathrm{WIN}_{\mathsf{sid}^*}] - (\mathsf{sid}^* - 1)\alpha$. The offset depending on the completeness error α is due to the fact that $G_{\mathsf{sid}^*}^A$ always accepts honest executions in sessions $\mathsf{sid} < \mathsf{sid}^*$, whereas this is not necessarily true in $\text{S-MIM}_{\mathsf{Auth}}^A$. Therefore,

$$\Pr\left[\text{S-MIM}_{\mathsf{Auth}}^A \Rightarrow \mathsf{true}\right] = \Pr[\mathrm{WIN}] \leq r^2\alpha + \sum_{\mathsf{sid}^*=1}^{r} \Pr[G_{\mathsf{sid}^*}^A \Rightarrow \mathsf{true}]. \quad (1)$$

In the remainder of this proof, for every $\mathsf{sid}^* \in \{1, \ldots, r\}$, we are going to prove an upper bound on $\Pr[G_{\mathsf{sid}^*}^A \Rightarrow \mathsf{true}]$. In particular, we now fix an arbitrary $\mathsf{sid}^* \in \{1, \ldots, r\}$, and let $H_0 = G_{\mathsf{sid}^*}$.

The proof now continues by transitioning from Game H_0 in turn to games H_1, H_2 and H_3. With respect to H_0, these games will only differ in the way in which queries to P are answered, but all games will otherwise inherit the **main**

Procedure P(c'): // H$_1$	**Procedure** P(c'): // H$_2$	**Procedure** P(c'): // H$_3$
If $c'[\mathsf{sid}] = \bot$ then	If $c'[\mathsf{sid}] = \bot$ then	If $\mathsf{sid} < \mathsf{sid}^*$ and $c' = c[\mathsf{sid}^*]$
$\quad c'[\mathsf{sid}] \leftarrow c'$	$\quad c'[\mathsf{sid}] \leftarrow c'$	$\quad$ then Ret $\bot$
Else Ret $\bot$	Else Ret $\bot$	If $c'[\mathsf{sid}] = \bot$ then $c'[\mathsf{sid}] \leftarrow c'$
$(\tau_1, \tau_2) \xleftarrow{\$} \mathsf{P}'_{K_{\mathsf{Auth}'}}(c')$	$(\tau_1, \tau_2) \xleftarrow{\$} \mathsf{P}'_{K_{\mathsf{Auth}'}}(c')$	Else Ret $\bot$
If $c' \neq c[\mathsf{sid}^*]$ then	If $c' = c[\mathsf{sid}^*]$ then	$(\tau_1, \tau_2) \xleftarrow{\$} \mathsf{P}'_{K_{\mathsf{Auth}'}}(c')$
$\quad \tau_2 \xleftarrow{\$} \mathcal{T}_2$	$\quad \sigma_2 \leftarrow \tau_2 \circ K_{\mathbb{F}} + \mathsf{H}_{K_{\mathsf{H}}}(\tau_1)$	If $c' = c[\mathsf{sid}^*]$ then
$\sigma_2 \leftarrow \tau_2 \circ K_{\mathbb{F}} + \mathsf{H}_{K_{\mathsf{H}}}(\tau_1)$	Else $\sigma_2 \xleftarrow{\$} \mathcal{T}_2$	$\quad \sigma_2 \leftarrow \tau_2 \circ K_{\mathbb{F}} + \mathsf{H}_{K_{\mathsf{H}}}(\tau_1)$
$\sigma[\mathsf{sid}] \leftarrow (\tau_1, \sigma_2)$	$\sigma[\mathsf{sid}] \leftarrow (\tau_1, \sigma_2)$	Else $\sigma_2 \xleftarrow{\$} \mathcal{T}_2$
Ret $\sigma[\mathsf{sid}]$	Ret $\sigma[\mathsf{sid}]$	$\sigma[\mathsf{sid}] \leftarrow (\tau_1, \sigma_2)$
		Ret $\sigma[\mathsf{sid}]$

Fig. 2. Modified prover oracles in the games H$_1$, H$_2$, and H$_3$.

procedure, as well as C and V, verbatim from $\mathsf{G}_{\mathsf{sid}^*} = \mathsf{H}_0$. A formal specification of the respective procedures is given in Fig. 2, and we now discuss them in detail.

We first transition to Game H$_1$, where we will use ror-cma security of Auth' to replace the right half of every tag computed by P to a random component whenever $c' \neq c[\mathsf{sid}^*]$, i.e., different from the random challenge used in the last round. The proof of the following lemma is given below.

Lemma 6. *There exists an attacker B such that*

$$\Pr\left[\mathsf{H}_0^A \Rightarrow \mathsf{true}\right] - \Pr\left[\mathsf{H}_1^A \Rightarrow \mathsf{true}\right] \leq \mathbf{Adv}_{\mathsf{Auth}'}^{\mathsf{ror\text{-}cma}}(B)\,,$$

where B has running time approximately equal to that of A, and makes at most r queries to its oracle.

Subsequently, in Game H$_2$, whenever $c' \neq c[\mathsf{sid}^*]$, instead of generating τ_2 at random, we directly generate σ_2 uniformly at random from the same set. Note that because $K_{\mathbb{F}} \neq 0$, we have that $\tau_2 \cdot K_{\mathbb{F}}$ is a fresh random value, and thus the two games H$_1$ and H$_2$ are identical,

In the next game, Game H$_3$, the procedure P replies to a query $c' = c[\mathsf{sid}^*]$ only if it is made in session sid^*, and otherwise returns $\bot$. As $c[\mathsf{sid}^*]$ is chosen uniformly at random, and independent of the interaction between the adversary and the oracles in the first $\mathsf{sid}^* - 1$ sessions, the "fundamental lemma" of game playing [4] yields

$$\Pr\left[\mathsf{H}_2^A \Rightarrow \mathsf{true}\right] - \Pr\left[\mathsf{H}_3^A \Rightarrow \mathsf{true}\right]$$
$$\leq \Pr\left[\,c[\mathsf{sid}^*] \in \{c'[1], c'[2], \ldots, c'[\mathsf{sid}^* - 1]\}\,\right] \leq \frac{r}{|\mathcal{C}|}\,. \quad (2)$$

Therefore, putting together Eq. (1), Lemma 6, and Eq. (2), we obtain that there exists an attacker B making at most r oracle queries and with time complexity close to the one of A such that

$$\Pr\left[\mathsf{S\text{-}MIM}_{\mathsf{Auth}}^A \Rightarrow \mathsf{true}\right] \leq r \cdot \left(\mathbf{Adv}_{\mathsf{Auth}'}^{\mathsf{ror\text{-}cma}}(B) + \frac{r}{|\mathcal{C}|} + \Pr\left[\mathsf{H}_3^A \Rightarrow \mathsf{true}\right]\right) + r^2\alpha\,.$$

In the rest of the proof, we give an upper bound on the probability that the game H_3 outputs true. The argument is going to rely on the almost pairwise-independence of H and the right-tag sparsity of Auth', and is from now on a purely information-theoretic argument. In particular, it does not rely on $K_{\mathsf{Auth}'}$ being hidden, but only on the fact that all the right tags in sessions prior to sid^* are random.

ANALYSIS OF WINNING PROBABILITY IN H_3. In the following, for notational convenience we let $\sigma[\mathsf{sid}^*] = (\sigma_1 = \tau_1, \sigma_2)$ and $\sigma'[\mathsf{sid}^*] = (\sigma'_1, \sigma'_2)$ be the original and modified values in the second-round of session sid^*. Similarly, we simply denote $c = c[\mathsf{sid}^*]$ and $c' = c'[\mathsf{sid}^*]$. Concretely, we are going to consider three different cases when analyzing the probability $\Pr\left[H_3^A \Rightarrow \mathsf{true}\right]$: (1) $c' \neq c$, (2) $c' = c$ and $\sigma_1 = \sigma'_1$, and (3) $c' = c$ and $\sigma_1 \neq \sigma'_1$. We now analyze the three individual cases.

CASE $c' \neq c$. Observe first that in session sid^*, the attacker obtains (τ_1, σ_2), where $(\tau_1, \tau_2) \xleftarrow{\$} \mathsf{P}'_K(c'_{\mathsf{sid}^*})$ and $\sigma_2 \xleftarrow{\$} T_2$, and inputs (σ'_1, σ'_2) to V. It wins if (σ'_1, τ'_2) is a valid tag, where $\tau'_2 = (\sigma'_2 - H_{K_H}(\sigma'_1)) \circ K_{\mathbb{F}}^{-1}$.

The crucial point is that $K_{\mathbb{F}}$ and K_H have *never* been used prior to the computation of τ'_2, as the oracle P has only returned random right tags. So we can equivalently think of generating these uniformly at random for the first time at this point independent of the rest of the game, and consider the probability that $\mathsf{V}'_K(c, (\sigma'_1, \tau'_2))$ verifies. Moreover, the value $Y := H_{K_H}(\sigma'_1)$ is going to be uniform (as we don't evaluate the function on any other point) by the δ-almost pairwise independence of H. Therefore, for every value $t \in \mathbb{F}$,

$$\Pr\left[\tau'_2 = t\right] = \Pr\left[(\sigma'_2 - Y) \circ K_{\mathbb{F}}^{-1} = t\right] = \Pr\left[Y = K_{\mathbb{F}} \circ t + \sigma'_2\right] = \frac{1}{|\mathbb{F}|}.$$

However, by ϵ-right tag sparsity, we know that there are at most $\epsilon|\mathbb{F}|$ possible values t for which (σ'_1, t) is a valid tag, and thus by the union bound

$$\Pr\left[H_3^A \Rightarrow \mathsf{true} \mid c' \neq c\right] = \Pr\left[\mathsf{V}'_K(c, (\sigma'_1, \tau'_2)) = \mathsf{accept}\right] \leq \epsilon \cdot |\mathbb{F}| \cdot \frac{1}{|\mathbb{F}|} = \epsilon. \quad (3)$$

CASE $c' = c$, $\sigma_1 = \sigma'_1 = \tau_1$ AND $\sigma_2 \neq \sigma'_2$. In this case, in session sid^*, the attacker obtains (τ_1, σ_2), where $(\tau_1, \tau_2) \xleftarrow{\$} \mathsf{P}'_K(c)$ and $\sigma_2 \xleftarrow{\$} \tau_2 \circ K_{\mathbb{F}} + H_{K_H}(\tau_1)$. Subsequently, it inputs (τ_1, σ'_2) to V. It wins if $\mathsf{V}'_K(c, (\tau_1, \tau'_2)) = \mathsf{accept}$, where $\tau'_2 = (\sigma'_2 - H_{K_H}(\tau_1)) \circ K_{\mathbb{F}}^{-1} \neq \tau_2$. Once again, we evaluate H only with one input, and as above $Y = H_{K_H}(\tau_1)$ is uniformly distributed.

Now, given σ_2, τ_2, and σ'_2, we want to upper bound the probability that $\tau'_2 = t \neq \tau_2$ for some value $t \in \mathbb{F}$, where the probability is over the choice of $K_{\mathbb{F}}$ and Y.

$$\Pr\left[\tau'_2 = t \mid \sigma_2 = \tau_2 \circ K_{\mathbb{F}} + Y\right] = \frac{\Pr\left[t \circ K_{\mathbb{F}} + Y = \sigma'_2 \wedge \tau_2 \circ K_{\mathbb{F}} + Y = \sigma_2\right]}{\Pr\left[\tau_2 \circ K_{\mathbb{F}} + Y = \sigma_2\right]}.$$

Since $\tau_2 \neq t$, there exists exactly one $K_{\mathbb{F}}$ such that $(\tau_2 - t) \cdot K_{\mathbb{F}} = \sigma_2 - \sigma'_2$, and moreover, this defines a unique value for Y, which is taken with probability at most $1/|\mathbb{F}|$, and thus the probability in the numerator is upper bounded by $1/(|\mathbb{F}|(|\mathbb{F}| - 1))$. Moreover, $\tau_2 \circ K_\mathbb{F} + Y$ is clearly uniform (because Y is uniform), and thus the denominator is $1/|\mathbb{F}|$. Putting these together gives us $\Pr[\tau_2' = t \mid \sigma_2 = \tau_2 \circ K_\mathbb{F} + Y] \le 1/(|\mathbb{F}| - 1)$. Now, due to ϵ-right tag sparsity, there are at most $\epsilon \cdot |\mathbb{F}|$ right tags τ_2' that verify, and thus

$$\Pr\left[H_3^A \Rightarrow \text{true} \mid c' = c \wedge \sigma_1 = \sigma_1' \right] \le \epsilon \cdot \frac{|\mathbb{F}|}{|\mathbb{F}| - 1} . \tag{4}$$

CASE $c' = c$ AND $\sigma_1 \neq \sigma_1'$. For the final case, the attacker obtains (τ_1, σ_2) as in the previous case, but inputs $(\sigma_1' \neq \tau_1, \sigma_2')$ to V, and the latter computes $\tau_2' = (\sigma_2' - H_{K_H}(\sigma_1')) \circ K_\mathbb{F}^{-1}$.

Here, we indeed evaluate H_{K_H} on *two* inputs. However, by Lemma 2, we see that for every possible values σ_1' and y', chosen adaptively depending on τ_1 and $H_{K_H}(\tau_1)$, $H_{K_H}(\sigma_1') = y'$ with probability at most $\delta/|\mathbb{F}|$. Therefore, for every possible t such that $V_K'(c, (\sigma_1', t)) = \text{accept}$, we have

$$\Pr[\tau_2' = t] = \Pr[H_{K_H}(\sigma') = K_\mathbb{F} \cdot t + \sigma_2'] \le \delta/|\mathbb{F}|.$$

Now, due to ϵ-right tag sparsity, there are at most $\epsilon \cdot |\mathbb{F}|$ such right tags, and thus

$$\Pr\left[H_3^A \Rightarrow \text{true} \mid c' = c \wedge \sigma_1 \neq \sigma_1' \right] \le \epsilon \cdot |\mathbb{F}| \cdot \frac{\delta}{|\mathbb{F}|} = \epsilon\delta . \tag{5}$$

PUTTING THINGS TOGETHER. To conclude the proof, we observe that all terms in Eqs. (3), (4) and (5) are upper bounded by $\epsilon \cdot \delta \cdot \frac{|\mathbb{F}|}{|\mathbb{F}|-1}$, and thus we also have $\Pr\left[H_3^A \Rightarrow \text{true} \right] \le \epsilon \cdot \delta \cdot \frac{|\mathbb{F}|}{|\mathbb{F}|-1}$. $\square$

Proof (Lemma 6). The attacker B for ROR-CMA(b) is very simple. It simulates the execution of H_b to the attacker A. Initially, B uses its input challenge c^* as c_{sid^*}. Then, when simulating queries to P on input c', it forwards them to its own oracle T, to obtain a pair (τ_1, τ_2). Finally, B uses the one available verification query to compute V's decision bit in session sid^*. Finally, B outputs the games H_b's output. By inspection, it is not hard to verify that $\Pr\left[\text{ROR-CMA}(b)_{\text{Auth}'}^B \Rightarrow \text{true} \right] = \Pr\left[H_b^A \Rightarrow \text{true} \right]$, which concludes the proof of the lemma. $\square$

4   Instantiations

In this section, we will provide examples of ror-cma-secure authentication protocols. All of them can be transformed to s-mim-secure authentication protocols using the transformation from Sect. 3. Table 2 summarizes the resulting protocols compactly.

Table 2. New s-mim-secure 2-round authentication protocols.

Scheme	Assump	Gen(1^n) / Response P(c) / Verify V(c, σ)
Auth$_{\mathsf{LPN}}$ Subsect. 4.1	LPN	Gen(1^n) : $(\mathbf{x}_1, \ldots, \mathbf{x}_5) \xleftarrow{\$} (\mathbb{F}_2^\ell)^5$
		P(c) = $(\mathbf{R}, \mathbf{R} \cdot (\mathbf{M_c} \cdot \mathbf{x}_1 + \mathbf{x}_2) + \mathbf{x}_3 \circ \mathbf{e} + \mathsf{H}_{\mathbf{x}_4, \mathbf{x}_5}(\mathbf{R}) \in \mathbb{F}_2^{\ell \times n} \times \mathbb{F}_2^\ell$
		V($c, (\mathbf{R}, \mathbf{z})$) : $\lvert (\mathbf{z} - \mathsf{H}_{\mathbf{x}_4, \mathbf{x}_5}(\mathbf{R}) - \mathbf{R} \cdot (\mathbf{M_c} \cdot \mathbf{x}_1 + \mathbf{x}_2)) \circ \mathbf{x}_3^{-1} \rvert$ small?
AuthT$_{\mathsf{LPN}}$ Subsect. 4.1	LPN	Gen(1^n) : $(\mathbf{X}_1, \mathbf{X}_2, \mathbf{x}_3, \mathbf{x}_4) \xleftarrow{\$} (\mathbb{F}_2^{n \times \ell})^2 \times (\mathbb{F}_2^\ell)^2$
		P(c) = $(\mathbf{r}, (\mathbf{X}_1 \cdot \mathbf{M_c} + \mathbf{X}_2) \cdot \mathbf{r} + \mathbf{x}_3 \circ \mathbf{e} + \mathbf{x}_4) \in \mathbb{F}_2^\ell \times \mathbb{F}_2^n$
		V($c, (\mathbf{r}, \mathbf{z})$) : $\lvert (\mathbf{z} - \mathbf{x}_4 - (\mathbf{X}_1 \cdot \mathbf{M_c} + \mathbf{X}_2) \cdot \mathbf{r}) \circ \mathbf{x}_3^{-1} \rvert$ small?
Auth$_{\mathsf{Field\text{-}LPN}}$ Subsect. 4.2	Field-LPN	Gen(1^n) : $(\mathbf{x}_1, \ldots, \mathbf{x}_4) \xleftarrow{\$} (\mathbb{F}_{2^\ell})^4$
		P(c) = $(\mathbf{r}, \mathbf{r} \circ (\mathbf{x}_1 \circ \mathbf{c} + \mathbf{x}_2) + \mathbf{x}_3 \circ \mathbf{e} + \mathbf{x}_4) \in \mathbb{F}_{2^\ell} \times \mathbb{F}_{2^\ell}$
		V($c, (\mathbf{R}, \mathbf{z})$) : $\lvert (\mathbf{z} - \mathbf{x}_4 - \mathbf{r} \circ (\mathbf{x}_1 \circ \mathbf{c} + \mathbf{x}_2)) \circ \mathbf{x}_3^{-1} + \mathbf{x}_4) \rvert$ small?
Auth$_{\mathsf{ddh}}$ Subsect. 4.4	ddh	Gen(1^n) : $(x_1, x_2, X) \xleftarrow{\$} \mathbb{F}_q^2 \times \mathbb{G}$
		P(c) = $(R, X \cdot R^{x_1 c + x_2}) \in \mathbb{G} \times \mathbb{G}$
		V($c, (r, z)$) : $X \cdot R^{x_1 c + x_2} = z$?
Auth$_{\mathsf{wprf}}$ Subsect. 4.3	wprf	$(x_{0,0}, \ldots, x_{\ell,1}, \mathbf{x}_1, \mathbf{x}_2) \xleftarrow{\$} \mathbb{D}^{2\ell} \times \mathbb{F}^2$
		P(c) = $(r, \sum_{i=1}^\ell F(x_{i,c_i}, r) + \mathsf{H}_{\mathbf{x}_1, \mathbf{x}_2}(r)) \in \mathbb{D} \times \mathbb{F}$
		V($c, (r, z)$) : $\sum_{i=1}^\ell F(x_{i,c_i}, r) + \mathsf{H}_{\mathbf{x}_1, \mathbf{x}_2}(r) = z$?

4.1   Instantiations from LPN

LEARNING PARITY WITH NOISE. For a parameter $0 < \gamma \leq 1/2$, we define the Bernoulli distribution $\mathcal{B}_\gamma$ that assigns $e \xleftarrow{\$} \mathcal{B}_\gamma$ the values 1 and 0 with probabilities γ and $1 - \gamma$, respectively. If $\mathcal{D}$ is a distribution over $\mathbb{D}$, then $\mathbf{x} \xleftarrow{\$} \mathcal{D}^n$ denotes the n-fold distribution where each component of $\mathbf{x} \in \mathbb{D}^n$ is chosen according to $\mathcal{D}$.

To define the $\mathsf{LPN}_{\ell,\gamma}$ problem in dimension $\ell \in \mathbb{N}$ and Bernoulli parameter $0 < \gamma \leq 1/2$ we introduce the LPN advantage as the quantity

$$\mathbf{Adv}^{\mathsf{LPN}}(A) = \Pr\left[A^{\mathrm{LPN}_{\mathbf{s},\gamma}()} \Rightarrow \text{true} \right] - \Pr\left[A^{\mathrm{LPN}_{\mathbf{s},1/2}()} \Rightarrow \text{true} \right],$$

where $\mathbf{s} \xleftarrow{\$} \mathbb{F}_2^\ell$ and $\mathrm{LPN}_{\mathbf{s},\alpha}$ ($\alpha \in \{\gamma, 1/2\}$) returns $(\mathbf{r}, \mathbf{r}^T \cdot \mathbf{s} + e)$ for $\mathbf{r} \xleftarrow{\$} \mathbb{F}_2^\ell$ and $e \xleftarrow{\$} \mathcal{B}_\alpha$. Note that oracle $\mathrm{LPN}_{\mathbf{s},1/2}$ always returns uniform $(\mathbf{r}, z) \xleftarrow{\$} \mathbb{F}_2^\ell \times \mathbb{F}_2$. We say that the $\mathsf{LPN}_{\ell,\gamma}$ is (t, q, ϵ)-hard if for all attackers A with time complexity t, making at most q oracle queries, we have $\mathbf{Adv}^{\mathsf{LPN}}(A) \leq \epsilon$.

ROR-CMA SECURE PROTOCOL. Let $n = O(\ell)$ denote the number of repetitions, γ the parameter of the Bernoulli distribution, and $\gamma' := 1/4 + \gamma/2$ controls the correctness error. The following authentication protocol $\mathsf{Auth}'_{\mathsf{LPN}} = \{\mathsf{Gen}', \mathsf{P}', \mathsf{V}'\}$ originates from [20]. It has associated key space $\mathcal{K} = (\mathbb{F}_2^\ell)^2$, tag space $\mathcal{T} = \mathcal{T}_1 \times \mathcal{T}_2 = \mathbb{F}_2^{\ell \times n} \times \mathbb{F}_2^n$, and challenge space $\mathcal{C} = \mathbb{F}_2^\ell$.

- Key Generation. The key-generation algorithm Gen' outputs a secret key $K = (\mathbf{k}_1, \mathbf{k}_2) \xleftarrow{\$} (\mathbb{F}_2^\ell)^2$.
- Challenge. The challenge is generated by the verifier V' as $\mathbf{c} \xleftarrow{\$} \mathbb{F}_2^\ell$.
- Response. The response $\tau = (\tau_1, \tau_2)$ to challenge $\mathbf{c} \in \mathbb{F}_2^\ell$ is computed by the prover P' by sampling $\mathbf{R} \xleftarrow{\$} \mathbb{F}_2^{\ell \times n}$ and computing $\tau = (\mathbf{R}, \mathbf{R}^T \cdot (\mathbf{M_c} \cdot \mathbf{k}_1 + \mathbf{k}_2) + \mathbf{e})$,

where $\mathbf{e} \xleftarrow{\$} \mathcal{B}_\gamma^n$. (Recall that $\mathbf{M_c}$ is the matrix representation of the finite field multiplication with $\mathbf{c}$.)
- Verification. Given challenge $\mathbf{c} \in \mathbb{F}_2^\ell$ and response $\tau = (\mathbf{R}, \mathbf{z}) \in \mathbb{F}_2^{\ell \times n} \times \mathbb{F}_2^n$, the verifier V' outputs accept iff: $rank(\mathbf{R}) = n$ and $|\mathbf{R}^T \cdot (\mathbf{M_c} \cdot \mathbf{k}_1 + \mathbf{k}_2) - \mathbf{z}| \leq \gamma' n$.

With the choice of $\gamma' = 1/4 + \gamma/2$, $\mathsf{Auth}'_{\mathsf{LPN}}$ has $2^{-O(n)}$ completeness error [20, Theorem 4]. Further, it has ϵ-sparse right tags, where $\epsilon = \Pr[\mathbf{z} \leq \gamma' n \mid \mathbf{z} \xleftarrow{\$} \mathbb{F}_2^n] \leq 2^{-O(n)}$, using the Hoeffding bound.

The proof of the following theorem is postponed to Appendix A.2.

Theorem 7. *If* $\mathsf{LPN}_{\ell,\gamma}$ *is* (t, nq, ϵ)*-hard, then* $\mathsf{Auth}'_{\mathsf{LPN}}$ *is* (t', q, ϵ)*-ror-cma-secure with* $t \approx t'$.

There exists an alternative ror-cma-secure authentication protocol [10,20] which defines $\tau_2 = \mathbf{R}^T \cdot \mathbf{k}_{\downarrow \mathbf{c}} + \mathbf{e}$, where $\mathbf{k}_{\downarrow \mathbf{c}}$ is the projection of $\mathbf{k}$ with respect to all ℓ non-zero bits of $\mathbf{c} \in \mathcal{C} := \{\mathbb{F}_2^{2\ell} : |\mathbf{c}| = \ell\}$.

MiM SECURE PROTOCOL. A s-mim-secure 2-round authentication protocol $\mathsf{Auth}_{\mathsf{LPN}}$ is obtained via the generic transformation from Sect. 3. An example instantiation using the almost pairwise independent hash function from Sect. 2 is given in Table 2.

TRADE-OFF. For all LPN-based protocols there exists a natural trade-off between key-size and communication complexity, as we will explain now. In the ror-cma-secure protocol $\mathsf{AuthT}'_{\mathsf{LPN}}$ we can chose the key as $(\mathbf{K}_1, \mathbf{K}_2) \xleftarrow{\$} (\mathbb{Z}_2^{\ell \times n})^2$ and define the response to a challenge $\mathbf{c} \in \mathbb{F}_2^\ell$ as $(\mathbf{r}, (\mathbf{M_c} \cdot \mathbf{K}_1 + \mathbf{K}_2) \cdot \mathbf{r} + \mathbf{e} \in \mathbb{F}_2^\ell \times \mathbb{F}_2^n$, where $\mathbf{r} \xleftarrow{\$} \mathbb{F}_2^\ell$. In the resulting s-mim-secure protocol we can use the specific pairwise independent hash function $\mathsf{H}_{\mathbf{S}_1, \mathbf{s}_2}(\mathbf{r}) := \mathbf{S}_1 \mathbf{r} + \mathbf{s}_2$, where $(\mathbf{S}_1, \mathbf{s}_2) \in \mathbb{F}_2^{\ell \times n} \times \mathbb{F}_2^n$. The response to a challenge $\mathbf{c}$ is computed as $\sigma = (\mathbf{r}, \mathbf{z})$, where

$$\mathbf{z} = ((\mathbf{M_c} \cdot \mathbf{K}_1 + \mathbf{K}_2) \cdot \mathbf{r} + \mathbf{e}) \circ K_\mathbb{F} + \mathbf{S}_1 \mathbf{r} + \mathbf{s}_2$$
$$= (\mathbf{M_c} \cdot \mathbf{K}_1 \cdot \mathbf{M}_{K_\mathbb{F}} + \mathbf{K}_2 \cdot \mathbf{M}_{K_\mathbb{F}} + \mathbf{S}_1) \cdot \mathbf{r} + \mathbf{M}_{K_\mathbb{F}} \cdot \mathbf{e} + \mathbf{s}_2.$$

This can be rewritten as $\mathbf{z} = (\mathbf{M_c} \cdot \mathbf{X}_1 + \mathbf{X}_2) \cdot \mathbf{r} + \mathbf{e} \circ \mathbf{x}_3 + \mathbf{x}_4$ using the substitutions

$$\mathbf{x}_1 := \mathbf{K}_1 \cdot \mathbf{M}_{K_\mathbb{F}}, \quad \mathbf{X}_2 := \mathbf{K}_2 \cdot \mathbf{M}_{K_\mathbb{F}} + \mathbf{S}_1, \quad \mathbf{x}_3 := K_\mathbb{F}, \quad \mathbf{x}_4 := \mathbf{s}_2.$$

The resulting protocol $\mathsf{AuthT}_{\mathsf{LPN}}$ is described in Table 2.

4.2 Instantiations from Field-LPN

FIELD LEARNING PARITY WITH NOISE. To define the Field-LPN$_{\ell,\gamma}$ problem over the extension field $(\mathbb{F}_{2^\ell}, \circ, +)$ and Bernoulli parameter $0 < \gamma \leq 1/2$, we introduce the Field-LPN advantage as the quantity

$$\mathbf{Adv}^{\mathsf{Field\text{-}LPN}}(A) = \Pr\left[A^{\mathsf{FLPN}_{\mathbf{s},\gamma}()} \Rightarrow \mathsf{true}\right] - \Pr\left[A^{\mathsf{FLPN}_{\mathbf{s},1/2}()} \Rightarrow \mathsf{true}\right],$$

where $\mathbf{s} \xleftarrow{\$} \mathbb{F}_{2^\ell}$ and $\text{FLPN}_{\mathbf{s},\alpha}$ returns $(\mathbf{r}, \mathbf{r} \circ \mathbf{s} + \mathbf{e})$ for $\mathbf{r} \xleftarrow{\$} \mathbb{F}_{2^\ell}$ and $\mathbf{e} \xleftarrow{\$} \mathcal{B}_\alpha^\ell$. Note that $\text{FLPN}_{\mathbf{s},1/2}$ always returns uniform $(\mathbf{r}, \mathbf{z}) \xleftarrow{\$} (\mathbb{F}_{2^\ell})^2$. We say that the Field-LPN$_{\ell,\gamma}$ is (t, q, ϵ)-hard if for all attackers A with time complexity t making at most q oracle queries, we have $\mathbf{Adv}^{\text{Field-LPN}}(A) \leq \epsilon$.

ROR-CMA SECURE PROTOCOL. Let γ the parameter of the Bernoulli distribution, and $\gamma' := 1/4 + \gamma/2$ controls the correctness error. We use $\mathbb{F} = \mathbb{F}_{2^\ell}$ to denote the finite field. The following authentication protocol $\text{Auth}'_{\text{Field-LPN}} = \{\text{Gen}', \text{P}', \text{V}'\}$ originates from [15]. It has associated key space $\mathcal{K} = \mathbb{F}^2$, split tag space $\mathcal{T} = \mathcal{T}_1 \times \mathcal{T}_2 = \mathbb{F} \times \mathbb{F}$, and challenge space $\mathcal{C} = \mathbb{F}$.

- Key Generation. The key-generation algorithm Gen$'$ outputs a secret key $K = (\mathbf{k}_1, \mathbf{k}_2) \xleftarrow{\$} \mathbb{F}^2$.
- Challenge. The challenge is generated by the verifier V$'$ as $\mathbf{c} \xleftarrow{\$} \mathbb{F}$.
- Response. The response $\tau = (\tau_1, \tau_2)$ to challenge $c \in \mathbb{F}$ is computed by the prover P$'$ as $\tau = (\mathbf{r}, \mathbf{r} \circ (\mathbf{k}_1 \circ \mathbf{c} + \mathbf{k}_2) + \mathbf{e})$, where $\mathbf{r} \xleftarrow{\$} \mathbb{F}$, $\mathbf{e} \xleftarrow{\$} \mathcal{B}_\gamma^\ell$.
- Verification. Given challenge $c \in \mathbb{F}$ and response $\tau = (\mathbf{r}, \mathbf{z}) \in \mathbb{F}^2$, the verifier V$'$ outputs accept iff $|\mathbf{r} \circ (\mathbf{k}_1 \circ \mathbf{c} + \mathbf{k}_2) - \mathbf{z}| \leq \gamma' n$.

As in the LPN case, $\text{Auth}'_{\text{Field-LPN}}$ has $2^{-O(\ell)}$ completeness error and $2^{-O(\ell)}$-sparse right tags. The proof of the following theorem is similar to that of Theorem 7 and is therefore omitted.

Theorem 8. *If* Field-LPN$_{\ell,\gamma}$ *is* (t, q, ϵ)-*hard, then* $\text{Auth}'_{\text{Field-LPN}}$ *is* (t', q, ϵ)-*ror-cma-secure with* $t' \approx t$.

MIM SECURE PROTOCOL. We now apply our generic transformation from Sect. 3 to $\text{Auth}'_{\text{Field-LPN}}$ to obtain a s-mim-secure protocol. The key consists of $(\mathbf{k}_1, \mathbf{k}_2, K_\mathbb{F}, \mathbf{s}_1, \mathbf{s}_2)$, where we use the concrete pairwise-independent hash function $\mathsf{H}_{\mathbf{s}_1,\mathbf{s}_2}(\mathbf{r}) = \mathbf{s}_1 \circ \mathbf{r} + \mathbf{s}_2$. The response to a challenge $\mathbf{c}$ is computed as $\sigma = (\mathbf{r}, \mathbf{z})$, where $\mathbf{z} = (\mathbf{r} \circ (\mathbf{k}'_1 \circ \mathbf{c} + \mathbf{k}'_2) + \mathbf{e}) \circ K_\mathbb{F} + \mathbf{s}_1 \circ \mathbf{r} + \mathbf{s}_2 = (\mathbf{r} \circ (\mathbf{k}'_1 \circ K_\mathbb{F} \circ \mathbf{c} + \mathbf{k}'_2 \circ K_\mathbb{F} + \mathbf{s}_1) + \mathbf{e} \circ K_\mathbb{F} + \mathbf{s}_2$. This can be written as $\mathbf{z} = (\mathbf{r} \circ (\mathbf{x}_1 \circ \mathbf{c} + \mathbf{x}_2) + \mathbf{e} \circ \mathbf{x}_3 + \mathbf{x}_4$ using the substitutions $\mathbf{x}_1 := \mathbf{k}_1 \circ K_\mathbb{F}$, $\mathbf{x}_2 := \mathbf{k}_2 \circ K_\mathbb{F} + \mathbf{s}_1$, $\mathbf{x}_3 := K_\mathbb{F}$, $\mathbf{x}_4 := \mathbf{s}_2$. The resulting simplified protocol $\text{Auth}_{\text{Field-LPN}}$ is given in Table 2.

4.3   Instantiations from Weak PRFs

WEAK PSEUDORANDOM FUNCTION. Let $\mathcal{F}$ be a function family $F : \mathbb{K} \times \mathbb{D} \to \mathbb{F}$. To define the $\text{wprf}_\mathcal{F}$ assumption over function family $\mathcal{F}$ we introduce the wprf advantage of an adversary A as the quantity

$$\mathbf{Adv}_\mathcal{F}^{\text{wprf}}(A) = \Pr[A^{F_x()} \Rightarrow \text{true}] - \Pr[A^{U()} \Rightarrow \text{true}],$$

where $x \xleftarrow{\$} \mathbb{K}$, F_x returns $(r, F(x, r))$ for $r \xleftarrow{\$} \mathbb{D}$, and U returns uniform $(r, z) \xleftarrow{\$} \mathbb{D} \times \mathbb{F}$. We say that $\mathcal{F}$ is a (t, q, ϵ)-weak PRF if for all attackers A with time complexity t, making at most q oracle queries, we have $\mathbf{Adv}_\mathcal{F}^{\text{wprf}}(A) \leq \epsilon$.

ROR-CMA SECURE PROTOCOL. We define an authentication protocols $\mathsf{Auth}'_{\mathsf{wprf}} = \{\mathsf{Gen}', \mathsf{P}', \mathsf{V}'\}$ with associated key space $\mathcal{K} = \mathbb{K}^\ell$, split tag space $\mathcal{T} = \mathcal{T}_1 \times \mathcal{T}_2 = \mathbb{D} \times \mathbb{F}$, and challenge space $\mathcal{C} = \{0,1\}^\ell$.

- Key Generation. The key-generation algorithm Gen' outputs a secret key $K = (x_{1,0}, \ldots, x_{\ell,0}, x_{1,1}, \ldots, x_{\ell,1}) \xleftarrow{\$} \mathbb{K}^{2 \times \ell}$.
- Challenge. The challenge is generated by the verifier V' as $c \xleftarrow{\$} \{0,1\}^\ell$.
- Response. The response $\tau = (\tau_1, \tau_2)$ to challenge $c \in \{0,1\}^\ell$ is computed by the prover P' as $\tau = (r, z = \sum_{i=1}^\ell F(x_{i,c_i}, r))$, where $r \xleftarrow{\$} \mathbb{D}$.
- Verification. Given challenge $c \in \{0,1\}^\ell$ and response $\tau = (r,z) \in \mathbb{D} \times \mathbb{F}$, the verifier V' outputs accept iff $\sum_{i=1}^\ell F(x_{i,c_i}, r) = z$.

The protocol has perfect completeness and $1/|\mathbb{F}|$-sparse right tags. It is easy to extend $\mathsf{Auth}'_{\mathsf{wprf}}$ to *randomized* weak PRFs (with additive noise), as defined in [22]. This way we obtain protocols from a more general class of assumptions, such as Toeplitz-LPN [22]. The proof of the following theorem is in Appendix A.2.

Theorem 9. *If $\mathcal{F}$ is a (t,q,ϵ)-weak PRF, then $\mathsf{Auth}'_{\mathsf{wprf}}$ is $(t',q,\epsilon/\ell)$-ror-cma-secure with $t' \approx t$.*

4.4 Instantiation from DDH

THE DDH PROBLEM. Let $\mathcal{G}$ be a family of groups with $\mathcal{G}_n = (\mathbb{G}, g, p)$, where $\mathbb{G}$ is a cyclic group of prime-order p with $\lceil \log p \rceil = n$ and g generates $\mathbb{G}$. To define the $\mathsf{ddh}_\mathcal{G}$ problem over group family $\mathcal{G}$ we introduce the ddh advantage as the quantity

$$\mathbf{Adv}^{\mathsf{ddh}}_{\mathcal{G}}(A) = \Pr\left[A^{\mathrm{DDH}_x()} \Rightarrow \mathsf{true} \right] - \Pr\left[A^{\mathrm{U}()} \Rightarrow \mathsf{true} \right],$$

where $x \xleftarrow{\$} \mathbb{Z}_p$ and DDH_x returns (R, R^x) for $R \xleftarrow{\$} \mathbb{Z}_p$, and U returns uniform $(R, Z) \xleftarrow{\$} \mathbb{G}^2$. We say that $\mathsf{ddh}_\mathcal{G}$ is (t,q,ϵ)-hard if for all attackers A with time complexity t making at most q oracle queries, we have $\mathbf{Adv}^{\mathsf{ddh}}(A) \le \epsilon$. Note that classical ddh hardness is exactly $(t',1,\epsilon')$-hardness of $\mathsf{ddh}_\mathcal{G}$ and by the random self-reducibility of ddh we have that $\mathsf{ddh}_\mathcal{G}$ is (t,q,ϵ)-hard iff it is $(t',1,\epsilon')$-hard with $t \approx t'$ and $\epsilon \approx \epsilon'$.

ROR-CMA SECURE PROTOCOL. We define an authentication protocol $\mathsf{Auth}'_{\mathsf{ddh}} = \{\mathsf{Gen}', \mathsf{P}', \mathsf{V}'\}$ with associated key space $\mathcal{K} = \mathbb{Z}_p^2$, split tag space $\mathcal{T} = \mathcal{T}_1 \times \mathcal{T}_2 = \mathbb{G} \times \mathbb{G}$, and challenge space $\mathcal{C} = \mathbb{Z}_p$.

- Key Generation. The key-generation algorithm Gen' outputs a secret key $K = (y_1, y_2) \xleftarrow{\$} \mathbb{Z}_p^2$.
- Challenge. The challenge is generated by the verifier V' as $\mathbf{c} \xleftarrow{\$} \mathbb{Z}_p$.
- Response. The response $\tau = (\tau_1, \tau_2)$ to challenge $c \in \mathbb{F}_p$ is computed by the prover P' as $\tau = (R, R^{y_1 \cdot c + y_2})$, where $R \xleftarrow{\$} \mathbb{G}$.

- <u>Verification.</u> Given challenge $c \in \mathbb{Z}_p$ and response $\tau = (R, Z) \in \mathbb{G}^2$, the verifier V' outputs accept iff $R^{y_1 \cdot c + y_2} = Z$.

The protocol $\mathsf{Auth}'_{\mathsf{ddh}}$ has perfect completeness and $1/p$-sparse right tags.

Theorem 10. *If* $\mathsf{ddh}_{\mathcal{G}}$ *is* (t, q, ϵ)-*hard, then* $\mathsf{Auth}'_{\mathsf{ddh}}$ *is* (t', q, ϵ)-ror-cma-*secure with* $t' \approx t$.

The proof is similar to the one of Theorem 7 and is omitted.

MiM SECURE PROTOCOL. We now apply our generic transformation from Sect. 3 to $\mathsf{Auth}'_{\mathsf{ddh}}$ to obtain a s-mim-secure protocol. By using the field structure of $\mathbb{Z}_p$ in the exponent, we can use the concrete pairwise-independent hash function $H_{s_1, s_2}(R) = R^{s_1} \cdot S_2 \in \mathbb{G}$, where $(s_1, S_2) \in \mathbb{Z}_p \times \mathbb{G}$. The key of $\mathsf{Auth}_{\mathsf{ddh}}$ consists of $(y_1, y_2, K_{\mathbb{F}}, s_1, S_2)$. We now show that the key of $\mathsf{Auth}_{\mathsf{ddh}}$ can be shrinked by two elements, see Table 2. The response to a challenge c is computed as $\sigma = (R, Z)$, where $Z = (R^{y_1 \cdot c + y_2})^{K_{\mathbb{F}}} \cdot R^{s_1} S_2 = R^{y_1 K_{\mathbb{F}} c + y_2 K_{\mathbb{F}} + s_1} S_2$. This can be written as $Z = R^{x_1 c + x_2} S_2$ using the substitutions $x_1 := y_1 K_{\mathbb{F}}$, $x_2 := y_2 K_{\mathbb{F}} + s_1$, $X := S_3$. The resulting simplified protocol $\mathsf{Auth}_{\mathsf{ddh}}$ is given in Table 2.

Acknowledgements. David Cash was partially supported by NSF grant CNS-1453132.

Eike Kiltz was supported by a Sofja Kovalevskaja Award of the Alexander von Humboldt Foundation and ERC Project ERCC (FP7/615074).

Stefano Tessaro was partially supported by NSF grants CNS-1423566 and the Glen and Susanne Culler Chair.

This work was done in part while David Cash and Stefano Tessaro were visiting the Simons Institute for the Theory of Computing, supported by the Simons Foundation and by the DIMACS/Simons Collaboration in Cryptography through NSF grant CNS-1523467.

A   Omitted Proofs

A.1   Proof of Theorem 7

Proof. Let A be an adversary in the $\mathsf{ROR\text{-}CMA}^A_{\mathsf{Auth}'}(b)$ security game. We define an adversary $B^{\mathsf{LPN}_{s,\alpha}}()$ against the $\mathsf{LPN}_{\ell, \gamma}$ problem, where $\alpha \in \{\gamma, \frac{1}{2}\}$ is unknown.

Adversary $B^{\mathsf{LPN}_{s,\alpha}}$:	**Procedure** $T(\mathbf{c})$:
$\mathbf{k}'_2 \xleftarrow{\$} \mathbb{Z}^{\ell}_2$	If $\mathbf{c} = \mathbf{c}^*$ then
$\mathbf{c}^* \xleftarrow{\$} \mathbb{F}^{\ell}_2$	$\quad \mathbf{z} \xleftarrow{\$} \mathcal{B}^n_{\gamma}$
$(\tau^*, state') \xleftarrow{\$} A^{T(\cdot)}(1^k, \mathbf{c}^*)$	$\quad \mathbf{R} \xleftarrow{\$} \mathbb{F}^{\ell \times n}_2$
Parse $\tau^* = (\mathbf{R}^*, \mathbf{z}^*) \in \mathbb{F}^{\ell \times n}_2 \times \mathbb{F}^n_2$	Else
If $|(\mathbf{R}^*)^T \cdot \mathbf{k}'_2 - \mathbf{z}^*| \leq \gamma' n$ and $rank(\mathbf{R}) = n$	$\quad (\tilde{\mathbf{R}}, \mathbf{z}) \xleftarrow{\$} \mathsf{LPN}^n_{s,\alpha}()$
$\quad d \leftarrow$ accept	$\quad \mathbf{R}^T \leftarrow \tilde{\mathbf{R}}^T \cdot (\mathbf{M}_{\mathbf{c}} - \mathbf{M}_{\mathbf{c}^*})^{-1}$
Else $d \leftarrow$ reject	$\quad \tau_1 \leftarrow \mathbf{R}$
Ret $A(state, d)$	$\quad \tau_2 \leftarrow \mathbf{z} + \mathbf{R}^T \mathbf{k}'_2$
	Ret (τ_1, τ_2)

Note that due to the finite field properties of the linear map $\mathbf{M_c}$, matrix $\mathbf{M_c} - \mathbf{M_{c^*}}$ is always invertible for $\mathbf{c} \neq \mathbf{c}^*$. Adversary B implicitly defines $\mathbf{k}_1 := \mathbf{s}$ and $\mathbf{k}_2 := -\mathbf{M_{c^*}} \cdot \mathbf{k}_1 + \mathbf{k}_2'$, where $\mathbf{s}$ is the LPN secret. As $\mathbf{k}_2'$ is uniform, the key $\mathbf{k} = (\mathbf{k}_1, \mathbf{k}_2)$ has the correct distribution. The definition of $K = (\mathbf{k}_1, \mathbf{k}_2)$ implies that

$$K(\mathbf{c}) := \mathbf{M_c} \cdot \mathbf{k}_1 + \mathbf{k}_2 = (\mathbf{M_c} - \mathbf{M_{c^*}}) \cdot \mathbf{k}_1 + \mathbf{k}_2'. \tag{6}$$

As $K(\mathbf{c}^*) = \mathbf{k}_2'$, the bit d is always computed correctly by B. We now consider the distribution of $\mathbf{T}(\mathbf{c})$. First note that τ_1 is always a uniform matrix in $\mathbb{F}_2^{\ell \times n}$. For $\mathbf{c} = \mathbf{c}^*$, $\mathbf{z}$ is Bernoulli distributed and, using Eq. (6), $\tau_2 = \mathbf{R}^T \mathbf{k}_2' + \mathbf{z}$ is distributed as computed by prover P'. Further, for $\mathbf{c} \neq \mathbf{c}^*$ we have

$$\begin{aligned}
\tau_2 &= \tilde{\mathbf{R}}^T \cdot \mathbf{s} + \mathbf{e} + \mathbf{R}^T \mathbf{k}_2' \\
&= \mathbf{R}^T \cdot (\mathbf{M_c} - \mathbf{M_{c^*}}) \cdot \mathbf{k}_1 + \mathbf{e} + \mathbf{R}^T \mathbf{k}_2' \\
&= \mathbf{R}^T \cdot (\mathbf{M_c} \cdot \mathbf{k}_1 + \mathbf{k}_2) + \mathbf{e},
\end{aligned}$$

where $\mathbf{e} \xleftarrow{\$} \mathcal{B}_\alpha^\ell$. If $\alpha = \frac{1}{2}$, then τ_1 and τ_2 are uniformly distributed and

$$\Pr[\,B^{\mathrm{LPN}_{\mathbf{s},1/2}()} \Rightarrow \mathrm{true}\,] = \Pr[\,\mathrm{ROR\text{-}CMA}_{\mathrm{Auth}'}^A(0) \Rightarrow \mathrm{true}\,].$$

If $\alpha = \gamma$, then $\tau = (\tau_1, \tau_2)$ is distributed as computed by prover P'. Hence $\Pr\left[\,B^{\mathrm{LPN}_{\mathbf{s},\gamma}()} \Rightarrow \mathrm{true}\,\right] = \Pr[\,\mathrm{ROR\text{-}CMA}_{\mathrm{Auth}'}^A(1) \Rightarrow \mathrm{true}\,]$. The last two equations provide $\mathbf{Adv}^{\mathrm{LPN}}(B) = \mathbf{Adv}_{\mathrm{Auth}'}^{\mathrm{ror\text{-}cma}}(A)$, where the running time of B is approximately that of A.   □

A.2   Proof of Theorem 9

Proof. Let A be an attacker in the ROR-CMA(1) game. We now describe games $\mathsf{G}_0, \ldots, \mathsf{G}_\ell$ that are exactly like the ROR-CMA(1) game, but with modified procedure $T(c)$. For $j \in \{0, \ldots, \ell - 1\}$, let $S_j : \{0,1\}^j \to \mathbb{F}$ be a random function, where $S_0(\varepsilon)$ is defined to be 0. Note that S_j can be efficiently simulated by lazy evaluation.

main G_j:	**Procedure $T(\mathbf{c})$:**	$/\!/\mathsf{G}_j$
$K \xleftarrow{\$} \mathrm{Gen}'(1^k)$	$r \xleftarrow{\$} \mathbb{D}$	
$c^* \xleftarrow{\$} \{0,1\}^\ell$	If $c_{|j} = c_{|j}^*$ then	
$(\tau^*, state) \xleftarrow{\$} A^{T(\cdot)}(c^*)$	$z = \sum_{i=1}^\ell F(x_{i,c_i}, r)$	
$d \leftarrow \mathsf{V}_K'(c^*, \tau^*)$	Else	
Ret $A(state, d)$	$z = S_j(c_{|j}) + \sum_{i=j+1}^\ell F(x_{i,c_i}, r)$	
	Ret $\tau = (\tau_1 \leftarrow r, \tau_2 \leftarrow z)$	

Note that in game G_0 all tags τ are computed correctly by T and hence $\mathsf{G}_0 = \mathrm{ROR\text{-}CMA}(1)$. Furthermore, in game G_ℓ, all tags except for challenge $\mathbf{c}^*$ are uniform and hence $\mathsf{G}_\ell = \mathrm{ROR\text{-}CMA}(0)$. The following lemma completes the proof of Theorem 9.

Lemma 11. *For any $j \in \{0, \ldots, \ell - 1\}$, there exists an attacker B_j such that*

$$\Pr\left[\, G_j^A \Rightarrow \mathsf{true} \,\right] - \Pr\left[\, G_{j+1}^A \Rightarrow \mathsf{true} \,\right] \leq \mathbf{Adv}_{\mathcal{F}}^{\mathsf{wprf}}(B).$$

To prove the lemma, we define an adversary $B = B_j^{O()}$ ($0 \leq j \leq \ell - 1$) against $\mathcal{F}$, where $O \in \{F_x, U\}$.

Adversary B^O:

$c^* \overset{\$}{\leftarrow} \{0,1\}^\ell$

$x_{i,k} = \begin{cases} \text{undefined} & i = j+1 \wedge k \neq c_{j+1}^* \\ \text{uniform in } \mathbb{K} & \text{otherwise} \end{cases}$

$(\tau^*, state) \overset{\$}{\leftarrow} A^{T(\cdot)}(c^*)$

Parse $\tau^* = (r^*, z^*) \in \mathbb{D} \times \mathbb{F}$

If $\sum_{i=1}^{\ell} F(x_{i,c_i^*}, r^*) = z^*$

$\quad d \leftarrow \mathsf{accept}$

Else $d \leftarrow \mathsf{reject}$

Ret $A(state, d)$

Procedure $T(\mathbf{c})$:

If $c_{|j+1} = c_{|j+1}^*$ then

$\quad r \overset{\$}{\leftarrow} \mathbb{D}; z = \sum_{i=1}^{\ell} F(x_{i,c_i}, r)$

Else

$\quad$ if $c_{j+1} \neq c_{j+1}^*$ then $(r, z') \overset{\$}{\leftarrow} O()$

$\quad$ Else $r \overset{\$}{\leftarrow} \mathbb{D}; z' = F(x_{j+1,c_{j+1}}, r)$

$\quad z = S_j(c_{|j}) + z' + \sum_{i=j+2}^{\ell} F(x_{i,c_i}, r)$

$\tau_1 \leftarrow r$

$\tau_2 \leftarrow z$

Ret (τ_1, τ_2)

Adversary B knows all secrets $x_{i,k}$ except $x_{j+1,1-c_j^*}$ which he defines implicitly as the secret x from the F_x oracle. In particular, he knows x_{i,c_i^*} and the bit d is always computed correctly. It remains to analyze the distribution of $T(\mathbf{c})$. If $c_{j+1} = c_{j+1}^*$, then the output of $T(\mathbf{c})$ in games G_j and G_{j+1} is identical. We now analyze the case $c_{j+1} \neq c_{j+1}^*$. If $O = F_x$, then $z = S_j(c_{|j}) + F(x, r) + \sum_{i=j}^{\ell} F(x_{i,c_i}, r) = S_j(c_{|j}) + \sum_{i=j+1}^{\ell} F(x_{i,c_i}, r)$ and hence $\Pr[B^{F_x()} \Rightarrow \mathsf{true}] = \Pr[G_j^A \Rightarrow \mathsf{true}]$. If $O = U$, then $z = S_j(c_{|j}) + z' + \sum_{i=j+1}^{\ell} F(x_{i,c_i}, r) = S_j(c_{|j+1}) + \sum_{i=j+1}^{\ell} F(x_{i,c_i}, r)$ and hence $\Pr[B^{U()} \Rightarrow \mathsf{true}] = \Pr[G_{j+1}^A \Rightarrow \mathsf{true}]$. $\qquad \square$

References

1. Akavia, A., Bogdanov, A., Guo, S., Kamath, A., Rosen, A.: Candidate weak pseudorandom functions in ac^0; mod$_2$. In: ITCS, pp. 251–260 (2014)
2. Applebaum, B., Cash, D., Peikert, C., Sahai, A.: Fast cryptographic primitives and circular-secure encryption based on hard learning problems. In: Halevi, S. (ed.) CRYPTO 2009. LNCS, vol. 5677, pp. 595–618. Springer, Heidelberg (2009)
3. Bellare, M., Rogaway, P.: Entity authentication and key distribution. In: Stinson, D.R. (ed.) CRYPTO 1993. LNCS, vol. 773, pp. 232–249. Springer, Heidelberg (1994)
4. Bellare, M., Rogaway, P.: The security of triple encryption and a framework for code-based game-playing proofs. In: Vaudenay, S. (ed.) EUROCRYPT 2006. LNCS, vol. 4004, pp. 409–426. Springer, Heidelberg (2006)
5. Bernstein, D.J., Lange, T.: Never trust a bunny. In: Hoepman, J.-H., Verbauwhede, I. (eds.) RFIDSec 2012. LNCS, vol. 7739, pp. 137–148. Springer, Heidelberg (2013)
6. Blum, A., Furst, M.L., Kearns, M., Lipton, R.J.: Cryptographic primitives based on hard learning problems. In: Stinson, D.R. (ed.) CRYPTO 1993. LNCS, vol. 773, pp. 278–291. Springer, Heidelberg (1994)

7. Blum, A., Kalai, A., Wasserman, H.: Noise-tolerant learning, the parity problem, and the statistical query model. In: 32nd ACM STOC, pp. 435–440, Portland, Oregon, USA, May 21–23, 2000. ACM Press (2000)
8. Bringer, J., Chabanne, H., Dottax, E.: HB^{++}: a lightweight authentication protocol secure against some attacks. In: SecPerU, pp. 28–33 (2006)
9. Damgård, I., Park, S.: Towards optimally efficient secret-key authentication from PRG. Cryptology ePrint Archive, Report 2014/426 (2014). http://eprint.iacr.org/2014/426
10. Dodis, Y., Kiltz, E., Pietrzak, K., Wichs, D.: Message authentication, revisited. In: Pointcheval, D., Johansson, T. (eds.) EUROCRYPT 2012. LNCS, vol. 7237, pp. 355–374. Springer, Heidelberg (2012)
11. Duc, D.N., Kim, K.: Securing HB$^+$ against GRS man-in-the-middle attack. In: SCIS (2007)
12. Fiat, A., Shamir, A.: How to prove yourself: practical solutions to identification and signature problems. In: Odlyzko, A.M. (ed.) CRYPTO 1986. LNCS, vol. 263, pp. 186–194. Springer, Heidelberg (1987)
13. Gilbert, H., Robshaw, M., Seurin, Y.: HB$^\sharp$: increasing the security and efficiency of HB$^+$. In: Smart, N.P. (ed.) EUROCRYPT 2008. LNCS, vol. 4965, pp. 361–378. Springer, Heidelberg (2008)
14. Gilbert, H., Robshaw, M., Seurin, Y.: How to encrypt with the LPN problem. In: Aceto, L., Damgård, I., Goldberg, L.A., Halldórsson, M.M., Ingólfsdóttir, A., Walukiewicz, I. (eds.) ICALP 2008, Part II. LNCS, vol. 5126, pp. 679–690. Springer, Heidelberg (2008)
15. Heyse, S., Kiltz, E., Lyubashevsky, V., Paar, C., Pietrzak, K.: Lapin: an efficient authentication protocol based on ring-LPN. In: Canteaut, A. (ed.) FSE 2012. LNCS, vol. 7549, pp. 346–365. Springer, Heidelberg (2012)
16. Hopper, N.J., Blum, M.: Secure human identification protocols. In: Boyd, C. (ed.) ASIACRYPT 2001. LNCS, vol. 2248, pp. 52–66. Springer, Heidelberg (2001)
17. Jain, A., Krenn, S., Pietrzak, K., Tentes, A.: Commitments and efficient zero-knowledge proofs from learning parity with noise. In: Wang, X., Sako, K. (eds.) ASIACRYPT 2012. LNCS, vol. 7658, pp. 663–680. Springer, Heidelberg (2012)
18. Juels, A., Weis, S.A.: Authenticating pervasive devices with human protocols. In: Shoup, V. (ed.) CRYPTO 2005. LNCS, vol. 3621, pp. 293–308. Springer, Heidelberg (2005)
19. Katz, J., Shin, J.S., Smith, A.: Parallel and concurrent security of the HB and HB+ protocols. J. Cryptol. 23(3), 402–421 (2010)
20. Kiltz, E., Pietrzak, K., Cash, D., Jain, A., Venturi, D.: Efficient authentication from hard learning problems. In: Paterson, K.G. (ed.) EUROCRYPT 2011. LNCS, vol. 6632, pp. 7–26. Springer, Heidelberg (2011)
21. Levieil, É., Fouque, P.-A.: An improved LPN algorithm. In: De Prisco, R., Yung, M. (eds.) SCN 2006. LNCS, vol. 4116, pp. 348–359. Springer, Heidelberg (2006)
22. Lyubashevsky, V., Masny, D.: Man-in-the-middle secure authentication schemes from LPN and weak PRFs. In: Canetti, R., Garay, J.A. (eds.) CRYPTO 2013, Part II. LNCS, vol. 8043, pp. 308–325. Springer, Heidelberg (2013)
23. Maurer, U.M., Sjödin, J.: A fast and key-efficient reduction of chosen-ciphertext to known-plaintext security. In: Naor, M. (ed.) EUROCRYPT 2007. LNCS, vol. 4515, pp. 498–516. Springer, Heidelberg (2007)
24. Maurer, U.M., Tessaro, S.: Basing PRFs on constant-query weak prfs: minimizing assumptions for efficient symmetric cryptography. In: Pieprzyk, J. (ed.) ASIACRYPT 2008. LNCS, vol. 5350, pp. 161–178. Springer, Heidelberg (2008)

248   D. Cash et al.

25. Munilla, J., Peinado, A.: HB-MP: a further step in the hb-family of lightweight authentication protocols. Computer Networks **51**(9), 2262–2267 (2007)
26. Ouafi, K., Overbeck, R., Vaudenay, S.: On the security of HB# against a man-in-the-middle attack. In: Pieprzyk, J. (ed.) ASIACRYPT 2008. LNCS, vol. 5350, pp. 108–124. Springer, Heidelberg (2008)
27. Regev, O.: On lattices, learning with errors, random linear codes, and cryptography. In: Gabow, H.N., Fagin, R. (eds.) 37th ACM STOC, Baltimore, Maryland, USA, May 22–24, 2005, pp. 84–93. ACM Press

Public Key Encryption, Signatures, and VRF

Algebraic Partitioning: Fully Compact and (almost) Tightly Secure Cryptography

Dennis Hofheinz[(✉)]

Karlsruhe Institute of Technology, Karlsruhe, Germany
Dennis.Hofheinz@kit.edu

Abstract. We describe a new technique for conducting "partitioning arguments". Partitioning arguments are a popular way to prove the security of a cryptographic scheme. For instance, to prove the security of a signature scheme, a partitioning argument could divide the set of messages into "signable" messages for which a signature can be simulated during the proof, and "unsignable" ones for which any signature would allow to solve a computational problem. During the security proof, we would then hope that an adversary only requests signatures for signable messages, and later forges a signature for an unsignable one.

In this work, we develop a new class of partitioning arguments from simple assumptions. Unlike previous partitioning strategies, ours is based upon an algebraic property of the partitioned elements (e.g., the signed messages), and not on their bit structure. This allows to perform the partitioning efficiently in a "hidden" way, such that already a single "slot" for a partitioning operation in the scheme can be used to implement many different partitionings sequentially, one after the other. As a consequence, we can construct complex partitionings out of simple basic (but algebraic) partitionings in a very space-efficient way.

As a demonstration of our technique, we provide the first signature and public-key encryption schemes that achieve the following properties simultaneously: they are (almost) tightly secure under a simple assumption, and they are fully compact (in the sense that parameters, keys, and signatures, resp. ciphertexts only comprise a constant number of group elements).

Keywords: Partitioning arguments · Tight security proofs · Digital signatures · Public-key encryption

1 Introduction

Partitioning Arguments. Many security reductions rely on a *partitioning argument*. Informally, a partitioning argument divides the parts of a large system into those parts that are under the control of the simulation, and those parts into which a computational challenge can be embedded. For instance, a partitioning

---

Supported by DFG grants HO 4534/2-2, HO 4534/4-1.

E. Kushilevitz and T. Malkin (Eds.): TCC 2016-A, Part I, LNCS 9562, pp. 251–281, 2015.
DOI: 10.1007/978-3-662-49096-9_11

argument for a signature scheme could divide the set of message into "signable messages" (for which a signature can be generated by the security reduction), and "unsignable messages" (for which any signature would solve an underlying problem). During the security reduction, we hope that an adversary only asks for the signatures of signable messages, but forges a signature for an unsignable one. Partitioning arguments are a popular means for proving the security of signature schemes (e.g., [17,29,35,38]), identity-based encryption schemes (e.g., [9,10,14,38]), or tightly secure cryptosystems (e.g., [6,15,32]).

The Complexity of Bit-based Partitioning. All of the above works (except for [10,17], which use a programmable random oracle to implement a partitioning) partition messages or identities according to their bit representation. For instance, in the signature scheme from [29], messages are signable precisely if they do not start with a particular bit prefix. This non-algebraic approach requires a certain preparation in the scheme itself: already the scheme must establish certain distinctions of messages based on their bit representation. For instance, the signature scheme of [38] uses a hash function of the form $H(M) = h_0 \prod_j h_{j,M_j}$, where M_j are the bits of the signed message M, and h_0 and the $h_{j,b}$ are public group elements. This leads to comparatively large public parameters or keys, in particular because all potential distinctions (based on the values of the M_j) are already present in the scheme.

Our Contribution. In this work, we develop an entirely different partitioning approach: instead of partitioning based on the bit representation, we partition according to a simple algebraic predicate. Namely, we view a message M as above as a $\mathbb{Z}_p$-element, and consider various Legendre symbols $L_j = \left(\frac{f_j(M)}{p}\right)$ for different affine functions f_j. Taken together, sufficiently many L_j uniquely determine M, but the computation of each L_j can be encoded as a series of $\mathbb{Z}_p$-operations.[1] Intuitively, this algebraic property allows to "internalize" and hide the computations of the L_j, e.g., by hiding the f_j inside a homomorphic commitment. As a consequence, only one "universal" partitioning (according to a single L_j) needs to be performed in the scheme itself; in the analysis, several simple partitionings can then be implemented sequentially, by varying the f_j.

Comparison with Previous Partitioning Techniques. Compared to previous, bit-based partitioning approaches, our new strategy has the advantage that it simultaneously leads to compact schemes and to a tight security reduction. Previous partitioning strategies were either based on more complex partitionings (such as [9,29,35,38]) that lead to a non-tight security reduction, or on a sequence of simple bit-based partitionings (such as [6,15,32]) that lead to large public parameters or keys. In contrast, we support many simple algebraic partitionings (and thus a tight security reduction), but we occupy only one "partitioning slot" in the public parameters. This leads to tightly secure and very compact applications, as we will detail next.

---

[1] Technically, we will not even need to explicitly compute L_j, but only prove that $L_j = 1$. This is possible using a quadratic equation over $\mathbb{Z}_p$.

Applications. Specifically, we demonstrate the usefulness of our partitioning technique by describing the first (almost) tightly secure signature and PKE schemes that are fully compact, in the sense that parameters, keys, and signatures (resp. ciphertexts) only contain a constant number of group elements. Our security reduction loses only a factor of $\mathbf{O}(k)$, where k is the security parameter. In particular, our security reduction does not degrade in the number of users or signatures, resp. ciphertexts. The security of our schemes is based upon the Decisional Diffie-Hellman (DDH) assumption in both preimage groups of a pairing. (This assumption is also called "Symmetric External Diffie-Hellman" or SXDH.) Tables 1 and 2 give a more detailed comparison with existing schemes.

In the following, we give more details on our techniques and results. To do so, we start with a little background concerning our applications.

Tight Security Reductions. To argue for the security of a given cryptographic scheme S, we usually employ a security reduction. That is, we try to argue that every hypothetical adversary $\mathcal{A}_S$ on S can be converted into an adversary $\mathcal{A}_P$ on an allegedly hard computational problem P. In that sense, the only way to break S is to solve P. Of course, we are mostly interested in reductions to well-investigated problems P. Furthermore, there are reasons to consider the *tightness* of the reduction: a tight reduction guarantees that $\mathcal{A}_P$'s success ε_P in solving P (in a reasonable metric) is about the same as $\mathcal{A}_S$'s success ε_S in attacking S.

To explain the impact of a (non-)tight reduction in more detail, consider a public-key encryption (PKE) scheme S that is deployed in a many-user environment. In this setting, an adversary $\mathcal{A}_S$ on S may observe, say, n_C ciphertexts generated for each of the, say, n_U users. Most known security reductions in this setting are non-tight, in the sense that $\varepsilon_P \leq \frac{\varepsilon_S}{n_U \cdot n_C}$. As a consequence, keylength recommendations should also take n_U and n_C into account; no "universal" keylength recommendation can be given for such a scheme. This is particularly problematic in settings that grow significantly beyond initial expectations.

Tightly Secure Encryption and Signature Schemes. The construction of tightly secure cryptographic schemes appears to be a nontrivial task. For instance, although already explicitly considered in 2000 [3], tightly secure PKE schemes have only been constructed very recently [2,6,15,28,32].[2,3] Moreover, the schemes from [2,28] have rather large ciphertexts, and the schemes induced by [6,15] and from [32] require large parameters (but offer small keys and ciphertexts).

The situation for tightly secure signature schemes is somewhat brighter, but results are still limited. There are efficient signature schemes that are tightly secure under "q-type" [8,16,36] or interactive [21] assumptions, or in the random

---

[2] Actually, [6,15] construct tightly secure identity-based encryption (IBE) schemes. However, those IBE schemes can be viewed as tightly secure signature schemes (using Naor's trick [11]), and then converted into tightly secure PKE schemes using the transformation from [28]. In fact, the PKE scheme of [32] can be viewed as a (modified and highly optimized) conversion of the IBE scheme from [15].

[3] We note that earlier PKE schemes achieve at least a certain form of tight security under "q-type" assumptions [22,23,27], or in the random oracle model [7,13,20].

Table 1. Comparison of different (at least almost) tightly EUF-CMA secure signature schemes from simple[4] assumptions in pairing-friendly groups. The **parameters, verification key**, and **signature** columns denote space complexity, measured in group elements. The **reduction loss** column denotes the (multiplicative) loss of the security reduction to the respective **assumption**. For the schemes from [6,15], we assume the signature scheme induced by the presented IBE scheme. Furthermore, $n = \Theta(k)$ denotes the bitlength of the signed message (if the signed message is a bitstring and not a group element or an exponent). We note that [32] mention that their scheme can be generalized to the d-LIN assumption (including 1-LIN=DDH). However, since they only give explicit complexities for the arising signatures (identical to the ones from [6]), we restrict to their DLIN-based scheme. Finally, we remark that all of these schemes (except for [12]) imply tightly secure PKE schemes (cf. Table 2).

Scheme	Parameters	Verification key	Signature	Reduction loss	Assumption
BMS03 [12]	0	$k+3$	$k+1$	$\mathbf{O}(k)$	CDH
HJ12 [28]	2	28	$8k+22$	$\mathbf{O}(1)$	DLIN
CW13 [15]	$2d^2(2n+1)$	d	$4d$	$\mathbf{O}(k)$	d-LIN
BKP14 [6]	d	$d^2(2n+1)$	$2d+1$	$\mathbf{O}(k)$	$\mathcal{D}_d$-MDDH
LJYP14 [32]	0	$\mathbf{O}(d^2n)$	$2d+1$	$\mathbf{O}(k)$	d-LIN
This work	14	6	25	$\mathbf{O}(k)$	DDH

oracle model [5,24,30]. There are also more recent and somewhat less efficient schemes tightly secure under simple[4] assumptions [6,12,15,28,32] (see also [1,2]). Some of these latter schemes can even be converted into tightly secure PKE schemes; however, all of the schemes [2,6,12,15,28,32] suffer from asymptotically large parameters, keys, or signatures (resp. ciphertexts).

The Scheme of Chen and Wee. Our technical ideas are best presented with our signature scheme. At a very high level, we follow the strategy of Chen and Wee [15] (see also [6]), where we interpret their IBE scheme as a signature scheme using Naor's trick [11]. In their scheme, signatures are of the form

$$\sigma = \left(h_0, \; sigk \cdot \prod_{i=1}^{n} h_{i,M_i} \right), \tag{1}$$

where $sigk$ is the secret key, $M = (M_i)_{i=1}^{n} \in \{0,1\}^n$ is the bit representation of the signed message, and $h_0, (h_{i,0}, h_{i,1})_{i=1}^{n}$ are group elements chosen from a joint public distribution.[5]

---

[4] With a "simple" assumption, we mean one in which the adversary gets a challenge whose size only depends on the security parameter, and is then supposed to output a unique solution without further interaction. Examples of simple assumptions are DLOG, DDH, d-LIN, or RSA, but not, say, Strong Diffie-Hellman [8] or q-ABDHE [22].

[5] We note that although their scheme can be viewed as a generalization of Waters signatures [38], their analysis is entirely different. Also, we omit here certain subtleties regarding the used distributions of group elements.

During their proof of existential unforgeability (EUF-CMA security), Chen and Wee gradually modify signatures generated by the security experiment for an adversary $\mathcal{A}$. This is done via a small hybrid argument over the bit indices of messages, and thus yields a security proof that loses a factor of $\mathbf{O}(n)$. Concretely, in the i-th hybrid, generated signatures are of the form $\sigma = (h_0, sigk_{M_1,\ldots,M_i} \cdot \prod_{j=1}^n h_{j,M_j})$, where $sigk_{M_1,\ldots,M_i} = \mathcal{R}(M_1,\ldots,M_i)$ for a truly random function $\mathcal{R}$. Similarly, a forged message-signature pair (M^*, σ^*) from $\mathcal{A}$ is only considered valid if it is consistent with $sigk_{M_1^*,\ldots,M_i^*}$ (instead of $sigk$). In other words, in the i-th hybrid, the secret key used in signatures depends on the first i bits of the signed message.

Table 2. Comparison of different (at least almost) tightly IND-CCA secure PKE schemes from simple[4] assumptions. As in Table 1, the **parameters, public key,** and **ciphertext** columns denote space complexity, measured in group elements, and the **reduction loss** column denotes the (multiplicative) loss of the security reduction to the respective **assumption**. For the schemes from [6,15], we assume the PKE scheme induced by the respective signature scheme when going through the construction of [28]. We note that [32] only describe a symmetric-pairing version of their scheme, so their DDH-based scheme is not explicit. However, we expect that their DDH-based scheme has slightly more compact ciphertexts than ours.

Scheme	Parameters	Public key	Ciphertext	Reduction loss	Assumption
HJ12 [28]	$\mathbf{O}(1)$	$\mathbf{O}(1)$	$\mathbf{O}(k)$	$\mathbf{O}(1)$	DLIN
AKDNO13 [2]	$\mathbf{O}(1)$	$\mathbf{O}(1)$	$\mathbf{O}(k)$	$\mathbf{O}(1)$	DLIN
CW13 [15]	$\mathbf{O}(d^2 k)$	$\mathbf{O}(d)$	$\mathbf{O}(d)$	$\mathbf{O}(k)$	d-LIN
BKP14 [6]	$\mathbf{O}(d)$	$\mathbf{O}(d^2 k)$	$\mathbf{O}(d)$	$\mathbf{O}(k)$	$\mathcal{D}_d$-MDDH
LJYP14 [32]	$\mathbf{O}(1)$	$\mathbf{O}(d^2 k)$	$\mathbf{O}(d)$	$\mathbf{O}(k)$	d-LIN
LJYP14 [32]	3	$24k + 30$	69	$\mathbf{O}(k)$	DLIN
This work	15	2	60	$\mathbf{O}(k)$	DDH

Thus, the difference between the $(i-1)$-th and the i-th hybrid is an additional dependency of used secret keys on the i-th message bit M_i. To progress from hybrid $i - 1$ to hybrid i, Chen and Wee first partition the message space in two halves (according to M_i). Then, using an elaborate argument, they consistently modify the secret keys used for messages from one half, and thus essentially decouple those keys from the keys used for messages from the other half. This creates an additional dependency on M_i. After $n = |M|$ such steps, each signature uses a different secret key (up to multiple signatures of the same message). In particular, $\mathcal{A}$ gets no information about the secret key $sigk_{M_1^*,\ldots,M_n^*}$ used to verify its own forgery, and existential unforgeability follows.

We would like to highlight the partitioning character of their analysis: in their proof, Chen and Wee introduce more and more dependencies of signatures on the corresponding messages, and each such dependency is based upon a different partitioning of the message space.[6] Now observe that already regular signatures (as in (1)) feature distinctions based on all bits of M. These distinctions provide the technical tool to introduce dependencies in the security proof. However, as a consequence, rather complex joint distributions need to be sampled during signature generation, which results in public parameters of $\mathbf{O}(n)$ group elements.

Algebraic Partitioning. In a nutshell, our main technical tool is a new way to partition the message space of a signature scheme. We call this tool "algebraic partitioning." Concretely, a signature for a message $M \in \mathbb{Z}_p$ in our scheme consists essentially of an encryption of the secret key X, along with a consistency proof:

$$\sigma = (C = \mathrm{Enc}(pk, X), \pi). \tag{2}$$

The corresponding encryption key pk is part of the verification key vk, and the consistency proof π proves the following statement:

> "**Either** C encrypts the secret key X, **or** $f(M) \in \mathbb{Z}_p$ is a quadratic residue (or both)."

Here, p is the order of the underlying group, and $f : \mathbb{Z}_p \to \mathbb{Z}_p$ is an affine function fixed (but hidden) in the verification key. Implicitly, this provides a *single* partitioning of messages into those for which $f(M)$ is a quadratic residue, and those for which $f(M)$ is not. However, since f is hidden, many partitionings can be induced (one after the other) by varying f during a proof.

In fact, during the security proof, this partitioning will fulfill the same role as the bit-based partitioning in the analysis of Chen and Wee. In particular, it will help to introduce additional dependencies of the signature on the message. More specifically, in the i-th hybrid of the security proof, C will not encrypt X, but a value X_M that depends on the i Legendre symbols $\left(\frac{f_j(M)}{p}\right)$ for randomly chosen (but fixed) affine functions $f_1, \ldots, f_i$. Each new such dependency is introduced by first refreshing the affine function f hidden in vk, and then modifying all values encrypted in signatures whenever possible (i.e., whenever $f(M)$ is a quadratic residue).[7] Observe that the single explicit partitioning in regular signatures is used several times (for different f_j) to introduce many dependencies of signatures on messages in the proof. The remaining strategy can then be implemented as in [15].

Our different strategy to partition the message space results in a very compact scheme. Namely, since only one explicit partitioning step is performed in the scheme, parameters, keys, and signatures comprise only a constant number of group elements. Specifically, parameters, keys, and signatures contain 14, 6, and 25 group elements, respectively. Besides, our scheme is compatible with Groth-Sahai proofs [26]. Hence, when used in the construction of [28], we immediately

---

[6] We note that a similar technique has also been used in the context of pseudorandom functions [25,33].

[7] This neglects a number of details. For instance, in the somewhat simplified scheme above, π always ties the ciphertexts in signatures for quadratic non-residues $f(M)$ to a single value X. In our actual proof, we will thus simulate a part of π, such that the encrypted values can be decoupled from the original secret key X.

get the first compact (in the above sense) PKE scheme that is tightly IND-CCA secure under a simple assumption.[8]

Different Perspective: Our Scheme as a MAC. So far our high-level discussion can be equally used to justify a similar message authentication code (MAC), in which verification is non-public. Such a MAC can then be converted into a signature scheme, e.g., using the technique of Bellare and Goldwasser [4].[9] One could hope that this yields a more modular construction, possibly with a MAC as a simpler basic building block. (In particular, this approach was suggested by a reviewer.)

In this work, we still present our idea directly in terms of a signature scheme. One reason is that a MAC following the strategy described above would actually not be significantly less complex than a full signature scheme. In particular, already a MAC would require Groth-Sahai proofs. Moreover, a modular approach in the spirit of [4] would require "algebraically compatible" building blocks (to allow for an efficient and tightly secure overall scheme), and would seem to lead to a more complex presentation.

Open Problems. Besides of course obtaining more efficient (and compact) schemes, it would be interesting to apply similar ideas in the identity-based setting. Specifically, currently there is no fully compact identity-based encryption (IBE) scheme whose security can be tightly based on a standard assumption.[10] However, it is not obvious how to use algebraic partitioning in the identity-based setting. Specifically, it is not clear how to "derive functionality" from valid signature proofs, in the following sense.

Namely, first note that IBE schemes can be interpreted as signature schemes, in a sense noted by Naor (cf. [11]): IBE user secret keys for an identity M correspond to signatures for message M, and verification simply checks whether the alleged signature works as a decryption key for identity M. It is natural to use the same interpretation to try to "upgrade" a signature scheme to an IBE scheme. For this strategy, however, one must find a way to make a signature σ act as a decryption trapdoor, and thus to "derive functionality from σ" (as opposed to just check σ for validity). In common discrete-log-based IBE schemes, this functionality property is achieved by the fact that a pairing operation is

---

[8] Actually, plugging our scheme directly into the construction of [28] yields an asymptotically compact, but not very efficient scheme. Thus, we provide a more direct and efficient explicit PKE construction with parameters, public keys, and ciphertexts comprised of 15, 2, and 60 group elements, respectively.

[9] In a signature scheme derived using the conversion of Bellare and Goldwasser, the verification key contains an encryption of the MAC secret key. A signature for a message M then consists of a MAC tag τ for M, along with a non-interactive zero-knowledge proof that τ is valid relative to the encrypted MAC key.

[10] The schemes of [22,23] are tightly secure and fully compact, but rely on a non-standard (q-type) assumption. On the other hand, IBE schemes obtained through the "dual systems" technique (e.g., [31,37]) are compact and secure under standard assumptions, but not known to be tightly secure.

used to pair IBE user secret keys with ciphertext elements. The result of this pairing operation is then a common secret that is shared between encryptor and decryptor.

Our strategy, however, crucially uses *quadratic* $\mathbb{Z}_p$-equations in signatures (to implement the algebraic partitioning of messages). In particular, our signature scheme uses a pairing operation already to implement these quadratic equations (even though signatures in our scheme consist solely of group elements in the source group of the pairing). As a consequence, the pairing operation cannot be used anymore to derive a common secret shared with the encryptor. Hence, at least a straightforward way to turn our signature scheme into an IBE scheme fails.[11]

Roadmap. After recalling some basic definitions, we present our signature scheme in Sect. 3. In Sect. 4, we give a direct construction of a PKE scheme derived from our signature scheme. In Sect. 5, we give more details on the exact Groth-Sahai equations arising from the consistency proofs of signatures and ciphertexts. In Appendix A, we provide additional illustrations for the proof of our signature scheme.

2 Preliminaries

Notation. Throughout the paper, $k \in \mathbb{N}$ denotes the security parameter. For $n \in \mathbb{N}$, let $[n] := \{1, \ldots, n\}$. For a finite set S, we denote with $s \leftarrow S$ the process of sampling s uniformly from S. For a probabilistic algorithm A, we denote with $y \leftarrow A(x; R)$ the process of running A on input x and with randomness R, and assigning y the result. We write $y \leftarrow A(x)$ for $y \leftarrow A(x; R)$ with uniformly chosen R, and we write $A(x) = y$ for the event that $A(x; R)$ (for uniform R) outputs y. If A's running time is polynomial in k, then A is called probabilistic polynomial-time (PPT). A function $f : \mathbb{N} \to \mathbb{R}$ is negligible if it vanishes faster than the inverse of any polynomial (i.e., if $\forall c \exists k_0 \forall k \geq k_0 : |f(x)| \leq 1/k^c$).

Collision-Resistant Hashing. A hash function generator is a PPT algorithm $\mathcal{H}$ that, on input 1^k, outputs (the description of) an efficiently computable function $\mathrm{H} : \{0,1\}^* \to \{0,1\}^k$.

Definition 1 (Collision-Resistance). *We say that a hash function generator $\mathcal{H}$ outputs collision-resistant functions H (or, when the reference to $\mathcal{H}$ is clear, that such an H is collision-resistant), if*

$$\mathrm{Adv}^{\mathrm{cr}}_{\mathcal{H},\mathcal{A}}(k) = \Pr\left[x \neq x' \wedge \mathrm{H}(x) = \mathrm{H}(x') \mid \mathrm{H} \leftarrow \mathcal{H}(1^k),\ (x, x') \leftarrow \mathcal{A}(1^k, \mathrm{H})\right]$$

is negligible for every PPT adversary $\mathcal{A}$.

---

[11] We realize that this explanation is somewhat technical and may not seem very compelling. We wish we had a better one.

Signature Schemes. A signature scheme SIG consists of four PPT algorithms SPars, SGen, Sig, Ver. Parameter generation $\text{SPars}(1^k)$ outputs public parameters spp that are shared among all users. Key generation $\text{SGen}(spp)$ takes public parameters spp, and outputs a verification key vk and a signing key $sigk$. The signature algorithm $\text{Sig}(spp, sigk, M)$ takes public parameters spp, a signing key $sigk$, and a message M, and outputs a signature σ. Verification $\text{Ver}(spp, vk, M, \sigma)$ takes public parameters spp, a verification key vk, a message M, and a potential signature σ, and outputs a verdict $b \in \{0, 1\}$. For correctness, we require that $1 \leftarrow \text{Ver}(spp, vk, M, \sigma) = 1$ always and for all M, all $(vk, sigk) \leftarrow \text{SGen}(1^k)$, and all $\sigma \leftarrow \text{Sig}(spp, sigk, M)$. For the sake of readability, we will omit the public parameters spp from invocations of Sig and Ver when the reference is clear.

Definition 2 (Multi-user (One-Time) Existential Unforgetability). *Let* SIG *be a signature scheme as above, and consider the following experiment for an adversary $\mathcal{A}$:*

1. *$\mathcal{A}$ specifies (in unary) the number $n_U \in \mathbb{N}$ of desired scheme instances.*
2. *The experiment then samples parameters $spp \leftarrow \text{SPars}(1^k)$ as well as n_U keypairs $(vk^{(\ell)}, sigk^{(\ell)}) \leftarrow \text{SGen}(spp)$.*
3. *$\mathcal{A}$ is invoked on input $(1^k, spp, (vk^{(\ell)})_{\ell=1}^{n_U})$, and gets access to signing oracles $\text{Sig}(sigk^{(\ell)}, \cdot)$ for all $\ell \in [n_U]$. Finally, $\mathcal{A}$ outputs an index $\ell^* \in [n_U]$ and a potential forgery (M^*, σ^*).*
4. *$\mathcal{A}$ wins iff $\text{Ver}(vk^{(\ell^*)}, M^*, \sigma^*) = 1$ and M^* was not queried to $\text{Sig}(sigk^{(\ell^*)}, \cdot)$.*

Let $\text{Adv}_{\text{SIG}, \mathcal{A}}^{\text{euf-mcma}}(k)$ denote the probability that $\mathcal{A}$ wins in the above experiment. We say that SIG is existentially unforgeable under chosen-message attacks in the multi-user setting (EUF-mCMA secure) iff $\text{Adv}_{\text{SIG}, \mathcal{A}}^{\text{euf-mcma}}(k)$ is negligible for every PPT $\mathcal{A}$. Let $\text{Adv}_{\text{SIG}, \mathcal{A}}^{\text{ot-euf-mcma}}(k)$ be the probability that $\mathcal{A}$ wins in the slightly modified experiment in which only one Sig-query to each scheme instance ℓ is allowed. We say that SIG is existentially unforgeable under one-time chosen-message attacks in the multi-user setting (OT-EUF-mCMA secure) iff $\text{Adv}_{\text{SIG}, \mathcal{A}}^{\text{ot-euf-mcma}}(k)$ is negligible for every PPT $\mathcal{A}$.

Public-key Encryption Schemes. A public-key encryption (PKE) scheme PKE consists of four PPT algorithms (EPars, EGen, Enc, Dec). The parameter generation algorithm $\text{EPars}(1^k)$ outputs public parameters epp. Key generation $\text{EGen}(epp)$ outputs a public key pk and a secret key sk. Encryption $\text{Enc}(epp, pk, M)$ takes parameters epp, a public key pk, and a message M, and outputs a ciphertext C. Decryption $\text{Dec}(epp, sk, C)$ takes public parameters epp, a secret key sk, and a ciphertext C, and outputs a message M. For correctness, we require $\text{Dec}(epp, sk, C) = M$ always and for all M, all $epp \leftarrow \text{EPars}(1^k)$, all $(pk, sk) \leftarrow \text{EGen}(epp)$, and all $C \leftarrow \text{Enc}(epp, pk, M)$. As with signatures, we usually omit the public parameters epp from invocations of Enc and Dec.

Definition 3 (Multi-user, Multi-challenge Indistinguishability of Ciphertexts). *For a public-key encryption scheme PKE and an adversary $\mathcal{A}$, consider the following security experiment $\text{Exp}_{\text{PKE}, \mathcal{A}}^{\text{ind-mcca}}(k)$:*

1. $\mathcal{A}$ specifies (in unary) the number $n_U \in \mathbb{N}$ of desired scheme instances.
2. The experiment samples parameters $epp \leftarrow \mathrm{EPars}(1^k)$, and n_U keypairs through $(pk^{(\ell)}, sk^{(\ell)}) \leftarrow \mathrm{EGen}(epp)$, and uniformly chooses a bit $b \leftarrow \{0, 1\}$.
3. $\mathcal{A}$ is invoked on input $(1^k, epp, (pk^{(\ell)})_{\ell=1}^{n_U})$, and gets access to challenge oracles $\mathcal{O}^{(\ell)}$ and decryption oracles $\mathrm{Dec}(sk^{(\ell)}, \cdot)$ for all $\ell \in [n_U]$. Here, challenge oracle $\mathcal{O}^{(\ell)}$, on input two messages M_0, M_1, outputs an encryption $C \leftarrow \mathrm{Enc}(pk^{(\ell)}, M_b)$ of M_b.
4. Finally, $\mathcal{A}$ outputs a bit b', and the experiment outputs 1 iff $b = b'$.

A PPT adversary $\mathcal{A}$ is valid if every pair (M_0, M_1) of messages submitted to an $\mathcal{O}^{(\ell)}$ by $\mathcal{A}$ satisfies $|M_0| = |M_1|$, and if $\mathcal{A}$ never submits any challenge ciphertext (previously received from an $\mathcal{O}^{(\ell)}$) to the corresponding decryption oracle $\mathrm{Dec}(sk^{(\ell)}, \cdot)$. Let

$$\mathrm{Adv}_{\mathrm{PKE},\mathcal{A}}^{\mathrm{ind\text{-}mcca}}(k) = \mathrm{Pr}\left[\mathrm{Exp}_{\mathrm{PKE},\mathcal{A}}^{\mathrm{ind\text{-}mcca}}(k) = 1\right] - 1/2.$$

We say that PKE has indistinguishable ciphertexts under chosen-ciphertext attacks in the multi-user, multi-challenge setting (short: is IND-mCCA secure) iff $\mathrm{Adv}_{\mathrm{PKE},\mathcal{A}}^{\mathrm{ind\text{-}mcca}}(k)$ is negligible for all valid $\mathcal{A}$. Let $\mathrm{Adv}_{\mathrm{PKE},\mathcal{A}}^{\mathrm{ind\text{-}mcpa}}$ be defined similarly, except that $\mathcal{A}$ has no access to any Dec oracles. PKE has indistinguishable ciphertexts under chosen-plaintext attacks in the multi-user, multi-challenge setting (short: is IND-mCPA secure) iff $\mathrm{Adv}_{\mathrm{PKE},\mathcal{A}}^{\mathrm{ind\text{-}mcpa}}(k)$ is negligible for all valid $\mathcal{A}$.

Quadratic Residues and Legendre Symbols. Let p be a prime. Then, $\mathrm{QR}_p \subseteq \mathbb{Z}_p^*$ is the set of quadratic residues modulo p, i.e., the set of all $x \in \mathbb{Z}_p^*$ for which an $r \in \mathbb{Z}_p^*$ with $r^2 = x \bmod p$ exists. Given p and an $x \in \mathrm{QR}_p$, such an r can be computed efficiently. For $x \in \mathbb{Z}_p$, we let $\left(\frac{x}{p}\right) = x^{\frac{p-1}{2}} \bmod p$ denote the Legendre of x modulo p. We have $\left(\frac{x}{p}\right) \in \{-1, 0, 1\}$, and in particular $\left(\frac{x}{p}\right) = 1 \Leftrightarrow x \in \mathrm{QR}_p$, as well as $\left(\frac{x}{p}\right) = 0 \Leftrightarrow x = 0$, and $\left(\frac{x}{p}\right) = -1 \Leftrightarrow x \in \mathbb{Z}_p^* \setminus \mathrm{QR}_p$.

Group and Pairing Generators. A group generator $\mathcal{G}$ is a PPT algorithm that, on input 1^k, outputs the description of a group $\mathbb{G}$, along with its (prime) order p, and a generator g of $\mathbb{G}$. A pairing generator $\mathcal{P}$ is a PPT algorithm that, on input 1^k, outputs descriptions of:

- three groups $\mathbb{G}, \hat{\mathbb{G}}, \mathbb{G}_T$ of the same prime order p, along with p, and generators $g, \hat{g}$ of $\mathbb{G}, \hat{\mathbb{G}}$,
- a bilinear map $e : \mathbb{G} \times \hat{\mathbb{G}} \to \mathbb{G}_T$ that is non-degenerate in the sense of $e(g, \hat{g}) \neq 1 \in \mathbb{G}_T$.

Occasionally, it will also be useful to consider a pairing generator $\mathcal{P}$ as a group generator (that only outputs $(\mathbb{G}, p, g)$ or $(\hat{\mathbb{G}}, p, \hat{g})$).

Assumption 1 (Decisional Diffie-Hellman). For a group generator $\mathcal{G}$ and an adversary $\mathcal{A}$, let $\mathrm{Adv}_{\mathcal{G},\mathcal{A}}^{\mathrm{ddh}}(k)$ be the following difference:

$$\mathrm{Pr}\left[\mathcal{A}(1^k, \mathbb{G}, p, g, g^x, g^y, g^{xy}) = 1\right] - \mathrm{Pr}\left[\mathcal{A}(1^k, \mathbb{G}, p, g, g^x, g^y, g^z) = 1\right].$$

Here, the probability is over $(\mathbb{G}, p, g) \leftarrow \mathcal{G}(1^k)$ *and uniformly chosen* $x, y, z \in \mathbb{Z}_p$. *We say that the Decisional Diffie-Hellman (DDH) assumption holds with respect to* $\mathcal{G}$ *iff* $\mathrm{Adv}^{\mathrm{ddh}}_{\mathcal{G}, \mathcal{A}}$ *is negligible for every PPT* $\mathcal{A}$. *When the reference to* $\mathcal{G}$ *is clear, we also say that the DDH assumption holds in* $\mathbb{G}$ *(and write* $\mathrm{Adv}^{\mathrm{ddh}}_{\mathbb{G}, \mathcal{A}}$). *On occasion, we might also say that the DDH assumption holds in groups* $\mathbb{G}$ *or* $\hat{\mathbb{G}}$ *sampled by a pairing generator, with the obvious meaning.*

ElGamal Encryption. The ElGamal encryption scheme $\mathrm{PKE}_{\mathrm{eg}}$ is defined as follows, where we assume a suitable group generator $\mathcal{G}$.

- $\mathrm{EPars}_{\mathrm{eg}}(1^k)$ runs $(\mathbb{G}, p, g) \leftarrow \mathcal{G}(1^k)$ and outputs $epp = (\mathbb{G}, p, g)$.
- $\mathrm{EGen}_{\mathrm{eg}}(epp)$ picks a uniform $sk \leftarrow \mathbb{Z}_p$, sets $pk = g^{sk}$, and outputs (pk, sk).
- $\mathrm{Enc}(pk, M)$, for $M \in \mathbb{G}$, picks an $R \leftarrow \mathbb{Z}_p$, and outputs $C = (g^R, pk^R \cdot M)$.
- $\mathrm{Dec}(sk, C)$, for $C = (C_1, C_2) \in \mathbb{G}^2$, outputs $M = C_2 / C_1^{sk}$.

The ElGamal scheme is tightly IND-mCPA secure under the DDH assumption in $\mathbb{G}$. Concretely, for every valid IND-mCPA adversary $\mathcal{A}$, there is a DDH adversary $\mathcal{B}$ (of roughly the same complexity as the IND-mCPA experiment with $\mathcal{A}$) with $\mathrm{Adv}^{\mathrm{ddh}}_{\mathbb{G}, \mathcal{B}}(k) = \mathrm{Adv}^{\mathrm{ind\text{-}mcpa}}_{\mathrm{PKE}_{\mathrm{eg}}, \mathcal{A}}(k)$.

Groth-Sahai Proofs. In a setting with a pairing generator, Groth-Sahai proofs [26] provide a very versatile and efficient way to prove the satisfiability of very general classes of equations over $\mathbb{G}$ and $\hat{\mathbb{G}}$. We will not need them in full generality, and the next definition only captures a number of abstract properties of Groth-Sahai proofs we will use. In particular, we will not formalize the exact classes of languages amenable to Groth-Sahai proofs. (For the exact languages used in our application, however, we give more details in Sect. 5.1.) Like [18,19], we formalize Groth-Sahai proofs as *commit-and-prove* systems:

Definition 4 (GS Proofs [26]). *The Groth-Sahai proof system for a given pairing generator* $\mathcal{P}$ *consists of the following PPT algorithms, where gpp denotes group parameters sampled by* $\mathcal{P}$.

Common Reference Strings. $\mathrm{HGen}(gpp)$ *and* $\mathrm{BGen}(gpp)$ *sample hiding, resp. binding common reference strings (CRSs) CRS.*

Commitments. *For a (hiding or binding) CRS CRS and a* $\mathbb{G}$-, $\hat{\mathbb{G}}$-, *or* $\mathbb{Z}_p$-*element* v, *the commitment algorithm* $\mathrm{Com}(gpp, \mathrm{CRS}, v; R)$ *outputs a commitment* C, *where* R *denotes the used random coins.*

Proofs. *Let CRS be a CRS, and let* $\mathcal{X}$ *be a system of equations. Each equation may be over* $\mathbb{G}$, $\hat{\mathbb{G}}$, *or* $\mathbb{Z}_p$, *and involve variables and constants. Let* $(v_i)_i$ *be a variable assignment that satisfies* $\mathcal{X}$, *and let* $(R_i)_i$ *be a vector of random coins for* Com. *Then* $\mathrm{Prove}(gpp, \mathrm{CRS}, \mathcal{X}, (v_i, R_i)_i)$ *outputs a proof* π.

Verification. *For a CRS CRS, a system* $\mathcal{X}$ *of equations, a commitment vector* $(C_i)_i$ *to an assignment of the variables in* $\mathcal{X}$, *and a proof* π, *the verification algorithm* $\mathrm{Verify}(gpp, \mathrm{CRS}, \mathcal{X}, (C_i)_i, \pi)$ *outputs a verdict* $b \in \{0, 1\}$.

Simulation. *For a hiding CRS generated as* $\mathrm{CRS} \leftarrow \mathrm{HGen}(gpp; R_{\mathrm{CRS}})$, *a system* $\mathcal{X}$ *of equations, and a vector* $(R_i)_i$ *of commitment random coins, we have that* $\mathrm{Sim}(gpp, R_{\mathrm{CRS}}, \mathcal{X}, (R_i)_i)$ *outputs a simulated proof* π.

As with signatures and encryption, we usually omit the group parameters gpp on invocations of C, Prove, Verify, Sim when the reference is clear.

Theorem 1 (Properties of GS Proofs [26]). *The algorithms from Definition 4 satisfy the following for all choices group parameters $gpp \leftarrow \mathcal{P}(1^k)$ (unless noted otherwise):*

Homomorphic Commitments. *For any (hiding or binding) CRS CRS, any two given commitments $\mathrm{Com}(\mathrm{CRS}, v; R)$ and $\mathrm{Com}(\mathrm{CRS}, v'; R')$ to $\mathbb{G}$-elements v, v' allow to efficiently compute a commitment $\mathrm{Com}(\mathrm{CRS}, v \cdot v'; R \cdot R')$ to $v \cdot v'$. (Note that the corresponding random coins $R \cdot R'$ can be efficiently computed from R and R'.) The same holds for two commitments to $\hat{\mathbb{G}}$-elements, and two commitments to $\mathbb{Z}_p$-elements (where the homomorphic operation on $\mathbb{Z}_p$-elements is addition).*

Dual-Mode Commitments. *Consider a commitment $C \leftarrow \mathrm{Com}(\mathrm{CRS}, v; R)$. If CRS is binding, then C uniquely determines v, and if CRS is hiding, then the distribution of C does not depend on v.*

CRS Indistinguishability. *For every PPT adversary $\mathcal{A}$, there are PPT adversaries $\mathcal{A}_1$ and $\mathcal{A}_2$ with*

$$\left| \Pr\left[\mathcal{A}(1^k, \mathrm{HGen}(gpp)) = 1 \right] - \Pr\left[\mathcal{A}(1^k, \mathrm{BGen}(gpp)) = 1 \right] \right|$$
$$\leq \left| \mathrm{Adv}^{\mathrm{ddh}}_{\mathbb{G}, \mathcal{A}_1}(k) \right| + \left| \mathrm{Adv}^{\mathrm{ddh}}_{\hat{\mathbb{G}}, \mathcal{A}_2}(k) \right|,$$

where the probability is over $gpp \leftarrow \mathcal{P}(1^k)$, and the random coins of HGen, BGen, and $\mathcal{A}$.

Perfect Completeness. *For every (hiding or binding) CRS CRS, every system $\mathcal{X}$ of equations, every satisfying assignment $(v_i)_i$ of $\mathcal{X}$, and every possible vector $(C_i)_i$ of commitments generated through $C_i \leftarrow \mathrm{Com}(\mathrm{CRS}, v_i; R_i)$, we have $\mathrm{Verify}(\mathrm{CRS}, \mathcal{X}, (C_i)_i, \mathrm{Prove}(\mathrm{CRS}, \mathcal{X}, (v_i, R_i)_i)) = 1$ with probability 1.*

Perfect Soundness. *For every binding CRS CRS, every system $\mathcal{X}$ of equations that is not satisfiable, and every $(C_i)_i$ and π, $\mathrm{Verify}(\mathrm{CRS}, \mathcal{X}, (C_i)_i, \pi) = 0$ always.*

Perfect Simulation. *For every hiding CRS CRS $\leftarrow \mathrm{HGen}(gpp; R_{\mathrm{CRS}})$, and every system $\mathcal{X}$ of equations that is satisfied by a variable assignment $(v_i)_i$, the following two distributions are identical:*

$$((C_i)_i, \mathrm{Prove}(\mathrm{CRS}, \mathcal{X}, (v_i, R_i)_i)) \quad \text{for } C_i \leftarrow \mathrm{Com}(\mathrm{CRS}, v_i; R_i) \text{ and fresh} R_i,$$
$$((C_i)_i, \mathrm{Sim}(R_{\mathrm{CRS}}, \mathcal{X}, (R_i)_i)) \quad \text{for } C_i \leftarrow \mathrm{Com}(\mathrm{CRS}, 1; R_i) \text{ and fresh} R_i.$$

(The probability space consists of the R_i and the coins of Prove and Sim.)

Since simulation is perfect (in the sense above), it also holds for reused commitments (i.e., when multiple adaptively chosen statements $\mathcal{X}$ that involve the same variables and commitments are proven, see also [18]). Besides, perfect simulation directly implies perfect witness-indistinguishability (under a hiding CRS): for any two vectors $(v_i)_i$ and $(v'_i)_i$ of satisfying assignments of a given system $\mathcal{X}$ of equations, the corresponding commitments and proofs $((C_i)_i, \pi)$ and $((C'_i)_i, \pi')$ are identically distributed. Again, this holds even if the same commitments are used in several proofs for adaptively generated statements $\mathcal{X}$.

3 The Signature Scheme

3.1 Scheme Description

Setting and Ingredients. We assume the following ingredients:

- A pairing generator $\mathcal{P}$ that outputs groups $\mathbb{G} = \langle g \rangle$ and $\hat{\mathbb{G}} = \langle \hat{g} \rangle$ of prime order $p > 2^k$ and an asymmetric pairing $e : \mathbb{G} \times \hat{\mathbb{G}} \to \mathbb{G}_T$. We make the DDH assumption in both $\mathbb{G}$ and $\hat{\mathbb{G}}$.
- The ElGamal encryption scheme (given by algorithms $\mathrm{EGen}_{eg}, \mathrm{Enc}_{eg}, \mathrm{Dec}_{eg}$) over $\mathbb{G}$. (That is, we will use $\mathcal{P}$ in place of EPars_{eg} to generate the group $\mathbb{G}$ for ElGamal.)
- A Groth-Sahai proof system for $\mathcal{P}$ (see Definition 4), given by algorithms $\mathrm{HGen}, \mathrm{BGen}, \mathrm{Com}, \mathrm{Prove}, \mathrm{Verify}, \mathrm{Sim}$.

Public Parameters. $\mathrm{SPars}(1^k)$ samples group parameters

$$gpp = (\mathbb{G}, \hat{\mathbb{G}}, \mathbb{G}_T, p, g, \hat{g}, e) \leftarrow \mathcal{P}(1^k)$$

and sets $epp_{eg} = (\mathbb{G}, p, g)$. Then, SPars generates two binding Groth-Sahai CRSs and two ElGamal keypairs:

$$\mathrm{CRS}_1 \leftarrow \mathrm{BGen}(gpp) \qquad (pk_0, sk_0) \leftarrow \mathrm{EGen}_{eg}(epp_{eg})$$
$$\mathrm{CRS}_2 \leftarrow \mathrm{BGen}(gpp) \qquad (pk_1, sk_1) \leftarrow \mathrm{EGen}_{eg}(epp_{eg}).$$

The public parameters are then defined as

$$spp = (gpp, \mathrm{CRS}_1, \mathrm{CRS}_2, pk_0, pk_1).$$

Key Generation. $\mathrm{SGen}(spp)$ first sets up the exponents

$$Z = X \leftarrow \mathbb{Z}_p^* \qquad \text{and} \qquad \alpha = \beta = 0,$$

and commits to them using fresh random coins R_Z, R_α, R_β:

$$C_\alpha \leftarrow \mathrm{Com}(\mathrm{CRS}_1, \alpha; R_\alpha), \qquad C_\beta \leftarrow \mathrm{Com}(\mathrm{CRS}_1, \beta; R_\beta),$$
$$C_Z \leftarrow \mathrm{Com}(\mathrm{CRS}_2, Z; R_Z).$$

We will use that α, β define an affine function $f : \mathbb{Z}_p \to \mathbb{Z}_p$ through $f(x) = \alpha \cdot x + \beta \bmod p$.

Verification and signing key are given by

$$vk = (C_Z, C_\alpha, C_\beta) \qquad\qquad sigk = (X, R_Z, R_\alpha, R_\beta).$$

Signature Generation. $\mathrm{Sig}(sigk, M)$, for $M \in \mathbb{Z}_p$, picks fresh random coins R and encrypts

$$C_0 = \mathrm{Enc}_{eg}(pk_0, g^{Z_0}; R) \qquad\qquad C_1 = \mathrm{Enc}_{eg}(pk_1, g^{Z_1}; R)$$

for $Z_0 = Z_1 = X \in \mathbb{Z}_p$, using the same coins R in both encryptions for efficiency. Then, Sig generates proofs π_1 and π_2 for the respective statements

$$\Big(\underbrace{Z_0 = Z_1}_{S1} \quad \vee \quad \underbrace{f(M) \in \mathrm{QR}_p \cup \{0\}}_{S2} \Big) \qquad \text{and} \qquad \underbrace{Z_0 = Z}_{S3}. \qquad (3)$$

Here, Z_0, Z_1, Z, f refer to the values encrypted (resp. committed to) in C_0, C_1, $C_Z, (C_\alpha, C_\beta)$. Concretely, Sig generates a proof π_1 for $S1 \vee S2$ under CRS_1, using as witness $Z_0 = Z_1 = X$ and the encryption coins R. Also, Sig computes a proof π_2 for $S3$ under CRS_2, using as witness X and R_Z, R. We stress that π_1 and π_2 are independently generated, with different (fresh) Groth-Sahai commitments to the respective witnesses. We describe the exact Groth-Sahai equations for these proofs in Sect. 5.1, and give some intuition on the meaning of the statements $S1$-$S3$ in Sect. 3.2 below.

The signature is then defined as

$$\sigma = (C_0, C_1, \pi_1, \pi_2).$$

Verification. $\mathrm{Ver}(spp, vk, M, \sigma)$ outputs 1 if and only if both proofs π_1 and π_2 in σ are valid with respect to $M, C_0, C_1, C_Z, C_\alpha, C_\beta$.

Correctness. The completeness of Groth-Sahai proofs implies the correctness of SIG.

Efficiency. SIG has the following efficiency characteristics (cf. Section 5.1):

- The public parameters consist of 8 $\mathbb{G}$- and 6 $\hat{\mathbb{G}}$-elements, plus the group parameters gpp.
- Each verification key contains 2 $\mathbb{G}$- and 4 $\hat{\mathbb{G}}$-elements.
- Each signing key contains 7 $\mathbb{Z}_p$-exponents.
- Each signature contains 11 $\mathbb{G}$- and 14 $\hat{\mathbb{G}}$-elements.

3.2   Security Analysis

More Details on the Role of π_1 and π_2 in Signatures. Before we proceed to the proof, we give some intuition on the proofs π_1 and π_2 published in signatures (and the statements $S1$-$S3$):

- π_1 proves that *either* C_0 and C_1 encrypt the same value *or* that the signed message satisfies a special property $S2$ (or both). In the scheme, all messages are special in this sense (because $f(M) = 0$ for all M). However, in the proof, we can adjust f and, e.g., partition the set of messages into special and non-special ones in a random and roughly balanced way. Intuitively, this provides a means to make the double encryption (C_0, C_1) inconsistent (and subsequently change the encrypted values) in signatures for special messages. At the same time, any valid adversarial forgery on a *non-special* message (that does not satisfy $S2$) must carry a consistent double encryption (C_0, C_1).

– In the scheme, π_2 ties the plaintext encrypted in C_0 to the master secret Z. In the simulation, we will remove that connection by simulating π_2. Specifically, recall that π_1 and π_2 are independently generated, using independently generated Groth-Sahai commitments to the respective witnesses. Thus, in the proof, we can simulate π_2 without witness (by choosing a hiding CRS_2 and using Sim), while preserving the soundness of π_1 (assuming CRS_1 is binding). This simulation of π_2 will be instrumental in changing the message encrypted in C_0 (when the signed message is special in the above sense).

Theorem 2 (Security of SIG). *Under the DDH assumptions in $\mathbb{G}$ and $\hat{\mathbb{G}}$, the signature scheme SIG from Sect. 3.1 is EUF-mCMA secure. Concretely, for every EUF-mCMA adversary $\mathcal{A}$ on SIG, there exist DDH adversaries $\mathcal{B}$ and $\mathcal{B}'$ (of roughly the same complexity as the EUF-mCMA experiment with $\mathcal{A}$ and SIG) with*

$$\mathrm{Adv}^{\mathrm{euf\text{-}mcma}}_{\mathrm{SIG},\mathcal{A}}(k) \leq (8n+1)\cdot\left|\mathrm{Adv}^{\mathrm{ddh}}_{\mathbb{G},\mathcal{B}}(k)\right| + (4n+1)\cdot\left|\mathrm{Adv}^{\mathrm{ddh}}_{\hat{\mathbb{G}},\mathcal{B}'}(k)\right| + \mathbf{O}(n/2^k) \quad (4)$$

for $n = 2\lceil\log_2(p)\rceil + k$, where p denotes the order of $\mathbb{G}$ and $\hat{\mathbb{G}}$, and k is the security parameter.

Proof Outline. The proof starts with a number of preparations for the core argument. Our main goal during this phase will be to implement an additional and explicit check of $\mathcal{A}$'s forgery $\sigma^* = (C_0^*, C_1^*, \pi_1^*, \pi_2^*)$ for $\mathrm{Dec}_{\mathrm{eg}}(sk_0, C_0^*) = g^{X^*}$. (Note that in the default key setup, this explicit check is redundant, since valid signatures *must* fulfill statement $S3$ from (3).)

In the core argument (from Game 4 to Game 5, detailed in Lemma 1), we replace the value X used in generated signatures and the additional forgery check with a value $\mathcal{H}(M)$ that depends on the signed message. We start with a constant function $\mathcal{H}(M) = X$ (which corresponds to Game 4), and then introduce more and more dependencies of $\mathcal{H}(M)$ on the Legendre symbols $\left(\frac{f_j(M)}{p}\right)$ for independently and randomly selected (invertible) affine functions f_j.

Each such dependency is introduced as follows. We start by committing to (the coefficients of) a new random function f^* in C_α, C_β. This change allows us to modify the messages Z_0, Z_1 encrypted in generated signatures for all M with $f^*(M) \in \mathrm{QR}_p \cup \{0\}$ (and only for those M), by proving $S2$ (and not $S1$) in signatures. We will also abort if $\mathcal{A}$'s forgery satisfies $f^*(M^*) \in \mathrm{QR}_p \cup \{0\}$, and we will keep enforcing our forgery check on C_0^*. Hence, from $\mathcal{A}$'s point of view, an additional dependency on $\left(\frac{f^*(M)}{p}\right)$ is consistently introduced on *all* signatures. More importantly, this dependency is also enforced during the additional forgery check.

After sufficiently many such dependencies are introduced (for several different f^*), all signatures are consistently generated with (or checked for) $Z_0 = Z_1 = \mathcal{R}(M)$ for a truly random function $\mathcal{R}$. At this point, $\mathcal{A}$ has to predict a truly random function $\mathcal{R}$ on a fresh input M^* in order to produce a valid forgery. Hence, $\mathcal{A}$'s forgery success must be negligible.

Figures 1 and 2 (on page 27 and page 28) give a more technical summary of the game transitions of the proof (also taking into account the notation for the multi-user case). The remainder of this section is devoted to a detailed proof.

Proof (Proof of Theorem 2) We proceed in games. Let out_i denote the output of Game i.

Game 1 is the original EUF-mCMA game with $\mathcal{A}$ and SIG. Of course,

$$\Pr[out_1 = 1] = \mathrm{Adv}_{\mathrm{SIG},\mathcal{A}}^{\mathrm{euf\text{-}mcma}}(k). \tag{5}$$

In the following, we apply a superscript to variables to denote to which SIG instance they belong. For instance, we denote with $X^{(\ell)}$ and $sk_0^{(\ell)}, sk_1^{(\ell)}$ the respective values from the ℓ-th used SIG instance. Furthermore, we write X^* for $X^{(\ell^*)}$ for the challenge instance ℓ^* selected by $\mathcal{A}$ for his forgery, and similarly for sk_0^* and sk_1^*.

Thus, in **Game 2**, we implement an additional "forgery check". Concretely, we only consider a forgery $\sigma^* = (C_0^*, C_1^*, \pi_1^*, \pi_2^*)$ from $\mathcal{A}$ as valid if π_1^* and π_2^* are valid *and* if $\mathrm{Dec}_{\mathrm{eg}}(sk_0^*, C_0^*) = g^{X^*}$. (Otherwise, the game outputs 0.) This change is purely conceptual: indeed, since CRS_2 is binding, we can use the soundness of Groth-Sahai proofs. Thus, any valid proof π_2^* guarantees that $S3$ (from (3)) holds, and so $\mathrm{Dec}_{\mathrm{eg}}(sk_0^*, C_0^*) = g^{X^*}$. We obtain

$$\Pr[out_2 = 1] = \Pr[out_1 = 1]. \tag{6}$$

In **Game 3**, we generate both CRS_1 and CRS_2 as hiding CRSs, using HGen. The CRS indistinguishability of Groth-Sahai proofs yields

$$\Pr[out_3 = 1] - \Pr[out_2 = 1] = \mathrm{Adv}_{\mathbb{G},\mathcal{B}_3}^{\mathrm{ddh}}(k) + \mathrm{Adv}_{\hat{\mathbb{G}},\mathcal{B}_3'}^{\mathrm{ddh}}(k) \tag{7}$$

for suitable DDH adversaries $\mathcal{B}_3$ and $\mathcal{B}_3'$. (Here, we use the re-randomizability of DDH tuples. This enables a reduction that loses only a factor of 1 instead of 2.)

In **Game 4**, we simulate all proofs π_2 in signatures generated for $\mathcal{A}$, using the Groth-Sahai simulator Sim (on input the random coins R_{CRS} used to prepare CRS). We also generate the corresponding commitments C_Z in all verification keys as $C_Z \leftarrow \mathrm{Com}(\mathrm{CRS}_2, 1)$. We stress that all $X^{(\ell)}$ are still chosen randomly, and all signatures are generated with encryptions C_0, C_1 of $X^{(\ell)}$. By the simulation property of Groth-Sahai proofs (see Theorem 1 and the following comment concerning the reuse of commitments), these changes do not affect $\mathcal{A}$'s view:

$$\Pr[out_4 = 1] = \Pr[out_3 = 1]. \tag{8}$$

In **Game 5**, we change the generation of signatures *and* the forgery check from Game 2 as follows. To describe these changes, let $\mathcal{R}^{(\ell)} : \mathbb{Z}_p \to \mathbb{Z}_p^*$ (for all scheme instances $\ell \in [n_U]$) be truly random functions. Our changes in Game 5 are then as follows:

– All signatures generated for $\mathcal{A}$ contain encryptions C_0, C_1 of exponents $Z_0 = Z_1 = \mathcal{R}^{(\ell)}(M)$ (encoded as g^{Z_0}, g^{Z_1}) instead of $Z_0 = Z_1 = X^{(\ell)}$, where M is the signed message. As in Game 4, the corresponding proof π is generated using witnesses for $S1$ and $S3$ from (3).

– Any forgery $\sigma^* = (C_0^*, C_1^*, \pi_1^*, \pi_2^*)$ for a (fresh) message M^* from $\mathcal{A}$ is considered valid only if π_1^* and π_2^* are valid *and* $\mathrm{Dec_{eg}}(sk_0^*, C_0^*) = \mathcal{R}^*(M^*)$ holds. Otherwise, the game outputs 0. (Again, we use the shorthand notation $\mathcal{R}^* = \mathcal{R}^{(\ell^*)}$ for the challenge instance ℓ^*.)

In particular, the second change implies that

$$\Pr[out_5 = 1] \leq 1/(p-1) \leq 1/2^k, \tag{9}$$

since $\mathcal{R}^*(M^*)$ is information-theoretically hidden from $\mathcal{A}$.

Hence, it remains to relate Game 4 and Game 5:

Lemma 1. *For $n = 2\lceil \log_2(p) \rceil + k$ and suitable DDH adversaries $\mathcal{B}_5$ and $\mathcal{B}_5'$, we have*

$$\left| \Pr[out_5 = 1] - \Pr[out_4 = 1] \right| \leq 8n \cdot \left| \mathrm{Adv}_{\mathbb{G}, \mathcal{B}_5}^{\mathrm{ddh}}(k) \right| + 4n \cdot \left| \mathrm{Adv}_{\hat{\mathbb{G}}, \mathcal{B}_5'}^{\mathrm{ddh}}(k) \right| + \mathbf{O}(n/2^k). \tag{10}$$

Before we prove Lemma 1, we remark that putting together (5–10), we obtain (4), which is sufficient to show Theorem 2.

Proof. (of Lemma 1) We will consider a series of hybrid games between Game 4 and Game 5. Concretely, Game 4.i (for $i \geq 0$) is defined like Game 4, except for the following changes:

– We initially uniformly and independently choose i invertible affine functions $f_j : \mathbb{Z}_p \to \mathbb{Z}_p$ (for $j \in [i]$). The f_j define a "partial fingerprint" function $\mathcal{L}_i : \mathbb{Z}_p \to \{-1, 0, 1\}^i$ through

$$\mathcal{L}_i(M) = \left(\left(\frac{f_1(M)}{p} \right), \dots, \left(\frac{f_i(M)}{p} \right) \right). \tag{11}$$

For every scheme instance $\ell \in [n_U]$, let $\mathcal{H}_i^{(\ell)} : \mathbb{Z}_p \to \mathbb{Z}_p^*$ be the composition of $\mathcal{L}_i$ with a truly random function $\mathcal{R}_i^{(\ell)} : \{-1, 0, 1\}^i \to \mathbb{Z}_p^*$ (so that $\mathcal{H}_i^{(\ell)}(M) = \mathcal{R}_i^{(\ell)}(\mathcal{L}_i(M))$).

– Signatures for $\mathcal{A}$ contain encryptions C_0, C_1 of exponents $Z_0 = Z_1 = \mathcal{H}_i^{(\ell)}(M)$.

– Any forgery $\sigma^* = (C_0^*, C_1^*, \pi_1^*, \pi_2^*)$ for a (fresh) message M^* from $\mathcal{A}$ is considered valid only if π_1^* and π_2^* are valid *and* $\mathrm{Dec_{eg}}(sk_0^*, C_0^*) = \mathcal{H}_i^{(\ell)}(M^*)$.

Note that every $\mathcal{H}_0^{(\ell)}$ is a constant function that maps every input M to the same random value. Hence, Game 4.0 is identical to Game 4:

$$\Pr[out_{4.0} = 1] = \Pr[out_4 = 1]. \tag{12}$$

Conversely, for large enough i and with high probability, the "fingerprint function" $\mathcal{L}_i$ becomes injective, so that all $\mathcal{H}_i^{(\ell)}$ become independent truly random functions from $\mathbb{Z}_p$ to $\mathbb{Z}_p^*$:

Lemma 2. For $n = 2\lceil \log_2(p)\rceil + k$, the function $\mathcal{L}_n$ from (11) is injective, except with probability $1/2^k$ (over the choice of the invertible affine functions $f_j : \mathbb{Z}_p \to \mathbb{Z}_p$).

We postpone a proof of Lemma 2 for now.

Hence, the functions $\mathcal{H}_n^{(\ell)} = \mathcal{R}_n^{(\ell)} \circ \mathcal{L}_n$ used in Game 4.n (for $n = 2\lceil \log_2(p)\rceil + k$) are statistically close to truly random functions $\mathcal{R}^{(\ell)}$ (as used in Game 5):

$$\left| \Pr\left[out_{4.n} = 1\right] - \Pr\left[out_5 = 1\right] \right| \leq 1/2^k. \tag{13}$$

The Algebraic Partitioning Step. Thus, we only need to show that there is no detectable difference between Game 4.i and Game 4.(i+1) for any i. We do so using a hybrid argument (i.e., a sequence of games) that interpolates between Game 4.i and Game 4.(i+1). (See Fig. 2 for an overview.) In short, we first refresh the affine function f from C_α, C_β to a fresh random (but invertible) affine function f^*. Next, we use f^* to implement a different treatment of signatures, depending on $\left(\frac{f(M)}{p}\right)$. We detail these steps in the following.

Concretely, **Game** 4.i.0 is identical to Game 4.i. Thus,

$$\Pr\left[out_{4.i.0} = 1\right] = \Pr\left[out_{4.i} = 1\right]. \tag{14}$$

Step 1: Refresh f. In **Game** 4.i.1, we initially choose an invertible affine function $f^* : \mathbb{Z}_p \to \mathbb{Z}_p$ uniformly, and we abort (with output 0) if the message M^* for which $\mathcal{A}$ finally prepares a forgery satisfies $f^*(M^*) \in QR_p \cup \{0\}$. We stress that f^* is not (yet) committed to in any C_α, C_β, and thus completely hidden from $\mathcal{A}$. Hence, an abort occurs with probability $\frac{p+1}{2p} = \frac{1}{2} + \frac{1}{2p}$, independently of $\mathcal{A}$'s view, so

$$\Pr\left[out_{4.i.1} = 1\right] = \left(\frac{1}{2} - \frac{1}{2p}\right) \cdot \Pr\left[out_{4.i.0} = 1\right] \geq \frac{1}{2} \cdot \Pr\left[out_{4.i.0} = 1\right] - \frac{1}{2p}. \tag{15}$$

In **Game** 4.i.2, we commit to the coefficients f_0^*, f_1^* of f^* from Game 4.i.1 in C_α, C_β for all verification keys (instead of the coefficients $\alpha = \beta = 0$). Accordingly, we generate all signatures for $\mathcal{A}$ by proving statement $S2$ (and not $S1$) from (3) whenever possible (i.e., upon all signature queries with $f^*(M) \in QR_p \cup \{0\}$). Since CRS_1 is hiding, we can use the witness-indistinguishability of Groth-Sahai proofs to obtain

$$\Pr\left[out_{4.i.2} = 1\right] = \Pr\left[out_{4.i.1} = 1\right]. \tag{16}$$

Step 2: Use f^* to Decouple Signatures. To describe our change in **Game** 4.i.3, recall that in Game 4.i.2, functions $\mathcal{H}_i^{(\ell)}$ is used to determine both the values $Z_0 = Z_1 = \mathcal{H}_i^{(\ell)}(M)$ encrypted in C_0, C_1 upon signature queries, and to implement the forgery check. In Game 4.i.3, we use *three* such functions $\mathcal{H}_i^{(\ell)}, \mathcal{Z}_i^{(\ell)}, \mathcal{Q}_i^{(\ell)} : \mathbb{Z}_p \to \mathbb{Z}_p^*$. Each of these functions is defined like $\mathcal{H}_i^{(\ell)}$, for the same fingerprint function $\mathcal{L}_i$, but with different (i.e., independently chosen) random functions $\mathcal{R}_i^{(\ell)}$. (In other words, we can write $\mathcal{H}_i^{(\ell)} = F \circ \mathcal{L}_i$, and $\mathcal{Z}_i^{(\ell)} = F' \circ \mathcal{L}_i$, and $\mathcal{Q}_i^{(\ell)} = F'' \circ \mathcal{L}_i$ for independently random functions $F, F', F'' : \{-1, 0, 1\}^i \to \mathbb{Z}_p^*$. Intuitively, thus, $\mathcal{Z}_i^{(\ell)}$ and $\mathcal{Q}_i^{(\ell)}$ are "decoupled copies" of $\mathcal{H}_i^{(\ell)}$.)

Our goal will be to use the functions $\mathcal{H}_i^{(\ell)}, \mathcal{Z}_i^{(\ell)}, \mathcal{Q}_i^{(\ell)}$ for messages M satisfying $f^*(M) \notin \mathrm{QR}_p$, $f^*(M) = 0$, and $f^*(M) \in \mathrm{QR}_p$, respectively. (Hence the symbols $\mathcal{Z}$ and $\mathcal{Q}$.) This will be conceptually identical to using a single function $\mathcal{H}_{i+1}^{(\ell)}$ for all messages of a given scheme instance ℓ. At this point, however, we can only partially implement this strategy, since we can only replace the messages encrypted in C_1, but not those from C_0. (Indeed, sk_0^* is still required to implement the additional forgery check in Game 4.i.3.)

Thus, in Game 4.i.3, for every scheme instance $\ell \in [n_U]$, we use the respective function $\mathcal{H}_i^{(\ell)}$ to generate all ciphertexts C_0, C_1 in signatures (as in Game 4.i.2), with the following exceptions:

- For signature queries with $f^*(M) = 0$, we encrypt $Z_1 = \mathcal{Z}_i^{(\ell)}(M)$ (instead of $Z_1 = \mathcal{H}_i^{(\ell)}(M)$) in the ciphertext C_1 of the generated signature.
- For signature queries with $f^*(M) \in \mathrm{QR}_p$, we encrypt $Z_1 = \mathcal{Q}_i^{(\ell)}(M)$ in C_1.

Note that for signatures with $f^*(M) \in \mathrm{QR}_p \cup \{0\}$, the random coins used to generate C_1 (or C_0) are not used as a witness in the process of constructing π. Furthermore, no secret key $sk_1^{(\ell)}$ has to be known to the game. A reduction to the (tight) IND-mCPA security of ElGamal yields

$$\sum_{i=0}^{n-1} \Pr\left[out_{4.i.3} = 1\right] - \Pr\left[out_{4.i.2} = 1\right] = n \cdot \mathrm{Adv}_{\mathbb{G}, \mathcal{B}_{4.i.3}}^{\mathrm{ddh}}(k) \qquad (17)$$

for a suitable DDH adversary $\mathcal{B}_{4.i.3}$. (We note that even though the random coins R of C_1 are not known explicitly to $\mathcal{B}_{4.i.3}$, a C_0 with reused R can be constructed from $sk_0^{(\ell)}$ and a given g^R.)

Our next step will be to replace the values encrypted in C_0 in a similar way. To do so, however, we need some preparations, since Game 4.i.3 still knows the secret keys $sk_0^{(\ell)}$ (to finally implement the forgery check). Fortunately, however, we can alternatively use the $sk_1^{(\ell)}$ to implement this check. (To see why this yields the same functionality, recall that by our abort rule from Game 1, we may restrict to forgeries with $f^*(M^*) \notin \mathrm{QR}_p \cup \{0\}$. However, by (3), a valid forgery for such a message must contain C_0^* and C_1^* that encrypt the same message.)

As a first step, in **Game** 4.i.4, we initially generate a binding CRS CRS_1 (using $\mathrm{CRS}_1 \leftarrow \mathrm{BGen}(gpp)$). The CRS indistinguishability of Groth-Sahai proofs ensures that

$$\sum_{i=0}^{n-1} \Pr\left[out_{4.i.4} = 1\right] - \Pr\left[out_{4.i.3} = 1\right] = n \cdot \left(\mathrm{Adv}_{\mathbb{G}, \mathcal{B}_{4.i.4}}^{\mathrm{ddh}}(k) + \mathrm{Adv}_{\mathbb{G}, \mathcal{B}'_{4.i.4}}^{\mathrm{ddh}}(k)\right)$$
$$(18)$$

for suitable DDH adversaries $\mathcal{B}_{4.i.4}$ and $\mathcal{B}'_{4.i.4}$.

Next, in **Game** 4.i.5, we implement the forgery check rule from Game 2 using sk_1^* (and not sk_0^*). That is, when $\mathcal{A}$ submits a forgery $\sigma^* = (C_0^*, C_1^*, \pi_1^*, \pi_2^*)$, we check if $\mathrm{Dec}_{\mathrm{eg}}(sk_1^*, C_1^*) = \mathcal{H}_i^*(M^*)$ holds (and reject the forgery if not). We may assume that $M^* \notin \mathrm{QR}_p \cup \{0\}$ (since otherwise, we trivially abort anyway). But for such M^*, a valid forgery *must* fulfill $S1$ from (3), since at this point, CRS_1 is binding. In other words, we have $\mathrm{Dec}_{\mathrm{eg}}(sk_1^*, C_1^*) = \mathcal{H}_i^*(M^*)$ if and only if $\mathrm{Dec}_{\mathrm{eg}}(sk_0^*, C_0^*) = \mathcal{H}_i^*(M^*)$. Hence, the change in Game 4.i.5 is purely conceptual, and we get:

$$\Pr[out_{4.i.5} = 1] = \Pr[out_{4.i.4} = 1]. \tag{19}$$

Since we no longer use sk_0^* (or the random coins from any C_1 generated upon a signature query), we can continue with our strategy. Specifically, in **Game** 4.i.6, we generate all ciphertexts C_0, C_1 in signatures as follows:

- For queries with $f^*(M) \notin \mathrm{QR}_p$, we encrypt $Z_0 = Z_1 = \mathcal{H}_i^{(\ell)}(M)$ in C_0 and C_1.
- For queries with $f^*(M) = 0$, we encrypt $Z_0 = Z_1 = \mathcal{Z}_i^{(\ell)}(M)$ in C_0 and C_1.
- For queries with $f^*(M) \in \mathrm{QR}_p$, we encrypt $Z_0 = Z_1 = \mathcal{Q}_i^{(\ell)}(M)$ in C_0 and C_1.

Observe that the only difference to Game 4.i.5 is that the messages Z_0 encrypted in ciphertexts C_0 in signatures with $f^*(M) \in \mathrm{QR}_p \cup \{0\}$ are changed. For such encryptions, neither secret key nor random coins are used by the game. Hence, a reduction to the (tight) IND-mCPA security of ElGamal yields

$$\sum_{i=0}^{n-1} \Pr[out_{4.i.6} = 1] - \Pr[out_{4.i.5} = 1] = n \cdot \mathrm{Adv}_{\mathbb{G}, \mathcal{B}_{4.i.6}}^{\mathrm{ddh}}(k) \tag{20}$$

for a suitable DDH adversary $\mathcal{B}_{4.i.6}$. (Again, a reuse of random coins between C_0 and C_1 is possible since the secret key sk_1 is known to $\mathcal{B}_{4.i.6}$ during the reduction.)

Step 3: Clean Up. Now in Game 4.i.6, we handle both signature queries and $\mathcal{A}$'s forgery with either $\mathcal{H}_i^{(\ell)}$, $\mathcal{Z}_i^{(\ell)}$, or $\mathcal{Q}_i^{(\ell)}$, depending on the Legendre symbol $\left(\frac{M}{p}\right)$ of M. This is equivalent to handling all messages with a single function $\mathcal{H}_{i+1}^{(\ell)}$ by the definition of $\mathcal{H}_i^{(\ell)}$ (see also (11)). Hence, we already "almost" implement the rules of Game 4.$(i+1)$, and we only need to clean up things a little.

Namely, in **Game** 4.i.7, we again implement the forgery check from Game 2 using sk_0^* (and not sk_1^*). With the same reasoning as in Game 5, we get:

$$\Pr[out_{4.i.7} = 1] = \Pr[out_{4.i.6} = 1]. \tag{21}$$

Next, in **Game** 4.i.8, we again set up CRS_1 as a hiding CRS (using HGen). Again, CRS indistinguishability guarantees

$$\sum_{i=0}^{n-1} \Pr[out_{4.i.8} = 1] - \Pr[out_{4.i.7} = 1] = n \cdot \left(\mathrm{Adv}_{\mathbb{G}, \mathcal{B}_{4.i.8}}^{\mathrm{ddh}}(k) + \mathrm{Adv}_{\mathbb{G}, \mathcal{B}'_{4.i.8}}^{\mathrm{ddh}}(k)\right) \tag{22}$$

for suitable DDH adversaries $\mathcal{B}_{4.i.8}$ and $\mathcal{B}'_{4.i.8}$.

In **Game** $4.i.9$, we again set up the commitments C_α, C_β in all verification keys as commitments to $\alpha = \beta = 0$. Accordingly, we generate all signatures for $\mathcal{A}$ by proving statement $S1$ from (3). (Note that this is possible again since all generated pairs (C_0, C_1) do encrypt the same message.) By the witness-indistinguishability of Groth-Sahai proofs,

$$\Pr[out_{4.i.9} = 1] = \Pr[out_{4.i.8} = 1]. \tag{23}$$

Finally, in **Game** $4.i.10$, we do not abort anymore. (That is, we take back the abort rule from Game 1.) To see how this change affects the game's output, we make a few observations. First, note that in both Game $4.i.9$ and Game $4.i.10$, $\mathcal{A}$'s view only depends on the way f^* partitions the set of messages depending on $\left(\frac{f^*(M)}{p}\right)$, but not on *which* messages M are mapped by f^* to squares, and which to non-squares. (Indeed, any partitioning of the M is invariant under multiplying f^* with an invertible non-square modulo p. However, multiplication with an invertible non-square inverts the Legendre symbol of $f^*(M)$.)

Thus, the probability for $\mathcal{A}$ to successfully forge a signature with $\left(\frac{f^*(M^*)}{p}\right) = 1$ is exactly the same as that to forge a signature with $\left(\frac{f^*(M^*)}{p}\right) = -1$. Hence, if we cease to abort upon $f^*(M^*) \in QR_p \cup \{0\}$, we *at least* double $\mathcal{A}$'s success probability:

$$\Pr[out_{4.i.10} = 1] \geq 2 \cdot \Pr[out_{4.i.9} = 1]. \tag{24}$$

At the same time, Game $4.i.10$ is identical to Game $4.(i+1)$. (As argued, the use of three functions $\mathcal{H}_i^{(\ell)}, \mathcal{Z}_i^{(\ell)}, \mathcal{Q}_i^{(\ell)}$ for each scheme instance ℓ is equivalent to the use of a single function $\mathcal{H}_{i+1}^{(\ell)}$ in Game $4.(i+1)$. Furthermore, CRS_1 is hiding, the C_α, C_β are set up as commitments to $\alpha = \beta = 0$, and the signatures use proofs of statement $S1$.) Thus,

$$\Pr[out_{4.i.10} = 1] = \Pr[out_{4.(i+1)} = 1]. \tag{25}$$

Collecting all differences of probabilities from (14–25), we obtain

$$\left| \Pr[out_{4.0} = 1] - \Pr[out_{4.n} = 1] \right| \leq \left| \sum_{i=0}^{n-1} \Pr[out_{4.i} = 1] - \Pr[out_{4.(i+1)} = 1] \right|$$
$$\leq 8n \cdot \left| \mathrm{Adv}_{\mathbb{G}, \mathcal{B}_5}^{\mathrm{ddh}}(k) \right| + 4n \cdot \left| \mathrm{Adv}_{\mathbb{G}, \mathcal{B}_5'}^{\mathrm{ddh}}(k) \right| + \mathbf{O}(n/2^k)$$

for DDH adversaries $\mathcal{B}_5$ and $\mathcal{B}_5'$ that combine all adversaries from the collected differences. Together with (12) and (13), we obtain (10).

It remains to prove Lemma 2:

Proof. (of Lemma 2) For any distinct $M_0, M_1 \in \mathbb{Z}_p$ and a uniformly chosen invertible affine function $f : \mathbb{Z}_p \to \mathbb{Z}_p$, we have $\Pr\left[\left(\frac{f(M_0)}{p}\right) = \left(\frac{f(M_1)}{p}\right)\right] \leq 1/2$, since f is pairwise independent. As all f_j from (11) are chosen independently, we get

$$\Pr[\mathcal{L}_n(M_0) = \mathcal{L}_n(M_1)] \leq 1/2^n$$

for any two distinct M_0, M_1. A union bound over all $\mathbf{O}(p^2)$ such pairs (M_0, M_1) shows the claim.

4   Compact and (almost) Tightly Secure Public-Key Encryption

Our signature scheme SIG from Sect. 3 is "almost" automorphic (in the sense of [1]). Namely, while its verification can be expressed as a system of equations that is compatible with Groth-Sahai proofs, its messages are exponents (as opposed to group elements). However, our scheme can still be used in the generic construction of [28]. This yields an (almost) tightly secure public-key encryption scheme with compact parameters, keys and ciphertexts. (Here, "compact" means "comprised of only a constant number of group elements or exponents.")

But although compact in the above sense, the resulting encryption scheme would be rather inefficient (in particular since it would use nested Groth-Sahai proofs). Thus, here we describe an optimized and more compact (almost) tightly secure public-key encryption scheme PKE.

Setting and Ingredients. The basis for our PKE construction is the signature scheme SIG from Sect. 3, and we assume similar ingredients. In particular, we assume groups $\mathbb{G}$ and $\hat{\mathbb{G}}$, along with the ElGamal encryption and Groth-Sahai proofs over $\mathbb{G}$. Additionally, we assume:

- An OT-EUF-mCMA secure signature scheme with message space $\mathbb{Z}_p$, given by algorithms OPars, OGen, OSig, OVer. For concreteness, in all of the following, we assume the one-time signature scheme TOTS from [28] in $\mathbb{G}$. Its OT-EUF-mCMA security can be tightly reduced to the discrete logarithm assumption in $\mathbb{G}$ (which is implied by the DDH assumption in $\mathbb{G}$).
- A generator $\mathcal{H}$ of collision-resistant hash functions H : $\{0,1\}^* \rightarrow \{0,1\}^k$. We will interpret H-outputs as $\mathbb{Z}_p$-elements in the natural way. (Recall that $p > 2^k$.)

All ingredients can be instantiated under the DDH assumptions in $\mathbb{G}$ and $\hat{\mathbb{G}}$.

Public Parameters. EPars(1^k) first proceeds like the parameter generation of SIG, and samples group parameters gpp, a hiding Groth-Sahai CRS, and two ElGamal public keys pk_0, pk_1. Then, EPars sets up exponents Z, α, β and *ciphertexts*

$$C_\alpha \leftarrow \text{Enc}_{\text{eg}}(pk_0, g^\alpha; R_\alpha), \ C_\beta \leftarrow \text{Enc}_{\text{eg}}(pk_0, g^\beta; R_\beta), \ C_Z \leftarrow \text{Enc}_{\text{eg}}(pk_0, g^Z; R_Z).$$

Note that here, we encrypt (and do not commit to) Z, α, β in order to be able to produce slightly more compact proofs involving Z, α, β later on. However, we note that conceptually, we could have as well committed to Z, α, β as with SIG.

Finally, EPars chooses parameters $opp \leftarrow$ OPars(1^k) and a hash function H, and outputs $epp = (gpp, \text{CRS}, pk_0, pk_1, opp, \text{H}, C_\alpha, C_\beta, C_Z)$.

Key Generation. EGen(epp) samples two ElGamal keypairs (pk_0', sk_0'), $(pk_1', sk_1') \leftarrow$ EGen$_{\text{eg}}(\mathbb{G}, p, g)$, and outputs a public and a secret key as

$$pk = (pk_0', pk_1') \qquad\qquad sk = (d, sk_d')$$

for a uniformly chosen bit $d \leftarrow \{0,1\}$.

Encryption. Intuitively, encryption corresponds to a Naor-Yung style double encryption with consistency proof [34]. The consistency proof itself proceeds as in [28], and essentially proves that either the double encryption is consistent, or a signature to a fresh value is known. (A suitable fresh value will be hash of a freshly sampled verification key of the one-time signature scheme.) Concretely, $\text{Enc}(pk, M)$, for $M \in \mathbb{G}$, chooses a one-time signature keypair $(ovk, osk) \leftarrow \text{OGen}(opp)$, and encrypts the values $Z_0' = Z_1' = M \in \mathbb{G}$ and $Z_0 = Z_1 = 0$ as

$$C_0' = \text{Enc}_{eg}(pk_0', Z_0'; R') \qquad C_0 = \text{Enc}_{eg}(pk_0, g^{Z_0}; R)$$
$$C_1' = \text{Enc}_{eg}(pk_1', Z_1'; R') \qquad C_1 = \text{Enc}_{eg}(pk_1, g^{Z_1}; R).$$

(Note that for efficiency and to simplify proofs involving these values, we reuse the encryption random coins R' and R.) Then, Enc generates a proof π (under CRS) of the statement

$$Z_0' = Z_1' \lor \left((Z_0 = Z_1 \lor f(\text{H}(ovk)) \in \text{QR}_p \cup \{0\}) \land (Z_0 = Z \lor Z = 0) \right). \quad (26)$$

Enc will prove the left branch $S1'$ of the outer $\lor$ clause, using as witness the encryption randomness R'. Hence, π essentially proves consistency of C_0', C_1', or the same statement as for a SIG-signature for $\text{H}(ovk)$. (There are some slight differences compared to a SIG-signature: first, we use only one CRS. Hence, we cannot simulate proofs for substatement $Z_0 = Z$ during the proof. Instead, however, we can set $Z = 0$ to be able to generate proofs for $S3'$ without knowledge of Z_0. Second, because the random coins used for C_α, C_β, C_Z are not known at encryption time, the proof of quadratic residuosity becomes somewhat less efficient than the one in SIG's signing algorithm. We refer to Sect. 5.2 for more details on the exact proof equations.)

Finally, Enc signs $\sigma \leftarrow \text{OSig}(osk, \text{H}(C_0', C_1', C_0, C_1, \pi))$ and outputs the ciphertext $C = (C_0', C_1', C_0, C_1, \pi, ovk, \sigma)$.

Decryption. $\text{Dec}(sk, C)$ checks the validity of σ and π. If both σ and π are valid, Dec outputs $M \leftarrow \text{Dec}_{eg}(sk_d', C_d')$; otherwise, Dec outputs $\bot$.

Efficiency. PKE has the following efficiency characteristics (cf. Section 5.2):

- The public parameters consist of 12 $\mathbb{G}$- and 3 $\hat{\mathbb{G}}$-elements, plus the group parameters gpp, and a description of the hash function H.
- Each public key contains 2 $\mathbb{G}$-elements.
- Each secret key contains one $\mathbb{Z}_p$-exponent and a bit.
- Each ciphertext contains 27 $\mathbb{G}$- and 30 $\hat{\mathbb{G}}$-elements, and 3 $\mathbb{Z}_p$-exponents.

Theorem 3. (Security of PKE). *Under the DDH assumptions in $\mathbb{G}$ and $\hat{\mathbb{G}}$, and assuming that H is collision-resistant, the PKE scheme PKE described above is IND-mCCA secure. Concretely, for every EUF-mCMA adversary $\mathcal{A}$ on SIG, there exist DDH adversaries $\mathcal{B}$ and $\mathcal{B}'$, and an adversary $\mathcal{C}$ on the collision-resistance of H (of roughly the same complexity as the EUF-mCMA experiment with $\mathcal{A}$ and SIG) with*

$$\text{Adv}_{\text{SIG},\mathcal{A}}^{\text{euf-mcma}}(k) \leq \mathbf{O}(k) \cdot \left| \text{Adv}_{\mathbb{G},\mathcal{B}}^{\text{ddh}}(k) \right| + \mathbf{O}(k) \cdot \left| \text{Adv}_{\hat{\mathbb{G}},\mathcal{B}'}^{\text{ddh}}(k) \right| + \text{Adv}_{\mathcal{H},\mathcal{C}}^{\text{cr}}(k) + \mathbf{O}(k/2^k).$$
$$(27)$$

Proof. (Proof sketch) The proof combines the strategy from [28] with our concrete signature scheme, and thus we outline only the main strategy. This strategy proceeds in games, and modifies an IND-mCCA attack with adversary $\mathcal{A}$ as follows:

- First, the consistency proofs in all ciphertexts are prepared with different witnesses. More specifically, instead of proving $Z_0' = Z_1'$, we prove the right branch of (26). (Note that this right branch corresponds to the validity of a SIG-signature for message $H(ovk)$.) Thanks to the witness-indistinguishability of Groth-Sahai proofs, this change is not detectable by $\mathcal{A}$.
- Next, all challenge ciphertexts generated for $\mathcal{A}$ are made inconsistent. (This is possible since the ciphertext consistency proofs are prepared from signature witnesses now.) Concretely, recall that so far we have encrypted the respective challenge message M_b^* (for the secret bit b chosen by the IND-mCCA experiment) in both C_0' and C_1' of all challenge ciphertexts. Now we encrypt M_b^* in C_d' and M_{1-b}^* in C_{1-d}', where d is the bit chosen for the respective PKE instance i. Hence, we change the encrypted message for all ElGamal instances whose secret key is not used. Since only the secret keys sk_d' (but not the sk_{1-d}') are used in the experiment, this game modification can be justified with the (tight) security of ElGamal.
- We now reject all inconsistent (in the sense $\mathrm{Dec}_{eg}(sk_0', C_0') \neq \mathrm{Dec}_{eg}(sk_1', C_1')$) decryption queries from $\mathcal{A}$. At this point in the proof, we know both sk_0' and sk_1' for all PKE-instances, and can thus recognize the first inconsistent (in the above sense) decryption query with a valid consistency proof. Note that any such query implies a valid SIG-signature for a message $H(ovk)$. The security of the one-time signature scheme guarantees that this message is fresh, so that $\mathcal{A}$ has essentially forged a SIG-signature. Any such forgery can be excluded with the same strategy as in the proof of Theorem 2 (with the differences described above). This step entails the dominant terms in (27) related to DDH reductions.

At this point, $\mathcal{A}$ gets no information about the IND-mCCA secret b anymore. Namely, each challenge ciphertext contains ElGamal encryptions of both M_0^* and M_1^*, in an order determined by $d \oplus b$, where d denotes which ElGamal secret key sk_d' the experiment uses to decrypt for this instance. Now since inconsistent ciphertexts are rejected, the game's answer to $\mathcal{A}$'s decryption queries does not depend on the any of the bits d. Moreover, unless (any) d is known, also b is hidden. Hence, $\mathcal{A}$'s view is now completely independent of b, and thus $\mathcal{A}$'s IND-mCCA success is zero.

5   Details on the Exact Groth-Sahai Equations in Our Schemes

5.1   The Exact Groth-Sahai Equations for the Proofs in Signatures

We now give details on the proofs π_1 and π_2 in signatures from SIG. Recall that π_1 and π_2 shall prove the respective statements

$$\left(\underbrace{Z_0 = Z_1}_{S1} \quad \vee \quad \underbrace{f(M) \in \mathrm{QR}_p \cup \{0\}}_{S2}\right) \quad \text{and} \quad \underbrace{Z_0 = Z}_{S3}. \tag{28}$$

The Statements $S1$-$S3$. We now discuss the three individual statements $S1$-$S3$ from (28) in more detail. To this end, let us write the ElGamal ciphertexts C_0, C_1 from a signature as

$$C_0 = (A, B_0) = (g^R, pk_0^R \cdot g^{Z_0}) \qquad C_1 = (A, B_1) = (g^R, pk_1^R \cdot g^{Z_1}).$$

(Of course, the reused value $A = g^R$ will only appear once in a signature.)

$S1$. The statement $Z_0 = Z_1$ holds if and only if $(g, pk_1/pk_0, A, B_1/B_0)$ is a Diffie-Hellman tuple. Thus, $S1$ is equivalent to the equations $A = g^R$ and $B_1/B_0 = (pk_1/pk_0)^R$, with witness R.

$S2$. The statement $f(M) \in \mathrm{QR}_p \cup \{0\}$ is equivalent to the existence of an exponent $W \in \mathbb{Z}_p$ with $f(M) = W^2 \bmod p$. (Recall that a commitment to $f(M)$ can be homomorphically computed from M and the commitments C_α, C_β.) Hence, a witness to $S2$ is given by (α, β, W).

$S3$. We can express $Z_0 = Z$ as an equation $B_0 = pk_0^R \cdot g^Z$ with witness (R, Z).

All involved commitment random coins are additionally required to construct a valid proof. Besides, so far we have neglected that in a setting with an asymmetric pairing, not all combinations of, e.g., $\mathbb{Z}_p$-products can be directly expressed. (For instance, a square W^2 needs to be rephrased as $W \cdot \widehat{W}$, with an additional proof that $W = \widehat{W}$.) Hence, in the rest of this section, we will decorate variables that correspond to a $\widehat{\mathbb{G}}$-commitment with a hat (e.g., $\widehat{W}$).

The Equations for π_1. Equations for the disjunction $S1 \vee S2$ can be derived using standard techniques. However, if we optimize a little, we obtain the following equations for $S1 \vee S2$:

$$A^{\widehat{U}} = g^{\widehat{V}} \quad (B_1/B_0)^{\widehat{U}} = (pk_1/pk_0)^{\widehat{V}} \quad \widehat{f(M)} = W \cdot \widehat{W} \quad W = \widehat{W} + \widehat{U}.$$

(For instance, if we want to prove $S2$, we can set $\widehat{U} = \widehat{V} = 0$ and $W = \widehat{W}$ such that $f(M) = W^2$.) The involved variables from the verification key are $\widehat{\alpha}$ and $\widehat{\beta}$ (used to homomorphically construct $\widehat{f(M)}$). The variables whose commitments are placed in the signature are $\widehat{U}, \widehat{V}, W, \widehat{W}$. All of these variables are committed to using CRS_1.

The Equations for π_2. Similarly, we obtain the following equations for $S3$:

$$A = g^{\widehat{S}} \qquad\qquad B_0 = pk_0^{\widehat{S}} \cdot g^Z.$$

The variables are Z (committed to in vk) and $\widehat{S}$ (from σ), both committed to using CRS_2.

Remarks and Efficiency Summary. We emphasize that hence, the proofs π_1 and π_2 are independent (and in particular do not share commitments). Furthermore, thanks to the composability of Groth-Sahai proofs, the commitments

C_α, C_β, C_Z to α, β, Z that are placed in the verification key can be directly (re-)used in proofs. Each commitment occupies 2 group elements. In total, the equations above comprise 4 linear equations over $\mathbb{G}$, and 2 quadratic equations over $\mathbb{Z}_p$. Thus, π_1 contains $4 \cdot 2 + 2 \cdot 1 + 2 \cdot 4 = 18$ group elements (12 of them from $\hat{\mathbb{G}}$), and π_2 contains $1 \cdot 2 + 2 \cdot 1 = 4$ group elements (2 of them from $\hat{\mathbb{G}}$).

5.2 The Exact Groth-Sahai Equations for the Proofs in Ciphertexts

We now detail the proof π in ciphertexts from PKE. Recall that π shall prove the statement

$$\underbrace{Z_0' = Z_1'}_{S1'} \vee \left(\left(\underbrace{Z_0 = Z_1}_{S2'} \vee \underbrace{f(\mathrm{H}(ovk)) \in \mathrm{QR}_p \cup \{0\}}_{S3'} \right) \wedge \left(\underbrace{Z_0 = Z}_{S4'} \vee \underbrace{Z = 0}_{S5'} \right) \right). \quad (29)$$

The variables in (29) refer to the messages encrypted in $\mathrm{PKE}_{\mathrm{eg}}$-ciphertexts from the public parameters and the PKE-ciphertext at hand. We make these $\mathrm{PKE}_{\mathrm{eg}}$-ciphertexts explicit as

$$C_0 = \mathrm{Enc}_{\mathrm{eg}}(pk_0, g^{Z_0}; R) = (A, B_0) \qquad C_0' = \mathrm{Enc}_{\mathrm{eg}}(pk_0', g^{Z_0'}; R') = (A', B_0')$$
$$C_1 = \mathrm{Enc}_{\mathrm{eg}}(pk_1, g^{Z_1}; R) = (A, B_1) \qquad C_1' = \mathrm{Enc}_{\mathrm{eg}}(pk_1', g^{Z_1'}; R') = (A', B_1')$$
$$C_Z = \mathrm{Enc}_{\mathrm{eg}}(pk_0, g^{Z}; R_Z) = (A_Z, B_Z).$$

Besides, a $\mathrm{PKE}_{\mathrm{eg}}$-ciphertext $C_f = \mathrm{Enc}_{\mathrm{eg}}(pk_0, g^{f(\mathrm{H}(ovk))}; R_f) = (A_f, B_f)$ that determines the variable $f(\mathrm{H}(ovk))$ can be homomorphically computed from the ciphertexts C_α, C_β, and $\mathrm{H}(ovk)$.

The Statements $S1'$-$S5'$. Let us take a closer look at the individual statements $S1'$-$S5'$:

$S1', S2'$. These statements can be formalized like statement $S1$ for SIG. For instance, $S1'$ holds if and only if $(g, pk_1'/pk_0', A', B_1'/B_0')$ is a Diffie-Hellman tuple; a suitable witness is R'.

$S4', S5'$. Similarly, $S4'$ holds precisely if $(g, pk_0, A/A_Z, B_0/B_Z)$ is a Diffie-Hellman tuple; a witness is $R - R_Z$. (Statement $S5'$ can be formalized analogously, with a witness R_Z.)

$S3'$. As with SIG, $S3'$ holds if and only if there is a $W \in \mathbb{Z}_p$ with $f(\mathrm{H}(ovk)) = W^2 \bmod p$. A suitable witness consists of W, and the encryption randomness R_f of C_f.

A Reformulation. The composed statement from (29) is equivalent to

$$\left(S1' \vee S2' \vee S3' \right) \quad \wedge \quad \left(S1' \vee S4' \vee S5' \right).$$

By the above, the first sub-statement $S1' \vee S2' \vee S3'$ is implied by the equations

$$A^{\widehat{U}} = g^{\widehat{V}} \qquad\qquad A'^{\widehat{U'}} = g^{\widehat{V'}} \qquad\qquad A_f^{\widehat{U_f}} = g^{\widehat{V_f}}$$
$$(B_1/B_0)^{\widehat{U}} = (pk_1/pk_0)^{\widehat{V}} \quad (B_1'/B_0')^{\widehat{U'}} = (pk_1'/pk_0')^{\widehat{V'}} \quad B_0^{\widehat{U_f}} = pk_0^{\widehat{V_f}} \cdot g^{\widehat{F}}$$
$$\widehat{F} = W \cdot \widehat{W} \qquad\qquad W = \widehat{W} \qquad\qquad 1 = \widehat{U} + \widehat{U'} + \widehat{U_f}$$
$$(30)$$

for new variables $\widehat{U}, \widehat{V}, \widehat{U'}, \widehat{V'}, \widehat{U_f}, \widehat{V_f}, \widehat{F}, W, \widehat{W}$. (We adopt the notation from Sect. 5.1 to decorate variables in $\hat{\mathbb{G}}$ with a hat.) Roughly, the last equation guarantees that one of $\widehat{U}, \widehat{U'}, \widehat{U_f}$ is nonzero, and in fact that $\widehat{U_f} = 1$ once $\widehat{U} = \widehat{U'} = 0$. Furthermore, we have $\widehat{U'} \neq 0 \Rightarrow S1'$, and $\widehat{U} \neq 0 \Rightarrow S2'$, and $\widehat{U_f} \neq 0 \Rightarrow S3'$. Finally, a witness for (30) can be produced from *either* a witness for $S1'$, or for $S2'$, or for $S3'$. (For instance, we can set $\widehat{U'} = \widehat{V'} = 0$ whenever a witness for $S1'$ is not available.)

Similarly, sub-statement $S1' \vee S4' \vee S5'$ yields additional equations

$$(A/A_Z)^{\widehat{U_0}} = g^{\widehat{V_0}} \qquad A_Z^{\widehat{U_Z}} = g^{\widehat{V_Z}} \qquad \widehat{U'} + \widehat{U_0} + \widehat{U_Z} = 1$$
$$(B_0/B_Z)^{\widehat{U_0}} = pk_0^{\widehat{V_0}} \qquad B_Z^{\widehat{U_Z}} = pk_0^{\widehat{V_Z}}$$

for new variables $\widehat{U_0}, \widehat{V_0}, \widehat{U_Z}, \widehat{V_Z}$.

Summary. Summing up, π contains commitments to 13 variables (12 of them from $\hat{\mathbb{G}}$), and proves 10 $\mathbb{G}$-linear, 2 $\mathbb{Z}_p$-linear, and 3 quadratic equations over $\mathbb{Z}_p$. This yields a proof of $13 \cdot 2 + 10 \cdot 1 + 3 \cdot 4 = 48$ group elements (30 of them from $\hat{\mathbb{G}}$) and $2 \cdot 1 = 2$ exponents from $\mathbb{Z}_p$.

Acknowledgments. The author would like to thank Eike Kiltz, Julia Hesse, Willi Geiselmann, and the anonymous reviewers for helpful feedback.

A   Illustration of proof strategy for Theorem 2

In this section, we give a brief overview over the steps used to prove Theorem 2.

#	CRS_2	Z	π_2	$Z_0 = Z_1$	forgery check	remark
1	binding	$X^{(\ell)}$	proof of $S3$	$X^{(\ell)}$	—	EUF-mCMA
2	binding	$X^{(\ell)}$	proof of $S3$	$X^{(\ell)}$	$\boxed{\mathrm{Dec}_{eg}(sk_0^*, C_0) = X^*}$	GS soundness
3	$\boxed{\text{hiding}}$	$X^{(\ell)}$	proof of $S3$	$X^{(\ell)}$	$\mathrm{Dec}_{eg}(sk_0^*, C_0) = X^*$	GS CRS indist.
4	hiding	$\boxed{1}$	$\boxed{\text{Sim-output}}$	$X^{(\ell)}$	$\mathrm{Dec}_{eg}(sk_0^*, C_0) = X^*$	GS simulation
5	hiding	1	Sim-output	$\boxed{\mathcal{R}^{(\ell)}(M)}$	$\mathrm{Dec}_{eg}(sk_0^*, C_0) = \boxed{\mathcal{R}^{(\ell)}(M^*)}$	see Fig. 2

Fig. 1. Outline of the main proof, see Theorem 2. $\boxed{\text{Boxes}}$ denote changes compared to the previous game. The first column denotes the game number, CRS_2 denotes the setup of the Groth-Sahai common reference string CRS_2, and Z denotes the value committed to in C_Z in verification keys. Column π_2 describes how proofs are prepared in signatures. Z_0, Z are the messages encrypted in C_0, C_1 in signatures generated for $\mathcal{A}$. **forgery check** describes an additional check required for a forgery to pass as valid (beyond being valid in the sense of Ver). The core of the proof is the transition from Game 4 to Game 5 (with the previous transitions preparing the ground), see also Fig. 2.

#	CRS_1	f	π_1	if $\left(\frac{f^*(M)}{p}\right)=1$		if $\left(\frac{f^*(M)}{p}\right)=-1$ $Z_0 = Z_1$	π_1	forgery check	abort condition	remark
				Z_0	Z_1					
4.i.0	hiding	0	$S1$	$\mathcal{H}_i^{(\ell)}(M)$	$\mathcal{H}_i^{(\ell)}(M)$	$\mathcal{H}_i^{(\ell)}(M)$	$S1$	$\mathrm{Dec}_{eg}(sk_0^*, C_0^*) = \mathcal{H}_i^{(\ell)}(M^*)$	—	same as 4.i
4.i.1	hiding	0	$S1$	$\mathcal{H}_i^{(\ell)}(M)$	$\mathcal{H}_i^{(\ell)}(M)$	$\mathcal{H}_i^{(\ell)}(M)$	$S1$	$\mathrm{Dec}_{eg}(sk_0^*, C_0^*) = \mathcal{H}_i^{(\ell)}(M^*)$	$\boxed{f^*(M^*) \in \mathrm{QR}_p \cup \{0\}}$	loses factor ≈ 2
4.i.2	hiding	$\boxed{f^*}$	$S1$	$\mathcal{H}_i^{(\ell)}(M)$	$\mathcal{H}_i^{(\ell)}(M)$	$\mathcal{H}_i^{(\ell)}(M)$	$S1$	$\mathrm{Dec}_{eg}(sk_0^*, C_0^*) = \mathcal{H}_i^{(\ell)}(M^*)$	$f^*(M^*) \in \mathrm{QR}_p \cup \{0\}$	GS witness-ind.
4.i.3	hiding	f^*	$\boxed{S2}$	$\boxed{\mathcal{Q}_i^{(\ell)}(M)}$	$\mathcal{H}_i^{(\ell)}(M)$	$\mathcal{H}_i^{(\ell)}(M)$	$S1$	$\mathrm{Dec}_{eg}(sk_0^*, C_0^*) = \mathcal{H}_i^{(\ell)}(M^*)$	$f^*(M^*) \in \mathrm{QR}_p \cup \{0\}$	ElGamal
4.i.4	binding	f^*	$S2$	$\mathcal{Q}_i^{(\ell)}(M)$	$\mathcal{H}_i^{(\ell)}(M)$	$\mathcal{H}_i^{(\ell)}(M)$	$S1$	$\mathrm{Dec}_{eg}(sk_0^*, C_0^*) = \mathcal{H}_i^{(\ell)}(M^*)$	$f^*(M^*) \in \mathrm{QR}_p \cup \{0\}$	GS CRS indist.
4.i.5	binding	f^*	$S2$	$\mathcal{Q}_i^{(\ell)}(M)$	$\mathcal{H}_i^{(\ell)}(M)$	$\mathcal{H}_i^{(\ell)}(M)$	$S1$	$\mathrm{Dec}_{eg}(\boxed{sk_1^*, C_1^*}) = \mathcal{H}_i^{(\ell)}(M^*)$	$f^*(M^*) \in \mathrm{QR}_p \cup \{0\}$	GS soundness
4.i.6	binding	f^*	$S2$	$\mathcal{Q}_i^{(\ell)}(M)$	$\mathcal{H}_i^{(\ell)}(M)$	$\mathcal{H}_i^{(\ell)}(M)$	$S1$	$\mathrm{Dec}_{eg}(sk_1^*, C_1^*) = \mathcal{H}_i^{(\ell)}(M^*)$	$f^*(M^*) \in \mathrm{QR}_p \cup \{0\}$	ElGamal
4.i.7	binding	f^*	$S2$	$\mathcal{Q}_i^{(\ell)}(M)$	$\boxed{\mathcal{Q}_i^{(\ell)}(M)}$	$\mathcal{H}_i^{(\ell)}(M)$	$S1$	$\mathrm{Dec}_{eg}(\boxed{sk_0^*, C_0^*}) = \mathcal{H}_i^{(\ell)}(M^*)$	$f^*(M^*) \in \mathrm{QR}_p \cup \{0\}$	GS soundness
4.i.8	$\boxed{\text{hiding}}$	f^*	$S2$	$\mathcal{Q}_i^{(\ell)}(M)$	$\mathcal{Q}_i^{(\ell)}(M)$	$\mathcal{H}_i^{(\ell)}(M)$	$S1$	$\mathrm{Dec}_{eg}(sk_0^*, C_0^*) = \mathcal{H}_i^{(\ell)}(M^*)$	$f^*(M^*) \in \mathrm{QR}_p \cup \{0\}$	GS CRS indist.
4.i.9	hiding	$\boxed{0}$	$\boxed{S1}$	$\mathcal{Q}_i^{(\ell)}(M)$	$\mathcal{Q}_i^{(\ell)}(M)$	$\mathcal{H}_i^{(\ell)}(M)$	$S1$	$\mathrm{Dec}_{eg}(sk_0^*, C_0^*) = \mathcal{H}_i^{(\ell)}(M^*)$	$f^*(M^*) \in \mathrm{QR}_p \cup \{0\}$	GS witness-ind.
4.i.10	hiding	0	$S1$	$\mathcal{Q}_i^{(\ell)}(M)$	$\mathcal{Q}_i^{(\ell)}(M)$	$\mathcal{H}_i^{(\ell)}(M)$	$S1$	$\mathrm{Dec}_{eg}(sk_0^*, C_0^*) = \mathcal{H}_i^{(\ell)}(M^*)$	$\boxed{\text{—}}$	gains factor ≈ 2 same as 4.$(i+1)$

Fig. 2. Transitions between two hybrids Game 4.i and Game 4.$(i+1)$ that in turn interpolate between Game 4 and Game 5 of the main proof. Again, boxes denote changes compared to the previous game. The notation follows Fig. 1: $\#$ denotes the game number, and CRS_1 and f denote the setup of these values in the public parameters. The column π_1 describes which sub-statement (i.e., $S1$ or $S2$) the proof π_1 actually proves, and the columns Z_0, Z_1 describe how the game prepares signatures for $\mathcal{A}$. In this, we distinguish the cases where the Legendre symbol $\left(\frac{f^*(M)}{p}\right)$ of the message to be signed is 1 and -1, respectively. (We neglect the unlikely case $\left(\frac{f^*(M)}{p}\right) = 0$ in this overview.) Also, as in Fig. 1, **forgery check** describes an additional check required for a forgery to pass as valid. Finally, the functions $\mathcal{H}_i^{(\ell)}$ are defined at the beginning of the proof of Lemma 1. (Intuitively, $\mathcal{H}_i^{(\ell)}$ is a random function that however does not depend on its full input M, but only on i values $\left(\frac{f_j(M)}{p}\right)$ for randomly chosen f_j.) We refer to Lemma 1 for a detailed proof and a justification for each game transition.

References

1. Abe, M., Fuchsbauer, G., Groth, J., Haralambiev, K., Ohkubo, M.: Structure-preserving signatures and commitments to group elements. In: Rabin, T. (ed.) CRYPTO 2010. LNCS, vol. 6223, pp. 209–236. Springer, Heidelberg (2010)
2. Abe, M., David, B., Kohlweiss, M., Nishimaki, R., Ohkubo, M.: Tagged one-time signatures: tight security and optimal tag size. In: Kurosawa, K., Hanaoka, G. (eds.) PKC 2013. LNCS, vol. 7778, pp. 312–331. Springer, Heidelberg (2013)
3. Bellare, M., Boldyreva, A., Micali, S.: Public-key encryption in a multi-user setting: security proofs and improvements. In: Preneel, B. (ed.) EUROCRYPT 2000. LNCS, vol. 1807, pp. 259–274. Springer, Heidelberg (2000)
4. Bellare, M., Goldwasser, S.: New paradigms for digital signatures and message authentication based on non-interactive zero knowledge proofs. In: Brassard, G. (ed.) CRYPTO 1989. LNCS, vol. 435, pp. 194–211. Springer, Heidelberg (1990)
5. Bernstein, D.J.: Proving tight security for Rabin-Williams signatures. In: Smart, N.P. (ed.) EUROCRYPT 2008. LNCS, vol. 4965, pp. 70–87. Springer, Heidelberg (2008)
6. Blazy, O., Kiltz, E., Pan, J.: (Hierarchical) Identity-based encryption from affine message authentication. In: Garay, J.A., Gennaro, R. (eds.) CRYPTO 2014, Part I. LNCS, vol. 8616, pp. 408–425. Springer, Heidelberg (2014)
7. Boldyreva, A.: Strengthening Security of RSA-OAEP. In: Fischlin, M. (ed.) CT-RSA 2009. LNCS, vol. 5473, pp. 399–413. Springer, Heidelberg (2009)
8. Boneh, D., Boyen, X.: Efficient selective-ID secure identity-based encryption without random oracles. In: Cachin, C., Camenisch, J.L. (eds.) EUROCRYPT 2004. LNCS, vol. 3027, pp. 223–238. Springer, Heidelberg (2004)
9. Boneh, D., Boyen, X.: Secure identity based encryption without random oracles. In: Franklin, M. (ed.) CRYPTO 2004. LNCS, vol. 3152, pp. 443–459. Springer, Heidelberg (2004)
10. Boneh, D., Franklin, M.: Identity-based encryption from the Weil pairing. In: Kilian, J. (ed.) CRYPTO 2001. LNCS, vol. 2139, p. 213. Springer, Heidelberg (2001)
11. Boneh, D., Franklin, M.K.: Identity-based encryption from the Weil pairing. SIAM J. Comput. 32(3), 586–615 (2003)
12. Boneh, D., Mironov, I., Shoup, V.: A secure signature scheme from Bilinear maps. In: Joye, M. (ed.) CT-RSA 2003. LNCS, vol. 2612, pp. 98–110. Springer, Heidelberg (2003)
13. Cash, D.M., Kiltz, E., Shoup, V.: The twin Diffie-Hellman problem and applications. In: Smart, N.P. (ed.) EUROCRYPT 2008. LNCS, vol. 4965, pp. 127–145. Springer, Heidelberg (2008)
14. Cash, D., Hofheinz, D., Kiltz, E., Peikert, C.: Bonsai trees, or how to delegate a lattice basis. In: Gilbert, H. (ed.) EUROCRYPT 2010. LNCS, vol. 6110, pp. 523–552. Springer, Heidelberg (2010)
15. Chen, J., Wee, H.: Fully, (Almost) tightly secure IBE and dual system groups. In: Canetti, R., Garay, J.A. (eds.) CRYPTO 2013, Part II. LNCS, vol. 8043, pp. 435–460. Springer, Heidelberg (2013)
16. Chevallier-Mames, B., Joye, M.: A practical and tightly secure signature scheme without hash function. In: Abe, M. (ed.) CT-RSA 2007. LNCS, vol. 4377, pp. 339–356. Springer, Heidelberg (2006)
17. Coron, J.-S.: On the exact security of full domain hash. In: Bellare, M. (ed.) CRYPTO 2000. LNCS, vol. 1880, pp. 229–235. Springer, Heidelberg (2000)

18. Escala, A., Groth, J.: Fine-tuning groth-sahai proofs. In: Krawczyk, H. (ed.) PKC 2014. LNCS, vol. 8383, pp. 630–649. Springer, Heidelberg (2014)
19. Fuchsbauer, G.: Commuting signatures and verifiable encryption. In: Paterson, K.G. (ed.) EUROCRYPT 2011. LNCS, vol. 6632, pp. 224–245. Springer, Heidelberg (2011)
20. Galindo, D., Martín, S., Morillo, P., Villar, J.L.: Easy verifiable primitives and practical public key cryptosystems. In: Boyd, C., Mao, W. (eds.) ISC 2003. LNCS, vol. 2851, pp. 69–83. Springer, Heidelberg (2003)
21. Gennaro, R., Halevi, S., Rabin, T.: Secure hash-and-sign signatures without the random oracle. In: Stern, J. (ed.) EUROCRYPT 1999. LNCS, vol. 1592, pp. 123–139. Springer, Heidelberg (1999)
22. Gentry, C.: Practical identity-based encryption without random oracles. In: Vaudenay, S. (ed.) EUROCRYPT 2006. LNCS, vol. 4004, pp. 445–464. Springer, Heidelberg (2006)
23. Gentry, C., Halevi, S.: Hierarchical identity based encryption with polynomially many levels. In: Reingold, O. (ed.) TCC 2009. LNCS, vol. 5444, pp. 437–456. Springer, Heidelberg (2009)
24. Goh, E.-J., Jarecki, S., Katz, J., Wang, N.: Efficient signature schemes with tight reductions to the Diffie-Hellman problems. J. Cryptology 20(4), 493–514 (2007)
25. Goldreich, O., Goldwasser, S., Micali, S.: On the cryptographic applications of random functions. In: Blakely, G.R., Chaum, D. (eds.) CRYPTO 1984. LNCS, vol. 196, pp. 276–288. Springer, Heidelberg (1985)
26. Groth, J., Sahai, A.: Efficient noninteractive proof systems for bilinear groups. SIAM J. Comput. 41(5), 1193–1232 (2012)
27. Hofheinz, D.: All-but-many lossy trapdoor functions. In: Pointcheval, D., Johansson, T. (eds.) EUROCRYPT 2012. LNCS, vol. 7237, pp. 209–227. Springer, Heidelberg (2012)
28. Hofheinz, D., Jager, T.: Tightly secure signatures and public-key encryption. In: Safavi-Naini, R., Canetti, R. (eds.) CRYPTO 2012. LNCS, vol. 7417, pp. 590–607. Springer, Heidelberg (2012)
29. Hohenberger, S., Waters, B.: Short and stateless signatures from the RSA assumption. In: Halevi, S. (ed.) CRYPTO 2009. LNCS, vol. 5677, pp. 654–670. Springer, Heidelberg (2009)
30. Kakvi, S.A., Kiltz, E.: Optimal security proofs for full domain hash, revisited. In: Pointcheval, D., Johansson, T. (eds.) EUROCRYPT 2012. LNCS, vol. 7237, pp. 537–553. Springer, Heidelberg (2012)
31. Lewko, A., Waters, B.: New techniques for dual system encryption and fully secure HIBE with short ciphertexts. In: Micciancio, D. (ed.) TCC 2010. LNCS, vol. 5978, pp. 455–479. Springer, Heidelberg (2010)
32. Libert, B., Joye, M., Yung, M., Peters, T.: Concise multi-challenge CCA-secure encryption and signatures with almost tight security. In: Sarkar, P., Iwata, T. (eds.) ASIACRYPT 2014, Part II. LNCS, vol. 8874, pp. 1–21. Springer, Heidelberg (2014)
33. Naor, M., Reingold, O.: Number-theoretic constructions of efficient pseudo random functions. In: Proceedings of the FOCS 1997, pp. 458–467. IEEE Computer Society (1997)
34. Naor, M., Yung, M.: Public-key cryptosystems provably secure against chosen ciphertext attacks. In: Proceedings of the STOC 1990, pp. 427–437. ACM (1990)
35. Naor, M., Yung, M.: Universal one-way hash functions and their cryptographic applications. In: Proceedings of the STOC 1989, pp. 33–43. ACM (1989)

36. Schäge, S.: Tight proofs for signature schemes without random oracles. In: Paterson, K.G. (ed.) EUROCRYPT 2011. LNCS, vol. 6632, pp. 189–206. Springer, Heidelberg (2011)

37. Waters, B.: Dual system encryption: realizing fully secure IBE and HIBE under simple assumptions. In: Halevi, S. (ed.) CRYPTO 2009. LNCS, vol. 5677, pp. 619–636. Springer, Heidelberg (2009)

38. Waters, B.: Efficient identity-based encryption without random oracles. In: Cramer, R. (ed.) EUROCRYPT 2005. LNCS, vol. 3494, pp. 114–127. Springer, Heidelberg (2005)

Standard Security Does Imply Security Against Selective Opening for Markov Distributions

Georg Fuchsbauer[1][(✉)], Felix Heuer[2], Eike Kiltz[2], and Krzysztof Pietrzak[1]

[1] Institute of Science and Technology Austria, Klosterneuburg, Austria
{gfuchsbauer,pietrzak}@ist.ac.at
[2] Horst Görtz Institute for IT-Security, Ruhr-University Bochum,
Bochum, Germany
{felix.heuer,eike.kiltz}@rub.de

Abstract. About three decades ago it was realized that implementing private channels between parties which can be adaptively corrupted requires an encryption scheme that is secure against *selective opening attacks*. Whether standard (IND-CPA) security implies security against selective opening attacks has been a major open question since. The only known reduction from selective opening to IND-CPA security loses an exponential factor. A polynomial reduction is only known for the very special case where the distribution considered in the selective opening security experiment is a product distribution, i.e., the messages are sampled independently from each other.

In this paper we give a reduction whose loss is quantified via the dependence graph (where message dependencies correspond to edges) of the underlying message distribution. In particular, for some concrete distributions including Markov distributions, our reduction is polynomial.

Keywords: Public-key encryption · Selective opening security · Markov · IND-CPA · IND-SO-CPA

1 Introduction

Security Under Selective Opening Attacks. Consider a scenario where many parties $1, \ldots, n$ send messages to one common receiver. To transmit a message $\mathbf{m}_i$, party i samples fresh randomness $\mathbf{r}_i$ and sends the ciphertext $\mathbf{c}_i = \mathsf{Enc}_{pk}(\mathbf{m}_i; \mathbf{r}_i)$ to the receiver. Consider an adversary $\mathcal{A}$ that does not only eavesdrop on the sent ciphertexts $(\mathbf{c}_1, \ldots, \mathbf{c}_n)$, but corrupts a set $\mathcal{I} \subseteq [n]$ of the sender's systems, thus learning the encrypted message $\mathbf{m}_i$ *and* the randomness $\mathbf{r}_i$ used to encrypt $\mathbf{m}_i$. The natural question to ask is whether the messages of uncorrupted parties remain confidential. Such attacks are referred to as *selective opening* (SO) *attacks* (under sender corruption).

Selective opening attacks naturally occur in multi-party computation where we assume secure channels between parties. Since a party might become corrupted, we would need the encryption on the channels to be selective opening

© International Association for Cryptologic Research 2016
E. Kushilevitz and T. Malkin (Eds.): TCC 2016-A, Part I, LNCS 9562, pp. 282–305, 2016.
DOI: 10.1007/978-3-662-49096-9_12

secure. In practice the same argument applies to a server that establishes secure connections that shall remain secure if users are corrupted.

Difficulty of Proving Security Under Selective Opening Attacks. The widely accepted standard notion for public-key encryption schemes is indistinguishability under chosen-plaintext attacks (IND-CPA security). At first sight one might consider a straight-forward hybrid argument to show that IND-CPA security already implies security against selective opening attacks since every party samples fresh randomness independently. However, so far nobody has been able to bring forward such a hybrid argument in general. Notice that revealing randomness r_i allows a selective opening adversary to verify that a corrupted ciphertext c_i is an encryption of m_i. The adversary's possibility to corrupt parties introduces a difficulty in proving that standard (IND-CPA) security already implies selective opening security. It seems that the reduction has to know (i.e. guess) the complete set $\mathcal{I}$ of all corruptions going to be made by $\mathcal{A}$ in order to serve its security game *before* $\mathcal{A}$ actually announces the senders it wishes to corrupt. Since $\mathcal{I}$ might be any subset of $\{1, \ldots, n\}$, a direct approach would lead to an exponential loss in the reduction. A main technical obstacle is that the encrypted messages may depend on each other. If, for example, they are encrypted and sent sequentially, message m_i may depend on m_{i-1} and all previous messages. Thus, corrupting some parties might already leak some information on messages sent by parties that have not been corrupted.

Until today, the only result in the standard model, given in [3,8], shows that IND-CPA implies selective opening security for the special case of a product distribution, i.e., when all messages $m_1, \ldots, m_n$ are sampled independently from each other. Intuitively, this holds since corrupting some ciphertext cannot reveal information on related messages if there are no related messages at all and the hybrid argument one might expect to work goes through. This leaves the following open question:

Does standard security imply selective opening security for any non-trivial message distribution?

1.1 Our Contributions

We present the first non-trivial positive results in the standard model, namely we show that IND-CPA security implies IND-SO-CPA security for a class of message distributions with few dependencies. Here IND-SO-CPA security refers to the indistinguishability-based definition of selective opening security sometimes referred to as *weak* IND-SO-CPA *security* [4].

IND-SO-CPA requires that a passive adversary that obtains a vector of ciphertexts $(c_1, \ldots, c_n)$ and has access to a ciphertext *opening* oracle, revealing the underlying message m_i of some ciphertext c_i and the randomness used to encrypt m_i, cannot distinguish the originally encrypted messages from freshly resampled messages that are *as likely as the original messages* given the messages of opened ciphertexts.

We consider *graph-induced* distributions where dependencies among messages correspond to edges in a graph and show that IND-CPA implies IND-SO-CPA security for all graph-induced distributions that satisfy a certain *low connectivity* property.

In particular, our result holds for the class of Markov distributions, i.e. distributions on message vectors $(\mathbf{m}_1, \ldots, \mathbf{m}_n)$ where all information relevant for the distribution of $\mathbf{m}_i$ is present in $\mathbf{m}_{i-1}$. We prove that any IND-CPA secure public-key encryption scheme is IND-SO-CPA secure if the messages are sampled from a Markov distribution. Our results cover for instance distributions where message $\mathbf{m}_i$ contains all previous messages (e.g. email conversations) or distributions where messages are increasing, i.e., $\mathbf{m}_1 \leq \mathbf{m}_2 \leq \ldots \leq \mathbf{m}_n$.

Note that a positive result on "weak" IND-SO-CPA security for all IND-CPA-secure encryption schemes for certain distributions is the best we can hope for due to the negative result of Bellare et al. [1] ruling out such an implication for SIM-SO-CPA security.

Details. Think of a vector of n messages sampled from some distribution $\mathfrak{D}$ as a graph G on n vertices $\{1, \ldots, n\}$ where we have an edge from message $\mathbf{m}_i$ to message $\mathbf{m}_j$ if the distribution of $\mathbf{m}_j$ depends on $\mathbf{m}_i$. Further, fix any subset $\mathcal{I} \subseteq \{1, \ldots, n\}$ of opening queries made by some adversary. The main observation is that removing $\mathcal{I}$ and all incident edges, G decomposes into connected components $C_1, \ldots, C_{n'}$ that can be resampled independently, since the distribution of messages on C_k solely depends on the messages in the neighborhood of C_k and $\mathfrak{D}$.

To argue that there is no efficient adversary $\mathcal{A}_{\mathsf{SO}}$ that distinguishes sampled and resampled messages in the selective opening experiment, we proceed in a sequence of hybrid games, starting in a game where after receiving encryptions of sampling messages and replies to opening queries, $\mathcal{A}_{\mathsf{SO}}$ obtains the sampled messages. In each hybrid step we use IND-CPA security to replace *sampled* messages on a connected component C_k with *resampled* messages without $\mathcal{A}_{\mathsf{SO}}$ noticing. To this end, the reduction from IND-CPA to the indistinguishability of two consecutive hybrids has to identify C_k to embed its own challenge before $\mathcal{A}_{\mathsf{SO}}$ makes any opening query.

We consider two approaches for guessing C_k. The first will consider graphs that have only polynomially many connected subgraphs; hence, the reduction can guess C_k right away. The second approach studies graphs for which every connected subgraph has a neighborhood of constant size; this allows the reduction to guess C_k by guessing its neighborhood. We show that the first approach ensures a reduction with polynomial loss for a strictly greater class of graphs than the second one.

Additionally, when the distribution is induced by an acyclic graph, we give a more sophisticated hybrid argument for the second approach, where in each hybrid transition only a single sampled message is replaced by a resampled message, allowing for a tighter reduction. Due to the definition of the hybrids, it will suffice to guess on fewer vertices of C_k's neighborhood.

1.2   Previous Work

There are three not polynomially equivalent definitions of SO-secure encryption [4]. Since messages in the IND-SO experiment have to be resampled conditioned on opened messages, there are two notions based on indistinguishability: *Weak* IND-SO restricts to distributions that support *efficient conditional* resampling. Bellare et al. [2] gave an indistinguishability-based notion for passive adversaries, usually referred to as IND-SO-CPA. *Full* IND-SO allows for arbitrary distributions on the messages and is due to Böhl et al. [4], who adopted a notion for commitment schemes from [2] to encryption.

SIM-SO captures semantic security and demands that everything an adversary can output can be computed by a simulator that only sees the messages of corrupted parties, whereas it does not see the public key, any ciphertext or any randomness. The notion dates back to Dwork et al. [8], who studied the *selective decommitment* problem, and does not suffer from a distribution restriction like *weak* IND-SO, since it does not involve resampling.

The first IND-SO-CPA-secure encryption scheme in the standard model was given in [2] based on lossy encryption. Selective opening secure encryption can be constructed from *deniable encryption* [6] as well as *non-committing encryption* [7]. Bellare et al. [1,3] separated SIM-SO-CPA from IND-CPA security and showed that IND-CPA security implies *weak* IND-SO-CPA security if the messages are (basically) sampled independently. The same result was already established for commitment schemes in [8].

To date, this is the only positive result that shows that IND-CPA implies *weak* IND-SO-CPA in the standard model. *Full* IND-SO-CPA and SIM-SO-CPA security were separated in [4]; neither of them implies the other. Hofheinz et al. [10] proved that IND-CPA implies *weak* IND-SO-CPA in the generic group model for a certain class of encryption schemes and separated IND-CCA from *weak* IND-SO-CCA security.

Recently, Hofheinz et al. [9] constructed the first (even IND-CCA-secure) PKE that is not *weakly* IND-SO-CPA secure. Their result relies on the existence of *public-coin differing-inputs obfuscation* and certain *correlation intractable hash functions*. Their scheme employs "secret-sharing message distributions" whose messages are evaluations of some polynomial. It is easily seen that such distributions have too many dependencies to be covered by our positive result. There is a gap between their result and ours, that is, distributions for which it is still open whether IND-CPA implies IND-SO-CPA.

2   Preliminaries

We denote by λ the security parameter. A function f is polynomial in n, $f(n) = \mathsf{poly}(n)$, if $f(n) = \mathcal{O}(n^c)$ for some $c > 0$. Let $0 < n := n(\lambda) = \mathsf{poly}(\lambda)$. A function $f(n)$ is negligible in n, $f(n) = \mathsf{negl}(n)$, if $f(n) = \mathcal{O}(n^{-c})$ for all $c > 0$. Any algorithm receives the unary representation 1^λ of the security parameter as first input. We say that an algorithm is a PPT algorithm if it runs in probabilistic polynomial time (in λ). For a finite set $\mathcal{S}$ we denote the sampling of a uniform random element a by $a \leftarrow_\$ \mathcal{S}$, and the sampling according to some distribution $\mathfrak{D}$ by $a \leftarrow \mathfrak{D}$. For $a, b \in \mathbb{N}$, $a \leq b$, let $[a, b] := \{a, a+1, \ldots, b\}$ and $[a] := [1, a]$. For $a < b$ let $[b, a] := \emptyset$. For $\mathcal{I} \subseteq [n]$ let $\bar{\mathcal{I}} := [n] \setminus \mathcal{I}$. We use boldface letters to denote vectors, which are of length n if not indicated otherwise. For a vector $\mathbf{m}$ and $i \in [n]$ let $\mathbf{m}_i$ denote the i-th entry of $\mathbf{m}$ and $|\mathbf{m}|$ the number of entries in $\mathbf{m}$. For a set $\mathcal{I} = \{i_1, \ldots, i_{|\mathcal{I}|}\}$, $i_1 < \ldots < i_{|\mathcal{I}|}$ let $\mathbf{m}_\mathcal{I}$ denote the projection of $\mathbf{m}$ to its $\mathcal{I}$-entries: $\mathbf{m}_\mathcal{I} := (\mathbf{m}_{i_1}, \ldots, \mathbf{m}_{i_{|\mathcal{I}|}})$. For an event E let $\bar{\mathsf{E}}$ denote the complementary event.

2.1   Games

A game G is a collection of procedures or oracles $\{\text{INITIALIZE}, \mathrm{P}_1, \mathrm{P}_2, \ldots, \mathrm{P}_t, \text{FINALIZE}\}$ for $t \geq 0$. Procedures P_1 to P_t and FINALIZE might require some input parameters. We implicitly assume that boolean flags are initialized to *false*, numerical types are initialized to 0, sets are initialized to $\emptyset$, while strings are initialized to the empty string ϵ. An adversary $\mathcal{A}$ is *run in game* G if $\mathcal{A}$ calls INITIALIZE. During the game $\mathcal{A}$ may run some procedure P_i as often as allowed by the game.

For each game in this paper, the "OPEN" procedure may be called an arbitrary number of times, while every other procedure is called once during the execution.

The interface of the game is provided by the *challenger*. If $\mathcal{A}$ calls P, the output of P is returned to $\mathcal{A}$, except for the FINALIZE procedure. On $\mathcal{A}$'s call of FINALIZE the game ends and outputs whatever FINALIZE returns. Let $\mathsf{G}^\mathcal{A} \Rightarrow \mathsf{out}$ denote the event that G runs $\mathcal{A}$ and outputs out. The *advantage* $\mathbf{Adv}(\mathsf{G}^\mathcal{A}, \mathsf{H}^\mathcal{A})$ of $\mathcal{A}$ in distinguishing games G and H is defined as $|\Pr[\mathsf{G}^\mathcal{A} \Rightarrow 1] - \Pr[\mathsf{H}^\mathcal{A} \Rightarrow 1]|$. We let Bad denote the event that a boolean flag Bad was set to *true* during the execution of some game.

2.2   Public-Key Encryption Schemes

A public-key encryption scheme consists of three PPT algorithms. Gen generates a key pair $(pk, sk) \leftarrow \mathsf{Gen}(1^\lambda)$ on input 1^λ. The public key pk implicitly contains 1^λ and defines three finite sets: the message space $\mathcal{M}$, the randomness space $\mathcal{R}$, and the ciphertext space $\mathcal{C}$. Given pk, a message $m \in \mathcal{M}$ and randomness $r \in \mathcal{R}$, Enc outputs an encryption $c = \mathsf{Enc}_{pk}(m; r) \in \mathcal{C}$ of m under pk. The decryption algorithm Dec takes a secret key sk and a ciphertext $c \in \mathcal{C}$ as input and outputs a message $m = \mathsf{Dec}_{sk}(c) \in \mathcal{M}$, or a special symbol $\perp \notin \mathcal{M}$ indicating that c is not a valid ciphertext. In the following we let $\mathsf{PKE} = (\mathsf{Gen}, \mathsf{Enc}, \mathsf{Dec})$ denote a public-key encryption scheme.

We require PKE to be correct: for all security parameters λ, for all $(pk, sk) \leftarrow \mathsf{Gen}(1^\lambda)$, and for all $m \in \mathcal{M}$ we have $\Pr[\mathsf{Dec}_{sk}(\mathsf{Enc}_{pk}(m; r)) = m] = 1$ where the probability is taken over the choice of r. We apply Enc and Dec to message vectors $\mathbf{m} = (\mathbf{m}_1, \ldots, \mathbf{m}_n)$ and randomness $\mathbf{r} = (\mathbf{r}_1, \ldots, \mathbf{r}_n)$ as $\mathsf{Enc}(\mathbf{m}; \mathbf{r}) := (\mathsf{Enc}(\mathbf{m}_1; \mathbf{r}_1), \ldots, \mathsf{Enc}(\mathbf{m}_n; \mathbf{r}_n))$.

Procedure INITIALIZE	**Procedure** CHALLENGE($\mathbf{m}^0, \mathbf{m}^1$)	**Procedure** FINALIZE(b')
$(pk, sk) \leftarrow \mathsf{Gen}(1^\lambda)$	$\mathbf{c} \leftarrow \mathsf{Enc}_{pk}(\mathbf{m}^b)$	Return b'
Return pk	Return $\mathbf{c}$	

Fig. 1. Game mult-IND-CPA$_{\mathsf{PKE},b}$; $\mathcal{B}_{\mathsf{mult}}$ must submit $\mathbf{m}^0, \mathbf{m}^1 \in \mathcal{M}^s$

2.3 IND-CPA and Mult-IND-CPA Security

We revise the standard notion of IND-CPA security and give a definition of indistinguishable ciphertext vectors under chosen-plaintext attacks that will allow for cleaner proofs of our results.

Definition 1 (mult-IND-CPA security). *For* PKE, *an adversary* $\mathcal{B}_{\mathsf{mult}}$, $s \in \mathbb{N}$ *and a bit* b *we consider game* mult-IND-CPA$_{\mathsf{PKE},b}^{\mathcal{B}_{\mathsf{mult}}}$ *as given in Fig. 1.* $\mathcal{B}_{\mathsf{mult}}$ *may only submit message vectors* $\mathbf{m}^0$, $\mathbf{m}^1 \in \mathcal{M}^s$. *To* PKE, $\mathcal{B}_{\mathsf{mult}}$ *and* λ *we associate the following advantage function*

$$\mathbf{Adv}_{\mathsf{PKE}}^{\mathsf{mult\text{-}IND\text{-}CPA}}(\mathcal{B}_{\mathsf{mult}}, \lambda) := \mathbf{Adv}\left(\mathsf{mult\text{-}IND\text{-}CPA}_{\mathsf{PKE},0}^{\mathcal{B}_{\mathsf{mult}}}, \mathsf{mult\text{-}IND\text{-}CPA}_{\mathsf{PKE},1}^{\mathcal{B}_{\mathsf{mult}}}\right).$$

PKE *is* mult-IND-CPA *secure if* $\mathbf{Adv}_{\mathsf{PKE}}^{\mathsf{mult\text{-}IND\text{-}CPA}}(\mathcal{B}_{\mathsf{mult}}, \lambda)$ *is negligible for all* PPT *adversaries* $\mathcal{B}_{\mathsf{mult}}$.

For an adversary $\mathcal{B}_{\mathsf{CPA}}$, we obtain the definition of IND-CPA security by letting $s := 1$ and write $\mathbf{Adv}_{\mathsf{PKE}}^{\mathsf{IND\text{-}CPA}}(\mathcal{B}_{\mathsf{CPA}}, \lambda)$ instead of $\mathbf{Adv}_{\mathsf{PKE}}^{\mathsf{mult\text{-}IND\text{-}CPA}}(\mathcal{B}_{\mathsf{CPA}}, \lambda)$. A standard hybrid argument proves the following lemma.

Lemma 2. *For any adversary* $\mathcal{B}_{\mathsf{mult}}$ *sending message vectors from* $\mathcal{M}^s$ *to the* mult-IND-CPA *game there exists an* IND-CPA *adversary* $\mathcal{B}_{\mathsf{CPA}}$ *with roughly the same running time as* $\mathcal{B}_{\mathsf{mult}}$ *such that*

$$\mathbf{Adv}_{\mathsf{PKE}}^{\mathsf{mult\text{-}IND\text{-}CPA}}(\mathcal{B}_{\mathsf{mult}}, \lambda) \leq s \cdot \mathbf{Adv}_{\mathsf{PKE}}^{\mathsf{IND\text{-}CPA}}(\mathcal{B}_{\mathsf{CPA}}, \lambda).$$

2.4 IND-SO-CPA Security

In this section we recall an indistinguishability-based definition for selective opening security under chosen-plaintext attacks and discuss the existing notions of SO security.

Definition 3 (Efficiently resamplable distribution). *Let* $\mathcal{M}$ *be a finite set. A family of distributions* $\{\mathfrak{D}_\lambda\}_{\lambda \in \mathbb{N}}$ *over* $\mathcal{M}^n = \mathcal{M}^{n(\lambda)}$ *is efficiently resamplable if the following properties hold for every* $\lambda \in \mathbb{N}$:

Length consistency. *For every* $i \in [n]$: $\Pr_{\mathbf{m}^1, \mathbf{m}^2 \leftarrow \mathfrak{D}_\lambda} \left[|\mathbf{m}_i^1| = |\mathbf{m}_i^2|\right] = 1$.
Resamplability. *There exists a* PPT *resampling algorithm* $\mathsf{Resamp}_{\mathfrak{D}_\lambda}(\cdot, \cdot)$ *that runs on* $(\mathbf{m}, \mathcal{I})$ *for* $\mathbf{m} \in \mathcal{M}^n$, $\mathcal{I} \subseteq [n]$ *and outputs a* $\mathfrak{D}_\lambda$-*distributed vector* $\mathbf{m}' \in \mathcal{M}^n$ *conditioned on* $\mathbf{m}'_\mathcal{I} = \mathbf{m}_\mathcal{I}$.

Procedure INITIALIZE	Procedure OPEN(i)
$(pk, sk) \leftarrow \mathsf{Gen}(1^\lambda)$	$\mathcal{I} := \mathcal{I} \cup \{i\}$
Return pk	Return $(\mathbf{m}_i^0, \mathbf{r}_i)$
Procedure $\mathsf{ENC}(\mathfrak{D}, \mathsf{Resamp}_{\mathfrak{D}})$	**Procedure** CHALLENGE
$\mathbf{m}^0 \leftarrow \mathfrak{D}$	$\mathbf{m}^1 \leftarrow \mathsf{Resamp}_{\mathfrak{D}}(\mathbf{m}^0, \mathcal{I})$
$\mathbf{r} \leftarrow_\$ \mathcal{R}^n$	Return $\mathbf{m}^b$
$\mathbf{c} = \mathsf{Enc}_{pk}(\mathbf{m}^0; \mathbf{r})$	**Procedure** FINALIZE(b')
Return $\mathbf{c}$	Return b'

Fig. 2. Game IND-SO-CPA$_{\mathsf{PKE}, b}$

A class of families of distributions $\mathcal{D}$ is efficiently resamplable if every family $\{\mathfrak{D}_\lambda\}_{\lambda \in \mathbb{N}} \in \mathcal{D}$ is efficiently resamplable.

Since the security parameter uniquely specifies an element of a family $\mathfrak{D}_\lambda$ we write $\mathfrak{D}$ instead of $\mathfrak{D}_\lambda$ whenever the security parameter is already fixed.

Definition 4. *For* PKE, *a bit b, an adversary $\mathcal{A}_{\mathsf{SO}}$ and a class of families of distributions $\mathcal{D}$ over $\mathcal{M}^n$ we consider game* IND-SO-CPA$_{\mathsf{PKE}, b}^{\mathcal{A}_{\mathsf{SO}}}$ *in Fig. 2. Run in the game, $\mathcal{A}_{\mathsf{SO}}$ calls* ENC *once right after* INITIALIZE *and has to submit $\mathfrak{D} \in \mathcal{D}$ along with a* PPT *resampling algorithm* $\mathsf{Resamp}_{\mathfrak{D}}$. *$\mathcal{A}_{\mathsf{SO}}$ may call* OPEN *multiple times and invokes* CHALLENGE *once after its last* OPEN *query before calling* FINALIZE. *We define the advantage of $\mathcal{A}_{\mathsf{SO}}$ run in game* IND-SO-CPA$_{\mathsf{PKE}, b}$ *as*

$$\mathbf{Adv}_{\mathsf{PKE}}^{\mathsf{IND\text{-}SO\text{-}CPA}}(\mathcal{A}_{\mathsf{SO}}, \mathfrak{D}_\lambda, \lambda) := \mathbf{Adv}\left(\mathsf{IND\text{-}SO\text{-}CPA}_{\mathsf{PKE}, 0}^{\mathcal{A}_{\mathsf{SO}}}, \mathsf{IND\text{-}SO\text{-}CPA}_{\mathsf{PKE}, 1}^{\mathcal{A}_{\mathsf{SO}}}\right).$$

PKE *is* IND-SO-CPA *secure w.r.t. $\mathcal{D}$ if* $\mathbf{Adv}_{\mathsf{PKE}}^{\mathsf{IND\text{-}SO\text{-}CPA}}(\mathcal{A}_{\mathsf{SO}}, \mathfrak{D}_\lambda, \lambda)$ *is negligible for all* PPT *$\mathcal{A}_{\mathsf{SO}}$.*

Notions of Selective Opening Security. Definition 4 is in the spirit of [2] but we allow for adaptive corruptions and let the adversary choose the distribution, as done by Böhl et al. [4]. The latter renamed IND-SO-CPA to *weak* IND-SO-CPA and introduced a strictly stronger notion, called *full* IND-SO-CPA, where $\mathcal{A}_{\mathsf{SO}}$ may submit any distribution (even one not efficiently resamplable) and need not provide a resampling algorithm.[1] We consider the name *weak* IND-SO-CPA unfortunate and simply refer to the security notion in Definition 4 as IND-SO-CPA security.

3   Selective Opening for Graph-Induced Distributions

This section considers graph-induced distributions and identifies connectivity properties so that IND-CPA entails IND-SO-CPA security. We introduce some

---

[1] E.g., for a one-way function OWF a distribution $(m, \mathsf{OWF}(m))$ may not support efficient resampling.

notation in Sect. 3.1. Sections 3.2 and 3.3 discuss a hybrid argument that considers the connected components of $G_{\overline{\mathcal{I}}}$, switching one of them from *sampled* to *resampled* in each transition. Section 3.4 discusses a different hybrid argument that will allow for tighter proofs if the distribution-inducing graph is acyclic.

3.1 Graphs

A *directed graph* G consists of a set of vertices V, identified with $[n]$ for $n > 0$ and a set of edges $E \subseteq V^2 \setminus \{(v,v) : v \in V\}$, i.e. we do not allow loops. G is *undirected* if $(v_2, v_1) \in E$ for each $(v_1, v_2) \in E$. For $V' \subseteq V$ let $G_{V'} := (V', E')$ denote the *induced subgraph* of G where $E' := E \cap (V')^2$. For $G = (V, E)$ we obtain its *undirected version*, $G^{\leftrightarrow} = (V, E^{\leftrightarrow})$ where $E^{\leftrightarrow} \supseteq E$ is obtained by adding the minimum number of edges to E so that the graph becomes undirected. For $V' \subseteq V$ let $N(V') := \{v \in V \setminus V' : \exists v' \in V' \text{ s.t. } (v, v') \in E^{\leftrightarrow}\}$ denote the *(open) neighborhood* of V' in G. For a vertex v, we denote by $P(v) = \{j : (j, v) \in E\}$ the set of its *parents*.

A *path* from v_1 to v_ℓ in G is a list of at least two vertices $(v_1, \ldots, v_\ell)$ where $v_i \in V$ for $i \in [\ell]$ and $(v_i, v_{i+1}) \in E$ for all $i \in [\ell - 1]$. If there is a path from u to v then u is a *predecessor* of v. Let $\text{pred}(v)$ denote the set of all predecessors of v. A *cycle* is a path where $v_\ell = v_1$. If G contains no cycles, it is *acyclic*. A directed, acyclic graph is called DAG.

A non-empty subset $V' \subseteq V$ is *connected* in G if for every pair of distinct vertices $(v_1, v_2) \in V'$ there exists a path form v_1 to v_2 in $G^{\leftrightarrow}$. G *is connected* if V is connected in G. G is *disconnected* if G is not connected. We assume G to be connected if not stated otherwise. A (set-)maximal connected set of vertices of G is called *connected component*.

Notational Convention. We do not distinguish between the i-th message of an n-message vector and vertex i in a graph on n vertices.

We start with defining Markov distributions, which are distributions on vectors of random variables reflecting processes, that is, variables with higher indices depend on ones with lower indices. A distribution is Markov if it is *memoryless* in the sense that all relevant information for the distribution of a value $\mathbf{M}_i$ is already present in $\mathbf{M}_{i-1}$, although the latter itself depends on its predecessor.

Definition 5. *Let* $\{\mathfrak{D}_\lambda\}_{\lambda \in \mathbb{N}}$ *be a family of distributions over* $\mathcal{M}^n$. *Let* $\mathbf{M} = (\mathbf{M}_1, \ldots, \mathbf{M}_n)$ *denote a vector of* $\mathcal{M}$-*valued random variables. We say* $\{\mathfrak{D}_\lambda\}_{\lambda \in \mathbb{N}}$ *is* Markov *if the following holds for all* $\lambda \in \mathbb{N}$ *and all* $\mathbf{m} \in \mathcal{M}^n$:

$$\Pr_{\mathbf{M} \leftarrow \mathfrak{D}_\lambda} \left[\mathbf{M}_i = \mathbf{m}_i \,\Big|\, \bigwedge_{j=1}^{i-1} \mathbf{M}_j = \mathbf{m}_j \right] = \Pr_{\mathbf{M} \leftarrow \mathfrak{D}_\lambda} \left[\mathbf{M}_i = \mathbf{m}_i \,\Big|\, \mathbf{M}_{i-1} = \mathbf{m}_{i-1} \right].$$

Markov distribution can be seen as "induced" by a chain graph $\mathbf{M}_1 \to \mathbf{M}_2 \to \ldots \to \mathbf{M}_n$, where edges represent dependencies. We will now generalize this to arbitrary graphs and still require (a generalization of) "memorylessness". We say that a graph G induces a distribution $\mathfrak{D}$ if whenever the distribution of $\mathbf{M}_j$ depends on $\mathbf{M}_i$ then there is a path from i to j in G. As for Markov distributions, we require that the distribution of a message only depends on its parents; in particular, for all $\lambda \in \mathbb{N}$, all $j \in [n]$ and $\mathbf{M} = (\mathbf{M}_1, \ldots, \mathbf{M}_n) \leftarrow \mathfrak{D}_\lambda$ the distribution of $\mathbf{M}_j$ only depends on its parents in G_λ, i.e. the set $P(j)$, rather than all its predecessors $\mathsf{pred}(j)$.

Definition 6 (Graph-induced distribution). *Let $\{\mathfrak{D}_\lambda\}_{\lambda \in \mathbb{N}}$ be a family of distributions over $\mathcal{M}^n$ and let $\{G_\lambda\}_{\lambda \in \mathbb{N}}$ be a family of graphs on n vertices. We say that $\{\mathfrak{D}_\lambda\}_{\lambda \in \mathbb{N}}$ is $\{G_\lambda\}_{\lambda \in \mathbb{N}}$-induced if the following holds for all $\lambda \in \mathbb{N}$:*

- *For all $i \neq j \in [n]$ if for $\mathfrak{D}_\lambda$ the distribution of $\mathbf{M}_j$ depends on $\mathbf{M}_i$ then there is a path from i to j in G_λ.*
- *For all $j \in [n]$ and all $\mathbf{m} \in \mathcal{M}^n$ we have*

$$\Pr_{\mathbf{M} \leftarrow \mathfrak{D}_\lambda}\left[\mathbf{M}_j = \mathbf{m}_j \,\middle|\, \bigwedge_{i \in \mathsf{pred}(j)} \mathbf{M}_i = \mathbf{m}_i\right] = \Pr_{\mathbf{M} \leftarrow \mathfrak{D}_\lambda}\left[\mathbf{M}_j = \mathbf{m}_j \,\middle|\, \bigwedge_{i \in P(j)} \mathbf{M}_i = \mathbf{m}_i\right].$$

We demand that for any $\lambda \in \mathbb{N}$ one can efficiently reconstruct G_λ from $\mathfrak{D}_\lambda$.

As with a family of distributions, we drop the security parameter and say that $\mathfrak{D}$ is *G-induced* whenever λ is already fixed. Note that G may contain cycles and may be undirected. Further note that Markov distributions can be seen as graph-induced distributions where the graph $G = (V, E)$ is a chain on n vertices, that is, $V = [n]$ and $E = \{(i-1, i) : i \in [n]\}$.

Although our proof ideas can be applied to disconnected graphs directly, Sects. 3.2, 3.3, and 3.4 consider *connected* graphs for simplicity. A hybrid argument over the connected components of a graph as given in Sect. 3.5 extends all our results to disconnected graphs.

3.2 A Bound Using Connected Subgraphs

Definition 7 (Number of connected subgraphs). *Let $G = (V, E)$. We define the number of connected subgraphs of G:*

$$S(G) := |\{V' \subseteq V : V' \text{ connected}\}| .$$

For example, for a chain graph on n vertices we have $S(G) = \frac{1}{2} \cdot n \cdot (n+1)$ and for the complete graph C_n on n vertices we have $S(C_n) = 2^n - 1$.

Theorem 8. *Let PKE be IND-CPA secure. Then PKE is IND-SO-CPA secure w.r.t. the class of efficiently resamplable and G-induced distribution families over $\mathcal{M}^n$ where $S(G) = \mathsf{poly}(n)$ and G is connected.*

Precisely, for any adversary $\mathcal{A}_{SO}$ run in game IND-SO-CPA_{PKE} there exists an IND-CPA_{PKE} adversary $\mathcal{B}_{CPA}$ with roughly the running time of $\mathcal{A}_{SO}$ plus two executions of Resamp such that

$$\mathbf{Adv}_{PKE}^{\text{IND-SO-CPA}}(\mathcal{A}_{SO}, \mathfrak{D}_\lambda, \lambda) \leq n \cdot (n-1) \cdot S(G_\lambda) \cdot \mathbf{Adv}_{PKE}^{\text{IND-CPA}}(\mathcal{B}_{CPA}, \lambda).$$

$$\boxed{\begin{array}{l} \textbf{Procedure } \text{CHALLENGE} \\[4pt] \mathbf{m}^1 \leftarrow \mathsf{Resamp}_{\mathfrak{D}}(\mathbf{m}^0, \mathcal{I}) \\[4pt] \mathbf{m}_i := \begin{cases} \mathbf{m}_i^1 & \text{for } i \in \bigcup_{j=1}^k C_j \\ \mathbf{m}_i^0 & \text{else} \end{cases} \\[4pt] \text{Return } \mathbf{m} = (\mathbf{m}_1, \ldots, \mathbf{m}_n) \end{array}}$$

Fig. 3. CHALLENGE procedure of hybrid game H_k. C_i denotes the i-th connected component of $G_{\overline{\mathcal{I}}}$. The challenge vector contains resampled messages in the first k batches $C_1, \ldots, C_k$ while the other messages remain sampled.

Proof Idea. Recall game $\mathsf{IND\text{-}SO\text{-}CPA}_{\mathsf{PKE},b}$ given in Fig. 2. During CHALLENGE the game sends $\mathbf{m}^b$, where $\mathbf{m}_{\overline{\mathcal{I}}}^0$ consists of messages sampled at the beginning, while $\mathbf{m}_{\overline{\mathcal{I}}}^1$ is resampled (conditioned on $\mathbf{m}_{\mathcal{I}}^1 = \mathbf{m}_{\mathcal{I}}^0$). We will define hybrid games $\mathsf{H}_0, \mathsf{H}_1, \ldots, \mathsf{H}_n$. For this, let $\mathcal{S} \subseteq 2^V$ denote all the connected subgraphs of G. We have $|\mathcal{S}| = S(G)$.

Note that $G_{\overline{\mathcal{I}}}$ consists of connected components $C_1, \ldots, C_{n'} \subseteq \mathcal{S}$ for some $n' \leq n - 1$. (This upper bound is attained by the star graph when I consists of the internal node.) We assume those components to be ordered, e.g. by the smallest vertex contained in each.

Thus, if $b = 1$ in game $\mathsf{IND\text{-}SO\text{-}CPA}$ then the challenger can resample $\mathbf{m}_{\overline{\mathcal{I}}}^1$ in n' batches $\mathbf{m}_{C_1}^1, \ldots, \mathbf{m}_{C_{n'}}^1$ (as $\overline{\mathcal{I}} = \bigcup_{i=1}^{n'} C_i$). Moreover, each batch $\mathbf{m}_{C_i}^1$ can be resampled *independently*, i.e., as a function of $\mathbf{m}_{\mathcal{I}}^0$ and $\mathfrak{D}$, but not $\mathbf{m}_{C_j}^1$, $j \neq i$.

Proof (Theorem 8). For $k = 0, \ldots, n$ we define hybrid game H_k as a modified game $\mathsf{IND\text{-}SO\text{-}CPA}_{\mathsf{PKE}}$, in which the messages of the first k batches $C_1, \ldots, C_k$ are resampled during CHALLENGE while the remaining batches stay sampled.

Every procedure except CHALLENGE remains as in Definition 4, and CHALLENGE is given in Fig. 3. Clearly, H_0 is the (real) game $\mathsf{IND\text{-}SO\text{-}CPA}_{\mathsf{PKE},0}$ and $\mathsf{H}_{n'}$ for some $n' \leq n - 1$ is the (random) game $\mathsf{IND\text{-}SO\text{-}CPA}_{\mathsf{PKE},1}$. Note that for $k, j \in [n', n]$ hybrids H_k and H_j are identical. We have

$$\mathbf{Adv}_{\mathsf{PKE}}^{\mathsf{IND\text{-}SO\text{-}CPA}}(\mathcal{A}_{\mathsf{SO}}, \mathfrak{D}_\lambda, \lambda) = \mathbf{Adv}\big(\mathsf{H}_0^{\mathcal{A}_{\mathsf{SO}}}, \mathsf{H}_{n'}^{\mathcal{A}_{\mathsf{SO}}}\big) \leq \sum_{k=0}^{n'-1} \mathbf{Adv}\big(\mathsf{H}_k^{\mathcal{A}_{\mathsf{SO}}}, \mathsf{H}_{k+1}^{\mathcal{A}_{\mathsf{SO}}}\big).$$

We now upper-bound the distance between two consecutive hybrids using the following lemma.

Lemma 9. *For every adversary $\mathcal{A}_{\mathsf{SO}}$ that distinguishes hybrids H_k and H_{k+1} there exists a* mult-IND-CPA *adversary $\mathcal{B}_{\mathsf{mult}}$ with roughly the running time of $\mathcal{A}_{\mathsf{SO}}$ plus two executions of* Resamp *such that*

$$\mathbf{Adv}\big(\mathsf{H}_k^{\mathcal{A}_{\mathsf{SO}}}, \mathsf{H}_{k+1}^{\mathcal{A}_{\mathsf{SO}}}\big) \leq S(G) \cdot \mathbf{Adv}_{\mathsf{PKE}}^{\mathsf{mult\text{-}IND\text{-}CPA}}(\mathcal{B}_{\mathsf{mult}}, \lambda).$$

Proof. We construct adversary $\mathcal{B}_{\mathsf{mult}}$ as follows (cf. Fig. 4):

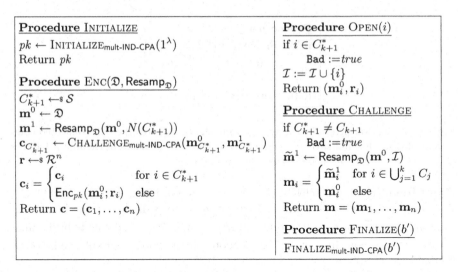

Fig. 4. $\mathcal{A}_{\mathsf{SO}}$'s game interface as provided by $\mathcal{B}_{\mathsf{mult}}$ run in game mult-IND-CPA. $\mathcal{B}_{\mathsf{mult}}$ interpolates between hybrids H_k, H_{k+1} for $k \in [0, n-1]$.

$\mathcal{B}_{\mathsf{mult}}$ forwards pk to $\mathcal{A}_{\mathsf{SO}}$ and picks $C^*_{k+1} \leftarrow^\$ \mathcal{S}$ uniformly at random (trying to guess C_{k+1}) after receiving $(\mathfrak{D}, \mathsf{Resamp}_\mathfrak{D})$. $\mathcal{B}_{\mathsf{mult}}$ samples $\mathbf{m}^0 \leftarrow \mathfrak{D}$ and resamples $\mathbf{m}^1$ keeping the neighborhood of C^*_{k+1} fixed. It submits $(\mathbf{m}^0_{C^*_{k+1}}, \mathbf{m}^1_{C^*_{k+1}})$ to its mult-IND-CPA challenger, obtains ciphertexts for positions in C^*_{k+1}, picks randomness and uses it to encrypt each message in $\overline{C^*_{k+1}}$. $\mathcal{B}_{\mathsf{mult}}$ sends $(\mathbf{c}_1, \ldots, \mathbf{c}_n)$ to $\mathcal{A}_{\mathsf{SO}}$, embedding its challenge at positions C^*_{k+1} and answers opening queries honestly if they do not occur on C^*_{k+1}. If $\mathcal{A}_{\mathsf{SO}}$ issues such a query, $\mathcal{B}_{\mathsf{mult}}$ cannot answer and sets $\mathsf{Bad} := true$ since it guessed C_{k+1} wrong. During CHALLENGE, $\mathcal{B}_{\mathsf{mult}}$ verifies that it guessed C_{k+1} correctly and sets $\mathsf{Bad} := true$ if not. $\mathcal{B}_{\mathsf{mult}}$ resamples messages $\widetilde{\mathbf{m}}^1$ that are sent in the first k batches while messages from $\mathbf{m}^0$ are sent in every other position. $\mathcal{B}_{\mathsf{mult}}$ outputs $\mathcal{A}_{\mathsf{SO}}$'s output.

In the following we use $\mathbf{m} \equiv \mathbf{m}'$ if $\mathbf{m}$ and $\mathbf{m}'$, interpreted as random variables, are identically distributed where the probability is taken over all choices in the computation of $\mathbf{m}$, $\mathbf{m}'$, respectively.

Assume, $\mathcal{B}_{\mathsf{mult}}$ guessed correctly, i.e. $C^*_{k+1} = C_{k+1}$. Clearly, $\mathcal{B}_{\mathsf{mult}}$ perfectly simulates hybrids H_k and H_{k+1} for messages and ciphertexts at positions in $\overline{C_{k+1}}$. Run in mult-IND-CPA$_{\mathsf{PKE},0}$, $\mathcal{B}_{\mathsf{mult}}$ obtains $\mathsf{Enc}_{pk}(\mathbf{m}^0_{C_{k+1}})$ and $\mathcal{A}_{\mathsf{SO}}$ therefore receives encryptions of sampled messages. During CHALLENGE the $(k+1)$-th batch contains sampled messages $\mathbf{m}^0_{C_{k+1}}$, thus $\mathcal{B}_{\mathsf{mult}}$ perfectly simulates hybrid H_k.

When $\mathcal{B}_{\mathsf{mult}}$ is run in mult-IND-CPA$_{\mathsf{PKE},1}$, $\mathcal{A}_{\mathsf{SO}}$ obtains encryptions of *resampled* messages $\mathsf{Enc}_{pk}(\mathbf{m}^1_{C_{k+1}})$ while it expects encrypted *sampled* messages: $\mathsf{Enc}_{pk}(\mathbf{m}^0_{C_{k+1}})$. During CHALLENGE $\mathcal{A}_{\mathsf{SO}}$ expects *resampled* messages $\widetilde{\mathbf{m}}^1_{C_{k+1}}$ but obtains *sampled* $\mathbf{m}^0_{C_{k+1}}$. Thus, the *sampled* and *resampled* messages change roles on C_{k+1}.

However, they are equally distributed, i.e., $\mathbf{m}^0_{C_{k+1}} \equiv \mathbf{m}^1_{C_{k+1}}$ since the messages in $N(C_{k+1})$ were fixed when resampling $\mathbf{m}^1$ and the distribution of messages in C_{k+1} depends on $\mathfrak{D}$ and messages in positions $N(C_{k+1})$ only. Likewise, $\mathbf{m}^1_{C_{k+1}} \equiv \widetilde{\mathbf{m}}^1_{C_{k+1}}$ for $\mathbf{m}^1 \leftarrow \mathsf{Resamp}_{\mathfrak{D}}(\mathbf{m}^0, N(C_{k+1}))$ and $\widetilde{\mathbf{m}}^1 \leftarrow \mathsf{Resamp}_{\mathfrak{D}}(\mathbf{m}^0, \mathcal{I})$ since the distribution of messages in C_{k+1} solely depends on $\mathfrak{D}$ and messages in $N(C_{k+1}) \subseteq \mathcal{I}$ and $\mathcal{A}_{\mathsf{SO}}$'s view is identical to hybrid H_{k+1}. We have

$$\Pr[\mathsf{mult\text{-}IND\text{-}CPA}^{\mathcal{B}_{\mathsf{mult}}}_{\mathsf{PKE},0} \Rightarrow 1] = \Pr[\mathsf{H}^{\mathcal{A}_{\mathsf{SO}}}_k \Rightarrow 1 \wedge \overline{\mathsf{Bad}}] \text{ and}$$

$$\Pr[\mathsf{mult\text{-}IND\text{-}CPA}^{\mathcal{B}_{\mathsf{mult}}}_{\mathsf{PKE},1} \Rightarrow 1] = \Pr[\mathsf{H}^{\mathcal{A}_{\mathsf{SO}}}_{k+1} \Rightarrow 1 \wedge \overline{\mathsf{Bad}}] \ .$$

Observe that Bad does not happen when $\mathcal{B}_{\mathsf{mult}}$ guessed C_{k+1} correctly. Since $\overline{\mathsf{Bad}}$ is independent of $\mathcal{A}_{\mathsf{SO}}$'s output in a hybrid and $|\mathcal{S}| = S(G)$, we have

$$\mathbf{Adv}^{\mathsf{mult\text{-}IND\text{-}CPA}}_{\mathsf{PKE}}(\mathcal{B}_{\mathsf{mult}}, \lambda) \geq \frac{1}{S(G)} \cdot \mathbf{Adv}\big(\mathsf{H}^{\mathcal{A}_{\mathsf{SO}}}_k, \mathsf{H}^{\mathcal{A}_{\mathsf{SO}}}_{k+1}\big),$$

which concludes the proof. $\square$

We proceed with the proof of Theorem 8. Using Lemma 9 we have

$$\mathbf{Adv}^{\mathsf{IND\text{-}SO\text{-}CPA}}_{\mathsf{PKE}}(\mathcal{A}_{\mathsf{SO}}, \mathfrak{D}_\lambda, \lambda) \leq \sum_{k=0}^{n'-1} \mathbf{Adv}\big(\mathsf{H}^{\mathcal{A}_{\mathsf{SO}}}_k, \mathsf{H}^{\mathcal{A}_{\mathsf{SO}}}_{k+1}\big)$$

$$\leq \sum_{k=0}^{n'-1} S(G_\lambda) \cdot \mathbf{Adv}^{\mathsf{mult\text{-}IND\text{-}CPA}}_{\mathsf{PKE}}(\mathcal{B}_{\mathsf{mult}}, \lambda).$$

$\mathcal{B}_{\mathsf{mult}}$ sends message vectors of length $|C^*_{k+1}| \leq n$ to its $\mathsf{mult\text{-}IND\text{-}CPA}$ challenger. Using Lemma 2, we have

$$\leq \sum_{k=0}^{n'-1} n \cdot S(G_\lambda) \cdot \mathbf{Adv}^{\mathsf{IND\text{-}CPA}}_{\mathsf{PKE}}(\mathcal{B}_{\mathsf{CPA}}, \lambda) \leq n \cdot (n-1) \cdot S(G_\lambda) \cdot \mathbf{Adv}^{\mathsf{IND\text{-}CPA}}_{\mathsf{PKE}}(\mathcal{B}_{\mathsf{CPA}}, \lambda),$$

since $n' \leq n - 1$, which completes the proof of Theorem 8. $\square$

Markov Distributions. Markov distributions (Definition 5) are induced by the chain graph $(V = [n], E = \{(i-1, i) : i \in [n]\})$, for which $S(G) = \frac{1}{2} \cdot n \cdot (n+1)$. We thus immediately obtain the following corollary from Theorem 8.

Corollary 10. *Let* PKE *be* IND-CPA *secure. Then* PKE *is* IND-SO-CPA *secure w.r.t. efficiently resamplable Markov distributions over* $\mathcal{M}^n$.

Precisely, for any adversary $\mathcal{A}_{\mathsf{SO}}$ *run in game* IND-SO-CPA$_{\mathsf{PKE}}$ *there exists an* IND-CPA$_{\mathsf{PKE}}$ *adversary* $\mathcal{B}_{\mathsf{CPA}}$ *with roughly the running time of* $\mathcal{A}_{\mathsf{SO}}$ *plus two executions of* Resamp *such that*

$$\mathbf{Adv}^{\mathsf{IND\text{-}SO\text{-}CPA}}_{\mathsf{PKE}}(\mathcal{A}_{\mathsf{SO}}, \mathfrak{D}_\lambda, \lambda) \leq \frac{1}{2} \cdot n^2 \cdot (n^2 - 1) \cdot \mathbf{Adv}^{\mathsf{IND\text{-}CPA}}_{\mathsf{PKE}}(\mathcal{B}_{\mathsf{CPA}}, \lambda).$$

3.3   A Bound Using the Maximum Border

Definition 11 (Maximum border). *Let $G = (V, E)$. We define the* maximum border *of G as the maximal size of the neighborhood of any connected subgraph in G.*

$$B(G) := \max\big\{\, |N(V')| \ : \ V' \subseteq V \text{ connected}\big\}.$$

For example, if G is an n-path for $n \geq 3$ then $B(G) = 2$. For the complete graph or star graph on n vertices we have $B(G) = n - 1$. Notice that $B(G) < n$.

In the reduction in Sect. 3.2 we guessed a connected component in $G_{\overline{\mathcal{I}}}$ that would be switched from sampled to resampled in a hybrid transition. Alternatively, we can guess a connected component in $G_{\overline{\mathcal{I}}}$ via its neighborhood. The following theorem expresses $S(G)$ in terms of $B(G)$.

Theorem 12. *Let G be a connected graph. Then the following bound on $S(G)$ holds:*

$$S(G) \leq \frac{2}{(B(G) - 1)!} \cdot n^{B(G)} \quad \text{for all} \quad 0 < B(G) \leq \frac{n - 2}{3} \ .$$

We begin with a simple observation before proving the theorem.

Lemma 13. *Let $G = (V, E)$ and $V_1 \neq V_2$ each of them connected in G such that $N(V_1) = N(V_2)$. Then $V_1 \cap V_2 = \emptyset$.*

Proof. Assume $V_1 \cap V_2 \neq \emptyset$. As $V_1 \neq V_2$ we have $V_1 \setminus V_2 \neq \emptyset$ without loss of generality. Because V_1 is connected, there exist vertices $v_\cap \in V_1 \cap V_2$ and $v_1 \in V_1 \setminus V_2$ such that $(v_1, v_\cap) \in E^{\leftrightarrow}$. Since $v_1 \notin V_2$, $v_\cap \in V_2$ and $(v_1, v_\cap) \in E^{\leftrightarrow}$, we see that $v_1 \in N(V_2)$. As $N(V_2) = N(V_1)$ it follows that $v_1 \in N(V_1)$; a contradiction since $v_1 \in V_1$. □

Proof (Theorem 12). Let $B := B(G)$. We have

$$S(G) = \sum_{i=0}^{B} \big|\{V' \subseteq V : V' \text{ connected} \wedge |N(V')| = i\}\big| .$$

For $i = 0$ we count the connected components of G.

$$= 1 + \sum_{i=1}^{B} \big|\{V' \subseteq V : V' \text{ connected} \wedge |N(V')| = i\}\big|$$

$$= 1 + \sum_{i=1}^{B} \sum_{\substack{V_i \subseteq V \\ |V_i| = i}} \big|\{V' \subseteq V : V' \text{ connected} \wedge N(V') = V_i\}\big| .$$

Let $V_i \subseteq V$ be non-empty and $\{V' \subseteq V : V' \text{ connected} \wedge N(V') = V_i\} = \{V'_1, \ldots, V'_k\}$ for appropriate k. Applying Lemma 13 to $V'_1, \ldots, V'_k$, we see that those sets V'_j are pairwise disjoint. Fix any vertex $v_i \in V_i$. Since $N(V'_j) = V_i$ for $j \in [k]$ and all V'_j are pairwise disjoint, there exists at least one vertex v'_j in each

V_j' such that $(v_j', v_i) \in E$ for all $j \in [k]$. Thus, $N(v_i) \geq k$, i.e. $B \geq k$. Hence, $k \leq B$ for given B and we obtain an upper bound for the number of possible sets V' for each fixed V_i. It follows

$$S(G) \leq 1 + \sum_{i=1}^{B} \sum_{\substack{V_i \subseteq V \\ |V_i| = i}} B = 1 + B \cdot \sum_{i=1}^{B} \binom{n}{i} \leq B \cdot \sum_{i=0}^{B} \binom{n}{i}. \tag{1}$$

To bound the sum in (1) we use the geometric series and upper-bound the quotient of two consecutive binomial coefficients by $\frac{1}{2}$:

$$\frac{\binom{n}{i}}{\binom{n}{i+1}} = \frac{i+1}{n-i} \leq \frac{1}{2} \Leftrightarrow i \leq \frac{n-2}{3}.$$

Hence

$$B \cdot \sum_{i=0}^{B} \binom{n}{i} \leq B \cdot \sum_{i=0}^{B} \frac{1}{2^i} \binom{n}{B} \leq B \cdot \binom{n}{B} \cdot \sum_{i=0}^{\infty} \frac{1}{2^i} \leq 2 \cdot B \cdot \frac{n^B}{B!} = \frac{2}{(B-1)!} \cdot n^B$$

for $B(G) \leq \frac{n-2}{3}$, which concludes the proof. $\qquad\qquad\qquad\qquad \square$

Theorems 8 and 12 together now yield the following corollary.

Corollary 14. *Let* PKE *be* IND-CPA *secure. Then* PKE *is* IND-SO-CPA *secure w.r.t. the class of efficiently resamplable and G-induced distribution families over $\mathcal{M}^n$ where $B(G) = $ const, $n \geq 3 \cdot B(G) + 2$ and G is connected.*

Concretely, for any adversary $\mathcal{A}_{SO}$ in game IND-SO-CPA$_{PKE}$ *there exists an* IND-CPA$_{PKE}$ *adversary $\mathcal{B}_{CPA}$ with roughly the running time of $\mathcal{A}_{SO}$ plus two executions of* Resamp *such that*

$$\mathbf{Adv}_{PKE}^{IND\text{-}SO\text{-}CPA}(\mathcal{A}_{SO}, \mathfrak{D}_\lambda, \lambda) \leq \frac{2 \cdot (n-1)}{(B(G_\lambda) - 1)!} \cdot n^{B(G_\lambda)+1} \cdot \mathbf{Adv}_{PKE}^{IND\text{-}CPA}(\mathcal{B}_{CPA}, \lambda).$$

Since Corollary 14 ensures a polynomial loss in the reduction for $B(G) = $ const and we are interested in asymptotic statements, we do not consider the restriction to $n \geq 3 \cdot B(G) + 2$ grave. One can easily obtain a version of Theorem 12 that is weaker by a factor of roughly $B(G)$ but holds for all $B(G) < n$. To this end one bounds the sum of binomial coefficients in (1) in terms of the incomplete upper gamma function Γ to get

$$\sum_{i=1}^{B} \binom{n}{i} \leq \sum_{i=1}^{B} \frac{n^i}{i!} = \frac{e^n \Gamma(B+1, n)}{B!} - 1.$$

Using a nice bound on Γ due to [11] that can be found in [5] we obtain a bound for $B(G) < n$.

> **Procedure** CHALLENGE
> $\mathbf{m}^1 \leftarrow \mathsf{Resamp}_{\mathfrak{D}}(\mathbf{m}^0, [k+1, n] \cup \mathcal{I})$
> Return $\mathbf{m}^b$

Fig. 5. CHALLENGE procedure of hybrid game H_k. For $k = n$ we have $[n + 1, n] = \emptyset$.

Think of a direct reduction for proving Corollary 14 as implicitly guessing C_{k+1} via guessing $N(C_{k+1})$ by picking up to $B(G)$ vertices in G and guessing one of at most $B(G)$ connected subgraphs that have the guessed neighborhood.

Note that Corollary 14 cannot provide a tighter bound on the loss than Theorem 8. In particular, there are (even connected) graphs for which Theorem 8 ensures an at most polynomial loss, while Corollary 14 does not. For instance, let G be the star graph on $\log n$ vertices attached to the chain graph of $n - \log n$ vertices, then $S(G) = \mathsf{poly}(n)$, but $B(G) > \mathsf{const}$.

3.4   A Tighter Reduction for Acyclic Graphs

While we considered graph-induced distributions for arbitrary graphs in Sects. 3.2 and 3.3, we now consider DAG-induced distributions for which we obtain a tighter reduction than what is guaranteed by Corollary 14.

For a DAG G we require that the vertices are semi-ordered in such a way that there is no directed path from i to j for $i < j$. Such an ordering always exists as G has no cycles. Note that the dependencies now go the other way as for Markov distributions, but this will allow us to replaced sampled messages by resampled ones from left to right as in the previous hybrids. We will traverse dependencies *backwards*, that is, if message $\mathbf{m}_i$ depends on $\mathbf{m}_j$ then $\mathbf{m}_i$ is switched from *sampled* to *resampled* before $\mathbf{m}_j$ is switched. So, as in the previous proofs, messages $\mathbf{m}_1, \ldots, \mathbf{m}_i$ will be resampled in the i-th hybrid.

Theorem 15. *Let* PKE *be* IND-CPA *secure. Then* PKE *is* IND-SO-CPA *secure w.r.t. the class of efficiently resamplable and G-induced distribution families over $\mathcal{M}^n$ where $B(G) = \mathsf{const}$ and G is a connected DAG.*

Precisely, for any adversary $\mathcal{A}_{\mathsf{SO}}$ run in game IND-SO-CPA$_{\mathsf{PKE}}$ *there exists an* IND-CPA$_{\mathsf{PKE}}$ *adversary $\mathcal{B}_{\mathsf{CPA}}$ with roughly the running time of $\mathcal{A}_{\mathsf{SO}}$ plus three executions of* Resamp *such that*

$$\mathbf{Adv}_{\mathsf{PKE}}^{\mathsf{IND\text{-}SO\text{-}CPA}}(\mathcal{A}_{\mathsf{SO}}, \mathfrak{D}_\lambda, \lambda) \leq 3 \cdot n^{B(G_\lambda)+1} \cdot \mathbf{Adv}_{\mathsf{PKE}}^{\mathsf{IND\text{-}CPA}}(\mathcal{B}_{\mathsf{CPA}}, \lambda).$$

Proof. We proceed in a sequence of hybrid games $\mathsf{H}_0, \mathsf{H}_1, \ldots, \mathsf{H}_n$ and switch message $\mathbf{m}_{k+1}$ from sampled to resampled in hybrid transition H_k to H_{k+1}. Hybrid H_k will return the sampled messages for all positions $[k + 1, n] \cup \mathcal{I}$, but resampled messages on all positions $[k] \setminus \mathcal{I}$ where the resampling is conditioned on *every message in* $[k + 1, n] \cup \mathcal{I}$. The code for CHALLENGE in given in Fig. 5, every other procedure stays as in Fig. 2.

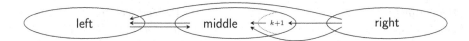

Fig. 6. Structure of G. Edges between particular sets cannot exist if there is no arrow depicted. If right $\neq \emptyset$, there is at least one edge from right to middle since G is connected. left and middle are disconnected in $G_{\overline{\mathcal{I}}}$.

Hybrid H_0 is identical to game IND-SO-CPA$_{\mathsf{PKE},0}$, and H_n is identical to IND-SO-CPA$_{\mathsf{PKE},1}$, hence

$$\mathbf{Adv}_{\mathsf{PKE}}^{\mathsf{IND\text{-}SO\text{-}CPA}}(\mathcal{A}_{\mathsf{SO}}, \mathfrak{D}_\lambda, \lambda) = \mathbf{Adv}(\mathsf{H}_0^{\mathcal{A}_{\mathsf{SO}}}, \mathsf{H}_n^{\mathcal{A}_{\mathsf{SO}}}) \leq \sum_{k=0}^{n-1} \mathbf{Adv}(\mathsf{H}_k^{\mathcal{A}_{\mathsf{SO}}}, \mathsf{H}_{k+1}^{\mathcal{A}_{\mathsf{SO}}}) . \quad (2)$$

We bound the distance between two consecutive hybrids H_k, H_{k+1} and proceed with the following lemma.

Lemma 16. *For every adversary $\mathcal{A}_{\mathsf{SO}}$ that distinguishes hybrids H_k and H_{k+1} there exists a* mult-IND-CPA *adversary $\mathcal{B}_{\mathsf{mult}}$ with roughly the running time of $\mathcal{A}_{\mathsf{SO}}$ plus three executions of* Resamp *such that*

$$\mathbf{Adv}(\mathsf{H}_k^{\mathcal{A}_{\mathsf{SO}}}, \mathsf{H}_{k+1}^{\mathcal{A}_{\mathsf{SO}}}) \leq \Pr[\overline{Bad}_k]^{-1} \cdot \mathbf{Adv}_{\mathsf{PKE}}^{\mathsf{mult\text{-}IND\text{-}CPA}}(\mathcal{B}_{\mathsf{mult}}, \lambda),$$

where $\Pr[\overline{Bad}_k]^{-1} = \sum_{i=0}^{B(G_\lambda)-1} \binom{k}{i}$ for $k < n-1$ and $\Pr[\overline{Bad}_k]^{-1} = \sum_{i=0}^{B(G_\lambda)} \binom{k}{i}$ for $k = n-1$.

Proof Idea: We construct a mult-IND-CPA adversary $\mathcal{B}_{\mathsf{mult}}$ that interpolates between hybrids H_k and H_{k+1}. Ideally, $\mathcal{B}_{\mathsf{mult}}$ embeds its own challenge at position $k+1$, but might have to resample some already resampled messages in $\mathbf{m}_{[k]}$ to avoid inconsistencies. Let middle denote the connected component in $G_{[k+1]\backslash\mathcal{I}}$ that contains $\mathbf{m}_{k+1}$. Let right $:= [k+2, n]$, and left $:= \overline{(\text{middle} \cup \text{right})}$. Observe that it is sufficient to resample middle again to obtain consistent resampled messages. In particular, there is no need to resample any right message due to the semi-order imposed on the vertices, as a message in right does not depend on any message in $\overline{\text{right}}$ (cf. Fig. 6). The reduction will guess middle to embed its mult-IND-CPA challenge, while it waits for all opening queries to happen to resample the left messages. Note that middle and left are disconnected in $G_{\overline{\mathcal{I}}}$, thus can be resampled independently of each other only depending on their respective neighborhood. Since right messages are fixed while resampling, it suffices to guess $N(\text{middle}) \cap [k]$. Further, G is connected, i.e. $N(\text{middle})$ contains at least one vertex from right $= [k+2, n]$ as long as $k < n-1$. Hence, for $k < n-1$, we have $|N(\text{middle}) \cap [k]| \leq B(G) - 1$.

Proof (Lemma 16). For $k \in [0, n]$ and $i \in [n]$ let $\mathsf{Open}_k(i)$ denote the event that $\mathcal{A}_{\mathsf{SO}}$ calls OPEN(i) in hybrid H_k. Two arbitrary hybrids only differ in the CHALLENGE procedure, hence $\Pr[\mathsf{Open}_s(i)] = \Pr[\mathsf{Open}_t(i)]$ for all $s, t \in [0, n]$, for all $i \in [n]$. Additionally, two consecutive hybrids H_k, H_{k+1} only differ in the

---

Procedure INITIALIZE

$pk \leftarrow$ INITIALIZE$_{\text{mult-IND-CPA}}(1^{\lambda})$
Return pk

Procedure ENC$(\mathfrak{D}, \text{Resamp}_{\mathfrak{D}})$

if $k < n - 1$
  $N^* \leftarrow_{\$} \{V' \subseteq [k] : |V'| \in [0, B(G) - 1]\}$
else
  $N^* \leftarrow_{\$} \{V' \subseteq [k] : |V'| \in [0, B(G)]\}$
Let middle* denote the connected component in $G_{[k+1] \backslash N^*}$ that contains $k + 1$.
$\mathbf{m}^0 \leftarrow \mathfrak{D}$
$\mathbf{m}^{1,0} \leftarrow \text{Resamp}_{\mathfrak{D}}(\mathbf{m}^0, N^* \cup \text{right})$
$\mathbf{m}^{1,1} \leftarrow \text{Resamp}_{\mathfrak{D}}(\mathbf{m}^0, N^* \cup \text{right} \cup \{k + 1\})$
$\mathbf{c}_{\text{middle}*} \leftarrow$ CHALLENGE$_{\text{mult-IND-CPA}}(\mathbf{m}^{1,0}_{\text{middle}*}, \mathbf{m}^{1,1}_{\text{middle}*})$
$\mathbf{r} \leftarrow_{\$} \mathcal{R}^n$
$\mathbf{c}_i = \begin{cases} \mathbf{c}_i & \text{for } i \in \text{middle}* \\ \text{Enc}_{pk}(\mathbf{m}^0_i; \mathbf{r}_i) & \text{else} \end{cases}$
Return $\mathbf{c} = (\mathbf{c}_1, \ldots, \mathbf{c}_n)$

Procedure OPEN(i)

if $i \in \text{middle}* \backslash \{k + 1\}$
  Bad $:= true$
$\mathcal{I} := \mathcal{I} \cup \{i\}$
Return $(\mathbf{m}^0_i, \mathbf{r}_i)$

Procedure CHALLENGE

if $N^* \not\subseteq \mathcal{I}$
  Bad $:= true$
$\mathbf{m}^1 \leftarrow \text{Resamp}_{\mathfrak{D}}(\mathbf{m}^0, \mathcal{I} \cup \text{right})$
$\mathbf{m}_i = \begin{cases} \mathbf{m}^1_i & \text{for } i \in \text{left} \\ \mathbf{m}^0_i & \text{else} \end{cases}$
Return $\mathbf{m} = (\mathbf{m}_1, \ldots, \mathbf{m}_n)$

Procedure FINALIZE(b')

FINALIZE$_{\text{mult-IND-CPA}}(b')$

Fig. 7. $\mathcal{A}_{\text{SO}}$'s game interface as provided by $\mathcal{B}_{\text{mult}}$ run in game mult-IND-CPA. $\mathcal{B}_{\text{mult}}$ interpolates between hybrids H_k, H_{k+1} for $k \in [0, n-1]$.

$(k+1)$-th message returned during CHALLENGE *unless* $\mathcal{A}_{\text{SO}}$ calls OPEN$(k+1)$ in game H_{k+1}. Thus, we have

$$\Pr[\mathsf{H}_k^{\mathcal{A}_{\text{SO}}} \Rightarrow 1 \wedge \mathsf{Open}_k(k+1)] = \Pr[\mathsf{H}_{k+1}^{\mathcal{A}_{\text{SO}}} \Rightarrow 1 \wedge \mathsf{Open}_{k+1}(k+1)]$$

and obtain

$$\mathbf{Adv}\left(\mathsf{H}_k^{\mathcal{A}_{\text{SO}}}, \mathsf{H}_{k+1}^{\mathcal{A}_{\text{SO}}}\right) =$$
$$\left| \Pr[\mathsf{H}_{k+1}^{\mathcal{A}_{\text{SO}}} \Rightarrow 1 \wedge \overline{\mathsf{Open}_{k+1}(k+1)}] - \Pr[\mathsf{H}_k^{\mathcal{A}_{\text{SO}}} \Rightarrow 1 \wedge \overline{\mathsf{Open}_k(k+1)}] \right|. \quad (3)$$

We describe $\mathcal{B}_{\text{mult}}$ (cf. Fig. 7): It passes pk on to $\mathcal{A}_{\text{SO}}$; obtaining $(\mathfrak{D}, \text{Resamp}_{\mathfrak{D}})$, $\mathcal{B}_{\text{mult}}$ makes a guess for middle (labeled middle*) by making a guess (labeled N^*) of middle's neighborhood in $G_{[k+1]}$ and samples $\mathbf{m}^0 \leftarrow \mathfrak{D}$. $\mathcal{B}_{\text{mult}}$ resamples $\mathbf{m}^{1,0}$ fixing $N^* \cup \text{right}$ and resamples $\mathbf{m}^{1,1}$ fixing $N^* \cup \text{right} \cup \{k + 1\}$. $\mathcal{B}_{\text{mult}}$ sends $(\mathbf{m}^{1,0}_{\text{middle}*}, \mathbf{m}^{1,1}_{\text{middle}*})$ to its mult-IND-CPA challenger, receives $\mathbf{c}_{\text{middle}*}$, samples fresh randomness to encrypt messages in middle* on its own and forwards $(\mathbf{c}_1, \ldots, \mathbf{c}_n)$ to $\mathcal{A}_{\text{SO}}$. $\mathcal{B}_{\text{mult}}$ sets Bad $:= true$ if $\mathcal{A}_{\text{SO}}$ calls OPEN(i) for some $i \in \text{middle}* \backslash \{k + 1\}$ since it cannot answer those queries.[2] Other opening queries are answered honestly. On $\mathcal{A}_{\text{SO}}$'s call of CHALLENGE, $\mathcal{B}_{\text{mult}}$ checks if $N^* \subseteq \mathcal{I}$. If not, $\mathcal{B}_{\text{mult}}$ guessed middle wrong and sets Bad to $true$. Otherwise, $\mathcal{B}_{\text{mult}}$ resamples messages fixing those at positions $\mathcal{I} \cup \text{right}$ to obtain resampled messages $\mathbf{m}^1$

---

[2] Equation (3) directly accounts for $\mathcal{A}_{\text{SO}}$ calling OPEN$(k + 1)$.

and sends $\mathbf{m}_i^1$ for all left positions and $\mathbf{m}_i^0$ for all remaining positions to $\mathcal{A}_{\mathsf{SO}}$. $\mathcal{B}_{\mathsf{mult}}$ outputs whatever $\mathcal{A}_{\mathsf{SO}}$ outputs.

Assume that $\mathcal{B}_{\mathsf{mult}}$ guessed correctly, i.e. N^* is the neighborhood of middle in $G_{[k]}$. Then $\mathsf{middle}^* = \mathsf{middle}$ holds and by definition of middle, Bad cannot happen.

Clearly, $\mathcal{B}_{\mathsf{mult}}$ correctly simulates $\mathcal{A}_{\mathsf{SO}}$'s hybrid view in all left and right positions. Note that $\mathcal{A}_{\mathsf{SO}}$ obtains *resampled* encryptions $\mathsf{Enc}_{pk}(\mathbf{m}_{\mathsf{middle}}^{1,b})$ during ENC, but expects *sampled* encryptions $\mathsf{Enc}_{pk}(\mathbf{m}_{\mathsf{middle}}^0)$, while receiving *sampled* $\mathbf{m}_{\mathsf{middle}}^0$ when calling CHALLENGE, expecting *resampled* $\mathbf{m}_{\mathsf{middle}}$. Thus, *sampled* middle messages become *resampled* middle messages from $\mathcal{A}_{\mathsf{SO}}$'s view and vice versa.

However, we have $\mathbf{m}_{\mathsf{middle}} \equiv \mathbf{m}_{\mathsf{middle}}^0$ since $N(\mathsf{middle}) \subseteq \mathcal{I} \cup \mathsf{right}$, whereby $\mathcal{I} \cup \mathsf{right}$ is fixed when resampling $\mathbf{m}_{\mathsf{middle}}$.

For $\mathcal{B}_{\mathsf{mult}}$ run in game $\mathsf{mult\text{-}IND\text{-}CPA}_{\mathsf{PKE},0}$, $\mathcal{A}_{\mathsf{SO}}$ receives $\mathsf{Enc}_{pk}(\mathbf{m}_{\mathsf{middle}}^{1,0})$ where $\mathbf{m}_{\mathsf{middle}}^{1,0} \equiv \mathbf{m}_{\mathsf{middle}}^0$ since $N^* \cup \mathsf{right} = N \cup \mathsf{right}$ is fixed when $\mathbf{m}^{1,0}$ is resampled. Hence, *all* middle messages sent during CHALLENGE look resampled and $\mathcal{A}_{\mathsf{SO}}$'s view is identical to hybrid H_{k+1}.

When $\mathcal{B}_{\mathsf{mult}}$ is run in $\mathsf{mult\text{-}IND\text{-}CPA}_{\mathsf{PKE},1}$, it forwards $\mathsf{Enc}_{pk}(\mathbf{m}_{\mathsf{middle}}^{1,1})$ to $\mathcal{A}_{\mathsf{SO}}$ where $\mathbf{m}_{\mathsf{middle}}^{1,1} \equiv \mathbf{m}_{\mathsf{middle}}^1$ for the same reason as for $b = 0$. In particular, we have $\mathbf{m}_{k+1}^0 = \mathbf{m}_{k+1}^{1,1}$ since $\mathbf{m}_{k+1}^0$ is fixed while resampling. Consequently, each message in middle *except* the $(k+1)$-th looks resampled during CHALLENGE and $\mathcal{A}_{\mathsf{SO}}$'s view is identical to hybrid H_k.

$\mathcal{B}_{\mathsf{mult}}$ outputs 1 in its game $\mathsf{mult\text{-}IND\text{-}CPA}$ if $\mathcal{A}_{\mathsf{SO}}$ outputs 1 in its respective hybrid and $\mathcal{A}_{\mathsf{SO}}$ does not open ciphertext $\mathbf{c}_{k+1}$ and Bad does not happen. We thus have

$$\mathbf{Adv}_{\mathsf{PKE}}^{\mathsf{mult\text{-}IND\text{-}CPA}}(\mathcal{B}_{\mathsf{mult}}, \lambda) \geq$$
$$\left| \Pr[\mathsf{mult\text{-}IND\text{-}CPA}_{\mathsf{PKE},0}^{\mathcal{B}_{\mathsf{mult}}} \Rightarrow 1] - \Pr[\mathsf{mult\text{-}IND\text{-}CPA}_{\mathsf{PKE},1}^{\mathcal{B}_{\mathsf{mult}}} \Rightarrow 1] \right|$$

$$= \left| \Pr[\mathsf{H}_{k+1}^{\mathcal{A}_{\mathsf{SO}}} \Rightarrow 1 \wedge \overline{\mathsf{Open}_{k+1}(k+1)} \wedge \overline{\mathsf{Bad}}] - \Pr[\mathsf{H}_k^{\mathcal{A}_{\mathsf{SO}}} \Rightarrow 1 \wedge \overline{\mathsf{Open}_k(k+1)} \wedge \overline{\mathsf{Bad}}] \right|.$$

Since $\overline{\mathsf{Bad}}$ is independent of $\mathsf{H}_i^{\mathcal{A}_{\mathsf{SO}}} \Rightarrow 1 \wedge \mathsf{Open}_i(k+1)$ for $i \in \{k, k+1\}$ we have

$$= \Pr[\overline{\mathsf{Bad}}] \cdot \left| \Pr[\mathsf{H}_{k+1}^{\mathcal{A}_{\mathsf{SO}}} \Rightarrow 1 \wedge \overline{\mathsf{Open}_{k+1}(k+1)}] - \Pr[\mathsf{H}_k^{\mathcal{A}_{\mathsf{SO}}} \Rightarrow 1 \wedge \overline{\mathsf{Open}_k(k+1)}] \right|$$
$$= \Pr[\overline{\mathsf{Bad}}] \cdot \mathbf{Adv}(\mathsf{H}_k^{\mathcal{A}_{\mathsf{SO}}}, \mathsf{H}_{k+1}^{\mathcal{A}_{\mathsf{SO}}}),$$

by Eq. (3). $\mathcal{B}_{\mathsf{mult}}$ picks N^* from a set of size $\sum_{i=0}^{B(G\lambda)-1} \binom{k}{i}$ for $k < n - 1$, and of size $\sum_{i=0}^{B(G\lambda)} \binom{k}{i}$ for $k = n - 1$, respectively, which proves Lemma 16.   □

The remaining proof consists of tedious computations. From Eq. (2) and Lemma 16 we have

$$\mathbf{Adv}_{\mathsf{PKE}}^{\mathsf{IND\text{-}SO\text{-}CPA}}(\mathcal{A}_{\mathsf{SO}}, \mathfrak{D}_\lambda, \lambda) \leq \sum_{k=0}^{n-1} \Pr[\overline{\mathsf{Bad}_k}]^{-1} \cdot \mathbf{Adv}_{\mathsf{PKE}}^{\mathsf{mult\text{-}IND\text{-}CPA}}(\mathcal{B}_{\mathsf{mult}}, \lambda).$$

Let $B := B(G)$. Since $\mathcal{B}_{\text{mult}}$ submits message vectors of length $|\text{middle}^*| \leq k+1$ to its mult-IND-CPA challenger and by Lemma 2:

$$\mathbf{Adv}_{\mathsf{PKE}}^{\mathsf{IND\text{-}SO\text{-}CPA}}(\mathcal{A}_{\mathsf{SO}}, \mathfrak{D}_\lambda, \lambda) \leq$$

$$\left(\sum_{k=0}^{n-2}(k+1) \cdot \sum_{i=0}^{B-1}\binom{k}{i} + n \cdot \sum_{i=0}^{B}\binom{n-1}{i} \right) \cdot \mathbf{Adv}_{\mathsf{PKE}}^{\mathsf{IND\text{-}CPA}}(\mathcal{B}_{\text{mult}}, \lambda). \qquad (4)$$

We upper-bound the loss in (4). Let $2 \leq B < n$.

$$\sum_{k=0}^{n-2}(k+1) \cdot \sum_{i=0}^{B-1}\binom{k}{i} + n \cdot \sum_{i=0}^{B}\binom{n-1}{i}$$

$$= \sum_{i=0}^{B-1}\binom{0}{i} + 2 \cdot \sum_{i=0}^{B-1}\binom{1}{i} + \sum_{k=2}^{n-2}(k+1) \cdot \sum_{i=0}^{B-1}\binom{k}{i} + n \cdot \sum_{i=0}^{B}\binom{n-1}{i}$$

$$\leq 5 + \sum_{k=2}^{n-2}(k+1) \cdot \sum_{i=0}^{B-1} k^i + n \cdot \sum_{i=0}^{B}\binom{n-1}{i}$$

$$= 5 + \sum_{k=2}^{n-2}(k+1) \cdot \frac{k^B - 1}{k - 1} + n \cdot \sum_{i=0}^{B}\binom{n-1}{i}$$

$$= 5 + \sum_{k=2}^{n-2}\underbrace{\frac{k+1}{k-1}}_{\leq 3} \cdot (k^B - 1) + n \cdot \sum_{i=0}^{B}\binom{n-1}{i}$$

$$\leq 5 + 3 \cdot \sum_{k=2}^{n-2}(k^B - 1) + n \cdot \sum_{i=0}^{B}\binom{n-1}{i}$$

$$= 5 + 3 \cdot \sum_{k=2}^{n-2}k^B - 3 \cdot (n-3) + n \cdot \sum_{i=0}^{B}\binom{n-1}{i}$$

$$= 14 - 3n + 3 \cdot \sum_{k=2}^{n-2}k^B + n \cdot \sum_{i=0}^{B}\binom{n-1}{i}$$

$$= 11 - 3n + 3 \cdot \sum_{k=0}^{n-2}k^B + n \cdot \sum_{i=0}^{B}\binom{n-1}{i} \qquad \text{since } B \geq 1$$

$$\leq 11 - 3n + 3 \cdot \sum_{k=0}^{n-2}k^B + n \cdot \sum_{i=0}^{B}n^i = 11 - 3n + 3 \cdot \sum_{k=0}^{n-2}k^B + n \cdot \frac{n^{B+1} - 1}{n - 1}$$

$$= 11 - 3n + 3 \cdot \sum_{k=0}^{n-2}k^B + \underbrace{\frac{n}{n-1}}_{\leq 2} \cdot (n^{B+1} - 1) \qquad \text{since } n \geq 2$$

$$\leq 9 - 3n + 3 \cdot \sum_{k=0}^{n-2}k^B + 2 \cdot n^{B+1} \leq 9 - 3n + 3 \cdot \int_0^n k^B \mathrm{d}k + 2 \cdot n^{B+1}$$

$$= 9 - 3n + 3 \cdot \frac{n^{B+1}}{B+1} + 2 \cdot n^{B+1} = 9 - 3n + \left(2 + \frac{3}{B+1}\right) \cdot n^{B+1}$$

$$\leq 9 - 3n + 3 \cdot n^{B+1} \qquad \text{since } B \geq 2$$

$$\leq 3 \cdot n^{B+1} \qquad \text{since } n \geq 3 \ .$$

Since G is connected we have $B = 0 \Leftrightarrow n = 1$, $B = 1 \Leftrightarrow n = 2$. Thus, it is easily verified that the bound holds for $(B, n) \in \{(0, 1), (1, 2)\}$ as well. □

Because Markov distributions are DAG-induced by chain graphs and the maximum border of a chain graph is at most 2 we immediately obtain a tighter version of Corollary 10 whose proof directly follows from Theorem 15.

Corollary 17. *Let* PKE *be* IND-CPA *secure. Then* PKE *is* IND-SO-CPA *secure with respect to efficiently resamplable Markov distributions over* $\mathcal{M}^n$.

In particular, for any adversary $\mathcal{A}_{\mathsf{SO}}$ *run in game* IND-SO-CPA$_{\mathsf{PKE}}$ *there exists an* IND-CPA$_{\mathsf{PKE}}$ *adversary* $\mathcal{B}_{\mathsf{CPA}}$ *with roughly the running time of* $\mathcal{A}_{\mathsf{SO}}$ *plus three executions of* Resamp *such that*

$$\mathbf{Adv}_{\mathsf{PKE}}^{\mathsf{IND\text{-}SO\text{-}CPA}}(\mathcal{A}_{\mathsf{SO}}, \mathfrak{D}_\lambda, \lambda) \leq 3 \cdot n^3 \cdot \mathbf{Adv}_{\mathsf{PKE}}^{\mathsf{IND\text{-}CPA}}(\mathcal{B}_{\mathsf{CPA}}, \lambda).$$

Applying the proof of Theorem 15 directly to the Markov case gives a slightly better bound on the loss, namely $n \cdot (n+1) \cdot (2n+1)/6$, since $N(\mathsf{middle}) \cap [n-1] = 1$ even for the last transition H_{n-1} to H_n. Hence, the loss in Eq. (4) decreases to $\sum_{k=0}^{n-1} (k+1)^2$.

Recall that the hybrids in the proof of Theorem 15 saved us a factor of n because it suffices to guess a set of size at most $B(G) - 1$ instead of $B(G)$ for $k < n - 1$ as at least one vertex of the neighborhood of middle is contained in right.

The same hybrids can be used to strengthen Theorem 8 as it suffices to guess a connected subgraph in $[k + 1]$ (instead of $[n]$) containing vertex $k + 1$.

Since G is connected, there is at least a path in $\{k + 1\} \cup$ right that contains $k + 1$, i.e. at least $n - k$ connected subgraphs in right $\cup \{k + 1\}$. Thus, there exist at least $n - k$ connected subgraphs in G that contain vertex $k + 1$ and are identical if restricted to $[k + 1]$. Hence the probability that the reduction guesses C_{k+1} correctly can be increased from $1/S(G)$ to $(n - k)/S(G)$, bringing the loss from $\mathcal{O}(n^2) \cdot S(G)$ down to $\mathcal{O}(n \cdot \log n) \cdot S(G)$.

3.5 A Hybrid Argument for Disconnected Graphs

Let G be a graph with z' connected components. Fix any semi-order on them, e.g. ordered by the smallest vertex in each component and let $V_1, \ldots, V_{z'}$ denote the sets of vertices of the connected components of G. For $j \in [z' + 1, n]$ let $V_j := \emptyset$. We define a security game where an adversary plays the IND-SO-CPA game on a connected component of the graph that induced the distribution chosen by the adversary.

Procedure INITIALIZE	**Procedure** OPEN(i)
$(pk, sk) \leftarrow \mathsf{Gen}(1^\lambda)$	$\mathcal{I} := \mathcal{I} \cup \{i\}$
Return pk	Return $(\mathbf{m}_i^0, \mathbf{r}_i)$
Procedure ENC($\mathfrak{D}$, Resamp$_\mathfrak{D}$)	**Procedure** CHALLENGE
$\mathbf{m}^0 \leftarrow \mathfrak{D}$	$\mathbf{m}^1 \leftarrow \mathsf{Resamp}_\mathfrak{D}(\mathbf{m}^0, \mathcal{I})$
$\mathbf{r} \leftarrow_\$ \mathcal{R}^n$	Return $\mathbf{m}_{V_z}^b$
$\mathbf{c} = \mathsf{Enc}_{pk}(\mathbf{m}_{V_z}^0 ; \mathbf{r}_{V_z})$	**Procedure** FINALIZE(b')
Return $\mathbf{c}$	Return b'

Fig. 8. $\mathcal{B}_{\mathsf{G\text{-}SO}}$'s interface in game G-IND-SO-CPA$_{\mathsf{PKE},b,z}$.

Definition 18. *For a public-key encryption scheme* $\mathsf{PKE} := (\mathsf{Gen}, \mathsf{Enc}, \mathsf{Dec})$, *a bit* b, *a family* $\mathcal{F}$ *of efficiently resamplable, G-induced distributions over* $\mathcal{M}^n$, $z \in [n]$ *and an adversary* $\mathcal{B}_{\mathsf{G\text{-}SO}}$ *we consider game* G-IND-SO-CPA$_{\mathsf{PKE},b,z}^{\mathcal{B}_{\mathsf{G\text{-}SO}}}$ *given in Fig. 8. Run in the game,* $\mathcal{B}_{\mathsf{G\text{-}SO}}$ *calls* ENC *once right after* INITIALIZE *and submits* $\mathfrak{D} \in \mathcal{F}$ *along with a* PPT *resampling algorithm* Resamp$_\mathfrak{D}$. $\mathcal{B}_{\mathsf{G\text{-}SO}}$ *may call* OPEN *multiple times but only for* $i \in V_z$ *and invokes* CHALLENGE *once after its last* OPEN *query before calling* FINALIZE. *We define the advantage of* $\mathcal{B}_{\mathsf{G\text{-}SO}}$ *run in* IND-SO-CPA$_{\mathsf{PKE},b,z}$ *as*

$$\mathbf{Adv}_{\mathsf{PKE},z}^{\mathsf{G\text{-}IND\text{-}SO\text{-}CPA}}(\mathcal{B}_{\mathsf{G\text{-}SO}}, \mathfrak{D}_\lambda, \lambda) :=$$
$$\mathbf{Adv}\left(\mathsf{G\text{-}IND\text{-}SO\text{-}CPA}_{\mathsf{PKE},0,z}^{\mathcal{B}_{\mathsf{G\text{-}SO}}}, \mathsf{G\text{-}IND\text{-}SO\text{-}CPA}_{\mathsf{PKE},1,z}^{\mathcal{B}_{\mathsf{G\text{-}SO}}}\right).$$

PKE *is* G-IND-SO-CPA$_z$ *secure w.r.t.* $\mathcal{F}$ *if* $\mathbf{Adv}_{\mathsf{PKE},z}^{\mathsf{IND\text{-}SO\text{-}CPA}}(\mathcal{B}_{\mathsf{G\text{-}SO}}, \mathfrak{D}_\lambda, \lambda)$ *is negligible for all* PPT *adversaries* $\mathcal{B}_{\mathsf{G\text{-}SO}}$. PKE *is* G-IND-SO-CPA *secure w.r.t.* $\mathcal{F}$ *if* PKE *is* G-IND-SO-CPA$_z$ *secure w.r.t.* $\mathcal{F}$ *for all* $z \in [n]$.

We have $\mathbf{Adv}_{\mathsf{PKE},z}^{\mathsf{G\text{-}IND\text{-}SO\text{-}CPA}}(\mathcal{B}_{\mathsf{G\text{-}SO}}, \mathfrak{D}_\lambda, \lambda) = 0$ *for* $z \in [z'+1, n]$.

Theorem 19. *Let* PKE *be* G-IND-SO-CPA *secure w.r.t. a family* $\mathcal{F}$ *of efficiently resamplable and G-induced distributions over* $\mathcal{M}^n$, *then* PKE *is* IND-SO-CPA *secure w.r.t* $\mathcal{F}$.

Proof. Again, the main idea is that connected components can be dealt with independently. We give a hybrid argument over the connected components of G_λ using G-IND-SO-CPA$_z$ security for switching connected component z from sampled to resampled. See Fig. 9 for code of CHALLENGE in hybrid H_z; every other procedure stays as in IND-SO-CPA$_{\mathsf{PKE},b}$ (cf. Fig. 2).

Note that H_0 is identical to game IND-SO-CPA$_{\mathsf{PKE},0}$ and $\mathsf{H}_{z'}$ is identical to IND-SO-CPA$_{\mathsf{PKE},1}$. Thus

$$\mathbf{Adv}_{\mathsf{PKE}}^{\mathsf{IND\text{-}SO\text{-}CPA}}(\mathcal{A}_{\mathsf{SO}}, \mathfrak{D}_\lambda, \lambda) = \mathbf{Adv}\left(\mathsf{H}_0^{\mathcal{A}_{\mathsf{SO}}}, \mathsf{H}_{z'}^{\mathcal{A}_{\mathsf{SO}}}\right) \leq \sum_{z=0}^{z'-1} \mathbf{Adv}\left(\mathsf{H}_z^{\mathcal{A}_{\mathsf{SO}}}, \mathsf{H}_{z+1}^{\mathcal{A}_{\mathsf{SO}}}\right) .$$

We proceed with the following Lemma.

Procedure CHALLENGE
$\mathbf{m}^1 \leftarrow \mathsf{Resamp}_{\mathfrak{D}}(\mathbf{m}^0, \mathcal{I})$
$$\mathbf{m}_i = \begin{cases} \mathbf{m}_i^1 & \text{for } i \in \bigcup_{j=1}^z V_j \\ \mathbf{m}_i^0 & \text{else} \end{cases}$$
Return $\mathbf{m} = (\mathbf{m}_1, \dots, \mathbf{m}_n)$

Fig. 9. Hybrid H_z. The first z connected components are already resampled conditioned on opening queries, while the rest remain sampled.

Lemma 20. *For every adversary $\mathcal{A}_{SO}$ distinguishing hybrids H_z and H_{z+1} there exists an adversary $\mathcal{B}_{G\text{-}SO}$ run in game $\mathsf{G\text{-}IND\text{-}SO\text{-}CPA}_{\mathsf{PKE},z+1}$ with roughly the running time plus one executions of Resamp such that*

$$\mathbf{Adv}\big(H_z^{\mathcal{A}_{SO}}, H_{z+1}^{\mathcal{A}_{SO}}\big) \leq \mathbf{Adv}_{\mathsf{PKE},z+1}^{\mathsf{G\text{-}IND\text{-}SO\text{-}CPA}}(\mathcal{B}_{G\text{-}SO}, \mathfrak{D}_\lambda, \lambda).$$

Proof. We construct an adversary $\mathcal{B}_{G\text{-}SO}$ that interpolates between hybrids H_z and H_{z+1} for $\mathcal{A}_{SO}$. $\mathcal{B}_{G\text{-}SO}$ proceeds as follows (cf. Fig. 10).

$\mathcal{B}_{G\text{-}SO}$ forwards pk to $\mathcal{A}_{SO}$. On $\mathcal{A}_{SO}$'s call of ENC, $\mathcal{B}_{G\text{-}SO}$ calls $\mathsf{ENC}_{\mathsf{G\text{-}IND\text{-}SO\text{-}CPA}_{z+1}}$ to obtain an encryption $\mathbf{c}_{V_{z+1}}$ of messages in the component V_{z+1}. $\mathcal{B}_{G\text{-}SO}$ samples messages $\mathbf{m}^0 \leftarrow \mathfrak{D}$ on its own and encrypts the messages in $\overline{V_{z+1}}$. $\mathcal{B}_{G\text{-}SO}$ sends $\mathbf{c} = (\mathbf{c}_1, \dots, \mathbf{c}_n)$ to $\mathcal{A}_{SO}$. $\mathcal{B}_{G\text{-}SO}$ answers opening queries on its own unless they occur on V_{z+1}, where it invokes its $\mathsf{OPEN}_{\mathsf{G\text{-}IND\text{-}SO\text{-}CPA}_{z+1}}$ oracle to answer. On CHALLENGE, $\mathcal{B}_{G\text{-}SO}$ receives a challenge message vector $\mathbf{m}_{V_{z+1}}$ by calling $\mathsf{CHALLENGE}_{\mathsf{G\text{-}IND\text{-}SO\text{-}CPA}_{z+1}}$ and resamples $\mathbf{m}^1$ conditioned on $\mathcal{I}$. $\mathcal{B}_{G\text{-}SO}$ returns resampled messages $\mathbf{m}^1$ on $\bigcup_{j=1}^z V_j$, its challenge messages $\mathbf{m}_{V_{z+1}}$ and sampled messages $\mathbf{m}^0$ for $\bigcup_{j=z+2}^n V_j$ to $\mathcal{A}_{SO}$. $\mathcal{B}_{G\text{-}SO}$ outputs whatever $\mathcal{A}_{SO}$ outputs.

Obviously $\mathcal{B}_{G\text{-}SO}$ simulates the hybrids correctly during ENC since it always returns encryptions of sampled messages. On $\mathcal{A}_{SO}$'s call of CHALLENGE the messages in the first z connected components are already resampled while the messages in the last $n-z-1$ connected components are sampled as in hybrids H_z and H_{z+1}. When $\mathcal{B}_{G\text{-}SO}$ is run in game $\mathsf{G\text{-}IND\text{-}SO\text{-}CPA}_{\mathsf{PKE},0,z+1}$, it obtains sampled messages for the $(z+1)$-th connected component; thus it runs $\mathcal{A}_{SO}$ in hybrid H_z. When run in $\mathsf{G\text{-}IND\text{-}SO\text{-}CPA}_{\mathsf{PKE},1,z+1}$, $\mathcal{B}_{G\text{-}SO}$ receives resampled messages for V_{z+1}; hence running $\mathcal{A}_{SO}$ in hybrid H_{z+1}. Thus

$$\Pr[\mathsf{G\text{-}IND\text{-}SO\text{-}CPA}_{\mathsf{PKE},0,z+1}^{\mathcal{B}_{G\text{-}SO}} \Rightarrow 1] = \Pr[H_z^{\mathcal{A}_{SO}} \Rightarrow 1] \text{ and}$$
$$\Pr[\mathsf{G\text{-}IND\text{-}SO\text{-}CPA}_{\mathsf{PKE},1,z+1}^{\mathcal{B}_{G\text{-}SO}} \Rightarrow 1] = \Pr[H_{z+1}^{\mathcal{A}_{SO}} \Rightarrow 1].$$

Lemma 20 follows.   □
We obtain

$$\mathbf{Adv}_{\mathsf{PKE}}^{\mathsf{IND\text{-}SO\text{-}CPA}}(\mathcal{A}_{SO}, \mathfrak{D}_\lambda, \lambda) \leq \sum_{z=1}^{z'} \mathbf{Adv}_{\mathsf{PKE},z}^{\mathsf{G\text{-}IND\text{-}SO\text{-}CPA}}(\mathcal{B}_{G\text{-}SO}, \mathfrak{D}_\lambda, \lambda)$$

and Theorem 19 follows immediately since $z' \leq n$.   □

Procedure INITIALIZE

$pk \leftarrow \text{Gen}_{\text{G-IND-SO-CPA}_{z+1}}(1^{\lambda})$

Return pk

Procedure $\text{ENC}(\mathfrak{D}, \text{Resamp}_{\mathfrak{D}})$

$\mathbf{c}_{V_{z+1}} \leftarrow \text{ENC}_{\text{G-IND-SO-CPA}_{z+1}}(\mathfrak{D}, \text{Resamp}_{\mathfrak{D}})$

$\mathbf{m}^0 \leftarrow \mathfrak{D}$

$\mathbf{r} \leftarrow^{\$} \mathcal{R}^n$

$$c_i = \begin{cases} \mathbf{c}_i & \text{for } i \in V_{z+1} \\ \text{Enc}_{pk}(\mathbf{m}_i^0; \mathbf{r}_i) & \text{else} \end{cases}$$

Return $\mathbf{c} = (\mathbf{c}_1, \ldots, \mathbf{c}_n)$

Procedure FINALIZE(b')

$\text{FINALIZE}_{\text{G-IND-SO-CPA}}(b')$

Procedure OPEN(i)

$\mathcal{I} := \mathcal{I} \cup \{i\}$

if $i \in V_{z+1}$

  Return $\text{OPEN}_{\text{G-IND-SO-CPA}_{z+1}}(i)$

else

  Return $(\mathbf{m}_i^0, \mathbf{r}_i)$

Procedure CHALLENGE

$\mathbf{m}_{V_{z+1}} \leftarrow \text{CHALLENGE}_{\text{G-IND-SO-CPA}_{z+1}}$

$\mathbf{m}^1 \leftarrow \text{Resamp}_{\mathfrak{D}}(\mathbf{m}^0, \mathcal{I})$

$$\mathbf{m}_i = \begin{cases} \mathbf{m}_i^1 & \text{for } i \in \bigcup_{j=1}^{z} V_j \\ \mathbf{m}_i & \text{for } i \in V_{z+1} \\ \mathbf{m}_i^0 & \text{else} \end{cases}$$

Return $\mathbf{m} = (\mathbf{m}_1, \ldots, \mathbf{m}_n)$

Fig. 10. Reduction run by $\mathcal{B}_{\text{G-SO}}$ to simulate H_z (or H_{z+1}) when $\mathcal{B}_{\text{G-SO}}$ is run in G-IND-SO-CPA$_{\text{PKE},0,z+1}$ (or G-IND-SO-CPA$_{\text{PKE},1,z+1}$).

In particular, we achieve versions of Theorem 8, Corollary 14 and Theorem 15 for disconnected graphs, where

$$S(G) = \sum_{i=1}^{z'} S(C_i) \quad \text{and} \quad B(G) = \max_{i \in [z']}\{B(C_i)\}$$

for a graph G consisting of connected components $C_1, \ldots, C_{z'}$.

Moreover, for $G = ([n], \emptyset)$, G-induced distributions become product distributions, i.e. the messages are sampled independently. Hence, the positive result of [3] can be seen as a special case of Theorem 19.

Acknowledgements. We would like to thank the anonymous reviewers for their helpful comments and remarks.

  G. Fuchsbauer and K. Pietrzak are supported by the European Research Council, ERC Starting Grant (259668-PSPC). F. Heuer is funded by a Sofja Kovalevskaja Award of the Alexander von Humboldt Foundation and DFG SPP 1736, Algorithms for BIG DATA. E. Kiltz is supported by a Sofja Kovalevskaja Award of the Alexander von Humboldt Foundation, the German Israel Foundation, and ERC Project ERCC (FP7/615074).

References

1. Bellare, M., Dowsley, R., Waters, B., Yilek, S.: Standard security does not imply security against selective-opening. In: Pointcheval, D., Johansson, T. (eds.) EURO-CRYPT 2012. LNCS, vol. 7237, pp. 645–662. Springer, Heidelberg (2012)

2. Bellare, M., Hofheinz, D., Yilek, S.: Possibility and impossibility results for encryption and commitment secure under selective opening. In: Joux, A. (ed.) EUROCRYPT 2009. LNCS, vol. 5479, pp. 1–35. Springer, Heidelberg (2009)
3. Bellare, M., Yilek, S.: Encryption schemes secure under selective opening attack. IACR Cryptology ePrint Archive 2009, p. 101 (2009)
4. Böhl, F., Hofheinz, D., Kraschewski, D.: On definitions of selective opening security. In: Fischlin, M., Buchmann, J., Manulis, M. (eds.) PKC 2012. LNCS, vol. 7293, pp. 522–539. Springer, Heidelberg (2012)
5. Borwein, J.M., Chan, O.-Y.: Uniform bounds for the complementary incomplete gamma function. Math. Inequal. Appl. **12**, 115–121 (2009)
6. Canetti, R., Dwork, C., Naor, M., Ostrovsky, R.: Deniable encryption. In: Kaliski Jr., B.S. (ed.) CRYPTO 1997. LNCS, vol. 1294, pp. 90–104. Springer, Heidelberg (1997)
7. Canetti, R., Feige, U., Goldreich, O., Naor, M.: Adaptively secure multi-party computation. In: 28th ACM STOC, 22–24 May 1996, Philadephia, Pennsylvania, USA, pp. 639–648 (1996)
8. Dwork, C., Naor, M., Reingold, O., Stockmeyer, L.J.: Magic functions. In: 40th FOCS, 17–19 October 1999, pp. 523–534. IEEE Computer Society Press, New York (1999)
9. Hofheinz, D., Rao, V., Wichs. D.: Standard security does not imply indistinguishability under selective opening. Cryptology ePrint Archive, Report 2015/792 (2015). http://eprint.iacr.org/
10. Hofheinz, D., Rupp, A.: Standard versus selective opening security: separation and equivalence results. In: Lindell, Y. (ed.) TCC 2014. LNCS, vol. 8349, pp. 591–615. Springer, Heidelberg (2014)
11. Natalini, P., Palumbo, B.: Inequalities for the incomplete gamma function. Math. Inequal. Appl. **3**, 69–77 (2000)

Non-Malleable Encryption: Simpler, Shorter, Stronger

Sandro Coretti[1]([✉]), Yevgeniy Dodis[2], Björn Tackmann[3],
and Daniele Venturi[4]

[1] Department of Computer Science, ETH Zürich, Zürich, Switzerland
corettis@inf.ethz.ch
[2] Department of Computer Science, New York University, New York, USA
dodis@cs.nyu.edu
[3] Department of Computer Science & Engineering, UC San Diego, La Jolla, USA
btackmann@eng.ucsd.edu
[4] Department of Computer Science, Sapienza University of Rome, Rome, Italy
venturi@di.uniroma1.it

Abstract. In a seminal paper, Dolev *et al.* [15] introduced the notion of
non-malleable encryption (NM-CPA). This notion is very intriguing since
it suffices for many applications of chosen-ciphertext secure encryption
(IND-CCA), and, yet, can be generically built from semantically secure
(IND-CPA) encryption, as was shown in the seminal works by Pass *et al.*
[29] and by Choi *et al.* [9], the latter of which provided a black-box
construction. In this paper we investigate three questions related to NM-
CPA security:

1. Can the rate of the construction by Choi *et al.* of NM-CPA from
   IND-CPA be improved?
2. Is it possible to achieve multi-bit NM-CPA security more efficiently
   from a single-bit NM-CPA scheme than from IND-CPA?
3. Is there a notion stronger than NM-CPA that has natural applications
   and can be achieved from IND-CPA security?

We answer all three questions in the positive. First, we improve the rate
in the scheme of Choi *et al.* by a factor $\mathcal{O}(\lambda)$, where λ is the security
parameter. Still, encrypting a message of size $\mathcal{O}(\lambda)$ would require ciphertext and keys of size $\mathcal{O}(\lambda^2)$ times that of the IND-CPA scheme, even in
our improved scheme. Therefore, we show a more efficient domain extension technique for building a λ-bit NM-CPA scheme from a single-bit
NM-CPA scheme with keys and ciphertext of size $\mathcal{O}(\lambda)$ times that of the
NM-CPA one-bit scheme. To achieve our goal, we define and construct a
novel type of continuous non-malleable code (NMC), called *secret-state
NMC*, as we show that standard continuous NMCs are *not enough* for
the natural "encode-then-encrypt-bit-by-bit" approach to work.

Finally, we introduce a new security notion for public-key encryption that we dub *non-malleability under (chosen-ciphertext) self-destruct
attacks* (NM-SDA). After showing that NM-SDA is a *strict* strengthening of NM-CPA and allows for more applications, we nevertheless show
that both of our results—(faster) construction from IND-CPA and domain

© International Association for Cryptologic Research 2016
E. Kushilevitz and T. Malkin (Eds.): TCC 2016-A, Part I, LNCS 9562, pp. 306–335, 2016.
DOI: 10.1007/978-3-662-49096-9_13

extension from one-bit scheme—also hold for our stronger NM-SDA security. In particular, the notions of IND-CPA, NM-CPA, and NM-SDA security are all equivalent, lying (plausibly, strictly?) below IND-CCA security.

1   Introduction

Several different security notions for public-key encryption (PKE) have been proposed. The most basic one is that of indistinguishability under chosen-plaintext attacks (IND-CPA) [21], which requires that an adversary with no decryption capabilities be unable to distinguish between the encryption of two messages. Although extremely important and useful for a number of applications, in many cases IND-CPA security is not sufficient. For example, consider the simple setting of an electronic auction, where the auctioneer U publishes a public key pk, and invites several participants $P_1, \ldots, P_q$ to encrypt their bids b_i under pk. As was observed in the seminal paper of Dolev *et al.* [15], although IND-CPA security of encryption ensures that P_1 cannot decrypt a bid of P_2 under the ciphertext e_2, it leaves open the possibility that P_1 can construct a special ciphertext e_1 which decrypts to a *related* bid b_1 (e.g., $b_1 = b_2 + 1$). Hence, to overcome such "malleability" problems, stronger forms of security are required.

The strongest such level of PKE security is indistinguishability under chosen-ciphertext attacks (IND-CCA), where the adversary is given unrestricted, adaptive access to a decryption oracle (modulo not being able to ask on the "challenge ciphertext"). This notion is sufficient for most natural applications of PKE, and several generic [5,15,25,28,31] and concrete [13,14,22,24] constructions of IND-CCA secure encryption schemes are known by now. Unfortunately, all these constructions either rely on specific number-theoretic assumptions, or use much more advanced machinery (such as non-interactive zero-knowledge proofs or identity-based encryption) than IND-CPA secure encryption. Indeed, despite numerous efforts (e.g., a partial negative result [20]), the relationship between IND-CPA and IND-CCA security remains unresolved until now. This motivates the study of various "middle-ground" security notions between IND-CPA and IND-CCA, which are sufficient for applications, and, yet, might be constructed from simpler basic primitives (e.g., any IND-CPA encryption).

One such influential notion is non-malleability under chosen-plaintext attacks (NM-CPA), originally introduced by Dolev *et al.* [15] with the goal of precisely addressing the auction example above, by demanding that an adversary not be able to maul ciphertexts to other ciphertexts encrypting related plaintexts. As was later shown by Bellare and Sahai [4] and by Pass *et al.* [30], NM-CPA is equivalent to security against adversaries with access to a *non-adaptive* decryption oracle, meaning that the adversary can only ask one "parallel" decryption query. Although NM-CPA appears much closer to IND-CCA than IND-CPA security, a seminal result by Pass *et al.* [29] showed that one can generically build NM-CPA encryption from any IND-CPA-secure scheme, and Choi *et al.* [9] later proved that this transformation can also be achieved via a black-box construction. Thus, NM-CPA schemes can be potentially based on weaker assumptions than IND-CCA schemes, and yet suffice for important applications.

Our Work. We investigate three questions related to NM-CPA security:

1. Can the efficiency of the construction by Choi *et al.* of NM-CPA from IND-CPA be improved?
2. Is it possible to achieve multi-bit NM-CPA security more efficiently from a single-bit NM-CPA scheme than from IND-CPA?
3. Is there a notion stronger than NM-CPA that has natural applications and can be achieved from IND-CPA security?

We answer all three questions positively. We start with Question 3, as it will also allow us to achieve stronger answers for Questions 1 and 2. In a recent paper, Coretti *et al.* [10] introduced a new middle-ground security notion for encryption—termed indistinguishability under (chosen-ciphertext) self-destruct attacks (IND-SDA) in this paper[1]—where the adversary gets access to an *adaptive* decryption oracle, which, however, stops decrypting after the first *invalid* ciphertext is submitted. Applying this notion to the auction example above, it means that the auctioneer can reuse the secret key for subsequent auctions, as long as all the encrypted bids are valid. Unfortunately, if an invalid ciphertext is submitted, even the results of the *current* auction should be discarded, as IND-SDA security is not powerful enough to argue that the decryptions of the remaining ciphertexts are unrelated w.r.t. prior plaintexts.

Motivated by the above, we introduce a new security notion that we dub *non-malleability under (chosen-ciphertext) self-destruct attacks* (NM-SDA). This notion (see Definition 3) naturally combines NM-CPA and IND-SDA, by allowing the adversary to ask many adaptive "parallel" decryption queries (i.e., a query consists of many ciphertexts) up to the point when the first invalid ciphertext is submitted. In such a case, the whole parallel decryption query containing an invalid ciphertext is still answered in full, but no future decryption queries are allowed. By being stronger (as we show below) than both NM-CPA and IND-SDA, NM-SDA security appears to be a strongest natural PKE security notion that is still weaker (as we give evidence below) than IND-CCA—together with q-bounded CCA-secure PKE [12], to which it seems incomparable. In particular, it seems to apply better to the auction example above: First, unlike with basic NM-CPA, the auctioneer can reuse the same public key pk, provided no invalid ciphertexts were submitted. Second, unlike IND-SDA, the current auction can be safely completed, even if some ciphertexts are invalid. Compared to IND-CCA, however, the auctioneer will still have to change its public key for *subsequent* auctions if some of the ciphertexts are invalid. Still, one can envision situations where parties are penalized for submitting such malformed ciphertexts, in which case NM-SDA security might be practically sufficient, leading to an implementation under (potentially) lesser computational assumptions as compared to using a full-blown IND-CCA PKE.

Having introduced and motivated NM-SDA security, we provide a comprehensive study of this notion, and its relationship to other PKE security notions. The prior notions of NM-CPA and IND-SDA are incomparable, meaning that

---

[1] The original name used in [10] is self-destruct chosen-ciphertext attacks security.

there are (albeit contrived) schemes that satisfy the former but not the latter notion and vice versa. This is shown in the full version of this work and implies that NM-SDA security is strictly stronger than either of the two other notions.

We turn to Question 2 above and answer it affirmatively even for our *stronger* notion of NM-SDA security; indeed, our security proof is easily seen to carry over to the simpler case of NM-CPA security. Finally, we also *simultaneously* answer Questions 1 and 3, by presenting a generalization of the Choi et al. [9] construction from IND-CPA encryption which: (a) allows us to improve the plaintext-length to ciphertext-length rate by a factor linear in the security parameter as compared to the construction of [9] (which is a special case of our abstraction, but with sub-optimal parameters); (b) generically achieves NM-SDA security (with or without the efficiency improvement). We detail these results below.

Domain Extension. For several security notions in public-key cryptography, is is known that single-bit public-key encryption implies multi-bit public-key encryption. For IND-CPA, this question is simple [21], since the parallel repetition of a single-bit scheme (i.e., encrypting every bit of a message separately) yields an IND-CPA secure multi-bit scheme. For the other notions considered in this paper, i.e., for NM-CPA, IND-SDA, and NM-SDA, as well as for IND-CCA, the parallel repetition (even using independent public keys) is not a scheme that achieves the same security level as the underlying single-bit scheme. However, Coretti et al. [10] provide a single-to-multi-bit transformation for IND-SDA security based on non-malleable codes [17] (see below), and Myers and Shelat [27], as well as Hohenberger et al. [23], provide (much) more complicated such transformations for IND-CCA security. To complement these works, we answer the question of domain extension for NM-SDA and NM-CPA in the affirmative. In particular we show the following result:

Theorem 1 (Informal). *Let λ be the security parameter. Then there is a black-box construction of a λ-bit NM-SDA (resp. NM-CPA) PKE scheme from a single-bit NM-SDA (resp. NM-CPA) PKE scheme, making $\mathcal{O}(\lambda)$ calls to the underlying single-bit scheme.*[2]

The proof of Theorem 1 can be found in Sect. 4. Our approach follows that for IND-SDA [10] and combines single-bit PKE with so-called *non-malleable codes (NMCs)*, introduced by Dziembowski et al. [17]. Intuitively, NMCs protect encoded messages against a tampering adversary, which tampers with the codeword by means of applying functions f from a particular function class $\mathcal{F}$ to it, in the sense that the decoding results in either the original message or a completely unrelated value.

Our construction has the following simple structure (see also Fig. 4): The plaintext m is first encoded using an appropriate non-malleable code into an encoding c, which is in turn encrypted bit-by-bit (under independent public keys) with the single-bit NM-SDA scheme.[3] The fact that NM-SDA security

---

[2] For longer than λ-bit messages, one can also use standard hybrid encryption.

[3] Technically, this scheme only achieves a relaxation of NM-SDA security, called *replayable* NM-SDA security, but the latter can be easily transformed into the former.

guarantees that an attacker can either leave a ciphertext intact or replace it, which results in an unrelated message, translates to the following capability of an adversary w.r.t. decryption queries: It can either leave a particular bit of the encoding unchanged, or fix it to 0 or to 1. Therefore, the tamper class against which the non-malleable code must be resilient is the class $\mathcal{F}_{\mathsf{set}}$ of functions that tamper with each bit of an encoding individually and can either leave it unchanged or set it to a fixed value.

The main new challenge for our construction is to deal with the *parallel* decryption queries: in order for the combined scheme to be NM-SDA secure, the NMC needs to be resilient against parallel tamper queries as well. Unfortunately, we show that no standard non-malleable code (as originally defined by Dziembowski *et al.* [17] and Faust *et al.* [18]) can achieve this notion (see Sect. 4.6). Fortunately, we observe that the NMC concept can be extended to allow the decoder to make use of (an initially generated) secret state, which simply becomes part of the secret key in the combined scheme. This modification of NMCs—called secret-state NMCs—allows us to achieve resilience against parallel tampering and may be of independent interest. This reduces our question to building a secret-state non-malleable code resilient against continuous parallel tampering attacks from $\mathcal{F}_{\mathsf{set}}$. We construct such a code in Sect. 4.3, by combining the notion of linear error-correcting secret sharing (see [17]) with the idea of a secret "trigger set" [9]. This construction forms one of the main technical contributions of our work.

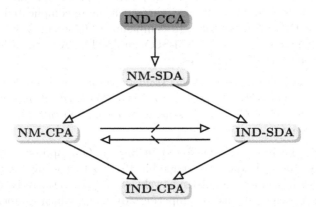

Fig. 1. Diagram of the main relationships between the security notions considered in this paper. $X \to Y$ means that X implies Y; $X \nrightarrow Y$ indicates a separation between X and Y. Notions with the same color are equivalent under black-box transformations; notions with different colors are not known to be equivalent.

NM-SDA from IND-CPA. Next, we show:

Theorem 2 (Informal). *There exists a black-box construction of an NM-SDA-secure PKE scheme from an IND-CPA-secure PKE.*

Hence, the notions of IND-CPA, NM-CPA, IND-SDA, and NM-SDA security are all equivalent, lying (plausibly, strictly?) below IND-CCA security. See Fig. 1.

The proof of Theorem 2 appears in Sect. 5. In fact, we show that a generalization of the construction by Choi *et al.* already achieves NM-SDA security (rather than only NM-CPA security). Our proof much follows the pattern of the original one, except for one key step in the proof, where a brand new proof technique is required. Intuitively, we need to argue that no sensitive information about the secret "trigger set" is leaked to the adversary, unless one of the ciphertexts is invalid. This rather general technique (for analyzing security of so called "parallel stateless self-destruct games") may be interesting in its own right (e.g., it is also used in the security proof of our non-malleable code in Sect. 4), and is detailed in Sect. 6.

Along the way, we also manage to slightly abstract the transformation of [9], and to re-phrase it in terms of certain linear error-correcting secret-sharing schemes (LECSSs) satisfying a special property (as opposed to using Reed-Solomon codes directly as an example of such a scheme). Aside from a more modular presentation (which gives a more intuitive explanation for the elegant scheme of Choi *et al.* [9]), this also allows us to instantiate the required LECSS more efficiently and thereby improve the rate of the transformation of [9] by a factor linear in the security parameter (while also arguing NM-SDA, instead of NM-CPA, security), giving us the positive answer to Question 1.[4]

2   Preliminaries

This section introduces notational conventions and basic concepts that we use throughout the work.

Bits and Symbols. Let $\ell \in \mathbb{N}$. For any multiple $m = t\ell$ of ℓ, an m-bit string $x = (x[1], \ldots, x[m]) = (x_1, \ldots, x_t)$ can be seen as composed of its *bits* $x[j]$ or its *symbols* $x_i \in \{0,1\}^\ell$. For two m-bit strings x and y, denote by $d_\mathsf{H}(x, y)$ their hamming distance as the number of *symbols* in which they differ.

Oracle Algorithms. Oracle algorithms are algorithms that can make special oracle calls. An algorithm A with an oracle O is denoted by $A(O)$. Note that oracle algorithms may make calls to other oracle algorithms (e.g., $A(B(O))$).

Distinguishers and Reductions. A *distinguisher* is an (possibly randomized) oracle algorithm $D(\cdot)$ that outputs a single bit. The distinguishing advantage on two (possibly stateful) oracles S and T is defined by

$$\Delta^D(S, T) \quad := \quad |\mathsf{P}[D(S) = 1] - \mathsf{P}[D(T) = 1]|,$$

---

[4] Note that Choi *et al.* [9] consider the ciphertext blow-up between the underlying IND-CPA scheme and the resulting scheme as quality measure of their construction, while we consider the rate (number of plaintext bits per ciphertext bit) of the resulting scheme.

where probabilities are over the randomness of D as well as S and T, respectively.

Reductions between distinguishing problems are modeled as oracle algorithms as well. Specifically, when reducing distinguishing two oracles U and V to distinguishing S and T, one exhibits an oracle algorithm $R(\cdot)$ such that $R(U)$ behaves as S and $R(V)$ as T; then, $\Delta^D(S,T) = \Delta^D(R(U), R(V)) = \Delta^{D(R(\cdot))}(U,V)$.

Linear Error-Correcting Secret Sharing. The following notion of a linear error-correcting secret sharing, introduced by Dziembowski *et al.* [17], is used in several places in this paper.

Definition 1 (Linear error-correcting sharing scheme). *Let $n \in \mathbb{N}$ be a security parameter and $\mathbb{F}$ a field of size $L = 2^\ell$ for some $\ell \in \mathbb{N}$. A (k, n, δ, τ) linear error-correcting secret sharing (LECSS) over $\mathbb{F}$ is a pair of algorithms (E, D), where $\mathsf{E} : \mathbb{F}^k \rightarrow \mathbb{F}^n$ is randomized and $\mathsf{D} : \mathbb{F}^n \times \mathbb{N} \rightarrow \mathbb{F}^k \cup \{\bot\}$ is deterministic, with the following properties:*

- Linearity: *For any vectors w output by E and any $c \in \mathbb{F}^n$,*

$$\mathsf{D}(w + c) = \begin{cases} \bot & \text{if } \mathsf{D}(c) = \bot, \text{ and} \\ \mathsf{D}(w) + \mathsf{D}(c) & \text{otherwise.} \end{cases}$$

- Minimum distance: *For any two codewords w, w' output by E, $d_\mathsf{H}(w, w') \geq \delta n$.*
- Error correction: *It is possible to efficiently correct up to $\delta n/2$ errors, i.e., for any $x \in \mathbb{F}^k$ and any w output by $\mathsf{E}(x)$, if $d_\mathsf{H}(c, w) \leq t$ for some $c \in \mathbb{F}^n$ and $t < \delta n/2$, then $\mathsf{D}(c,t) = x$.*
- Secrecy: *The symbols of a codeword are individually uniform over $\mathbb{F}$ and and τn-wise independent (over the randomness of E).*

This paper considers various instantiations of LECSSs, which are described in Sects. 4.5 and 5.3, where they are used.

One-time Signatures. A *digital signature scheme* (DSS) is a triple of algorithms $\Sigma = (KG, S, V)$, where the key-generation algorithm KG outputs a key pair $(\mathsf{sk}, \mathsf{vk})$, the (probabilistic) signing algorithm S takes a message m and a signing key sk and outputs a signature $s \leftarrow S_\mathsf{sk}(m)$, and the verification algorithm takes a verification key vk, a message m, and a signature s and outputs a single bit $V_\mathsf{vk}(m, s)$. A *(strong) one-time signature (OTS) scheme* is a digital signature scheme that is secure as long as an adversary only observes a single signature. More precisely, OTS security is defined using the following game $G^{\Sigma, \mathsf{ots}}$ played by an adversary A: Initially, the game generates a key pair $(\mathsf{sk}, \mathsf{vk})$ and hands the verification key vk to A. Then, A can specify a single message m for which he obtains a signature $s \leftarrow S_\mathsf{vk}(m)$. Then, the adversary outputs a pair (m', s'). The adversary wins the game if $(m', s') \neq (m, s)$ and $V_\mathsf{vk}(m', s') = 1$. The *advantage* of A is the probability (over all involved randomness) that A wins the game, and is denoted by $\Gamma^A(G^{\Sigma, \mathsf{ots}})$.

Definition 2. *A DSS scheme Σ is a (t, ε)-strong one-time signature scheme if for all adversaries A with running time at most t, $\Gamma^A(G^{\Sigma, \mathsf{ots}}) \leq \varepsilon$.*

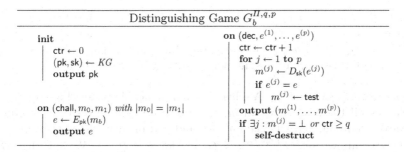

Fig. 2. Distinguishing game $G_b^{\Pi,q,p}$, where $b \in \{0,1\}$, used to define security of a PKE scheme $\Pi = (KG, E, D)$. The numbers $q, p \in \mathbb{N}$ specify the maximum number of decryption queries and their size, respectively. The command **self − destruct** results in all future decryption queries being answered by $\perp$.

3   Non-Malleability Under Self-Destruct Attacks

A *public-key encryption (PKE) scheme* with message space $\mathcal{M} \subseteq \{0,1\}^*$ and ciphertext space $\mathcal{C}$ is defined as three algorithms $\Pi = (KG, E, D)$, where the key-generation algorithm KG outputs a key pair $(\mathsf{pk}, \mathsf{sk})$, the (probabilistic) encryption algorithm E takes a message $m \in \mathcal{M}$ and a public key pk and outputs a ciphertext $c \leftarrow E_{\mathsf{pk}}(m)$, and the decryption algorithm takes a ciphertext $c \in \mathcal{C}$ and a secret key sk and outputs a plaintext $m \leftarrow D_{\mathsf{sk}}(c)$. The output of the decryption algorithm can be the special symbol $\perp$, indicating an invalid ciphertext. A PKE scheme is correct if $m = D_{\mathsf{sk}}(E_{\mathsf{pk}}(m))$ (with probability 1 over the randomness in the encryption algorithm) for all messages m and all key pairs $(\mathsf{pk}, \mathsf{sk})$ generated by KG.

Security notions for PKE schemes in this paper are formalized using the distinguishing game $G_b^{\Pi,q,p}$, depicted in Fig. 2: The distinguisher (adversary) is initially given a public key and then specifies two messages m_0 and m_1. One of these, namely m_b, is encrypted and the adversary is given the resulting challenge ciphertext. During the entire game, the distinguisher has access to a decryption oracle that allows him to make at most q decryption queries, each consisting of at most p ciphertexts. Once the distinguisher specifies an invalid ciphertext, the decryption oracle self-destructs, i.e., no further decryption queries are answered.

The general case is obtained when both q and p are arbitrary (denoted by $q = p = *$), which leads to our main definition of non-malleability under (chosen-ciphertext) self-destruct attacks (NM-SDA). For readability, set $G_b^{\Pi,\mathsf{nm\text{-}sda}} := G_b^{\Pi,*,*}$ for $b \in \{0,1\}$. Formally, NM-SDA is defined as follows:

Definition 3 (Non-malleability under self-destruct attacks). *A public-key encryption scheme Π is (t, q, p, ε)-NM-SDA-secure if for all distinguishers D with running time at most t and making at most q decryption queries of size at most p each, $\Delta^D(G_0^{\Pi,\mathsf{nm\text{-}sda}}, G_1^{\Pi,\mathsf{nm\text{-}sda}}) \leq \varepsilon$.*

All other relevant security notions in this paper can be derived as special cases of the above definition, by setting the parameters q and p appropriately.

Chosen-Plaintext Security (IND-CPA). In this variant, the distinguisher is not given access to a decryption oracle, i.e., $q = p = 0$. For readability, set $G_b^{\Pi,\text{ind-sda}} := G_b^{\Pi,0,0}$ for $b \in \{0,1\}$ in the remainder of this paper. We say that Π is (t,ε)-IND-CPA-secure if it is, in fact, $(t,0,0,\varepsilon)$-NM-SDA-secure.

Non-malleability (NM-CPA). A scheme is non-malleable under chosen-plaintext attacks [29], if the adversary can make a single decryption query consisting of arbitrarily many ciphertexts, i.e., $q = 1$ and p arbitrary (denoted by $p = *$). Similarly to above, set $G_b^{\Pi,\text{nm-cpa}} := G_b^{\Pi,1,*}$ for $b \in \{0,1\}$. We say that Π is (t,p,ε)-NM-CPA-secure if it is, in fact, $(t,1,p,\varepsilon)$-NM-SDA-secure.[5]

Indistinguishability Under Self-Destruct Attacks (IND-SDA). This variant, introduced in [10], allows arbitrarily many queries to the decryption oracle, but each of them may consist of a single ciphertext only, i.e., q arbitrary (denoted by $q = *$) and $p = 1$. Once more, set $G_b^{\Pi,\text{ind-sda}} := G_b^{\Pi,*,1}$. We say that Π is (t,q,ε)-IND-SDA-secure if it is, in fact, $(t,q,1,\varepsilon)$-NM-SDA-secure.

Chosen-Ciphertext Security (IND-CCA). The standard notion of IND-CCA security can be obtained as a strengthening of NM-SDA where $q = *$, $p = 1$, and the decryption oracle never self-destructs. We do not define this notion formally, as it is not the main focus of this paper.

Asymptotic Formulation. To allow for concise statements, sometimes we prefer to use an asymptotic formulation instead of stating concrete parameters. More precisely, we will say that a PKE scheme Π is X-secure for X $\in$ {IND-CPA, NM-CPA, IND-SDA, NM-SDA} if for all efficient adversaries the advantage ε in the distinguishing game is negligible in the security parameter.

Non-malleable CPA vs. Indistinguishable SDA. We provide a separation between the notions of NM-CPA and IND-SDA security; a corresponding theorem and proof can be found in the full version of this work. Given such a separation, our notion of NM-SDA security (see Definition 3) is strictly stronger than either of the two other notions.

4   Domain Extension

This section contains one of our main technical results. We show how single-bit NM-SDA PKE can be combined with so-called *secret-state non-malleable codes* resilient against *continuous parallel tampering*, which we believe is an interesting notion in its own right, to achieve multi-bit NM-SDA-secure PKE. We construct such a code and prove its security. In the full version of this paper, we additionally

---

[5] Note that the way NM-CPA is defined here is stronger than usual. This is due to the adversary's ability to ask a parallel decryption query at any time—as opposed to only after receiving the challenge ciphertext in earlier definitions (cf., e.g., [29]).

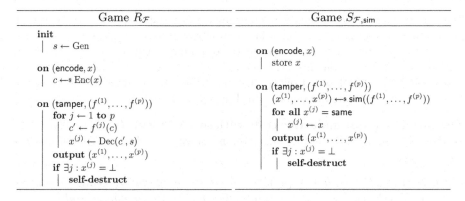

Fig. 3. Distinguishing game $(R_{\mathcal{F}}, S_{\mathcal{F},\text{sim}})$ used to define non-malleability of a secret-state coding scheme (Gen, Enc, Dec). The command **self − destruct** has the effect that all *future* queries are answered by $\bot$.

show that no code without secret state can achieve security against parallel tampering unconditionally.[6]

4.1 A New Flavor of Non-Malleable Codes

Non-malleable codes were introduced by Dziembowski *et al.* [17]. Intuitively, they protect encoded messages in such a way that any tampering with the codeword causes the decoding to either output the original message or a completely unrelated value. The original notion can be extended to include the aforementioned secret state in the decoder as follows:

Definition 4 (Code with secret state). *A (k, n)-code with secret state (CSS) is a triple of algorithms* (Gen, Enc, Dec), *where the (randomized) state-generation algorithm* Gen *outputs a secret state s from some set $\mathcal{S}$, the (randomized) encoding algorithm* Enc *takes a k-bit plaintext x and outputs an n-bit encoding $c \leftarrow \text{Enc}(x)$, and the (deterministic) decoding algorithm* Dec *takes an encoding as well as some secret state $s \in \mathcal{S}$ and outputs a plaintext $x \leftarrow \text{Dec}(c, s)$ or the special symbol $\bot$, indicating an invalid encoding.*

Tampering attacks are captured by functions f, from a certain function class $\mathcal{F}$, that are applied to an encoding. The original definition by [17] allows an attacker to apply only a single tamper function. In order to capture continuous parallel attacks, the definition below permits the attacker to repeatedly specify parallel tamper queries, each consisting of several tamper functions. The process ends as soon as one of the tamper queries leads to an invalid codeword.

The non-malleability requirement is captured by considering a real and an ideal experiment. In both experiments, an attacker is allowed to encode a message of his choice. In the real experiment, he may tamper with an actual encoding of

---

[6] The question whether the notion is achievable by a computationally-secure code remains open for future work.

that message, whereas in the ideal experiment, the tamper queries are answered by a (stateful) simulator. The simulator is allowed to output the special symbol same, which the experiment replaces by the originally encoded message. In either experiment, if a component of the answer vector to a parallel tamper query is the symbol $\perp$, a self-destruct occurs, i.e., all *future* tamper queries are answered by $\perp$. The experiments are depicted in Fig. 3.

Definition 5 (Non-malleable code with secret state). *Let* $q, p \in \mathbb{N}$ *and* $\varepsilon > 0$. *A CSS* (Gen, Enc, Dec) *is* $(\mathcal{F}, q, p, \varepsilon)$-non-malleable *if the following properties are satisfied:*

- *Correctness: For each* $x \in \{0,1\}^k$ *and all* $s \in \mathcal{S}$ *output by* Gen, *correctness means* $\text{Dec}(\text{Enc}(x), s) = x$ *with probability 1 over the randomness of* Enc.
- *Non-Malleability: There exists a (possibly stateful) simulator* sim *such that for any distinguisher* D *asking at most* q *parallel queries, each of size at most* p, $\Delta^D(R_\mathcal{F}, S_{\mathcal{F},\text{sim}}) \le \varepsilon$.

We remark that for codes without secret state (as the ones considered in [17]), one obtains the standard notion of non-malleability [17] by setting $q = p = 1$, and continuous non-malleability [18] by letting $p = 1$ and q arbitrary (i.e., $q = *$).

4.2   Combining Single-Bit PKE and Non-Malleable Codes

Our construction of a multi-bit NM-SDA-secure PKE scheme Π' from a single-bit NM-SDA-secure scheme Π and a secret-state non-malleable (k, n)-code follows the approach of [10]: It encrypts a k-bit message m by first computing an encoding $c = (c[1], \ldots, c[n])$ of m and then encrypting each bit $c[j]$ under an independent public key of Π; it decrypts by first decrypting the individual components and then decoding the resulting codeword using the secret state of the non-malleable code; the secret state is part of the secret key. The scheme is depicted in detail in Fig. 4.

Intuitively, NM-SDA security (or CCA security in general) guarantees that an attacker can either leave a message intact or replace it by an independently created one. For our construction, which separately encrypts every bit of an encoding of the plaintext, this translates to the following capability of an adversary w.r.t. decryption queries: It can either leave a particular bit of the encoding unchanged or fix it to 0 or to 1. Therefore, the tamper class against which the non-malleable code must be resilient is the class $\mathcal{F}_{\text{set}} \subseteq \{f \mid f : \{0,1\}^n \to \{0,1\}^n\}$ of functions that tamper with each bit of an encoding individually and can either leave it unchanged or set it to a fixed value. More formally, $f \in \mathcal{F}_{\text{set}}$ can be characterized by $(f[1], \ldots, f[n])$, where $f[j] : \{0,1\} \to \{0,1\}$ is the action of f on the jth bit and $f[j] \in \{\text{zero}, \text{one}, \text{keep}\}$ with the meaning that it either sets the jth bit to 0 (zero) or to 1 (one) or leaves it unchanged (keep).

Before stating the theorem about the security of our construction Π', it needs to be pointed out that it achieves only the so-called *replayable* variant of NM-SDA security. The notion of replayable CCA (RCCA) security (in general) was

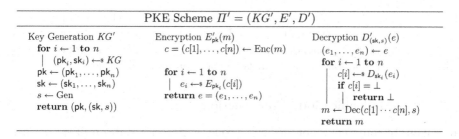

Fig. 4. The k-bit PKE scheme $\Pi' = (KG', E', D')$ built from a 1-bit PKE scheme $\Pi = (KG, E, D)$ and a (k, n)-coding scheme with secret state $(\mathrm{Gen}, \mathrm{Enc}, \mathrm{Dec})$.

introduced by Canetti *et al.* [6] to deal with the fact that for many applications (full) CCA security is unnecessarily strict. Among other things, they provide a MAC-based generic transformation of RCCA-secure schemes into CCA-secure ones, which we can also apply in our setting (as we show) to obtain a fully NM-SDA-secure scheme Π''.

Theorem 3. *Let* $q, p \in \mathbb{N}$ *and* Π *be a* $(t + t_{\mathsf{1bit}}, q, p, \varepsilon_{\mathsf{1bit}})$-*NM-SDA-secure 1-bit PKE scheme,* (T, V) *a* $(t + t_{\mathsf{mac}}, 1, qp, \varepsilon_{\mathsf{mac}})$-*MAC, and* $(\mathrm{Gen}, \mathrm{Enc}, \mathrm{Dec})$ *a* $(\mathcal{F}_{\mathsf{set}}, q, p, \varepsilon_{\mathsf{nmc}})$-*non-malleable* (k, n)-*code with secret state. Then,* Π'' *is* (t, q, p, ε)-*NM-SDA-secure PKE scheme with* $\varepsilon = 2(3(n\varepsilon_{\mathsf{1bit}} + \varepsilon_{\mathsf{nmc}}) + qp \cdot 2^{-\ell} + \varepsilon_{\mathsf{mac}})$, *where* t_{1bit} *and* t_{mac} *are the overheads incurred by the corresponding reductions and* ℓ *is the length of a verification key for the MAC.*

The full proof of Theorem 3 can be found in the full version; here we only provide a sketch. We stress that an analogous statement as the one of the above theorem works for domain extension of NM-CPA, i.e., for constructing a multi-bit NM-CPA scheme out of a single-bit NM-CPA scheme. The proof is very similar to the one of Theorem 3 and therefore omitted.

Proof (Sketch). The proof considers a series of n hybrid experiments. In very rough terms, the ith hybrid generates the challenge ciphertext by computing an encoding $c = (c[1], \dots, c[n])$ of the challenge plaintext and by replacing the first i bits $c[i]$ of c by random values $\tilde{c}[i]$ before encrypting the encoding bit-wise, leading to the challenge $(c_1^*, \dots, c_n^*)$. Moreover, when answering decryption queries $(c_1', \dots, c_n')$, if $c_j' = c_j^*$ for $j \leq i$, the ith hybrid sets the outcome of c_j''s decryption to be the corresponding bit $c[j]$ of the original encoding c, whereas if $c_j' \neq c_j^*$, it decrypts normally (then it decodes the resulting n-bit string normally). This follows the above intuition that a CCA-secure PKE scheme guarantees that if a decryption query is different from the challenge ciphertext, then the plaintext contained in it must have been created independently of the challenge plaintext. The indistinguishability of the hybrids follows from the security of the underlying single-bit scheme Π.

In the nth hybrid, the challenge consists of n encryptions of random values. Thus, the only information about the encoding of the challenge plaintext that an attacker gets is that leaked through decryption queries. But in the nth hybrid there is a 1-to-1 correspondence between decryption queries and the tamper function $f = (f[1], \ldots, f[n])$ applied to the encoding of the challenge plaintext: The case $c'_j = c^*_j$ corresponds to $f[j] = \mathsf{keep}$, and the case $c'_j \neq c^*_j$ corresponds to $f[j] = \mathsf{zero}$ or $f[j] = \mathsf{one}$, depending on whether c'_j decrypts to zero or to one. This allows a reduction to the security of the non-malleable code.   □

4.3 Non-Malleable Code Construction

It remains to construct a non-malleable code (with secret state) resilient against parallel tampering. The intuition behind our construction is the following: If a code has the property (as has been the case with previous schemes secure against (non-parallel) bit-wise tampering) that changing a single bit of a valid encoding results in an invalid codeword, then the tamper function that fixes a particular bit of the encoding and leaves the remaining positions unchanged can be used to determine the value of that bit; this attack is parallelizable, and thus a code of this type cannot provide security against parallel tampering. A similar attack is also possible if the code corrects a fixed (known) number of errors. To circumvent this issue, our construction uses a—for the lack of a better word—"dynamic" error-correction bound: The secret state (initially chosen at random) determines the positions of the encoding in which a certain amount of errors is tolerated.

Construction. Let $\mathbb{F} = \mathrm{GF}(2)$ and $\alpha > 0$. Let (E, D) be a (k, n, δ, τ)-LECSS (cf. Definition 1) with minimum distance δ and secrecy τ over $\mathbb{F}$ such that:[7]

- *Minimum distance:* $\delta > 1/4 + 2\alpha$ and $\delta/2 > 2\alpha$.
- *Constant rate:* $k/n = \Omega(1)$.
- *Constant secrecy:* $\tau = \Omega(1)$.

In the following, we assume that $\alpha \geq \tau$, an assumption that can always be made by ignoring some of the secrecy. Consider the following (k, n)-code with secret state $(\mathsf{Gen}, \mathsf{Enc}, \mathsf{Dec})$:

- Gen: Choose a subset T of $[n]$ of size τn uniformly at random and output it.
- Enc(x) for $x \in \{0, 1\}^k$: Compute $c = \mathsf{E}(x)$ and output it.
- Dec(c, T) for $c \in \{0, 1\}^n$: Find codeword $w = (w[1], \ldots, w[n])$ with $d_\mathsf{H}(w, c) \leq \alpha n$. If no such w exists, output $\bot$. If $w[j] \neq c[j]$ for some $j \in T$, output $\bot$ as well. Otherwise, decode w to its corresponding plaintext x and output it.

We prove the following theorem:

Theorem 4. *For all $q, p \in \mathbb{N}$, (k, n)-code $(\mathsf{Gen}, \mathsf{Enc}, \mathsf{Dec})$ based on a (k, n, δ, τ)-LECSS satisfying the three conditions above is $(\mathcal{F}_{\mathsf{set}}, q, p, \varepsilon_{\mathsf{nmc}})$-non-malleable with $\varepsilon_{\mathsf{nmc}} = p(\mathcal{O}(1) \cdot e^{-\tau n/16} + e^{-\tau^2 n/4}) + p e^{-\tau^2 n}$.*

---

[7] The reasons for these restrictions become apparent in the proof; of course, α must be chosen small enough in order for these constraints to be satisfiable.

Instantiating the Construction. Section 4.5 details how a LECSS satisfying the above properties can be constructed by combining high-distance binary codes with a recent result by Cramer *et al.* [11] in order to "add" secrecy. The resulting LECSS has secrecy $\tau = \Omega(1)$ and rate $\rho = \Omega(1)$ (cf. Corollary 1 in Sect. 4.5). The secrecy property depends on the random choice of a universal hash function. Thus, the instantiated code can be seen as a construction in the CRS model. When combined with the single-bit PKE as described above, the description of the hash function can be made part of the public key.

By combining Theorems 3, and 4, and Corollary 1, we obtain a 1-to-k-bit black-box domain extension for NM-SDA (and NM-CPA) making $\mathcal{O}(k)$ calls to the underlying 1-bit scheme, therefore establishing Theorem 1.[8]

4.4 Proof of the Non-Malleable Code Construction

For the proof of Theorem 4, fix $q, p \in \mathbb{N}$ and a distinguisher D making at most q tamper queries of size p each. Set $\mathcal{F} := \mathcal{F}_{\mathsf{set}}$ for the rest of the proof. In the following, we assume that $\alpha \geq \tau$, an assumption that can always be made by ignoring some of the secrecy. The goal is to show $\Delta^D(R_{\mathcal{F}}, S_{\mathcal{F},\mathsf{sim}}) \leq \varepsilon_{\mathsf{nmc}} = p(\mathcal{O}(1) \cdot e^{-\tau n/16} + e^{-\tau^2 n/4}) + p e^{-\tau^2 n}$ for a simulator sim to be determined.

On a high level, the proof proceeds as follows: First, it shows that queries that interfere with too many bits of an encoding and at the same time do not fix enough bits (called *middle* queries below) are rejected with high probability. The effect of the remaining query types (called *low* and *high* queries) on the decoding process can always be determined from the query itself and the bits of the encoding at the positions indexed by the secret trigger set T. Since the size of T is τn, these symbols are uniformly random and independent of the encoded message, which immediately implies a simulation strategy for sim.

Tamper-Query Types. Recall that $f \in \mathcal{F}_{\mathsf{set}}$ is characterized by $(f[1], \ldots, f[n])$, where $f[j] : \{0,1\} \to \{0,1\}$ is the action of f on the jth bit, for $f[j] \in \{\mathsf{zero}, \mathsf{one}, \mathsf{keep}\}$, with the meaning that it either sets the jth bit to 0 (zero) or to 1 (one) or leaves it unchanged (keep). Define $A(f)$ to be the set of all indices j such that $f[j] \in \{\mathsf{zero}, \mathsf{one}\}$, and let $q(f) := |A(f)|$. Moreover, let $\mathsf{val}(\mathsf{zero}) := 0$ and $\mathsf{val}(\mathsf{one}) := 1$.

A tamper query f is a *low query* if $q(f) \leq \tau n$, a *middle query* if $\tau n < q(f) < (1 - \tau)n$, and a *high query* if $q(f) \geq (1 - \tau)n$.

Analyzing Query Types. The following lemma states that an isolated middle query is rejected with high probability.

Lemma 1. *Let $f \in \mathcal{F}_{\mathsf{set}}$ be a middle query. Then, for any $x \in \{0,1\}^k$,*

$$\mathsf{P}[\mathsf{Dec}(f(\mathsf{Enc}(x))) \neq \bot] \quad \leq \quad \mathcal{O}(1) \cdot e^{-\tau n/16} + e^{-\tau^2 n/4}$$

---

[8] For the construction to be secure, it is necessary that $n = \Omega(\lambda)$ and, therefore, due to the constant rate of the LECSS, the plaintext length is $k = \Omega(\lambda)$ as well.

where the probability is over the randomness of Enc *and the choice of the secret trigger set* T.

Proof. Fix $x \in \{0,1\}^k$ and a middle query $f = (f[1], \ldots, f[n])$. Suppose first that $q(f) \geq n/2$. Define $\mathcal{W} := \{w \in \mathbb{F}^n \mid w \text{ is codeword} \wedge \exists r : d_H(f(\mathsf{E}(x;r)), w) \leq \alpha n\}$, where r is the randomness of E. That is, $\mathcal{W}$ is the set of all codewords that could possibly be considered while decoding an encoding of x tampered with via f. Consider two distinct codewords $w, w' \in \mathcal{W}$. From the definition of $\mathcal{W}$ it is apparent that $w[j] \neq \mathsf{val}(f[j])$ for at most αn positions $j \in A(f)$ (and similarly for w'), which implies that w and w' differ in at most $2\alpha n$ positions $j \in A(f)$. Therefore, w and w' differ in at least $(\delta - 2\alpha)n$ positions $j \notin A(f)$.

For $w \in \mathcal{W}$, let $\tilde{w}$ be the projection of w onto the unfixed positions $j \notin A(f)$ and set $\tilde{\mathcal{W}} := \{\tilde{w} \mid w \in \mathcal{W}\}$. The above distance argument implies that $|\mathcal{W}| = |\tilde{\mathcal{W}}|$. Moreover, $\tilde{\mathcal{W}}$ is a binary code with block length $n - q(f)$ and relative distance at least

$$\frac{(\delta - 2\alpha)n}{n - q(f)} \geq \frac{(\delta - 2\alpha)n}{n/2} = 2\delta - 4\alpha > 1/2,$$

where the last inequality follows from the fact that δ and α are such that $\delta - 2\alpha > 1/4$. Therefore, by the Plotkin bound (a proof can, e.g., be found in [26, p. 41]),[9]

$$|\mathcal{W}| = |\tilde{\mathcal{W}}| \leq \mathcal{O}(1).$$

Denote by $c = (c[1], \ldots, c[n])$ and $\tilde{c} = (\tilde{c}[1], \ldots, \tilde{c}[n])$ the (random variables corresponding to the) encoding $c = \mathsf{Enc}(x)$ and the tampered encoding $\tilde{c} = f(c)$, respectively. For an arbitrary (n-bit) codeword $w \in \mathcal{W}$,

$$\mathbf{E}[d_H(\tilde{c}, w)] = \sum_{j=1}^{n} \mathbf{E}[d_H(\tilde{c}[j], w[j])] \geq \sum_{j \in J} \mathbf{E}[d_H(\tilde{c}[j], w[j])],$$

where $J \subseteq [n]$ is the set containing the indices of the first τn bits *not* fixed by f. Note that by the definition of middle queries, there are at least that many, i.e., $|J| = \tau n$.

Observe that for $j \in J$, $d_H(\tilde{c}[j], w[j])$ is an indicator variable with expectation $\mathbf{E}[d_H(\tilde{c}[j], w[j])] \geq \frac{1}{2}$, since $c[j]$ is a uniform bit. Thus, $\mathbf{E}[d_H(\tilde{c}, w)] \geq \frac{\tau n}{2}$.

Additionally, $(d_H(\tilde{c}[j], w[j]))_{j \in J}$ are independent. Therefore, using a standard Chernoff bound, for $\varepsilon > 0$

$$\mathsf{P}[d_H(\tilde{c}, w) < (1 - \varepsilon)\tau n/2] \leq e^{-\tau \varepsilon^2 n/4}.$$

Therefore, the probability that there exists $w \in \mathcal{W}$ for which the above does not hold is at most $|\mathcal{W}| \cdot e^{-\tau \varepsilon^2 n/4} \leq \mathcal{O}(1) \cdot e^{-\tau \varepsilon^2 n/4}$, by a union bound.

Suppose now that $d_H(\tilde{c}, w) \geq (1 - \varepsilon)\tau n/2$ for all codewords $w \in \mathcal{W}$. Then, over the choice of T,[10]

$$\mathsf{P}[\forall j \in T : d_H(\tilde{c}[j], w[j]) = 0] \leq (1 - (1 - \varepsilon)\tau/2)^{\tau n} \leq e^{-(1-\varepsilon)\tau^2 n/2}.$$

---

[9] The size constant absorbed by $\mathcal{O}(1)$ here depends on how close $2\delta - 4\alpha$ is to $1/2$.
[10] Recall that $|T| = \tau n$.

The lemma now follows by setting $\varepsilon := \frac{1}{2}$.

If $q(f) < n/2$ an analogous argument can be made for the difference $d := c - \tilde{c}$ between the encoding and the tampered codeword, as such a query f fixes at least half of the bits of d (to 0, in fact) and $\mathsf{D}(d) \neq \bot$ implies $\mathsf{D}(\tilde{c}) \neq \bot$.   □

It turns out that low and high queries always result in $\bot$ or one other value.

Lemma 2. *Low queries $f \in \mathcal{F}_{\mathsf{set}}$ can result only in $\bot$ or the originally encoded message $x \in \{0,1\}^k$. High queries $f \in \mathcal{F}_{\mathsf{set}}$ can result only in $\bot$ or one other value $x_f \in \{0,1\}^k$, which solely depends on f. Furthermore, x_f, if existent, can be found efficiently given f.*

Proof. The statement for low queries is trivial, since a low query f cannot change the encoding beyond the error correction bound αn.

Consider now a high query f and the following efficient procedure:

1. Compute $\tilde{c}_f \leftarrow f(0^n)$.
2. Find codeword w_f with $d_{\mathsf{H}}(w_f, \tilde{c}_f) \leq 2\alpha n$ (this is possible since $2\alpha < \delta/2$).
3. Output w_f or $\bot$ if none exists.

Consider an arbitrary encoding c and let $\tilde{c} \leftarrow f(c)$ be the tampered encoding. Assume there exists w with $d_{\mathsf{H}}(w, \tilde{c}) \leq \alpha n$. Since a high query f fixes all but τn bits, $d_{\mathsf{H}}(\tilde{c}, \tilde{c}_f) \leq \tau n \leq \alpha n$, and, thus, $d_{\mathsf{H}}(w, \tilde{c}_f) \leq 2\alpha n$, by the triangle inequality. Hence, $w = w_f$.

In other words, if the decoding algorithm Dec on $\tilde{c}$ finds a codeword $w = w_f$, one can find it using the above procedure, which also implies that high queries can only result in $\bot$ or one other message $x_f = \mathsf{D}(w_f)$.   □

Handling Middle Queries. Consider the hybrid game H_1 that behaves as $R_{\mathcal{F}}$, except that it answers all middle queries by $\bot$.

Lemma 3. $\Delta^D(R_{\mathcal{F}}, H_1) \quad \leq \quad p(\mathcal{O}(1) \cdot e^{-\tau n/16} + e^{-\tau^2 n/4}).$

The proof of Lemma 3 follows a generic paradigm, at whose core is the so-called *self-destruct lemma*, which deals with the indistinguishability of hybrids with the self-destruct property and is explained in detail in Sect. 6. Roughly, this lemma applies whenever the first hybrid (in this case $R_{\mathcal{F}}$) can be turned into the second one (in this case H_1) by changing ("bending") the answers to a subset (the "bending set") of the possible queries to always be $\bot$, and when additionally non-bent queries have a unique answer (cf. the statement of Lemma 10). Intuitively, the lemma states that parallelism and adaptivity do not help distinguish (much) in such cases, which allows using Lemma 1.

Proof. The lemma is proved conditioned on the message x encoded by D. To use the self-destruct lemma, note first that both $R_{\mathcal{F}}$ and H_1 answer parallel tamper queries in which each component is from the set $\mathcal{X} := \mathcal{F}$ by vectors whose components are in $\mathcal{Y} := \{0,1\}^k \cup \{\bot\}$. Moreover, both hybrids use as internal randomness a uniformly chosen element from $\mathcal{R} := \{0,1\}^\rho \times \mathcal{S}$, where ρ

is an upper bound on the number of random bits used by Enc and S is the set of all τn-subsets T of $[n]$. $R_{\mathcal{F}}$ answers each component of a query $f \in \mathcal{X}$ by

$$g(f, (r, T)) \quad := \quad \text{Dec}(f(\text{Enc}(x; r)), T).$$

Define $\mathcal{B} \subseteq \mathcal{X}$ to be the set of all middle queries; H_1 is the $\mathcal{B}$-bending of $R_{\mathcal{F}}$ (cf. Definition 7).

Observe that queries $f \notin \mathcal{B}$ are either low or high queries. For low queries f, the unique answer is $y_f = x$, and for high queries f, $y_f = x_f$ (cf. Lemma 2). Thus, by Lemmas 10 and 1,

$$\Delta^D(R_{\mathcal{F}}, H_1) \quad \leq \quad p \cdot \max_{f \in \mathcal{B}} \mathsf{P}[g(f, (r, T)) \neq \bot] \quad \leq \quad p(\mathcal{O}(1) \cdot e^{-\tau n/16} + e^{-\tau^2 n/4}),$$

where the probability is over the choice of (r, T).   □

Handling High Queries. Consider the following hybrid game H_2: It differs from H_1 in the way it decodes high queries f. Instead of applying the normal decoding algorithm to the tampered codeword $\tilde{c}$, it proceeds as follows:

1. Find w_f (as in the proof of Lemma 2).
2. If w_f does not exist, return $\bot$.
3. If $\tilde{c}[j] = w_f[j]$ for all $j \in T$, return $\text{Dec}(w)$. Otherwise, return $\bot$.

Lemma 4. $\Delta^D(H_1, H_2) \quad \leq \quad pe^{-\tau^2 n}.$

Proof. The lemma is proved conditioned on the message x encoded by D *and the randomness r of the encoding.* For the remainder of the proof, r is therefore considered fixed inside H_1 and H_2. The proof, similarly to that of Lemma 3, again uses the self-destruct lemma.

Set $\mathcal{X} := \mathcal{F}$ and $\mathcal{Y} := \{0, 1\}^k \cup \{\bot\}$. However, this time, let $\mathcal{R} := \mathcal{S}$. For $f \in \mathcal{X}$ and $T \in \mathcal{R}$, define

$$g(f, T) \quad := \quad \text{Dec}(\tilde{c}, T),$$

where $\tilde{c} := f(\text{Enc}(x; r))$. The bending set $\mathcal{B} \subseteq \mathcal{X}$ is the set of all high queries f such that w_f exists and $d_H(w_f, \tilde{c}) > \alpha n$.[11] It is readily verified that H_2 is a parallel stateless self-destruct game (cf. Definition 6) that behaves according to g, and that H_1 is its $\mathcal{B}$-bending.

Consider a query $f \notin \mathcal{B}$. If f is a low query, the unique answer is $y_f = x$; if it is a middle query, $y_f = \bot$; if it is a high query, $y_f = x_f$ (cf. Lemma 2). Therefore,

$$\Delta^D(H_1, H_2) \quad \leq \quad \max_{f \in \mathcal{B}} \mathsf{P}[g(f, T) \neq \bot] \quad \leq \quad pe^{-\tau^2 n},$$

where the first inequality follows from Lemma 10 and the second one from the fact that $d_H(x_f, \tilde{c}) > \tau n$ for queries $f \in \mathcal{B}$, and therefore the probability over the choice of T that it is accepted is at most $(1 - \tau)^{\tau n} \leq e^{-\tau^2 n}$.   □

---

[11] These are queries potentially accepted by H_2 but not by H_1.

Simulation. By analyzing hybrid H_2, one observes that low and high queries can now be answered knowing only the query itself and the symbols of the encoding indexed by the secret trigger set $T \in \mathcal{S}$.

Lemma 5. *Consider the random experiment of distinguisher D interacting with H_2. There is an efficiently computable function* $\mathrm{Dec}' : \mathcal{F}_{\mathsf{set}} \times \mathcal{S} \times \{0,1\}^{\tau n} \to \{0,1\}^k \cup \{\mathsf{same}, \bot\}$ *such that for any low or high query f, any fixed message x, any fixed encoding c thereof, and any output T of Gen,*

$$\left[\mathrm{Dec}'(f, T, (c[j])_{j \in T})\right]_{\mathsf{same}/x} = \mathrm{Dec}(f(c)),$$

where $[\cdot]_{\mathsf{same}/x}$ *is the identity function except that* same *is replaced by x and where* $(c[j])_{j \in T}$ *are the symbols of c specified by T.*

Proof. Consider a low query f. Due to the error correction, $\mathrm{Dec}(f(c))$ is the message originally encoded if no bit indexed by T is changed and $\bot$ otherwise. Which one is the case can clearly be efficiently computed from f, T, and $(c[j])_{j \in T}$.

For high queries f the statement follows by inspecting the definition of H_2 and Lemma 2. $\qquad\square$

In H_2, by the τn-secrecy of the LECSS, the distribution of the symbols indexed by T is independent of the message x encoded by D. Moreover, the distribution of T is trivially independent of x. This suggests the following simulator sim: Initially, it chooses a random subset T from $\binom{[n]}{\tau n}$ and chooses τn random symbols $(c[j])_{j \in T}$. Every component f of any tamper query is handled as follows: If f is a low or a high query, the answer is $\mathrm{Dec}'(f, T, (c[j])_{j \in T})$; if f is a middle query, the answer is $\bot$. This implies:

Lemma 6. $H_2 \equiv S_{\mathcal{F}, \mathsf{sim}}$.

Proof (Theorem 4). From Lemmas 3, 4, and 6 and a triangle inequality. $\qquad\square$

4.5   LECSS for the Non-Malleable Code

Let $\mathbb{F} = \mathrm{GF}(2)$ and $\alpha > 0$. In this section we show how to construct a (k, n, δ, τ)-LECSS (E, D) (cf. Definition 1 in Sect. 2) with minimum distance δ and secrecy τ over $\mathbb{F}$ and the following properties (as required in Sect. 4.3):

- *Minimum distance:* $\delta > 1/4 + 2\alpha$ and $\delta/2 > 2\alpha$.
- *Constant rate:* $k/n = \Omega(1)$.
- *Constant secrecy:* $\tau = \Omega(1)$.

The construction combines high-distance binary codes with a recent result by Cramer *et al.* [11], which essentially allows to "add" secrecy to any code of sufficient rate.

Let $\mathcal{C}$ be a (n, l)-code with rate $R = \frac{l}{n}$ over $\mathbb{F}$. In the following we write $\mathcal{C}(x)$ for the codeword corresponding to $x \in \mathbb{F}^l$ and $\mathcal{C}^{-1}(c, e)$ for the output of the efficient error-correction algorithm attempting to correct up to e errors on c, provided that $e < \delta n/2$;[12] the output is $\perp$ if there is no codeword within distance e of c.

Adding Secrecy. Let l be such that $k < l < n$. The construction by [11] combines a surjective linear universal hash function $\mathsf{h} : \mathbb{F}^l \to \mathbb{F}^k$ with $\mathcal{C}$ to obtain a LECSS (E, D) as follows:[13]

- $\mathsf{E}(x)$ for $x \in \{0,1\}^k$: Choose $s \in \{0,1\}^l$ randomly such that $\mathsf{h}(s) = x$ and output $c = \mathcal{C}(s)$.
- $\mathsf{D}(c, e)$ for $c \in \{0,1\}^n$ and $e < \delta n/2$: Compute $s = \mathcal{C}^{-1}(c, e)$. If $s = \perp$, output $\perp$. Otherwise, output $x = \mathsf{h}(s)$.

The resulting LECSS has rate $\rho = \frac{k}{ln}$ and retains all distance and error-correction properties of $\mathcal{C}$. Additionally, if R is not too low, the LECSS has secrecy. More precisely, Cramer *et al.* prove the following theorem:

Theorem 5 ([11]). *Let $\tau > 0$ and $\eta > 0$ be constants and $\mathcal{H}$ be a family of linear universal hash functions $\mathsf{h} : \mathbb{F}^l \to \mathbb{F}^k$. Given that $R \geq \rho + \eta + \tau + h(\tau)$, there exists a function $\mathsf{h} \in \mathcal{H}$ such that (E, D) achieves secrecy τ. Moreover, such a function h can be chosen randomly with success probability $1 - 2^{-\eta n}$.*

The version of the above theorem presented in [11] does not claim that any τn bits of an encoding are uniform and independent but merely that they are independent of the message encoded. Yet, by inspecting their proof, it can be seen that uniformity is guaranteed if $\tau n \leq l - k$, which is the case if and only if $\tau \leq \frac{l}{n} - \frac{k}{n} = R - \rho$, which is clearly implied by the precondition of the theorem.

Zyablov Bound. For code $\mathcal{C}$, we use concatenated codes reaching the Zyablov bound:

Theorem 6. *For every $\delta < 1/2$ and all sufficiently large n, there exists a code $\mathcal{C}$ that is linear, efficiently encodable, of distance at least δn, allows to efficiently correct up to $\delta n/2$ errors, and has rate*

$$R \geq \max_{0 \leq r \leq 1 - h(\delta + \varepsilon)} r \left(1 - \frac{\delta}{h^{-1}(1 - r) - \varepsilon} \right),$$

for $\varepsilon > 0$ and where $h(\cdot)$ is the binary entropy function.

The Zyablov bound is achieved by concatenating Reed-Solomon codes with linear codes reaching the Gilbert-Varshamov bound (which can be found by brute-force search in this case). Alternatively, Shen [32] showed that the bound is also reached by an explicit construction using algebraic geometric codes.

---

[12] This assumes that $\mathcal{C}$ is efficiently decodable up to relative distance $\delta/2$. However, while the codes we consider here have this property, for our non-malleable code construction, it would be sufficient to have efficient error correction up to distance 2α for whatever particular choice of the constant α.

[13] Note that we switched the roles of l and k here in order to remain consistent with the notation in this paper.

Choice of Parameters. Set $\alpha := 1/200$ and $\delta := 1/4 + 2\alpha + \varepsilon$ for $\varepsilon := 1/500$, say. Then, $\delta - 2\alpha > 1/4$, as required. Moreover, the rate of the Zyablov code with said distance δ can be approximated to be $R \geq 0.0175$. Setting, $\tau := 1/1000$ yields $\tau + h(\tau) \leq 0.0125$, leaving a possible rate for the LECSS of up to $\rho \approx 0.005 - \eta$. Hence:

Corollary 1. *For any $\alpha > 0$ there exists a (k, n, δ, τ)-LECSS* (E, D) *with the following properties:*

- Minimum distance: $\delta > 1/4 + 2\alpha$ *and* $\delta/2 > 2\alpha$.
- Constant rate: $k/n = \Omega(1)$.
- Constant secrecy: $\tau = \Omega(1)$.

4.6 Impossibility for Codes Without State

We show that codes without secret state (as, e.g., the ones in [1, 2, 7, 10, 16, 17, 19]) cannot achieve (unconditional) non-malleability against parallel tampering. Specifically, we prove the following theorem:

Theorem 7. *Let $\mathcal{F} := \mathcal{F}_{\mathsf{set}}$. Let* (Enc, Dec) *be a (k, n)-code without secret state and noticeable rate. There exists a distinguisher D asking a single parallel tampering query of size n^6 such that, for all simulators* sim *and all n large enough,* $\Delta^D(R_{\mathcal{F}}, S_{\mathcal{F},\mathsf{sim}}) \geq 1/2$.

The above impossibility result requires that the rate of the code not be too small (in fact $n = o(2^{k/6})$ suffices, see the full version for the exact parameters). The distinguisher D is inefficient, so it might still be possible to construct a non-malleable code against parallel tampering with only computational security. We leave this as an interesting open question for future research.

Here, we outline an attack for the case where Dec is deterministic. A full proof and a generalization to the setting where Dec uses (independent) randomness for (each) decoding is in the full version.

Proof (Sketch). A possible attack works as follows: There exists an (inefficient) extraction algorithm that, by suitably tampering with an encoding in the real experiment $R_{\mathcal{F}}$, is able to recover the original plaintext with high probability. Since (modulo some technicalities) this is not possible in the ideal experiment $S_{\mathcal{F},\mathsf{sim}}$ (for any simulator sim), this constitutes a distinguishing attack.

For simplicity, suppose that the decoding algorithm Dec is deterministic. The extraction relies on the fact that for any position $i \in [n]$ with relevance in the decoding, there exist two codewords c_i' and c_i'' with $\mathrm{Dec}(c_i') \neq \mathrm{Dec}(c_i'')$ and differing in position i only. From the result of a tamper query fixing all but the ith position to correspond with the bits of c_i' (or c_i'') one can therefore infer the value of the ith bit of the encoding. This extraction is an independent process for every (relevant) position and thus parallelizable. In other words, a single parallel tamper query can be used to recover every relevant position of an encoding (from which the original message can be computed by filling the non-relevant positions with arbitrary values and applying the decoding algorithm). □

5   Construction from CPA Security

In this section we show that NM-SDA security can be achieved in a black-box fashion from IND-CPA security. Specifically, we prove that a generalization using LECSS (cf. Sect. 2) of the scheme by Choi et al. [9] (dubbed the CDMW construction in the remainder of this section) is NM-SDA secure. Using a constant-rate LECSS allows to improve the rate of the CDMW construction from $\Omega(1/\lambda^2)$ to $\Omega(1/\lambda)$, where λ is the security parameter. This abstraction might also give a deeper understanding of the result of [9]. The main difficulty in the analysis is to extend their proof to deal with adaptively chosen parallel decryption queries (with self-destruct).

5.1   The CDMW Construction

The CDMW construction uses a randomized Reed-Solomon code, which is captured as a special case by the notion of a linear error-correcting secret sharing (LECSS) (E, D) (cf. Sect. 2). For ease of description, we assume that the decoding algorithm returns not only the plaintext x but also the corresponding codeword w, i.e., $(x, w) \leftarrow \mathsf{D}(c, e)$, where $e \in \mathbb{N}$ specifies the number of errors to correct; moreover, the output is $(x, w) = (\bot, \bot)$ if c is not within distance e of any codeword.

The LECSS has to satisfy an additional property, which is that given a certain number of symbols chosen uniformly at random and independently and a plaintext x, one can efficiently produce an encoding that matches the given symbols and has the same distribution as $\mathsf{E}(x)$. It is described in more detail in the proof of Lemma 9, where it is needed.[14]

Let $\Pi = (KG, E, D)$ be a PKE scheme with message space $\mathcal{M} = \{0, 1\}^\ell$ (we assume $\ell = \Omega(\lambda)$), and let $\Sigma = (KG^{\mathsf{ots}}, S, V)$ be a one-time signature scheme with verification keys of length $\kappa = \mathcal{O}(\lambda)$. Moreover, let $\alpha > 0$ be any constant and (E, D) a (k, n, δ, τ)-LECSS over $\mathrm{GF}(2^\ell)$ with $\delta > 2\alpha$.

The CDMW construction (cf. Fig. 5), to encrypt a plaintext $m \in \{0, 1\}^{k\ell}$, first computes an encoding $(c_1, \ldots, c_n) \leftarrow \mathsf{E}(m)$ and then creates the $(\kappa \times n)$-matrix $\mathbf{C}$ in which this encoding is repeated in every row. For every entry $\mathbf{C}_{ij}$ of this matrix, there are two possible public keys $\mathsf{pk}_{i,j}^b$; which of them is used to encrypt the entry is determined by the ith bit $v[i]$ of the verification key $\mathsf{verk} = (v[1], \ldots, v[\kappa])$ of a freshly generated key pair for Σ. In the end, the encrypted matrix $\mathbf{E}$ is signed using verk, producing a signature σ. The ciphertext is $(\mathbf{E}, \mathsf{verk}, \sigma)$.

The decryption first verifies the signature. Then, it decrypts all columns indexed by a set $T \subset [n]$, chosen as part of the secret key, and checks that each column consists of a single value only. Finally, it decrypts the first row and tries to find a codeword with relative distance at most α. If so, it checks whether the codeword matches the first row in the positions indexed by T. If all checks pass, it outputs the plaintext corresponding to the codeword; otherwise it outputs $\bot$.

---

[14] Of course, the Reed-Solomon-based LECSS from [9] has this property.

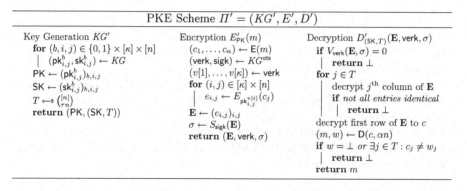

PKE Scheme $\Pi' = (KG', E', D')$		
Key Generation KG'	Encryption $E'_{PK}(m)$	Decryption $D'_{(SK,T)}(\mathbf{E}, \text{verk}, \sigma)$
for $(b,i,j) \in \{0,1\} \times [\kappa] \times [n]$	$(c_1,\dots,c_n) \leftarrow E(m)$	**if** $V_{\text{verk}}(\mathbf{E}, \sigma) = 0$
$\quad | \quad (pk^b_{i,j}, sk^b_{i,j}) \leftarrow KG$	$(\text{verk}, \text{sigk}) \leftarrow KG^{\text{ots}}$	$\quad | \quad$ **return** $\perp$
$PK \leftarrow (pk^b_{i,j})_{b,i,j}$	$(v[1],\dots,v[\kappa]) \leftarrow \text{verk}$	**for** $j \in T$
$SK \leftarrow (sk^b_{i,j})_{b,i,j}$	**for** $(i,j) \in [\kappa] \times [n]$	$\quad | \quad$ decrypt j^{th} column of $\mathbf{E}$
$T \leftarrow^{\$} \binom{[n]}{\tau n}$	$\quad | \quad e_{i,j} \leftarrow E_{pk^{v[i]}_{i,j}}(c_j)$	$\quad | \quad$ **if** *not all entries identical*
return $(PK, (SK, T))$	$\mathbf{E} \leftarrow (e_{i,j})_{i,j}$	$\quad | \quad | \quad$ **return** $\perp$
	$\sigma \leftarrow S_{\text{sigk}}(\mathbf{E})$	decrypt first row of $\mathbf{E}$ to c
	return $(\mathbf{E}, \text{verk}, \sigma)$	$(m, w) \leftarrow D(c, \alpha n)$
		if $w = \perp$ *or* $\exists j \in T : c_j \neq w_j$
		$\quad | \quad$ **return** $\perp$
		return m

Fig. 5. The CDMW PKE scheme Π' based on a CPA-secure scheme Π [9].

In the remainder of this section, we sketch the proof of the following theorem, which implies Theorem 2.

Theorem 8. *Let* $t \in \mathbb{N}$ *and* Π *be a* $(t + t_{\text{cpa}}, \varepsilon_{\text{cpa}})$-*IND-CPA-secure PKE scheme,* $\alpha > 0$, (E, D) *a* (k, n, δ, τ)-*LECSS with* $\delta > 2\alpha$, *and* Σ *a* $(t + t_{\text{ots}}, \varepsilon_{\text{ots}})$-*secure OTS scheme with verification-key length* κ. *Then, for any* $q, p \in \mathbb{N}$, *PKE scheme* Π' *is* (t, q, p, ε)-*NM-SDA-secure with*

$$\varepsilon = (1 - \tau)\kappa n \cdot \varepsilon_{\text{cpa}} + 2 \cdot \varepsilon_{\text{ots}} + 4 \cdot p(1 - \tau)^{\alpha n},$$

where t_{cpa} *and* t_{ots} *represent the overhead incurred by corresponding reductions.*

Instantiating the Construction. Note that the security proof below does not use the linearity of the LECSS. The CDMW construction can be seen as using a Reed-Solomon-based LECSS with rate $\mathcal{O}(1/\kappa)$. If the construction is instantiated with a constant-rate LECSS, the final rate improves over CDMW by a factor of $\Omega(\kappa) = \Omega(\lambda)$. More concretely, assuming a constant-rate CPA encryption, a ciphertext of length $\mathcal{O}(\lambda^3)$ can encrypt a plaintext of length $\Omega(\lambda^2)$ as compared to $\Omega(\lambda)$ for plain CDMW. As shown in Sect. 5.3, the LECSS can be instantiated with constructions based on Reed-Solomon or algebraic geometric codes (which also satisfy the additional property mentioned above), both with constant rate. Among the constant-rate codes, algebraic geometric codes allow to choose the parameters optimally also for shorter plaintexts.

5.2 Security Proof of the CDMW Construction

The proof follows the original one [9]. The main change is that one needs to argue that, unless they contain invalid ciphertexts, adaptively chosen parallel queries do not allow the attacker to obtain useful information, in particular on the secret set T. This is facilitated by using the *self-destruct lemma* (cf. Sect. 6). The proof proceeds in three steps using two hybrid games H_b and H'_b:

- The first hybrid H_b gets rid of signature forgeries for the verification key used to create the challenge ciphertext. The indistinguishability of the hybrid from $G_b^{\Pi',\text{nm-sda}}$ follows from the security of the OTS scheme and requires only minor modifications compared to the original proof.
- The second hybrid H_b' uses an alternative decryption algorithm. The indistinguishability of H_b' and H_b holds unconditionally; this step requires new techniques compared to the original proof.
- Finally, the distinguishing advantage between H_0' and H_1' is bounded by a reduction to the IND-CPA security of the underlying scheme Π; the reduction again resembles the one in [9].

Dealing with Forgeries. For $b \in \{0,1\}$, hybrid H_b behaves as $G_b^{\Pi',\text{nm-sda}}$ but generates the signature key pair $(\text{sigk}^*, \text{verk}^*)$ used for the challenge ciphertext initially and rejects any decryption query $(\mathbf{E}', \sigma', \text{verk}')$ if $\text{verk}' = \text{verk}^*$.

Lemma 7. *For $b \in \{0,1\}$, there exists a reduction $R_b'(\cdot)$ such that for all distinguishers D, $\Delta^D(G_b^{\Pi',\text{nm-sda}}, H_b) \leq \Gamma^{R_b'(D)}(G^{\Sigma,\text{ots}})$.*

Proof. $R_b'(\cdot)$ is a standard reduction to the unforgeability of Σ. $\qquad\square$

Alternative Decryption Algorithm. For $b \in \{0,1\}$, hybrid H_b' behaves as H_b but for the way it answers decryption queries $(\mathbf{E}', \sigma', \text{verk}')$: As before, it first verifies the signature σ' and checks that each column of $\mathbf{E}'$ consists of encryptions of a single value. Then, it determines the first position i at which verk' and verk^* differ, i.e., where $v'[i] \neq v^*[i]$. It decrypts the ith row of $\mathbf{E}$ and checks if there is a codeword w within distance $2\alpha n$.[15] If such w does not exist or else if w does not match the *first* row in a position indexed by T, the check fails. Otherwise, the plaintext corresponding to w is output.

Lemma 8. *For $b \in \{0,1\}$ and all distinguishers D, $\Delta^D(H_b, H_b') \leq 2 \cdot p(1-\tau)^{\alpha n}$.*

The proof of Lemma 8 shows that the original and alternative decryption algorithms are indistinguishable not just for a single parallel query (as is sufficient for NM-CPA) but even against adaptively chosen parallel queries (with self-destruct). It is the main technical contribution of this section.

At the core of the proof is an analysis of how different types of encoding matrices $\mathbf{C}$ are handled inside the two decryption algorithms. To that end, one can define two games B and B' (below) that capture the behaviors of the original and the alternative decryption algorithms, respectively. The proof is completed by bounding $\Delta(B, B')$ (for all distinguishers) and showing the existence of a wrapper W_b such that $W_b(B)$ behaves as H_b and $W_b(B')$ as H_b' (also below). This proves the lemma since $\Delta^D(H_b, H_b') = \Delta^D(W_b(B), W_b(B')) = \Delta^{D(W_b(\cdot))}(B, B')$.

The games B and B' behave as follows: Both initially choose a random size-τ subset of $[n]$. Then, they accept parallel queries with components $(\mathbf{C}, i)$ for $\mathbf{C} \in \mathbb{F}^{\kappa \times n}$ and $i \in [\kappa]$. The answer to each component is computed as follows:

---

[15] Recall that the actual decryption algorithm always decrypts the first row and tries to find w within distance αn.

1. Both games check that all columns indexed by T consist of identical entries.
2. Game B tries to find a codeword w with distance less than αn from the *first* row (regardless of i), whereas B' tries to find w within $2\alpha n$ of row i. Then, if such a w is found, *both* games check that it matches the *first* row of $\mathbf{C}$ in the positions indexed by T.
3. If all checks succeed, the answer to the (component) query is w; otherwise, it is $\bot$.

Both games then output the answer vector and implement the self-destruct, i.e., if any of the answers is $\bot$, all *future* queries are answered by $\bot$.

Claim. For $b \in \{0,1\}$ and all distinguishers D, $\Delta^D(B,B') \leq 2 \cdot p(1-\tau)^{\alpha n}$.

Encoding Matrices. Towards a proof of Claim 5.2, consider the following partition of the set of encoding matrices $\mathbf{C}$ (based on the classification in [9]):

1. There exists a codeword w within αn of the first row of $\mathbf{C}$, and all rows have distance at most αn.
2. (a) There exist two rows in $\mathbf{C}$ with distance greater than αn.
   (b) The rest; in this case the first row differs in more than αn positions from any codeword.

Observe that queries $(\mathbf{C}, i)$ with $\mathbf{C}$ of type 1 are treated identically by both B and B': A codeword w within αn of the first row of $\mathbf{C}$ is certainly found by B; since all rows have distance at most αn, w is within $2\alpha n$ of row i and thus also found by B'. Furthermore, note that if $\mathbf{C}$ is of type 2b, it is always rejected by B (but not necessarily by B').

Consider the hybrids C and C' that behave as B and B', respectively, but always reject *all* type-2 queries. Since type-1 queries are treated identically, C and C' are indistinguishable. Moreover:

Claim. For all distinguishers D, $\Delta^D(B,C) \leq p(1-\tau)^{\alpha n}$ and $\Delta^D(C',B') \leq p(1-\tau)^{\alpha n}$.

The proof of Claim 5.2 follows a generic paradigm, at whose core is the so-called *self-destruct lemma*, which deals with the indistinguishability of hybrids with the self-destruct property and is explained in detail in Sect. 6. Roughly, this lemma applies whenever the first hybrid (in this case B resp. B') can be turned into the second one (in this case C resp. C') by changing ("bending") the answers to a subset (the "bending set") of the possible queries to always be $\bot$, and when additionally non-bent queries have a unique answer (cf. the statement of Lemma 10). Intuitively, the lemma states that parallelism and adaptivity do not help distinguish (much) in such cases.

Proof. To use the self-destruct lemma, note that B, C, C', and B' all answer queries from $\mathcal{X} := \mathbb{F}^{\kappa \times n} \times [\kappa]$ by values from $\mathcal{Y} := \mathbb{F}^n$. Moreover, note that they use as internal randomness a uniformly chosen element T from the set $\mathcal{R} := \binom{[n]}{\tau n}$ of size-τn subsets of $[n]$.

Consider first B and C. Let $g : \mathcal{X} \times \mathcal{R} \to \mathcal{Y}$ correspond to how B answers queries $(\mathbf{C}, i)$ (see above). Let $\mathcal{B}$ be the set $\mathcal{B}$ of all type-2a-queries. Then, C is its $\mathcal{B}$-bending (cf. Definition 7). Observe that queries $x = (\mathbf{C}, i) \notin \mathcal{B}$ are either of type 1 or 2b. For the former, the unique answer y_x is the codeword w within αn of the first row of $\mathbf{C}$; for the latter, y_x is $\perp$. Therefore, using the self-destruct lemma (Lemma 10), for all distinguishers D, $\Delta^D(B, C) \leq p \cdot \max_{(\mathbf{C}, i) \in \mathcal{B}} \mathsf{P}[g((\mathbf{C}, i), T) \neq \perp]$, where the probability is over the choice of T. Since type-2a queries have two rows with distance greater than αn, the probability over the choice of T that this remains unnoticed is at most $(1 - \tau)^{\alpha n}$.

For the second part of the claim, consider B' and C'. Now, let $g : \mathcal{X} \times \mathcal{R} \to \mathcal{Y}$ correspond to how B' answers queries $(\mathbf{C}, i)$ (see above again), and let $\mathcal{B}$ be the set $\mathcal{B}$ of all type-2-queries. Then, C' is the $\mathcal{B}$-bending of B'.

Note that all queries $x = (\mathbf{C}, i) \notin \mathcal{B}'$ are of type 1, and the unique answer y_x is the codeword w within $2\alpha n$ of row i of $\mathbf{C}$. Therefore, using Lemma 10 again, for all distinguishers D, $\Delta^D(B', C') \leq p \cdot \max_{(\mathbf{C}, i) \in \mathcal{B}'} \mathsf{P}[g'((\mathbf{C}, i), T) \neq \perp]$, where the probability is again over the choice of T. Since type-2a queries have two rows with distance greater than αn and in type-2b queries the first row differs in more than αn positions from any codeword, the probability over the choice of T that this remains unnoticed is at most $(1 - \tau)^{\alpha n}$. □

Proof (Claim 5.2). The proof follows using the triangle inequality:

$$\Delta^D(B, B') \leq \Delta^D(B, C) + \Delta^D(C, C') + \Delta^D(C', B') \leq 2 \cdot p(1 - \tau)^{\alpha n}. \qquad \square$$

Wrapper. It remains to show that there exists a wrapper W_b such that $W_b(B)$ behaves as H_b and $W_b(B')$ as H_b'. The construction of W_b is straight forward: H_b and H_b' generate all keys and the challenge in the identical fashion; therefore, W_b can do it the same way. W_b answers decryption queries $(\mathbf{E}', \mathsf{verk}', \sigma')$ by first verifying the signature σ' and rejecting queries if σ' is invalid or if verk' is identical to the verification key verk^* chosen for the challenge, decrypting the entire matrix $\mathbf{E}'$ to $\mathbf{C}'$ and submitting $(\mathbf{C}', i)$ to the oracle (either B or B'), where i is the first position at which verk' and verk^* differ, and decoding the answer w and outputting the result or simply forwarding it if it is $\perp$. Moreover, W_b implements the self-destruct. By inspection it can be seen that $W_b(B)$ implements the original decryption algorithm and $W_b(B')$ the alternative one.

Reduction to IND-CPA Security. We prove:

Lemma 9. *There exists a reduction $R(\cdot)$ such that for all distinguishers D,*

$$\Delta^D(H_0', H_1') \quad = \quad (1 - \tau)\kappa n \cdot \Delta^{D(R(\cdot))}(G_{\Pi}^{\text{ind-cpa}} 0, G_{\Pi}^{\text{ind-cpa}} 1).$$

Proof (Sketch). The proof is a straight-forward generalization of the original proof by [9]; the only difference is that it needs to process multiple parallel decryption queries and implement the self-destruct feature appropriately. For ease of exposition, we describe the reduction to a many-public-key version of the CPA game for Π.[16]

Reduction $R(\cdot)$ initially chooses the secret set T and creates the challenge OTS key pair with verification key $\mathsf{verk}^* = (v^*[1], \ldots, v^*[\kappa])$ and all key pairs $(\mathsf{pk}_{i,j}^b, \mathsf{sk}_{i,j}^b)$ with $j \in T$ or $b \neq v^*[i]$. The remaining $(1 - \tau)\kappa n$ key pairs are generated by the CPA game.

Recall that the LECSS is assumed to satisfy the following property: Given τn symbols $(c_i)_{i \in T}$ chosen uniformly at random and independently and any plaintext $x \in \mathbb{F}^k$, one can efficiently sample symbols $(c_i)_{i \notin T}$ such that $(c_1, \ldots, c_n)$ has the same distribution as $\mathsf{E}(x)$. Using this fact, $R(\cdot)$ creates the challenge for m_0 and m_1 as follows: It picks the random symbols $(c_i)_{i \in T}$ and completes them to two full encodings c_{m_0} and c_{m_1} with the above procedure, once using m_0 and once using m_1 as the plaintext. Let $\mathbf{C}_{m_0}$ and $\mathbf{C}_{m_1}$ be the corresponding matrices (obtained by copying the encodings κ times). Observe that the two matrices match in the columns indexed by T. These entries are encrypted by $R(\cdot)$, using the public key $\mathsf{pk}_{i,j}^b$ for entry (i, j) for which $b \neq v^*[i]$. Denote by $\mathbf{C}'_{m_0}$ and $\mathbf{C}'_{m_1}$ the matrices $\mathbf{C}_{m_0}$ and $\mathbf{C}_{m_1}$ with the columns in T removed. The reduction outputs $(\mathsf{chall}, \mathbf{C}'_{m_0}, \mathbf{C}'_{m_1})$ to its oracle and obtains the corresponding ciphertexts, which it combines appropriately with the ones it created itself to form the challenge ciphertext.

Finally, since the reduction knows all the secret keys $\mathsf{pk}_{i,j}^b$ with $b \neq v^*[i]$, it can implement the alternative decryption algorithm (and the self-destruct). $\square$

Overall Proof. Finally, one obtains:

Proof (Theorem 8). Let t_{cpa} be the overhead caused by reduction $R(\cdot)$ and t_{ots} the larger of the overheads caused by $R_0'(\cdot)$ and $R_1'(\cdot)$. Moreover, let D be a distinguisher with running time at most t. Using the triangle inequality, and Lemmas 7, 8, and 9,

$$
\begin{aligned}
\Delta^D(G_0^{\Pi', \mathsf{nm\text{-}sda}}, G_1^{\Pi', \mathsf{nm\text{-}sda}}) \quad \leq \quad & \Delta^D(G_0^{\Pi', \mathsf{nm\text{-}sda}}, H_0) + \Delta^D(H_0, H_0') \\
& + \Delta^D(H_0', H_1') + \Delta^D(H_1', H_1) \\
& + \Delta^D(H_1, G_1^{\Pi', \mathsf{nm\text{-}sda}}) \\
\leq \quad & \Gamma^{D(R_0'(\cdot))}(G^{\Sigma, \mathsf{ots}}) + 2 \cdot p(1 - \tau)^{\alpha n} \\
& + (1 - \tau)\kappa n \cdot \Delta^{D(R(\cdot))}(G_\Pi^{\mathsf{ind\text{-}cpa}} 0, G_\Pi^{\mathsf{ind\text{-}cpa}} 1) \\
& + 2 \cdot p(1 - \tau)^{\alpha n} + \Gamma^{D(R_1'(\cdot))}(G^{\Sigma, \mathsf{ots}}) \\
\leq \quad & \varepsilon_{\mathsf{ots}} + 2 \cdot p(1 - \tau)^{\alpha n} \\
& + (1 - \tau)\kappa n \cdot \varepsilon_{\mathsf{cpa}} + 2 \cdot p(1 - \tau)^{\alpha n} + \varepsilon_{\mathsf{ots}}.
\end{aligned}
$$

$\square$

---

[16] In the many-public-key version of the CPA game, an attacker can play the CPA game for several independently generated public keys simultaneously; this is equivalent to the normal formulation by a standard hybrid argument [3].

5.3   LECSS for the CDMW Construction

In this section we show how to instantiate the LECSS used for the CDMW construction in Sect. 5. Let $\mathbb{F}$ be a finite field of size $L = 2^\ell$, where ℓ is the plaintext length of the IND-CPA scheme used in the construction. Then, there are the following variants of a (k, n, δ, τ)-LECSS:

– *CDMW Reed-Solomon codes:* The original CDMW construction can be seen as using a Reed-Solomon-based LECSS with rate $\Theta(1/\lambda)$, which is suboptimal (see next item).
– *Constant-Rate Reed-Solomon codes:* Cheraghchi and Guruswami [8] provide a LECSS based on a construction by Dziembowski *et al.* [17] and on Reed-Solomon (RS) codes with $\ell = \Theta(\log n)$. One can show that it achieves the following parameters (not optimized): $\alpha = 1/8$, $\tau = 1/8$ and rate $k/n \geq 1/4$ (i.e., all constant).
– *Algebraic geometric codes:* Using algebraic geometric (AG) codes, Cramer *et al.* [12] provide a LECSS with $\ell = \mathcal{O}(1)$ and still constant error correction, secrecy, and rate (but with worse concrete constants than Reed-Solomon codes).

Note that asymptotically, RS and AG codes are equally good: both have constant rate, distance, and secrecy. However, since with AG codes ℓ is constant (i.e., they work over an alphabet of constant size), the minimal plaintext length can be shorter than with RS codes.

6   A General Indistinguishability Paradigm

A recurring issue in this paper are proofs that certain self-destruct games answering successive parallel decryption/tampering queries are indistinguishable. We formalize such games as *parallel stateless self-destruct games*.

Definition 6. *An oracle U is a* parallel stateless self-destruct (PSSD) game *if*

– *it accepts parallel queries in which each component is from some set $\mathcal{X}$ and answers them by vectors with components from some set $\mathcal{Y}$,*
– $\bot \in \mathcal{Y}$,
– *there is a function $g : \mathcal{X} \times \mathcal{R} \to \mathcal{Y}$ such that every query component $x \in \mathcal{X}$ is answered by $g(x, r)$, where $r \in \mathcal{R}$ is the internal randomness of U, and*
– *the game self-destructs, i.e., after the first occurrence of $\bot$ in an answer vector all further outputs are $\bot$.*

A PSSD game can be transformed into a related one by "bending" the answers to some of the queries $x \in \mathcal{X}$ to the value $\bot$. This is captured by the following definition:

Definition 7. *Let U be a PSSD game that behaves according to g and let $\mathcal{B} \subseteq \mathcal{X}$. The $\mathcal{B}$-bending of U, denoted by U', is the PSSD game that behaves according to g', where*

$$g'(x, r) = \begin{cases} \bot & \text{if } x \in \mathcal{B}, \\ g(x, r) & \text{otherwise.} \end{cases}$$

The *self-destruct lemma* below states that in order to bound the distinguishing advantage between a PSSD and its bending, one merely needs to analyze a single, non-parallel query, provided that all non-bent queries x can only be answered by a unique value y_x or $\perp$.

Lemma 10. *Let U be a PSSD game and U' its $\mathcal{B}$-bending for some $\mathcal{B} \subseteq \mathcal{X}$. If for all $x \notin \mathcal{B}$ there exists $y_x \in \mathcal{Y}$ such that $\{g(x, r) \mid r \in \mathcal{R}\} = \{y_x, \perp\}$, then, for all distinguishers D, $\Delta^D(U, U') \leq p \cdot \max_{x \in \mathcal{B}} \mathsf{P}[g(x, R) \neq \perp]$, where the probability is over the choice of R.*

Proof. Fix a distinguisher D and denote by R and R' the random variables corresponding to the internal randomness of U and U', respectively. Call a value $x \in \mathcal{X}$ *dangerous* if $x \in \mathcal{B}$ and a query dangerous if it contains a dangerous value.

In the random experiment corresponding to the interaction between D and U, define the event E that the first dangerous query contains a dangerous value X with $g(X, R) \neq \perp$ and that the self-destruct has not been provoked yet. Similarly, define the event E' for the interaction between D and U' that the first dangerous query contains a dangerous value X' with $g(X', R') \neq \perp$ and that the self-destruct has not been provoked yet.[17]

Clearly, U and U' behave identically unless E resp. E' occur. Thus, it remains to bound $\mathsf{P}[E] = \mathsf{P}[E']$. To that end, note that adaptivity does not help in provoking E. For any distinguisher D, there exists a *non-adaptive* distinguisher $\tilde{D}$ such that whenever D provokes E, so does D'. D' proceeds as follows: First, it interacts with D only. Whenever D asks a non-dangerous query, D' answers every component $x \notin \mathcal{B}$ by y_x. As soon as D specifies a dangerous query, D' stops its interaction with D and sends all queries to U.

Fix all randomness in experiment $D'(U)$, i.e., the coins of D (inside D') and the randomness r of U. Suppose D would provoke E in the direct interaction with U. In such a case, all the answers by D' are equal to the answers by U, since, by assumption, the answers to components $x \notin \mathcal{B}$ in non-dangerous queries are y_x or $\perp$ and the latter is excluded if E is provoked. Thus, whenever D provokes E, D' provokes it as well.

The success probability of non-adaptive distinguishers D is upper bounded by the probability over R that their first dangerous query provokes E, which is at most $p \cdot \max_{x \in \mathcal{B}} \mathsf{P}[g(x, R) \neq \perp]$.   □

Acknowledgements. Sandro Coretti was supported by SNF project no. 200020-132794. Björn Tackmann was supported by the SNF Fellowship P2EZP2_155566 and NSF grants CNS-1228890 and CNS-1116800. Daniele Venturi was partially supported by the European Commission (Directorate-General Home Affairs) under the GAINS project HOME/2013/CIPS/AG/4000005057, and by the European Union's Horizon 2020 research and innovation programme under grant agreement No 644666.

---

[17] Note that the function g is the *same* in the definitions of either event.

References

1. Aggarwal, D., Dodis, Y., Lovett, S.: Non-malleable codes from additive combinatorics. In: STOC, pp. 774–783. ACM (2014)
2. Agrawal, S., Gupta, D., Maji, H.K., Pandey, O., Prabhakaran, M.: Explicit non-malleable codes resistant to permutations and perturbations. In: Gennaro, R., Robshaw, M. (eds.) CRYPTO 2015. LNCS, vol. 9215, pp. 538–557. Springer, Heidelberg (2015)
3. Bellare, M., Boldyreva, A., Micali, S.: Public-key encryption in a multi-user setting: security proofs and improvements. In: Preneel, B. (ed.) EUROCRYPT 2000. LNCS, vol. 1807, pp. 259–274. Springer, Heidelberg (2000)
4. Bellare, M., Sahai, A.: Non-malleable encryption: equivalence between two notions, and an indistinguishability-based characterization. In: Wiener, M. (ed.) CRYPTO 1999. LNCS, vol. 1666, pp. 519–536. Springer, Heidelberg (1999)
5. Canetti, R., Halevi, S., Katz, J.: Chosen-ciphertext security from identity-based encryption. In: Cachin, C., Camenisch, J.L. (eds.) EUROCRYPT 2004. LNCS, vol. 3027, pp. 207–222. Springer, Heidelberg (2004)
6. Canetti, R., Krawczyk, H., Nielsen, J.B.: Relaxing chosen-ciphertext security. In: Boneh, D. (ed.) CRYPTO 2003. LNCS, vol. 2729, pp. 565–582. Springer, Heidelberg (2003)
7. Chattopadhyay, E., Zuckerman, D.: Non-malleable codes against constant split-state tampering. Electron. Colloq. Comput. Complex. (ECCC) **21**, 102 (2014)
8. Cheraghchi, M., Guruswami, V.: Non-malleable coding against bit-wise and split-state tampering. In: Lindell, Y. (ed.) TCC 2014. LNCS, vol. 8349, pp. 440–464. Springer, Heidelberg (2014)
9. Choi, S.G., Dachman-Soled, D., Malkin, T., Wee, H.M.: Black-box construction of a non-malleable encryption scheme from any semantically secure one. In: Canetti, R. (ed.) TCC 2008. LNCS, vol. 4948, pp. 427–444. Springer, Heidelberg (2008)
10. Coretti, S., Maurer, U., Tackmann, B., Venturi, D.: From single-bit to multi-bit public-key encryption via non-malleable codes. In: Dodis, Y., Nielsen, J.B. (eds.) TCC 2015, Part I. LNCS, vol. 9014, pp. 532–560. Springer, Heidelberg (2015)
11. Cramer, R., Damgård, I.B., Döttling, N., Fehr, S., Spini, G.: Linear secret sharing schemes from error correcting codes and universal hash functions. In: Oswald, E., Fischlin, M. (eds.) EUROCRYPT 2015. LNCS, vol. 9057, pp. 313–336. Springer, Heidelberg (2015)
12. Cramer, R., Hanaoka, G., Hofheinz, D., Imai, H., Kiltz, E., Pass, R., Shelat, A., Vaikuntanathan, V.: Bounded CCA2-secure encryption. In: Kurosawa, K. (ed.) ASIACRYPT 2007. LNCS, vol. 4833, pp. 502–518. Springer, Heidelberg (2007)
13. Cramer, R., Shoup, V.: A practical public key cryptosystem provably secure against adaptive chosen ciphertext attack. In: Krawczyk, H. (ed.) CRYPTO 1998. LNCS, vol. 1462, pp. 13–25. Springer, Heidelberg (1998)
14. Cramer, R., Shoup, V.: Universal hash proofs and a paradigm for adaptive chosen ciphertext secure public-key encryption. In: Knudsen, L.R. (ed.) EUROCRYPT 2002. LNCS, vol. 2332, pp. 45–64. Springer, Heidelberg (2002)
15. Dolev, D., Dwork, C., Naor, M.: Nonmalleable cryptography. SIAM J. Comput. **30**(2), 391–437 (2000)
16. Dziembowski, S., Kazana, T., Obremski, M.: Non-malleable Codes from two-source extractors. In: Canetti, R., Garay, J.A. (eds.) CRYPTO 2013, Part II. LNCS, vol. 8043, pp. 239–257. Springer, Heidelberg (2013)

17. Dziembowski, S., Pietrzak, K., Wichs, D.: Non-malleable codes. In: ICS, pp. 434–452 (2010)
18. Faust, S., Mukherjee, P., Nielsen, J.B., Venturi, D.: Continuous non-malleable codes. In: Lindell, Y. (ed.) TCC 2014. LNCS, vol. 8349, pp. 465–488. Springer, Heidelberg (2014)
19. Faust, S., Mukherjee, P., Venturi, D., Wichs, D.: Efficient non-malleable codes and key-derivation for poly-size tampering circuits. In: Nguyen, P.Q., Oswald, E. (eds.) EUROCRYPT 2014. LNCS, vol. 8441, pp. 111–128. Springer, Heidelberg (2014)
20. Gertner, Y., Malkin, T., Myers, S.: Towards a separation of semantic and CCA security for public key encryption. In: Vadhan, S.P. (ed.) TCC 2007. LNCS, vol. 4392, pp. 434–455. Springer, Heidelberg (2007)
21. Goldwasser, S., Micali, S.: Probabilistic encryption. J. Comput. Syst. Sci. 28(2), 270–299 (1984)
22. Hofheinz, D., Kiltz, E.: Practical chosen ciphertext secure encryption from factoring. In: Joux, A. (ed.) EUROCRYPT 2009. LNCS, vol. 5479, pp. 313–332. Springer, Heidelberg (2009)
23. Hohenberger, S., Lewko, A., Waters, B.: Detecting dangerous queries: a new approach for chosen ciphertext security. In: Pointcheval, D., Johansson, T. (eds.) EUROCRYPT 2012. LNCS, vol. 7237, pp. 663–681. Springer, Heidelberg (2012)
24. Kurosawa, K., Desmedt, Y.G.: A new paradigm of hybrid encryption scheme. In: Franklin, M. (ed.) CRYPTO 2004. LNCS, vol. 3152, pp. 426–442. Springer, Heidelberg (2004)
25. Lindell, Y.: A simpler construction of CCA2-secure public-key encryption under general assumptions. In: Biham, E. (ed.) EUROCRYPT 2003. LNCS, vol. 2656, pp. 241–254. Springer, Heidelberg (2003)
26. MacWilliams, F., Sloane, N.: The Theory of Error-Correcting Codes, 2nd edn. North-holland Publishing Company, Amsterdam (1978)
27. Myers, S., Shelat, A.: Bit encryption is complete. In: FOCS, pp. 607–616 (2009)
28. Naor, M., Yung, M.: Public-key cryptosystems provably secure against chosen ciphertext attacks. In: STOC, pp. 427–437 (1990)
29. Pass, R., Shelat, A., Vaikuntanathan, V.: Construction of a non-malleable encryption scheme from any semantically secure one. In: Dwork, C. (ed.) CRYPTO 2006. LNCS, vol. 4117, pp. 271–289. Springer, Heidelberg (2006)
30. Pass, R., Shelat, A., Vaikuntanathan, V.: Relations among notions of non-malleability for encryption. In: Kurosawa, K. (ed.) ASIACRYPT 2007. LNCS, vol. 4833, pp. 519–535. Springer, Heidelberg (2007)
31. Sahai, A.: Non-malleable non-interactive zero knowledge and adaptive chosen-ciphertext security. In: FOCS, pp. 543–553 (1999)
32. Shen, B.: A Justesen construction of binary concatenated codes that asymptotically meet the Zyablov bound for low rate. IEEE Trans. Inf. Theory 39(1), 239–242 (1993)

Verifiable Random Functions from Standard Assumptions

Dennis Hofheinz[1]([⊠]) and Tibor Jager[2]

[1] Karlsruhe Institute of Technology, Karlsruhe, Germany
`dennis.hofheinz@kit.edu`
[2] Ruhr-University Bochum, Bochum, Germany
`tibor.jager@rub.de`

Abstract. The question whether there exist verifiable random functions with exponential-sized input space and full adaptive security based on a non-interactive, *constant-size* assumption is a long-standing open problem. We construct the first verifiable random functions which achieve all these properties simultaneously.

Our construction can securely be instantiated in groups with symmetric bilinear map, based on any member of the $(n-1)$-*linear* assumption family with $n \geq 3$. This includes, for example, the 2-linear assumption, which is also known as the *decision linear* (DLIN) assumption.

1 Introduction

A verifiable random function (VRF) V_{sk} is essentially a pseudorandom function, but with the additional feature that it is possible to create a *non-interactive* and *publicly verifiable* proof π that a given function value Y was computed correctly as $Y = V_{sk}(X)$. VRFs are useful ingredients for applications as various as resettable zero-knowledge proofs [37], lottery systems [38], transaction escrow schemes [31], updatable zero-knowledge databases [34], or e-cash [4,5].

Desired Properties of VRFs. The standard security properties required from VRFs are *pseudorandomness* (when no proof is given, of course) and *unique provability*. The latter means that for each X there is only one *unique* value Y such that a proof for the statement "$Y = V_{sk}(X)$" exists. Unique provability is a very strong requirement, because *not even the party that creates sk* (possibly maliciously) may be able to create fake proofs. For example, the natural attempt of constructing a VRF by combining a pseudorandom function with a non-interactive zero-knowledge proof system fails, because zero-knowledge proofs are *simulatable*, which contradicts uniqueness.

Most known constructions of verifiable random functions allow an only *polynomially bounded* input space, or do not achieve *full adaptive security*, or are based on an *interactive* complexity assumption. In the sequel, we will say that a VRF has *all desired properties*, if is has an *exponential-sized input space* and a proof of *full adaptive security* under a *non-interactive* complexity assumption.

Supported by DFG grants HO 4534/2-2 and HO 4534/4-1.

E. Kushilevitz and T. Malkin (Eds.): TCC 2016-A, Part I, LNCS 9562, pp. 336–362, 2016.
DOI: 10.1007/978-3-662-49096-9_14

VRFs with all Desired Properties. All known examples of VRFs that possess all desired properties are based on so-called Q-*type* complexity assumptions. For example, the VRF of Hohenberger and Waters [28] relies on the assumption that, given a list of group elements

$$(g, h, g^x, \ldots, g^{x^{Q-1}}, g^{x^{Q+1}}, \ldots, g^{x^{2Q}}, t) \in \mathbb{G}^{2Q+1} \times \mathbb{G}_T$$

and a bilinear map $e : \mathbb{G} \times \mathbb{G} \rightarrow \mathbb{G}_T$, it is computationally infeasible to distinguish $t = e(g, h)^{x^Q}$ from a random element of $\mathbb{G}_T$ with probability significantly better than $1/2$. Note that the assumption is parametrized by an integer Q, which determines the number of group elements in a given problem instance.

The main issue with Q-type assumptions is that they get stronger with increasing Q, as demonstrated by Cheon [18]. For example, the VRF described in [28] is based on a Q-type assumption with $Q = \Theta(q \cdot k)$, where k is the security parameter and q is the number of function evaluations queried by the attacker in the security experiment. Constructions from weaker Q-type assumptions were described by Boneh *et al.* [11] and Abdalla *et al.* [2], both require $Q = \Theta(k)$. A VRF-security proof for the classical verifiable *unpredictable* function of Lysyanskaya [35], which requires a Q-type assumption with only $Q = O(\log k)$, was recently given in [29]. Even though this is complexity assumption is relatively weak, it is still Q-type.

In summary, the construction of a VRF with all desired security properties, which is based on a standard, *constant-size* assumption (like the *decision-linear* assumption, for example) is a long-standing open problem, posed for example in [28,29]. Some authors even asked if it is possible to prove that a Q-type assumption is *inherently necessary* to construct such VRFs [28]. Indeed, by adopting the techniques of [30] to the setting of VRFs, one can prove [33] that some known VRF-constructions are *equivalent* to certain Q-type assumptions, which means that a security proof under a strictly weaker assumption is impossible. This includes the VRFs of Dodis-Yampolskiy [20] and Boneh *et al.* [11]. It is also known that it is impossible to construct verifiable random functions from one-way permutations [13] or even trapdoor permutations in a black-box manner [23].

Our Contribution. We construct the first verifiable random functions with exponential-sized input space, and give a proof of full adaptive security under any member of the $(n-1)$-*linear* assumption family with $n \geq 3$ in symmetric bilinear groups. The $(n - 1)$-linear assumption is a family of non-interactive, *constant-size* complexity assumptions, which get progressively weaker with larger n [42]. A widely-used special case is the 2-linear assumption, which is also known as the *decision-linear* (DLIN) assumption [10].

Recently, a lot of progress has been made in proving the security of cryptosystems which previously required a Q-type assumption, see [16,26,43], for example. Verifiable random functions with all desired properties were one of the last cryptographic applications that required Q-type assumptions. Our work eliminates VRFs from this list.

The New Construction and Proof Idea. The starting point for our construction is the VRF of Lysyanskaya [35]. Her function is in fact the Naor-Reingold pseudorandom function [40] with

$$V_{sk}(X) \; = \; g^{\prod_{i=1}^{k} a_{i,x_i}},$$

where $X = (x_1, \ldots, x_k)$, and the $a_{i,b}$ are randomly chosen exponents. However, unlike [40], Lysyanskaya considers this function in a "Diffie-Hellman gap group".[1] The corresponding verification key consists of all $g^{a_{i,x_i}}$. Relative to this verification key, an image y can be proven to be of the form $g^{\prod_{i=1}^{k} a_{i,x_i}}$ by publishing all "partial products in the exponent", that is, all values $\pi_\ell := g^{\prod_{i=1}^{\ell} a_{i,x_i}}$ for $\ell \in \{2, \ldots, k-1\}$. (Since the Decisional Diffie-Hellman problem is assumed to be easy, these partial products can be checked for consistency with the $g^{a_{i,b}}$ one after the other.)

Note that pseudorandomness of this construction is not obvious. Indeed, Lysyanskaya's analysis requires a computational assumption that offers k group elements in a computational challenge. (This "size-k" assumption could be reduced to a "size-$(\log(k))$" assumption recently [29].) One reason for this apparent difficulty lies in the verifiability property of a VRF. For instance, the original Naor-Reingold analysis of [40] (that shows that this V_{sk} is a *PRF*) can afford to gradually substitute images given to the adversary by random images, using a hybrid argument. Such a proof is not possible in a setting in which the adversary can ask for "validity proofs" for some of these images. (Note that by the uniqueness property of a VRF, we cannot expect to be able to simulate such validity proofs for non-images.) As a result, so far security proofs for VRFs have used "one-shot reductions" to suitable computational assumptions (which then turned out to be rather complex).

We circumvent this problem by a more complex function (with more complex public parameters) that can be modified *gradually*, using simpler computational assumptions. Following [22], in the sequel we will write $g^{\mathbf{u}}$, where $\mathbf{u} = (u_1, \ldots, u_n)^\top \in \mathbb{Z}_p^n$ is a vector, to denote the vector $g^{\mathbf{u}} := (g^{u_1}, \ldots, g^{u_n})$. We will also extend this notation to matrices in the obvious way. To explain our approach, consider the function

$$G_{sk}(X) \; = \; g^{\mathbf{u}^\top \cdot \prod_{i=1}^{k} \mathbf{M}_{i,x_i}}$$

for random (quadratic) *matrices* $\mathbf{M}_{i,x_i}$ and a random vector $\mathbf{u}$. The function G_{sk} will not be the VRF V_{sk} we seek, but it will form the basis for it. (In fact, V_{sk} will only postprocess G_{sk}'s output, in a way we will explain below.) V_{sk}'s verification key will include $g^{\mathbf{u}}$ and the $g^{\mathbf{M}_{i,b}}$. As in the VRF described above, validity proofs of images contain all partial products $g^{\mathbf{u}^\top \cdot \prod_{i=1}^{\ell} \mathbf{M}_{i,x_i}}$. (However, note that to *check* proofs, we now need a bilinear map, and not only an efficient DDH-solver, as with Lysyanskaya's VRF.)

---

[1] In a Diffie-Hellman gap group, the Decisional Diffie-Hellman problem is easy, but the Computational Diffie-Hellman is hard. A prominent candidate of such groups are pairing-friendly groups.

To show pseudorandomness, let us first consider the case of *selective security* in which the adversary $\mathcal{A}$ first commits to a challenge preimage X^*. Then, $\mathcal{A}$ receives the verification key and may ask for arbitrary images $V_{sk}(X)$ *and proofs* for $X \neq X^*$. Additionally, $\mathcal{A}$ gets either $V_{sk}(X^*)$ (without proof), or a random image, and has to decide which it is.

In this setting, we can gradually adapt the $g^{\mathbf{M}_{i,b}}$ given to $\mathcal{A}$ such that $\prod_{i=1}^{k} \mathbf{M}_{i,x_i}$ has full rank if and only if $X = X^*$. To this end, we choose $\mathbf{M}_{i,b}$ as a full-rank matrix exactly for $b = x_i^*$. (This change can be split up in a number of local changes, each of which changes only one $\mathbf{M}_{i,b}$ and can be justified with the $(n-1)$-linear assumption, where n is the dimension of $\mathbf{M}_{i,b}$.) Even more: we show that if we perform these changes carefully, and in a "coordinated" way, we can achieve that $\mathbf{v}^{\top} := \mathbf{u}^{\top} \prod_{i=1}^{k} \mathbf{M}_{i,x_i}$ lies in a fixed subspace $\mathfrak{U}^{\top}$ if and only if $X \neq X^*$. In other words, if we write $\mathbf{v} = \sum_{i=1}^{n} \beta_i \mathbf{b}_i$ for a basis $\{\mathbf{b}_i\}_{i=1}^{n}$ such that $\{\mathbf{b}_i\}_{i=1}^{n-1}$ is a basis of $\mathfrak{U}$, then we have that $\beta_n = 0$ if and only if $X \neq X^*$. Put differently: $\mathbf{v}$ has a $\mathbf{b}_n$-component if and only if $X = X^*$.

Hence, we could hope to embed (part of) a challenge from a computational hardness assumption into $\mathbf{b}_n$. For instance, to obtain a VRF secure under the Bilinear Decisional Diffie-Hellman (BDDH) assumption, one could set $V_{sk}(X) = e(G_{sk}(X), g^\alpha)^\beta$ for a pairing e and random α, β. A BDDH challenge can then be embedded into $\mathbf{b}_n$, α, and β. (Of course, also validity proofs need to be adapted suitably.)

In the main part of the paper, we show how to generalize this idea simultaneously to adaptive security (with a semi-generic approach that employs admissible hash functions), and based on the $(n-1)$-linear assumption for arbitrary $n \geq 3$ (instead of the BDDH assumption).

We note that we pay a price for a reduction to a standard assumption: since our construction relies on *matrix* multiplication (instead of multiplication of exponents), it is less efficient than previous constructions. For instance, compared to Lysyanskaya's VRF, our VRF has less compact proofs (by a factor of about n, when building on the $(n-1)$-linear assumption), and requires more pairing operations (by a factor of about n^2) for verification.

Programmable Vector Hash Functions. The proof strategy sketched above is implemented by a new tool that we call *programmable vector hash functions* (PVHFs). Essentially, PVHFs can be seen as a variant of programmable hash functions of Hofheinz and Kiltz [27], which captures the "coordinated" setup of G_{sk} described above in a modular building block. We hope that this building block will be useful for other cryptographic constructions.

More Related Work. VRFs were introduced by Micali, Rabin, and Vadhan [36]. Number-theoretic constructions of VRFs were described in [1,2,11,19,20,28,29, 35,36]. Abdalla *et al.* [1,2] also gave a generic construction from a special type of identity-based key encapsulation mechanisms. Most of these either do not achieve full adaptive security for large input spaces, or are based on *interactive* complexity assumptions, the exceptions [2,11,28,29] were mentioned above. We wish to avoid interactive assumptions to prevent circular arguments, as explained by Naor [39].

The notion of *weak* VRFs was proposed by Brakerski *et al.* [13], along with simple and efficient constructions, and proofs that neither VRFs, nor weak VRFs can be constructed (in a black-box way) from one-way permutations. Several works introduced related primitives, like simulatable VRFs [15] and constrained VRFs [25].

Other Approaches to Avoid Q-type Assumptions. One may ask whether the techniques presented by Chase and Meiklejohn [16], which in certain applications allow to replace Q-type assumption with *constant-size* subgroup hiding assumptions, give rise to alternative constructions of VRFs from constant-size assumptions. This technique is based on the idea of using the *dual-systems* approach of Waters [43], and requires to add *randomization* to group elements. This randomization makes it difficult to construct VRFs that meet the unique provability requirement. Consequently, Chase and Meiklejohn were able to prove that the VRF of Dodis and Yampolski [20] forms a secure pseudorandom function under a static assumption, but not that it is a secure VRF.

Open Problems. The verifiable random functions constructed in this paper are relatively inefficient, when compared to the q-type-based constructions of [2,11,28,29], for example. An interesting open problem is therefore the construction of more efficient VRFs from standard assumptions. In particular, it is not clear whether the constructions in this paper can also be instantiated from the SXDH assumption in asymmetric bilinear groups. This would potentially yield a construction with smaller matrices, and thus shorter proofs.

2   Certified Bilinear Group Generators

In order to be able to prove formally that a given verifiable random function satisfies *uniqueness* in the sense of Definition 8, we extend the notion of *certified trapdoor permutations* [7,8,32] to *certified bilinear group generators*. Previous works on verifiable random functions were more informal in this aspect, e.g., by requiring that group membership can be tested efficiently and that each group element has a unique representation.

Definition 1. *A* bilinear group generator *is a probabilistic polynomial-time algorithm* GrpGen *that takes as input a security parameter k (in unary) and outputs $\Pi = (p, \mathbb{G}, \mathbb{G}_T, \circ, \circ_T, e, \phi(1)) \xleftarrow{\$} \mathsf{GrpGen}(1^k)$ such that the following requirements are satisfied.*

1. *p is prime and $\log(p) \in \Omega(k)$.*
2. *$\mathbb{G}$ and $\mathbb{G}_T$ are subsets of $\{0,1\}^*$, defined by algorithmic descriptions of maps $\phi : \mathbb{Z}_p \to \mathbb{G}$ and $\phi_T : \mathbb{Z}_p \to \mathbb{G}_T$.*
3. *$\circ$ and $\circ_T$ are algorithmic descriptions of efficiently computable (in the security parameter) maps $\circ : \mathbb{G} \times \mathbb{G} \to \mathbb{G}$ and $\circ_T : \mathbb{G}_T \times \mathbb{G}_T \to \mathbb{G}_T$, such that*
   (a) *$(\mathbb{G}, \circ)$ and $(\mathbb{G}_T, \circ_T)$ form algebraic groups and*
   (b) *ϕ is a group isomorphism from $(\mathbb{Z}_p, +)$ to $(\mathbb{G}, \circ)$ and*
   (c) *ϕ_T is a group isomorphism from $(\mathbb{Z}_p, +)$ to $(\mathbb{G}_T, \circ_T)$.*

4. e is an algorithmic description of an efficiently computable (in the security parameter) bilinear map $e : \mathbb{G} \times \mathbb{G} \to \mathbb{G}_T$. We require that e is non-degenerate, *that is,*

$$x \neq 0 \implies e(\phi(x), \phi(x)) \neq \phi_T(0)$$

Definition 2. *We say that group generator* GrpGen *is* certified, *if there exists a deterministic polynomial-time algorithm* GrpVfy *with the following properties.*

Parameter Validation. *Given a string Π (which is not necessarily generated by* GrpGen*), algorithm* GrpVfy(Π) *outputs 1 if and only if Π has the form*

$$\Pi = (p, \mathbb{G}, \mathbb{G}_T, \circ, \circ_T, e, \phi(1))$$

and all requirements from Definition 1 are satsified.

Recognition and Unique Representation Elements of $\mathbb{G}$. *Furthermore, we require that each element in $\mathbb{G}$ has a* unique *representation, which can be efficiently recognized. That is, on input two strings Π and s,* GrpVfy(Π, s) *outputs 1 if and only if* GrpVfy$(\Pi) = 1$ *and it holds that $s = \phi(x)$ for some $x \in \mathbb{Z}_p$. Here $\phi : \mathbb{Z}_p \to \mathbb{G}$ denotes the fixed group isomorphism contained in Π to specify the representation of elements of $\mathbb{G}$ (see Definition 1).*

3   Programmable Vector Hash Functions

Notation. As explained in the introduction, for a vector $\mathbf{u} = (u_1, \dots, u_n)^\top \in \mathbb{Z}_p^n$ we will write $g^{\mathbf{u}}$ to denote the vector $g^{\mathbf{u}} := (g^{u_1}, \dots, g^{u_n})$, and we will generalize this notation to matrices in the obvious way. Moreover, whenever the reference to a group generator $g \in \mathbb{G}$ is clear (note that a generator $g = \phi(1)$ is always contained in the group parameters Π generated by GrpGen), we will henceforth follow [22] and simplify our notation by writing $[x] := g^x \in \mathbb{G}$ for an integer $x \in \mathbb{Z}_p$, $[\mathbf{u}] := g^{\mathbf{u}} \in \mathbb{G}^n$ for a vector $\mathbf{u} \in \mathbb{Z}_p^n$, and $[\mathbf{M}] := g^{\mathbf{M}} \in \mathbb{G}^{n \times n}$ for a matrix $\mathbf{M} \in \mathbb{Z}_p^{n \times n}$. We also extend our notation for bilinear maps: we write $e([\mathbf{A}], [\mathbf{B}])$ (for matrices $\mathbf{A} = (a_{i,j})_{i,j} \in \mathbb{Z}_p^{n_1 \times n_2}$ and $\mathbf{B} = (b_{i,j})_{i,j} \in \mathbb{Z}_p^{n_2 \times n_3}$) for the matrix whose (i,j)-th entry is $\prod_{\ell=1}^{n_2} e([a_{i,\ell}], [b_{\ell,j}])$. In other words, we have $e([\mathbf{A}], [\mathbf{B}]) = e(g, g)^{\mathbf{AB}}$.

For a vector space $\mathfrak{U} \subseteq \mathbb{Z}_p^{n \times n}$ of column vectors, we write $\mathfrak{U}^\top := \{\mathbf{u}^\top \mid \mathbf{u} \in \mathfrak{U}\}$ for the respective set of row vectors. Furthermore, we write $\mathfrak{U}^\top \cdot \mathbf{M} := \{\mathbf{u}^\top \cdot \mathbf{M} \mid \mathbf{u}^\top \in \mathfrak{U}^\top\}$ for an element-wise vector-matrix multiplication. Finally, we denote with $\mathrm{GL}_n(\mathbb{Z}_p) \subset \mathbb{Z}_p^{n \times n}$ the set of invertible n-by-n matrices over $\mathbb{Z}_p$. Recall that a uniformly random $\mathbf{M} \in \mathbb{Z}_p$ is invertible except with probability at most n/p. (Hence, the uniform distributions on $\mathrm{GL}_n(\mathbb{Z}_p)$ and $\mathbb{Z}_p^{n \times n}$ are statistically close.)

3.1   Vector Hash Functions

Definition 3. *Let* GrpGen *be group generator algorithm and let $n \in \mathbb{N}$ be a positive integer. A* verifiable vector hash function (VHF) *for* GrpGen *with domain $\{0,1\}^k$ and range $\mathbb{G}^n$ consists of algorithms (*Gen$_{\mathsf{VHF}}$, Eval$_{\mathsf{VHF}}$, Vfy$_{\mathsf{VHF}}$*) with the following properties.*

- $\mathsf{Gen_{VHF}}$ *takes as input parameters* $\Pi \xleftarrow{\$} \mathsf{GrpGen}(1^k)$ *and outputs a verification key* vk *and an evaluation key* ek *as* $(vk, ek) \xleftarrow{\$} \mathsf{Gen_{VHF}}(\Pi)$.
- $\mathsf{Eval_{VHF}}$ *takes as input an evaluation key* ek *and a string* $X \in \{0,1\}^k$. *It outputs* $([\mathbf{v}], \pi) \leftarrow \mathsf{Eval_{VHF}}(ek, X)$, *where* $[\mathbf{v}] = ([v_1], \ldots, [v_n])^\top \in \mathbb{G}^n$ *is the function value and* $\pi \in \{0,1\}^*$ *is a corresponding proof of correctness.*
- $\mathsf{Vfy_{VHF}}$ *takes as input a verification key* vk, *vector* $[\mathbf{v}] \in \mathbb{G}^n$, *proof* $\pi \in \{0,1\}^*$, *and* $X \in \{0,1\}^k$, *and outputs a bit:* $\mathsf{Vfy_{VHF}}(vk, [\mathbf{v}], \pi, X) \in \{0,1\}$.

We require correctness *and* unique provability *in the following sense.*

Correctness. *We say that* $(\mathsf{Gen_{VHF}}, \mathsf{Eval_{VHF}}, \mathsf{Vfy_{VHF}})$ *is* correct, *if for all* $\Pi \xleftarrow{\$} \mathsf{GrpGen}(1^k)$, *all* $(vk, ek) \xleftarrow{\$} \mathsf{Gen_{VHF}}(\Pi)$, *and all* $X \in \{0,1\}^k$ *holds that*

$$\Pr\left[\mathsf{Vfy_{VHF}}(vk, [\mathbf{v}], \pi, X) = 1 : \begin{array}{l} (vk, ek) \xleftarrow{\$} \mathsf{Gen_{VHF}}(\Pi), \\ ([\mathbf{v}], \pi) \leftarrow \mathsf{Eval_{VHF}}(ek, X) \end{array}\right] = 1$$

Unique Provability. *We say that a VHF has* unique provability, *if for all strings* $vk \in \{0,1\}^*$ *(not necessarily generated by* $\mathsf{Gen_{VHF}}$*) and all* $X \in \{0,1\}^k$ *there does not exist any tuple* $([\mathbf{v}_0], \pi_0, [\mathbf{v}_1], \pi_1)$ *with* $[\mathbf{v}_0] \neq [\mathbf{v}_1]$ *and* $[\mathbf{v}_0], [\mathbf{v}_1] \in \mathbb{G}^n$ *such that*

$$\mathsf{Vfy_{VHF}}(vk, [\mathbf{v}_0], \pi_0, X) = \mathsf{Vfy_{VHF}}(vk, [\mathbf{v}_1], \pi_1, X) = 1$$

3.2 Selective Programmability

Definition 4. *We say that VHF* $(\mathsf{Gen_{VHF}}, \mathsf{Eval_{VHF}}, \mathsf{Vfy_{VHF}})$ *is* selectively programmable, *if additional algorithms* $\mathsf{Trap_{VHF}} = (\mathsf{TrapGen_{VHF}}, \mathsf{TrapEval_{VHF}})$ *exist, with the following properties.*

- $\mathsf{TrapGen_{VHF}}$ *takes group parameters* $\Pi \xleftarrow{\$} \mathsf{GrpGen}(1^k)$, *matrix* $[\mathbf{B}] \in \mathbb{G}^{n \times n}$, *and* $X^{(0)} \in \{0,1\}^k$. *It computes* $(vk, td) \xleftarrow{\$} \mathsf{TrapGen_{VHF}}(\Pi, [\mathbf{B}], X^{(0)})$, *where* vk *is a verification key with corresponding trapdoor evaluation key* td.
- $\mathsf{TrapEval_{VHF}}$ *takes as input a trapdoor evaluation key* td *and a string* $X \in \{0,1\}^k$. *It outputs a vector* $\boldsymbol{\beta} \leftarrow \mathsf{TrapEval_{VHF}}(td, X)$ *with* $\boldsymbol{\beta} \in \mathbb{Z}_p^n$ *and a proof* $\pi \in \{0,1\}^k$.

We furthermore have the following requirements.

Correctness. *For all* $\Pi \xleftarrow{\$} \mathsf{GrpGen}(1^k)$, *all* $[\mathbf{B}] \in \mathbb{G}^{n \times n}$, *and all* $X, X^{(0)} \in \{0,1\}^k$ *we have*

$$\Pr\left[\mathsf{Vfy_{VHF}}(vk, [\mathbf{v}], X) = 1 : \begin{array}{l} (vk, td) \xleftarrow{\$} \mathsf{TrapGen_{VHF}}(\Pi, [\mathbf{B}], X^{(0)}) \\ (\boldsymbol{\beta}, \pi) \leftarrow \mathsf{TrapEval_{VHF}}(td, X) \\ [\mathbf{v}] := [\mathbf{B}] \cdot \boldsymbol{\beta} \end{array}\right] = 1$$

$\mathcal{O}_0(X):$	$\mathcal{O}_1(X):$	$\mathcal{O}_{\mathsf{check}}(X):$
$(\mathbf{v}, \pi) \leftarrow \mathsf{Eval}_{\mathsf{VHF}}(ek, X)$	$(\boldsymbol{\beta}, \pi) \leftarrow \mathsf{TrapEval}_{\mathsf{VHF}}(td, X)$	$(\boldsymbol{\beta}, \pi) \leftarrow \mathsf{TrapEval}_{\mathsf{VHF}}(td, X)$
Return $([\mathbf{v}], \pi)$	$[\mathbf{v}] := [\mathbf{B}] \cdot \boldsymbol{\beta}$	$(\beta_1, \ldots, \beta_n) := \boldsymbol{\beta}$
	Return $([\mathbf{v}], \pi)$	**If** $\beta_n \neq 0$ **then Return** 1
		Else Return 0

Fig. 1. Definition of oracles $\mathcal{O}_0$, $\mathcal{O}_1$, and $\mathcal{O}_{\mathsf{check}}$.

Indistinguishability. *Verification keys generated by* $\mathsf{TrapGen}_{\mathsf{VHF}}$ *are computationally indistinguishable from keys generated by* $\mathsf{Gen}_{\mathsf{VHF}}$. *More precisely, we require that for all PPT algorithms* $\mathcal{A} = (\mathcal{A}_0, \mathcal{A}_1)$ *holds that*

$$\mathsf{Adv}^{\mathsf{vhf-sel-ind}}_{\mathsf{VHF},\mathsf{Trap}_{\mathsf{VHF}}}(k) := 2 \cdot \Pr \left[\begin{array}{l} \Pi \xleftarrow{\$} \mathsf{GrpGen}(1^k); \ (X^{(0)}, st) \xleftarrow{\$} \mathcal{A}_0(1^k) \\ (vk_0, ek) \xleftarrow{\$} \mathsf{Gen}_{\mathsf{VHF}}(\Pi); \ \mathbf{B} \xleftarrow{\$} \mathrm{GL}_n(\mathbb{Z}_p) \\ (vk_1, td) \xleftarrow{\$} \mathsf{TrapGen}_{\mathsf{VHF}}(\Pi, [\mathbf{B}], X^{(0)}) \\ \overline{b} \xleftarrow{\$} \{0,1\}; \ \mathcal{A}_1^{\mathcal{O}_{\overline{b}}}(st, vk_{\overline{b}}) = \overline{b} \end{array} \right] - 1$$

is negligible, where oracles $\mathcal{O}_0$ *and* $\mathcal{O}_1$ *are defined in Fig. 1.*

Well-distributed Outputs. *Let* $q = q(k) \in \mathbb{N}$ *be a polynomial, and let* $\beta_n^{(i)}$ *denote the* n-*th coordinate of vector* $\boldsymbol{\beta}^{(i)} \in \mathbb{Z}_p^n$. *There exists a polynomial* poly *such that for all* $(X^{(0)}, \ldots, X^{(q)}) \in (\{0,1\}^k)^{q+1}$ *with* $X^{(0)} \neq X^{(i)}$ *for* $i \geq 1$ *holds that*

$$\Pr \left[\begin{array}{l} \beta_n^{(0)} \neq 0 \wedge \beta_n^{(i)} = 0 \\ \forall i \in \{1, \ldots, q\} \end{array} : \begin{array}{l} \Pi \xleftarrow{\$} \mathsf{GrpGen}(1^k) \\ \mathbf{B} \xleftarrow{\$} \mathrm{GL}_n(\mathbb{Z}_p) \\ (vk, td) \xleftarrow{\$} \mathsf{TrapGen}_{\mathsf{VHF}}(\Pi, [\mathbf{B}], X^{(0)}) \\ (\boldsymbol{\beta}^{(i)}, \pi) \leftarrow \mathsf{TrapEval}_{\mathsf{VHF}}(td, X^{(i)}) \ \forall i \end{array} \right] \geq \frac{1}{\mathsf{poly}(k)}$$

We note that in our security definitions, $\mathbf{B}$ is always a random *invertible* matrix, although $\mathsf{TrapGen}_{\mathsf{VHF}}$ would also work on arbitrary $\mathbf{B}$.

Furthermore, note that we only require a noticeable "success probability" in our "well-distributed outputs" requirement above. This is sufficient for our application; however, our (selectively secure) PVHF construction achieves a success probability of 1. (On the other hand, our adaptively secure construction only achieves well-distributedness in the sense above, with a significantly lower – but of course still noticeable – success probability.)

3.3 Adaptive Programmability

Definition 5. *We say that VHF* $(\mathsf{Gen}_{\mathsf{VHF}}, \mathsf{Eval}_{\mathsf{VHF}}, \mathsf{Vfy}_{\mathsf{VHF}})$ *is (adaptively) programmable, if algorithms* $\mathsf{Trap}_{\mathsf{VHF}} = (\mathsf{TrapGen}_{\mathsf{VHF}}, \mathsf{TrapEval}_{\mathsf{VHF}})$ *exist, which have exactly the same syntax and requirements on correctness, indistinguishability, and* well-formedness *as in Definition 4, with the following differences:*

- $\mathsf{TrapGen}_{\mathsf{VHF}}(\Pi, [\mathbf{B}])$ *does not take an additional string* $X^{(0)}$ *as input.*
- *In the indistinguishability experiment,* $\mathcal{A}_0$ *is the trivial algorithm, which outputs the empty string* $\emptyset$, *while* $\mathcal{A}_1$ *additionally gets access to oracle* $\mathcal{O}_{\mathsf{check}}$ *(see Fig. 1). We stress that this oracle always uses td to compute its output, independently of* $\overline{b}$. *We denote with* $\mathsf{Adv}^{\mathsf{vhf-ad-ind}}_{\mathsf{VHF},\mathsf{Trap}_{\mathsf{VHF}}}(k)$ *the corresponding advantage function.*

4   A PVHF Based on the Matrix-DDH Assumption

Overview. In this section, we present a programmable vector hash function, whose security is based upon the "Matrix-DDH" assumption introduced in [22] (which generalizes the matrix-DDH assumption of Boneh *et al.* [12] and the matrix d-linear assumption of Naor and Segev [41]). This assumption can be viewed as a relaxation of the $(n-1)$-linear assumption, so that in particular our construction will be secure under the $(n-1)$-linear assumption with $n \geq 3$.

Assumption 6. *The* n-rank assumption *states that* $[\mathbf{M}_{n-1}] \stackrel{c}{\approx} [\mathbf{M}_n]$, *where* $\mathbf{M}_i \in \mathbb{Z}_p^{n \times n}$ *is a uniformly distributed rank-i matrix, i.e., that*

$$\mathsf{Adv}^{n-\mathsf{rank}}_{\mathcal{A}}(k) := \Pr\left[\mathcal{A}([\mathbf{M}_{n-1}]) = 1\right] - \Pr\left[\mathcal{A}([\mathbf{M}_n]) = 1\right]$$

is negligible for every PPT adversary $\mathcal{A}$.

4.1   The Construction

Assume a bilinear group generator GrpGen and an integer $n \in \mathbb{N}$ as above. Consider the following vector hash function VHF:

- $\mathsf{Gen}_{\mathsf{VHF}}(\mathsf{GrpGen})$ uniformly chooses $2k$ invertible matrices $\mathbf{M}_{i,b} \in \mathbb{Z}_p^{n \times n}$ (for $1 \leq i \leq k$ and $b \in \{0,1\}$) and a nonzero vector $\mathbf{u}^\top \in \mathbb{Z}_p^n \setminus \{0\}$. The output is (vk, ek) with

$$vk = \left(([\mathbf{M}_{i,b}])_{1 \leq i \leq k, b \in \{0,1\}}, [\mathbf{u}]\right) \qquad ek = \left((\mathbf{M}_{i,b})_{1 \leq i \leq k, b \in \{0,1\}}, \mathbf{u}\right).$$

- $\mathsf{Eval}_{\mathsf{VHF}}(ek, X)$ (for $X = (x_1, \ldots, x_k)$) computes and outputs an image $[\mathbf{v}] = [\mathbf{v}_k] \in \mathbb{G}^n$ and a proof $\pi = ([\mathbf{v}_1], \ldots, [\mathbf{v}_{k-1}]) \in (\mathbb{G}^n)^{k-1}$, where

$$\mathbf{v}_i^\top = \mathbf{u}^\top \cdot \prod_{j=1}^{i} \mathbf{M}_{j,x_j}. \tag{1}$$

- $\mathsf{Vfy}_{\mathsf{VHF}}(vk, [\mathbf{v}], \pi, X)$ outputs 1 if and only if

$$e([\mathbf{v}_i^\top], [1]) = e([\mathbf{v}_{i-1}^\top], [\mathbf{M}_{i,x_i}]), \tag{2}$$

holds for all i with $1 \leq i \leq k$, where we set $[\mathbf{v}_0] := [\mathbf{u}]$ and $[\mathbf{v}_k] := [\mathbf{v}]$.

Theorem 1 (Correctness and Uniqueness of VHF). VHF *is a vector hash function. In particular,* VHF *satisfies the correctness and uniqueness conditions from Definition 3.*

Proof. First, note that (2) is equivalent to $\mathbf{v}_i^\top = \mathbf{v}_{i-1}^\top \cdot \mathbf{M}_{i,x_i}$. By induction, it follows that (2) holds for all i if and only if $\mathbf{v}_i^\top = \mathbf{u}^\top \cdot \prod_{j=1}^i \mathbf{M}_{j,x_j}$ for all i. By definition of $\mathsf{Eval}_{\mathsf{VHF}}$, this yields correctness. Furthermore, we get that $\mathsf{Vfy}_{\mathsf{VHF}}$ outputs 1 for precisely one value $[\mathbf{v}] = [\mathbf{v}_k]$ (even if the $\mathbf{M}_{i,b}$ are not invertible). In fact, the proof π is uniquely determined by vk and X.

4.2 Selective Security

We proceed to show the selective security of VHF:

Theorem 2 (Selectively Programmabability of VHF). VHF *is selectively programmable in the sense of Definition 4.*

The Trapdoor Algorithms. We split up the proof of Theorem 2 into three lemmas (that show correctness, well-distributed outputs, and indistinguishability of VHF). But first, we define the corresponding algorithms $\mathsf{TrapGen}_{\mathsf{VHF}}$ and $\mathsf{TrapEval}_{\mathsf{VHF}}$.

– $\mathsf{TrapGen}_{\mathsf{VHF}}(\Pi, [\mathbf{B}], X^{(0)})$ first chooses $k+1$ subspaces $\mathfrak{U}_i$ of $\mathbb{Z}_p^n$ (for $0 \le i \le k$), each of dimension $n-1$. Specifically,
  - the first k subspaces $\mathfrak{U}_i$ (for $0 \le i \le k-1$) are chosen independently and uniformly,
  - the last subspace $\mathfrak{U}_k$ is the subspace spanned by the first $n-1$ unit vectors. (That is, $\mathfrak{U}_k$ contains all vectors whose last component is 0.)
  Next, $\mathsf{TrapGen}_{\mathsf{VHF}}$ uniformly chooses $\mathbf{u} \in \mathbb{Z}_p^n \setminus \mathfrak{U}_0$ and $2k$ matrices $\mathbf{R}_{i,b}$ (for $1 \le i \le k$ and $b \in \{0,1\}$), as follows:

$$
\begin{aligned}
\mathbf{R}_{i,1-x_i^{(0)}} \text{ uniformly of rank } n-1 \text{ subject to} \quad & \mathfrak{U}_{i-1}^\top \cdot \mathbf{R}_{i,1-x_i^{(0)}} = \mathfrak{U}_i^\top \\
\mathbf{R}_{i,x_i^{(0)}} \text{ uniformly of rank } n \text{ subject to} \quad & \mathfrak{U}_{i-1}^\top \cdot \mathbf{R}_{i,x_i^{(0)}} = \mathfrak{U}_i^\top .
\end{aligned}
\tag{3}
$$

Finally, $\mathsf{TrapGen}_{\mathsf{VHF}}$ sets

$$
\mathbf{M}_{i,b} = \mathbf{R}_{i,b} \quad \text{for } 1 \le i \le k-1
$$
$$
[\mathbf{M}_{k,0}] = [\mathbf{R}_{k,0} \cdot \mathbf{B}^\top] \qquad [\mathbf{M}_{k,1}] = [\mathbf{R}_{k,1} \cdot \mathbf{B}^\top],
\tag{4}
$$

and outputs

$$
td = \left((\mathbf{R}_{i,b})_{i \in [k-1], b \in \{0,1\}}, \mathbf{u}, [\mathbf{B}]\right) \qquad vk = \left(([\mathbf{M}_{i,b}])_{i \in [k], b \in \{0,1\}}, [\mathbf{u}]\right).
$$

– $\mathsf{TrapEval}_{\mathsf{VHF}}(td, X)$ first computes an image $[\mathbf{v}] = [\mathbf{v}_k]$, along with a corresponding proof $\pi = [\mathbf{v}_1, \ldots, \mathbf{v}_{k-1}]$ exactly like $\mathsf{Eval}_{\mathsf{VHF}}$, using (1). (Note that

TrapEval$_{\mathsf{VHF}}$ can compute all $[\mathbf{v}_i]$ from its knowledge of $\mathbf{u}$, all $\mathbf{R}_{i,b}$, and $[\mathbf{B}]$.) Next, observe that the image $[\mathbf{v}]$ satisfies

$$\mathbf{v}^\top = \mathbf{v}_k^\top = \mathbf{u}^\top \cdot \prod_{j=1}^k \mathbf{M}_{j,x_j} = \underbrace{\left(\mathbf{u}^\top \cdot \prod_{j=1}^k \mathbf{R}_{j,x_j} \right)}_{=:\boldsymbol{\beta}^\top} \cdot \mathbf{B}^\top. \tag{5}$$

Hence, TrapEval$_{\mathsf{VHF}}$ outputs $(\boldsymbol{\beta}, \pi)$.

Lemma 1 (Correctness of Trap$_{\mathsf{VHF}}$). *The trapdoor algorithms* TrapGen$_{\mathsf{VHF}}$ *and* TrapEval$_{\mathsf{VHF}}$ *above satisfy correctness in the sense of Definition 4.*

Proof. This follows directly from (5).

Lemma 2. (Well-distributedness of Trap$_{\mathsf{VHF}}$). *The above trapdoor algorithms* TrapGen$_{\mathsf{VHF}}$ *and* TrapEval$_{\mathsf{VHF}}$ *enjoy well-distributed outputs in the sense of Definition 4.*

Proof. Fix any preimage $X^{(0)} = (x_i^{(0)})_{i=1}^k \in \{0,1\}^k$, matrix $[\mathbf{B}] \in \mathbb{G}^{n \times n}$, and corresponding keypair $(td, vk) \xleftarrow{\$} \mathsf{TrapGen}_{\mathsf{VHF}}(\Pi, [\mathbf{B}], X^{(0)})$. We will show first that for all $X = (x_i)_{i=1}^k \in \{0,1\}^k$, the corresponding vectors $\mathbf{v}_i^\top$ computed during evaluation satisfy

$$\mathbf{u}^\top \cdot \prod_{j=1}^i \mathbf{R}_{i,x_i} \in \mathfrak{U}_i^\top \quad \Longleftrightarrow \quad x_j \neq x_j^{(0)} \text{ for some } j \leq i. \tag{6}$$

Equation (6) can be proven by induction over i. The case $i = 0$ follows from the setup of $\mathbf{u} \notin \mathfrak{U}_0$. For the induction step, assume (6) holds for $i - 1$. To show (6) for i, we distinguish two cases:

- If $x_i = x_i^{(0)}$, then $\mathbf{R}_{i,x_i}$ has full rank, and maps $\mathfrak{U}_{i-1}^\top$ to $\mathfrak{U}_i^\top$. Thus, $\mathbf{u}^\top \cdot \prod_{j=1}^i \mathbf{R}_{j,x_j} \in \mathfrak{U}_i^\top$ if and only if $\mathbf{u}^\top \cdot \prod_{j=1}^{i-1} \mathbf{R}_{j,x_j} \in \mathfrak{U}_{i-1}^\top$.[2] By the induction hypothesis, and using $x_i = x_i^{(0)}$, hence, $\mathbf{u}^\top \cdot \prod_{j=1}^i \mathbf{R}_{j,x_j} \in \mathfrak{U}_i^\top$ if and only if $x_j \neq x_j^{(0)}$ for some $j \leq i$. This shows (6).
- If $x_i \neq x_i^{(0)}$, then $\mathbf{R}_{i,x_i}$ has rank $n - 1$. Together with $\mathfrak{U}_{i-1}^\top \cdot \mathbf{R}_{i,x_i} = \mathfrak{U}_i^\top$, this implies that in fact $(\mathbb{Z}_p^n)^\top \cdot \mathbf{R}_{i,x_i} = \mathfrak{U}_i^\top$. Hence, both directions of (6) hold.

This shows that (6) holds for all i. In particular, if we write

$$\boldsymbol{\beta}^\top = (\beta_1, \ldots, \beta_n) = \mathbf{u}^\top \cdot \prod_{j=1}^k \mathbf{R}_{i,x_i}$$

(as in (5)), then $\boldsymbol{\beta} \in \mathfrak{U}_k$ if and only if $X \neq X^{(0)}$. By definition of $\mathfrak{U}_k$, this means that $\beta_n = 0 \Leftrightarrow X \neq X^{(0)}$. Well-distributedness as in Definition 4 follows.

---

[2] Recall our notation from Sect. 3.

Lemma 3 (Indistinguishability of $\mathsf{Trap}_{\mathsf{VHF}}$). *If the n-rank assumption holds relative to GrpGen, then the above algorithms $\mathsf{TrapGen}_{\mathsf{VHF}}$ and $\mathsf{TrapEval}_{\mathsf{VHF}}$ satisfy the indistinguishability property from Definition 4. Specifically, for every adversary $\mathcal{A}$, there exists an adversary $\mathcal{B}$ (of roughly the same complexity) with*

$$\mathsf{Adv}^{\mathsf{vhf-sel-ind}}_{\mathsf{VHF},\mathsf{TrapGen}_{\mathsf{VHF}},\mathsf{TrapEval}_{\mathsf{VHF}},\mathcal{A}}(k) \;=\; k \cdot \mathsf{Adv}^{n\text{-}\mathsf{rank}}_{\mathbb{G},n,\mathcal{B}}(k) + \mathbf{O}(kn/p).$$

Proof. Fix an adversary $\mathcal{A}$. We proceed in games.

Game 0. Game 0 is identical to the indistinguishability game with $\bar{b} = 0$. In this game, $\mathcal{A}$ first selects a "target preimage" $X^{(0)}$, and then gets a verification key vk generated by $\mathsf{Gen}_{\mathsf{VHF}}$, and oracle access to an evaluation oracle $\mathcal{O}$. Let G_0 denote $\mathcal{A}$'s output in this game. (More generally, let G_i denote $\mathcal{A}$'s output in Game i.) Our goal will be to gradually change this setting such that finally, vk is generated by $\mathsf{TrapGen}_{\mathsf{VHF}}(\mathsf{GrpGen}, [\mathbf{B}], X^{(0)})$ (for an independently uniform invertible $\mathbf{B}$), and $\mathcal{O}$ uses the corresponding trapdoor to generate images and proofs. Of course, $\mathcal{A}$'s output must remain the same (or change only negligibly) during these transitions.

Game 1.ℓ (for $0 \leq \ell \leq k$). In Game 1.ℓ (for $0 \leq \ell \leq k$), vk is generated in part as by $\mathsf{TrapGen}_{\mathsf{VHF}}$, and in part as by $\mathsf{Gen}_{\mathsf{VHF}}$. ($\mathcal{O}$ is adapted accordingly.) Specifically, Game 1.ℓ proceeds like Game 0, except for the following changes:

- Initially, the game chooses $\ell + 1$ subspaces $\mathfrak{U}_i$ (for $0 \leq i \leq \ell$) of dimension $n - 1$ independently and uniformly, and picks $\mathbf{u} \in \mathbb{Z}_p^n \setminus \mathfrak{U}_0$. (Note that unlike in an execution of $\mathsf{TrapGen}_{\mathsf{VHF}}$, also $\mathfrak{U}_k$ is chosen uniformly when $\ell = k$.)
- Next, the game chooses $2k$ matrices $\mathbf{R}_{i,b}$ (for $1 \leq i \leq k$ and $b \in \{0,1\}$), as follows. For $i \leq \ell$, the $\mathbf{R}_{i,b}$ are chosen as by $\mathsf{TrapGen}_{\mathsf{VHF}}$, and thus conform to (3). For $i > \ell$, the $\mathbf{R}_{i,b}$ are chosen uniformly and independently (but invertible).
- Finally, the game sets up $\mathbf{M}_{i,b} := \mathbf{R}_{i,b}$ for all i, b. (Again, note the slight difference to $\mathsf{TrapGen}_{\mathsf{VHF}}$, which follows (4).)

The game hands the resulting verification key vk to $\mathcal{A}$; since all $\mathbf{M}_{i,b}$ are known over $\mathbb{Z}_p$, oracle $\mathcal{O}$ can be implemented as $\mathsf{Eval}_{\mathsf{VHF}}$.

Now let us take a closer look at the individual games Game 1.ℓ. First, observe that Game 1.0 is essentially Game 0: all $\mathbf{M}_{i,b}$ are chosen independently and uniformly, and $\mathcal{O}$ calls are answered in the only possible way (given vk). The only difference is that $\mathbf{u}$ is chosen independently uniformly from $\mathbb{Z}_p^n \setminus \{0\}$ in Game 0, and from $\mathbb{Z}_p^n \setminus \mathfrak{U}_0$ (for a uniform dimension-$(n-1)$ subspace $\mathfrak{U}_0$) in Game 1.0. However, both choices lead to the same distribution of $\mathbf{u}$, so we obtain

$$\Pr[G_0 = 1] \;=\; \Pr[G_{1.0} = 1]. \tag{7}$$

Next, we investigate the change from Game 1.$(\ell - 1)$ to Game 1.ℓ. We claim the following:

Lemma 4. *There is an adversary $\mathcal{B}$ on the n-rank problem with*

$$\sum_{\ell=1}^{k} \Pr[G_{1.\ell} = 1] - \Pr[G_{1.(\ell-1)} = 1] \;=\; k \cdot \mathsf{Adv}^{n\text{-}\mathsf{rank}}_{\mathbb{G},n,\mathcal{B}}(k) + \mathbf{O}(kn/p). \tag{8}$$

We postpone a proof of Lemma 4 until after the main proof.

Game 2. Finally, in Game 2, we slightly change the way $\mathfrak{U}_k$ and the matrices $\mathbf{M}_{k,b}$ (for $b \in \{0,1\}$) are set up:

- Instead of setting up $\mathfrak{U}_k$ uniformly (like all other $\mathfrak{U}_i$), we set up $\mathfrak{U}_k$ like $\mathsf{TrapGen}_{\mathsf{VHF}}$ would (i.e., as the subspace spanned by the first $n-1$ unit vectors).
- Instead of setting up $\mathbf{M}_{k,b} = \mathbf{R}_{k,b}$, we set $\mathbf{M}_{k,b} = \mathbf{R}_{k,b} \cdot \mathbf{B}^\top$ for an independently and uniformly chosen invertible $\mathbf{B}$, exactly like $\mathsf{TrapGen}_{\mathsf{VHF}}$ would.

Observe that since $\mathbf{B}$ is invertible, these modifications do not alter the distribution of the matrices $\mathbf{M}_{k,b}$ (compared to Game 1.ℓ). Indeed, in both cases, both $\mathbf{M}_{k,b}$ map $\mathfrak{U}_{k-1}^\top$ to the same uniformly chosen $(n-1)$-dimension subspace. In Game 1.ℓ, this subspace is $\mathfrak{U}_k$, while in Game 2, this subspace is the subspace spanned by the first $n-1$ columns of $\mathbf{B}$. We obtain:

$$\Pr\left[G_{1.\ell} = 1\right] = \Pr\left[G_2 = 1\right]. \tag{9}$$

Finally, it is left to observe that Game 2 is identical to the indistinguishability experiment with $\bar{b} = 1$: vk is prepared exactly as with $\mathsf{TrapGen}_{\mathsf{VHF}}(\Pi, [\mathbf{B}], X^{(0)})$ for a random $\mathbf{B}$, and $\mathcal{O}$ outputs the images and proofs uniquely determined by vk. Hence,

$$\begin{aligned}
\mathsf{Adv}_{\mathsf{VHF},\mathsf{Trap}_{\mathsf{VHF}},\mathcal{A}}^{\mathsf{vhf-sel-ind}}(k) &= \Pr\left[G_2 = 1\right] - \Pr\left[G_0 = 1\right] \\
&\overset{(7),(9)}{=} \Pr\left[G_{1.k} = 1\right] - \Pr\left[G_{1.0} = 1\right] \\
&= \sum_{\ell=1}^{k} \Pr\left[G_{1.\ell} = 1\right] - \Pr\left[G_{1.(\ell-1)} = 1\right] \\
&\overset{(8)}{=} k \cdot \mathsf{Adv}_{\mathbb{G},n,\mathcal{A}}^{n-\mathsf{rank}}(k) + \mathbf{O}(kn/p)
\end{aligned}$$

as desired.

It remains to prove Lemma 4.

Proof (Proof of Lemma 4). We describe an adversary $\mathcal{B}$ on the n-rank problem. $\mathcal{B}$ gets as input a matrix $[\mathbf{A}]$ "in the exponent," such that $\mathbf{A}$ is either of rank n, or of rank $n-1$. Initially, $\mathcal{B}$ uniformly picks $\ell \in \{1, \ldots, k\}$. Our goal is construct $\mathcal{B}$ such that it internally simulates Game 1.$(\ell-1)$ or Game 1.ℓ, depending on $\mathbf{A}$'s rank. To this end, $\mathcal{B}$ sets up vk as follows:

- Like Game 1.$(\ell-1)$, $\mathcal{B}$ chooses ℓ subspaces $\mathfrak{U}_0, \ldots, \mathfrak{U}_{\ell-1}$, and $\mathbf{u} \in \mathbb{Z}_p^n \setminus \mathfrak{U}_0$ uniformly.
- For $i < \ell$, $\mathcal{B}$ chooses matrices $\mathbf{R}_{i,b}$ like $\mathsf{TrapGen}_{\mathsf{VHF}}$ does, ensuring (3). For $i > \ell$, all $\mathbf{R}_{i,b}$ are chosen independently and uniformly but invertible. The case $i = \ell$ is more complicated and will be described next.
- To set up $\mathbf{M}_{\ell,0}$ and $\mathbf{M}_{\ell,1}$, $\mathcal{B}$ first asks $\mathcal{A}$ for its challenge input $X^{(0)} = (x_i^{(0)})_{i=1}^{k}$. Next, $\mathcal{B}$ embeds its own challenge $[\mathbf{A}]$ as $[\mathbf{R}_{\ell,1-x_\ell^{(0)}}] := [\mathbf{A}]$. To construct an $[\mathbf{R}_{\ell,x_\ell^{(0)}}]$ that achieves (3) (for $i = \ell$), $\mathcal{B}$ first uniformly chooses a basis

$\{c_1, \ldots, c_n\}$ of $\mathbb{Z}_p^n$, such that $\{c_1, \ldots, c_{n-1}\}$ forms a basis of $\mathfrak{U}_{\ell-1}$. (Note that $\mathcal{B}$ chooses the subspace $\mathfrak{U}_{\ell-1}$ on its own and over $\mathbb{Z}_p$, so this is possible efficiently for $\mathcal{B}$.) In the sequel, let $\mathbf{C}$ be the matrix whose i-th row is $c_i^\top$, and let $\mathbf{C}^{-1}$ be the inverse of $\mathbf{C}$. Jumping ahead, the purpose of $\mathbf{C}^{-1}$ is to help translate vectors from $\mathfrak{U}_{\ell-1}$ (as obtained through a partial product $\mathbf{u}^\top \prod_{j=1}^{\ell-1} \mathbf{M}_{j,x_j}$) to a "more accessible" form.

Next, $\mathcal{B}$ samples $n-1$ random vectors $[c_i']$ (for $1 \le i \le n-1$) in the image of $[\mathbf{A}]$ (e.g., by choosing random $r_i^\top$ and setting $[c_i'] = r_i^\top \cdot [\mathbf{A}]$). Furthermore, $\mathcal{B}$ samples $c_n' \in \mathbb{Z}_p^n$ randomly. Let $[\mathbf{C}']$ be the matrix whose i-th row is $[c_i'^\top]$. The purpose of $\mathbf{C}'$ is to define the image of $\mathbf{R}_{\ell, x_\ell^{(0)}}$. Specifically, $\mathcal{B}$ computes

$$[\mathbf{R}_{\ell, x_\ell^{(0)}}] := \mathbf{C}^{-1} \cdot [\mathbf{C}'].$$

(Note that $\mathcal{B}$ can compute $[\mathbf{R}_{\ell, x_\ell^{(0)}}]$ efficiently, since $\mathbf{C}^{-1}$ is known "in the clear.")
We will show below that, depending on the rank of $\mathbf{A}$, either $\mathfrak{U}_{\ell-1}^\top \cdot \mathbf{R}_{\ell, x_\ell^{(0)}} = \mathfrak{U}_{\ell-1}^\top \cdot \mathbf{A}$, or $\mathfrak{U}_{\ell-1}^\top \cdot \mathbf{R}_{\ell, x_\ell^{(0)}}$ is an independently random subspace of dimension $n-1$.
– Finally, $\mathcal{B}$ sets $[\mathbf{M}_{i,b}] = [\mathbf{R}_{i,b}]$ for all i, b, and hands $\mathcal{A}$ the resulting verification key vk.

Furthermore, $\mathcal{B}$ implements oracle $\mathcal{O}$ as follows: if $\mathcal{A}$ queries $\mathcal{O}$ with some $X = (x_i)_{i=1}^k \in \{0,1\}^k$, then $\mathcal{B}$ can produce the (uniquely determined) image and proof from the values

$$[v_i^\top] = [\mathbf{u}^\top \cdot \prod_{j=1}^i \mathbf{M}_{j,x_j}]. \tag{10}$$

On the other hand, $\mathcal{B}$ can compute all $[v_i^\top]$ efficiently, since it knows all factors in (10) over $\mathbb{Z}_p$, except for (at most) one factor $[\mathbf{M}_{\ell, x_\ell}]$.
Finally, $\mathcal{B}$ outputs whatever $\mathcal{A}$ outputs.
We now analyze this simulation. First, note that vk and $\mathcal{O}$ are simulated exactly as in both Game 1.$(\ell-1)$ and Game 1.ℓ, except for the definition of $[\mathbf{R}_{\ell, x_i^{(0)}}]$ and $[\mathbf{R}_{\ell, 1-x_i^{(0)}}]$. Now consider how these matrices are set up depending on the rank of $\mathcal{B}$'s challenge $\mathbf{A}$.

– If $\mathbf{A}$ is of rank n, then $\mathbf{R}_{\ell, x_\ell^{(0)}}$ and $\mathbf{R}_{\ell, 1-x_\ell^{(0)}}$ are (statistically close to) independently and uniformly random invertible matrices. Indeed, then each row $c_i'^\top$ of $\mathbf{C}'$ is independently and uniformly random: $c_1', \ldots, c_{n-1}'$ are independently random elements of the image of $\mathbf{A}$ (which is $\mathbb{Z}_p^n$), and c_n' is independently random by construction. Hence, $\mathbf{C}'$ is independently and uniformly random (and thus invertible, except with probability n/p). On the other hand, $\mathbf{R}_{\ell, x_\ell^{(0)}} = \mathbf{A}$ is uniformly random and invertible by assumption.
– If $\mathbf{A}$ is of rank $n-1$, then $\mathbf{R}_{\ell, x_\ell^{(0)}}$ and $\mathbf{R}_{\ell, 1-x_\ell^{(0)}}$ are (statistically close to) distributed as in (3). Indeed, then the rank of $\mathbf{R}_{\ell, 1-x_\ell^{(0)}} = \mathbf{A}$ is $n-1$, and the rank

of $\mathbf{R}_{\ell,x_{\ell}^{(0)}} = \mathbf{C}^{-1} \cdot \mathbf{C}'$ is n, except with probability at most n/p.[3] Moreover, if we write $\mathfrak{U}_{\ell}^{\top} := (\mathbb{Z}_p^n)^{\top} \cdot \mathbf{A}$, then by construction $(\mathbb{Z}_p^n)^{\top} \cdot \mathbf{R}_{\ell,1-x_{\ell}^{(0)}} = \mathfrak{U}_{\ell}^{\top}$, but also

$$(\mathbb{Z}_p^n)^{\top} \cdot \mathbf{R}_{\ell,x_{\ell}^{(0)}} = \mathfrak{W}^{\top} \cdot \mathbf{C}' = \mathfrak{U}_{\ell}^{\top},$$

where $\mathfrak{W}$ is the vector space spanned by the first $n-1$ unit vectors. Furthermore, $\mathbf{R}_{\ell,x_{\ell}^{(0)}}$ and $\mathbf{R}_{\ell,1-x_{\ell}^{(0)}}$ are distributed uniformly with these properties.

Hence, summarizing, up to a statistical defect of at most $2/p$, $\mathcal{B}$ simulates Game 1.($\ell - 1$) if $\mathbf{A}$ is of rank n, and Game 1.ℓ if $\mathbf{A}$ is of rank $n - 1$. This shows (8).

4.3 Adaptive Security

The idea behind the adaptively-secure construction is very similar to the selective case. Both the construction and the security proof are essentially identical, except that we apply an *admissible hash function* (AHF) AHF : $\{0,1\}^k \rightarrow \{0,1\}^{\ell_{\mathsf{AHF}}}$ (cf. Definition 7) to the inputs X of $\mathsf{Eval}_{\mathsf{VHF}}$ and $\mathsf{TrapEval}_{\mathsf{VHF}}$ before computing the matrix products. (We mention that suitable AHFs with $\ell_h = \mathbf{O}(k)$ exist [24,35].) Correctness and unique provability follow immediately. In order to prove well-distributedness, we rely on the properties of the admissible hash function. By a slightly more careful, AHF-dependent embedding of low-rank matrices in the verification key, these properties ensure that, for any sequence of queries issued by the adversary, it holds with non-negligible probability that the vector $[\mathbf{v}^{(0)}]$ assigned to input $X^{(0)}$ does not lie in the subspace generated by $(\mathbf{b_1}, \ldots, \mathbf{b_{n-1}})$, while all vectors $[\mathbf{v}^{(i)}]$ assigned to input $X^{(i)}$ do, which then yields the required well-distributedness property.

Admissible Hash Functions. To obtain adaptive security, we rely on a semi-blackbox technique based on admissible hash functions (AHFs, [3,9,14,24,35]). In the following, we use the formalization of AHFs from [24]:

Definition 7 (AHF). *For a function* AHF : $\{0,1\}^k \rightarrow \{0,1\}^{\ell_{\mathsf{AHF}}}$ *(for a polynomial $\ell_{\mathsf{AHF}} = \ell_{\mathsf{AHF}}(k)$) and $K \in (\{0,1,\perp\})^{\ell_{\mathsf{AHF}}}$, define the function $F_K : \{0,1\}^k \rightarrow \{\mathsf{CO},\mathsf{UN}\}$ through*

$$F_K(X) = \mathsf{UN} \iff \forall i : K_i = \mathsf{AHF}(X)_i \vee K_i = \perp,$$

where $\mathsf{AHF}(X)_i$ *denotes the i-th component of* $\mathsf{AHF}(X)$. *We say that* AHF *is q-admissible if there exists a PPT algorithm* KGen *and a polynomial* poly(k), *such that for all $X^{(0)}, \ldots, X^{(q)} \in \{0,1\}^k$ with $X^{(0)} \notin \{X^{(i)}\}$,*

$$\Pr\left[F_K(X^{(0)}) = \mathsf{UN} \wedge F_K(X^{(1)}) = \cdots = F_K(X^{(q)}) = \mathsf{CO}\right] \geq 1/\mathsf{poly}(k), \quad (11)$$

---

[3] To see this, observe that except with probability $(n-1)/p$, the first $n-1$ columns of $\mathbf{C}'$ are linearly independent (as they are random elements in the image of the rank-$(n-1)$ matrix $\mathbf{A}$). Further, the last row (which is independently and uniformly random) does not lie in the span of the first $n-1$ rows except with probability at most $1/p$.

where the probability is over $K \overset{\$}{\leftarrow} \mathsf{KGen}(1^k)$. We say that AHF is an admissible hash function (AHF) if AHF is q-admissible for all polynomials $q = q(k)$.

There are efficient constructions of admissible hash functions [24,35] with $\ell_{\mathsf{AHF}} = \mathbf{O}(k)$ from error-correcting codes.

A Hashed Variant of VHF. Fix an AHF $\mathsf{AHF} : \{0,1\}^k \to \{0,1\}^{\ell_{\mathsf{AHF}}}$ and a corresponding KGen algorithm. Essentially, we will hash preimages (using AHF) before feeding them into VHF to obtain a slight variant VHF' of VHF that we can prove adaptively secure. More specifically, let VHF' be the verifiable hash function that is defined like VHF, except for the following differences:

- $\mathsf{Gen}'_{\mathsf{VHF}}(\mathsf{GrpGen})$ proceeds like $\mathsf{Gen}_{\mathsf{VHF}}(\mathsf{GrpGen})$, but samples $2\ell_{\mathsf{AHF}}$ (not $2k$) matrices $\mathbf{M}_{i,b}$.
- $\mathsf{Eval}'_{\mathsf{VHF}}(ek, X)$ (for $X \in \{0,1\}^k$), first computes $X' = (x'_i)_{i=1}^{\ell_{\mathsf{AHF}}} = \mathsf{AHF}(X) \in \{0,1\}^{\ell_{\mathsf{AHF}}}$, and then outputs an image $[\mathbf{v}] = [\mathbf{v}_{\ell_{\mathsf{AHF}}}]$ and a proof

$$\pi = ([\mathbf{v}_1], \ldots, [\mathbf{v}_{\ell_{\mathsf{AHF}}-1}])$$

where $\mathbf{v}_i^\top = \mathbf{u}^\top \cdot \prod_{j=1}^{i} \mathbf{M}_{j,x'_j}$.
- $\mathsf{Vfy}'_{\mathsf{VHF}}(vk, [\mathbf{v}], \pi, X)$ computes $X' = (x'_i)_{i=1}^{\ell_{\mathsf{AHF}}} = \mathsf{AHF}(X) \in \{0,1\}^{\ell_{\mathsf{AHF}}}$ and outputs 1 if and only if $e([\mathbf{v}_i^\top], [1]) = e([\mathbf{v}_{i-1}^\top], [\mathbf{M}_{i,x'_i}])$ holds for all i with $1 \leq i \leq \ell_{\mathsf{AHF}}$, where $[\mathbf{v}_0] := [\mathbf{u}]$ and $[\mathbf{v}_{\ell_{\mathsf{AHF}}}] := [\mathbf{v}]$.

Theorem 3 (Adaptive Programmability of VHF'). VHF' *is adaptively programmable in the sense of Definition 5.*

The Trapdoor Algorithms. We proceed similarly to the selective case and start with a description of the algorithms $\mathsf{TrapGen}'_{\mathsf{VHF}}$ and $\mathsf{TrapEval}'_{\mathsf{VHF}}$.

- $\mathsf{TrapGen}'_{\mathsf{VHF}}(\Pi, [\mathbf{B}])$ proceeds like algorithm $\mathsf{TrapGen}_{\mathsf{VHF}}$ from Sect. 4.2, except that
  - $\mathsf{TrapGen}'_{\mathsf{VHF}}$ initializes $K \overset{\$}{\leftarrow} \mathsf{KGen}(1^k)$ and includes K in td.
  - $\mathsf{TrapGen}'_{\mathsf{VHF}}$ chooses $\ell_{\mathsf{AHF}} + 1$ (and not $k+1$) subspaces $\mathfrak{U}_i$ (for $0 \leq i \leq \ell_{\mathsf{AHF}}$). (The last subspace $\mathfrak{U}_{\ell_{\mathsf{AHF}}}$ is chosen in a special way, exactly like $\mathfrak{U}_k$ is chosen by $\mathsf{TrapGen}_{\mathsf{VHF}}$.)
  - $\mathsf{TrapGen}'_{\mathsf{VHF}}$ chooses $2\ell_{\mathsf{AHF}}$ (and not $2k$) matrices $\mathbf{R}_{i,b}$ (for $1 \leq i \leq \ell_{\mathsf{AHF}}$ and $b \in \{0,1\}$), as follows:
    If $K_i = b$, then $\mathbf{R}_{i,b}$ is chosen uniformly of rank $n-1$, subject to

$$\mathfrak{U}_{i-1}^\top \cdot \mathbf{R}_{i,1-x_i^{(0)}} = \mathfrak{U}_i^\top$$

If $K_i \neq b$, then $\mathbf{R}_{i,b}$ is chosen uniformly of rank n, subject to

$$\mathfrak{U}_{i-1}^\top \cdot \mathbf{R}_{i,x_i^{(0)}} = \mathfrak{U}_i^\top$$

- $\mathsf{TrapEval}'_{\mathsf{VHF}}(td, X)$ proceeds like algorithm $\mathsf{TrapEval}_{\mathsf{VHF}}$ on input a preimage $\mathsf{AHF}(X) \in \{0,1\}^{\ell_{\mathsf{AHF}}}$. Specifically, $\mathsf{TrapEval}'_{\mathsf{VHF}}$ computes $[\mathbf{v}] = [\mathbf{v}_k]$, along with a corresponding proof $\pi = [\mathbf{v}_1, \ldots, \mathbf{v}_{k-1}]$ exactly like $\mathsf{Eval}'_{\mathsf{VHF}}$. Finally, and analogously to $\mathsf{TrapEval}_{\mathsf{VHF}}$, $\mathsf{TrapEval}'_{\mathsf{VHF}}$ outputs $(\boldsymbol{\beta}, \pi)$ for $\boldsymbol{\beta}^\top := \mathbf{u}^\top \cdot \prod_{j=1}^{k} \mathbf{R}_{j,x_j}$.

Correctness and indistinguishability follow as for $\mathsf{TrapGen_{VHF}}$ and $\mathsf{TrapEval_{VHF}}$, so we state without proof:

Lemma 5 (Correctness of $\mathsf{Trap'_{VHF}}$). *The trapdoor algorithms $\mathsf{TrapGen'_{VHF}}$ and $\mathsf{TrapEval'_{VHF}}$ above satisfy correctness in the sense of Definition 5.*

Lemma 6 (Indistinguishability of $\mathsf{Trap'_{VHF}}$). *If the n-rank assumption holds relative to GrpGen, then the above algorithms $\mathsf{TrapGen'_{VHF}}$ and $\mathsf{TrapEval'_{VHF}}$ satisfy the indistinguishability property from Definition 5. Specifically, for every adversary $\mathcal{A}$, there exists an adversary $\mathcal{B}$ (of roughly the same complexity) with*

$$\mathsf{Adv}^{\mathsf{vhf-sel-ind}}_{\mathsf{VHF'},\mathsf{Trap'_{VHF}},\mathcal{A}}(k) = \ell_{\mathsf{AHF}} \cdot \mathsf{Adv}^{n-\mathsf{rank}}_{\mathbb{G},n,\mathcal{B}}(k) + \mathbf{O}(\ell_{\mathsf{AHF}}n/p).$$

The (omitted) proof of Lemma 6 proceeds exactly like that Lemma 3, only adapted to AHF-hashed inputs. Note that the additional oracle $\mathcal{O}_{\mathsf{check}}$ an adversary gets in the adaptive indistinguishability game can be readily implemented with the key K generated by $\mathsf{TrapGen'_{VHF}}$. (The argument from the proof of Lemma 3 does not rely on a secret $X^{(0)}$, and so its straightforward adaptation could even expose the full key K to an adversary.)

Lemma 7 (Well-distributedness of $\mathsf{Trap'_{VHF}}$). *The above trapdoor algorithms $\mathsf{TrapGen'_{VHF}}$ and $\mathsf{TrapEval'_{VHF}}$ have well-distributed outputs in the sense of Definition 5.*

Proof. First, we make an observation. Fix a matrix $[\mathbf{B}]$, and a corresponding key-pair $(td, vk) \xleftarrow{\$} \mathsf{TrapGen_{VHF}}(\Pi, [\mathbf{B}])$. Like (6), we can show that for all $X' = (x'_i)_{i=1}^{\ell_{\mathsf{AHF}}}$, the corresponding vectors $\mathbf{v}_i^\top$ computed during evaluation satisfy

$$\mathbf{u}^\top \cdot \prod_{j=1}^{i} \mathbf{R}_{i,x_i} \in \mathfrak{U}_i^\top \quad \Longleftrightarrow \quad x_j = K_j \text{ for some } j \leq i.$$

Hence, $\boldsymbol{\beta} \in \mathfrak{U}_{\ell_{\mathsf{AHF}}}$ (and thus $\beta_n = 0$) for the value $\boldsymbol{\beta}$ that is computed by $\mathsf{TrapEval'_{VHF}}(td, X)$ if and only if $F_K(X) = \mathsf{C0}$. By property (11) of AHF, the lemma follows.

5 VRFs from Verifiable PVHFs

Let $(\mathsf{Gen_{VRF}}, \mathsf{Eval_{VRF}}, \mathsf{Vfy_{VRF}})$ be the following algorithms.

– Algorithm $(vk, sk) \xleftarrow{\$} \mathsf{Gen_{VRF}}(1^k)$ takes as input a security parameter k and outputs a key pair (vk, sk). We say that sk is the *secret key* and vk is the *verification key*.
– Algorithm $(Y, \pi) \xleftarrow{\$} \mathsf{Eval_{VRF}}(sk, X)$ takes as input secret key sk and $X \in \{0,1\}^k$, and outputs a function value $Y \in \mathcal{Y}$, where $\mathcal{Y}$ is a finite set, and a proof π. We write $V_{sk}(X)$ to denote the function value Y computed by $\mathsf{Eval_{VRF}}$ on input (sk, X).

Initialize(1^k) :	Challenge(X^*) :	Exp$_{\mathcal{A}}^{\mathsf{VRF}}(1^k)$:
$b \xleftarrow{\$} \{0,1\}$	$(Y_0, \pi) \xleftarrow{\$} \mathsf{Eval}_{\mathsf{VRF}}(sk, X^*)$	$vk \xleftarrow{\$} \mathbf{Initialize}(1^k)$
$(vk, sk) \xleftarrow{\$} \mathsf{Gen}_{\mathsf{VRF}}(1^k)$	$Y_1 \xleftarrow{\$} \mathcal{Y}$	$X^* \xleftarrow{\$} \mathcal{A}^{\mathbf{Evaluate}}$
Return vk	**Return** Y_b	$Y_b \xleftarrow{\$} \mathbf{Challenge}(X^*)$
		$B \xleftarrow{\$} \mathcal{A}^{\mathbf{Evaluate}}$
Evaluate(X) :		**Return** $(B = b)$
$(Y, \pi) \xleftarrow{\$} \mathsf{Eval}_{\mathsf{VRF}}(sk, X)$		
Return (Y, π)		

Fig. 2. The VRF security experiment.

- Algorithm $\mathsf{Vfy}_{\mathsf{VRF}}(vk, X, Y, \pi) \in \{0,1\}$ takes as input verification key vk, $X \in \{0,1\}^k$, $Y \in \mathcal{Y}$, and proof π, and outputs a bit.

Definition 8. *We say that a tuple of algorithms* $(\mathsf{Gen}_{\mathsf{VRF}}, \mathsf{Eval}_{\mathsf{VRF}}, \mathsf{Vfy}_{\mathsf{VRF}})$ *is a verifiable random function (VRF), if all the following properties hold.*

Correctness. *Algorithms* $\mathsf{Gen}_{\mathsf{VRF}}$, $\mathsf{Eval}_{\mathsf{VRF}}$, $\mathsf{Vfy}_{\mathsf{VRF}}$ *are polynomial-time algorithms, and for all* $(vk, sk) \xleftarrow{\$} \mathsf{Gen}_{\mathsf{VRF}}(1^k)$ *and all* $X \in \{0,1\}^k$ *holds: if* $(Y, \pi) \xleftarrow{\$} \mathsf{Eval}_{\mathsf{VRF}}(sk, X)$, *then we have* $\mathsf{Vfy}_{\mathsf{VRF}}(vk, X, Y, \pi) = 1$.
Unique Provability. *For all strings* (vk, sk) *(which are not necessarily generated by* $\mathsf{Gen}_{\mathsf{VRF}}$*) and all* $X \in \{0,1\}^k$, *there does not exist any* (Y_0, π_0, Y_1, π_1) *such that* $Y_0 \neq Y_1$ *and* $\mathsf{Vfy}_{\mathsf{VRF}}(vk, X, Y_0, \pi_0) = \mathsf{Vfy}_{\mathsf{VRF}}(vk, X, Y_1, \pi_1) = 1$.
Pseudorandomness. *Let* $\mathsf{Exp}_{\mathcal{B}}^{\mathsf{VRF}}$ *be the security experiment defined in Fig. 2, played with adversary* $\mathcal{B}$. *We require that the advantage function*

$$\mathsf{Adv}_{\mathcal{B}}^{\mathsf{VRF}}(k) := 2 \cdot \Pr\left[\mathsf{Exp}_{\mathcal{B}}^{\mathsf{VRF}}(1^k) = 1\right] - 1$$

is negligible for all PPT $\mathcal{B}$ *that never query* **Evaluate** *on input* X^*.

5.1 A Generic Construction from Verifiable PVHFs

Let $(\mathsf{Gen}_{\mathsf{VHF}}, \mathsf{Eval}_{\mathsf{VHF}}, \mathsf{Vfy}_{\mathsf{VHF}})$ be a vector hash function according to Definition 3, and let $(\mathsf{GrpGen}, \mathsf{GrpVfy})$ be a certified bilinear group generator according to Definitions 1 and 2. Let $(\mathsf{Gen}_{\mathsf{VRF}}, \mathsf{Eval}_{\mathsf{VRF}}, \mathsf{Vfy}_{\mathsf{VRF}})$ be the following algorithms.

Key Generation. $\mathsf{Gen}_{\mathsf{VRF}}(1^k)$ runs $\Pi \xleftarrow{\$} \mathsf{GrpGen}(1^k)$ to generate bilinear group parameters, and then $(ek, vk') \xleftarrow{\$} \mathsf{Gen}_{\mathsf{VHF}}(\Pi)$. Then it chooses a random vector $\mathbf{w} \xleftarrow{\$} (\mathbb{Z}_p^*)^n$, defines $sk := (\Pi, ek, \mathbf{w})$ and $vk := (\Pi, vk', [\mathbf{w}])$, and outputs (vk, sk).
Function Evaluation. On input $sk := (\Pi, ek, \mathbf{w})$ with $\mathbf{w} = (w_1, \ldots, w_n)^\top \in (\mathbb{Z}_p^*)^n$ and $X \in \{0,1\}^k$, algorithm $\mathsf{Eval}_{\mathsf{VRF}}(sk, X)$ first runs

$$([\mathbf{v}], \pi') \leftarrow \mathsf{Eval}_{\mathsf{VHF}}(ek, X).$$

Then it computes the function value Y and an additional proof $[\mathbf{z}] \in \mathbb{G}^n$ as

$$Y := \prod_{i=1}^{n} \left[\frac{v_i}{w_i} \right] \qquad \text{and} \qquad [\mathbf{z}] := [(z_1, \ldots, z_n)^\top] := \left[\left(\frac{v_1}{w_1}, \ldots, \frac{v_n}{w_n} \right)^\top \right]$$

Finally, it sets $\pi := ([\mathbf{v}], \pi', [\mathbf{z}])$ and outputs (Y, π).

Proof Verification. On input (vk, X, Y, π), $\mathsf{Vfy}_{\mathsf{VRF}}$ outputs 0 if any of the following properties is not satisfied.

1. vk has the form $vk = (\Pi, vk', [\mathbf{w}])$, such that $[\mathbf{w}] = ([w_1], \ldots, [w_n])$ and the bilinear group parameters and group elements contained in vk are valid. That is, it holds that $\mathsf{GrpVfy}(\Pi) = 1$ and $\mathsf{GrpVfy}(\Pi, [w_i]) = 1$ for all $i \in \{1, \ldots, n\}$.
2. $X \in \{0, 1\}^k$.
3. π has the form $\pi = ([\mathbf{v}], \pi', [\mathbf{z}])$ with $\mathsf{Vfy}_{\mathsf{VHF}}(vk', [\mathbf{v}], \pi', X) = 1$ and both vectors $[\mathbf{v}]$ and $[\mathbf{z}]$ contain only validly-encoded group elements, which can be checked by running GrpVfy.
4. It holds that and $[z_i] = [v_i/w_i]$ for all $i \in \{1, \ldots, n\}$ and $Y = [\sum_{i=1}^n v_i/w_i]$. This can be checked by testing

$$e([z_i], [w_i]) \overset{?}{=} e([v_i], [1]) \quad \forall i \in \{1, \ldots, n\} \qquad \text{and} \qquad Y \overset{?}{=} \prod_{i=1}^{n} [z_i]$$

If all the above checks are passed, then $\mathsf{Vfy}_{\mathsf{VRF}}$ outputs 1.

5.2   Correctness, Unique Provability, and Pseudorandomness

Theorem 4 (Correctness and Unique Provability). *The triplet of algorithms* $(\mathsf{Gen}_{\mathsf{VRF}}, \mathsf{Eval}_{\mathsf{VRF}}, \mathsf{Vfy}_{\mathsf{VRF}})$ *forms a correct verifiable random function, and it satisfies the* unique provability *requirement in the sense of Definition 8.*

Proof. Correctness is straightforward to verify, therefore we turn directly to unique provability. We have to show that there does not *exist* any (Y_0, π_0, Y_1, π_1) such that $Y_0 \neq Y_1$ and $\mathsf{Vfy}_{\mathsf{VRF}}(vk, X, Y_0, \pi_0) = \mathsf{Vfy}_{\mathsf{VRF}}(vk, X, Y_1, \pi_1) = 1$. Let us first make the following observations.

– First of all, note that $\mathsf{Vfy}_{\mathsf{VRF}}$ on input $((\Pi, vk', [\mathbf{w}]), X, Y, ([\mathbf{v}], \pi', (\mathbf{z}))$ checks whether Π contains valid certified bilinear group parameters by running $\mathsf{GrpVfy}(\Pi)$. Moreover, it checks whether all group elements contained in $[\mathbf{w}]$, $[\mathbf{v}]$, and $[\mathbf{z}]$ are valid group elements with respect to Π. Thus, we may assume in the sequel that all these group elements are valid and have a unique encoding. In particular, $[\mathbf{w}]$ is uniquely determined by vk.
– Furthermore, it is checked that $X \in \{0, 1\}^k$. The unique provability property of the vector hash function $(\mathsf{Gen}_{\mathsf{VHF}}, \mathsf{Eval}_{\mathsf{VHF}}, \mathsf{Vfy}_{\mathsf{VHF}})$ guarantees that for all strings $vk' \in \{0, 1\}^*$ and all $X \in \{0, 1\}^k$ there does not exist any tuple $([\mathbf{v}_0], \pi_0, [\mathbf{v}_1], \pi_1)$ with $[\mathbf{v}_0] \neq [\mathbf{v}_1]$ and $[\mathbf{v}_0], [\mathbf{v}_1] \in \mathbb{G}^n$ such that

$$\mathsf{Vfy}_{\mathsf{VHF}}(vk', [\mathbf{v}_0], \pi_0, X) = \mathsf{Vfy}_{\mathsf{VHF}}(vk', [\mathbf{v}_1], \pi_1, X) = 1$$

Thus, we may henceforth use that there is only one unique vector of group elements $[\mathbf{v}]$ which passes the test $\mathsf{Vfy}_{\mathsf{VHF}}(vk', [\mathbf{v}], \pi', X) = 1$ performed by $\mathsf{Vfy}_{\mathsf{VRF}}$. Thus, $[\mathbf{v}]$ is uniquely determined by X and the values Π and vk' contained in vk.

- Finally, note that $\mathsf{Vfy}_{\mathsf{VRF}}$ tests whether $[z_i] = [v_i/w_i]$ holds. Due to the fact that the bilinear group is certified, which guarantees that each group element has a unique encoding and that the bilinear map is non-degenerate, for each $i \in \{1, \ldots, n\}$ there exists only one unique group element encoding $[z_i]$ such that the equality $[z_i] = [v_i/w_i]$ holds.

Therefore the value $Y = \prod_{i=1}^{n} [z_i]$ is uniquely determined by $[\mathbf{v}]$ and $[\mathbf{w}]$, which in turn are uniquely determined by X and the verification key vk.

Assumption 9. *The $(n-1)$-linear assumption states that* $[\mathbf{c}, \mathbf{d}, \sum_{i=1}^{n} d_i/c_i] \stackrel{c}{\approx} [\mathbf{c}, \mathbf{d}, r]$, *where* $\mathbf{c} = (c_1, \ldots, c_n)^\top \in (\mathbb{Z}_p^*)^n$, $\mathbf{d} = (d_1, \ldots, d_n)^\top \in \mathbb{Z}_p^n$, *and* $r \in \mathbb{Z}_p$ *are uniformly random. That is, we require that*

$$\mathsf{Adv}_{\mathcal{A}}^{(n-1)-\mathsf{lin}}(k) := \Pr\left[\mathcal{A}([\mathbf{c}, \mathbf{d}, \sum_{i=1}^{n} d_i/c_i]) = 1\right] - \Pr\left[\mathcal{A}([\mathbf{c}, \mathbf{d}, r]) = 1\right]$$

is negligible for every PPT adversary $\mathcal{A}$.

We remark that the above formulation is an equivalent formulation of the standard $(n-1)$-linear assumption, cf. [21, Page 9], for instance.

Theorem 5 (Pseudorandomness). *If* $(\mathsf{Gen}_{\mathsf{VHF}}, \mathsf{Eval}_{\mathsf{VHF}}, \mathsf{Vfy}_{\mathsf{VHF}})$ *is an adaptivly programmable VHF in the sense of Definition 3 and the $(n-1)$-linear assumption holds relative to* GrpGen, *then algorithms* $(\mathsf{Gen}_{\mathsf{VRF}}, \mathsf{Eval}_{\mathsf{VRF}}, \mathsf{Vfy}_{\mathsf{VRF}})$ *form a verifiable random function which satisfies the* pseudorandomness *requirement in the sense of Definition 8.*

Proof Sketch. The proof is based on a reduction to the indistinguishability and well-distributedness of the programmable vector hash function. The well-distributedness yields a leverage to embed the given instance of the $(n-1)$-linear assumption in the view of the adversary, following the approach sketched already in the introduction. Given the PVHF as a powerful building block, the remaining main difficulty of the proof lies in dealing with the fact that the "partitioning" proof technique provided by PVHFs is incompatible with "decisional" complexity assumptions. This is a well-known difficulty, which appeared in many previous works. It stems from the fact that different sequences of queries of the VRF-adversary may lead to different abort probabilities in the security proof. We can overcome this issue by employing the standard *artificial abort* technique [44], which has also been used to prove security of Waters' IBE scheme [44] and the VRF of Hohenberger and Waters [28], for example.

Proof. Let $\mathcal{A}$ be an adversary in the VRF security experiment from Definition 8. We will construct an adversary $\mathcal{B}$ on the $(n-1)$-linear assumption, which simulates the VRF pseudorandomness security experiment for $\mathcal{A}$. However, before we can construct this adversary, we have to make some changes to the security experiment. Consider the following sequence of games, where we let $\mathsf{Exp}_{\mathcal{A}}^i(1^k)$ denote the experiment executed in Game i and we write $\mathsf{Adv}_{\mathcal{A}}^i(k) := \Pr\left[\mathsf{Exp}_{\mathcal{A}}^i(1^k) = 1\right]$ to denote the advantage of $\mathcal{A}$ in Game i.

Game 0. This is the original VRF security experiment, executed with algorithms $(\mathsf{Gen}_{\mathsf{VRF}}, \mathsf{Eval}_{\mathsf{VRF}}, \mathsf{Vfy}_{\mathsf{VRF}})$ as constructed above. Clearly, we have

$$\mathsf{Adv}_{\mathcal{A}}^0(k) = \mathsf{Adv}_{\mathcal{A}}^{\mathsf{VRF}}(k)$$

In the sequel we write vk_0' to denote the VHF-key generated by $(vk_0', ek) \xleftarrow{\$} \mathsf{Gen}_{\mathsf{VHF}}(\Pi)$ in the experiment.

Game 1. This game proceeds exactly as before, except that it additionally samples a uniformly random invertible matrix $\mathbf{B} \xleftarrow{\$} \mathrm{GL}_n(\mathbb{Z}_p)$ and generates an *additional* key for the vector hash function as $(vk_1', td) \xleftarrow{\$} \mathsf{TrapGen}_{\mathsf{VHF}}(\Pi, [\mathbf{B}])$, which is *not* given to the adversary. That is, the adversary in Game 1 receives as input a VRF verification key $vk = (\Pi, vk_0', [\mathbf{w}])$, where vk_0' is generated by $\mathsf{Gen}_{\mathsf{VHF}}$, exactly as in Game 0.

Whenever $\mathcal{A}$ issues an **Evaluate**$(X^{(i)})$-query on some input $X^{(i)}$, the experiment proceeds as in Game 0, and additionally computes $((\beta_1, \ldots, \beta_n), \pi) \xleftarrow{\$} \mathsf{TrapEval}_{\mathsf{VHF}}(td, X^{(i)})$. If $\beta_n \neq 0$, then the experiment aborts and outputs a random bit. Moreover, when $\mathcal{A}$ issues a **Challenge**$(X^{(0)})$-query, then the experiment computes $((\beta_1, \ldots, \beta_n), \pi) \xleftarrow{\$} \mathsf{TrapEval}_{\mathsf{VHF}}(td, X^{(0)})$. If $\beta_n = 0$, then the experiment aborts and outputs a random bit.

The well-distributedness of the PVHF guarantees that there is a polynomial poly such that for all possible queries $X^{(0)}, X^{(1)}, \ldots, X^{(q)}$ the probability that the experiment is not aborted is at least

$$\Pr\left[\beta_n^{(0)} = 0 \wedge \beta_n^{(i)} \neq 0 \; \forall i \in \{1, \ldots, q\}\right] \geq 1/\mathsf{poly}(k) \geq \lambda$$

where λ is a non-negligible lower bound on the probability of not aborting.

Artificial Abort. Note that the probability that the experiment aborts depends on the particular sequence of queries issued by $\mathcal{A}$. This is problematic, because different sequences of queries may have different abort probabilities (cf. Appendix A). Therefore the experiment in Game 1 performs an additional *artificial abort* step, which ensures that the experiment is aborted with always the (almost) same probability $1 - \lambda$, independent of the particular sequence of queries issued by $\mathcal{A}$. To this end, the experiment proceeds as follows.

After $\mathcal{A}$ terminates and outputs a bit B, the experiment estimates the concrete abort probability $\eta(\mathbf{X})$ for the sequence of queries $\mathbf{X} := (X^{(0)}, \ldots, X^{(q)})$ issued by $\mathcal{A}$. To this end, the experiment:

1. Computes an estimate η' of $\eta(\mathbf{X})$, by R-times repeatedly sampling trapdoors $(vk_j', td_j) \xleftarrow{\$} \mathsf{TrapGen}_{\mathsf{VHF}}(\Pi, [\mathbf{B}])$ and checking whether $\beta_n^{(0)} = 0$ or $\beta_n^{(i)} \neq 0$, where

$$((\beta_1^{(i)}, \ldots, \beta_n^{(i)}), \pi) \leftarrow \mathsf{TrapEval}_{\mathsf{VHF}}(td_j, X^{(i)}) \quad \text{for} \quad i \in \{0, \ldots, q\}$$

for sufficiently large R. Here ϵ is defined such that $2 \cdot \epsilon$ is a lower bound on the advantage of $\mathcal{A}$ in the original security experiment.

2. If $\eta' \geq \lambda$, then the experiment aborts artificially with probability $(\eta' - \lambda)/\eta'$, and outputs a random bit.

Note that if η' was *exact*, that is, $\eta' = \eta(\mathbf{X})$, then the total probability of *not* aborting would *always* be $\eta(\mathbf{X}) \cdot (1 - (\eta' - \lambda)/\eta') = \lambda$, independent of the particular sequence of queries issued by $\mathcal{A}$. In this case we would have $\mathsf{Adv}^1_{\mathcal{A}}(k) = \lambda \cdot \mathsf{Adv}^0_{\mathcal{A}}(k)$. However, the estimate η' of $\eta(\mathbf{X})$ is not necessarily exact. By applying the standard analysis technique from [44] (see also [17, 28]), one can show that setting $R := \mathbf{O}(\epsilon^{-2} \ln(1/\epsilon) \lambda^{-1} \ln(1/\lambda))$ is sufficient to obtain

$$\mathsf{Adv}^1_{\mathcal{A}}(k) \geq \mathbf{O}(\epsilon \cdot \lambda) \cdot \mathsf{Adv}^0_{\mathcal{A}}(k).$$

Game 2. The experiment now provides the adversary with the *trapdoor* VHF verification key vk'_1, by including it in the VRF verification key $vk = (\Pi, vk'_1, [\mathbf{z}])$ in place of vk'_0. Moreover, the experiment now evaluates the VHF on inputs X by running $(\boldsymbol{\beta}, \pi) \xleftarrow{\$} \mathsf{TrapEval}_{\mathsf{VHF}}(td, X)$ and then computing $[\mathbf{v}] := [\mathbf{B}] \cdot \boldsymbol{\beta}$. The rest of the experiment proceeds exactly as before.

We claim that any adversary $\mathcal{A}$ distinguishing Game 2 from Game 1 implies an adversary $\mathcal{B}$ breaking the indistinguishability of the VHF according to Definition 5. Adversary $\mathcal{B}^{\mathcal{O}_b, \mathcal{O}_{\mathsf{check}}}(vk'_b)$ receives as input a verification key vk'_b, which is either generated by $\mathsf{Gen}_{\mathsf{VHF}}(\Pi)$ or $\mathsf{TrapGen}_{\mathsf{VHF}}(\Pi, [\mathbf{B}])$ for a uniformly invertible random matrix $\mathbf{B}$. It simulates the security experiment from Game 2 for $\mathcal{A}$ as follows.

- The given VHF verification key vk'_b is embedded in the VRF verification key $vk = (\Pi, vk'_b, [\mathbf{z}])$, where b is the random bit chosen by the indistinguishability security experiment played by $\mathcal{B}$. All other values are computed exactly as before.
- In order to evaluate the VHF on input X, $\mathcal{B}$ is able to query its oracle $\mathcal{O}_b$, which either computes and returns $([\mathbf{v}], \pi) \leftarrow \mathsf{Eval}_{\mathsf{VHF}}(ek, X)$ (in case $b = 0$), or it computes $(\boldsymbol{\beta}, \pi) \leftarrow \mathsf{TrapEval}_{\mathsf{VHF}}(td, X)$ and returns $[\mathbf{v}] := [\mathbf{B}] \cdot \boldsymbol{\beta}$ (in case $b = 1$).
- To test whether a given value X requires a (non-artificial) abort, $\mathcal{B}$ queries $\mathcal{O}_{\mathsf{check}}(X)$, which returns 1 if and only if $((\beta_1, \ldots, \beta_n), \pi) \leftarrow \mathsf{TrapEval}_{\mathsf{VHF}}(td)$ with $\beta_n \neq 0$.
- The artificial abort step is performed by $\mathcal{B}$ exactly as in Game 1.

Note that if $b = 0$, then the view of $\mathcal{A}$ is identical to Game 1, while if $b = 1$ then it is identical to Game 2. Thus, by the adaptive indistinguishability of the VHF, we have

$$\mathsf{Adv}^2_{\mathcal{A}}(k) \geq \mathsf{Adv}^1_{\mathcal{A}}(k) - \mathsf{negl}(k)$$

for some negligible function $\mathsf{negl}(k)$.

Game 3. Finally, we have to make one last technical modification before we are able to describe our reduction to the $(n-1)$-linear assumption. Game 3 proceeds exactly as before, except that matrix $[\mathbf{B}]$ has a slightly different distribution. In Game 2, $\mathbf{B} \xleftarrow{\$} \mathrm{GL}_n(\mathbb{Z}_p)$ is chosen uniformly random (and invertible). In Game 3, we instead choose matrix $\mathbf{B}$ by sampling $\mathbf{b_1}, \ldots, \mathbf{b_{n-1}} \xleftarrow{\$} (\mathbb{Z}_p^*)^n$ and $\mathbf{b_n} \xleftarrow{\$} \mathbb{Z}_p^n$, defining $\mathbf{B} := (\mathbf{b_1}, \ldots, \mathbf{b_n})$, and then computing $[\mathbf{B}]$. Thus, we ensure that the first $n-1$ vectors do not have any component which equals the identity element. This is done to adjust the distribution of $[\mathbf{B}]$ to the distribution chosen by our reduction algorithm.

By applying the union bound, we have $\left| \mathsf{Adv}_{\mathcal{A}}^3(k) - \mathsf{Adv}_{\mathcal{A}}^2(k) \right| \leq n^2/p$. Since n is polynomially bounded and $\log(p) \in \Omega(k)$, we have

$$\mathsf{Adv}_{\mathcal{A}}^3(k) \geq \mathsf{Adv}_{\mathcal{A}}^2(k) - \mathsf{negl}(k)$$

for some negligible function $\mathsf{negl}(k)$.

The Reduction to the $(n-1)$-Linear Assumption. In this game, we describe our actual reduction algorithm $\mathcal{B}$. Adversary $\mathcal{B}$ receives as input a $(n-1)$-linear challenge $[\mathbf{c}, \mathbf{d}, t]$, where $\mathbf{c} = (c_1, \ldots, c_n)^\top \xleftarrow{\$} (\mathbb{Z}_p^*)^n$, $\mathbf{d} \xleftarrow{\$} \mathbb{Z}_p^n$, and either $t = \sum_{i=1}^n d_i/c_i$ or $t \xleftarrow{\$} \mathbb{Z}_p$. It simulates the VRF security experiment exactly as in Game 3, with the following differences.

Initialization and Set-up of Parameters. Matrix $[\mathbf{B}]$ is computed as follows. First, $\mathcal{B}$ chooses $n(n-1)$ random integers $\alpha_{i,j} \xleftarrow{\$} \mathbb{Z}_p^*$ for $i \in \{1, \ldots, n-1\}$ and $j \in \{1, \ldots, n\}$. Then it sets $[\mathbf{b_i}] := (\alpha_{i,1}c_1, \ldots, \alpha_{i,n}c_n)^\top$ and $[\mathbf{b_n}] := [\mathbf{d}]$, and finally $[\mathbf{B}] := [\mathbf{b_1}, \ldots, \mathbf{b_n}]$. Vector $[\mathbf{w}]$ is set to $[\mathbf{w}] := [\mathbf{c}]$.

Note that matrix $[\mathbf{B}]$ and vector $[\mathbf{w}]$ are distributed exactly as in Game 3. Observe also that the first $n-1$ column vectors of $[\mathbf{B}]$ depend on $\mathbf{c}$, while the last vector is equal to $\mathbf{d}$.

Answering Evaluate-Queries. Whenever $\mathcal{A}$ issues an **Evaluate**-query on input $X^{(j)}$, then $\mathcal{B}$ computes $\boldsymbol{\beta} = (\beta_1, \ldots, \beta_n)^\top \leftarrow \mathsf{TrapEval}_{\mathsf{VHF}}(td, X^{(j)})$. If $\beta_n \neq 0$, then $\mathcal{B}$ aborts and outputs a random bit. Otherwise it computes

$$[\mathbf{v}] := [\mathbf{B} \cdot \boldsymbol{\beta}] = \left[(\mathbf{b_1}, \ldots, \mathbf{b_{n-1}}) \cdot (\beta_1, \ldots, \beta_{n-1})^\top \right] = \left[(\gamma_1 c_1, \ldots, \gamma_n c_n)^\top \right]$$

for integers $\gamma_1, \ldots, \gamma_n$, which are efficiently computable from $\boldsymbol{\beta}$ and the $\alpha_{i,j}$-values chosen by $\mathcal{B}$ above. Here we use that $\beta_n = 0$ holds for all **Evaluate**-queries that do not cause an abort.

Next, $\mathcal{B}$ computes the proof elements in $[\mathbf{z}]$ by setting $[z_i] := [\gamma_i]$ for all $i \in \{1, \ldots, n\}$. Note that, due to our setup of $[\mathbf{w}]$, it holds that

$$[\gamma_i] = \left[\frac{\gamma_i c_i}{c_i} \right] = \left[\frac{v_i}{w_i} \right]$$

thus all proof elements can be computed correctly by $\mathcal{B}$. Finally, $\mathcal{B}$ sets

$$Y := \prod_{i=1}^n [z_i] = \left[\sum_{i=1}^n z_i \right]$$

which yields the correct function value. Thus, all **Evaluate**-queries can be answered by $\mathcal{B}$ exactly as in Game 3.

Answering the Challenge-Query. When $\mathcal{A}$ issues a **Challenge**-query on input $X^{(0)}$, then $\mathcal{B}$ computes $\boldsymbol{\beta} = (\beta_1, \ldots, \beta_n)^\top \leftarrow \mathsf{TrapEval}_{\mathsf{VHF}}(td, X^{(0)})$. If $\beta_n = 0$, then $\mathcal{B}$ aborts and outputs a random bit. Otherwise again it computes the γ_i-values in

$$[\mathbf{v}] := [\mathbf{B} \cdot \beta] = \left[(\mathbf{b_1}, \ldots, \mathbf{b_{n-1}}) \cdot (\beta_1, \ldots, \beta_{n-1})^\top + \mathbf{b_n} \cdot \beta_n\right]$$
$$= \left[(\gamma_1 c_1, \ldots, \gamma_n c_n)^\top + \mathbf{d} \cdot \beta_n\right]$$

Writing v_i and d_i to denote the i-th component of $\mathbf{v}$ and $\mathbf{d}$, respectively, it thus holds that $v_i = \gamma_i c_i + d_i \beta_n$. Observe that then the function value is

$$Y = \left[\sum_{i=1}^n \frac{v_i}{c_i}\right] = \left[\sum_{i=1}^n \frac{\gamma_i c_i + d_i \beta_n}{c_i}\right]$$

$\mathcal{B}$ computes and outputs $[t \cdot \beta_n] \cdot [\sum_{i=1}^n \gamma_i] = [t \cdot \beta_n + \sum_{i=1}^n \gamma_i]$. Observe that if $[t] = [\sum_{i=1}^n d_i/c_i]$, then it holds that

$$\left[t \cdot \beta_n + \sum_{i=1}^n \gamma_i\right] = \left[\beta_n \cdot \sum_{i=1}^n \frac{d_i}{c_i} + \sum_{i=1}^n \frac{\gamma_i c_i}{c_i}\right] = \left[\sum_{i=1}^n \frac{\gamma_i c_i + d_i \beta_n}{c_i}\right] = Y$$

Thus, if $[t] = [\sum_{i=1}^n d_i/c_i]$, then $\mathcal{B}$ outputs the correct function value Y. However, if $[t]$ is uniformly random, then $\mathcal{B}$ outputs a uniformly random group element.

Finally, $\mathcal{B}$ performs an artificial abort step exactly as in Game 2. Note that $\mathcal{B}$ provides a perfect simulation of the experiment in Game 3, which implies that

$$\mathsf{Adv}_{\mathcal{B}}^{(n-1)-\mathsf{lin}}(k) = \mathsf{Adv}_{\mathcal{A}}^3(k)$$

which is non-negligible, if $\mathsf{Adv}_{\mathcal{A}}^{\mathsf{VRF}}(k)$ is.

A   The Need for an Artificial Abort

The "artificial abort" technique of Waters [44] has become standard for security proofs that combine a "partitioning" proof technique with a "decisional" complexity assumption. For example, it is also used to analyze Waters' IBE scheme [44], the verifiable random function of Hohenberger and Waters [28], and many other works.

Unfortunately, the artificial abort is necessary, because our $(n-1)$-linear reduction algorithm $\mathcal{B}$ is not able to use the output of $\mathcal{A}$ directly in case the experiment it not aborted. This is because the abort probability may depend on the particular sequence of queries issued by $\mathcal{A}$. For example, it may hold that $\Pr[B = b] = 1/2 + \epsilon$ for some non-negligible ϵ, which means that $\mathcal{A}$ has a non-trivial advantage in breaking the VRF-security, while $\Pr[B = b \mid \neg\mathsf{abort}] = 1/2$, which means that $\mathcal{B}$ does not have any non-trivial advantage in breaking the $(n-1)$-linear assumption. Essentially, the artificial abort ensures that $\mathcal{B}$ aborts for all sequences of queries made by $\mathcal{A}$ with approximately the same probability.

Alternatively, we could avoid the artificial abort by following the approach of Bellare and Ristenpart [6], which yields a tighter (but more complex) reduction. To this end, we would have to define and construct a PVHF which guarantees sufficiently close upper and lower bounds on the abort probability. This is possible by adopting the idea of balanced admissible hash functions (AHFs) from [29] to "balanced PVHFs". Indeed, instantiating our adaptively-secure PVHF with the balanced AHF from [29] yields such a balanced PVHF. However, this would have made the definition of PVHFs much more complex. We preferred to keep this novel definition as simple as possible, thus used the artificial abort approach of Waters [44].

References

1. Abdalla, M., Catalano, D., Fiore, D.: Verifiable random functions from identity-based key encapsulation. In: Joux, A. (ed.) EUROCRYPT 2009. LNCS, vol. 5479, pp. 554–571. Springer, Heidelberg (2009)
2. Abdalla, M., Catalano, D., Fiore, D.: Verifiable random functions: relations to identity-based key encapsulation and new constructions. J. Cryptology $27(3)$, 544–593 (2014)
3. Abdalla, M., Fiore, D., Lyubashevsky, V.: From selective to full security: semi-generic transformations in the standard model. In: Fischlin, M., Buchmann, J., Manulis, M. (eds.) PKC 2012. LNCS, vol. 7293, pp. 316–333. Springer, Heidelberg (2012)
4. Au, M.H., Susilo, W., Mu, Y.: Practical compact E-cash. In: Pieprzyk, J., Ghodosi, H., Dawson, E. (eds.) ACISP 2007. LNCS, vol. 4586, pp. 431–445. Springer, Heidelberg (2007)
5. Belenkiy, M., Chase, M., Kohlweiss, M., Lysyanskaya, A.: Compact E-cash and simulatable VRFs revisited. In: Shacham, H., Waters, B. (eds.) Pairing 2009. LNCS, vol. 5671, pp. 114–131. Springer, Heidelberg (2009)
6. Bellare, M., Ristenpart, T.: Simulation without the artificial abort: simplified proof and improved concrete security for waters' IBE scheme. In: Joux, A. (ed.) EUROCRYPT 2009. LNCS, vol. 5479, pp. 407–424. Springer, Heidelberg (2009)
7. Bellare, M., Yung, M.: Certifying cryptographic tools: the case of trapdoor permutations. In: Brickell, E.F. (ed.) CRYPTO 1992. LNCS, vol. 740, pp. 442–460. Springer, Heidelberg (1993)
8. Mihir Bellare and Moti Yung: Certifying Permutations: Noninteractive Zero-Knowledge Based on Any Trapdoor Permutation. J. Cryptology $9(3)$, 149–166 (1996)
9. Boneh, D., Boyen, X.: Secure identity based encryption without random oracles. In: Franklin, M. (ed.) CRYPTO 2004. LNCS, vol. 3152, pp. 443–459. Springer, Heidelberg (2004)
10. Boneh, D., Boyen, X., Shacham, H.: Short group signatures. In: Franklin, M. (ed.) CRYPTO 2004. LNCS, vol. 3152, pp. 41–55. Springer, Heidelberg (2004)
11. Boneh, D., Montgomery, H.W., Raghunathan, A.: Algebraic pseudorandom functions with improved efficiency from the augmented cascade. In: Proceedings of the ACM Conference on Computer and Communications Security 2010, pp. 131–140. ACM (2010)
12. Boneh, D., Halevi, S., Hamburg, M., Ostrovsky, R.: Circular-secure encryption from decision Diffie-Hellman. In: Wagner, D. (ed.) CRYPTO 2008. LNCS, vol. 5157, pp. 108–125. Springer, Heidelberg (2008)

13. Brakerski, Z., Goldwasser, S., Rothblum, G.N., Vaikuntanathan, V.: Weak verifiable random functions. In: Reingold, O. (ed.) TCC 2009. LNCS, vol. 5444, pp. 558–576. Springer, Heidelberg (2009)

14. Cash, D., Hofheinz, D., Kiltz, E., Peikert, C.: Bonsai trees, or how to delegate a lattice basis. In: Gilbert, H. (ed.) EUROCRYPT 2010. LNCS, vol. 6110, pp. 523–552. Springer, Heidelberg (2010)

15. Chase, M., Lysyanskaya, A.: Simulatable VRFs with applications to multi-theorem NIZK. In: Menezes, A. (ed.) CRYPTO 2007. LNCS, vol. 4622, pp. 303–322. Springer, Heidelberg (2007)

16. Chase, M., Meiklejohn, S.: Déjà Q: using dual systems to revisit q-type assumptions. In: Nguyen, P.Q., Oswald, E. (eds.) EUROCRYPT 2014. LNCS, vol. 8441, pp. 622–639. Springer, Heidelberg (2014)

17. Chatterjee, S., Sarkar, P.: On (Hierarchical) identity based en- cryption protocols with short public parameters (With an Exposition of Waters' Artificial Abort Technique). Cryptology ePrint Archive, Report 2006/279 (2006). http://eprint.iacr.org/

18. Cheon, J.H.: Security analysis of the strong Diffie-Hellman problem. In: Vaudenay, S. (ed.) EUROCRYPT 2006. LNCS, vol. 4004, pp. 1–11. Springer, Heidelberg (2006)

19. Dodis, Y.: Efficient construction of (distributed) verifiable random functions. In: Desmedt, Y.G. (ed.) PKC 2003. LNCS, vol. 2567, pp. 1–17. Springer, Heidelberg (2002)

20. Dodis, Y., Yampolskiy, A.: A verifiable random function with short proofs and keys. In: Vaudenay, S. (ed.) PKC 2005. LNCS, vol. 3386, pp. 416–431. Springer, Heidelberg (2005)

21. Escala, A., Herold, G., Kiltz, E., Ráfols, C., Villar, J.: An algebraic framework for Diffie-Hellman assumptions. Cryptology ePrint Archive, Report 2013/377 (2013). http://eprint.iacr.org/

22. Escala, A., Herold, G., Kiltz, E., Ràfols, C., Villar, J.: An algebraic framework for Diffie-Hellman assumptions. In: Canetti, R., Garay, J.A. (eds.) CRYPTO 2013, Part II. LNCS, vol. 8043, pp. 129–147. Springer, Heidelberg (2013)

23. Fiore, D., Schröder, D.: Uniqueness is a different story: impossibility of verifiable random functions from trapdoor permutations. In: Cramer, R. (ed.) TCC 2012. LNCS, vol. 7194, pp. 636–653. Springer, Heidelberg (2012)

24. Freire, E.S.V., Hofheinz, D., Paterson, K.G., Striecks, C.: Programmable hash functions in the multilinear setting. In: Canetti, R., Garay, J.A. (eds.) CRYPTO 2013, Part I. LNCS, vol. 8042, pp. 513–530. Springer, Heidelberg (2013)

25. Fuchsbauer, G.: Constrained verifiable random functions. In: Abdalla, M., De Prisco, R. (eds.) SCN 2014. LNCS, vol. 8642, pp. 95–114. Springer, Heidelberg (2014)

26. Gerbush, M., Lewko, A., O'Neill, A., Waters, B.: Dual form signatures: an approach for proving security from static assumptions. In: Wang, X., Sako, K. (eds.) ASIACRYPT 2012. LNCS, vol. 7658, pp. 25–42. Springer, Heidelberg (2012)

27. Hofheinz, D., Kiltz, E.: Programmable hash functions and their applications. In: Wagner, D. (ed.) CRYPTO 2008. LNCS, vol. 5157, pp. 21–38. Springer, Heidelberg (2008)

28. Hohenberger, S., Waters, B.: Constructing verifiable random functions with large input spaces. In: Gilbert, H. (ed.) EUROCRYPT 2010. LNCS, vol. 6110, pp. 656–672. Springer, Heidelberg (2010)

29. Jager, T.: Verifiable random functions from weaker assumptions. In: Dodis, Y., Nielsen, J.B. (eds.) TCC 2015, Part II. LNCS, vol. 9015, pp. 121–143. Springer, Heidelberg (2015)

30. Jao, D., Yoshida, K.: Boneh-Boyen signatures and the strong Diffie-Hellman problem. In: Shacham, H., Waters, B. (eds.) Pairing 2009. LNCS, vol. 5671, pp. 1–16. Springer, Heidelberg (2009)

31. Jarecki, S.: Handcuffing big brother: an abuse-resilient transaction escrow scheme. In: Cachin, C., Camenisch, J.L. (eds.) EUROCRYPT 2004. LNCS, vol. 3027, pp. 590–608. Springer, Heidelberg (2004)

32. Kakvi, S.A., Kiltz, E., May, A.: Certifying RSA. In: Wang, X., Sako, K. (eds.) ASIACRYPT 2012. LNCS, vol. 7658, pp. 404–414. Springer, Heidelberg (2012)

33. Lauer, S.: Verifiable random functions. Master Thesis, Ruhr- University Bochum (2015)

34. Liskov, M.: Updatable zero-knowledge databases. In: Roy, B. (ed.) ASIACRYPT 2005. LNCS, vol. 3788, pp. 174–198. Springer, Heidelberg (2005)

35. Lysyanskaya, A.: Unique signatures and verifiable random functions from the DH-DDH separation. In: Yung, M. (ed.) CRYPTO 2002. LNCS, vol. 2442, pp. 597–612. Springer, Heidelberg (2002)

36. Micali, S., Rabin, M.O., Vadhan, S.P.: Verifiable random functions. In: Proceedings of the FOCS 1999, pp. 120–130. IEEE Computer Society (1999)

37. Micali, S., Reyzin, L.: Soundness in the public-key model. In: Kilian, J. (ed.) CRYPTO 2001. LNCS, vol. 2139, pp. 542–565. Springer, Heidelberg (2001)

38. Micali, S., Rivest, R.L.: Micropayments revisited. In: Preneel, B. (ed.) CT-RSA 2002. LNCS, vol. 2271, pp. 149–163. Springer, Heidelberg (2002)

39. Naor, M.: On cryptographic assumptions and challenges. In: Boneh, D. (ed.) CRYPTO 2003. LNCS, vol. 2729, pp. 96–109. Springer, Heidelberg (2003)

40. Moni Naor and Omer Reingold: Number-theoretic constructions of efficient pseudo-random functions. J. ACM **51**(2), 231–262 (2004)

41. Naor, M., Segev, G.: Public-key cryptosystems resilient to key leakage. In: Halevi, S. (ed.) CRYPTO 2009. LNCS, vol. 5677, pp. 18–35. Springer, Heidelberg (2009)

42. Shacham, H.: A Cramer-shoup encryption scheme from the linear assumption and from progressively weaker linear variants. Cryptology ePrint Archive, Report 2007/074 (2007). http://eprint.iacr.org/

43. Waters, B.: Dual system encryption: realizing fully secure IBE and HIBE under simple assumptions. In: Halevi, S. (ed.) CRYPTO 2009. LNCS, vol. 5677, pp. 619–636. Springer, Heidelberg (2009)

44. Waters, B.: Efficient identity-based encryption without random oracles. In: Cramer, R. (ed.) EUROCRYPT 2005. LNCS, vol. 3494, pp. 114–127. Springer, Heidelberg (2005)

Complexity of Cryptographic Primitives

Homomorphic Evaluation Requires Depth

Andrej Bogdanov[1][(✉)] and Chin Ho Lee[2]

[1] Department of Computer Science and Engineering
and Institute for Theoretical Computer Science and Communications,
Chinese University of Hong Kong, Hong Kong, China
andrejb@cse.cuhk.edu.hk
[2] College of Computer and Information Science, Northeastern University,
Boston, USA
chlee@ccs.neu.edu

Abstract. We show that homomorphic evaluation of any non-trivial functionality of sufficiently many inputs with respect to any CPA secure homomorphic encryption scheme cannot be implemented by circuits of polynomial size and constant depth, i.e., in the class AC^0. In contrast, we observe that there exist ordinary public-key encryption schemes of quasipolynomial security in AC^0 assuming noisy parities are exponentially hard to learn. We view this as evidence that homomorphic evaluation is inherently more complex than basic operations in encryption schemes.

1 Introduction

A central objective in the theory of cryptography is to classify the relative complexity of various cryptographic tasks. One common way of arguing that task B is of comparable easiness to task A is to give a black-box implementation of B using A as a primitive. Notable examples include the construction of pseudorandom generators from one-way permutations [GL89] and one-way functions [HILL99, HRV10].

But how should we argue that task B is "more complex" than task A? In the generic setting, one looks for the existence of a black-box separation [IR89, RTV04], or a lower bound on the query complexity of a black-box reduction [GT00]. However such black box impossibility results are not always a good indicator of the relative complexity of the two tasks in the real world (under suitable complexity assumptions). For example, although collision-resistant hash functions cannot be constructed from one-way functions in a black-box manner [Sim98], both objects have simple, local (NC^0) implementations under standard assumptions [AIK07].

An alternative way to argue that task B is more complex than task A is to provide a concrete complexity model in which one can implement A (under plausible assumptions), but not B. For example, Applebaum et al. [AIK07] show that under plausible complexity assumptions, nontrivial pseudorandom generators

---

A. Bogdanov—Work supported by grant RGC GRF CUHK410113.
C.H. Lee—Work done at the Chinese University of Hong Kong.

E. Kushilevitz and T. Malkin (Eds.): TCC 2016-A, Part I, LNCS 9562, pp. 365–371, 2016.
DOI: 10.1007/978-3-662-49096-9_15

can be implemented in the complexity class NC^0. However, it is not difficult to see that this class does not contain pseudorandom functions; in fact, Linial, Mansour, and Nisan [LMN93] show that pseudorandom functions cannot be implemented even in AC^0. Taken together, these results may be viewed as concrete evidence that pseudorandom functions are more complex than pseudorandom generators, despite the existence of a black-box reduction [GGM86] and the lack of lower bounds on the complexity of such reductions [MV11].

In this work we give concrete complexity-theoretic evidence that homomorphic evaluation of essentially any non-trivial functionality is more complex than the basic cryptographic operations of key generation, encryption, and decryption. Our main result (Theorem 2) shows that homomorphic evaluation of any non-trivial functionality (for example the AND function) that depends on sufficiently many inputs cannot be implemented by circuits of constant depth and subexponential size with respect to any CPA secure encryption scheme. In Sect. 4 we show that encryption schemes in AC^0 of super-polynomial CPA security exist assuming Learning Noisy Parities is exponentially hard.

Thus constant-depth circuits provide sufficient computational power for implementing operations in both ordinary private and public-key encryption schemes (under a previously studied assumption), but not for realizing homomorphic evaluation of any non-trivial functionality.

2   Definitions

In this section we give a definition of what it means for an algorithm E to homomorphically evaluate a given functionality f. A fairly weak requirement is that a homomorphic evaluator for $f(m_1, \ldots, m_k)$ should take as inputs encryptions of $m_1, \ldots, m_k$ and output a ciphertext that decrypts to $f(m_1, \ldots, m_k)$.

We will allow for the evaluation algorithm to err on some fraction of the encryptions. This takes into account the possibility that the encryption scheme itself may produce incorrect encryptions with some probability.

Definition 1. *Let* (**Gen, Enc, Dec**) *be a private-key encryption scheme over message set* Σ *with ciphertexts in* $\{0,1\}^n$. *We say a circuit* E *is a* homomorphic evaluator *of* $f : \Sigma^k \to \Sigma$ *with error* δ *if for all* $m_1, \ldots, m_k \in \Sigma$,

$$\Pr[\mathbf{Dec}_{SK}(E(\mathbf{Enc}_{SK}(m_1, R_1), \ldots, \mathbf{Enc}_{SK}(m_k, R_k))) = f(m_1, \ldots, m_k)] \geq 1 - \delta,$$

where $SK \sim$ **Gen** *is a uniformly chosen secret key and* $R_1, \ldots, R_k$ *are independent random seeds.*

In the public-key setting, we are given an encryption scheme (**Gen, Enc, Dec**) and require that

$$\Pr[\mathbf{Dec}_{SK}(E(PK, \mathbf{Enc}_{PK}(m_1, R_1), \ldots, \mathbf{Enc}_{PK}(m_k, R_k)))$$
$$= f(m_1, \ldots, m_k)] \geq 1 - \delta.$$

where $(PK, SK) \sim$ **Gen** *is a random key pair.*

We point out one challenge that this natural definition poses in the context of ruling out the existence of homomorphic evaluators. When k is much smaller than n, the definition allows for plausible encryption schemes that admit trivial homomorphic evaluators, by "outsourcing" the homomorphic evaluation to the decryption algorithm. For example suppose that the meaningful portion of an encryption is only captured in the first n/k bits of the ciphertext. Then the homomorphic evaluator can simply copy the meaningful portion of its k encryptions in non-overlapping parts of the output. Upon seeing a ciphertext of this form, the decryption algorithm can easily compute the value $f(m_1, \ldots, m_k)$ by first decrypting the ciphertext corresponding to each of the k encryptions and then evaluating f.

Thus our negative result will only apply to functions whose number of relevant inputs k is sufficiently large in terms of n. Beyond this requirement, we do not make any assumption on f.

The requirement we make on the encryption scheme is CPA message indistinguishability. A private-key encryption scheme is (s, d, ε) CPA message indistinguishable if for every pair of messages $m, m' \in \Sigma$ and every distinguishing oracle circuit $D^?$ of size s and depth d,

$$|\Pr_{SK,R}[D^{\mathbf{Enc}(SK,\cdot)}(\mathbf{Enc}_{SK}(m, R)) = 1]$$
$$- \Pr_{SK,R}[D^{\mathbf{Enc}(SK,\cdot)}(\mathbf{Enc}_{SK}(m', R)) = 1]| \leq \varepsilon.$$

In the public key setting CPA security follows from ordinary message indistinguishability:

$$|\Pr_{PK,R}[D(PK, \mathbf{Enc}_{PK}(m, R)) = 1] - \Pr_{PK,R}[D(PK, \mathbf{Enc}_{PK}(m', R)) = 1]| \leq \varepsilon.$$

3   Homomorphic Evaluation Requires Depth

Theorem 2. *Suppose* $(\mathbf{Gen}, \mathbf{Enc}, \mathbf{Dec})$ *is an* $(2s + k + O(1), d + 1, 1/6(k+1))$ *CPA message indistinguishable private-key (resp. public-key) encryption scheme. Let E be a homomorphic evaluator of size s and depth d with error at most $1/3$ for some $f : \Sigma^k \to \Sigma$ that depends on all of its inputs with respect to this scheme. Then $s > 2^{\Omega((k/6n)^{1/(d-1)})}$.*

For notational simplicity, we present the proof for the private key variant. Since f depends on all its inputs, for every $i \in [k]$ there is a pair of messages m and m' that differ only in coordinate i such that $f(m) \neq f(m')$. Now suppose E is a homomorphic evaluator for f with error $1/3$. Then

$$\Pr[\mathbf{Dec}(E(\mathbf{Enc}(m_1, R_1), \ldots, \mathbf{Enc}(m_i, R_i), \ldots, \mathbf{Enc}(m_k, R_k))) \neq f(m)] \leq 1/3$$

and

$$\Pr[\mathbf{Dec}(E(\mathbf{Enc}(m_1, R_1), \ldots, \mathbf{Enc}(m_i', R_i'), \ldots, \mathbf{Enc}(m_k, R_k))) \neq f(m')] \leq 1/3,$$

where the probability is taken over the choice of secret key SK (which we omit to simplify notation) and the randomness $R_1, \ldots, R_i, R_i', \ldots, R_k$ used in the encryption. Since $f(m) \neq f(m')$, it follows that

$$\Pr[\mathbf{Dec}(E(\mathbf{Enc}(m_1, R_1), \ldots, \mathbf{Enc}(m_i, R_i), \ldots, \mathbf{Enc}(m_k, R_k)))$$
$$\neq \mathbf{Dec}(E(\mathbf{Enc}(m_1, R_1), \ldots, \mathbf{Enc}(m_i', R_i'), \ldots, \mathbf{Enc}(m_k, R_k)))] \geq 1/3.$$

Therefore it must be that

$$\Pr[E(\mathbf{Enc}(m_1, R_1), \ldots, \mathbf{Enc}(m_i, R_i), \ldots, \mathbf{Enc}(m_k, R_k))$$
$$\neq E(\mathbf{Enc}(m_1, R_1), \ldots, \mathbf{Enc}(m_i', R_i'), \ldots, \mathbf{Enc}(m_k, R_k))] \geq 1/3.$$

By CPA message indistinguishability and a hybrid argument, we can replace $m_1, \ldots, m_i, m_i', \ldots, m_k$ by 0 to obtain

$$\Pr[E(\mathbf{Enc}(0, R_1), \ldots, \mathbf{Enc}(0, R_i), \ldots, \mathbf{Enc}(0, R_k))$$
$$\neq E(\mathbf{Enc}(0, R_1), \ldots, \mathbf{Enc}(0, R_i'), \ldots, \mathbf{Enc}(0, R_k))] \geq 1/6. \quad (1)$$

Lemma 3. *Let* $D_1, \ldots, D_k$ *be any distributions over* $\{0,1\}^n$. *Let* $g : (\{0,1\}^n)^k \to \{0,1\}$ *be a circuit of size* s *and depth* d *where* $s \leq 2^{(\varepsilon k)^{1/(d-1)}/K}$ *for some absolute constant* K. *Then*

$$\Pr[g(X_1, \ldots, X_i, \ldots, X_k) \neq g(X_1, \ldots, X_i', \ldots, X_k)] < \varepsilon$$

where the randomness is taken over the choice of $i \sim [k]$ *and independent samples* $X_1 \sim D_1, \ldots, X_i, X_i' \sim D_i, \ldots, X_k \sim D_k$.

We apply this lemma with D_i equal to the distribution of encryptions of 0 and $\varepsilon = 1/6n$ to each of the n outputs of E and take a union bound to conclude that (1) is violated unless $s > 2^{\Omega((k/6n)^{1/(d-1)})}$.

Proof (of Lemma 3). Fix any pair $Z, Z' \in (\{0,1\}^n)^k$. For any $w \in \{0,1\}^k$, let $Z_w \in (\{0,1\}^n)^k$ be the string such that

$$\text{the } i\text{-th block of } Z_w = \begin{cases} \text{the } i\text{-th block of } Z, & \text{if } w_i = 0 \\ \text{the } i\text{-th block of } Z', & \text{if } w_i = 1. \end{cases}$$

Let $h_{Z,Z'}(w) = g(Z_w)$. Then h is of size at most s and depth at most d. By Boppana [Bop97], for every Z and Z' we have

$$\Pr_{W,i}[h_{Z,Z'}(W) \neq h_{Z,Z'}(W + e_i)] \leq (K \log s)^{d-1}/k$$

for some constant K, where W and i are uniform over $\{0,1\}^k$ and $[k]$ respectively, and e_i is the i-th indicator vector. Therefore for Z, Z' sampled independently from $D_1 \times \cdots \times D_k$ we can rewrite $\Pr[g(X_1, \ldots, X_i, \ldots, X_k) \neq g(X_1, \ldots, X_i', \ldots, X_k)]$ as

$$\mathrm{E}_{Z,Z'}[\Pr_{W,i}[h_{Z,Z'}(W) \neq h_{Z,Z'}(W + e_i)]] \leq \mathrm{E}_{Z,Z'}[(K \log s)^{d-1}/k]$$
$$= (K \log s)^{d-1}/k.$$

It follows that if this probability is at most ε, then $s \leq 2^{(\varepsilon k)^{1/(d-1)}/K}$.

Lemma 3 bounds the total influence of shallow circuits under independent inputs chosen from an arbitrary distribution. Our proof is based on ideas of Blais, O'Donnell, and Wimmer [BOW10], who bound the noise sensitivity of such circuits.

4   On CPA-Secure Encryption Schemes in AC^0

In this section we show that encryption schemes in AC^0 of super-polynomial CPA security exist assuming Learning Noisy Parities over $\{0,1\}^n$ requires time $2^{\Omega(n^\delta)}$ for some constant $\delta > 0$.

To begin with, we observe that asymptotically super-polynomial security cannot be achieved by NC^0 decryption circuits: If every output of the decryption circuit depends on at most d bits of the ciphertext, then for any message m the decryption circuit on the distribution of encryptions of m can be PAC-learned in time $O_d(n^d)$, violating CPA security.

We obtain candidate encryption schemes in AC^0 by applying the following reduction:

Lemma 4. *For every $d > 0$, every (public or private key) encryption scheme of size S and depth D can be implemented in size $S2^D \cdot 2^{d \cdot D \cdot S^{1/d}}$ and depth $2d + 1$.*

In particular, encryption schemes in the class NC^2 can be simulated by constant-depth circuit families of size $2^{O(n^\varepsilon)}$ for any constant $\varepsilon > 0$.

Two such schemes are the private-key one of Gilbert et al. [GRS08] and the public-key one of Alekhnovich [Ale11, Cryptosystem1]. The key generation, encryption, and decryption algorithms for these schemes apply linear algebra over $\mathbb{F}_2$ and thus admit NC^2 implementations [Ber84]. The security of these two schemes is based on the hardness of Learning Noisy Parities.

Noisy Parities over $\mathbb{F}_2^n$ with noise rate η can be learned by brute force in time $\text{poly}(n) \cdot \binom{n}{\eta n}$. A slight improvement in the exponent is achievable for high noise rates using the algorithm of Blum, Kalai, and Wasserman [BKW03]. Its running time is $2^{\Theta(n/\log n)}$. Assuming noisy parities are hard to learn in time $2^{\Omega(n^\delta)}$ for some constant $\delta > 0$, it follows from Lemma 4 that the above schemes have constant-depth implementations whose security is super-polynomial in their size. The error rate can be assumed constant in the cryptosystem of Gilbert et al. and $1/\sqrt{n}$ in the cryptosystem of Alekhnovich.

The cryptosystems of Gilbert et al. and Alekhnovich have noticeable encryption error. The error can be reduced to negligible by encrypting the message independently multiple times. While some of the multiple encryptions may be erroneous, with all but negligible probability at least $2/3$ of them will be correct. The errors can be corrected by taking approximate majority at the decryption stage, which can be implemented using circuits of depth 3 [Ajt83], thereby preserving the constant depth complexity of the implementation.

Proof (of Lemma 4). We show that the conclusion holds for every circuit of size S and depth D, so in particular it holds for the key generation, encryption, and decryption circuits (where the circuits are viewed as functions of both their input and their randomness). This is folklore and was recently used in [LV15]. We sketch the proof for completeness.

First, every circuit of size S and depth D can be simulated by a branching program of length S and width 2^D by traversing the circuit in depth first order while maintaining the value of the evaluated subtree at each level.

Second, for every k, every branching program of length S and width W can be written as an OR of W^k ANDs of k branching programs of length S/k and width W. This representation is obtained by factoring the branching program over its states at time S/k, $2S/k$, up to $(k-1)S/k$.

Applying this transformation recursively d times, we obtain a simulation of a size S, depth D circuit by a size $(kW^k)^d$, depth $2d$ circuit whose inputs are branching programs of length S/k^d and width w. Each such branching program can be trivially simulated by a CNF of size W^{S/k^d}. Putting this together, we obtain a simulation of size S, depth D circuits by size $k^d W^{dk+S/k^d}$, depth $2d+1$ circuits. Setting $k = S^{1/d}$ proves the lemma.

Acknowledgment. We thank Yuval Ishai for sharing his insights on encryption schemes in AC^0.

References

[AIK07] Applebaum, B., Ishai, Y., Kushilevitz, E.: Cryptography with constant input locality. In: Menezes, A. (ed.) CRYPTO 2007. LNCS, vol. 4622, pp. 92–110. Springer, Heidelberg (2007)

[Ajt83] Ajtai, M.: Σ_1^1-formulae on finite structures. Ann. Pure Appl. Logic **24**, 607–620 (1983)

[Ale11] Alekhnovich, M.: More on average case vs approximation complexity. Comput. Complex. **20**(4), 755–786 (2011)

[Ber84] Berkowitz, S.J.: On computing the determinant in small parallel time using a small number of processors. Inform. Process. Lett. **18**(3), 147–150 (1984)

[BKW03] Blum, A., Kalai, A., Wasserman, H.: Noise-tolerant learning, the parity problem, and the statistical query model. J. ACM **50**(4), 506–519 (2003)

[Bop97] Boppana, R.B.: The average sensitivity of bounded-depth circuits. Inf. Process. Lett. **63**(5), 257–261 (1997)

[BOW10] Blais, E., O'Donnell, R., Wimmer, K.: Polynomial regression under arbitrary product distributions. Mach. Learn. **80**, 273–294 (2010)

[GGM86] Goldreich, O., Goldwasser, S., Micali, S.: How to construct random functions. J. ACM **33**(4), 792–807 (1986)

[GL89] Goldreich, O., Levin, L.A.: A hard-core predicate for all one-way functions. In: Proceedings of the Twenty-First Annual ACM Symposium on Theory of computing, STOC 1989, pp. 25–32. ACM, New York (1989)

[GRS08] Gilbert, H., Robshaw, M., Seurin, Y.: How to encrypt with the LPN problem. In: Aceto, L., Damgård, I., Goldberg, L.A., Halldórsson, M.M., Ingólfsdóttir, A., Walukiewicz, I. (eds.) ICALP 2008, Part II. LNCS, vol. 5126, pp. 679–690. Springer, Heidelberg (2008)

[GT00]   Gennaro, R., Trevisan, L.: Lower bounds on the efficiency of generic cryptographic constructions. In: Proceedings of the 41st Annual Symposium on Foundations of Computer Science, FOCS 2000, p. 305. IEEE Computer Society, Washington (2000)

[HILL99] Håstad, J., Impagliazzo, R., Levin, L.A., Luby, M.: A pseudorandom generator from any one-way function. SIAM J. Comput. **28**(4), 1364–1396 (1999)

[HRV10] Haitner, I., Reingold, O., Vadhan, S.: Efficiency improvements in constructing pseudorandom generators from one-way functions. In: Proceedings of the 42nd ACM Symposium on Theory of Computing, STOC 2010, pp. 437–446. ACM, New York (2010)

[IR89]   Impagliazzo, R., Rudich, S.: Limits on the provable consequences of one-way permutations. In: Proceedings of the Twenty-First Annual ACM Symposium on Theory of computing, STOC 1989, pp. 44–61. ACM, New York (1989)

[LMN93] Linial, N., Mansour, Y., Nisan, N.: Constant depth circuits, fourier transform, and learnability. J. ACM **40**(3), 607–620 (1993)

[LV15]   Lee, C.H., Viola, E.: Some limitations of the sum of small-bias distributions. Electron. Colloquium Comput. Complex. (ECCC) **22**, 5 (2015)

[MV11]   Miles, E., Viola, E.: On the complexity of non-adaptively increasing the stretch of pseudorandom generators. In: Ishai, Y. (ed.) TCC 2011. LNCS, vol. 6597, pp. 522–539. Springer, Heidelberg (2011)

[RTV04] Reingold, O., Trevisan, L., Vadhan, S.P.: Notions of reducibility between cryptographic primitives. In: Naor, M. (ed.) TCC 2004. LNCS, vol. 2951, pp. 1–20. Springer, Heidelberg (2004)

[Sim98] Simon, D.R.: Findings collisions on a one-way street: can secure hash functions be based on general assumptions? In: Nyberg, K. (ed.) EUROCRYPT 1998. LNCS, vol. 1403, pp. 334–345. Springer, Heidelberg (1998)

On Basing Private Information Retrieval on NP-Hardness

Tianren Liu$^{(\boxtimes)}$ and Vinod Vaikuntanathan

MIT CSAIL, Cambridge, USA
liutr@mit.edu, vinodv@csail.mit.edu

Abstract. The possibility of basing the security of cryptographic objects on the (minimal) assumption that **NP** $\not\subseteq$ **BPP** is at the very heart of complexity-theoretic cryptography. Most known results along these lines are negative, showing that assuming widely believed complexity-theoretic conjectures, there are no reductions from an **NP**-hard problem to the task of breaking certain cryptographic schemes. We make progress along this line of inquiry by showing that the security of single-server single-round private information retrieval schemes cannot be based on **NP**-hardness, unless the polynomial hierarchy collapses. Our main technical contribution is in showing how to break the security of a PIR protocol given an **SZK** oracle. Our result is tight in terms of both the correctness and the privacy parameter of the PIR scheme.

1 Introduction

The possibility of basing the security of cryptographic objects on the (minimal) assumption that **NP** $\not\subseteq$ **BPP** is at the very heart of complexity-theoretic cryptography. Somewhat more precisely, "basing primitive X on **NP**-hardness" means that there is a construction of primitive X and a probabilistic polynomial-time oracle algorithm (a reduction) R such that for every oracle A that "breaks the security of X", $\Pr[R^A(\phi) = 1] \geq 2/3$ if $\phi \in \mathsf{SAT}$ and $\Pr[R^A(\phi) = 1] \leq 1/3$ otherwise.

There are a handful of impossibility results which show that, assuming widely believed complexity-theoretic conjectures, the security of various cryptographic objects cannot be based on **NP**-hardness. We discuss these results in detail in Sect. 1.2. In this work, we make progress along these lines of inquiry by showing that (single server) private information retrieval (PIR) schemes cannot be based on **NP**-hardness, unless the polynomial hierarchy collapses.

Main Theorem 1 (Informal). *If there is a probabilistic polynomial time reduction from solving* SAT *to breaking a single-server, one round, private information retrieval scheme, then* **NP** $\subseteq$ **coAM**.

T. Liu—Supported by NSF Grants CNS-1350619 and CNS-1414119.

V. Vaikuntanathan—Research supported in part by NSF Grants CNS-1350619 and CNS-1414119, Alfred P. Sloan Research Fellowship, Microsoft Faculty Fellowship, the NEC Corporation, and a Steven and Renee Finn Career Development Chair from MIT.

E. Kushilevitz and T. Malkin (Eds.): TCC 2016-A, Part I, LNCS 9562, pp. 372–386, 2016.
DOI: 10.1007/978-3-662-49096-9_16

Our result rules out security reductions from SAT that make black-box use of the adversary that breaks a PIR scheme. Other than being black-box in the adversary, the security reduction can be very general, in particular, it is allowed to make polynomially many adaptively chosen calls to the PIR-breaking adversary.

Our result is tight in terms of both the correctness and the privacy parameter of the PIR scheme. Namely, information-theoretically secure PIR schemes exist for those choice of parameters that are not ruled out by our result. We refer the reader to Sect. 3 for a formal statement of our result.

Private Information Retrieval. Private information retrieval (PIR) is a protocol between a database D holding a string $x \in \{0,1\}^n$, and a user holding an index $i \in [n]$. The user wishes to retrieve the i-th bit x_i from the database, without revealing any information about i. Clearly, the database can rather inefficiently accomplish this by sending the entire string x to the user. The objective of PIR, then, is to achieve this goal while communicating (significantly) less than n bits.

Chor, Goldreich, Kushilevitz and Sudan [CKGS98], who first defined PIR, also showed that non-trivial PIR schemes (with communication less than n bits) require computational assumptions. Subsequently, PIR has been shown to imply one-way functions [BIKM99], oblivious transfer [CMO00] and collision-resistant hashing [IKO05], placing it in cryptomania proper.

On the other hand, there have been several constructions of PIR with decreasing communication complexity under various cryptographic assumptions [KO97, CMS99, Lip05, BGN05, GR05, Gen09, BV11].

In particular, Kushilevitz and Ostrovsky [KO97] were the first to show a construction of PIR with $O(n^\epsilon)$ communication (for any constant $\epsilon > 0$) assuming the existence of additively homomorphic encryption schemes. Some of the later constructions of PIR [CMS99, Lip05, GR05, BV11] achieve polylog(n) communication under number-theoretic assumptions such as the Phi-hiding assumption and the LWE assumption. Notably, all of them are single-round protocols, involving one message from the user to the server and one message back.

1.1 Our Techniques

The core of our proof is an attack against any single-server one-round PIR protocol given access to an **SZK** oracle. In particular, we show that given an oracle to the *entropy difference* (ED) problem, which is complete for **SZK**, one can break any single-server one-round PIR protocol. Once we have this result, the rest follows from a beautiful work of Mahmoody and Xiao [MX10] who show that **BPP^{SZK}** $\subseteq$ **AM $\cap$ coAM**. That is, if there is a reduction from deciding SAT to breaking single-server one-round PIR, then SAT $\in$ **BPP^{SZK}** and therefore, by [MX10], SAT $\in$ **AM $\cap$ coAM**. In turn, from the work of Boppana, Håstad and Zachos [BHZ87], this means that the polynomial hierarchy collapses to the second level.

The intuition behind the attack against PIR protocols is simple. Assume that the database is uniformly random and the user's query is fixed. Let X be a random variable that denotes the database, and let A be a random variable that denotes the PIR answer (on input a query q from a user trying to retrieve the i-th bit). We have two observations.

1. *The answer enables the user to learn the i-th bit.* In other words, the mutual information between the i-th database bit X_i and the answer A has to be large. Indeed, we show that if the PIR protocol is correct with probability $1 - \varepsilon$, then this mutual information is at least $1 - h(\varepsilon)$, where h is the binary entropy function.
2. *The answer does not contain a large amount of information about all the database entries.* Indeed, the entropy of the answer is limited by its length which is much shorter than the size of the database. We show that for most indices j, the answer contains little information about the j-th bit, that is the mutual information between A and X_j is small.

We then proceed as follows. Given the user's query q, an efficient adversary can construct a circuit sampling from joint distribution $(X; A)$. Armed with the entropy difference ED oracle, the adversary can estimate $I(X_j; A)$ for any index j. Since $I(X_i; A)$ is close to 1 (where i is the index underlying the query q) and $I(X_j; A)$ is small for most indices j, the adversary can predict i much better than random guessing. This breaks the security of PIR.

We refer the reader to Theorem 3.1 for the formal statement, and to Proposition 2.8 which shows that the parameters of Theorem 3.1 are tight.

1.2   Related Work

Brassard [Bra79] showed that one-way permutations cannot be based on **NP**-hardness. Subsequently, Goldreich and Goldwasser [GG98], in the process of clarifying Brassard's work, showed that public-key encryption schemes that satisfy certain very special properties cannot be based on **NP**-hardness. In particular, one of their conditions require that it should be easy to certifying an invalid key as such.

Akavia, Goldreich, Goldwasser and Moshkovitz [AGGM06], and later Bogdanov and Brzuska [BB15], showed that a special class of one-way functions called *size-verifiable one-way functions* cannot be based on **NP**-hardness. A size-verifiable one-way function, roughly speaking, is one in which the size of the set of pre-images can be efficiently approximated via an **AM** protocol.

Most recently, Bogdanov and Lee [BL13a] showed that (even simple) homomorphic encryption schemes cannot be based on **NP**-hardness. This includes additively homomorphic encryption as well as homomorphic encryption schemes that only support the majority function, as special cases. While PIR schemes can be constructed from additively homomorphic encryption, we are not aware of a way to use PIR to obtain any type of non-trivial homomorphic encryption scheme.

Several works have also explored the problem of basing average-case hardness on (worst case) **NP**-hardness, via restricted types of reductions, most notably non-adaptive reductions that make all its queries to the oracle simultaneously.

The work of Feigenbaum and Fortnow, subsequently strengthened by Bogdanov and Trevisan [BT06], show that there cannot be a *non-adaptive* reduction from (worst-case) SAT to the average-case hardness of any problem in **NP**, unless **PH** $\subseteq \Sigma_2$ (that is, the polynomial hierarchy collapses to the second level). In contrast, our results rule out even adaptive reductions (to much stronger primitives).

2 Definitions

2.1 Information Theory Background

A *random variable* X over a finite set S is defined by its probability mass function $p_X : S \to [0,1]$ such that $\sum_{x \in S} p_X(x) = 1$. We use uppercase letters to denote random variables. The *Shannon entropy* of a random variable X, denoted $H(X)$, is defined as

$$H(X) = \sum_x p_X(x) \log_2 \frac{1}{p_X(x)}.$$

Let $\mathrm{Bern}(p)$ denote the Bernoulli distribution on $\{0,1\}$ which assigns a probability of p to 1 and $1 - p$ to 0. We will denote by $h(p) = H(\mathrm{Bern}(p)) = p \log_2 \frac{1}{p} + (1 - p) \log_2 \frac{1}{1-p}$ the Shannon entropy of the distribution $\mathrm{Bern}(p)$.

Let X and Y be two (possibly dependent) random variables. The *conditional entropy* of Y given X, denoted $H(Y|X)$, is defined as $H(Y|X) = H(XY) - H(X)$, where XY denotes the joint distribution of X and Y. Informally, $H(Y|X)$ measures the (residual) uncertainty of Y when X is known.

The *mutual information* between random variables X and Y is

$$I(X;Y) = H(X) + H(Y) - H(XY) = H(Y) - H(Y|X) = H(X) - H(X|Y)$$

which measures the information that X reveals about Y (and vice versa). In particular, if two random variables X, Y are independent, their mutual information is zero.

The *conditional mutual information* between random variables X and Y given Z, denoted $I(X;Y|Z)$, is defined as

$$I(X;Y|Z) = H(X|Z) + H(Y|Z) - H(XY|Z).$$

We will use without proof that entropy, conditional entropy, mutual information, conditional mutual information are non-negative.

We will need the following simple propositions.

Proposition 2.1. *Let* $X \sim \mathrm{Bern}(\frac{1}{2})$ *be a random variable uniformly distributed in* $\{0,1\}$, *let* $N \sim \mathrm{Bern}(\varepsilon)$ *be a noise that is independent from* X, *and let* $\hat{X} = X \oplus N$ *be the noisy version of* X. *Then* $I(\hat{X}; X) = 1 - h(\varepsilon)$. *Moreover, for any random variable* X' *satisfying* $\Pr[X' = X] \geq 1 - \varepsilon$,

$$I(X'; X) \geq 1 - h(\varepsilon).$$

Proof. Clearly, $I(\hat{X};X) = H(X) - H(X|\hat{X}) = 1 - h(\varepsilon)$. Furthermore, the random variable $\hat{X} = X \oplus N$ minimizes the mutual information $I(\hat{X};X)$ under the constraint that $\Pr[\hat{X} = X] \geq 1 - \varepsilon$. In particular, we have

$$I(X';X) = H(X) - H(X|X') = 1 - H(X \oplus X'|X') \geq 1 - H(X \oplus X') \geq 1 - h(\varepsilon)$$

for any random variable X' satisfying $\Pr[X' = X] \geq 1 - \varepsilon$.   □

Proposition 2.2 (Conditioning Decreases Entropy). *For any random variables X, Y, Z, it holds that $H(X) \geq H(X|Y) \geq H(X|YZ)$.*

In general, conditioning can increase or decrease mutual information, but when conditioning on an *independent* variable, mutual information increases.

Proposition 2.3 (Conditioning on Independent Variables Increases Mutual Information). *For random variables X, Y, Z such that Y and Z are independent, $I(X;Y|Z) \geq I(X;Y)$.*

Proof. As Y, Z are independent, $H(Y|Z) = H(Y)$.

$$I(X;Y|Z) = H(Y|Z) - H(Y|XZ) \geq H(Y) - H(Y|X) = I(X;Y).   □$$

Proposition 2.4 (Data Processing for Mutual Information). *Assume random variables X, Y, Z satisfies $X \to Y \to Z$, i.e. X and Z are independent conditional on Y, then $I(X;Y) \geq I(X;Z)$.*

Proof. Since X and Z are independent conditional on Y (meaning $I(X;Z|Y) = 0$), we have $H(X|YZ) = H(X|Y)$. Thus

$$I(X;Y) = H(X) - H(X|Y) = H(X) - H(X|YZ) \geq H(X) - H(X|Z) = I(X;Z).   □$$

Proposition 2.5 (Chain Rule for Mutual Information). *For random variables $X_1, \ldots, X_n, Y$, it holds that*

$$I(X_1 \ldots X_n; Y) = \sum_{i=1}^{n} I(X_i; Y | X_1 \ldots X_{i-1}).$$

2.2 Single-Server One-Round Private Information Retrieval

In a single-server private information retrieval (PIR) protocol, the database holds n bits of data $x \in \{0,1\}^n$. The user, given an index $i \in [n]$, would like to retrieve the i-th bit from the server, without revealing any information about i. The user does so by generating a query based on i using a randomized algorithm; the server responds to the query with an answer. The user, given the answer and the randomness used to generate the query, should be able to learn the i-th bit x_i.

We specialize our definitions to the case of single round protocols.

Definition 2.6 (Private Information Retrieval). *A single-server one round private information retrieval (PIR) scheme is a tuple $(\mathbf{Qry}, \mathbf{Ans}, \mathbf{Rec})$ of algorithms such that*

- *The query algorithm* **Qry** *is a probabilistic polynomial-time algorithm such that* $\mathbf{Qry}(1^n, i) \to (q, \sigma)$, *where* $i \in [n]$. *Here,* q *is the PIR query and* σ *is the secret state of the user (which, without loss of generality, is the randomness used by the algorithm).*
- *The answer algorithm* **Ans** *is a probabilistic polynomial-time algorithm such that* $\mathbf{Ans}(x, q) \to a$, *where* $x \in \{0, 1\}^n$. *Let* ℓ *denote the length of the answer, i.e.* $a \in \{0, 1\}^\ell$.
- *The reconstruction algorithm* **Rec** *is a probabilistic polynomial-time algorithm such that* $\mathbf{Rec}(a, \sigma) \to b$ *where* $b \in \{0, 1\}$.

Correctness. A PIR scheme $(\mathbf{Qry}, \mathbf{Ans}, \mathbf{Rec})$ is $(1 - \varepsilon)$-correct if for any $x \in \{0, 1\}^n$ and for any i,

$$\Pr\Big[\mathbf{Qry}(1^n, i) \to (q, \sigma), \mathbf{Ans}(x, q) \to a : \mathbf{Rec}(a, \sigma) = x_i\Big] \geq 1 - \varepsilon(n)$$

where the probability is taken over the random tapes of $\mathbf{Qry}, \mathbf{Ans}, \mathbf{Rec}$. We call ϵ the error probability of the PIR scheme.

Privacy. The standard definition of computational privacy for PIR requires that the database cannot efficiently distinguish between queries for different indices. Formally, a PIR scheme is δ-IND-secure (for some $\delta = \delta(n)$) if for any probabilistic polynomial-time algorithm $\mathcal{A} = (\mathcal{A}_1, \mathcal{A}_2)$, there exists a negligible function δ such that

$$\Pr\left[\begin{array}{c} \mathcal{A}_1(1^n) \to (i_0, i_1, \tau) \\ b \overset{\$}{\leftarrow} \{0, 1\} \\ \mathbf{Qry}(1^n, i_b) \to (q, \sigma) \\ \mathcal{A}_2(1^n, q, \tau) \to b' \end{array} : b' = b\right] < \frac{1}{2} + \delta(n) \tag{1}$$

(Here and in the sequel, τ will denote the state that $\mathcal{A}_1$ passes on to $\mathcal{A}_2$).

The adversary in this privacy definition is interactive, which introduces difficulties in defining an oracle that breaks PIR. To make our task easier, we consider an alternative, non-interactive definition which is equivalent to (1).

We call a PIR scheme δ-GUESS-secure if for any probabilistic polynomial-time algorithm $\mathcal{A}$, there exists a negligible function δ such that

$$\Pr\left[\begin{array}{c} j \overset{\$}{\leftarrow} [n] \\ \mathbf{Qry}(1^n, j) \to (q, \sigma) \\ \mathcal{A}(1^n, q) \to j' \end{array} : j' = j\right] < \frac{1}{n}\left(1 + \delta(n)\right) \tag{2}$$

These two definitions of privacy are equivalent up to a polynomial factor in n, as we show in the next proposition.

Proposition 2.7. *If a PIR scheme is δ_1-IND-secure (according to Definition (1)), then it is δ_2-GUESS-secure (according to Definition (2)) where $\delta_2 = n\delta_1$. Similarly, if a PIR scheme is δ_2-GUESS-secure, then it is δ_1-IND-secure where $\delta_1 = \delta_2/2$.*

Proof. Assume that a probabilistic polynomial-time (p.p.t.) adversary algorithm $\mathcal{A}$ breaks δ_2-privacy according to Definition (2). We construct an adversary $\mathcal{B} = (\mathcal{B}_1, \mathcal{B}_2)$ that breaks Definition (1).

The algorithm $\mathcal{B}_1(1^n)$ picks two random indices i_0 and i_1 and outputs i_0, i_1 and $\tau = (i_0, i_1)$, algorithm $\mathcal{B}_2(1^n, q, \tau = (i_0, i_1))$ calls $\mathcal{A}(1^n, q)$ to get an index i, and outputs 0 if and only if $i = i_0$. Then,

$$\Pr\left[\begin{matrix} \mathcal{B}_1(1^n) \to (i_0, i_1, \tau) \\ b \xleftarrow{\$} \{0, 1\} \\ \mathbf{Qry}(1^n, i_b) \to (q, \sigma) \\ \mathcal{B}_2(1^n, q, \tau) \to b' \end{matrix} : b' = b\right] = \Pr\left[\begin{matrix} i_0, i_1 \xleftarrow{\$} [n] \\ b \xleftarrow{\$} \{0, 1\} \\ \mathbf{Qry}(1^n, i_b) \to (q, \sigma) \\ \mathcal{A}(1^n, q) \to i \end{matrix} : \begin{matrix} i = i_0, b = 0 \\ \text{or} \\ i \neq i_0, b \neq 0 \end{matrix}\right]$$

$$= \frac{1}{2}\Pr\left[\begin{matrix} i_0, i_1 \xleftarrow{\$} [n] \\ \mathbf{Qry}(1^n, i_0) \to (q, \sigma) \\ \mathcal{A}(1^n, q) \to i \end{matrix} : i = i_0\right] + \frac{1}{2}\Pr\left[\begin{matrix} i_0, i_1 \xleftarrow{\$} [n] \\ \mathbf{Qry}(1^n, i_1) \to (q, \sigma) \\ \mathcal{A}(1^n, q) \to i \end{matrix} : i \neq i_0\right]$$

$$\geq \frac{1}{2}\frac{1}{n}(1 + \delta_2(n)) + \frac{1}{2}\left(1 - \frac{1}{n}\right) = \frac{1}{2}\left(1 + \frac{\delta_2(n)}{n}\right)$$

Thus, $(\mathcal{B}_1, \mathcal{B}_2)$ breaks $\frac{\delta_2}{n}$-privacy according to Definition (1).

In the other direction, assume that a p.p.t. adversary algorithm $\mathcal{A} = (\mathcal{A}_1, \mathcal{A}_2)$ breaks δ_1-privacy according to Definition (1). We construct an adversary $\mathcal{B}$ that works as follows. $\mathcal{B}$ runs $\mathcal{A}_1$ to get $(i_0, i_1, \tau) \leftarrow \mathcal{A}_1(1^n)$, gets a challenge query q and runs $\mathcal{A}_2$ to get $b \leftarrow \mathcal{A}_2(1^n, q, \tau)$. $\mathcal{B}$ simply outputs i_b. Then, we have:

$$\Pr\left[\begin{matrix} j \xleftarrow{\$} [n] \\ \mathbf{Qry}(1^n, j) \to (q, \sigma) \\ \mathcal{B}(1^n, q) \to j' \end{matrix} : j' = j\right] = \Pr\left[\begin{matrix} \mathcal{A}_1(1^n) \to (i_0, i_1, \tau) \\ j \xleftarrow{\$} [n] \\ \mathbf{Qry}(1^n, j) \to (q, \sigma) \\ \mathcal{A}_2(1^n, q, \tau) \to b \end{matrix} : j = i_b\right]$$

$$= \frac{2}{n}\Pr\left[\begin{matrix} \mathcal{A}_1(1^n) \to (i_0, i_1, \tau) \\ j \xleftarrow{\$} \{i_0, i_1\} \\ \mathbf{Qry}(1^n, j) \to (q, \sigma) \\ \mathcal{A}_2(1^n, q, \tau) \to b \end{matrix} : j = i_b\right] \geq \frac{2}{n}\left(\frac{1}{2} + \delta_1(n)\right) = \frac{1}{n}\left(1 + 2\delta_1(n)\right)$$

Thus, $\mathcal{B}$ breaks $2\delta_1$-privacy according to Definition (2). $\qquad\square$

Answer Communication Complexity. We define the answer communication complexity of the PIR scheme to be the number of bits in the server's response to a PIR query. (This is denoted by ℓ in Definition 2.6). Similarly, we call the bit-length of the query as the query communication complexity, and their sum as the total communication complexity. In this work, we are interested in PIR protocols with a "small" answer communication complexity (regardless of their query communication complexity). Since our main result is a lower bound, this only makes it stronger.

Typically, we are interested in PIR schemes with answer communication complexity $\ell = o(n)$. Otherwise, e.g. when $\ell = n$, there is a trivial PIR protocol with perfect privacy, where the user sends nothing and the server sends the whole database x. The following proposition shows a tradeoff between the correctness error and answer communication complexity of perfectly private PIR schemes.

Proposition 2.8. *There exists a PIR scheme with perfect information-theoretic privacy, error probability ε, and answer communication complexity $\ell = n \cdot (1 - h(\varepsilon) + O(n^{-1/4}))$.*

Consider a PIR scheme where the user sends nothing and the server sends the whole database to the user, incurring an answer communication complexity of n bits. The query contains no information about the index i, and this achieves perfect privacy and correctness. The idea is that given the possibility of a correctness error of ε, the server can compress the database into $\ell < n$ bits, such that the user can still recover the database with at most ε error.

This is a fundamental problem in information theory, called "lossy source coding" [Sha59]. Let X be a uniform random Bernoulli variable. Proposition 2.1 says that for any random variable $\hat{X}$ such that $\Pr[\hat{X} = X] \geq 1 - \varepsilon$, $I(\hat{X}, X) \geq 1 - h(\varepsilon)$. Therefore, to compress a random binary string and to recover the string from the lossy compression with $(1 - \varepsilon)$ accuracy, the compression ratio need to be at least $1 - h(\varepsilon)$.

There exists a lossy source coding scheme almost achieves the information theoretical bound [Ari09, KU10], i.e., when $\ell = n \cdot (1 - h(\varepsilon) + O(n^{-1/4}))$, there exists efficient algorithms $E : \{0,1\}^n \rightarrow \{0,1\}^\ell$ and $D : \{0,1\}^\ell \rightarrow \{0,1\}^n$, such that for randomly chosen $X \in \{0,1\}^n$ and for any index $i \in [n]$,

$$\Pr_X[\hat{X} = D(E(X)) : \hat{X}_i = X_i] \geq 1 - \varepsilon.$$

Therefore, if the server sends $E(x)$ as the answer, then the PIR scheme achieves $(1 - \varepsilon)$ correctness on a random database. Moreover, we can extend this to work for any database by the following scheme which has a query communication complexity of n bits and an answer communication complexity of ℓ bits.

- User sends a query m, which is a random string in $\{0,1\}^n$;
- Server answers by $a = E(m \oplus x)$;
- User retrieves the whole database by $\hat{x} = D(a) \oplus m$.

Then for any database and any index $i \in [n]$, $\Pr[\hat{x}_i = x_i] \geq 1 - \varepsilon$.

Reduction to Breaking PIR. What does it mean for a reduction to decide a language L assuming that there is a p.p.t. adversary that breaks PIR? For any language L, we say L can be reduced to breaking the δ-GUESS-security of PIR scheme (**Qry, Ans, Rec**) if there exists a probabilistic polynomial-time oracle Turing machine (OTM) M such that for all x and for all "legal" oracles $\mathcal{O}_\delta^{\text{PIR}}$,

$$\Pr[M^{\mathcal{O}_\delta^{\text{PIR}}}(x) = 1] \geq 2/3 \quad \text{if } x \in L$$
$$\Pr[M^{\mathcal{O}_\delta^{\text{PIR}}}(x) = 1] \leq 1/3 \quad \text{if } x \notin L$$

where the probability is taken over the coins of the machine M and the oracle $\mathcal{O}_\delta^{\text{PIR}}$. We stress that M is allowed to make adaptive queries to the oracle.

By a legal δ-breaking oracle $\mathcal{O}_\delta^{\mathsf{PIR}}$, we mean one that satifies

$$\Pr\left[\begin{array}{c} j \leftarrow [n] \\ \mathbf{Qry}(1^n, j) \rightarrow (q, \sigma) : j = j' \\ \mathcal{O}_\delta^{\mathsf{PIR}}(q) \rightarrow j' \end{array}\right] \geq \frac{1}{n}(1 + \delta) \tag{3}$$

where the probability is taken over the coins used in the experiment, including those of $\mathbf{Qry}$ and $\mathcal{O}_\delta^{\mathsf{PIR}}$.

2.3   Entropy Difference

Entropy Difference (ED) is a promise problem that is complete for **SZK** [GV99]. Entropy Difference is a promise problem defined as

- YES instances: (X, Y) such that $H(X) \geq H(Y) + 1$
- NO instances: (X, Y) such that $H(Y) \geq H(X) + 1$

where X and Y are distributions encoded as circuits which sample from them.

We list a few elementary observations regarding the power of an oracle that decides the entropy difference problem.

First, given an entropy difference oracle, a polynomial-time algorithm can distinguish between two distributions X and Y such that either $H(X) \geq H(Y) + \frac{1}{s}$ or $H(Y) \geq H(X) + \frac{1}{s}$ for any polynomial function s. That is, one can solve the entropy difference problem up to any inverse-polynomial precision. This can be done as follows: For distributions X, Y, we query the Entropy Difference oracle with $(X_1 \ldots X_s, Y_1 \ldots Y_s)$, where $X_i \sim X, Y_i \sim Y$ and $X_1, \ldots, X_s$ are i.i.d. and $Y_1, \ldots, Y_s$ are i.i.d. Then we would be able to distinguish between $H(X) \geq H(Y) + \frac{1}{s}$ and $H(Y) \geq H(X) + \frac{1}{s}$.

Similarly, a polynomial-time algorithm can use the Entropy Difference oracle to distinguish between $H(X) \geq \hat{h} + \frac{1}{s}$ and $H(X) \leq \hat{h} - \frac{1}{s}$ for a given $\hat{h}$. This can be done as follows: construct a distribution Y that $2s\hat{h} - 1 < H(Y) < 2s\hat{h} + 1$ and query the Entropy Difference oracle with the distributions $X_1 \ldots X_{2s}$ and Y, where $X_1, \ldots, X_{2s}$ are independent copies of X. Therefore, a polynomial-time algorithm given Entropy Difference oracle can estimate $H(X)$ to within any additive inverse-polynomial precision by binary search.

Finally, assume that X and Y are random variables encoded as a circuit which samples from their joint distributions. Then, a polynomial-time algorithm given an Entropy Difference oracle can also estimate the conditional entropy $H(X|Y)$, mutual information $I(X;Y)$ to any inverse-polynomial precision. Here the precision is measured by absolute additive error.

3   PIR and NP-Hardness

Theorem 3.1 (Main Theorem). *Let $\Pi = (\mathbf{Qry}, \mathbf{Ans}, \mathbf{Rec})$ be any $(1 - \epsilon)$-correct PIR scheme with n-bit databases and answer communication complexity ℓ. Let L be any language. If*

1. *there exists a reduction from L to breaking the δ-privacy of Π in the sense of Equation (2); and*
2. *there is a polynomial $p(n)$ such that*

$$\ell \cdot (1 + \delta) \leq n \cdot (1 - h(\varepsilon)) - 1/p(n)$$

then $L \in \mathbf{AM} \cap \mathbf{coAM}$.

In particular, using the result of [BHZ87], this tells us that unless the polynomial hierarchy collapses, there is no reduction from SAT to breaking the privacy of a PIR scheme with parameters as above.

We note that the bound in the lemma is tight. As Proposition 2.8 shows, there is in fact a perfectly (information-theoretically) private PIR protocol with a matching answer communication complexity of $n \cdot (1 - h(\varepsilon)) + o(n)$.

We prove our main theorem by combining the following two lemmas. The first lemma is our main ingredient, and says that if there is a reduction from deciding a language L to breaking a PIR scheme, and the PIR scheme has a low answer communication complexity, then L can be reduced to the entropy difference problem (defined in Sect. 2.3).

Lemma 3.2 ($\mathbf{BPP}^{\mathcal{O}_\delta^{PIR}} \subseteq \mathbf{BPP}^{ED}$). *Let $\Pi = (\mathbf{Qry}, \mathbf{Ans}, \mathbf{Rec})$ be any $(1 - \epsilon)$-correct PIR scheme with answer communication complexity ℓ and let L be any language. If there exists a reduction from L to δ-breaking the privacy of a PIR protocol such that*

$$\frac{1 - h(\varepsilon)}{\ell} - \frac{1 + \delta}{n} \geq \frac{1}{p(n)}$$

for some polynomial function $p(n)$, then there exists a probabilistic polynomial time reduction from L to ED.

As noted in Proposition 2.8, this condition is tight as there exists a PIR scheme achieving perfect privacy ($\delta = 0$) if $\ell \approx n \cdot (1 - h(\varepsilon))$.

The next lemma, originally shown in [MX10] and used in [BL13b], states that any language decidable by a randomized oracle machine with access to an entropy difference oracle is in $\mathbf{AM} \cap \mathbf{coAM}$.

Lemma 3.3 ($\mathbf{BPP}^{ED} \subseteq \mathbf{AM} \cap \mathbf{coAM}$ [MX10]). *For any language L, if there exists an OTM M such that for any oracle $\mathcal{O}$ solving entropy difference*

$$\Pr[M^{\mathcal{O}}(x) = 1] \geq 2/3 \quad \text{if } x \in L$$
$$\Pr[M^{\mathcal{O}}(x) = 1] \leq 1/3 \quad \text{if } x \notin L,$$

then $L \in \mathbf{AM} \cap \mathbf{coAM}$.

3.1   Proof of the Main Theorem

Assume that there exists a reduction from deciding a language L to breaking PIR with parameters as stated in Theorem 3.1. In other words, there is a reduction from L to δ-breaking PIR where

$$\frac{1}{n}(1+\delta) \leq \frac{1-h(\varepsilon)}{\ell} - \frac{1}{n \cdot \ell \cdot p(n)}.$$

where the inequality is using the hypothesis in Theorem 3.1 that $\ell \cdot (1+\delta) \leq n \cdot (1-h(\varepsilon)) - 1/p(n)$.

Then, by Lemma 3.2, there is a reduction from deciding L to solving the entropy difference problem ED. Combined with Lemma 3.3, we deduce that $L \in$ AM $\cap$ coAM.

3.2   Proof of Lemma 3.2

We start with two claims that are central to our proof. The first claim says that because of $(1-\varepsilon)$-correctness of the PIR scheme, the PIR answer a on a query $q \leftarrow \mathbf{Qry}(1^n, i)$ has to contain information about the ith bit of the database x_i.

Claim. Let $\Pi = (\mathbf{Qry}, \mathbf{Ans}, \mathbf{Rec})$ be a PIR scheme which is $(1-\varepsilon)$-correct. Fix any index $i \in [n]$. Let X denote a random n-bit database; $(Q, \Sigma) \leftarrow \mathbf{Qry}(1^n, i)$; and $A \leftarrow \mathbf{Ans}(X, Q)$. Then,

$$I(A; X_i | Q) \geq 1 - h(\varepsilon). \tag{4}$$

Proof. Define the random variable $\hat{X}_i \leftarrow \mathbf{Rec}(A, \Sigma)$. Since the PIR scheme is $(1-\varepsilon)$-correct, $\Pr[\hat{X}_i = X_i] \geq 1 - \varepsilon$. Since X_i is a uniform Bernoulli variable, we know from Proposition 2.1 that $I(\hat{X}_i; X_i) \geq 1 - h(\varepsilon)$.

As X_i is independent from Q, we know from Proposition 2.3 that

$$I(\hat{X}_i; X_i | Q) \geq I(\hat{X}_i; X_i).$$

Next, we claim that conditioning on Q, we have $X_i \to A \to \hat{X}_i$, in other words, $I(X_i; \hat{X}_i | A, Q) = 0$. This is because when A and Q are given, one can sample a random Σ consistent with Q, then compute $\hat{X}_i$ from Σ and A, with no knowledge of X_i. Now, Proposition 2.4 (data processing inequality for mutual information) shows that $I(A; X_i | Q) \geq I(\hat{X}_i; X_i | Q)$.

Combining what we have,

$$I(A; X_i | Q) \geq I(\hat{X}_i; X_i | Q) \geq I(\hat{X}_i; X_i) \geq 1 - h(\varepsilon).$$

This completes the proof.   □

Claim. Let $\Pi = (\mathbf{Qry}, \mathbf{Ans}, \mathbf{Rec})$ be a PIR scheme with an answer communication complexity of ℓ bits. Let X denote a random n-bit database; $(Q, \Sigma) \leftarrow \mathbf{Qry}(1^n, i)$; and $A \leftarrow \mathbf{Ans}(X, Q)$. Then, for any potential query q,

$$\sum_{j=1}^{n} I(A; X_j | Q = q) \leq \ell. \tag{5}$$

Proof. Recall that, by definition,

$$I(A; X_i | Q) = \mathbb{E}_Q\left[I(A; X_i | Q)\right] = \sum_q I(A; X_i | Q = q) \Pr[Q = q]$$

For any potential query q, the event $Q = q$ is independent from X. In particular, for any index j, random variable X_j is independent from $X_1 \ldots X_{j-1}$ given $Q = q$. So for any q,

$$\sum_{j=1}^{n} I(A; X_j | Q = q) \leq \sum_{j=1}^{n} I(A; X_j | X_1 \ldots X_{j-1}, Q = q)$$
$$= I(A; X_1 \ldots X_n | Q = q)$$
$$\leq H(A | Q = q) \leq \ell$$

where the first inequality is implied by the Proposition 2.3 and the second equality is Proposition 2.5 (chain rule for mutual information). $\quad\square$

Equations (4) and (5) are the core of the proof of Lemma 3.2. Equation (4) shows that, when retrieving the i-th bit, the mutual information between X_i and server's answer A is large. Equation (5) shows that, the sum of mutual information between each bit X_j and server's answer A is bounded by the answer communication complexity. Therefore, if we could measure the mutual information by an Entropy Difference oracle, we would have a pretty good knowledge of i.

In particular, we proceed as follows. Assume language L can be solved by a probabilistic polynomial-time oracle Turing machine $\mathcal{M}$ given any oracle $\mathcal{O}_\delta^{\mathsf{PIR}}$ that breaks the δ-GUESS-security of the PIR scheme $(\mathbf{Qry}, \mathbf{Ans}, \mathbf{Rec})$ where

$$\frac{1+\delta}{n} \leq \frac{1 - h(\varepsilon)}{\ell} - \frac{1}{p(n)} \tag{6}$$

where $p(\cdot)$ is a fixed polynomial. We construct an efficient oracle algorithm (see Algorithm 1) that solves L given an Entropy Difference oracle $\mathcal{O}^{\mathsf{ED}}$.

For any query q and index i, when $\mathcal{O}_\delta^{\mathsf{PIR}}(q)$ is simulated,

$$\Pr\left[\hat{i} \leftarrow \mathcal{O}_\delta^{\mathsf{PIR}}(q) : \hat{i} = i\right] = \frac{\hat{\mu}_i}{\sum_j \hat{\mu}_j} \geq \frac{\mu_i - \frac{1}{2n \cdot p(n)}}{\sum_j \mu_j + \frac{1}{2p(n)}}$$
$$\geq \frac{\mu_i - \frac{1}{2p(n)}}{\ell + \frac{1}{2p(n)}} \geq \frac{\mu_i}{\ell} \frac{1 - \frac{1}{2p(n)}}{1 + \frac{1}{2p(n)}} \geq \frac{\mu_i}{\ell}\left(1 - \frac{1}{p(n)}\right) \geq \frac{\mu_i}{\ell} - \frac{1}{p(n)}$$

Assuming q is generated from $q \leftarrow \mathbf{Qry}(1^n, i)$, then $\mathbb{E}[\mu_i] = I(X_i; A | Q) \geq 1 - h(\varepsilon)$. So

$$\Pr\left[q \leftarrow \mathbf{Qry}(1^n, i), \hat{i} \leftarrow \mathcal{O}_\delta^{\mathsf{PIR}}(q) : \hat{i} = i\right]$$
$$= \underset{q \leftarrow \mathbf{Qry}(1^n, i)}{\mathbb{E}} \left[\Pr[\hat{i} = i | Q = q]\right]$$
$$\geq \underset{q \leftarrow \mathbf{Qry}(1^n, i)}{\mathbb{E}} \left[\frac{\mu_i}{\ell} - \frac{1}{p(n)}\right]$$
$$= \frac{\mathbb{E}_{q \leftarrow \mathbf{Qry}(1^n, i)}[\mu_i]}{\ell} - \frac{1}{p(n)}$$

Algorithm 1. Solving L given ED oracle on input x

---

1. Simulate $\mathcal{M}^{\mathcal{O}_\delta^{\mathsf{PIR}}}(x)$
2. Whenever $\mathcal{M}$ queries $\mathcal{O}_\delta^{\mathsf{PIR}}(q)$, do the following:
   (a) For each index $j = 1, \ldots, n$, use the entropy difference oracle to estimate

   $$\mu_j = I(A; X_j | Q = q)$$

   to $\frac{1}{2n \cdot p(n)}$ precision. More precisely, construct a circuit $C = C_{q,j}$ such that

   $$C_{q,j}(x, r) = (x_j, \mathbf{Ans}(x, q, r))$$

   and estimate the mutual information between the two components of C's output. Let $\hat{\mu}_j \in [0, 1]$ denote the estimation.
   (b) Sample a random value $\hat{i} \in [n]$ according to probability distribution $p(\hat{i}) = \hat{\mu}_{\hat{i}} / \sum_j \hat{\mu}_j$
   (c) Answer $\mathcal{M}$'s query by $\hat{i}$
3. Output what $\mathcal{M}$ output

$$\geq \frac{1 - h(\varepsilon)}{\ell} - \frac{1}{p(n)}$$

$$\geq \frac{1}{n}(1 + \delta)$$

4   Discussion and Open Questions

We show that any non-trivial single-server single-round PIR scheme can be broken in **SZK**. Since languages that can be decided with (adaptive) oracle access to **SZK** live in **AM** ∩ **coAM**, this shows that there cannot be a reduction from SAT to **SZK**, and therefore also from SAT to breaking single-server single-round PIR.

The crucial underlying feature of single-round PIR schemes that we use is the ability to "re-randomize". By this, we mean that given a user query q for an index i, one can generate not just a single transcript, but the distribution over all transcripts where the database is uniformly random and the prefix of the transcript is q. This ability to generate a transcript distribution of the same index and random database allows the adversary to break a PIR scheme with an **SZK** oracle.

Indeed, this is reminiscent of the work of Bogdanov and Lee who show that breaking homomorphic encryption is not **NP**-hard [BL13b]. Their main contribution is to show that any homomorphic encryption (whose homomorphic evaluation process produces a ciphertext that is statistically close to a fresh encryption) can be turned into a (weakly) re-randomizable encryption scheme. Once this is done, an **SZK** oracle can be used to break the scheme in much the same way as we do.

A natural question arising from our work is to extend our results to multi-round PIR. The key technical difficulty that arises is in sampling a random "continuation" of a partial transcript. We conjecture that our lower bound can nevertheless be extended to the multi-round case, and leave this as an interesting open problem.

Acknowledgments. We would like to thank the anonymous TCC reviewers for their careful reading and excellent suggestions, and Jayadev Acharya for valuable comments about lossy source coding and polar codes.

References

[AGGM06] Akavia, A., Goldreich, O., Goldwasser, S., Moshkovitz, D.: On basing one-way functions on NP-hardness. In: Kleinberg, J.M. (ed.) Proceedings of the 38th Annual ACM Symposium on Theory of Computing, Seattle, WA, USA, 21–23 May 2006, pp. 701–710. ACM (2006)

[Ari09] Arikan, E.: Channel polarization: a method for constructing capacity-achieving codes for symmetric binary-input memoryless channels. IEEE Trans. Inf. Theory **55**(7), 3051–3073 (2009)

[BB15] Bogdanov, A., Brzuska, C.: On basing size-verifiable one-way functions on NP-hardness. In: Dodis, Y., Nielsen, J.B. (eds.) TCC 2015, Part I. LNCS, vol. 9014, pp. 1–6. Springer, Heidelberg (2015)

[BGN05] Boneh, D., Goh, E.-J., Nissim, K.: Evaluating 2-DNF formulas oncipher-texts. In: Kilian [Kil05], pages 325–341

[BHZ87] Boppana, R.B., Håstad, J., Zachos, S.: Does co-NP have short interactive proofs? Inf. Process. Lett. **25**(2), 127–132 (1987)

[BIKM99] Beimel, A., Ishai, Y., Kushilevitz, E., Malkin, T.: One-way functions are essential for single-server private information retrieval. In: Vitter, J.S., Larmore, L.L., Leighton, F.T. (eds.) Proceedings of the Thirty-First Annual ACM Symposium on Theory of Computing, 1–4 May 1999, Atlanta, Georgia, USA, pp. 89–98. ACM (1999)

[BL13a] Bogdanov, A., Lee, C.H.: Limits of provable security for homomorphic encryption. In: Canetti, R., Garay, J.A. (eds.) CRYPTO 2013, Part I. LNCS, vol. 8042, pp. 111–128. Springer, Heidelberg (2013)

[BL13b] Bogdanov, A., Lee, C.H.: Limits of provable security for homomorphic encryption. In: Canetti, R., Garay, J.A. (eds.) CRYPTO 2013, Part I. LNCS, vol. 8042, pp. 111–128. Springer, Heidelberg (2013)

[Bra79] Brassard, G.: Relativized cryptography. In: 20th Annual Symposium on Foundations of Computer Science, San Juan, Puerto Rico, 29–31 October 1979, pp. 383–391. IEEE Computer Society (1979)

[BT06] Bogdanov, A., Trevisan, L.: On worst-case to average-case reductions for NP problems. SIAM J. Comput. **36**(4), 1119–1159 (2006)

[BV11] Brakerski, Z., Vaikuntanathan, V.: Efficient fully homomorphic encryption from (standard) LWE. In: Ostrovsky, R. (ed.) IEEE 52nd Annual Symposium on Foundations of Computer Science, FOCS 2011, Palm Springs, CA, USA, 22–25 October 2011, pages 97–106. IEEE Computer Society (2011)

[CKGS98] Chor, B., Kushilevitz, E., Goldreich, O., Sudan, M.: Private information retrieval. J. ACM **45**(6), 965–981 (1998)

[CMO00] Di Crescenzo, G., Malkin, T., Ostrovsky, R.: Single database private information retrieval implies oblivious transfer. In: Preneel, B. (ed.) EUROCRYPT 2000. LNCS, vol. 1807, pp. 122–138. Springer, Heidelberg (2000)

[CMS99] Cachin, C., Micali, S., Stadler, M.A.: Computationally private information retrieval with polylogarithmic communication. In: Stern, J. (ed.) EUROCRYPT 1999. LNCS, vol. 1592, pp. 402–414. Springer, Heidelberg (1999)

[Gen09] Gentry, C.: Fully homomorphic encryption using ideal lattices. In: Mitzenmacher, M. (ed.) Proceedings of the 41st Annual ACM Symposium on Theory of Computing, STOC 2009, Bethesda, MD, USA, 31 May–2 June 2009, pp. 169–178. ACM (2009)

[GG98] Goldreich, O., Goldwasser, S.: On the possibility of basing cryptography on the assumption that $p \neq np$. IACR Cryptology ePrint Archive, 1998, 5 (1998)

[GR05] Gentry, C., Ramzan, Z.: Single-database private information retrieval with constant communication rate. In: Caires, L., Italiano, G.F., Monteiro, L., Palamidessi, C., Yung, M. (eds.) ICALP 2005. LNCS, vol. 3580, pp. 803–815. Springer, Heidelberg (2005)

[GV99] Goldreich, O., Vadhan, S.: Comparing entropies in statistical zero knowledge with applications to the structure of SZK. In: 1999 Proceedings of the Fourteenth Annual IEEE Conference on Computational Complexity, pp. 54–73. IEEE (1999)

[IKO05] Ishai, Y., Kushilevitz, E., Ostrovsky, R.: Sufficient conditionsfor collision-resistant hashing. In: Kilian [Kil05], pp. 445–456

[Kil05] Kilian, J. (ed.): TCC 2005. LNCS, vol. 3378. Springer, Heidelberg (2005)

[KO97] Kushilevitz, E., Ostrovsky, R.: Replication is NOT needed: SINGLE database, computationally-private information retrieval. In: 38th Annual Symposium on Foundations of Computer Science, FOCS 1997, Miami Beach, Florida, USA, 19–22 October 1997, pp. 364–373. IEEE Computer Society (1997)

[KU10] Korada, S.B., Urbanke, R.L.: Polar codes are optimal for lossy source coding. IEEE Trans. Inf. Theor. **56**(4), 1751–1768 (2010)

[Lip05] Lipmaa, H.: An oblivious transfer protocol with log-squared communication. In: Zhou, J., López, J., Deng, R.H., Bao, F. (eds.) ISC 2005. LNCS, vol. 3650, pp. 314–328. Springer, Heidelberg (2005)

[MX10] Mahmoody, M., Xiao, D.: On the power of randomized reductions and the checkability of sat. In: 2010 IEEE 25th Annual Conference on Computational Complexity (CCC), pp. 64–75. IEEE (2010)

[Sha59] Shannon, C.E.: Coding theorems for a discrete source with a fidelity criterion. IRE Nat. Conv. Rec. **4**(142–163), 1 (1959)

Obfuscation-Based Cryptographic Constructions

On the Correlation Intractability of Obfuscated Pseudorandom Functions

Ran Canetti[1,2]([⊠]), Yilei Chen[1], and Leonid Reyzin[1]

[1] Boston University, Boston, USA
{canetti,chenyl,reyzin}@bu.edu
[2] Tel Aviv University, Tel Aviv, Israel
canetti@tau.ac.il

Abstract. A family of hash functions is called "correlation intractable" if it is hard to find, given a random function in the family, an input-output pair that satisfies any "sparse" relation, namely any relation that is hard to satisfy for truly random functions. Indeed, correlation intractability is a strong and natural random-oracle-like property. However, it was widely considered unobtainable. In fact for some parameter settings, unobtainability has been demonstrated [26]. We construct a correlation intractable function ensemble that withstands all relations with a priori bounded polynomial complexity. We assume the existence of sub-exponentially secure indistinguishability obfuscators, puncturable pseudorandom functions, and input-hiding obfuscators for evasive circuits. The existence of the latter is implied by Virtual-Grey-Box obfuscation for evasive circuits [13].

1 Introduction

To what extent can we construct efficient function families that "behave like random functions"? This is an intriguing question in cryptography. One of the most elusive properties of random functions is correlation intractability, proposed by Canetti, Goldreich and Halevi [26]. Roughly speaking, correlation intractable functions guarantee that it is infeasible to find input-output pairs that satisfy some "rare" relation. A bit more precisely, a binary relation R is called *sparse*, if for each value x, only a negligible fraction of y values satisfy $(x, y) \in R$. A function family F is *correlation intractable* if, for any sparse relation R, it is infeasible for the adversary to find, given the full description of a random function f in F, a value x such that $(x, f(x))$ is in the relation.

The only known results regarding the existence of correlation intractable functions are negative. Specifically, for some settings of the parameters (e.g. when the key is shorter than the input), correlation intractable functions were shown not to exist. This observation was used in [26] to demonstrate the uninstantiability of the random oracle model [9]. However, whether correlation intractable functions exist for other settings of the parameters, and based on what assumptions, remains open.

© International Association for Cryptologic Research 2016
E. Kushilevitz and T. Malkin (Eds.): TCC 2016-A, Part I, LNCS 9562, pp. 389–415, 2016.
DOI: 10.1007/978-3-662-49096-9_17

Beyond the foundational appeal, correlation intractability is desirable in real world applications. For example, consider the hash function used to build the block chain in the Bitcoin protocol [47]. Its main security property, needed to obtain proofs of work, can be stated as correlation intractability with respect to a specific set of relations, which come from protocol-defined constraints on the input and the output. (Specifically, the input needs to contain appropriate transaction information and the output needs to begin with the correct number of zeros.) It should be noted that we do not claim that our result directly applies to the Bitcoin protocol: in this paper we consider only relations that are negligibly sparse, while for Bitcoin and other proof-of-work applications, it is necessary to consider relations that are moderately sparse and to define a more precise analog of correlation intractability (in which the difficulty of finding $(x, f(x)) \in R$ is closely related to the density of R).

More generally, consider a multi-party game which uses the value returned by a random oracle, applied to the previous moves of players, as a substitute for public randomness. Correlation intractable functions can potentially be used to instantiate the random oracle in such a game without significant change in the properties of the game.

Alternative Approaches to Obtaining Hash Functions with Random Oracle Like Properties. Several alternative notions have been proposed in attempt to capture random-oracle-like properties of hash functions. These notions include entropy preservation [7], seed incompressibility [41], perfect one-wayness [23,28], non-malleability [16], correlation robustness [43], correlated input security [38], and universal computational extractors [8]. Their relations to correlation intractability will be discussed later in Sect. 1.4. Still, to the best of our knowledge, none of the known results regarding these notions shed light on the question of the existence of correlation intractable functions.

Obfuscated Pseudorandom Functions. A natural approach to constructing functions with random-oracle-like properties is to obfuscate pseudorandom functions (PRFs). Indeed, if the obfuscation was perfect, then the adversary would be unable to take advantage of the code any more than by merely having oracle access to the function. This would render the function random-oracle-like. Strong security definitions of obfuscation are formalized in the work of Hada [39] and Barak et al. [6], e.g. *Virtual-black-box* (VBB) Obfuscation. However, they also show that VBB obfuscation is impossible for many function families. In particular, Barak et al. [6] explicitly construct a PRF such that given any program (no matter how obfuscated) that computes the PRF, the adversary can find an input which evaluates to a fixed value. This certainly breaks correlation intractability.

We also know that *no* pseudorandom function family can be VBB obfuscated with respect to auxiliary inputs [12,37]. However, these results do not rule out the possibility that there exist pseudorandom functions whose obfuscated version is correlation intractable.

A reasonable next step may thus be to consider PRFs with additional properties, such as constrained or puncturable PRFs [18,19,44]. Indeed, as demonstrated by multiple works, starting with the ingenious work of Sahai and Waters [51], puncturable PRFs are an extremely powerful tool when combined with obfuscation of general programs. In particular, puncturable PRFs have been used together with iO to instantiate some random-oracle-like hash functions, including universal hardcore functions [10], universal computational extractors [22], and functions used for the full-domain-hash construction [42]. Furthermore, the constructions of [10,22] are simply obfuscating puncturable PRFs. It is thus natural to ask:

Are obfuscated puncturable PRFs correlation intractable?
If so, under what assumptions?

1.1   Our Results

We make progress towards answering the above questions. Specifically, we show that puncturable pseudorandom functions, obfuscated using an indistinguishability obfuscator, satisfy *bounded* correlation intractability. Here "bounded" means that there is a polynomial upper bound on the computational complexity of the sparse relations considered, and the complexity of the function family depends on that bound. (We stress that this bound applies only to the relation. The adversary runs in arbitrary polynomial time.) Bounded correlation intractability is indeed a qualitatively weaker property than full correlation intractability (see definitions in Sect. 3). Still, even in its bounded form, correlation intractability is a very strong notion that has not been constructed before. In particular, in many specific applications, such as Bitcoin, an upper bound on the complexity of the sparse relation is known.

Our result holds under the assumption of sub-exponentially secure general iO and puncturable PRFs, and also requires the existence of *Input-Hiding Obfuscation* (IHO) for evasive circuit families, which we now explain. Recall that a boolean circuit family is evasive if for any input, only negligibly many circuits in the family evaluate to a non-zero value. An obfuscator on evasive circuits achieves the "input-hiding" property, if it is infeasible for a polytime adversary to find, given an obfuscated version of a random function in the family, a preimage of non-zero output for that function. (Note that no subexponential hardness is assumed here.) Candidate IHOs for general evasive circuits are proposed by Bitansky et al. [13] and Badrinarayanan et al. [3] (see Sect. 1.3). Our main theorem is thus the following:

Theorem 1 (Bounded correlation intractable function ensembles, informal). *Assume existence of input-hiding obfuscation for evasive circuits, subexponentially secure indistinguishability obfuscation, and subexponentially secure puncturable pseudorandom functions. Then there is a $p(n)$-bounded correlation intractable function ensemble for any polynomial $p(n)$.*

Note that if we only consider relations R where for any x, there are only very few y values in the range satisfy $R(x, y)$, and allow the range to be larger than the domain, then correlation intractability becomes easy to obtain. Indeed, for such a R and a 1-universal function f there will with high probability not *exist* inputs x such that $R(x, f(x))$ holds. However, we argue that this case is of less interest. Rather, we are interested in general sparse relations where the "bad inputs" exist, but are hard to find. Our solution is able to handle the general case. For further discussions of the parameters and other special relations, we refer the readers to the end of Sect. 3.

1.2   Our Techniques

Our goal is to prove correlation intractability of certain function family. At a high level, our approach is to show, given a relation R, that a function f sampled randomly from the initial function family is indistinguishable from another function, f^R, that is constructed specifically so as to make it hard to find "bad inputs" with respect to the given relation R.

However, the definition of this function f^R, and moreover showing that it is indistinguishable from the original function f, needs to be done with care. In particular, the "naive" methodology of simply puncturing f at all the bad points, so as to obtain a function where no bad points for relation R exist, fails. We start by briefly explaining this failure.

Failure of the "Standard" Puncturing Methodology. Recall that a PRF is puncturable if for any key K and input value x it is possible to generate a key $K\{x\}$ that is "punctured" at x, such that $F_K(x)$ remains pseudorandom even given $K\{x\}$, and yet $K\{x\}$ allows evaluating F_K at all points other than x. To prove security of constructions that use puncturable PRFs obfuscated with iO, the "standard" methodology proceeds in two steps to get an indistinguishable game that an adversary cannot win (thus showing, by indistinguishability, that the adversary also fails in the original game). In the first step (whose indistinguishability is proven via iO), one typically punctures the key at the bad inputs that threaten the security of the scheme, and hardwires the output values for the punctured inputs. In the second step (whose indistinguishability is proven via the puncturable PRF), the output values at the punctured inputs are changed to ensure the adversary can't exploit them.

In our scenario, given a relation R, the "bad" inputs are those x values that satisfy $R(x, F_K(x)) = 1$, where K is randomly sampled after R is fixed. However, it is not clear how puncturing at these bad points helps here, since it is not clear how to argue that changing the output values so as to avoid R is indistinguishable. (In fact, it can be seen from our analysis that such change may well be distinguishable overall.)

Said otherwise, the "standard" puncturing technique is geared toward the case where the bad input values are fixed before the PRF key K is chosen, whereas for correlation intractability, the bad points are determined by K.

A "Counterintuitive" Puncturing Strategy. To get around this difficulty, we start from the following observation: for any sparse relation, the "bad" inputs x (i.e., those for which $R(x, F_K(x)) = 1$) are rare—in fact, they can be recognized by a circuit from an evasive circuit family. All we need to do in order to prove correlation intractability is show an indistinguishable function in which those rare inputs are hidden from the adversary. We do so by decomposing the PRF into two branches: one defined on the bad inputs, which form an evasive set, the other defined on the "innocent" inputs. Then we apply an input-hiding obfuscator to the bad branch. However, the input-hiding obfuscator cannot work in the presence of auxiliary information given by the innocent branch: the value of the function on the innocent inputs may permit the adversary to find the evasive inputs. We therefore puncture the key and change the function at every input that belongs to the innocent branch. To avoid increasing the circuit size beyond polynomial as we puncture at exponentially many points, we build an alternative function family $\mathcal{F}^R$ that is designed to avoid R. The details of the key-switching strategy form the technical heart of the proof.

The Proof in a Nutshell. To better illustrate the main idea, we present an overview of the proof. The analysis goes through 3 hybrids, as will be presented by the games between the adversary and the challenger. Hybrid 0 represents the original game. Hybrid 1, 2, and 3 are intermediate games that are indistinguishable by the adversary. Finally we will show that the adversary cannot break correlation intractability in hybrid 3, therefore concluding that the adversary also fails in hybrid 0, since hybrids 0 and 3 are indistinguishable.

We note that the circuits being iOed shall be padded to the same size, which is possible in our construction if an a priori bound on the size of the relation is given. Under this limitation, our techniques suffice to prove only a bounded version of correlation intractability. For the simplicity of the overview, we postpone the details of padding to the formal proof and now present the hybrids.

For any sparse relation R that is recognizable by some bounded polynomial sized circuit:

0. The challenger samples a key K of puncturable PRF $\mathcal{F}$ and obfuscates it:

$$h_k^0(\cdot) = \mathrm{iO}(F_K(\cdot))$$

The adversary wins if it outputs x such that $(x, h_k^0(x)) \in R$. This is the original game. The only thing that changes in subsequent games is the circuit obfuscated iO.

1. The challenger samples a key K of puncturable PRF $\mathcal{F}$, and embeds the relation R into the description of the function:

$$h_k^1(x) = \mathrm{iO}\left(\begin{matrix} \text{if } R(x, F_K(x)) = 1, \text{return } F_K(x) & ; \text{ the "bad" branch} \\ \text{else,} \hspace{3.2em} \text{return } F_K(x) & ; \text{ the "innocent" branch} \end{matrix}\right)$$

Note that h^1 has the same functionality as h^0, and therefore it is indistinguishable from the original function by iO. (Recall that an iO scheme iO guarantees that $\mathrm{iO}(C) \approx \mathrm{iO}(C')$ for any two circuits C, C' that have the same size and functionality.) This is a preparation step, which enables us to partition the function as described above.

2. Replace the key that is evaluated on the innocent branch with a freshly generated key K' for a different puncturable PRF $\mathcal{F}^R$ parameterized by R:

$$h_k^2(x) = \text{iO} \left(\begin{array}{ll} \text{if } R(x, F_K(x)) = 1, \text{ return } F_K(x) & ; \text{ the "bad" branch} \\ \text{else,} \hspace{3.2em} \text{return } F_{K'}^R(x) & ; \text{ the "innocent" branch} \end{array} \right)$$

where $\mathcal{F}^R$ is designed such that there is no x such that $(x, F_{K'}^R(x)) \in R$ with high probability. To generate a key K' for $\mathcal{F}^R$, we sample a set of independent puncturable PRF keys $K_1, ..., K_{T(n)}$ from $\mathcal{F}$. The function $F_{K'}^R$ executes in a "rejection sampling" fashion, such that for input x, it goes through the keys $K_1, ..., K_{T(n)}$ one by one, evaluates on the first key K_i for which $(x, F_{K_i}(x))$ is not in the relation. Setting T to be linear in l (in fact, even slightly sublinear) is enough to make sure that x not in the relation is found except with exponentially small probability. A similar construction was proposed in [49] (the results are included in [26]) to achieve "relation-specific" correlation intractable functions.

To prove the indistinguishability of h^1 and h^2, we show that both of them are subexponentially secure puncturable PRFs, based on the subexponential security assumption on the underlying puncturable PRF $\mathcal{F}$. We then use the following lemma (derived from the proof methodology in the work of Canetti et al. [27]) to show that, h^1 and h^2 are indistinguishable after being obfuscated by subexponentially secure iO.

Lemma 1 (Informal). *If h_1 and h_2 are subexponentially secure punctured PRFs and iO is subexponentially secure, then $\text{iO}(h_1)$ and $\text{iO}(h_2)$ are indistinguishable.*

3. Wrap the first "if-trigger", together with the underlying evasive function, by input-hiding obfuscation. The function h_k^3 is then generated as:

$$h_k^3(x) = \text{iO} \left(\begin{array}{ll} y \leftarrow \text{IHO} \left(\begin{array}{ll} \text{if } R(x, F_K(x)) = 1, \text{ return } F_K(x) \\ \text{else,} \hspace{3em} \text{return } \bot \end{array} \right) ; \text{"bad"} \\ \text{if } y = \bot, \ y \leftarrow F_{K'}^R(x) \hspace{6.5em} ; \text{"innocent"} \\ \text{return } y \end{array} \right)$$

h^3 is indistinguishable from h^2 because they are functionally equivalent and obfuscated by iO.

Finally, we note that finding the x values that trigger the non-zero values on the "input-hiding-box" is hard, given R and an "innocent" function $F_{K'}^R$ generated independently (even if not obfuscated). Since the adversary cannot distinguish whether she is given the original function h^0 or the function h^3, and finding an input on h^3 that satisfies the relation is hard, it should also be infeasible for the adversary to break correlation intractability on the original function.

1.3   More on Input-Hiding Obfuscation for Evasive Functions

Our result depends on the existence of input-hiding obfuscation (IHO) for evasive circuits. In this section we survey the state of the art regaring the existence of such obfuscation.

IHO for the class NC^1 can be obtained as follows. Start with a primitive called *strong indistinguishability obfuscation* (siO), which guarantees that if two circuits C_0 and C_1 are drawn from two distributions that are *concentrated* on the same function, then $\mathsf{siO}(C_0)$ is indistinguishable from $\mathsf{siO}(C_1)$. We show in Sect. 2.1 that siO for evasive circuit class $\mathcal{C}$ implies input-hiding obfuscation for $\mathcal{C}$. Thus, it is enough get siO for NC^1. Bitansky et al. [13] show that siO is equivalent to worst-case VGB obfuscation, and that siO/VGB for NC^1 circuits can be obtained under the assumptions that certain graded encoding schemes satisfy a strong form of semantic security [50]. Therefore, under the same assumption as made in [13] plus the assumption that puncturable PRFs exist in NC^1 [17], we obtain correlation intractable functions w.r.t. relations recognizable by NC^1 circuits.

IHO for larger circuit classes is currently is not known to follow from simpler primitives. Still, one can simply assume (similarly to [13]) that existing candidate obfuscators for $\mathsf{P/poly}$ are IHO. This assumption is not contradicted by known impossibility results: for evasive (as opposed to general [6]) circuits, there are no impossibility results known even for such a strong notion as average-case VBB [4].

Alternatively, IHO can be built in idealized models. In fact, both VBB obfuscation and IHO for $\mathsf{P/poly}$ were shown possible in a model with idealized graded encodings [2,5,20,54]. Furthermore, IHO for $\mathsf{P/poly}$ was shown possible by Badrinarayanan et al. [3] in a more relaxed idealized model, which avoids the devastating zeroing attack [29] on the candidate graded encodings [30,34].

Proposing simpler constructions of IHO without going through the full-fledged VGB, or basing IHO on simpler assumptions is an interesting open problem.

1.4   More on Related Work

Correlation Intractability and Constant-Round Public-Coin Zero-Knowledge Proofs. Hada and Tanaka show that the existence of correlation intractable hash functions (w.r.t. relations that are not necessarily efficient) implies 3 round public-coin auxiliary-input zero-knowledge proofs exist only for languages in BPP [40]. The key observation is based on the relation $R_{\notin\mathcal{L}}$ defined as

$$(x||\alpha,\beta) \in R_{\notin\mathcal{L}} \Leftrightarrow x \notin \mathcal{L} \wedge \exists\gamma, \Pr[\mathsf{Ver}(x,\alpha,\beta,\gamma) = \mathsf{Accept}] \geq \text{non.negl}.$$

where x is the instance, α,β,γ are the 3 messages in the protocol. The relation is sparse due to the statistical soundness of the underlying proof. Given the fact that the bounded simulator cannot break the correlation intractability, it should be able to decide the membership of the instance.

However, deciding the membership in the relation $R_{\notin\mathcal{L}}$ requires (at least) an auxiliary string γ in addition to the instance x, input α, and output β, whereas the construction of correlation intractable function proposed in this paper can only handle relations that takes exactly one input and one output. An alternative way of describing the relation is proposed by Halevi et al. [41] who define the relation with multiple invocations, and set γ as part of the inputs of the additional invocations. Our construction hasn't been proved to work for relations with multiple invocations.

Entropy-Preserving Hashing. The notion of "entropy-preserving hashing", formalized by Barak, Lindell and Vadhan [7] as being sufficient to achieve Fiat-Shamir heuristics for proofs [32], is closely related to correlation intractability. Roughly speaking, the definition requires that after the adversary is given the key and chooses the input, the output conditioned on the input has high entropy.

We show (in Appendix A) that entropy preservation and correlation intractability implies each other. However, the connections are shown w.r.t. relations that are not necessarily decidable by poly-size circuits. Therefore, our construction is not necessarily entropy-preserving. The existence of entropy-preserving hash functions remains open. In fact Bitansky et al. show that entropy preservation is impossible to prove from black-box reduction to falsifiable assumptions [14]. As a corollary, correlation intractability w.r.t. possibly inefficient relations is impossible to obtain from black-box reduction to falsifiable assumptions. We don't know if the same impossibility holds for CI w.r.t. efficiently recognizable relations.

Alternative Approaches to Instantiating Random Oracles. Several alternative definitions have been proposed in order to capture the random-oracle-like properties. These notions include perfect one-wayness [23,28], non-malleability [16], seed incompressibility (SI) [41], correlation robustness [43], correlated input security (CIH) [38], and universal computational extractors (UCE) [8]. These definitions are quite different from correlation intractability. In particular, SI, CIH and UCE model the security game in two stages, where the adversary in the first stage doesn't get full access to the description of the function, to avoid the impossibility results in [26]. It turns out that one can separate correlation intractability and each of these notions. An example is given in Appendix A that separates CIH/UCE and correlation intractability.

Separations, of course, do not show incompatibility: indeed, a construction may naturally satisfy many security definitions simultaneously. For example, essentially the same construction as in this paper (obfuscated puncturable PRFs) was shown to also satisfy a subclass of UCE by Brzuska and Mittelbach [22]. Further exploring constructions that satisfy multiple definitions simultaneously (and, in particular, gaining a better understanding of puncturable PRFs) is an interesting future direction.

Additional Related Work. A canonical construction of a PRF from a pseudorandom generator (PRG), now known as the GGM PRF, was given by Goldreich, Goldwasser and Micali [36]. Suppose we simply publish a GGM PRF seed in the clear to allow public evaluation, without any obfuscation. Is such a function correlation intractable? This questions was posed in the 1990s and answered negatively by Goldreich [35]. He constructed a specialized PRG, such that the GGM PRF built on this PRG is not correlation intractable. In fact one can find a preimage of $0^{m(n)}$ with non-negligible probability.

Correlation intractability is a natural criterion for designing efficient ciphers and hash functions. For example, it is used by Mandal et al. [46] to analyze the 6-round Feistel construction. In particular, they show that the 6-round Feistel construction is sequentially indifferentiable from a random invertible permutation, which implies that it is correlation intractable under an idealized assumption on the Feistel round function.

2   Preliminaries

Many experiments and probability statements in this paper contain randomized algorithms (such as obfuscators or adversaries) within them. The probability of success of an experiment is always taken over the random coins used by the relevant randomized algorithms; therefore, we do not mention these coins explicitly.

A function ensemble $\mathcal{F}$ has a key generation function $g : S \to K$; on seeds s of length $\sigma(n)$, g produces a key k of length $\kappa(n)$ for a function with input length $l(n)$ and output length $m(n)$:

$$\mathcal{F} = \{f_k : \{0,1\}^{l(n)} \to \{0,1\}^{m(n)}, k = g(s), s \in \{0,1\}^{\sigma(n)}\}_{n\in\mathbb{N}}$$

By default we denote $k \xleftarrow{\$} \mathcal{F}_n$ (sometimes abbreviated as k in the equations) as sampling a key k uniformly random from $\mathcal{F}_n$.

For any definition based on computational indistinguishability, we will say that the relevant security notion is *subexponential* if for every distinguisher there exists $\epsilon > 0$ such that the distinguisher's advantage is 2^{-n^ϵ}, where n is the security parameter.

2.1   Obfuscation

In this work we use indistinguishability obfuscation for all circuits, and input-hiding obfuscation for all evasive circuit collections. Both obfuscators considered in this paper perfectly preserve the functionality, and cause a polynomial blow-up on the size of the function description. To be precise, for the circuit family $\mathcal{F} = \{f : \{0,1\}^{l(n)} \to \{0,1\}^{m(n)}\}_{f\in\mathcal{F}_n}$, a probabilistic algorithm Obf is an obfuscator, if

1. The string Obf(f) describes a circuit that computes the same function as f;
2. There is a polynomial $B(\cdot)$ such that $|\mathsf{Obf}(f)| \le B(|f|)$.

The difference lies in the security properties: indistinguishability obfuscation guarantees that the obfuscation of any functionally equivalent circuits cannot be distinguished; whereas input-hiding obfuscation only applies on evasive circuits, and promises to hide all the inputs which lead to non-zero outputs.

Definition 1 (Indistinguishability Obfuscation [6]). *Obf is an indistinguishability Obfuscator (iO) for $\mathcal{F}$ if for any feasible adversary A, there is a negligible function negl($\cdot$) such that for all circuits f_0 and f_1 that have identical functionalities, and are of the same size, it holds that*

$$|\Pr[A(\text{iO}(f_0)) = 1] - \Pr[A(\text{iO}(f_1)) = 1]| \leq negl(n)$$

Definition 2 (Evasive circuit collections). *Let $\mathcal{F} = \{f_k : \{0,1\}^{l(n)} \rightarrow \{0,1\}^{m(n)}\}_{n \in \mathbb{N}}$ be a circuit collection, we say $\mathcal{F}_n$ is evasive if there is a negligible function negl($\cdot$) such that for all $x \in \{0,1\}^{l(n)}$:*

$$\Pr_k[f_k(x) \neq 0^{m(n)}] \leq negl(n)$$

Definition 3 (Input-hiding Obfuscation for evasive circuits [4]). *An obfuscator for a evasive circuit collection $\mathcal{F}$ is input-hiding (IHO) if for every p.p.t. adversary A there exist a negligible function negl($\cdot$) s.t. for every auxiliary input $z \in \{0,1\}^{\text{poly}(n)}$:*

$$\Pr_k[f_k(A(\text{IHO}(f_k), z)) \neq 0^{m(n)}] \leq negl(n)$$

The notion of IHO (unlike iO) is inherently average-case, i.e., the function f_k is random and independent of the auxiliary input z (see [4, Sect. 2] for a discussion of this issue). In particular, impossibility results, such as [21], for notions of obfuscation that allow a related auxiliary input, do not apply.

Remark 1. The original definitions of evasive circuit collections and the corresponding obfuscators proposed by Barak et al. [4] are stated for circuits with 1-bit output; whereas our definition of evasive circuit collections is for multi-bit output. For the case of input-hiding obfuscation, the existence of IHO for *all* evasive circuits with 1-bit output implies the existence of IHO for *all* evasive circuits with multi-bit output: for circuit $C(x)$ with m-bit output, we can obfuscate the circuit $C(x;i) = C(x)^{(i)}$ that returns the i-th output bit, and run $\text{IHO}(C(x;i))$ with $i \in [m]$. This transformation is mentioned by Bitansky et al. [13] for VGB obfuscation for all circuits. We note that the transformation also works for certain restricted circuit classes including NC^1.

Throughout this paper, we will assume the existence of IHO for all evasive circuits with 1-bit output, and use IHO for evasive circuits with possibly multi-bit output without loss of generality.

Input-Hiding Obfuscation from VGB Obfuscation. We introduce one of the known approaches to designing input-hiding obfuscation for evasive circuits. As a corollary of the result from [13], IHO is implied by Virtual-Grey-Box (VGB) obfuscation, or equivalently, strong indistinguishability obfuscation (siO).

Definition 4 (Concentrated/Evasive function distribution). *Let $\mathcal{F} = \{f_k : \{0,1\}^{l(n)} \rightarrow \{0,1\}\}_{n \in \mathbb{N}}$ be a function ensemble, $\tilde{\mathcal{F}}_n$ be a distribution on $\mathcal{F}_n$. Let $\text{maj}_{\tilde{\mathcal{F}}_n}(x) = \mathbb{E}_{f \leftarrow \tilde{\mathcal{F}}_n} f(x)$ be the common output on x for functions drawn from $\tilde{\mathcal{F}}_n$.*

1. $\tilde{\mathcal{F}}_n$ is concentrated if there is a negligible function $\mathsf{negl}(\cdot)$ that

$$\max_{x \in \{0,1\}^{l(n)}} \Pr_{f \leftarrow \tilde{\mathcal{F}}_n} [f(x) \neq \mathsf{maj}_{\tilde{\mathcal{F}}_n}(x)] \leq \mathsf{negl}(n)$$

2. (Rephrasing Definition 2 for 1-bit output) $\tilde{\mathcal{F}}_n$ is evasive if it is concentrated, and $\forall x \in \{0,1\}^{l(n)}$, $\mathsf{maj}_{\tilde{\mathcal{F}}_n}(x) = 0$.

Definition 5 (Strong indistinguishability Obfuscator [13]). *An obfuscator is a strong indistinguishability Obfuscator (siO) for $\mathcal{F}$ if for any two concentrated distribution ensembles $\tilde{\mathcal{F}}_n^0$, $\tilde{\mathcal{F}}_n^1$ on $\mathcal{F}_n$ s.t. $\mathsf{maj}_{\tilde{\mathcal{F}}_n^0} \equiv \mathsf{maj}_{\tilde{\mathcal{F}}_n^1}$, and for any p.p.t. adversary A, there is a negligible function $\mathsf{negl}(\cdot)$:*

$$\left| \Pr_{f_0 \leftarrow \tilde{\mathcal{F}}_n^0} [A(\mathsf{siO}(f_0)) = 1] - \Pr_{f_1 \leftarrow \tilde{\mathcal{F}}_n^1} [A(\mathsf{siO}(f_1)) = 1] \right| \leq \mathsf{negl}(n)$$

Definition 6 (Virtual-Grey-Box Obfuscation [11]). *Obf is a Virtual-Grey-Box (VGB) Obfuscator for $\mathcal{F}$ if for any feasible adversary A, there is a simulator S, and a negligible function $\mathsf{negl}(\cdot)$ such that for all $f \in \mathcal{F}$:*

$$| \Pr[A(\mathsf{Obf}(f)) = 1] - \Pr[S^f(1^{|f|}) = 1]| \leq \mathsf{negl}(|f|)$$

where the running time of S is computationally unbounded, but only sends polynomially many queries to f (such a simulator is usually called "semi-bounded").

Theorem 2 ([13]). *An obfuscator is siO for $\mathcal{F}$ iff it is worst-case VGB obfuscator for $\mathcal{F}$.*

Theorem 3 (SiO implies IHO for evasive functions). *Let $\mathcal{F} = \{f_k : \{0,1\}^{l(n)} \rightarrow \{0,1\}\}_{n \in \mathbb{N}}$ be an evasive function ensemble, Obf be a strong iO for $\mathcal{F}$, then Obf is an input-hiding obfuscator for $\mathcal{F}$.*

Proof. Let $\tilde{\mathcal{F}}_n^0$ be the uniform distribution on $\mathcal{F}$ and $\tilde{\mathcal{F}}_n^1$ be the one-element distribution consisting of the zero function. Then $\mathsf{maj}_{\tilde{\mathcal{F}}_n^0} \equiv \mathsf{maj}_{\tilde{\mathcal{F}}_n^1} \equiv 0$. Therefore

$$\Pr_{f_0 \leftarrow \tilde{\mathcal{F}}_n^0} [f_0(A(\mathsf{siO}(f_0), z)) = 1] \leq \Pr_{f_1 \leftarrow \tilde{\mathcal{F}}_n^1} [f_1(A(\mathsf{siO}(f_1), z)) = 1] + \mathsf{negl}(n) = \mathsf{negl}(n).$$

2.2 Puncturable Pseudorandom Functions

Definition 7 (Puncturable PRF [18,19,44,51]). *Let $l(n)$ and $m(n)$ be the input and output lengths. A family of puncturable pseudorandom functions $\mathcal{F} = \{F_K\}$ is given by a triple of efficient functions (Gen, Eval, Puncture), where $\mathsf{Gen}(1^n)$ generates the key K, such that F_K maps from $\{0,1\}^{l(n)}$ to $\{0,1\}^{m(n)}$; $\mathsf{Eval}(K,x)$ takes a key K, an input x, outputs $F_K(x)$; $\mathsf{Puncture}(K,x^*)$ takes a key and an input x^*, outputs a punctured key $K\{x^*\}$.*
  It satisfies the following conditions:

Functionality Preserved Over Unpunctured Points: *For all x^* and keys K, if $K\{x^*\} = \mathsf{Puncture}(K, x^*)$, then for all $x \neq x^*$, $\mathsf{Eval}(K, x) = \mathsf{Eval}(K\{x^*\}, x)$.*

Pseudorandom on the Punctured Points: *For every input x^*, the value of F on x^* is indistinguishable from random in the presence of the key punctured at x^*. That is, the following two distributions are indistinguishable for every x^*:*

$$(x^*, K\{x^*\}, F_K(x^*)) \text{ and } (x^*, K\{x^*\}, r^*),$$

where K is output by $\mathsf{Gen}(1^n)$, $K\{x^*\}$ is output by $\mathsf{Puncture}(K, x^*)$, and r^* is uniform in $\{0, 1\}^{m(n)}$.

Theorem 4 ([18, 19, 36, 44]). *If one-way function exists, then for all length parameters $l(n)$, $m(n)$, there is a puncturable PRF family that maps from $l(n)$ bits to $m(n)$ bits.*

3   Correlation Intractability

We recall the definitions of correlation intractability, initially proposed in [25, 26].

Definition 8 (Sparse relations[1]). *A binary relation R is sparse with respect to length parameters $l(n)$, $m(n)$, if there is a negligible function $\delta(\cdot)$ such that for every $x \in \{0, 1\}^{l(n)}$:*

$$\Pr_{y \in \{0,1\}^{m(n)}} [R(x, y) = 1] \leq \delta(n)$$

In some cases, we quantitatively describes the relations as $\delta(n)$-sparse, and even more precisely, $\delta_x(n)$-sparse when specifying the density on the input x.

Definition 9 (Correlation intractability). *A family of functions $\mathcal{H} = \{h_k : \{0, 1\}^{l(n)} \to \{0, 1\}^{m(n)}\}_{n \in \mathbb{N}}$ is correlation intractable (CI) if for all (nonuniform, p.p.t.) adversary A, for all sparse relations R, there's a negligible function $\mathsf{negl}(\cdot)$ such that:*

$$\Pr_{k \xleftarrow{\$} \mathcal{H}_n} [x \leftarrow A(k) : R(x, h_k(x)) = 1] < \mathsf{negl}(n)$$

---

[1] This is called $(l(n), m(n))$-restricted sparse relation in [26], as opposed to the "unrestricted" version where the input length is not prescribed. In this paper we remove the "restriction" in the term, since the case where the input length is unbounded is shown to be impossible (cf. Claim 3), and the "restricted" definition is indeed a natural and interesting setting. Also, in [26] and subsequently in [40, 41, 46], they also define "evasive" relations, which is equivalent to sparse for relations with 1-invocation, and with non-uniform adversaries. Throughout this paper, we only define and use "sparse" relations, since we focus on 1-invocation relations. The term "evasive" only serves the definition of "evasive circuit collections" [4] (cf. Definition 2) to avoid confusion.

In the definition above, the sparse relations may not be efficiently recognizable. A reasonable weakening on Definition 9 is to restrict the relations to be recognizable by poly-size circuits:

Definition 10 (CI-P/poly[2]). *The definition is same as Definition 9 except that we restrict the relations to be recognizable by poly-size circuits*

$$C : \{0,1\}^{l(n)+m(n)} \to \{0,1\}$$

s.t. $C(x,y) = 1$ *iff* $R(x,y) = 1$.

This definition can be further weakened by giving an a priori bound $p(n)$ on the size of the circuit that defines the relation, instead of allowing circuits of arbitrary polynomial size.

Definition 11 (Bounded correlation intractability). *Given a polynomial* $p(\cdot)$. *A family of functions* $\mathcal{H} = \{h_k : \{0,1\}^{l(n)} \to \{0,1\}^{m(n)}\}_{n \in \mathbb{N}}$ *is* $p(n)$-*bounded correlation intractable (bounded CI, or* $p(\cdot)$-*CI) if for all (non-uniform, p.p.t.) adversary A, for all sparse relations R that can be recorgnized by a circuit of size smaller or equal to* $p(n)$, *there's a negligible function* negl$(\cdot)$ *such that:*

$$\Pr_{k \xleftarrow{\$} \mathcal{H}_n} [x \leftarrow A(k) : R(x, h_k(x)) = 1] < \mathsf{negl}(n)$$

On the Length Parameters. It is shown in [26] that a function family cannot be correlation intractable when the key length $\kappa(n)$ of the function is short compared to the input length $l(n)$:

Claim ([26]). $\mathcal{H}_n$ is not correlation intractable w.r.t. poly-size relations when $\kappa(n) \leq l(n)$.

Proof. Consider the diagonalization relation $R = \{(k, h_k(k)) | k \in K\}$ (pad k with 0s to get length $l(n)$ if $\kappa(n) < l(n)$). The attacker outputs k (padded with 0s to length $l(n)$ as the x).

If $\kappa(n) > l(n)$, then there is no way to pad k to get x. However, some extensions of the impossibility result are still possible; we refer the readers to [26] for the details.

As opposed to the relation between input and key lengths, the relation between input and outputs lengths is not restricted. The only requirement is that the output length $m(n)$ shall be super-logarithmic, i.e. $m(n) \geq \omega(\log(n))$. Although CI is meant to model cryptographic hash functions (which have short outputs), the definition of CI is also meaningful for the functions whose output is longer than their input. In fact, our construction works for both cases.

We note that a function family that is correlation intractable against a more general class of sparse relations captures an essential feature of random oracles better. However, if one is interested in defending against certain restricted types

---

[2] This notion is called "weak correlation intractability" in [26].

of sparse relations, we may have simpler constructions based on standard cryptographic assumptions. For example, Ajtai's function [1], based on the hardness of approximating the Short Independent Vector Problem for Lattice in the worst case, suffices to prevent the adversary from finding the preimage of any fixed output. We also note that any 1-universal hash function family is correlation intractable, if one only considers very sparse relations — more specifically relations where, for any x, the number of y's that stand in the relation with x is at most a negligible fraction of the ratio between the size of the range and the size of the domain of functions in the family. Indeed, in this case with high probability a random function from the 1-universal hashing family has no input-output pairs in the relation. (We note that in this case the output is inherently longer than the input.)

4   Bounded Correlation Intractability from Obfuscating Puncturable PRF

In this section we give the construction of correlation intractable function ensembles with respect to all the sparse relations recognizable by circuits of size up to a given polynomial $p(\cdot)$.

Construction 5 (Bounded CI). *Let* $\mathcal{F} = \{F_K : \{0,1\}^{l(n)} \rightarrow \{0,1\}^{m(n)}\}_{n \in \mathbb{N}}$ *be a puncturable pseudorandom function. Let the function ensemble* $\mathcal{H} = \{h_k : \{0,1\}^{l(n)} \rightarrow \{0,1\}^{m(n)}\}_{n \in \mathbb{N}}$ *be constructed as*

$$h_k(\cdot) = \text{iO}(F_K(\cdot), \text{padding}(n))$$

where $K \xleftarrow{\$} \mathcal{F}_n$, *for some length of* **padding**.

Theorem 6 (Bounded CI). *Let* $p(n)$ *be a polynomial in the security parameter* n. *Assuming the existence of input-hiding obfuscation for all evasive circuits, sub-exponentially secure indistinguishability obfuscation for* P/poly, *and sub-exponentially secure puncturable PRF, there is an appropriate polynomial size of* **padding** *such that the family* $\mathcal{H}$ *is* $p(n)$-*bounded correlation intractable.*

The size of padding (which represents arbitrary gates that do not change the functionality of the circuit) will be discussed at the end of the proof (see Remark 2). In short, it depends on p and the blow-up due to input-hiding obfuscation. In the proof below, we drop the explicit mention of padding from the construction in order to simplify notation.

Proof of Theorem 6: The proof in this section follows the outline presented in Sect. 1.2. The proof goes through 3 hybrids. From the original game which captures the security definition of correlation intractability, we move to intermediate games 1, 2, and 3 that are indistinguishable by the adversary. Finally we will show that the adversary cannot win in game 3 except for negligible probability. We conclude that the adversary also fails in game 0, since the adversary cannot distinguish game 0 and game 3.

More specifically, fix an adversary and a $\delta(n)$-sparse relation R. Then:

Game 0: The Original Game. The adversary receives the key of the function h_k^0 constructed by the challenger:

$$h_k^0(\cdot) = \mathsf{iO}(F_K(\cdot)) \tag{0}$$

The adversary wins if he outputs an x such that $R(x, h_k^0(x)) = 1$. The winning condition is the same in each subsequent game; what changes is that h^0 is replaced by h^1, h^2, and h^3, which are computed as obfuscations of different circuits, each described in the corresponding game below.

Game 1: Embed the Relation into the Description Without Changing the Functionality. The challenger samples a puncturable key K, then generates h_k^1 which has the relation R embedded:

$$\bullet \quad h_k^1(x) = \mathsf{iO} \left(\begin{array}{ll} \text{if } R(x, F_K(x)) = 1, \text{ return } F_K(x) \\ \text{else,} \hspace{3.2cm} \text{return } F_K(x) \end{array} \right) \tag{1}$$

The hybrids h_k^0 and h_k^1 have identical functionality. Therefore, because both h_k^0 and h_k^1 are obfuscated by iO, they are indistinguishable for any p.p.t. adversary.

Game 2: Switch to a Function Where the "Innocent" Branch is Generated Independently from the "Bad" Branch and Avoids R. The challenger constructs a new function family $\mathcal{F}^R$ that always avoids R, as described below, and generates h_k^2 as:

$$h_k^2(x) = \mathsf{iO} \left(\begin{array}{ll} \text{if } R(x, F_K(x)) = 1, \text{ return } F_K(x) \\ \text{else,} \hspace{3.2cm} \text{return } F_{K'}^R(x) \end{array} \right) \tag{2}$$

where $F_K \xleftarrow{\$} \mathcal{F}_n$ and $F_{K'}^R \xleftarrow{\$} \mathcal{F}^R$. The function family $\mathcal{F}^R$ is constructed as follows:

Construction 7 ($\mathcal{F}^R$). *Let $\mathcal{F}^R = \{F_{K'}^R : \{0,1\}^{l(n)} \to \{0,1\}^{m(n)}\}_n$ be a function family, where each $F_{K'}^R$ is constructed as follows:*

$$F_{K'}^R(x) = \left(\begin{array}{l} K' = (K_1, K_2, \ldots, K_{T(n)}) \\ \hline \text{for } i = 1 \text{ to } T(n): \\ \quad \text{if } R(x, F_{K_i}(x)) = 0, \text{ return } F_{K_i}(x) \\ \text{return } \perp \end{array} \right) \tag{2.else}$$

where $T(n) = \frac{l(n)}{\log(n)}$. The functions $F_{K_1}, \ldots, F_{K_{T(n)}}$ are sampled independently from any puncturable PRF family $\mathcal{F}$.

The functionality of $F_{K'}^R$ is to output, given an input x, the pseudorandom value $F_{K_i}(x)$, where K_i is the first key among $K_1, \ldots, K_{T(n)}$ s.t. $R(x, F_{K_i}(x)) = 0$ (if no such K_i exists, output $\perp$). The iteration bound $T(n)$ is set large enough to make sure that $F_{K'}^R$ outputs $\perp$ with probability less than $2^{-l(n)} \cdot \mathsf{negl}(n)$ (we prove and use this fact in Lemma 2).

To prove that h_k^2 is indistinguishable from h_k^1, let g_k^2 be the same as h_k^2 but without the iO:

$$g_k^2(x) = \begin{cases} \text{if } R(x, F_K(x)) = 1, \text{ return } F_K(x) \\ \text{else,} \qquad\qquad\qquad \text{return } F_{K'}^R(x) \end{cases} \quad (2.\text{inner})$$

First, using subexponential security of F_K, we show in Lemma 2 that the g_k^2 is also a subexponentially secure puncturable PRF. Then, in Lemma 3 (whose proof methodology is derived from the work of Canetti et al. [27]), we show that any two subexponentially secure puncturable PRFs are indistinguishable after being obfuscated by subexponentially secure iO. This makes $h_2^k = \text{iO}(g_k^2)$ indistinguishable from $h_0^k = \text{iO}(F_K)$, and therefore also indistinguishable from h_1^k. (Note that technically h_1^k is not needed at all—we can move directly from h_0^k to h_2^k; but we believe that moving to h_1^k first clarifies presentation.)

Lemmas 2 and 3 below are based on the sub-exponential hardness of puncturability and iO, respectively. Let $\epsilon_{\text{Puncture}}$ be the adversary's advantage of winning the puncturability game of $\mathcal{F}$ and ϵ_{iO} be the advantage of distinguishing the iO of two identical functions. We need to set

$$\epsilon_{\text{Puncture}} = \epsilon_{\text{iO}} = 2^{-l(n)} \cdot \text{negl}(n)$$

This level of security can always be achieved from subexponential hardness by setting the security parameter λ for the puncturable PRF and for iO sufficiently high, but still polynomial in n: if the security of these two objects is $2^{-\lambda^\epsilon}$ for security parameter λ, then setting $\lambda = (2l(n))^{1/\epsilon}$ is sufficient.

Lemma 2. *Assume that $\mathcal{F}$ is a subexponentially secure puncturable PRF with the advantage of distinguishing being $\epsilon_{\text{Puncture}} = 2^{-l(n)} \cdot \text{negl}(n)$. Then the function g_k^2 (i.e., the function being obfuscated in h_k^2) is also a subexponentially secure puncturable PRF with the advantage of distinguishing at most $2^{-l(n)} \cdot \text{negl}(n)$.*

Proof. To puncture g_k^2 on input x^*, we puncture all the inner PRF keys $K, K_1, \ldots, K_{T(n)}$ on x^*, and construct the punctured function as follows:

$$k\{x^*\} = (R, K\{x^*\}, K'\{x^*\} = (K_1\{x^*\}, \ldots, K_{T(n)}\{x^*\}))$$
$$g_{k\{x^*\}}(x) = \begin{pmatrix} \text{if } R(x, F_{K\{x^*\}}(x)) = 1, \text{ return } F_{K\{x^*\}}(x) \\ \text{else,} \qquad\qquad\qquad\quad \text{return } F_{K'\{x^*\}}^R(x) \end{pmatrix} \quad (2.\text{p})$$

where $F_{K'\{x^*\}}^R$ is constructed as:

$$F_{K'\{x^*\}}^R(x) = \begin{pmatrix} K'\{x^*\} = (K_1\{x^*\}, \ldots, K_{T(n)}\{x^*\}) \\ \hline \text{for } i = 1 \text{ to } T(n) : \\ \quad \text{if } R(x, F_{K_i\{x^*\}}(x)) = 0, \text{ return } F_{K_i\{x^*\}}(x) \\ \text{return } \perp \end{pmatrix} \quad (2.\text{else.p})$$

By the puncturability of $\mathcal{F}$, the outputs of $F_{K\{x^*\}}$ and $F_{K_i\{x^*\}}$ on the punctured points are indistinguishable from random even given $k\{x^*\}$. More precisely,

$$(k\{x^*\}, F_K(x^*), F_{K_1}(x^*), \ldots, F_{K_{T(n)}}(x^*)) \approx (k\{x^*\}, U_0, U_1, \ldots, U_{T(n)})$$

(where $(U_0, U_1, ..., U_{T(n)}) \overset{\$}{\leftarrow} \{0,1\}^{(T(n)+1)\cdot m(n)}$). The advantage of any p.p.t. adversary to distinguish these two tuples is

$$(T(n)+1) \cdot \epsilon_{\mathsf{Puncture}} = (T(n)+1) \cdot 2^{-l(n)} \cdot \mathsf{negl}(n) = 2^{-l(n)} \cdot \mathsf{negl}(n)$$

Construct the distribution V_{x^*} by sampling random $U_0, \ldots, U_{T(n)}$ and computing

$$V_{x^*} = \begin{pmatrix} \text{if } R(x^*, U_0) = 1, \text{ return } U_0 \\ \text{else}: \text{ for } i = 1 \text{ to } T(n): \\ \qquad\qquad \text{if } R(x^*, U_i) = 0, \text{ return } U_i \\ \text{return } \bot \end{pmatrix}$$

From the indistinguishability of $F_K(x^*)$ and $F_{K_i}(x^*)$ from uniform, it follows that V_{x^*} is indistinguishable from $g_k^2(x^*)$:

$$\left(k\{x^*\}, g_k^2(x^*)\right) \approx \left(k\{x^*\}, V_{x^*}\right)$$

and the advantage of any p.p.t. adversary to distinguish these two pairs is $2^{-l(n)} \cdot \mathsf{negl}(n)$. To complete the proof, we will show that V_{x^*} is very close to uniform over $\{0,1\}^{m(n)}$: it differs from uniform by the probability that $V_{x^*} = \bot$. Indeed,

- For all $y \in \{0,1\}^{m(n)}$ such that $R(x^*, y) = 1$,

$$\Pr[V_{x^*} = y] = \Pr[U_0 = y] = 2^{-m(n)}$$

- $\Pr[V_{x^*} = \bot] = (1 - \delta_{x^*}(n))\delta_{x^*}(n)^{T(n)}$
- For all $y \in \{0,1\}^{m(n)}$ such that $R(x^*, y) = 0$ (note that there are $2^{m(n)}(1 - \delta_{x^*}(n))$ such values)

$$\begin{aligned}
&\Pr[V_{x^*} = y] \\
&= \Pr[V_{x^*} = y | R(x^*, V_{x^*}) \neq 1 \wedge V_{x^*} \neq \bot] \Pr[R(x^*, V_{x^*}) \neq 1 \wedge V_{x^*} \neq \bot] \\
&= \frac{1}{2^{m(n)}(1 - \delta_{x^*}(n))}(1 - \Pr[V_{x^*} \neq \bot \wedge R(x^*, V_{x^*}) = 1] - \Pr[V_{x^*} = \bot]) \\
&= \frac{1}{2^{m(n)}(1 - \delta_{x^*}(n))}(1 - \delta_{x^*}(n) - (1 - \delta_{x^*}(n))\delta_{x^*}(n)^{T(n)}) \\
&= 2^{-m(n)} \cdot \left(1 - \frac{(1 - \delta_{x^*}(n))\delta_{x^*}(n)^{T(n)}}{1 - \delta_{x^*}(n)}\right) = 2^{-m(n)} \cdot \left(1 - \delta_{x^*}(n)^{T(n)}\right)
\end{aligned}$$

Thus, the statistical difference between V_{x^*} and the uniform distribution on $\{0,1\}^{m(n)}$ (which is a bound on any distinguisher's advantage) is

$$\frac{1}{2} \sum_{y \in \{\bot\} \cup \{0,1\}^n} |\Pr[V_{x^*} = y] - \Pr[U = y]| \quad (U \text{ is uniform over } \{0,1\}^{m(n)})$$

$$= \frac{1}{2}\left((1 - \delta_{x^*}(n))\delta_{x^*}(n)^{T(n)}\right.$$

$$\left. + \sum_{y \text{ s.t. } R(x^*, y) = 0} \left(2^{-m(n)} - 2^{-m(n)} \cdot \left(1 - \delta_{x^*}(n)^{T(n)}\right)\right)\right)$$

$$= (1 - \delta_{x^*}(n))\delta_{x^*}(n)^{T(n)} \leq \delta_{x^*}(n)^{T(n)}$$

We thus obtain that V_{x^*} can be distinguished from uniform with advantage at most $\delta_{x^*}(n)^{T(n)} = 2^{-l(n)} \cdot \mathsf{negl}(n)$, because $T(n) = \frac{l(n)}{\log(n)}$ and $\delta_x(n)$ is a negligible function.

V_{x^*} is independent of $k\{x^*\}$. Therefore, the advantage of any adversary in distinguishing $(k\{x^*\}, V_{x^*})$ from $(k\{x^*\}, U)$ is $2^{-l(n)} \cdot \mathsf{negl}(n)$. And we already know the same is true for distinguishing $(k\{x^*\}, g_k^2(x^*))$ from $(k\{x^*\}, V_{x^*})$. Thus, even given $k\{x^*\}$, g_k^2 cannot be distinguished from uniform with advantage better than $2^{-l(n)} \cdot \mathsf{negl}(n)$, which concludes the proof.

Next we show that for arbitrary puncturable PRF families $\mathcal{F}_1, \mathcal{F}_2 : \{0,1\}^{l(n)} \to \{0,1\}^{m(n)}$ that are $2^{-l(n)} \cdot \mathsf{negl}(n)$-secure, the pseudorandom functions sampled independently from these families are indistinguishable after being obfuscated by $2^{-l(n)} \cdot \mathsf{negl}(n)$-secure indistinguishability obfuscation. The following lemma is derived from the "piO" proof methodology developed in the work of Canetti et al. [27].

Lemma 3. *Let $\mathcal{F}_1, \mathcal{F}_2 : \{0,1\}^{l(n)} \to \{0,1\}^{m(n)}$ be $2^{-l(n)} \cdot \mathsf{negl}(n)$-secure puncturable PRF families, iO be $\epsilon_{\mathsf{iO}} = 2^{-l(n)} \cdot \mathsf{negl}(n)$-secure indistinguishability obfuscation. Let $F_{K_1} \xleftarrow{\$} \mathcal{F}_1, F_{K_2} \xleftarrow{\$} \mathcal{F}_2$, then $\mathsf{iO}(F_{K_1})$ and $\mathsf{iO}(F_{K_2})$ are indistinguishable.*

Proof. We prove the indistinguishability via $2^{l(n)} + 1$ intermediate hybrids, one for each input. More precisely, for $z^* \in \{0, 1, ..., 2^{l(n)} - 1, 2^{l(n)}\}$, we construct f_{z^*} as

$$f_{z^*}(x) = \mathsf{iO}\left(\begin{array}{ll} \text{if } x = z^*, \text{ return } F_{K_1}(x) \\ \text{else,} \qquad \text{return } \left(\begin{array}{ll} \text{if } x > z^*, \text{ return } F_{K_1}(x) \\ \text{else,} \qquad \text{return } F_{K_2}(x) \end{array}\right) \end{array}\right)$$

Note that f_0 is functionally equivalent to F_{K_1}, therefore, they are $2^{-l(n)} \cdot \mathsf{negl}(n)$ indistinguishable after being obfuscated by iO. Likewise, $f_{2^{l(n)}}$ is functionally equivalent to F_{K_2}, hence being $2^{-l(n)} \cdot \mathsf{negl}(n)$-indistinguishable following iO.

Next we show that each intermediate pairs f_{z^*} and $f_{z^*+1}, z^* \in \{0, 1, ..., 2^{l(n)} - 1\}$, are $2^{-l(n)} \cdot \mathsf{negl}(n)$-indistinguishable. We introduce 3 more sub-hybrids:

$$f_{z^*,y^*}(x) = \mathsf{iO}\left(\begin{array}{ll} \text{if } x = z^*, \text{ return } y^* \\ \text{else,} \qquad \text{return } \left(\begin{array}{ll} \text{if } x > z^*, \text{ return } F_{K_1\{z^*\}}(x) \\ \text{else,} \qquad \text{return } F_{K_2\{z^*\}}(x) \end{array}\right) \end{array}\right)$$

where y^* equals to $F_{K_1}(z^*)$, $U \xleftarrow{\$} \{0,1\}^{m(n)}$, and $F_{K_2}(z^*)$ respectively.

Note that $f_{z^*, F_{K_1}(z^*)}$ is functionally equivalent to f_{z^*}; $f_{z^*, F_{K_2}(z^*)}$ is functionally equivalent to f_{z^*+1}. They are $2^{-l(n)} \cdot \mathsf{negl}(n)$-indistinguishable following iO. In between, $f_{z^*, F_{K_1}(z^*)}$ is indistinguishable from $f_{z^*,U}$ and $f_{z^*,U}$ is indistinguishable from $f_{z^*, F_{K_2}(z^*)}$, following the $2^{-l(n)} \cdot \mathsf{negl}(n)$-puncturability of K_1 and K_2.

To conclude, f_{z^*} and f_{z^*+1} are $4 \cdot 2^{-l(n)} \cdot \mathsf{negl}(n)$-indistinguishable following the $2^{-l(n)} \cdot \mathsf{negl}(n)$ security of $\mathcal{F}_1, \mathcal{F}_2$, and iO. Summing up all the $2^{l(n)} + 1$

intermediate hybrids, the total advantage of distinguishing $\mathsf{iO}(F_{K_1})$ and $\mathsf{iO}(F_{K_2})$ is negligible.

Combining Lemmas 2 and 3, h_k^1 is indistinguishable from h_k^2.

Game 3: Wrap the "Bad" Branch by Input-Hiding Obfuscation, Without Changing the Functionality. The challenger generates h_k^3 that is functionally equivalent to h_k^2 but is computed differently. The difference is that in game 3, the challenger first wraps the if statement together with the true branch with input-hiding obfuscation (the challenger also applies iO to the entire function, just like in the previous games, which ensures that h_k^2 is indistinguishable from h_k^3):

$$h_k^3(x) = \mathsf{iO}\left(\begin{array}{l} y \leftarrow \mathsf{IHO}\left(\begin{array}{ll} \text{if } R(x, F_K(x)) = 1, \text{return } F_K(x) \\ \text{else,} \qquad\qquad\qquad\quad \text{return } \perp \end{array}\right) \\ \text{if } y = \perp, \; y \leftarrow F_{K'}^R(x) \\ \text{return } y \end{array}\right) \tag{3}$$

Let $E_K^R(x)$ denote $\left(\begin{array}{ll} \text{if } R(x, F_K(x)) = 1, \text{return } F_K(x) \\ \text{else,} \qquad\qquad\qquad\quad \text{return } \perp \end{array}\right)$.

Proposition 1. $\mathcal{E}^R = \{E_K^R : \{0,1\}^{l(n)} \to \{0,1\}^{m(n)}\}_{n\in\mathbb{N}}$ *is an evasive circuit family.*

Proof. Assume, for contradiction, that there is an input $x' \in \{0,1\}^{l(n)}$ on which there are non-negligibly many keys that evaluate to a value other than $\perp$. We can then build a (non-uniform) adversary that distinguishes the PRF $F_K(x)$ from a truly random function with non-neglible advantage. The adversary simply queries input x' to the function and checks if the output y satisfies $R(x', y)$.

Note that h_k^2 and h_k^3 are functionally equivalent. Therefore, by indistinguishability obfuscation, the adversary cannot distinguish game 2 and game 3.

Finally, in Game 3: Suppose that there is a p.p.t. adversary A who gets h_k^3, finds an input x such that $R(x, h_k^3(x)) = 1$ with non-negligible probability $\eta(n)$, we build an adversary A' that breaks IHO for evasive circuit family $\mathcal{E}^R$: A' gets $\mathsf{IHO}(E_K^R(\cdot))$, samples $F_{K'}^R$ independently, and creates h_k^3 as described in construction (3), sends it to A. For adversary A, finding an input x to h^3 such that $R(x, h_k^3(x)) = 1$ is equivalent to finding such an input to $\mathsf{IHO}(E_K^R(\cdot))$ that evaluates to an non-bottom value, because $F_{K'}^R$ is independently generated and always avoids R ($F_{K'}^R$ outputs $\perp$ rather than hit R).

The advantage of adversary A' is thus the following:

$$\Pr_K[A'(\mathsf{IHO}(E_{R,K}(\cdot))) \to x : \; E_{R,K}(x) \neq \perp]$$

$$= \Pr_{K,K'}[A(\mathsf{IHO}(E_{R,K}(\cdot)), R, F_{K'}^R) \to x : \; E_{R,K}(x) \neq \perp]$$

$$\geq \Pr_k[A(h_k^3(\cdot)) \to x : R(x, h_k^3(x)) = 1] \geq \eta(n)$$

which forms the contradiction.

If a p.p.t. adversary could find x such $R(x, h_k^0(x)) = 1$, then she could distinguish h^0 from h^3 (because testing R is polynomial-time). Thus, we complete the proof that $\mathcal{H}$ is correlation intractable. □

Remark 2 (The size of padding). Let $\kappa_{\mathcal{F}}(n)$ be the key size of $\mathcal{F}_n$, $\kappa_{\mathcal{F}}^*(n)$ be the punctured key size of $\mathcal{F}_n$, $B(\cdot)$ be the maximum blow-up of the input-hiding obfuscation. The size of $F_{K'}^R$ is $T(n) \cdot (p(n) + 2 \cdot \kappa_{\mathcal{F}}(n))$. The maximum size of $\mathsf{IHO}(E_{R,K})$ is $B(p(n) + 2 \cdot \kappa_{\mathcal{F}}(n))$. The size of padding is bounded by

$$|\mathsf{padding}(n)|$$
$$\leq B(p(n) + 2 \cdot \kappa_{\mathcal{F}}(n)) + T(n) \cdot (p(n) + 2 \cdot \kappa_{\mathcal{F}}(n)) + (T(n) + 2) \cdot \kappa_{\mathcal{F}}^*(n)$$
$$= \mathsf{poly}(n)$$

As the analysis suggests, the key size of the function inherently exceeds the maximum size of R. The existence of correlation intractable functions with a prescribed description size that works for all poly-size relations (i.e. CI-P/poly) remains an open problem.

Acknowledgments. We are grateful to Nir Bitansky, Cheng Chen, Omer Paneth, and Oxana Poburinnaya for their enlightening discussions in the early stage of this work. We thank Ethan Heilman for discussions on the Bitcoin protocol, and Stefano Tessaro for pointing out "the piO proof methodology", which simplified the proof compared to earlier versions of this paper. We also thank the anonymous reviewers for their helpful comments.

This work is supported by US NSF grants 1012798, 1012910, 1218461, 1413920, 1422965, ISF grant 1523/14, and the Check Point Institute for Information Security. Part of the research by Y.C. was conducted while at Tel Aviv University funded by the Check Point Institute for Information Security.

Appendix

A   Correlation Intractability Versus Other Notions

We explore the relation between correlation intractability and other security definitions for cryptographic hash functions.

A.1   Relations with Entropy-Preserving Hashing

Recall the definition of Entropy Preserving (EP) from [7]:

Definition 12 (Entropy preservation). A family of hash function $\mathcal{H} = \{h_k : \{0,1\}^{l(n)} \to \{0,1\}^{m(n)}, k = g(s), s \in \{0,1\}^{\sigma(n)}\}_{n \in \mathbb{N}}$ ensures conditional entropy[3] greater than $\delta(n)$ if for all (non-uniform, p.p.t.) adversary A:

$$H(h_k(A(k))|A(k)) > \delta(n)$$

---

[3] The entropy of a random variable X is defined as $H(X) = \mathbb{E}_{x \xleftarrow{\$} X}[\log \frac{1}{\Pr[X=x]}]$. For jointly distributed random variables (X, Y), the conditional entropy of X given Y is defined to be $\mathbb{E}_{y \xleftarrow{\$} Y}[H(X|_{Y=y})]$, where $X|_{Y=y}$ denotes the conditional distribution of X given that $Y = y$.

Equivalently:

$$\mathbb{E}_{k,A}[H(h_k(X)|_{X=A(k)})] > \delta(n)$$

Notice that in order to get meaningful (i.e. non-zero) conditional entropy, the length of the key $\kappa(n)$ must be bigger then the length of the input $l(n)$, otherwise the adversary could always output the key (i.e. $A(k) \to k$) so that the conditional entropy will be zero (same to the diagonalization attack of correlation intractability [26]). In other words, we hope that there are multiple choices of keys that could lead the adversary to return the same input, and $h_k(x)$ on these candidate seeds and fixed input has different values.

[7] proposed 3 bounds for $\delta(n)$, each being interested on its own:

- (Best possible) $\delta(n) > m(n) - O(\log n)$. If achievable, would imply that constant-round public-coin auxiliary-input zero-knowledge proofs exist only for languages in BPP.
- (Somewhat) $\delta(n) > 1/\mathsf{poly}(n)$, also interesting. If achievable, would imply that 3-round public-coin auxiliary-input optimally sound zero-knowledge proofs exist only for languages in BPP.
- (Minimum/Weakest) $\delta(n) > 0$, still interesting. Even the existance of the weakest entropy-preserving hash functions implies that the parallel composition of some classic protocols (e.g. Blum's protocol [15]) is not auxiliary-input zero-knowledge.

An equivalent formalization of the minimum/weakest notion:

Conjecture 1 ([7]). There is a polynomial $p(\cdot)$ such that the following holds: For every *non-uniform deterministic* polynomial-time algorithm A and all sufficiently large n, there are circuits C_1, C_2 of size at most $p(n)$ such that $\alpha = A(C_1) = A(C_2)$ but $C_1(\alpha) \neq C_2(\alpha)$.

Note that even the construction of the weakest notion of entropy-preservation is unknown. In fact it is shown by Bitansky et al. to be impossible to obtain from black-box reduction to falsifiable assumptions [14].

Connections with CI. We show that correlation intractability (where the sparse relations are not necessarily efficiently recognizable) impies entropy preservation; and entropy preservation implies a weaker variant of correlation intractability in which if the adversary exists, it breaks correlation intractability with probability 1.

Theorem 8. *If a function family $\mathcal{H}$ is correlation intractable, then it is also entropy-preserving, i.e. for all p.p.t. adversary A:*

$$H(h_k(A(k))|A(k)) > m(n) - O(\log(n))$$

Proof. Assume by contradiction that $\mathcal{H}$ is not entropy-preserving, then there's an Adv A, such that

$$H(h_k(A(k))|A(k)) < m(n) - \omega(\log(n))$$

We define a relation by enumerating the keys, and query A on each key to get x, and the corresponding $y = h_k(x)$, then adding (x, y) into the relation. Formally, let R be:

$$R = \{(x, h_k(x)) \mid x = A(k),\ k = g(s),\ s \in \{0, 1\}^{\sigma(n)}\}$$

R is sparse since the adversary can always break entropy-preservation, which means the portion of the possible outputs conditioned on the adversary's choice of the input is negligible.

Notice that this relation is not likely to be efficiently recognizable, which means our construction of bounded correlation intractable functions is not necessarily entropy-preserving.

Definition 13 (Weak correlation intractability[4]). *A family of functions* $\mathcal{H} = \{h_k : \{0, 1\}^{l(n)} \to \{0, 1\}^{m(n)}\}_{n \in \mathbb{N}}$ *is weak correlation intractable (wCI) if for all (non-uniform, p.p.t.) adversary A, for all sparse relations R, there's a non-negligible function* non.negl$(\cdot)$ *such that:*

$$\Pr_{k \overset{\$}{\leftarrow} \mathcal{H}_n} [x \leftarrow A(k) : R(x, h_k(x)) = 1] < 1 - \textsf{non.negl}(n)$$

Theorem 9. *If a function family $\mathcal{H}$ guarantees the best possible entropy preservation, i.e. for all p.p.t. adversary A:*

$$H(h_k(A(k))|A(k)) > m(n) - O(\log(n))$$

then it is weakly correlation intractable.

Proof. If $\mathcal{H}$ is not weakly correlation intractable, which means there is a sparse relation R, an adversary A that:

$$\Pr_k[x \leftarrow A(k) : (x, h_k(x)) \in R] = 1$$

Since R is sparse, which means for all x, the possible y values form a negligibly small subset of the range. Therefore the conditional entropy is:

$$H(h_k(A(k))|A(k)) < m(n) - \omega(\log(n))$$

which forms a contradiction.

A.2   Separations Between Correlation Intractability and Other Notions

Several random-oracle-like notions are defined in an "indistinguishability" fashion. These definitions attempt to capture the intuition that, given only limited

---

[4] This notion is different from the "weak correlation intractability" in [26]. The "weak correlation intractability" in [26] is redefined as CI-P/poly in this article, cf. Definition 10.

access to or partial information from the function, it is hard for the adversary to distinguish whether the information is obtained from the hash function or a truly random function. The notions defined in this way include correlation robustness[5] [43], seed-incompressibility[6] [41], correlated input security (CIH) [38], and universal computational extractor (UCE) [8].

These notions are quite different from correlation intractability. In the next few paragraphs, we demonstrate the difference by showing that a simple version of correlated-input hash function (defined by [38], rephrased by [8] as a subclass of UCE and by [22] as q-CIH) is separated from correlation intractability. We emphasize that the purpose of showing separations is to demonstrate the properties of these definitions on their own, rather than showing incompatibility. In fact, there is evidence that these notions are compatible with correlation intractability: the same construction that we show to be correlation intractable (iO of puncturable PRFs with appropriate padding) was shown to satisfy a subclass of UCE by Brzuska and Mittelbach [22].

Definition 14 *(q-CIH [8, 22, 38]). Let q be a polynomial. For a hash function family $\mathcal{H} = \{h_k : \{0,1\}^{l(n)} \to \{0,1\}^{m(n)}\}_{n \in \mathbb{N}}$, consider the following game between the p.p.t. adversary $A = (A_1, A_2)$ and the challenger:*

1. *The challenger samples a hash function from the family $h_k \xleftarrow{\$} \mathcal{H}_n$.*
2. *A_1 samples $q(n)$ (possibly correlated) inputs x_i, $i \in [q(n)]$.*
3. *The challenger tosses a coin b. If $b = 0$, then let $y_i = h_k(x_i)$, $i \in [q(n)]$; if $b = 1$, then let $y_i \xleftarrow{\$} \{0,1\}^{m(n)}$, $i \in [q(n)]$.*
4. *A_2 gets h_k, y_i, $i \in [q(n)]$, outputs $b' \in \{0,1\}$, and wins if $b' = b$.*

$\mathcal{H}$ is called q-CIH if any p.p.t. adversary $A = (A_1, A_2)$ wins with probability less than $1/2 + \mathsf{negl}(n)$.

Theorem 10. *If q-CIH exists, then there is a function ensemble that is q-CIH but not correlation intractable. If correlation intractable function ensemble exists, then there is a function ensemble that is correlation intractable but not q-CIH.*

Proof. The constructions that demonstrate the separation of CIH and correlation intractability are very similar to the ones in ([8], Sect. 4.4) where they are used to separate UCE from other notions including collision resistance.

Consider the following constructions:

Construction 11. *Let $\mathcal{H} = \{h_k : \{0,1\}^{l(n)} \to \{0,1\}^{m(n)}\}_{n \in \mathbb{N}}$ be q-CIH. We construct $\mathcal{H}'$ by adding a uniformly random string $u \in \{0,1\}^{l(n)}$ as the prefix of the key, and define $h'_{k'} = h'_{u||k}$ as:*

$$h'_{u||k}(x) = \begin{cases} \text{if } x = u, & \text{return } 0^{m(n)}\,; \\ \text{else,} & \text{return } h_k(x)\,. \end{cases}$$

---

[5] Correlation robustness is defined for keyless hash functions, unlike the other notions in this article.

[6] [41] discussed both indistinguishability-style and correlation intractability-style definitions, when the adversary is only given partial information of the key (e.g. with an a priori bound on the length).

Lemma 4. $\mathcal{H}'$ *is q-CIH but not correlation intractable.*

Proof. To break correlation intractability, the adversary outputs u which is a preimage of $0^{m(n)}$.

To show $\mathcal{H}'$ is q-CIH, assume by contradiction that there is an adversary $A' = (A'_1, A'_2)$ that wins the q-CIH game with probability $1/2 + \eta(n)$ where η is non-negligible. We use the exact same adversary to break the q-CIH of $\mathcal{H}$: note that with probability $(1 - 2^{-l(n)})^{q(n)}$, A'_1 won't sample an input that equals to u, beyond which the view of A'_2 will be exactly the same for $\mathcal{H}$ and $\mathcal{H}'$. Therefore, A' wins the q-CIH game for $\mathcal{H}$ with probability no less than

$$(1 - 2^{-l(n)})^{q(n)} \cdot (1/2 + \eta(n)) \geq 1/2 + \eta(n) - q(n) \cdot 2^{-l(n)}$$

where $\eta(n) - q(n) \cdot 2^{-l(n)}$ is non-negligible, thus forming a contradiction.

Construction 12. *Let $\mathcal{H} = \{h_k : \{0,1\}^{l(n)} \rightarrow \{0,1\}^{m(n)-1}, k = g(s), s \in \{0,1\}^{\sigma(n)}\}_{n \in \mathbb{N}}$ be a correlation intractable function ensemble, we construct $\mathcal{H}'$ by padding an 1-bit at the end of the output:*

$$h'_{k'}(x) = h_k(x)||1$$

Lemma 5. $\mathcal{H}'$ *is correlation intractable but not q-CIH.*

Proof. To break q-CIH, the adversary outputs 0 if all the y_i, $i \in [q(n)]$ end with 1; otherwise, the adversary outputs 1.

To show $\mathcal{H}'$ is correlation intractable, assume by contradiction that there is an attacker A', a sparse relation $R' : \{0,1\}^{l(n)+m(n)} \rightarrow \{0,1\}$, and a non-negligible function $\eta(\cdot)$ such that

$$\Pr_{k'}[x \leftarrow A'(k') : R'(x, h'_{k'}(x)) = 1] > \eta(n)$$

Then we build an adversary A and a sparse relation $R : \{0,1\}^{l(n)+m(n)-1} \rightarrow \{0,1\}$ against $\mathcal{H}$: the relation R is defined as

$$R = \{(x, y) \mid R'(x, y||1) = 1, \ x \in \{0,1\}^{l(n)}, \ y \in \{0,1\}^{m(n)-1}\}$$

The density of R is at most twice as much as the density of R', so it is sparse. Given the key k, A constructs $h'_{k'}$ by padding a bit '1' at the end of the output of h_k, then sends $h'_{k'}$ to A' and outputs the answer of A'. The probability that A breaks R is exactly the probability that A' breaks R', which contradicts the assumption that $\mathcal{H}$ is correlation intractable.

Note that this transformation works regardless of the efficiency of checking the relation.

The proof completes by combining Constructions 11 and 12 and Lemmas 4 and 5.

References

1. Ajtai, M.: Generating hard instances of lattice problems (extended abstract). In: STOC, pp. 99–108 (1996)
2. Applebaum, B., Brakerski, Z.: Obfuscating circuits via composite-order graded encoding. In: Dodis and Nielsen [31], pp. 528–556
3. Badrinarayanan, S., Miles, E., Sahai, A., Zhandry, M.: Post-zeroizing obfuscation: the case of evasive circuits. Cryptology ePrint Archive, Report 2015/167 (2015). http://eprint.iacr.org/
4. Barak, B., Bitansky, N., Canetti, R., Kalai, Y.T., Paneth, O., Sahai, A.: Obfuscation for evasive functions. In: Lindell [45], pp. 26–51
5. Barak, B., Garg, S., Kalai, Y.T., Paneth, O., Sahai, A.: Protecting obfuscation against algebraic attacks. In: Nguyen and Oswald [48], pp. 221–238
6. Barak, B., Goldreich, O., Impagliazzo, R., Rudich, S., Sahai, A., Vadhan, S.P., Yang, K.: On the (im)possibility of obfuscating programs. J. ACM 59(2), 6 (2012)
7. Barak, B., Lindell, Y., Vadhan, S.P.: Lower bounds for non-black-box zero knowledge. J. Comput. Syst. Sci. 72(2), 321–391 (2006)
8. Bellare, M., Hoang, V.T., Keelveedhi, S.: Instantiating random oracles via UCEs. In: Canetti, R., Garay, J.A. (eds.) CRYPTO 2013, Part II. LNCS, vol. 8043, pp. 398–415. Springer, Heidelberg (2013)
9. Bellare, M., Rogaway, P.: Random oracles are practical: a paradigm for designing efficient protocols. In: ACM Conference on Computer and Communications Security, pp. 62–73 (1993)
10. Bellare, M., Stepanovs, I., Tessaro, S.: Poly-many hardcore bits for any one-way function and a framework for differing-inputs obfuscation. In: Sarkar and Iwata [52], pp. 102–121
11. Bitansky, N., Canetti, R.: On strong simulation and composable point obfuscation. In: Rabin, T. (ed.) CRYPTO 2010. LNCS, vol. 6223, pp. 520–537. Springer, Heidelberg (2010)
12. Bitansky, N., Canetti, R., Cohn, H., Goldwasser, S., Kalai, Y.T., Paneth, O., Rosen, A.: The impossibility of obfuscation with auxiliary input or a universal simulator. In: Garay and Gennaro [33], pp. 71–89
13. Bitansky, N., Canetti, R., Kalai, Y.T., Paneth, O.: On virtual grey box obfuscation for general circuits. In: Garay and Gennaro [33], pp. 108–125
14. Bitansky, N., Dachman-Soled, D., Garg, S., Jain, A., Kalai, Y.T., López-Alt, A., Wichs, D.: Why "Fiat-shamir for proofs" lacks a proof. In: Sahai, A. (ed.) TCC 2013. LNCS, vol. 7785, pp. 182–201. Springer, Heidelberg (2013)
15. Blum, M.: How to prove a theorem so no one else can claim it. In: The Proceedings of ICM 86, August 1986. Invited 45 Minute Address to the International Congress of Mathematicians (1986, to appear)
16. Boldyreva, A., Cash, D., Fischlin, M., Warinschi, B.: Foundations of non-malleable hash and one-way functions. In: Matsui, M. (ed.) ASIACRYPT 2009. LNCS, vol. 5912, pp. 524–541. Springer, Heidelberg (2009)
17. Boneh, D., Lewi, K., Montgomery, H.W., Raghunathan, A.: Key homomorphic PRFs and their applications. In: Canetti and Garay [24], pp. 410–428
18. Boneh, D., Waters, B.: Constrained pseudorandom functions and their applications. In: Sako, K., Sarkar, P. (eds.) ASIACRYPT 2013, Part II. LNCS, vol. 8270, pp. 280–300. Springer, Heidelberg (2013)
19. Boyle, E., Goldwasser, S., Ivan, I.: Functional signatures and pseudorandom functions. In: Krawczyk, H. (ed.) PKC 2014. LNCS, vol. 8383, pp. 501–519. Springer, Heidelberg (2014)

20. Brakerski, Z., Rothblum, G.N.: Virtual black-box obfuscation for all circuits via generic graded encoding. In: Lindell [45], pp. 1–25

21. Brzuska, C., Mittelbach, A.: Indistinguishability obfuscation versus multi-bit point obfuscation with auxiliary input. In: Sarkar and Iwata [52], pp. 142–161

22. Brzuska, C., Mittelbach, A.: Using indistinguishability obfuscation via UCEs. In: Sarkar and Iwata [52], pp. 122–141

23. Canetti, R.: Towards realizing random oracles: hash functions that hide all partial information. In: Kaliski Jr., B.S. (ed.) CRYPTO 1997. LNCS, vol. 1294, pp. 455–469. Springer, Heidelberg (1997)

24. Canetti, R., Garay, J.A. (eds.): CRYPTO 2013, Part I. LNCS, vol. 8042. Springer, Heidelberg (2013)

25. Canetti, R., Goldreich, O., Halevi, S.: The random oracle methodology, revisited (preliminary version). In: Vitter [53], pp. 209–218

26. Canetti, R., Goldreich, O., Halevi, S.: The random oracle methodology, revisited. J. ACM **51**(4), 557–594 (2004)

27. Canetti, R., Lin, H., Tessaro, S., Vaikuntanathan, V.: Obfuscation of probabilistic circuits and applications. In: Dodis and Nielsen [31], pp. 468–497

28. Canetti, R., Micciancio, D., Reingold, O.: Perfectly one-way probabilistic hash functions (preliminary version). In: Vitter [53], pp. 131–140

29. Cheon, J.H., Han, K., Lee, C., Ryu, H., Stehlé, D.: Cryptanalysis of the multilinear map over the integers. In: Oswald, E., Fischlin, M. (eds.) EUROCRYPT 2015. LNCS, vol. 9056, pp. 3–12. Springer, Heidelberg (2015)

30. Coron, J.-S., Lepoint, T., Tibouchi, M.: Practical multilinear maps over the integers. In: Canetti and Garay [24], pp. 476–493

31. Dodis, Y., Nielsen, J.B. (eds.): TCC 2015, Part II. LNCS, vol. 9015. Springer, Heidelberg (2015)

32. Fiat, A., Shamir, A.: How to prove yourself: practical solutions to identification and signature problems. In: Odlyzko, A.M. (ed.) CRYPTO 1986. LNCS, vol. 263, pp. 186–194. Springer, Heidelberg (1987)

33. Garay, J.A., Gennaro, R. (eds.): CRYPTO 2014, Part II. LNCS, vol. 8617. Springer, Heidelberg (2014)

34. Garg, S., Gentry, C., Halevi, S.: Candidate multilinear maps from ideal lattices. In: Johansson, T., Nguyen, P.Q. (eds.) EUROCRYPT 2013. LNCS, vol. 7881, pp. 1–17. Springer, Heidelberg (2013)

35. Goldreich, O.: The GGM construction does not yield correlation intractable function ensembles. IACR Cryptol. ePrint Arch. **2002**, 110 (2002)

36. Goldreich, O., Goldwasser, S., Micali, S.: How to construct random functions. J. ACM **33**(4), 792–807 (1986)

37. Goldwasser, S., Kalai, Y.T.: On the impossibility of obfuscation with auxiliary input. In: FOCS, pp. 553–562. IEEE Computer Society (2005)

38. Goyal, V., O'Neill, A., Rao, V.: Correlated-input secure hash functions. In: Ishai, Y. (ed.) TCC 2011. LNCS, vol. 6597, pp. 182–200. Springer, Heidelberg (2011)

39. Hada, S.: Zero-knowledge and code obfuscation. In: Okamoto, T. (ed.) ASIACRYPT 2000. LNCS, vol. 1976, pp. 443–457. Springer, Heidelberg (2000)

40. Hada, S., Tanaka, T.: Zero-knowledge and correlation intractability. IEICE Trans. **89-A**(10), 2894–2905 (2006)

41. Halevi, S., Myers, S., Rackoff, C.: On seed-incompressible functions. In: Canetti, R. (ed.) TCC 2008. LNCS, vol. 4948, pp. 19–36. Springer, Heidelberg (2008)

42. Hohenberger, S., Sahai, A., Waters, B.: Replacing a random oracle: full domain hash from indistinguishability obfuscation. In: Nguyen and Oswald [48], pp. 201–220

43. Ishai, Y., Kilian, J., Nissim, K., Petrank, E.: Extending oblivious transfers efficiently. In: Boneh, D. (ed.) CRYPTO 2003. LNCS, vol. 2729, pp. 145–161. Springer, Heidelberg (2003)

44. Kiayias, A., Papadopoulos, S., Triandopoulos, N., Zacharias, T.: Delegatable pseudorandom functions and applications. In: Sadeghi, A.-R., Gligor, V.D., Yung, M. (eds.) ACM Conference on Computer and Communications Security, pp. 669–684. ACM (2013)

45. Lindell, Y. (ed.): TCC 2014. LNCS, vol. 8349. Springer, Heidelberg (2014)

46. Mandal, A., Patarin, J., Seurin, Y.: On the public indifferentiability and correlation intractability of the 6-round feistel construction. In: Cramer, R. (ed.) TCC 2012. LNCS, vol. 7194, pp. 285–302. Springer, Heidelberg (2012)

47. Nakamoto, S.: Bitcoin: a peer-to-peer electronic cash system. Consulted 1(2012), 28 (2008)

48. Nguyen, P.Q., Oswald, E. (eds.): EUROCRYPT 2014. LNCS, vol. 8441. Springer, Heidelberg (2014)

49. Nissim, K.: Two results regarding correlation intractability. Manuscript (1999)

50. Pass, R., Seth, K., Telang, S.: Indistinguishability obfuscation from semantically-secure multilinear encodings. In: Garay, J.A., Gennaro, R. (eds.) CRYPTO 2014, Part I. LNCS, vol. 8616, pp. 500–517. Springer, Heidelberg (2014)

51. Sahai, A., Waters, B.: How to use indistinguishability obfuscation: deniable encryption, and more. In: Shmoys, D.B. (ed) STOC, pp. 475–484. ACM (2014)

52. Sarkar, P., Iwata, T. (eds.): ASIACRYPT 2014, Part II. LNCS, vol. 8874. Springer, Heidelberg (2014)

53. Vitter, J.S. (ed.) Proceedings of the Thirtieth Annual ACM Symposium on the Theory of Computing, Dallas, Texas, USA, 23–26 May 1998. ACM (1998)

54. Zimmerman, J.: How to obfuscate programs directly. In: Oswald, E., Fischlin, M. (eds.) EUROCRYPT 2015. LNCS, vol. 9057, pp. 439–467. Springer, Heidelberg (2015)

Reconfigurable Cryptography: A Flexible Approach to Long-Term Security

Julia Hesse[✉], Dennis Hofheinz, and Andy Rupp

Karlsruhe Institute of Technology, Karlsruhe, Germany
{julia.hesse,dennis.hofheinz,andy.rupp}@kit.edu

Abstract. We put forward the concept of a *reconfigurable* cryptosystem. Intuitively, a reconfigurable cryptosystem allows to increase the security of the system at runtime, by changing a single central parameter we call common reference string (CRS). In particular, e.g., a cryptanalytic advance does not necessarily entail a full update of a large public-key infrastructure; only the CRS needs to be updated. In this paper we focus on the reconfigurability of encryption and signature schemes, but we believe that this concept and the developed techniques can also be applied to other kind of cryptosystems.

Besides a security definition, we offer two reconfigurable encryption schemes, and one reconfigurable signature scheme. Our first reconfigurable encryption scheme uses indistinguishability obfuscation (however only in the CRS) to adaptively derive short-term keys from long-term keys. The security of long-term keys can be based on a one-way function, and the security of both the indistinguishability obfuscation and the actual encryption scheme can be increased on-the-fly, by changing the CRS. We stress that our scheme remains secure even if previous short-term secret keys are leaked.

Our second reconfigurable encryption scheme has a similar structure (and similar security properties), but relies on a pairing-friendly group instead of obfuscation. Its security is based on the recently introduced hierarchy of k-SCasc assumptions. Similar to the k-Linear assumption, it is known that k-SCasc implies $(k + 1)$-SCasc, and that this implication is proper in the generic group model. Our system allows to increase k on-the-fly, just by changing the CRS. In that sense, security can be increased without changing any long-term keys.

We also offer a reconfigurable signature scheme based on the same hierarchy of assumptions.

Keywords: Long-term security · Security definitions · Public-key cryptography

1 Introduction

Motivation. Public-key cryptography plays an essential role in security and privacy in wide networks such as the internet. Secure channels are usually

Supported by DFG grants HO 4534/2-2 and HO 4534/4-1.

E. Kushilevitz and T. Malkin (Eds.): TCC 2016-A, Part I, LNCS 9562, pp. 416–445, 2016.
DOI: 10.1007/978-3-662-49096-9_18

established using hybrid encryption, where the exchange of session keys for fast symmetric encryption algorithms relies on a public key infrastructure (PKI). These PKIs incorporate public keys from large groups of users. For instance, the PKI used by OpenPGP for encrypting and signing emails consists of roughly four million public keys. This PKI is continuously growing, especially so since the Snowden leaks multiplied the amount of newly registered public keys.

One drawback of large PKIs is that they are slow to react to security incidents. For instance, consider a PKI that predominantly stores 2048-bit RSA keys, and imagine a sudden cryptanalytic advance that renders 2048-bit RSA keys insecure. In order to change all keys to, say, 4096-bit keys, every user would have to generate new keypairs and register the new public key. Similarly, expensive key refresh processes are necessary in case, e.g., a widely deployed piece of encryption software turns out to leak secret keys, the assumed adversarial resources the system should protect from suddenly increase (e.g., from the computing resources of a small group of hackers to that of an intelligence agency), etc.

In this paper, we consider a scenario where key updates are triggered by a central authority for all users/devices participating in a PKI (and not by the individuals themselves), e.g., such as a large company maintaining a PKI for its employees who wants the employees to update their keys every year or when new recommendations on minimal key lengths are released. Other conceivable examples include operators of a PKI for wireless-sensor networks or for other IoT devices. We do not consider the problem of making individually initiated key updates more efficient.

Reconfigurable Cryptography. This paper introduces the concept of reconfigurable cryptography. In a nutshell, in a reconfigurable cryptographic scheme, there are long-term and short-term public and secret keys. Long-term public and secret keys are generated once for each user, and the long-term public key is publicized, e.g., in a PKI. Using a central and public piece of information (the common reference string or CRS), long-term keys allow to derive short-term keys, which are then used to perform the actual operation. If the short-term keys become insecure (or leak), only the central CRS (but not the long-term keys) needs to be updated (and certified). Note that the long-term secret keys are only needed for the process of deriving new short-term secret keys and not for the actual decryption process. Thus, they can be kept "offline" at a secure place.

We call the process of updating the CRS *reconfiguration.* An attack model for a reconfigurable cryptography scheme is given by an adversary who can ask for short-term secret keys derived from the PKI and any deprecated CRSs. After that, the adversary is challenged on a fresh short-term key pair. This models the fact that short-term key pairs should not reveal any information about the long-term secret keys of the PKI and thus, after their leakage, the whole system can be rescued by updating only the central CRS. Note that for most such schemes (except some trivial ones described below), the entity setting up the CRS needs to be trusted not to keep a trapdoor allowing to derive short-term secret keys for all users and security levels. In order to mitigate this risk however, a CRS could also be computed in a distributed fashion using MPC techniques.

Related Concepts and First Examples. An objection to our approach that might come to mind when first thinking about long-term secure encryption is the following: why do we not follow a much simpler approach like letting users exchange sufficiently long symmetric encryption keys once (which allow for fast encryption/decryption), using a (slow) public key scheme with comparable security? Unfortunately, it quickly turns out that there are multiple drawbacks with this approach: advanced encryption features known only for public-key encryption (e.g., homomorphic encryption) are excluded; each user needs to maintain a secure database containing the shared symmetric keys with his communication partners; the long-term secret key of the PKE scheme needs to be kept "online" in order to be able to decrypt symmetric keys from new communication partners, etc. Hence, we do not consider this a satisfying approach to long-term security.

A first attempt to create a scheme which better complies with our concept of reconfigurable encryption could be the following: simply define the long-term keys as a sequence of short-term keys. For instance, a long-term public key could consist of RSA keys of different lengths, say, of 2048, 4096, and 8192 bits. The CRS could be an index that selects which key (or, keylength) to use as a short-term key. If a keylength must be considered broken, simply take the next. This approach is perfectly viable, but does not scale well: only an a-priori fixed number (and type) of keys can be stored in a long-term key, and the size of such a long-term key grows (at least) linearly in the number of possible short-term keys.

A second attempt might be to use identity-based techniques: for instance, the long-term public and secret key of a user of a reconfigurable encryption scheme could be the master public and secret key of an identity-based encryption (IBE [6,17,21]) scheme. The CRS selects an IBE identity (used by all users), and the short-term secret key is the IBE user secret key for the identity specified by the CRS. Encryptions are always performed to the current identity (as specified by the CRS), such that the short-term secret key can be used to decrypt. In case (some of) the current short-term secret keys are revealed, simply change the identity specified in the CRS. This scheme scales much better to large numbers of reconfigurations than the trivial scheme above. Yet, security does not increase after a reconfiguration. (For instance, unlike in the trivial example above, there is no obvious way to increase keylengths through reconfiguration.)

Finally, we note that our security requirements are somewhat orthogonal to the ones found in forward security [4,9,10]. Namely, in a forward-secure scheme, we would achieve that revealing a current (short-term) secret key does not harm the security of *previous* instances of the scheme. In contrast, we would like to achieve that revealing the current (and previous) short-term secret keys does not harm the security of *future* instances of the scheme. Furthermore, we are interested in *increasing* the security of the scheme gradually, through reconfigurations (perhaps at the cost of decreased efficiency).

Our Contribution. We introduce the concept of reconfigurable cryptography. For this purpose, it is necessary to give a security definition for a cryptographic scheme defined in *two* security parameters, a long-term and a short-term security parameter. This definition needs to capture the property that security can be increased by varying the short-term security parameter. As it turns out, finding a reasonable definition which captures our notion and is satisfiable at the same time is highly non-trivial. Ultimately, here we present a non-uniform security definition based on an asymptotic version of *concrete security* introduced by Bellare et al. in [2,3]. The given definition is intuitive and leads to relatively simple proofs. Consequently, also our building blocks need to be secure against non-uniform adversaries (what can be assumed when building on non-uniform complexity assumptions). Alternatively, also a uniform security definition is conceivable which, however, would lead to more intricate proofs.

Besides a security definition, we offer three constructions: two reconfigurable public-key encryption schemes (one based on indistinguishability obfuscation [1,12,20], the other based on the family of SCasc assumptions [11] in pairing-friendly groups), and a reconfigurable signature scheme based on arbitrary families of matrix assumptions (also in pairing-friendly groups).

To get a taste of our solutions, we now sketch our schemes.

Some Notation. We call $\lambda \in \mathbb{N}$ the long-term security parameter, and $k \in \mathbb{N}$ the short-term security parameter. λ has to be fixed at setup time, and intuitively determines how hard it should be to retrieve the long-term secret key from the long-term public key. (As such, λ gives an an upper bound of the security of the whole system. In particular, we should be interested in systems in which breaking the long-term public key should be qualitatively harder than breaking short-term keys.) In contrast, k can (and should) increase with each reconfiguration. Intuitively, a larger value of k should make it harder to retrieve short-term keys.

Our Obfuscation-Based Reconfigurable Encryption Scheme. Our first scheme uses indistinguishability obfuscation [1,12,20], a pseudorandom generator PRG, and an arbitrary public-key encryption scheme PKE. As a long-term secret key, we use a value $x \in \{0,1\}^{\lambda}$; the long-term public key is PRG(x). A CRS consists of the obfuscation of an algorithm Gen, that inputs either a long-term public key PRG(x) or a long-term secret key x, and proceeds as follows:

- Gen(PRG(x)) generates a PKE public key, using random coins derived from PRG(x) for PKE key generation,
- Gen(x) generates a PKE secret key, using random coins derived from PRG(x).

Note that Gen(x) outputs the matching PKE secret key to the public key output by Gen(PRG(x)). Furthermore, we use $\lambda + k$ as a security parameter for the indistinguishability obfuscation, and k for the PKE key generation. (Hence, with larger k, the keys produced by Gen become more secure.)

We note that the long-term security of our scheme relies *only* on the security of PRG. Moreover, the short-term security (which relies on the obfuscator and PKE) can be increased (by increasing k and replacing the CRS) without changing the PKI. Furthermore, we show that releasing short-term secret keys for previous CRSs does not harm the security of the current instance of the scheme. (We remark that a similar setup and technique has been used by [7] for a different purpose, in the context of non-interactive key exchange.)

Reconfigurable Encryption in Pairing-Friendly Groups. We also present a reconfigurable encryption scheme in a cyclic group $G = \langle g \rangle$ that admits a symmetric pairing $e : G \times G \to G_T$ into some target group $G_T = \langle g_T \rangle$. Both groups are of prime order $p > 2^\lambda$. The long-term assumption is the hardness of computing discrete logarithms in G, while the short-term assumption is the k-SCasc assumption from [11] over G (with a pairing).[1] To explain our scheme in a bit more detail, we adopt the notation of [11] and write $[x] \in G$ (resp. $[x]_T \in G_T$) for the group element g^x (resp. g_T^x), and similarly for vectors $[\vec{u}]$ and matrices $[\mathbf{A}]$ of group elements.

A long-term secret key is an exponent x, and the corresponding long-term public key is $[x]$. A CRS for a certain value $k \in \mathbb{N}$ is a uniform vector $[\vec{y}] \in G^k$ of group elements. The induced short-term public key is a matrix $[\mathbf{A}_x] \in G^{(k+1) \times k}$ derived from $[x]$, and the short-term secret key is a vector $[\vec{r}] \in G^{k+1}$ satisfying $\vec{r}^\top \cdot \mathbf{A}_x = \vec{y}$. An encryption of a message $m \in G_T$ is of the form

$$c = ([\mathbf{A}_x \cdot \vec{s}], [\vec{y}^\top \cdot \vec{s}]_T \cdot m)$$

for a uniformly chosen $[\vec{s}] \in G^k$. Intuitively, the k-SCasc assumption states that $[\mathbf{A}_x \cdot \vec{s}]$ is computationally indistinguishable from a random vector of group elements. This enables a security proof very similar to that for (dual) Regev encryption [13,18] (see also [8]).

Hence, the long-term security of the above scheme is based on the discrete logarithm problem. Its short-term security relies on the k-SCasc assumption, where k can be adapted at runtime, without changing keys in the underlying PKI. Furthermore, we show that revealing previous short-term keys $[\vec{r}]$ does not harm the security of the current instance.[2]

We remark that [11] also present a less complex generalization of ElGamal to the k-SCasc assumption. Although they do not emphasize this property, their scheme allows to dynamically choose k at encryption time. However, their scheme does not in any obvious way allow to derive a short-term secret key that would be restricted to a given value of k. In other words, after, e.g., a key leakage, their scheme becomes insecure for all k, without the possibility of a reconfiguration.

---

[1] The k-SCasc assumption states that it is hard to distinguish vectors of group elements from a certain linear subspace from vectors of independently uniform group elements. Here, the parameter k determines the size of vectors, and – similar to the k-Linear assumption –, it is known that the k-SCasc assumption implies the $(k+1)$-SCasc assumption. In the generic group model, the $(k+1)$-SCasc assumption is also *strictly* weaker than the k-SCasc assumption [11]. Hence, increasing k leads to (at least generically) weaker assumptions.

[2] Currently, the best way to solve most problems in cyclic groups (such as k-SCasc or k-Linear instances) appears to be to compute discrete logarithms. In that sense, it would seem that the long-term and short-term security of our scheme are in a practical sense equivalent. Still, we believe that it is useful to offer solutions that give progressively stronger *provable* security guarantees (such as in our case with the k-SCasc assumption), if only to have fallback solutions in case of algorithmic advances, say, concerning the Decisional Diffie-Hellman problem.

Our Reconfigurable Signature Scheme. We also construct a reconfigurable signature scheme in pairing-friendly groups. Its long-term security is based on the Computational Diffie-Hellman (CDH) assumption, and its short-term security can be based on any matrix assumption (e.g., on k-SCasc). Of course, efficient (non-reconfigurable) signature schemes from the CDH assumption already exist (e.g., Waters' signature scheme [23]). Compared to such schemes, our scheme still offers reconfigurability in case, e.g., short-term secret keys are leaked.

Roadmap. We start with some preliminaries in Sect. 2, followed by the definition of a reconfigurable encryption scheme and the security experiment in Sect. 3. In Sect. 4, we give the details of our two constructions for reconfigurable encryption. Finally, we treat reconfigurable signature schemes in Sect. 5.

2   Preliminaries

Notation. Throughout the paper, $\lambda, k, \ell \in \mathbb{N}$ denote security parameters. For a finite set $\mathcal{S}$, we denote by $s \leftarrow \mathcal{S}$ the process of sampling s uniformly from $\mathcal{S}$. For a probabilistic algorithm $\mathcal{A}$, we denote with $\mathcal{R}_{\mathcal{A}}$ the space of $\mathcal{A}$'s random coins. $y \leftarrow \mathcal{A}(x; r)$ denotes the process of running $\mathcal{A}$ on input x and with uniform randomness $r \in \mathcal{R}_{\mathcal{A}}$, and assigning y the result. We write $y \leftarrow \mathcal{A}(x)$ for $y \leftarrow \mathcal{A}(x; r)$ with uniform r. If $\mathcal{A}$'s running time, denoted by $\mathbf{T}(\mathcal{A})$, is polynomial in λ, then $\mathcal{A}$ is called probabilistic polynomial-time (PPT). We call a function η negligible if for every polynomial p there exists λ_0 such that for all $\lambda \geq \lambda_0$ holds $|\eta(\lambda)| \leq \frac{1}{p(\lambda)}$.

Concrete Security. To formalize security of reconfigurable encryption schemes, we make use of the concept of concrete security as introduced in [2,3]. Here one considers an explicit function for the adversarial advantage in breaking an assumption, a primitive, a protocol, etc. which is parameterized in the adversarial resources. More precisely, as usual let $\mathsf{Adv}^{\mathsf{x}}_{\mathcal{P},\mathcal{A}}(\lambda)$ denote the advantage function of an adversary $\mathcal{A}$ in winning some security experiment $\mathsf{Exp}^{\mathsf{x}}_{\mathcal{P},\mathcal{A}}(\lambda)$ defined for some cryptographic object $\mathcal{P}$ (e.g., a PKE scheme, the DDH problem, etc.) in the security parameter λ. For an integer $t \in \mathbb{N}$, we define the *concrete advantage* $\mathsf{CAdv}^{\mathsf{x}}_{\mathcal{P}}(t, \lambda)$ of breaking $\mathcal{P}$ with runtime t by

$$\mathsf{CAdv}^{\mathsf{x}}_{\mathcal{P}}(t, \lambda) := \max_{\mathcal{A}}\{\mathsf{Adv}^{\mathsf{x}}_{\mathcal{P},\mathcal{A}}(\lambda)\}, \tag{1}$$

where the maximum is over all $\mathcal{A}$ with time complexity t. It is straightforward to extend this definition to cryptographic objects defined in two security parameters which we introduce in this paper. In the following, if we are given an advantage function $\mathsf{Adv}^{\mathsf{x}}_{\mathcal{P},\mathcal{A}}(\lambda)$ for a cryptographic primitive $\mathcal{P}$ that we consider, the definition of the concrete advantage can then be derived as in (1). Asymptotic security (against non-uniform adversaries and when only one security parameter is considered) then means that $\mathsf{CAdv}^{\mathsf{x}}_{\mathcal{P}}(t(\lambda), \lambda)$ is negligible for all polynomials t in λ. Hence, if we only give the usual security definition for a cryptographic building block in the following its concrete security is also defined implicitly as described above.

Implicit Representation. Let G be a cyclic group of order p generated by g. Then by $[a] := g^a$ we denote the *implicit representation* of $a \in \mathbb{Z}_p$ in G. To distinguish between implicit representations in two groups G and G_T, we use $[\cdot]$ and $[\cdot]_T$, respectively. The notation naturally extends to vectors and matrices of group elements.

Matrix-Vector Products. Sometimes, we will need to perform simple operations from linear algebra "in the exponent", aided by a pairing operation as necessary. Concretely, we will use the following operations: If a matrix $[\mathbf{A}] = [(a_{i,j})_{i,j}] \in G^{m \times n}$ is known "in the exponent", and a vector $\vec{u} = (u_i)_i \in \mathbb{Z}_p^n$ is known "in plain", then the product $[\mathbf{A} \cdot \vec{u}] \in G^m$ can be efficiently computed as $[(v_i)_i]$ for $[v_i] = \sum_{j=1}^n u_j \cdot [a_{i,j}]$. Similarly, inner products $[\vec{u}^\top \cdot \vec{v}]$ can be computed from $[\vec{u}]$ and $\vec{v}$ (or from $\vec{u}$ and $[\vec{v}]$). Finally, if only $[\mathbf{A}]$ and $[\vec{u}]$ are known (i.e., only "in the exponent"), still $[\mathbf{A} \cdot \vec{u}]_T$ can be computed in the target group, as $[(v_i)_i]_T$ for $[v_i]_T = \sum_{j=1}^n e([a_{i,j}], [u_j])$.

Symmetric Pairing-Friendly Group Generator. A symmetric pairing-friendly group generator is a probabilistic polynomial time algorithm $\mathcal{G}$ that takes as input a security parameter 1^λ and outputs a tuple $\mathbb{G} := (p, G, g, G_T, e)$ where

- G and G_T are cyclic groups of prime order p, $\lceil \log_2(p) \rceil = \lambda$ and $\langle g \rangle = G$
- $e : G \times G \longrightarrow G_T$ is an efficiently computable non-degenerate bilinear map.

The Matrix Diffie-Hellman Assumption ([11]). Let $k, q \in \mathbb{N}$ and $\mathcal{D}_k$ be an efficiently samplable matrix distribution over $\mathbb{Z}_q^{(k+1) \times k}$. The $\mathcal{D}_k$-Diffie-Hellman assumption ($\mathcal{D}_k$-MDDH) relative to a pairing-friendly group generator $\mathcal{G}$ states that for all PPT adversaries $\mathcal{A}$ it holds that

$$\mathsf{Adv}_{\mathcal{G},\mathcal{A}}^{\mathcal{D}_k\text{-MDDH}}(\lambda) := |\Pr[\mathcal{A}(\mathbb{G}, [\mathbf{A}, \mathbf{A}\vec{w}]) = 1] - \Pr[\mathcal{A}(\mathbb{G}, [\mathbf{A}, \vec{u}]) = 1]|$$

is negligible in λ, where the probability is over the random choices $\mathbf{A} \leftarrow \mathcal{D}_k, \vec{w} \leftarrow \mathbb{Z}_q^k$ and $\vec{u} \leftarrow \mathbb{Z}_q^{k+1}$, $\mathbb{G} := (p, G, g, G_T, e) \leftarrow \mathcal{G}$ and the random coins of $\mathcal{A}$. Examples of $\mathcal{D}_k$-MDDH assumptions are the k-Lin assumption and the compact symmetric k-cascade assumption (k-SCasc or $\mathcal{SC}_k$-MDDH). For the latter the matrix distribution $\mathcal{SC}_k$ samples matrices of the form

$$\mathbf{A}_x := \begin{pmatrix} x & 0 & \cdots & 0 & 0 \\ 1 & x & \cdots & 0 & 0 \\ 0 & 1 & \cdots & 0 & 0 \\ \vdots & & \ddots & & \vdots \\ 0 & 0 & \cdots & 1 & x \\ 0 & 0 & \cdots & 0 & 1 \end{pmatrix} \in \mathbb{Z}_n^{(k+1) \times k} \tag{2}$$

for uniformly random $x \leftarrow \mathbb{Z}_n$. In Sect. 4.2, we will consider a version of the SCasc assumption defined in two security parameters.

PKE Schemes. A public-key encryption (PKE) scheme PKE with message space $\mathcal{M}$ consists of three PPT algorithms Gen, Enc, Dec. Key generation Gen(1^ℓ) outputs a public key pk and a secret key sk. Encryption Enc(pk, m) takes pk and a message $m \in \mathcal{M}$, and outputs a ciphertext c. Decryption Dec(sk, c) takes

sk and a ciphertext c, and outputs a message m. For correctness, we want $\mathsf{Dec}(sk, c) = m$ for all $m \in \mathcal{M}$, all $(pk, sk) \leftarrow \mathsf{Gen}(1^\ell)$, and all $c \leftarrow \mathsf{Enc}(pk, m)$.

IND-CPA and IND-CCA Security. Let PKE be a PKE scheme as above. For an adversary $\mathcal{A}$, consider the following experiment: first, the experiment samples $(pk, sk) \leftarrow \mathsf{Gen}(1^k)$ and runs $\mathcal{A}$ on input pk. Once $\mathcal{A}$ outputs two messages m_0, m_1, the experiment flips a coin $b \leftarrow \{0, 1\}$ and runs $\mathcal{A}$ on input $c^* \leftarrow \mathsf{Enc}(pk, m_b)$. We say that $\mathcal{A}$ wins the experiment iff $b' = b$ for $\mathcal{A}$'s final output b'. We denote $\mathcal{A}$'s advantage with $\mathsf{Adv}^{\mathsf{ind\text{-}cpa}}_{\mathsf{PKE}, \mathcal{A}}(k) := |\Pr[\mathcal{A} \text{ wins}] - 1/2|$ and say that PKE is IND-CPA secure iff $\mathsf{Adv}^{\mathsf{ind\text{-}cpa}}_{\mathsf{PKE}, \mathcal{A}}(k)$ is negligible for all PPT $\mathcal{A}$. Similarly, write $\mathsf{Adv}^{\mathsf{ind\text{-}cca}}_{\mathsf{PKE}, \mathcal{A}}(k) := |\Pr[\mathcal{A} \text{ wins}] - 1/2|$ for $\mathcal{A}$'s winning probability when $\mathcal{A}$ additionally gets access to a decryption oracle $\mathsf{Dec}(sk, \cdot)$ at all times. (To avoid trivialities, $\mathcal{A}$ may not query Dec on c^*, though.)

PRGs. Informally, a pseudorandom generator (PRG) is a deterministic algorithm that maps a short random bit string (called seed) to a longer pseudorandom bitstring. More formally, let $p(\cdot)$ be a polynomial such that $p(\lambda) > \lambda$ for all $\lambda \in \mathbb{N}$ and let PRG be a deterministic polynomial-time algorithm which on input of a bit string in $\{0, 1\}^\lambda$ returns a bit string in $\{0, 1\}^{p(\lambda)}$ (also denoted by $\mathsf{PRG} : \{0, 1\}^\lambda \to \{0, 1\}^{p(\lambda)}$). The security of PRG is defined through

$$\mathsf{Adv}^{\mathsf{prg}}_{\mathsf{PRG}, D}(\lambda) := |\Pr[1 \leftarrow D(\mathsf{PRG}(x))] - \Pr[1 \leftarrow D(r)]|,$$

where D is a distinguisher, $x \leftarrow \{0, 1\}^\lambda$ and $r \leftarrow \{0, 1\}^{p(\lambda)}$.

Indistinguishability Obfuscation ($i\mathcal{O}$). For our construction in Sect. 4.1, we make use of indistinguishability obfuscators for polynomial-size circuits. Intuitively, such an algorithm is able to obfuscate two equivalent circuits in a way such that a PPT adversary who receives the two obfuscated circuits as input is not able to distinguish them. The following definition is taken from [12].

Definition 1 (Indistinguishability Obfuscator). *A uniform PPT machine $i\mathcal{O}$ is called an indistinguishability obfuscator for a circuit class $\{\mathcal{C}_\ell\}$ if the following conditions are satisfied:*

- *For all security parameters $\ell \in \mathbb{N}$, for all $C \in \mathcal{C}_\ell$, for all inputs x, we have that*

$$\Pr[C'(x) = C(x) : C' \leftarrow i\mathcal{O}(\ell, C)] = 1$$

- *For any (not necessarily uniform) PPT distinguisher D, there exists a negligible function α such that the following holds: For all security parameters $\ell \in \mathbb{N}$, for all pairs of circuits $C_0, C_1 \in \mathcal{C}_\ell$, we have that if $C_0(x) = C_1(x)$ for all inputs x, then*

$$\mathsf{Adv}^{\mathsf{io}}_{i\mathcal{O}, D}(\ell) := |\Pr[1 \leftarrow D(i\mathcal{O}(\ell, C_0))] - \Pr[1 \leftarrow D(i\mathcal{O}(\ell, C_1))]| \leq \alpha(\ell)$$

Note that an $i\mathcal{O}$ candidate for circuit classes $\{\mathcal{C}_\ell\}$, where the input size as well as the maximum circuit size are polynomials in ℓ has been proposed in [12].

Puncturable PRF. Informally speaking, a puncturable (or constrained) PRF $F_K : \{0, 1\}^{n(\ell)} \to \{0, 1\}^{p(\ell)}$ is a PRF for which it is possible to constrain the key

K (i.e., derive a new key K_S) in order to exclude a certain subset $S \subset \{0,1\}^{n(\ell)}$ of the domain of the PRF. (Note that this means that $F_{K_S}(x)$ is not defined for $x \in S$ and equal to $F_K(x)$ for $x \notin S$.) Given the punctured key K_S, an adversary may not be able to distinguish $F_K(x)$ from a random $y \in \{0,1\}^{p(\ell)}$ for $x \in S$. The following definition adapted from [19] formalizes this notion.

Definition 2. *A puncturable family of PRFs F is given by three PPT algorithms* Gen_F, $\mathsf{Puncture}_F$, *and* Eval_F, *and a pair of computable functions* $(n(\cdot), p(\cdot))$, *satisfying the following conditions:*

– *For every* $S \subset \{0,1\}^{n(\ell)}$, *for all* $x \in \{0,1\}^{n(\ell)}$ *where* $x \notin S$, *we have that:*

$$\Pr[\mathsf{Eval}_F(K,x) = \mathsf{Eval}_F(K_S,x) : K \leftarrow \mathsf{Gen}_F(1^\ell), K_S \leftarrow \mathsf{Puncture}_F(K,S)] = 1$$

– *For every PPT adversary* $\mathcal{A}$ *such that* $\mathcal{A}(1^\ell)$ *outputs a set* $S \subset \{0,1\}^{n(\ell)}$ *and a state* state, *consider an experiment where* $K \leftarrow \mathsf{Gen}_F(1^\ell)$ *and* $K_S = \mathsf{Puncture}_F(K,S)$. *Then the advantage* $\mathsf{Adv}_{F,\mathcal{A}}^{\mathsf{pprf}}(\ell)$ *of* $\mathcal{A}$ *defined by*

$$\left| \Pr[1 \leftarrow \mathcal{A}(\mathsf{state}, K_S, \mathsf{Eval}_F(K,S))] - \Pr[1 \leftarrow \mathcal{A}(\mathsf{state}, K_S, U_{p(\ell)\cdot|S|})] \right|$$

is negligible, where $\mathsf{Eval}_F(K,S)$ *denotes the concatenation of* $\mathsf{Eval}_F(K,x_i)$, $i = 1, ..., m$, *where* $S = \{x_1, ..., x_m\}$ *is the enumeration of the elements in* S *in lexicographic order, and* U_t *denotes the uniform distribution over* t *bits.*

To simplify notation, we write $F_K(x)$ instead of $\mathsf{Eval}_F(K,x)$. Note that if one-way functions exist, then there also exist a puncturable PRF family for any efficiently computable functions $n(\ell)$ and $p(\ell)$.

3   Definitions

The idea behind our concept of a reconfigurable public key cryptosystem is very simple: instead of directly feeding a PKI into the algorithms of the cryptosystem, we add some precomputation routines to derive a temporary short-term PKI. This PKI is then used by the cryptosystem. Instructions on how to derive and when to update the short-term PKI are given by a trusted entity. Our concept is quite modular and, thus, is applicable to other cryptosystems as well. In this section, we consider the case of reconfigurable encryption.

In Definition 3, we give a formal description of a reconfigurable public key encryption (RPKE) scheme. An RPKE scheme is a multi-user system which is setup (once) by some trusted entity generating public system parameters given a long-term security parameter 1^λ. Based on these public parameters, each user generates his long-term key pair. Moreover, the entity uses the public parameters to generate a common reference string defining a certain (short-term) security level k. Note that only this CRS is being updated when a new short-term security level for the system should be established. The current CRS is distributed to all users, who derive their short-term secret and public keys for the corresponding security level from their long-term secret and public keys and the CRS. Encryption and decryption of messages works as in a standard PKE using the short-term key pair of a user.

Definition 3. *A reconfigurable public-key encryption (RPKE) scheme* RPKE *consists of the following PPT algorithms:*

- Setup(1^λ) *receives a long-term security parameter* 1^λ *as input, and returns (global) long-term public parameters* $\mathcal{PP}$.
- MKGen$(\mathcal{PP})$ *takes the long-term public parameters* $\mathcal{PP}$ *as input and returns the long-term public and private key* (mpk, msk) *of a user.*
- CRSGen$(\mathcal{PP}, 1^k)$ *is given the long-term public parameters* $\mathcal{PP}$, *a short-term security parameter* 1^k, *and returns a (global) short-term common reference string CRS. We assume that the message space* $\mathcal{M}$ *is defined as part of CRS.*
- PKGen(CRS, mpk) *takes the CRS CRS as well as the long-term public key* mpk *of a user as input and returns a short-term public key* pk *for this user.*
- SKGen(CRS, msk) *takes the CRS CRS as well as the long-term secret key* msk *of a user as input and returns a short-term secret key* sk *for this user.*
- Enc(pk, m) *receives a user's short-term public key* pk *and a message* $m \in \mathcal{M}$ *as input and returns a ciphertext* c.
- Dec(sk, c) *receives a user's short-term secret key* sk *and a ciphertext* c *as input and returns* $m \in \mathcal{M} \cup \{\bot\}$.

We call RPKE *correct if for all values of* $\lambda, k \in \mathbb{N}$, $\mathcal{PP} \leftarrow$ Setup(1^λ), $(mpk, msk) \leftarrow$ MKGen$(\mathcal{PP})$, *CRS* $\leftarrow$ CRSGen$(\mathcal{PP}, 1^k)$, $m \in \mathcal{M}$, $pk \leftarrow$ PKGen(CRS, mpk), $sk \leftarrow$ SKGen(CRS, msk), *and all* $c \leftarrow$ Enc(pk, m), *it holds that* Dec$(sk, c) = m$.

Security. Our security experiment for RPKE systems given in Fig. 1 is inspired by the notion of IND-CCA (IND-CPA) security, extended to the more involved key generation phase of a reconfigurable encryption scheme. Note that we provide the adversary with a secret key oracle for deprecated short-term keys. The intuition behind our security definition is that we can split the advantage of an adversary into three parts. One part (called f_1 in Definition 4) reflects its advantage in attacking the subsystem of an RPKE that is only responsible for long-term security (λ). Another part (f_2) represents its advantage in attacking the subsystem that is only responsible for short-term security (k). The remaining part (f_3) stands for its advantage in attacking the subsystem that links the long-term with the short-term security subsystem (e.g., short-term key derivation). We demand that all these advantages are negligible in the corresponding security parameter, i.e., part one in λ, part two in k, and part three in both λ (where k is fixed) and in k (where λ is fixed).

Note that it is not reasonable to demand that the overall advantage is negligible in λ and in k. For instance, consider the advantage function CAdv $(t(\lambda, k), \lambda, k) \leq 2^{-\lambda} + 2^{-k} + 2^{-(\lambda+k)}$. Intuitively, we would like to call an RPKE exhibiting this bound secure. Unfortunately, it is neither negligible in λ nor in k.

Definition 4. *Let* RPKE *be an RPKE scheme according to Definition 3. Then we define the advantage of an adversary* $\mathcal{A}$ *as*

$$\mathsf{Adv}^{\mathsf{r\text{-}ind\text{-}cca}}_{\mathsf{RPKE},\mathcal{A}}(\lambda, k) := \left| \Pr[\mathsf{Exp}^{\mathsf{r\text{-}ind\text{-}cca}}_{\mathsf{RPKE},\mathcal{A}}(\lambda, k) = 1] - \frac{1}{2} \right|$$

where $\mathsf{Exp}_{\mathsf{RPKE},\mathcal{A}}^{\text{r-ind-cca}}$ is the experiment given in Fig. 1. The concrete advantage $\mathsf{CAdv}_{\mathsf{RPKE}}^{\text{r-ind-cca}}(t, \lambda, k)$ of adversaries against RPKE with time complexity t follows canonically (cf. Sect. 2).

An RPKE scheme RPKE is then called R-IND-CCA secure if for all polynomials $t(\lambda, k)$, there exist positive functions $f_1 : \mathbb{N}^2 \to \mathbb{R}_0^+$, $f_2 : \mathbb{N}^2 \to \mathbb{R}_0^+$, and $f_3 : \mathbb{N}^3 \to \mathbb{R}_0^+$ as well as polynomials $t_1(\lambda, k)$, $t_2(\lambda, k)$, and $t_3(\lambda, k)$ such that

$$\mathsf{CAdv}_{\mathsf{RPKE}}^{\text{r-ind-cca}}(t(\lambda, k), \lambda, k) \le f_1(t_1(\lambda, k), \lambda) + f_2(t_2(\lambda, k), k) + f_3(t_3(\lambda, k), \lambda, k)$$

for all λ, k, and the following conditions are satisfied for f_1, f_2, f_3:

- For all $k \in \mathbb{N}$ it holds that $f_1(t_1(\lambda, k), \lambda)$ is negligible in λ
- For all $\lambda \in \mathbb{N}$ it holds that $f_2(t_2(\lambda, k), k)$ is negligible in k
- For all $k \in \mathbb{N}$ it holds that $f_3(t_3(\lambda, k), \lambda, k)$ is negligible in λ
- For all $\lambda \in \mathbb{N}$ it holds that $f_3(t_3(\lambda, k), \lambda, k)$ is negligible in k

We define R-IND-CPA security analogously with respect to the modified experiment $\mathsf{Exp}_{\mathsf{RPKE},\mathcal{A}}^{\text{r-ind-cpa}}(\lambda, k)$, which is identical to $\mathsf{Exp}_{\mathsf{RPKE},\mathcal{A}}^{\text{r-ind-cca}}(\lambda, k)$ except that $\mathcal{A}$ has no access to an Dec-Oracle.

In Sect. 1 we already sketched an IBE-based RPKE scheme that would be secure in the sense of Definition 4. However, for this RPKE f_2 and f_3 can be set to be the zero function, meaning that the adversarial advantage cannot be decreased by increasing k. In this paper we are not interested in such schemes.

Of course, one can think of several reasonable modifications to the security definition given above. For instance, one may want to omit the "learn" stage in the experiment and instead give the algorithm access to the Break-Oracle during

---

Experiment $\mathsf{Exp}_{\mathsf{RPKE},\mathcal{A}}^{\text{r-ind-cca}}(\lambda, k)$

$\mathcal{PP} \leftarrow \mathsf{Setup}(1^\lambda)$

$(mpk, msk) \leftarrow \mathsf{MKGen}(\mathcal{PP})$

$state \leftarrow \mathcal{A}^{\mathsf{Break}(\mathcal{PP}, msk, \cdot)}(1^\lambda, 1^k, \mathcal{PP}, mpk, \text{"learn"})$

$CRS^* \leftarrow \mathsf{CRSGen}(\mathcal{PP}, 1^k)$

$sk^* \leftarrow \mathsf{SKGen}(CRS^*, msk)$

$pk^* \leftarrow \mathsf{PKGen}(CRS^*, mpk)$

$(m_0, m_1, state') \leftarrow \mathcal{A}^{\mathsf{Dec}(sk^*, \cdot)}(CRS^*, state, \text{"select"})$

$b \leftarrow \{0, 1\}$

$c^* \leftarrow \mathsf{Enc}(pk^*, m_b)$

$out_\mathcal{A} \leftarrow \mathcal{A}^{\mathsf{Dec}(sk^*, \cdot)}(c^*, state', \text{"guess"})$

Let $k_1, \ldots, k_\ell$ be the inputs sent to the Break-Oracle by $\mathcal{A}$. On input k_i, the Oracle returns $CRS_{k_i} \leftarrow \mathsf{CRSGen}(\mathcal{PP}, 1^{k_i})$ as well as $sk_{k_i} \leftarrow \mathsf{SKGen}(CRS_{k_i}, msk)$ to $\mathcal{A}$.

Return 1 if $k_i < k$ for all i, $|m_0| = |m_1|$, $out_\mathcal{A} = b$, and c^* has never been sent as input to the Dec-Oracle. Otherwise, return 0.

Fig. 1. R-IND-CCA experiment for reconfigurable PKE.

the "select" and "guess" stages. Fortunately, it turned out that most of these reasonable, slight modifications lead to a definition which is equivalent to the simple version we chose.

4 Constructions

4.1 Reconfigurable Encryption from Indistinguishability Obfuscation

We can build a R-IND-CCA (R-IND-CPA) secure reconfigurable encryption scheme from any IND-CCA (IND-CPA) secure PKE using indistinguishable obfuscation and puncturable PRFs. The basic idea is simple: We obfuscate a circuit which on input of the long-term public or secret key, where the public key is simply the output of a PRG on input of the secret key, calls the key generator of the PKE scheme using random coins derived by means of the PRF. It outputs the public key of the PKE scheme if the input to the circuit was the long-term public key and the secret key if the input was the long-term secret key.

Ingredients. Let $\mathsf{PKE_{CCA}} = (\mathsf{Gen_{CCA}}, \mathsf{Enc_{CCA}}, \mathsf{Dec_{CCA}})$ be an IND-CCA secure encryption scheme. Assuming the first component of the key pair that $\mathsf{Gen_{CCA}}(1^\ell)$ outputs is the public key, we define the PPT algorithms $\mathsf{PKGen_{CCA}}(1^\ell) := \#1(\mathsf{Gen_{CCA}}(1^\ell))$ and $\mathsf{SKGen_{CCA}}(1^k) := \#2(\mathsf{Gen_{CCA}}(1^k))$ which run $\mathsf{Gen_{CCA}}(1^\ell)$ and output only the public key or the secret key, respectively. By writing $\mathsf{Gen_{CCA}}(1^k; r)$, $\mathsf{PKGen_{CCA}}(1^k; r)$, $\mathsf{SKGen_{CCA}}(1^k; r)$ we will denote the act of fixing the randomness used by $\mathsf{Gen_{CCA}}$ for key generation to be r, a random bit string of sufficient length. For instance, r could be of polynomial length $p(k)$, where p equals the runtime complexity of $\mathsf{Gen_{CCA}}$. We allow r to be longer than needed and assume that any additional bits are simply ignored by $\mathsf{Gen_{CCA}}$.[3] Furthermore, let $\mathsf{PRG} : \{0,1\}^\lambda \to \{0,1\}^{2\lambda}$ be a pseudo-random generator and F be a family of puncturable PRFs mapping $n(\ell) := 2\ell$ bits to $p(\ell)$ bits. For $i \in \mathbb{N}$ we define $\mathsf{pad}_i : \{0,1\}^* \to \{0,1\}^*$ as the function which appends i zeroes to a given bit string. As a last ingredient, we need an indistinguishability obfuscator $i\mathcal{O}(\ell, C)$ for a class of circuits of size at most $q(\ell)$, where q is a suitable polynomial in $\ell = \lambda + k$ which upper bounds the size of the circuit $\mathsf{Gen}(a, b)$ to be defined as part of CRSGen.[4]

Our Scheme. With the ingredients described above our RPKE $\mathsf{RPKE}_{i\mathcal{O}}$ can be defined as in Fig. 2. Note that the security parameter ℓ used in the components for deriving short-term keys from long-term keys, i.e., F and $i\mathcal{O}$, is set to $\lambda + k$. That means, it increases (and the adversarial advantage becomes negligible) with both, the long-term and the short-term security parameter. (Alternative choices with the same effect like $\ell = \frac{\lambda}{2} + k$ are also possible.) Since the components which generate and use the short-term secrets depend on k, the security of the

---

[3] Equivalently, we could always apply a truncate function $\mathsf{trunc}_{p(k)} : \{0,1\}^* \to \{0,1\}^{p(k)}$ which outputs the $p(k)$ most significant bits of a given input.

[4] Note that actually q must be chosen as an upper bound of both Gen and Gen', where the latter is defined in the security proof.

Setup(1^λ)	MKGen($\mathcal{PP}$)
return $\mathcal{PP} := (1^\lambda)$	$x \leftarrow \{0,1\}^\lambda$
	set $mpk := \mathsf{PRG}(x)$, $msk := x$
	return (mpk, msk)

CRSGen($\mathcal{PP}, 1^k$)
$K \leftarrow \mathsf{Gen}_\mathsf{F}(1^{\lambda+k})$
$\mathsf{Gen}(a,b) := \begin{cases} \mathsf{PKGen}_{\mathsf{CCA}}(1^k; \mathsf{F}_K(\mathsf{pad}_{2k}(a))), & b = 0 \wedge a \in \{0,1\}^{2\lambda} \\ \mathsf{SKGen}_{\mathsf{CCA}}(1^k; \mathsf{F}_K(\mathsf{pad}_{2k}(\mathsf{PRG}(a)))), & b = 1 \wedge a \in \{0,1\}^\lambda \\ \bot, & \text{else} \end{cases}$
$\mathsf{iOGen} \leftarrow i\mathcal{O}(\lambda + k, \mathsf{Gen}(a,b))$
return $CRS := (\mathsf{iOGen})$

PKGen(CRS, mpk)	SKGen(CRS, msk)
parse $\mathsf{iOGen} := CRS$	parse $\mathsf{iOGen} := CRS$
return $\mathsf{iOGen}(mpk, 0)$	return $\mathsf{iOGen}(msk, 1)$

Enc(pk, m)	Dec(sk, c)
return $\mathsf{Enc}_{\mathsf{CCA}}(pk, m)$	return $\mathsf{Dec}_{\mathsf{CCA}}(sk, c)$

Fig. 2. Our $i\mathcal{O}$-based RPKE scheme $\mathsf{RPKE}_{i\mathcal{O}}$

scheme can be increased by raising k. As a somewhat disturbing side-effect of our choice of ℓ, the domain of F, which is used to map the long-term public key $mpk \in \{0,1\}^{2\lambda}$ to a pseudo-random string to be used by $\mathsf{Gen}_{\mathsf{CCA}}$, is actually too large. Hence, we have to embed 2λ-bit strings into $2(\lambda + k)$-bit strings by applying pad_{2k}.

Security. R-IND-CCA security of $\mathsf{RPKE}_{i\mathcal{O}}$ follows from the following Lemma.

Lemma 1. *Let a $t \in \mathbb{N}$ be given and let t' denote the maximal runtime of the experiment $\mathsf{Exp}^{\text{r-ind-cca}}_{\mathsf{RPKE}_{i\mathcal{O}},\cdot}(\lambda, k)$ involving arbitrary adversaries with runtime t. Then it holds that*

$$\mathsf{CAdv}^{\text{r-ind-cca}}_{\mathsf{RPKE}_{i\mathcal{O}}}(t, \lambda, k) \leq \tfrac{1}{2^\lambda} + \mathsf{CAdv}^{\text{prg}}_{\mathsf{PRG}}(s_1, \lambda) + \mathsf{CAdv}^{\text{ind-cca}}_{\mathsf{PKE}_{\mathsf{CCA}}}(s_2, k) \\ + \mathsf{CAdv}^{\text{pprf}}_{\mathsf{F}}(s_3, \lambda + k) + \mathsf{CAdv}^{\text{io}}_{i\mathcal{O}}(s_4, \lambda + k) \quad (3)$$

where $t' \approx s_1 \approx s_2 \approx s_3 \approx s_4$.

Proof. The following reduction will be in the non-uniform adversary setting. Consider an adversary $\mathcal{A}$ against $\mathsf{RPKE}_{i\mathcal{O}}$ for fixed security parameters λ and k who has an advantage denoted by $\mathsf{Adv}^{\text{r-ind-cca}}_{\mathsf{RPKE}_{i\mathcal{O}},\mathcal{A}}(\lambda, k)$. We will first show that $\mathcal{A}$ can be turned into adversaries

- $\mathcal{B}$ against PRG for fixed security parameter λ with advantage $\mathsf{Adv}^{\mathsf{prg}}_{\mathsf{PRG},\mathcal{B}_k}(\lambda)$,
- $\mathcal{C}$ against $i\mathcal{O}$ for fixed security parameter $\lambda+k$ with advantage $\mathsf{Adv}^{\mathsf{io}}_{i\mathcal{O},\mathcal{C}}(\lambda+k)$,
- $\mathcal{D}$ against F for fixed security parameter $\lambda+k$ with advantage $\mathsf{Adv}^{\mathsf{pprf}}_{\mathsf{F},\mathcal{D}}(\lambda+k)$,
- $\mathcal{E}$ against $\mathsf{PKE}_{\mathsf{CCA}}$ for fixed security parameter k with advantage $\mathsf{Adv}^{\mathsf{ind\text{-}cca}}_{\mathsf{PKE}_{\mathsf{CCA}},\mathcal{E}}(k)$

such that the advantage $\mathsf{Adv}^{\mathsf{r\text{-}ind\text{-}cca}}_{\mathsf{RPKE}_{i\mathcal{O}},\mathcal{A}}(\lambda,k)$ is upper bounded by

$$\frac{1}{2^{\lambda}} + \mathsf{Adv}^{\mathsf{prg}}_{\mathsf{PRG},\mathcal{B}}(\lambda) + \mathsf{Adv}^{\mathsf{ind\text{-}cca}}_{\mathsf{PKE}_{\mathsf{CCA}},\mathcal{E}}(k) + \mathsf{Adv}^{\mathsf{io}}_{i\mathcal{O},\mathcal{C}}(\lambda+k) + \mathsf{Adv}^{\mathsf{pprf}}_{\mathsf{F},\mathcal{D}}(\lambda+k). \quad (4)$$

After that, we will argue that from Eq. 4 the upper bound on the concrete advantage stated in Eq. 3 from our Lemma follows.

Throughout the reduction proof, let $\mathsf{Adv}^{\mathsf{Game}_i}_{\mathsf{RPKE}_{i\mathcal{O}},\mathcal{A}}(\lambda,k)$ denote the advantage of $\mathcal{A}$ in winning Game i for fixed λ, k.

Game 1 is the real experiment $\mathsf{Exp}^{\mathsf{r\text{-}ind\text{-}cca}}_{\mathsf{RPKE}_{i\mathcal{O}},\mathcal{A}}$. So we have

$$\mathsf{Adv}^{\mathsf{r\text{-}ind\text{-}cca}}_{\mathsf{RPKE}_{i\mathcal{O}},\mathcal{A}}(\lambda,k) = \mathsf{Adv}^{\mathsf{Game}_1}_{\mathsf{RPKE}_{i\mathcal{O}},\mathcal{A}}(\lambda,k). \quad (5)$$

Game 2 is identical to Game 1 except that a short-term secret key returned by the Break-Oracle on input $k' < k$ is computed by executing

$$\mathsf{SKGen}_{\mathsf{CCA}}(1^{k'}; F_K(\mathsf{pad}_{2k'}(mpk)))$$

instead of calling $\mathsf{SKGen}(CRS_{k'}, msk)$, where $CRS_{k'} \leftarrow \mathsf{CRSGen}(\mathcal{PP}, 1^{k'})$ and $K \leftarrow \mathsf{Gen}_\mathsf{F}(1^{\lambda+k'})$ is the corresponding PRF key generated in the scope of $\mathsf{CRSGen}(\mathcal{PP}, 1^{k'})$. Similarly, the challenge secret key sk^* is computed by the challenger by executing

$$\mathsf{SKGen}_{\mathsf{CCA}}(1^k; F_{K^*}(\mathsf{pad}_{2k}(mpk))),$$

and not by calling $\mathsf{SKGen}(CRS^*, msk)$, where CRS^* denotes the challenge CRS and K^* the PRF key used in the process of generating CRS^* by applying $\mathsf{CRSGen}(\mathcal{PP}, 1^k)$. In this way, msk is not used in the game anymore after $mpk = \mathsf{PRG}(msk)$ has been generated. Obviously, this change cannot be noticed by $\mathcal{A}$ and so we have

$$\mathsf{Adv}^{\mathsf{Game}_2}_{\mathsf{RPKE}_{i\mathcal{O}},\mathcal{A}}(\lambda,k) = \mathsf{Adv}^{\mathsf{Game}_1}_{\mathsf{RPKE}_{i\mathcal{O}},\mathcal{A}}(\lambda,k). \quad (6)$$

Game 3 is identical to Game 2 except that the challenge long-term public key is no longer computed as $mpk = \mathsf{PRG}(msk)$ but set to be a random bit string $r \leftarrow \{0,1\}^{2\lambda}$. Note with the change introduced in Game 2, we achieved that this game only depended on $\mathsf{PRG}(msk)$ but not on msk itself. Hence, we can immediately build an adversary $\mathcal{B}$ against PRG for (fixed) security parameter λ from a distinguisher between Games 1 and 2 with advantage

$$\mathsf{Adv}^{\mathsf{prg}}_{\mathsf{PRG},\mathcal{B}_k}(\lambda) = \left|\mathsf{Adv}^{\mathsf{Game}_2}_{\mathsf{RPKE}_{i\mathcal{O}},\mathcal{A}}(\lambda,k) - \mathsf{Adv}^{\mathsf{Game}_3}_{\mathsf{RPKE}_{i\mathcal{O}},\mathcal{A}}(\lambda,k)\right|. \quad (7)$$

As a consequence, in Game 3 nothing at all is leaked about msk.

The PRG adversary $\mathcal{B}$ receives a bit string $y \in \{0,1\}^{2\lambda}$ from the PRG challenger which is either random (as in Game 3) or the output of $\mathsf{PRG}(x)$ for

$x \leftarrow \{0,1\}^{\lambda}$ (as in Game 3). It computes $\mathcal{PP} \leftarrow \mathsf{Setup}(1^{\lambda})$, $CRS^* \leftarrow \mathsf{CRSGen}$ $(\mathcal{PP}, 1^k)$, and sets $mpk := y$. Note that due to the changes in Game 2 the key msk (which would be the unknown x) is not needed to execute the experiment. Then it runs $\mathcal{A}$ on input $\mathcal{PP}$ and mpk. A Break-Query is handled as described in Game 2, i.e., sk is computed by $\mathcal{B}$ based on mpk. The challenge short-term key sk^* is computed in the same way from mpk. In this way, $\mathcal{B}$ can perfectly simulate the Dec-Oracle when it runs $\mathcal{A}$ on input CRS^*. When receiving two messages m_0 and m_1 from the adversary, $\mathcal{B}$ returns $c^* \leftarrow \mathsf{Enc}(pk^*, m_b)$ for random b where pk^* has been generated as usual from mpk. Then $\mathcal{B}$ forwards the final output of $\mathcal{A}$. Clearly, if y was random $\mathcal{B}$ perfectly simulated Game 3, otherwise it simulated Game 2.

To introduce the changes in **Game 4**, let

$$K^*_{\{\mathsf{pad}_{2k}(mpk)\}} := \mathsf{Puncture}_F(K^*, \{\mathsf{pad}_{2k}(mpk)\})$$

denote the key K^* (used in the construction of CRS^*) where we punctured out mpk (represented as an element of $\{0,1\}^{2(\lambda+k)}$). This implies that $F_{K^*_{\{\mathsf{pad}_{2k}(mpk)\}}}(a)$ is no longer defined for $a = \mathsf{pad}_{2k}(mpk)$. Now, we set $r := F_{K^*}(\mathsf{pad}_{2k}(mpk))$ and the challenge short-term keys $pk^* := \mathsf{PKGen}_{\mathsf{CCA}}(1^k; r)$ and $sk^* := \mathsf{SKGen}_{\mathsf{CCA}}$ $(1^k; r)$. Those keys are computed in the experiment immediately after the generation of the long-term key pair (mpk, msk). This is equivalent to the way these keys have been computed in Game 2. Additionally, we replace $\mathsf{Gen}(a, b)$ in CRSGen for the challenge security level k by

$$\mathsf{Gen}'(a,b) := \begin{cases} pk^*, & b = 0 \wedge a = mpk \\ \mathsf{PKGen}_{\mathsf{CCA}}(1^k; F_{K^*_{\{\mathsf{pad}_{2k}(mpk)\}}}(\mathsf{pad}_{2k}(a))), & b = 0 \wedge a \in \{0,1\}^{2\lambda} \setminus \{mpk\} \\ \mathsf{SKGen}_{\mathsf{CCA}}(1^k; F_{K^*_{\{\mathsf{pad}_{2k}(mpk)\}}}(\mathsf{pad}_{2k}(\mathsf{PRG}(a)))), & b = 1 \wedge a \in \{0,1\}^{\lambda} \\ \bot, & \text{else} \end{cases}$$

CRS^* will now include the obfuscated circuit $\mathsf{iOGen}' \leftarrow i\mathcal{O}(\lambda + k, \mathsf{Gen}'(a, b))$.

We now verify that the circuits Gen and Gen' are indeed equivalent (most of the time). Obviously, it holds that $\mathsf{Gen}(a, 0) = \mathsf{Gen}'(a, 0)$ for all $a \in \{0,1\}^{2\lambda}$: the precomputed value pk^* results from running $\mathsf{PKGen}_{\mathsf{CCA}}(1^{\lambda+k}; F_{K^*}(\mathsf{pad}_{2k}(mpk)))$ which is exactly what $\mathsf{Gen}(mpk, 0)$ would run too. Moreover, we have

$$F_{K^*}(\mathsf{pad}_{2k}(a)) = F_{K^*_{\{\mathsf{pad}_{2k}(mpk)\}}}(\mathsf{pad}_{2k}(a))$$

for all $a \in \{0,1\}^{2\lambda} \setminus \{mpk\}$. Let us now consider $\mathsf{Gen}'(a, 1)$ for $a \in \{0,1\}^{\lambda}$. Remember that starting with Game 3, mpk is a random element from $\{0,1\}^{2\lambda}$. That means, with probability at least $1 - \frac{1}{2^{\lambda}}$ we have that mpk is not in the image of PRG and, thus,

$$F_{K^*}(\mathsf{pad}_{2k}(\mathsf{PRG}(a))) = F_{K^*_{\{\mathsf{pad}_{2k}(mpk)\}}}(\mathsf{pad}_{2k}(\mathsf{PRG}(a)))$$

for all $a \in \{0,1\}^{\lambda}$. Hence, with probability $1 - \frac{1}{2^{\lambda}}$ the circuits Gen and Gen' are equivalent for all inputs. So a distinguisher between Game 4 and Game 3 can be turned into an adversary $\mathcal{C}$ against $i\mathcal{O}$ for security parameter $\lambda + k$ with advantage

$$\mathsf{Adv}^{\mathsf{io}}_{i\mathcal{O},\mathcal{C}}(\lambda + k) \geq \left| \mathsf{Adv}^{\mathsf{Game}_3}_{\mathsf{RPKE}_{i\mathcal{O}},\mathcal{A}}(\lambda, k) - \mathsf{Adv}^{\mathsf{Game}_4}_{\mathsf{RPKE}_{i\mathcal{O}},\mathcal{A}}(\lambda, k) \right| - \frac{1}{2^\lambda}. \qquad (8)$$

$\mathcal{C}$ computes $\mathcal{PP} \leftarrow \mathsf{Setup}(1^\lambda)$ and $mpk \leftarrow \{0,1\}^{2\lambda}$. Then it chooses a PPRF $\mathsf{F} : \{0,1\}^{2(\lambda+k)} \rightarrow \{0,1\}^{p(\lambda+k)}$ and a corresponding key $K^* \leftarrow \mathsf{Gen}_\mathsf{F}(1^{\lambda+k})$. Using these ingredients it sets up circuits $C_0 := \mathsf{Gen}$ according to the definition from Game 3 and $C_1 := \mathsf{Gen}'$ according to the definition from Game 4. As explained above, with probability $1 - \frac{1}{2^\lambda}$ these circuits are equivalent for all inputs. CRS^* is then set as the output of the $i\mathcal{O}$ challenger for security parameter $\lambda + k$ on input of the circuits C_0 and C_1.[5] sk^* and pk^* can either be computed as defined in Game 3 or as in Game 4. As both ways are equivalent, it does not matter for the reduction. The remaining parts of Game 3 and Game 4 are identical. In particular, Break-Queries of $\mathcal{A}$ can be handled without knowing msk. The output bit of the third and final execution of $\mathcal{A}$ is simply forwarded by $\mathcal{C}$ to the $i\mathcal{O}$ challenger.

Game 5 is identical to Game 4 except that the value r is chosen as a truly random string from $\{0,1\}^{p(\lambda+k)}$ and not set to $\mathsf{F}_{K^*}(\mathsf{pad}_{2k}(mpk))$. As besides r, Game 4 did not depend on K^* anymore but only on $K^*_{\{\mathsf{pad}_{2k}(mpk)\}}$, a distinguisher between Game 4 and Game 5 can directly be turned into an adversary $\mathcal{D}$ against the pseudorandomness of the puncturable PRF family for security parameter $\lambda + k$. Thus, we have

$$\mathsf{Adv}^{\mathsf{pprf}}_{\mathsf{F},\mathcal{D}}(\lambda + k) = \left| \mathsf{Adv}^{\mathsf{Game}_4}_{\mathsf{RPKE}_{i\mathcal{O}},\mathcal{A}}(\lambda, k) - \mathsf{Adv}^{\mathsf{Game}_5}_{\mathsf{RPKE}_{i\mathcal{O}},\mathcal{A}}(\lambda, k) \right|. \qquad (9)$$

$\mathcal{D}$ computes $\mathcal{PP} \leftarrow \mathsf{Setup}(1^\lambda)$, $mpk \leftarrow \{0,1\}^{2\lambda}$, and chooses a PPRF $\mathsf{F} : \{0,1\}^{2(\lambda+k)} \rightarrow \{0,1\}^{p(\lambda+k)}$. Then it sends $\mathsf{pad}_{2k}(mpk)$ to its challenger who chooses a key $K^* \leftarrow \mathsf{Gen}_\mathsf{F}(1^{\lambda+k})$ and computes the punctured key $K^*_{\{\mathsf{pad}_{2k}(mpk)\}}$. Furthermore, the challenger sets $r_0 := \mathsf{F}_{K^*}(\mathsf{pad}_{2k}(mpk))$ and $r_1 \leftarrow \{0,1\}^{p(\lambda+k)}$. It chooses $b \leftarrow \{0,1\}$ and sends r_b along with $K^*_{\{\mathsf{pad}_{2k}(mpk)\}}$ to $\mathcal{D}$. $\mathcal{D}$ sets $r := t_b$, $pk^* := \mathsf{PKGen}_{\mathsf{CCA}}(1^k; r)$ and $sk^* := \mathsf{SKGen}_{\mathsf{CCA}}(1^k; r)$. Using the given punctured key $K^*_{\{\mathsf{pad}_{2k}(mpk)\}}$, $\mathcal{D}$ can also generate CRS^* as described in Game 4. The rest of the reduction is straightforward. The output bit of the final execution of $\mathcal{A}$ is simply forwarded by $\mathcal{C}$ to its challenger. If $b = 0$, $\mathcal{D}$ perfectly simulates Game 4, otherwise it simulates Game 5.

Now, observe that in Game 5, the keys pk^* and sk^* are generated using $\mathsf{Gen}_{\mathsf{CCA}}$ with a uniformly chosen random string r on its random tape. In particular, pk^* and sk^* are completely independent of the choice of mpk and msk. After the generation of these short-term keys, the adversary has access to the Break-Oracle, which, of course, will also not yield any additional information about them since the output of this oracle only depends on independent random choices like mpk and the PRF keys K. The remaining steps of Game 5

---

[5] C_0 and C_1 are assumed to be of the same size, otherwise the smaller one is padded accordingly.

correspond to the regular IND-CCA game for $\mathsf{PKE_{CCA}}$ except that the adversary is given the additional input CRS^*, which however only depends on pk^*, and the independent choices mpk and K^*. So except for pk^* (which is the output of $\mathsf{PKGen}(CRS^*, mpk)$), the adversary does not get any additional useful information from CRS^* (which he could not have computed by himself). Hence, it is easy to construct an IND-CCA adversary $\mathcal{E}$ against $\mathsf{PKE_{CCA}}$ for security parameter k from $\mathcal{A}$ which has the same advantage as $\mathcal{A}$ in winning Game 5, i.e.,

$$\mathsf{Adv}^{\mathsf{ind\text{-}cca}}_{\mathsf{PKE_{CCA}},\mathcal{E}}(k) = \mathsf{Adv}^{\mathsf{Game5}}_{\mathsf{RPKE}_{i\mathcal{O}},\mathcal{A}}(\lambda, k). \tag{10}$$

$\mathcal{E}$ computes $\mathcal{PP} \leftarrow \mathsf{Setup}(1^\lambda)$ and $mpk \leftarrow \{0,1\}^{2\lambda}$. Break-Queries from $\mathcal{A}$ can be answered by $\mathcal{E}$ only based on mpk (as described in Game 2). Then $\mathcal{E}$ receives pk^* generated using $\mathsf{Gen_{CCA}}(1^k)$ from the IND-CCA challenger. To compute CRS^*, $\mathcal{E}$ chooses a PPRF $\mathsf{F} : \{0,1\}^{2(\lambda+k)} \rightarrow \{0,1\}^{p(\lambda+k)}$, the corresponding key $K^* \leftarrow \mathsf{Gen_F}(1^{\lambda+k})$ and sets the punctured key $K^*_{\{\mathsf{pad}_{2k}(mpk)\}}$. Using these ingredients, Gen' can be specified as in Game 4 and its obfuscation equals CRS^*. When $\mathcal{E}$ runs $\mathcal{A}$ on input CRS^*, $\mathcal{A}$'s queries to the Dec-Oracle are forwarded to the IND-CCA challenger. Similarly, the messages m_0 and m_1 that $\mathcal{A}$ outputs are sent to $\mathcal{E}$'s challenger. When $\mathcal{E}$ receives c^* from its challenger, it runs $\mathcal{A}$ on this input, where Dec-Oracle calls are again forwarded, and outputs the output bit of $\mathcal{A}$.

Putting Eqs. 5–10 together, we obtain Eq. 4.

From Eq. 4 to Eq. 3. Let t denote the runtime of $\mathcal{A}$ and t' the maximal runtime of the experiment $\mathsf{Exp}^{\mathsf{r\text{-}ind\text{-}cca}}_{\mathsf{RPKE}_{i\mathcal{O}},\cdot}(\lambda, k)$ involving an arbitrary adversary with runtime t. Furthermore, note that the reduction algorithms $\mathcal{B}, \mathcal{C}, \mathcal{D}, \mathcal{E}$ are uniform in the sense that they perform the same operations for any given adversary $\mathcal{A}$ of runtime t. Let $s_1, s_2, s_3,$ and s_4 denote the maximal runtime of our PRG, IND-CCA, PPRF, and $i\mathcal{O}$ reduction algorithm, respectively, for an RPKE adversary with runtime t. As all these reduction algorithms basically execute the R-IND-CCA experiment (including minor modifications) with the RPKE adversary, we have that $t' \approx s_1 \approx s_2 \approx s_3 \approx s_4$. Clearly, the runtime of our reduction algorithms are upper bounded by the corresponding values t_i and thus it follows

$$\mathsf{Adv}^{\mathsf{r\text{-}ind\text{-}cca}}_{\mathsf{RPKE}_{i\mathcal{O}},\mathcal{A}}(\lambda, k) \leq \tfrac{1}{2^\lambda} + \mathsf{CAdv}^{\mathsf{prg}}_{\mathsf{PRG}}(s_1, \lambda) + \mathsf{CAdv}^{\mathsf{ind\text{-}cca}}_{\mathsf{PKE_{CCA}}}(s_2(\lambda, k), k)$$
$$+ \mathsf{CAdv}^{\mathsf{pprf}}_{\mathsf{F}}(s_3(\lambda, k), \lambda, k) + \mathsf{CAdv}^{\mathsf{io}}_{i\mathcal{O}}(s_4, \lambda + k). \tag{11}$$

Finally, since the same upper bound (on the right-hand side of Eq. 11) on the advantage holds for any adversary $\mathcal{A}$ with runtime t, this is also an upper bound for $\mathsf{CAdv}^{\mathsf{r\text{-}ind\text{-}cca}}_{\mathsf{RPKE}_{i\mathcal{O}}}(t, \lambda, k)$.

Theorem 1. *Let us assume that for any polynomial $s(\ell)$, the concrete advantages* $\mathsf{CAdv}^{\mathsf{prg}}_{\mathsf{PRG}}(s(\ell), \ell)$, $\mathsf{CAdv}^{\mathsf{io}}_{i\mathcal{O}}(s(\ell), \ell)$, $\mathsf{CAdv}^{\mathsf{pprf}}_{\mathsf{F}}(s(\ell), \ell)$ *and* $\mathsf{CAdv}^{\mathsf{ind\text{-}cca}}_{\mathsf{PKE_{CCA}}}(s(\ell), \ell)$ *are negligible. Then* $\mathsf{RPKE}_{i\mathcal{O}}$ *is R-IND-CCA secure.*

Proof. Let $t(\lambda, k)$ be a polynomial and let us consider the upper bound on $\mathsf{CAdv}^{\mathsf{r\text{-}ind\text{-}cca}}_{\mathsf{RPKE}_{i\mathcal{O}}}(t(\lambda, k), \lambda, k)$ given by Lemma 1. First, note that since RPKE is efficient there is also a polynomial bound $t'(\lambda, k)$ on the runtime complexity of the experiment and thus $s_1(\lambda, k)$, $s_2(\lambda, k)$, $s_3(\lambda, k)$, and $s_4(\lambda, k)$ will be polynomial as $t'(\lambda, k) \approx s_1(\lambda, k) \approx s_2(\lambda, k) \approx s_3(\lambda, k) \approx s_4(\lambda, k)$ for all $\lambda, k \in \mathbb{N}$. Furthermore, let $t_1(\lambda, k) := s_1(\lambda, k)$, $t_2(\lambda, k) := s_2(\lambda, k)$, and $t_3(\lambda, k)$ be a polynomial upper bound on $s_3(\lambda, k)$ and $s_4(\lambda, k)$. Now, consider the following partition of $\mathsf{CAdv}_{\mathsf{RPKE}_{i\mathcal{O}}}^{\mathsf{r\text{-}ind\text{-}cca}}(t(\lambda, k), \lambda, k)$ as demanded in Definition 4: $f_1(t_1(\lambda, k), \lambda) := \frac{1}{2^\lambda} + \mathsf{CAdv}_{\mathsf{PRG}}^{\mathsf{prg}}(t_1(\lambda, k), \lambda)$, $f_2(t_2(\lambda, k), k) := \mathsf{CAdv}_{\mathsf{PKE}_{\mathsf{CCA}}}^{\mathsf{ind\text{-}cca}}(t_2(\lambda, k), k)$, and

$$f_3(t_3(\lambda, k), \lambda, k) := \mathsf{CAdv}_{i\mathcal{O}}^{\mathsf{io}}(t_3(\lambda, k), \lambda + k) + \mathsf{CAdv}_{\mathsf{F}}^{\mathsf{pprf}}(t_3(\lambda, k), \lambda + k)$$

Obviously, for all fixed $k \in \mathbb{N}$, $t_1(\lambda, k)$ is a polynomial in a single variable, namely λ, and thus $f_1(t_1(\lambda, k), \lambda)$ is negligible in λ by assumption. Similarly, for all fixed $\lambda \in \mathbb{N}$, $f_2(t_2(\lambda, k), k)$ is negligible in k by assumption. Moreover, for all fixed $k \in \mathbb{N}$ and for all fixed $\lambda \in \mathbb{N}$, $t_3(\lambda, k)$ becomes a polynomial in λ and in k, respectively, and the advantages $\mathsf{CAdv}_{i\mathcal{O}}^{\mathsf{io}}(t_3(\lambda, k), \lambda+k)$ and $\mathsf{CAdv}_{\mathsf{F}}^{\mathsf{pprf}}(t_3(\lambda, k), \lambda+ k)$ are negligible in λ and in k by assumption.

Versatility of Our $i\mathcal{O}$-Based Construction. As one can easily see, the $i\mathcal{O}$-based construction of an RPKE we presented above is very modular and generic: there was no need to modify the standard cryptosystem (the IND-CCA secure PKE) itself to make it reconfigurable but we just added a component "in front" which fed its key generator with independently-looking randomness. Thus, the same component may be used to make other types of cryptosystems reconfigurable in this sense. Immediate applications would be the construction of an $i\mathcal{O}$-based R-IND-CPA secure RPKE from an IND-CPA secure PKE or of an R-EUF-CMA secure reconfigurable signature scheme (cf. Definition 6) from an EUF-CMA secure signature scheme. The construction is also very flexible in the sense that it allows to switch to a completely different IND-CCA secure PKE (or at least to a more secure algebraic structure for the PKE) on-the-fly when the short-term security level k gets increased. One may even use the same long-term keys to generate short-term PKIs for multiple different cryptosystems (e.g., a signature and an encryption scheme) used in parallel. We leave the security analysis of such extended approaches as an open problem.

4.2 Reconfigurable Encryption from SCasc

Our second construction of a R-IND-CPA secure reconfigurable encryption scheme makes less strong assumptions than our construction using $i\mathcal{O}$. Namely, it uses a pairing-friendly group generator $\mathcal{G}$ as introduced in Sect. 2 and the only assumption is (a suitable variant of) the $\mathcal{SC}_k$-MDDH assumption with respect to $\mathcal{G}$. Our construction is heavily inspired by Regev's lattice-based encryption scheme [18] (in its "dual variant" [13]). However, instead of computing with noisy integers, we perform similar computations "in the exponent". (A similar adaptation of lattice-based constructions to a group setting was already undertaken in [8], although with different constructions and for a different purpose.)

A Two-Parameter Variant of the $\mathcal{SC}_k$-MDDH Assumption. For our purposes, it will be useful to consider the $\mathcal{SC}_k$-MDDH assumption as an assumption in *two* security parameters, λ and k. Namely, let

$$\mathsf{Adv}^{\mathsf{SC}}_{\mathcal{G},\mathcal{B}}(\lambda, k) := \mathsf{Adv}^{\mathcal{D}_k\text{-MDDH}}_{\mathcal{G},\mathcal{A}}(\lambda)$$

where $\mathcal{D}_k = \mathcal{SC}_k$ as defined by Eq. 2 in Sect. 2. Note that this also defines the concrete advantage $\mathsf{CAdv}^{\mathsf{SC}}_{\mathcal{G}}(t, \lambda, k)$ (generically defined in Sect. 2).

It is not immediately clear how to define asymptotic security with this two-parameter advantage function. To do so, we follow the path taken for our reconfigurable security definition, with λ as a long-term, and k as a short term security parameter: We say that the SCasc assumption holds relative to $\mathcal{G}$ iff $\mathsf{CAdv}^{\mathsf{SC}}_{\mathcal{G}}(t, \lambda, k)$ can be split up into three components, as follows. We require that for every polynomial $t = t(\lambda, k)$, there exist nonnegatively-valued functions $f_1 : \mathrm{N}^2 \to \mathrm{R}^+_0, f_2 : \mathrm{N}^2 \to \mathrm{R}^+_0, f_3 : \mathrm{N}^3 \to \mathrm{R}^+_0$ and polynomials $t_1(\lambda, k), t_2(\lambda, k), t_3(\lambda, k)$ such that

$$\mathsf{CAdv}^{\mathsf{SC}}_{\mathcal{G}}(t(\lambda, k), \lambda, k) \le f_1(t_1(\lambda, k), \lambda) + f_2(t_2(\lambda, k), k) + f_3(t_3(\lambda, k), \lambda, k)$$

and the following conditions are satisfied for f_1, f_2, f_3:

- For all $k \in \mathrm{N}$ it holds that $f_1(t_1(\lambda, k), \lambda)$ is negligible in λ
- For all $\lambda \in \mathrm{N}$ it holds that $f_2(t_2(\lambda, k), k)$ is negligible in k
- For all $k \in \mathrm{N}$ it holds that $f_3(t_3(\lambda, k), \lambda, k)$ is negligible in λ
- For all $\lambda \in \mathrm{N}$ it holds that $f_3(t_3(\lambda, k), \lambda, k)$ is negligible in k.

The interpretation is quite similar to reconfigurable security: we view λ (which determines, e.g., the group order) as a long-term security parameter. On the other hand, k determines the concrete computational problem considered in this group, and we thus view k as a short-term security parameter. (For instance, it is conceivable that an adversary may successfully break one computational problem in a given group, but not a potentially harder problem. Hence, increasing k may be viewed as increasing the security of the system.) It is not hard to show that $\mathsf{CAdv}^{\mathsf{SC}}_{\mathcal{G}}(t, \lambda, k)$ holds in the generic group model, although, the usual proof technique only allows for a trivial splitting of the adversarial advantage into the f_1, f_2 and f_3.

Choosing Subspace Elements. We will face the problem of sampling a vector $[\vec{r}] \in G^{k+1}$ satisfying $\vec{r}^\top \cdot \mathbf{A}_x = \vec{y}^\top$ for given $\mathbf{A}_x \in \mathbb{Z}_p^{(k+1) \times k}$ (of the form of Eq. 2) and $[\vec{y}] \in G^k$. One efficient way to choose a uniform solution $[\vec{r}] = [(r_i)_i]$ is as follows: choose r_1 uniformly, and set $[r_{i+1}] = [y_i] - x \cdot [r_i]$ for $2 \le i \le k + 1$.

Our Scheme RPKE_{SC}. Now our encryption scheme has message space G_T and is given by the following algorithms:

Setup(1^λ): sample $(p, G, g, G_T, e) \leftarrow \mathcal{G}(1^\lambda)$ and return $\mathcal{PP} := (p, G, g, G_T, e)$.
MKGen($\mathcal{PP}$): sample $x \leftarrow \mathbb{Z}_p$ and return $mpk := [x] \in G$ and $msk := x$.
CRSGen($\mathcal{PP}, 1^k$): sample $\vec{y} \leftarrow \mathbb{Z}_p^k$ and return $CRS := (1^k, \mathcal{PP}, [\vec{y}^\top] \in G^k)$.
PKGen(CRS, mpk): compute $[\mathbf{A}_x]$ from $mpk = [x]$, return $pk := (CRS, [\mathbf{A}_x])$.
SKGen(CRS, msk): compute $\mathbf{A}_x$ from $msk = x$ and sample a uniform solution $[\vec{r}] \in G^{k+1}$ of $\vec{r}^\top \cdot \mathbf{A}_x = \vec{y}^\top$, and return $sk := (CRS, [\vec{r}])$.

$\mathsf{Enc}(pk, m)$: sample $\vec{s} \leftarrow \mathbb{Z}_p^k$, return $c = ([\vec{R}], [S]_T) = ([\mathbf{A}_x \cdot \vec{s}], [\vec{y}^\top \cdot \vec{s}]_T \cdot m) \in G^{k+1} \times G_T$

$\mathsf{Dec}(sk, c)$: return $m = [S]_T - [\vec{r}^\top \cdot \vec{R}]_T \in G_T$.

Correctness and Security. Correctness follows from

$$\mathsf{Dec}(sk, c) = [S]_T - [\vec{r}^\top \cdot \vec{R}]_T = ([\vec{y}^\top \cdot \vec{s}]_T - [\vec{r}^\top \cdot \mathbf{A}_x \cdot \vec{s}]_T) \cdot m,$$

since $\vec{y}^\top = \vec{r}^\top \cdot \mathbf{A}_x$ by definition. For security, consider

Lemma 2. *Let $t \in \mathbb{N}$ be given and let t' denote the maximal runtime of the experiment $\mathsf{Exp}^{\mathsf{r\text{-}ind\text{-}cca}}_{\mathsf{RPKE}_{SC},\cdot}(\lambda, k)$ involving arbitrary adversaries with runtime t. Then it holds that*

$$\mathsf{CAdv}^{\mathsf{r\text{-}ind\text{-}cpa}}_{\mathsf{RPKE}_{SC}}(t, \lambda, k) \leq \frac{1}{2^\lambda} + \mathsf{CAdv}^{\mathsf{SC}}_{\mathcal{G}}(s, \lambda, k) \tag{12}$$

where $t' \approx s$.

Proof. Similar to the proof of Lemma 1, the following reduction will be in the non-uniform setting, where we consider an adversary $\mathcal{A}$ against RPKE_{SC} for fixed security parameters λ and k. We show that $\mathcal{A}$ can be turned into an algorithm $\mathcal{B}$ solving SCasc for fixed λ and k with advantage $\mathsf{Adv}^{\mathsf{SC}}_{\mathcal{G},\mathcal{B}}(\lambda, k)$ such that

$$\mathsf{Adv}^{\mathsf{r\text{-}ind\text{-}cpa}}_{\mathsf{RPKE}_{SC},\mathcal{A}}(\lambda, k) \leq \frac{1}{2^\lambda} + \mathsf{Adv}^{\mathsf{SC}}_{\mathcal{G},\mathcal{B}}(\lambda, k). \tag{13}$$

We proceed in games, with **Game** 1 being the $\mathsf{Exp}^{\mathsf{r\text{-}ind\text{-}cpa}}_{\mathsf{RPKE}_{SC},\mathcal{A}}$ experiment. Let $\mathsf{Adv}^{\mathsf{Game}_i}_{\mathsf{RPKE}_{SC},\mathcal{A}}(\lambda, k)$ denote the advantage of $\mathcal{A}$ in Game i. Thus, by definition,

$$\mathsf{Adv}^{\mathsf{r\text{-}ind\text{-}cca}}_{\mathsf{RPKE}_{SC},\mathcal{A}}(\lambda, k) = \mathsf{Adv}^{\mathsf{Game}_1}_{\mathsf{RPKE}_{SC},\mathcal{A}}(\lambda, k). \tag{14}$$

In **Game** 2, we implement the $\mathsf{Break}(\mathcal{PP}, msk, \cdot)$ oracle differently for $\mathcal{A}$. Namely, recall that in Game 1, upon input $k' < k$, the Break-Oracle chooses a CRS $CRS_{k'} = (1^{k'}, \mathcal{PP}, [\vec{y}^\top] \leftarrow G^{k'})$, then computes a secret key $sk_{k'} = [\vec{r}] \in G^{k'+1}$ with $\vec{r}^\top \mathbf{A}_x = \vec{y}^\top$, and finally returns $CRS_{k'}$ and $sk_{k'}$ to $\mathcal{A}$.

Instead, we will now let Break first choose $\vec{r} \in \mathbb{Z}_p^{k'+1}$ uniformly, and then compute $[\vec{y}^\top] = [\vec{r}^\top \mathbf{A}_x]$ from $\vec{r}$ and set $CRS_{k'} = (1^{k'}, \mathcal{PP}, [\vec{y}^\top])$. This yields exactly the same distribution for $sk_{k'}$ and $CRS_{k'}$, but only requires knowledge about $[\mathbf{A}_x]$ (and not $\mathbf{A}_x$). Hence, we have

$$\mathsf{Adv}^{\mathsf{Game}_1}_{\mathsf{RPKE}_{SC},\mathcal{A}}(\lambda, k) = \mathsf{Adv}^{\mathsf{Game}_2}_{\mathsf{RPKE}_{SC},\mathcal{A}}(\lambda, k). \tag{15}$$

In **Game** 3, we prepare the challenge ciphertext c^* differently for $\mathcal{A}$. As a prerequisite, we let the game also choose CRS^* like the Break oracle from Game 2 chooses the $CRS_{k'}$. In other words, we set up $CRS^* = [\vec{y}^\top] = [\vec{r}^{*\top} \mathbf{A}_x]$ for uniformly chosen $\vec{r}^*$. This way, we can assume that $sk^* = (CRS^*, [\vec{r}^*])$ is known to the game, even for an externally given $[\mathbf{A}_x]$.

Next, recall that in Game 2, we have first chosen $\vec{s} \leftarrow \mathbb{Z}_p^k$ and then computed $c^* = ([\vec{R}], [S]_T) = ([\mathbf{A}_x \cdot \vec{s}], [\vec{y}^\top \cdot \vec{s}]_T \cdot m_b)$. In Game 3, we still first choose $\vec{s}$ and compute $[\vec{R}] = [\mathbf{A}_x \cdot \vec{s}]$. However, we then compute $[S]_T = [\vec{r^*}^\top \cdot R]_T \cdot m_b$ in a black-box way from $[\vec{R}]$, without using $\vec{s}$ again.

These changes are again purely conceptual, and we get

$$\mathsf{Adv}^{\mathsf{Game}_2}_{\mathsf{RPKE}_{SC}, \mathcal{A}}(\lambda, k) = \mathsf{Adv}^{\mathsf{Game}_3}_{\mathsf{RPKE}_{SC}, \mathcal{A}}(\lambda, k). \tag{16}$$

Now, in **Game** 4, we are finally ready to use the SCasc assumption. Specifically, instead of computing the value $[\vec{R}]$ of c^* as $[\vec{R}] = [\mathbf{A}_x \cdot \vec{s}]$ for a uniformly chosen $\vec{s} \in \mathbb{Z}_p^k$, we sample $[\vec{R}] \in G^{k+1}$ independently and uniformly. (By our change from Game 3, then $[S]_T$ is computed from $[\vec{R}]$ using sk^*.)

Our change hence consists in replacing an element of the form $[\mathbf{A}_x \cdot \vec{s}]$ by a random vector of group elements. Besides, at this point, our game only requires knowledge of $[\mathbf{A}_x]$ (but not of $\mathbf{A}_x$). Hence, a straightforward reduction to the SCasc assumption yields an adversary $\mathcal{B}$ with

$$\mathsf{Adv}^{\mathsf{SC}}_{\mathcal{G}, \mathcal{B}}(\lambda, k) = \left| \mathsf{Adv}^{\mathsf{Game}_4}_{\mathsf{RPKE}_{SC}, \mathcal{A}}(\lambda, k) - \mathsf{Adv}^{\mathsf{Game}_3}_{\mathsf{RPKE}_{SC}, \mathcal{A}}(\lambda, k) \right|. \tag{17}$$

Finally, it is left to observe that in Game 4, the challenge ciphertext is (statistically close to) independently random. Indeed, recall that the challenge ciphertext is chosen as $c^* = ([\vec{R}], [S]_T)$ for uniform $\vec{R} \in \mathbb{Z}_p^{k+1}$, and $[S]_T = [\vec{r^*}^\top \cdot R]_T \cdot m_b$. Suppose now that $\vec{R}$ does not lie in the image of $\mathbf{A}_x$. (That is, $\vec{R}$ cannot be explained as a combination of columns of $\mathbf{A}_x$.) Then, for random $\vec{r}$, the values $\vec{r^*}^\top \mathbf{A}_x$ and $\vec{r^*}^\top \cdot R$ are independently random. In particular, even given $[\mathbf{A}_x]$ and CRS^*, the value $[\vec{r^*}^\top \cdot R]_T$ looks independently random to $\mathcal{A}$.

Hence, $\mathcal{A}$'s view is independent of the encrypted message m_b (at least when conditioned on $\vec{R}$ not being in the image of $\mathbf{A}_x$). On the other hand, since $\vec{R}$ is uniformly random in Game 4, it lies in the image of $\mathbf{A}_x$ only with probability $1/p$. Thus, we get

$$\mathsf{Adv}^{\mathsf{Game}_4}_{\mathsf{RPKE}_{SC}, \mathcal{A}}(\lambda, k) \le \frac{1}{p}. \tag{18}$$

Putting Eqs. 14–18 together (and using that $p \ge 2^\lambda$), we obtain Eq. 13.

From Eq. 13 to Eq. 12. Let t denote the runtime of $\mathcal{A}$ and t' the maximal runtime of the experiment $\mathsf{Exp}^{\mathsf{r\text{-}ind\text{-}cca}}_{\mathsf{RPKE}_{SC}, \cdot}(\lambda, k)$ involving an arbitrary adversary with runtime t. Note that the reduction algorithm $\mathcal{B}$ is uniform in the sense that it performs the same operations for any given adversary $\mathcal{A}$ of runtime t. Let s denote the maximal runtime of our SCasc algorithm for an RPKE adversary with runtime t. As the SCasc algorithm basically executes the R-IND-CCA experiment (including minor modifications) with the RPKE adversary, we have that $t' \approx s$. Clearly, the runtime of $\mathcal{B}$ is upper bounded by s and thus it follows

$$\mathsf{Adv}^{\mathsf{r\text{-}ind\text{-}cca}}_{\mathsf{RPKE}_{SC}, \mathcal{A}}(\lambda, k) \le \frac{1}{2^\lambda} + \mathsf{CAdv}^{\mathsf{SC}}_{\mathcal{G}}(s, \lambda, k). \tag{19}$$

Finally, since the same upper bound (on the right-hand side of Eq. 19) on the advantage holds for any adversary $\mathcal{A}$ with runtime t, this is also an upper bound for $\mathsf{CAdv}^{\mathsf{r\text{-}ind\text{-}cca}}_{\mathsf{RPKE}_{SC}}(t, \lambda, k)$.

Theorem 2. *If the two-parameter variant of the SCasc assumption holds, then* RPKE$_{SC}$ *is R-IND-CPA secure.*

Proof. Let $t(\lambda, k)$ be a polynomial. Since RPKE$_{SC}$ is efficient, $t'(\lambda, k)$ will be polynomial and so $s(\lambda, k)$. As $s(\lambda, k)$ is polynomial, according to the SCasc assumption there exist functions g_1, g_2, and g_3 as well as polynomials $s_1(\lambda, k)$, $s_2(\lambda, k)$, and $s_3(\lambda, k)$ such that

$$\mathsf{CAdv}_{\mathcal{G}}^{\mathsf{SC}}(s(\lambda, k), \lambda, k) \leq g_1(s_1(\lambda, k), \lambda) + g_2(s_2(\lambda, k), k) + g_3(s_3(\lambda, k), \lambda, k).$$

Now consider the following partition of $\mathsf{CAdv}_{\mathsf{RPKE}_{SC}}^{\mathsf{r\text{-}ind\text{-}cca}}(t(\lambda, k), \lambda, k)$: $f_1(s_1(\lambda, k), \lambda) := \frac{1}{2^\lambda} + g_1(s_1(\lambda, k), \lambda, k)$, $f_2(s_2(\lambda, k), k) := g_2(s_2(\lambda, k), \lambda, k)$, and $f_3(s_3(\lambda, k), \lambda, k) = g_3(s_3(\lambda, k), \lambda, k)$. The properties demanded for f_1, f_2, f_3 by Definition 4 immediately follow from the SCasc assumption.

5   Reconfigurable Signatures

The concept of reconfiguration is not restricted to encryption schemes. In this section, we consider the case of reconfigurable signatures. We start with some preliminaries, define reconfigurable signatures and a security experiment (both in line with the encryption case) and finally give a construction.

5.1   Preliminaries

Signature Schemes. A signature scheme SIG with message space $\mathcal{M}$ consists of three PPT algorithms Setup, Gen, Sig, Ver. Setup(1^λ) outputs public parameters $\mathcal{PP}$ for the scheme. Key generation Gen($\mathcal{PP}$) outputs a verification key vk and a signing key sk. The signing algorithm Sig(sk, m) takes the signing key and a message $m \in \mathcal{M}$, and outputs a signature σ. Verification Ver(vk, σ, m) takes the verification key, a signature and a message m and outputs 1 or $\perp$. For correctness, we require that for all $m \in \mathcal{M}$ and all $(vk, sk) \leftarrow$ Gen(1^k) we have Ver(sk, Sig(sk, m), m) = 1.

EUF-CMA Security. The *EUF-CMA-advantage* of an adversary $\mathcal{A}$ on SIG is defined by $\mathsf{Adv}_{\mathsf{SIG},\mathcal{A}}^{\mathsf{euf\text{-}cma}}(\lambda) := \Pr[\mathsf{Exp}_{\mathsf{SIG},\mathcal{A}}^{\mathsf{euf\text{-}cma}}(\lambda) = 1]$ for the experiment $\mathsf{Exp}_{\mathsf{SIG},\mathcal{A}}^{\mathsf{euf\text{-}cma}}$ described below. In $\mathsf{Exp}_{\mathsf{SIG},\mathcal{A}}^{\mathsf{euf\text{-}cma}}$, first, $\mathcal{PP} \leftarrow$ Setup(1^λ) and $(pk, sk) \leftarrow$ Gen($\mathcal{PP}$) is sampled. The we run $\mathcal{A}$ on input pk, where $\mathcal{A}$ also has access to a signature oracle. The experiment returns 1 if for $\mathcal{A}$'s output (σ^*, m^*) it holds that Ver(pk, σ^*, m^*) = 1 and m^* was not sent to the signature oracle. A signature scheme SIG is called EUF-CMA-secure if for all PPT algorithms $\mathcal{A}$ the advantage $\mathsf{Adv}_{\mathsf{SIG},\mathcal{A}}^{\mathsf{euf\text{-}cma}}(\lambda)$ is negligible.

Non-Interactive Proof Systems. A non-interactive proof system for a language $\mathcal{L}$ consists of three PPT algorithms (CRSGen, Prove, Ver). CRSGen($\mathcal{L}$) gets as input information about the language and outputs a *common reference string* (CRS). Prove(CRS, x, w) with statement x and witness w outputs a proof π, and Ver(CRS, π, x) outputs 1 if π is a valid proof for $x \in \mathcal{L}$, and $\perp$ otherwise.

The proof system is *complete* if Ver always accepts proofs if x is contained in $\mathcal{L}$, and it is *perfectly sound* if Ver always rejects proofs if x is not in $\mathcal{L}$.

Witness Indistinguishability (WI). Suppose a statement $x \in \mathcal{L}$ has more than one witness. A proof of membership can be generated using any of the witnesses. If a proof $\pi \leftarrow$ Prove(CRS, x, w) information theoretically hides the choice of the witness, it is called *perfectly witness indistinguishable*.

Groth-Sahai (GS) Proofs. In [15], Groth and Sahai introduced efficient non-interactive proof systems in pairing-friendly groups. We will only give a high level overview of the properties that are needed for our reconfigurable signature scheme and refer to the full version [15] for the details of their construction.

In GS proof systems, the algorithm CRSGen takes as input a pairing-friendly group $\mathbb{G} := (p, G, g, G_T, e)$ and outputs a CRS suitable for proving satisfiability of various types of equations in these groups. Furthermore, CRSGen has two different modes of operation, producing a CRS that leads to either perfectly witness indistinguishable or perfectly sound proofs. The two types of CRS can be shown to be computationally indistinguishable under different security assumptions such as subgroup decision, SXDH and 2-Linear.

In both modes, CRSGen additionally outputs a trapdoor. In the WI mode, this trapdoor can be used to produce proofs of false statements[6]. In the sound mode, the trapdoor can be used to extract the witness from the proof. To easily distinguish the two operating modes, we equip CRSGen with an additional parameter $mode \in \{\texttt{wi}, \texttt{sound}\}$.

Statements provable with GS proofs have to be formulated in terms of satisfiability of equations in pairing-friendly groups. For example, it is possible to prove the statement $\mathcal{X} := \text{``}\exists s \in \mathbb{Z}_n : [s]_1 = S\text{''}$ for an element $S \in G_1$. A witness for this statement is a value s satisfying the equation $[s] = S$, i.e., the DL of S to the basis g_1. Furthermore, GS proofs are nestable and thus admit proving statements about proofs, e.g., $\mathcal{Y} := \text{``}\exists \pi : \text{Ver}(CRS, \pi, \mathcal{X}) = 1\text{''}$.

5.2 Definitions

Similar to the case of RPKE, we can define reconfigurable signatures.

Definition 5. *A reconfigurable signature (RSIG) scheme* RSIG *consists of algorithms* Setup, MKGen, CRSGen, PKGen, SKGen, Sig *and* Ver. *The first five algorithms are defined as in Definition 3.* Sig *and* Ver *are the signature generation and verification algorithms and are defined as in a regular signature scheme.* RSIG *is called correct if for all* $\lambda, k \in \mathbb{N}$, $\mathcal{PP} \leftarrow$ Setup(1^λ), $(mpk, msk) \leftarrow$ MKGen($\mathcal{PP}$), $CRS \leftarrow$ CRSGen($\mathcal{PP}, 1^k$), *messages* $m \in \mathcal{M}$, $sk \leftarrow$ SKGen(CRS, msk) *and* $pk \leftarrow$ PKGen(CRS, mpk) *we have that* Ver(pk, Sig(sk, m), m) = 1.

---

[6] Actually, the original paper only describes a method for generating proofs for specific false statements. Arbitrary statements can be proven at the cost of slightly larger proofs and CRSs, using known methods that apply to WI proofs [14].

Experiment $\mathsf{Exp}_{\mathsf{RSIG},\mathcal{A}}^{\mathsf{r\text{-}euf\text{-}cma}}(\lambda,k)$

$\mathcal{PP} \leftarrow \mathsf{Setup}(1^\lambda)$

$(mpk, msk) \leftarrow \mathsf{MKGen}(\mathcal{PP})$

$state \leftarrow \mathcal{A}^{\mathsf{Break}(\mathcal{PP}, msk, \cdot)}(1^\lambda, 1^k, \mathcal{PP}, mpk, \text{"learn"})$

$CRS^* \leftarrow \mathsf{CRSGen}(\mathcal{PP}, 1^k)$

$sk^* \leftarrow \mathsf{SKGen}(CRS^*, msk)$

$pk^* \leftarrow \mathsf{PKGen}(CRS^*, mpk)$

$(m^*, \sigma^*) \leftarrow \mathcal{A}^{\mathsf{Sig}(sk^*, \cdot)}(CRS^*, state)$

Let $k_1, \ldots, k_\ell$ be the inputs sent to the Break-Oracle by $\mathcal{A}$. On input k_i, the Break-Oracle returns $CRS_{k_i} \leftarrow \mathsf{CRSGen}(\mathcal{PP}, 1^{k_i})$ as well as $sk_{k_i} \leftarrow \mathsf{SKGen}(CRS_{k_i}, msk)$ to $\mathcal{A}$. Return 1 if $k_i < k$ for all i, $\mathsf{Ver}(pk^*, \sigma^*, m^*) = 1$, and m^* was not an input to the Sig-Oracle. Otherwise, return 0.

Fig. 3. R-EUF-CMA experiment for a reconfigurable signature scheme RSIG.

We define R-EUF-CMA security for an RSIG scheme RSIG analogously to R-IND-CCA security for RPKE, where the security experiment $\mathsf{Exp}_{\mathsf{RSIG},\mathcal{A}}^{\mathsf{r\text{-}euf\text{-}cma}}(\lambda,k)$ is defined in Fig. 3.

Definition 6. *Let* RSIG *be an RSIG scheme according to Definition 5. Then we define the advantage of an adversary $\mathcal{A}$ as*

$$\mathsf{Adv}_{\mathsf{RSIG},\mathcal{A}}^{\mathsf{r\text{-}euf\text{-}cma}}(\lambda, k) := \Pr[\mathsf{Exp}_{\mathsf{RSIG},\mathcal{A}}^{\mathsf{r\text{-}euf\text{-}cma}}(\lambda, k) = 1]$$

where $\mathsf{Exp}_{\mathsf{RSIG},\mathcal{A}}^{\mathsf{r\text{-}euf\text{-}cma}}(\lambda, k)$ *is the experiment given in Fig. 3. The concrete advantage* $\mathsf{CAdv}_{\mathsf{RSIG}}^{\mathsf{r\text{-}euf\text{-}cma}}(t, \lambda, k)$ *of adversaries against* RSIG *with time complexity t follows canonically (cf. Sect. 2).*

An RSIG scheme RSIG *is then called R-EUF-CMA secure if for all polynomials $t(\lambda, k)$, there exist positive functions $f_1 : \mathbb{N}^2 \to \mathbb{R}_0^+$, $f_2 : \mathbb{N}^2 \to \mathbb{R}_0^+$, and $f_3 : \mathbb{N}^3 \to \mathbb{R}_0^+$ as well as polynomials $t_1(\lambda, k)$, $t_2(\lambda, k)$, and $t_3(\lambda, k)$ such that*

$$\mathsf{CAdv}_{\mathsf{RSIG}}^{\mathsf{r\text{-}euf\text{-}cma}}(t(\lambda, k), \lambda, k) \leq f_1(t_1(\lambda, k), \lambda) + f_2(t_2(\lambda, k), k) + f_3(t_3(\lambda, k), \lambda, k)$$

for all λ, k, and the following conditions are satisfied for f_1, f_2, f_3:

- *For all $k \in \mathbb{N}$ it holds that $f_1(t_1(\lambda, k), \lambda)$ is negligible in λ*
- *For all $\lambda \in \mathbb{N}$ it holds that $f_2(t_2(\lambda, k), k)$ is negligible in k*
- *For all $k \in \mathbb{N}$ it holds that $f_3(t_3(\lambda, k), \lambda, k)$ is negligible in λ*
- *For all $\lambda \in \mathbb{N}$ it holds that $f_3(t_3(\lambda, k), \lambda, k)$ is negligible in k*

5.3 Reconfigurable Signatures from Groth-Sahai Proofs

The intuition behind our scheme is as follows. Each user of the system has a long-term key pair, consisting of a public instance of a hard problem and a private solution of this instance. A valid signature is a proof of knowledge of *either* knowledge of the long-term secret key *or* a valid signature of the message under another signature scheme. The proof system and signature scheme for generating the proofs of knowledge are published, e.g. using a CRS. We are now able to reconfigure the scheme by updating the CRS with a new proof system and a new signature scheme. This way, old short-term secret keys of a user (i.e., valid proofs of knowledge of the user's long-term secret key under deprecated proof systems) become useless and can thus be leaked to the adversary.

Our reconfigurable signature scheme RSIG with message space $\mathcal{M} = \{0,1\}^m$ is depicted in Fig. 4. It makes use of a symmetric pairing-friendly group generator $\mathcal{G}$, a family of GS proof systems $\mathsf{PS} := \{\mathsf{PS}_k := (\mathsf{CRSGen}_{\mathsf{PS}_k}, \mathsf{Prove}_{\mathsf{PS}_k}, \mathsf{Ver}_{\mathsf{PS}_k})\}_{k\in\mathbb{N}}$ for proving equations in the groups generated by $\mathcal{G}(1^\lambda)$ and a family of EUF-CMA-secure signature schemes $\mathsf{SIG} := \{\mathsf{SIG}_k := (\mathsf{Setup}_{\mathsf{SIG}_k}, \mathsf{Gen}_{\mathsf{SIG}_k}, \mathsf{Sig}_{\mathsf{SIG}_k}, \mathsf{Ver}_{\mathsf{SIG}_k})\}_{k\in\mathbb{N}}$ with message space $\mathcal{M}$, where $\mathsf{Setup}_{\mathsf{SIG}_k}(1^\lambda)$ outputs $\mathbb{G}$ with $\mathbb{G} \leftarrow \mathcal{G}(1^\lambda)$ for all $k \in \mathbb{N}$ (i.e., each SIG_k can be instantiated using the same symmetric pairing-friendly groups $\mathbb{G}$).

Two-Parameter Families of GS Proofs and EUF-CMA-Secure Signatures. Let us view PS as a family of GS proof systems and SIG a family of EUF-CMA-secure signature schemes defined in *two* security parameters λ and k. Such families may be constructed based on the (two parameters variant) of the SCasc assumption or other matrix assumptions. Consequently, we consider a security experiment where the adversary receives two security parameters and has advantage $\mathsf{Adv}_{\mathsf{PS},\mathcal{A}}^{\mathsf{ind\text{-}crs}}(\lambda, k)$ and $\mathsf{Adv}_{\mathsf{SIG},\mathcal{B}}^{\mathsf{euf\text{-}cma}}(\lambda, k)$, respectively. Note that this also defines the concrete advantages $\mathsf{CAdv}_{\mathsf{PS}}^{\mathsf{ind\text{-}crs}}(t, \lambda, k)$ and $\mathsf{CAdv}_{\mathsf{SIG}}^{\mathsf{euf\text{-}cma}}(t, \lambda, k)$ (as generically defined in Sect. 2). We define asymptotic security for these families following the approach taken for our reconfigurable security definition. That means, we call PS (SIG) secure if for every polynomial $t(\lambda, k)$ the advantage $\mathsf{CAdv}_{\mathsf{PS}}^{\mathsf{ind\text{-}crs}}(t(\lambda, k), \lambda, k)$ ($\mathsf{CAdv}_{\mathsf{SIG}}^{\mathsf{euf\text{-}cma}}(t(\lambda, k), \lambda, k)$) can be split up into nonnegatively-valued functions $f_1 : \mathbb{N}^2 \to \mathbb{R}_0^+, f_2 : \mathbb{N}^2 \to \mathbb{R}_0^+, f_3 : \mathbb{N}^3 \to \mathbb{R}_0^+$ such that for some polynomials $t_1(\lambda, k), t_2(\lambda, k), t_3(\lambda, k)$ the sum $f_1(t_1(\lambda, k), \lambda) + f_2(t_2(\lambda, k), k) + f_3(t_3(\lambda, k), \lambda, k)$ is an upper bound on the advantage. Furthermore, the following conditions need to be satisfied for f_1, f_2, f_3:

- For all $k \in \mathbb{N}$ it holds that $f_1(t_1(\lambda, k), \lambda)$ is negligible in λ
- For all $\lambda \in \mathbb{N}$ it holds that $f_2(t_2(\lambda, k), k)$ is negligible in k
- For all $k \in \mathbb{N}$ it holds that $f_3(t_3(\lambda, k), \lambda, k)$ is negligible in λ
- For all $\lambda \in \mathbb{N}$ it holds that $f_3(t_3(\lambda, k), \lambda, k)$ is negligible in k.

Correctness of RSIG, in terms of Definition 5, directly follows from the completeness of the underlying proof system.

Lemma 3. *Let a $t \in \mathbb{N}$ be given and let t' denote the maximal runtime of the experiment* $\mathsf{Exp}_{\mathsf{RSIG},\cdot}^{\mathsf{r\text{-}euf\text{-}cma}}(\lambda, k)$ *involving arbitrary adversaries with runtime t. Then it holds that*

$$\mathsf{CAdv}_{\mathsf{RSIG}}^{\mathsf{r\text{-}euf\text{-}cma}}(t, \lambda, k) \leq 2 \cdot \mathsf{CAdv}_{\mathsf{PS}}^{\mathsf{ind\text{-}crs}}(s_1, \lambda, k) + \mathsf{CAdv}_{\mathcal{G}}^{\mathsf{cdh}}(s_2, \lambda) + \mathsf{CAdv}_{\mathsf{SIG}}^{\mathsf{euf\text{-}cma}}(s_3, \lambda, k) \quad (20)$$

where $t' \approx s_1 \approx s_2 \approx s_3$.

Setup(1^λ)	MKGen$(\mathcal{PP})$
$(p, G, g, G_T, e) \leftarrow \mathcal{G}(1^\lambda)$ return $\mathcal{PP} := (p, G, g, G_T, e)$	parse $\mathcal{PP} := (p, G, g, G_T, e)$ $x, y \leftarrow \mathbb{Z}_n$ return $mpk := ([x], [y])$, $msk := [xy]$

CRSGen$(\mathcal{PP}, 1^k)$	SKGen(CRS_k, msk)
$(CRS_{\mathsf{PS}_k}, td_k) \leftarrow \mathsf{CRSGen}_{\mathsf{PS}_k}(\texttt{wi}, \mathcal{PP})$ $(\widetilde{sk}_k, \widetilde{vk}_k) \leftarrow \mathsf{Gen}_{\mathsf{SIG}_k}(\mathcal{PP})$ return $CRS_k := (CRS_{\mathsf{PS}_k}, \widetilde{vk}_k, \mathcal{PP}, k)$	parse CRS_k as $(CRS_{\mathsf{PS}_k}, \widetilde{vk}_k, \mathcal{PP}, k)$ set $\mathcal{X} := "\exists z : e(mpk_1, mpk_2) = e(z, [1])"$ $\pi_k \leftarrow \mathsf{Prove}_{\mathsf{PS}_k}(CRS_{\mathsf{PS}_k}, \mathcal{X}, msk)$ return $sk_k := (CRS_k, \pi_k)$

PKGen(CRS_k, mpk)	
return $pk_k := (CRS_k, mpk)$	

Sig(m, sk_k)	Ver(pk_k, σ, m)
parse sk_k as (CRS_k, π_k) and CRS_k as $(CRS_{\mathsf{PS}_k}, \widetilde{vk}_k, \mathcal{PP}, k)$ set $\mathcal{Y}_k := "\exists(\pi_k, \Sigma_k) :$ $\mathsf{Ver}_{\mathsf{PS}_k}(CRS_{\mathsf{PS}_k}, \pi_k, \mathcal{X}) = 1 \lor \mathsf{Ver}_{\mathsf{SIG}_k}(\widetilde{vk}_k, \Sigma_k, m) = 1"$ $\pi_m \leftarrow \mathsf{Prove}_{\mathsf{PS}_k}(CRS_{\mathsf{PS}_k}, \mathcal{Y}_k, sk_k)$ return $\sigma := (\pi_m, \mathcal{Y}_k)$	parse pk_k as (CRS_k, mpk) and CRS_k as $(CRS_{\mathsf{PS}_k}, \widetilde{vk}_k, \mathcal{PP}, k)$ parse $\sigma := (\pi_m, \mathcal{Y}_k)$ verify that $\mathcal{Y}_k$ contains m and $\mathcal{X}$ and $\mathcal{X}$ contains mpk return $\mathsf{Ver}_{\mathsf{PS}_k}(CRS_{\mathsf{PS}_k}, \pi_m, \mathcal{Y}_k)$

Fig. 4. Our reconfigurable signature scheme

Theorem 3. *Let us assume that* PS *is a secure two-parameter family of Groth-Sahai proof systems,* SIG *a secure two-parameter family of EUF-CMA secure signature schemes and the CDH assumption holds with respect to* $\mathcal{G}$. *Then* RSIG *is R-EUF-CMA secure.*

We omit the proof of Theorem 3 as it is analogous to the proof of Lemma 2. In the remainder of this section, we sketch a proof for Lemma 3.

Proof Sketch: We use a hybrid argument to prove our theorem. Starting with the R-EUF-CMA security game, we end up with a game in which the adversary has no chance of winning. It follows that $\mathsf{Adv}^{\mathsf{r\text{-}euf\text{-}cma}}_{\mathsf{RSIG}, \mathcal{A}}(\lambda, k)$ is smaller than the sum of advantages of adversaries distinguishing between all subsequent intermediate games. Throughout the proof, $\mathsf{Adv}^{Gi}_{\mathcal{A}}(\lambda, k)$ denotes the winning probability of $\mathcal{A}$ when running in game i.

Game 0: This is the original security game $\mathsf{Exp}^{\mathsf{r\text{-}euf\text{-}cma}}_{\mathsf{RSIG}, \mathcal{A}}$. Note that the signature oracle of $\mathcal{A}$ is implemented using sk_k and thus, implicitly, msk as a witness. We have that $\mathsf{Adv}^{\mathsf{r\text{-}euf\text{-}cma}}_{\mathsf{RSIG}, \mathcal{A}}(\lambda, k) = \mathsf{Adv}^{G0}_{\mathcal{A}}(\lambda, k)$.

Game 1: Here we modify the implementation of the signature oracle by letting the experiment use the formerly unused signing key of the signature scheme SIG_k. More formally, let $state$ denote the output of $\mathcal{A}^{\mathsf{Break}}(\mathcal{PP}, mpk, \text{"learn"})$. While running $(CRS^*, \widetilde{vk}^*, \mathcal{PP}, k) \leftarrow \mathsf{CRSGen}(\mathcal{PP}, 1^k)$, the experiment learns $\widetilde{sk}^*$, the signing key corresponding to $\widetilde{vk}^*$. We now let the experiment answer $\mathcal{A}$'s oracle queries $\mathsf{Sig}_k(sk^*, m)$ for $m \in \mathcal{M}$ with signatures $\mathsf{Prove}_{\mathsf{PS}_k}(CRS^*, \mathcal{Y}^*, \tau)$, where $\tau \leftarrow \mathsf{Sig}_{\mathsf{SIG}_k}(\widetilde{sk}^*, m)$ and $\mathcal{Y}^* := \text{"}\exists(\pi^*, \Sigma^*) : \mathsf{Ver}_{\mathsf{PS}_k}(CRS^*, \pi^*, \mathcal{X}) = 1 \vee \mathsf{Ver}_{\mathsf{SIG}_k}(\widetilde{vk}^*, \Sigma^*, m) = 1\text{"}$.

Since the proofs generated by PS_k are perfectly WI, the $\mathcal{A}$'s view in game 0 and game 1 is exactly the same and thus we have $\mathsf{Adv}_{\mathcal{A}}^{G1}(\lambda, k) = \mathsf{Adv}_{\mathcal{A}}^{G0}(\lambda, k)$.

Game 2: In this game, we want to switch the CRS for which $\mathcal{A}$ forges a message from witness indistinguishable to sound mode. For this, the experiment runs $(CRS_{\mathsf{PS}_k}, td_k) \leftarrow \mathsf{CRSGen}_{\mathsf{PS}_k}(\mathsf{sound}, \mathcal{PP})$ and $(\widetilde{sk}^*, \widetilde{vk}^*) \leftarrow \mathsf{Gen}_{\mathsf{SIG}_k}(\mathcal{PP})$ and sets $CRS^* := (CRS_{\mathsf{PS}_k}, \widetilde{vk}^*, \mathcal{PP}, k)$.

Claim. For every λ, k and $\mathcal{A}$, there is an adversary $\mathcal{B}$ with $\mathbf{T}(\mathcal{A}) \approx \mathbf{T}(\mathcal{B})$ and $\mathsf{Adv}_{\mathsf{PS},\mathcal{B}}^{\mathsf{ind\text{-}crs}}(\lambda, k) := \left|\frac{1}{2} - \Pr\left[\mathcal{B}(CRS_{\mathsf{PS}_k}) \to \mathsf{mode}\right]\right| = \left|\frac{\mathsf{Adv}_{\mathcal{A}}^{G1}(\lambda, k) - \mathsf{Adv}_{\mathcal{A}}^{G2}(\lambda, k)}{2}\right|$, where $(CRS_{\mathsf{PS}_k}, td_k) \leftarrow \mathsf{CRSGen}_{\mathsf{PS}_k}(\mathsf{mode}, \mathcal{PP})$ and $\mathsf{mode} \in \{\mathtt{wi}, \mathtt{sound}\}$.

Proof. Note that $\mathcal{A}$'s view in game 1 and 2 is exactly the same until he sees CRS^*. We construct $\mathcal{B}$ as follows. $\mathcal{B}$ gets CRS_{PS_k} and then plays game 1 with $\mathcal{A}$ until $\mathcal{A}$ outputs $state$. Now $\mathcal{B}$ sets $CRS^* := (CRS_{\mathsf{PS}_k}, \widetilde{vk}^*, \mathcal{PP}, k)$ and proceeds the game. Note that this is possible since $\mathcal{B}$ does not make use of a trapdoor for CRS_{PS_k}. $\mathcal{B}$ finally outputs $\mathtt{wi}$ if $\mathcal{A}$ wins the game. If $\mathcal{A}$ loses, $\mathcal{B}$ outputs $\mathtt{sound}$.

We now analyze the advantage of $\mathcal{B}$ in guessing the CRS mode. For this, note that if $\mathsf{mode} = \mathtt{wi}$, then $\mathcal{A}$'s view is as in game 1, and if $\mathsf{mode} = \mathtt{sound}$, then $\mathcal{A}$'s view is as in game 2. Let X_i denote the event that $\mathcal{A}$ wins game i, and thus $\mathsf{Adv}_{\mathcal{A}}^{Gi}(\lambda, k) = \Pr[X_i]$. We have that

$$\Pr[\mathcal{B}\text{wins}] = \Pr[\mathcal{B} \text{ wins} \mid \mathsf{mode} = \mathtt{wi}] + \Pr[\mathcal{B} \text{ wins} \mid \mathsf{mode} = \mathtt{sound}]$$

$$= \frac{1}{2}\sum_{i=1}^{2}(\Pr[\mathcal{B} \text{ wins} \mid X_i] + \Pr[\mathcal{B} \text{ wins} \mid \neg X_i])$$

$$= \frac{1}{2}(1 \cdot \mathsf{Adv}_{\mathcal{A}}^{G1}(\lambda, k) + 0 \cdot (1 - \mathsf{Adv}_{\mathcal{A}}^{G1}(\lambda, k)) + 0 \cdot \mathsf{Adv}_{\mathcal{A}}^{G2}(\lambda, k) + 1 - \mathsf{Adv}_{\mathcal{A}}^{G2}(\lambda, k)$$

$$= \frac{1}{2}(\mathsf{Adv}_{\mathcal{A}}^{G1}(\lambda, k) + 1 - \mathsf{Adv}_{\mathcal{A}}^{G2}(\lambda, k)) = \frac{1}{2} + \frac{\mathsf{Adv}_{\mathcal{A}}^{G1}(\lambda, k) - \mathsf{Adv}_{\mathcal{A}}^{G2}(\lambda, k)}{2}$$

$$\Rightarrow \left|\Pr[\mathcal{B}\text{wins}] - \frac{1}{2}\right| = \left|\frac{\mathsf{Adv}_{\mathcal{A}}^{G1}(\lambda, k) - \mathsf{Adv}_{\mathcal{A}}^{G2}(\lambda, k)}{2}\right|$$

Game 3: Now, the experiment no longer uses knowledge of msk to produce answers $sk_k \leftarrow \mathsf{SKGen}(CRS_{\mathsf{PS}_k}, msk)$ to Break-queries. Instead, we let the experiment use the trapdoor of the CRS to generate the proofs. This can be done since the experiment always answers Break-oracle queries by running $(CRS_{\mathsf{PS}_k}, td_k) \leftarrow \mathsf{CRSGen}_{\mathsf{PS}_k}(\mathsf{wi}, \mathcal{PP})$ and, since in wi mode, td_k can be used to simulate a proof sk_k without actually using msk. Moreover, the proofs are perfectly indistinguishable from the proofs in Game 2 and thus $\mathcal{A}$'s view in Games 2 and 3 are identical and we have $\mathsf{Adv}_{\mathcal{A}}^{G3}(\lambda, k) = \mathsf{Adv}_{\mathcal{A}}^{G2}(\lambda, k)$.

Game 4: We modify the winning conditions of the experiment: $\mathcal{A}$ loses if sk^*, i.e., a solution to a CDH instance, can be extracted from the forgery.

Claim. For every λ and k, and every adversary $\mathcal{A}$, there exists an adversary $\mathcal{C}$ with $\mathbf{T}(\mathcal{A}) \approx \mathbf{T}(\mathcal{C})$ and

$$\mathsf{Adv}_{\mathcal{G},\mathcal{C}}^{\mathsf{cdh}}(\lambda) := \Pr\left[\mathcal{C}(\mathbb{G}, [x], [y]) = [xy]\right] \geq \left|\mathsf{Adv}_{\mathcal{A}}^{G3}(\lambda, k) - \mathsf{Adv}_{\mathcal{A}}^{G4}(\lambda, k)\right| \quad (21)$$

where $\mathbb{G} \leftarrow \mathcal{G}(1^\lambda)$ and the probability is over the random coins of $\mathcal{G}$ and $\mathcal{C}$.

Proof. First note that $\mathcal{A}$'s view is identical in both games, since we only modified the winning condition. Let E denote the event that sk^* can be extracted from the forgery produced by $\mathcal{A}$. Let X_3, X_4 denote the random variables describing the output of the experiment in Game 3 and Game 4, respectively. From the definition of the winning conditions of both games it follows that

$$\Pr\left[X_3 = 1 | \neg E\right] = \Pr\left[X_4 = 1 | \neg E\right] \implies |\Pr\left[X_3 = 1\right] - \Pr\left[X_4 = 1\right]| \leq \Pr\left[E\right]$$
$$\leq \Pr\left[\mathcal{C}(\mathbb{G}, mpk) = msk\right]$$

where the first inequality follows from the difference lemma [22] and the latter holds because, since msk is not needed to run the experiment, $\mathcal{C}$ can run $\mathcal{A}$ and, since E happened, extract the CDH solution from the forgery.

Game 5: We again modify the winning conditions of $\mathcal{A}$ by: $\mathcal{A}$ loses the game if a valid signature under SIG_k can be extracted from the forgery.

Claim. For every λ and k, and every adversary $\mathcal{A}$, there exists a $\mathcal{D}$ with $\mathbf{T}(\mathcal{A}) \approx \mathbf{T}(\mathcal{D})$ and

$$\mathsf{Adv}_{\mathsf{SIG}_k,\mathcal{D}}^{\mathsf{euf\text{-}cma}}(\lambda) := \Pr\left[\mathsf{Exp}_{\mathsf{SIG}_k,\mathcal{D}}^{\mathsf{euf\text{-}cma}}(\lambda) = 1\right] \geq \mathsf{Adv}_{\mathcal{A}}^{G4}(\lambda, k) - \mathsf{Adv}_{\mathcal{A}}^{G5}(\lambda, k) \quad (22)$$

Proof. The proof proceeds similar to the proof of the last claim. Note that the signature oracle provided by the EUF-CMA experiment can be used to answer $\mathcal{A}$'s queries to the oracle $\mathsf{Sig}_k(sk^*, \cdot)$.

Now let us determine the chances of $\mathcal{A}$ in winning game 5. If $\mathcal{A}$ does not know any of the two witnesses, it follows from the perfect soundness of CRS^* that $\mathcal{A}$ can not output a valid proof and therefore never wins game 5. Collecting advantages over all games concludes our proof sketch of Theorem 3.

Instantiation Based on SCasc. Towards an instantiation of our scheme, we need to choose a concrete family PS_k of NIWI proof systems and a family SIG_k of EUF-CMA signature schemes. We seek an interesting instantiation where reconfiguration of the PKI using a higher value of k (i.e., publishing a new CRS) leads to a system with increased security.

For this purpose, PS_k and SIG_k should be based on a family of assumptions that (presumably) become weaker as k grows such as the $\mathcal{D}_k$-MDDH assumption families from [11]. The k-SCasc assumption family seen in Sect. 2 is one interesting member of this class.

In the uniform adversary setting, [11,16] shows that any $\mathcal{D}_k$-MDDH assumption family is enough to obtain a family of GS proof system $PS_k := (\mathsf{CRSGen}_{PS_k}, \mathsf{Prove}_{PS_k}, \mathsf{Ver}_{PS_k})$ with computationally indistinguishable CRS modes. More formally, one can show for any k that if $\mathcal{D}_k$-MDDH holds w.r.t. $\mathcal{G}$, then for all PPT adversaries $\mathcal{A}$, the advantage $\mathsf{Adv}_{PS_k,\mathcal{A}}^{\mathsf{ind\text{-}crs}}(\lambda) := |\Pr[\mathcal{A}(CRS_{PS_k}) = \mathsf{mode}] - \frac{1}{2}|$ is negligible in λ, where $CRS_{PS_k} \leftarrow \mathsf{CRSGen}_{PS_k}(\mathbb{G})$ and $\mathbb{G} \leftarrow \mathcal{G}(1^\lambda)$. If we base the construction in [11,16] on the two-parameter variant of SCasc as defined in Sect. 4.2 (or of any other $\mathcal{D}_k$-MDDH assumption, which can be defined in a straightforward manner), we obtain a family of GS proof systems as required by our RSIG scheme.

Very recently, the concept of affine MACs was introduced in [5]. Basing their construction on the Naor-Reingold PRF, whose security follows from any $\mathcal{D}_k$-MDDH assumption, we can now construct a family of signature schemes SIG_k, where for each k we have that SIG_k is is EUF-CMA secure under $\mathcal{D}_k$-MDDH using the well-known fact that every PR-ID-CPA-secure IBE system implies an EUF-CMA-secure signature system.[7] Furthermore, we claim that using the same construction we can obtain a family of signature schemes as required by using the two-parameter variant of SCasc (or of any other $\mathcal{D}_k$-MDDH assumption) as the underlying assumption.

References

1. Barak, B., Goldreich, O., Impagliazzo, R., Rudich, S., Sahai, A., Vadhan, S.P., Yang, K.: On the (im)possibility of obfuscating programs. J. ACM **59**(2), 6 (2012)
2. Bellare, M., Desai, A., Jokipii, E., Rogaway, P.: A concrete security treatment of symmetric encryption. In: Proceedings of FOCS 1997, pp. 394–403. IEEE Computer Society (1997)
3. Bellare, M., Kilian, J., Rogaway, P.: The security of the cipher block chaining message authentication code. J. Comput. Syst. Sci. **61**(3), 362–399 (2000)
4. Bellare, M., Miner, S.K.: A forward-secure digital signature scheme. In: Wiener, M. (ed.) CRYPTO 1999. LNCS, vol. 1666, pp. 431–448. Springer, Heidelberg (1999)
5. Blazy, O., Kiltz, E., Pan, J.: (Hierarchical) identity-based encryption from affine message authentication. In: Garay, J.A., Gennaro, R. (eds.) CRYPTO 2014, Part I. LNCS, vol. 8616, pp. 408–425. Springer, Heidelberg (2014)

---

[7] In fact, [5] constructs an IB-KEM. It is straightforward to verify that a PR-IDKEM-CPA secure IB-KEM scheme also implies an EUF-CMA-secure signature scheme.

6. Boneh, D., Franklin, M.: Identity-based encryption from the weil pairing. In: Kilian, J. (ed.) CRYPTO 2001. LNCS, vol. 2139, pp. 213–229. Springer, Heidelberg (2001)
7. Boneh, D., Zhandry, M.: Multiparty key exchange, efficient traitor tracing, and more from indistinguishability obfuscation. In: Garay, J.A., Gennaro, R. (eds.) CRYPTO 2014, Part I. LNCS, vol. 8616, pp. 480–499. Springer, Heidelberg (2014)
8. Brakerski, Z., Kalai, Y.T., Katz, J., Vaikuntanathan, V.: Overcoming the hole in the bucket: public-key cryptography resilient to continual memory leakage. In: Proceedings of FOCS 2010, pp. 501–510. IEEE Computer Society (2010)
9. Canetti, R., Halevi, S., Katz, J.: A forward-secure public-key encryption scheme. J. Cryptology **20**(3), 265–294 (2007)
10. Diffie, W., van Oorschot, P.C., Wiener, M.J.: Authentication and authenticated key exchanges. Des. Codes Crypt. **2**(2), 107–125 (1992)
11. Escala, A., Herold, G., Kiltz, E., Ràfols, C., Villar, J.: An algebraic framework for Diffie-Hellman assumptions. In: Canetti, R., Garay, J.A. (eds.) CRYPTO 2013, Part II. LNCS, vol. 8043, pp. 129–147. Springer, Heidelberg (2013)
12. Garg, S., Gentry, C., Halevi, S., Raykova, M., Sahai, A., Waters, B.: Candidate indistinguishability obfuscation and functional encryption for all circuits. In: Proceedings of FOCS 2013, pp. 40–49. IEEE Computer Society (2013)
13. Gentry, C., Peikert, C., Vaikuntanathan, V.: Trapdoors for hard lattices and new cryptographic constructions. In: Proceedings of STOC 2008, pp. 197–206. ACM (2008)
14. Groth, J.: Simulation-sound NIZK proofs for a practical language and constant size group signatures. In: Lai, X., Chen, K. (eds.) ASIACRYPT 2006. LNCS, vol. 4284, pp. 444–459. Springer, Heidelberg (2006)
15. Groth, J., Sahai, A.: Efficient non-interactive proof systems for bilinear groups. In: Smart, N.P. (ed.) EUROCRYPT 2008. LNCS, vol. 4965, pp. 415–432. Springer, Heidelberg (2008)
16. Herold, G., Hesse, J., Hofheinz, D., Ràfols, C., Rupp, A.: Polynomial spaces: a new framework for composite-to-prime-order transformations. In: Garay, J.A., Gennaro, R. (eds.) CRYPTO 2014, Part I. LNCS, vol. 8616, pp. 261–279. Springer, Heidelberg (2014)
17. Maurer, U.M., Yacobi, Y.: A non-interactive public-key distribution system. Des. Codes Cryptograph. **9**(3), 305–316 (1996)
18. Regev, O.: On lattices, learning with errors, random linear codes, and cryptography. In: Proceedings of STOC 2005, pp. 84–93. ACM (2005)
19. Sahai, A., Waters, B.: How to use indistinguishability obfuscation: deniable encryption, and more. Cryptology ePrint Archive, Report 2013/454 (2013). http://eprint.iacr.org/2013/454
20. Sahai, A., Waters, B.: How to use indistinguishability obfuscation: deniable encryption, and more. In: Proceedings of STOC 2014, pp. 475–484. ACM (2014)
21. Shamir, A.: Identity-based cryptosystems and signature schemes. In: Blakely, G.R., Chaum, D. (eds.) CRYPTO 1984. LNCS, vol. 196, pp. 47–53. Springer, Heidelberg (1985)
22. Shoup, V.: Sequences of games: a tool for taming complexity in security proofs. IACR Cryptology ePrint Archive 2004, 332 (2004). http://eprint.iacr.org/2004/332
23. Waters, B.: Efficient identity-based encryption without random oracles. In: Cramer, R. (ed.) EUROCRYPT 2005. LNCS, vol. 3494, pp. 114–127. Springer, Heidelberg (2005)

Multilinear Maps from Obfuscation

Martin R. Albrecht[1][(✉)], Pooya Farshim[2], Dennis Hofheinz[3],
Enrique Larraia[1], and Kenneth G. Paterson[1]

[1] Royal Holloway, University of London, Egham, UK
martin.albrecht@rhul.ac.uk
[2] Queen's University Belfast, Belfast, UK
[3] Karlsruhe Institute of Technology, Karlsruhe, Germany

Abstract. We provide constructions of multilinear groups equipped with natural hard problems from indistinguishability obfuscation, homomorphic encryption, and NIZKs. This complements known results on the constructions of indistinguishability obfuscators from multilinear maps in the reverse direction.

We provide two distinct, but closely related constructions and show that multilinear analogues of the DDH assumption hold for them. Our first construction is *symmetric* and comes with a κ-linear map $\mathbf{e} : \mathbb{G}^\kappa \longrightarrow \mathbb{G}_T$ for prime-order groups $\mathbb{G}$ and $\mathbb{G}_T$. To establish the hardness of the κ-linear DDH problem, we rely on the existence of a base group for which the $(\kappa-1)$-strong DDH assumption holds. Our second construction is for the *asymmetric* setting, where $\mathbf{e} : \mathbb{G}_1 \times \cdots \times \mathbb{G}_\kappa \longrightarrow \mathbb{G}_T$ for a collection of $\kappa + 1$ prime-order groups $\mathbb{G}_i$ and $\mathbb{G}_T$, and relies only on the standard DDH assumption in its base group. In both constructions the linearity κ can be set to any arbitrary but a priori fixed polynomial value in the security parameter.

We rely on a number of powerful tools in our constructions: (probabilistic) indistinguishability obfuscation, dual-mode NIZK proof systems (with perfect soundness, witness indistinguishability and zero knowledge), and additively homomorphic encryption for the group $\mathbb{Z}_N^+$. At a high level, we enable "bootstrapping" multilinear assumptions from their simpler counterparts in standard cryptographic groups, and show the equivalence of IO and multilinear maps under the existence of the aforementioned primitives.

Keywords: Multilinear map · Indistinguishability obfuscation · Homomorphic encryption · Decisional Diffie–Hellman · Groth–Sahai proofs

1 Introduction

1.1 Main Contribution

In this paper, we explore the relationship between multilinear maps and obfuscation. Our main contribution is a construction of multilinear maps for groups of prime order equipped with natural hard problems, using indistinguishability obfuscation (IO) in combination with other tools, namely NIZK proofs,

© International Association for Cryptologic Research 2016
E. Kushilevitz and T. Malkin (Eds.): TCC 2016-A, Part I, LNCS 9562, pp. 446–473, 2016.
DOI: 10.1007/978-3-662-49096-9_19

homomorphic encryption, and a base group $\mathbb{G}_0$ satisfying a mild cryptographic assumption. This complements known results in the reverse direction, showing that various forms of indistinguishability obfuscation can be constructed from multilinear maps [GGH+13b, CLTV15, Zim15]. The relationship between IO and multilinear maps is a very natural question to study, given the rich diversity of cryptographic constructions that have been obtained from both multilinear maps and obfuscation, and the apparent fragility of current constructions for multilinear maps. More on this below.

We provide two distinct but closely related constructions. One is for multilinear maps in the *symmetric* setting, that is non-degenerate multilinear maps $\mathbf{e} : \mathbb{G}_1{}^\kappa \longrightarrow \mathbb{G}_T$ for groups $\mathbb{G}_1$ and $\mathbb{G}_T$ of prime order N. Our construction relies on the existence of a base group $\mathbb{G}_0$ in which the $(\kappa - 1)$-SDDH assumption holds—this states that, given a κ-tuple of $\mathbb{G}_0$-elements $(g, g^\omega, \ldots, g^{\omega^{\kappa-1}})$, we cannot efficiently distinguish g^{ω^κ} from a random element of $\mathbb{G}_0$. Under this assumption, we prove that the κ-MDDH problem, a natural analogue of the DDH problem as stated below, is hard.

(**The κ-MDDH problem, informal**). Given a generator g_1 of $\mathbb{G}_1$ and $\kappa +$ 1 group elements $g_1^{a_i}$ in $\mathbb{G}$ with $a_i \leftarrow_\$ \mathbb{Z}_N$, distinguish $\mathbf{e}(g_1, \ldots, g_1)^{\prod_{i=1}^{\kappa+1} a_i}$ from a random element of $\mathbb{G}_T$.

This problem can be used as the basis for several cryptographic constructions [BS03] including, as the by now the classic example of multiparty non-interactive key exchange (NIKE) [GGH13a].

Our other construction is for the *asymmetric* setting, that is multilinear maps $\mathbf{e} : \mathbb{G}_1 \times \cdots \times \mathbb{G}_\kappa \longrightarrow \mathbb{G}_T$ for a collection of κ groups $\mathbb{G}_i$ and $\mathbb{G}_T$ all of prime order N. It uses a base group $\mathbb{G}_0$ in which we require only that the standard DDH assumption holds. For this construction, we show that a natural asymmetric analogue of the κ-MDDH assumption holds (wherein all but two of the $\kappa + 1$ group elements input to $\mathbf{e}$ come from distinct groups).

In Sect. 7, we also show the intractability of the *rank problem* for our construction for multilinear maps in the symmetric setting; this is a generalization of DDH-like problems to matrices that has proven to be useful in cryptographic constructions [BHHO08, NS09, GHV12, BLMR13, EHK+13].

At a high level, then, our constructions are able to "bootstrap" from rather mild assumptions in a standard cryptographic group to much stronger multilinear assumptions in a group (or groups, in the asymmetric setting) equipped with a κ-linear map. Here κ is fixed up-front at construction time, but is otherwise unrestricted. Of course, such constructions cannot be expected to come "for free," and we need to make use of powerful tools including probabilistic IO (PIO) for obfuscating randomized circuits [CLTV15], dual-mode NIZK proofs enjoying perfect soundness (for a binding CRS), perfect witness indistinguishability (for a hiding CRS), and perfect zero knowledge, and additive homomorphic encryption for the group $(\mathbb{Z}_N, +)$ (or alternatively, a perfectly correct FHE scheme). It is an important open problem arising from our work to weaken the requirements on, or remove altogether, these additional tools.

1.2   General Approach

Our approach to obtaining multilinear maps in the symmetric setting is as follows (with many details to follow in the main body). Let $\mathbb{G}_0$ with generator g_0 be a group of prime order N in which the $(\kappa - 1)$-SDDH assumption holds.

We work with redundant encodings of elements h of the base group $\mathbb{G}_0$ of the form $h = g_0^{x_0}(g_0^\omega)^{x_1}$ where g_0^ω comes from a $(\kappa - 1)$-SDDH instance; we write $\mathbf{x} = (x_0, x_1)$ for the vector of exponents *representing* h. Then $\mathbb{G}_1$ consists of all strings of the form $(h, \mathbf{c}_1, \mathbf{c}_2, \pi)$ where $h \in \mathbb{G}_0$, ciphertext $\mathbf{c}_1$ is a homomorphic encryption under public key pk_1 of a vector $\mathbf{x}$ representing h, ciphertext $\mathbf{c}_2$ is a homomorphic encryption under a second public key pk_2 of another vector $\mathbf{y}$ also representing h, and π is a NIZK proof showing consistency of the two vectors $\mathbf{x}$ and $\mathbf{y}$, i.e., a proof that the plaintexts $\mathbf{x}, \mathbf{y}$ underlying $\mathbf{c}_1, \mathbf{c}_2$ encode the *same* group element h. Note that each element of the base group $\mathbb{G}_0$ is multiply represented when forming elements in $\mathbb{G}_1$, but that equality of group elements in $\mathbb{G}_1$ is easy to test. An alternative viewpoint is to consider $(\mathbf{c}_1, \mathbf{c}_2, \pi)$ as being *auxiliary information* accompanying element $h \in \mathbb{G}_0$; we prefer the perspective of redundant encodings, and our abstraction in Sect. 3 is stated in such terms. When viewed in this way, our approach can be seen as closely related to the Naor–Yung paradigm for constructing CCA-secure PKE [NY90].

Addition of two elements in $\mathbb{G}_1$ is carried out by an obfuscation of a circuit C_{Add} that is published along with the groups. It has the secret keys sk_1, sk_2 hard-coded in; it first checks the respective proofs, then uses the additive homomorphic property of the encryption scheme to combine ciphertexts, and finally uses the secret keys sk_1, sk_2 as witnesses to generate a new NIZK proof showing equality of encodings. Note that the new encoding is as compact as that of the two input elements.

The multilinear map on inputs $(h_i, \mathbf{c}_{i,1}, \mathbf{c}_{i,2}, \pi_i)$ for $1 \le i \le \kappa$ is computed using the obfuscation of a circuit C_{Map} that has sk_1 and ω hard-coded in. This allows C_{Map} to "extract" full exponents of h_i in the form $(x_{i,1} + \omega \cdot x_{i,2})$ from $\mathbf{c}_{i,1}$, and thereby compute the element $g_0^{\prod_i (x_{i,1} + \omega \cdot x_{i,2})}$. This is defined to be the output of our multilinear map $\mathbf{e}$, and so our target group $\mathbb{G}_T$ is in fact $\mathbb{G}_0$, the base group. The multilinearity of $\mathbf{e}$ follows immediately from the form of the exponent.

In the asymmetric case, the main difference is that we work with different values ω_i in each of our input groups $\mathbb{G}_i$. However, the groups are all constructed via redundant encodings, just as above.

This provides a high-level view of our approach, but no insight into why the approach achieves our aim of building multilinear maps with associated hard problems. Let us give some intuition on why the κ-MDDH problem is hard in our setting. We transform a κ-MDDH tuple $\mathbf{h} = ((g_1^{a_i})_{i \le \kappa+1}, g_T^d)$, where d is the product of the $a_i \in \mathbb{Z}_N$, g_1 is in the "encoded" form above, thus $g_1 = (h_1, \mathbf{c}_1, \mathbf{c}_2, \pi)$, and g_T is a generator of $\mathbb{G}_T = \mathbb{G}_0$, into another κ-MDDH tuple $\mathbf{h}'$ with exponents $a_i' = a_i + \omega$ for $i \le \kappa$. This means that the exponent of the challenge element in the target group $d' = \prod_1^\kappa (a_i + \omega) a_{\kappa+1}$ can be seen as a degree κ polynomial in ω. Therefore, with the knowledge of the a_i and a $(\kappa - 1)$-SDDH challenge, with ω implicit in the exponent, we are able to randomize $g_T^{d'}$ replacing $g_T^{\omega^\kappa}$ with a uniform value.

Nevertheless, in the preceding simplistic argument we have made two assumptions. The first is that we are able to provide an obfuscation of a circuit C'_{Map} that has the same functionality as C_{Map} over $\mathbb{G}_1$ *without* the explicit knowledge of ω. We resolve this by showing a way of evaluating the κ-linear map on any elements of $\mathbb{G}_1$ using only the powers $g_0^{\omega^i}$ for $1 \leq i \leq \kappa-1$, and vectors extracted from the accompanying ciphertexts, and then applying IO to the two circuits.[1]

The second assumption we made is that we can indeed switch from $\mathbf{h}$ to $\mathbf{h}'$ without being noticed. In other words, that the vectors $\mathbf{x}_i$, $\mathbf{y}_i$ representing g^{a_i} can be replaced (without being noticed) with vectors $\mathbf{h}_i'$ whose second coordinate is always fixed. Intuitively this is based on the IND-CPA security of the FHE scheme, but in order to give a successful reduction we also have to change the circuit C_{Add} (since C_{Add} uses both decryption keys). We show two ways to do this: one is based on probabilistic indistinguishability obfuscation [CLTV15], and the other uses only (deterministic) indistinguishability obfuscation, and additionally exploits the specific structure of a particular (pairing-based) NIZK implementation due to Groth and Sahai [GS08].

We note that in this work we do not construct graded encoding schemes as in [GGH13a]. That is, we do not construct maps from $\mathbb{G}_i \times \mathbb{G}_j$ to $\mathbb{G}_{i+j}$. On the other hand, our construction is noiseless and is closer to multilinear maps as defined by Boneh and Silverberg [BS03].

1.3   Attacks on Multilinear Maps

Multilinear maps have been in a state of turmoil, with the discovery of attacks [CHL+15, HJ15, CLR15, MF15, Cor15] against the GGH13 [GGH13a], CLT [CLT13, CLT15] and GGH15 [GGH15] proposals. Hence, our confidence in constructions for graded encoding schemes (and thereby multilinear maps) has been shaken. On the other hand, when IO is constructed from graded encoding schemes via Barrington's theorem [GGH+13b] or dual-input straddling sets [AB15, Zim15], then none of the known attacks on graded encoding schemes seem to apply [CGH+15]. Indeed, when building IO from multilinear maps one restricts the pool of available operations to an attacker by fixing a circuit a priori which means that certain "interesting" elements cannot be (easily) constructed. Hence, currently it is perhaps more plausible to assume that IO exists than it is to assume that secure multilinear maps exist. However, we stress that more cryptanalysis of IO constructions is required to investigate what security they provide.

Moreover, even though current constructions for IO rely on graded encoding schemes, it is not implausible that alternative routes to achieving IO without relying on multilinear maps will emerge in due course. And setting aside the novel applications obtained directly from IO, multilinear maps, and more generally graded encoding schemes, have proven to be very fruitful as constructive tools

---

[1] This is not trivial since the new method should not lead to an exponential blow-up in κ.

in their own right (cf. [BS03, PTT10], resp., [FHPS13, GGH+13c, HSW13] and [GGSW13, BWZ14, TLL14, BLR+15]). This rich set of applications coupled with the current uncertainty over the status of graded encoding schemes and multilinear maps provides additional motivation to ask what additional tools are needed in order to upgrade IO to multilinear maps. As an additional benefit, we upgrade (via IO) noisy graded encoding schemes to clean multilinear maps—sometimes now informally called "dream" or "ideal" multilinear maps.

1.4 Related Work

The closest related work to ours is that of Yamakawa et al. [YYHK14, YYHK15]; indeed, their work was the starting point for ours. Yamakawa et al. construct a *self-pairing map*, that is a bilinear map from $\mathbb{G} \times \mathbb{G}$ to $\mathbb{G}$; multilinear maps can be obtained by iterating their self-pairing. Their work is limited to the RSA setting. It uses the group of signed quadratic residues modulo a Blum integer N, denoted QR_N^+, to define a pairing function that, on input elements g^x, g^y in QR_N^+, outputs g^{2xy}. In their construction, elements of QR_N^+ are augmented with auxiliary information to enable the pairing computation—in fact, the auxiliary information for an element g^x is simply an obfuscation of a circuit for computing the $2x$th power modulo $\mathrm{ord}(QR_N^+)$, and the pairing is computed by evaluating this circuit on an input g^y (say). The main contribution of [YYHK14] is in showing that these obfuscated circuits leak nothing about x or the group order.

A nice feature of their scheme is that the degree of linearity κ that can be accommodated is not limited up-front in the sense that the pairing output is also a group element to which further pairing operations (derived from auxiliary information for other group elements) can be applied. However, the construction has several drawbacks. First, the element output by the pairing does not come with auxiliary information.[2] Second, the size of the auxiliary information for a product of group elements grows exponentially with the length of the product, as each single product involves computing the obfuscation of a circuit for multiplying, with its inputs already being obfuscated circuits. Third, the main construction in [YYHK14] only builds hard problems for the self-pairing of the computational type (in fact, they show the hardness of the computational version of the κ-MDDH problem in QR_N^+ assuming that factoring is hard). Still, this is sufficient for several cryptographic applications.

In contrast, our construction is *generic* with respect to its platform group. Furthermore, the equivalent of the auxiliary information in our approach does not itself involve any obfuscation. Consequently, the description of a product

---

[2] The authors of [YYHK14] state that such information can be added in their construction, but what would be needed is the obfuscation of a circuit for computing $4xy$th powers. The information available for building this would be obfuscations of circuits for computing $2x$th and $2y$th powers, so an obfuscation of a *composition* of *already* obfuscated circuits would be required. Strictly speaking then, the auxiliary information associated with elements output by their pairing is of a different type to that belonging to the inputs, making it questionable whether "self-pairing" is the right description of what is constructed in [YYHK14].

of group elements stays compact. Indeed, given perfect additive homomorphic encryption for $(\mathbb{Z}_p, +)$, we can perform arbitrary numbers of group operations in each component group $\mathbb{G}_i$. It is an open problem to find a means of augmenting our construction with the equivalent of auxiliary information in the *target* group $\mathbb{G}_T$, to make our multilinear maps amenable to iteration and thereby achieve graded maps as per [GGH13a, CLT13].

2   Background

The security parameter is denoted by $\lambda \in \mathbb{N}$. We assume that λ is an implicit input given in unary to all algorithms. Given a randomized algorithm $\mathcal{A}$ we denote the action of running $\mathcal{A}$ on inputs $(x_1, \ldots)$ with fresh random coins r and assigning the output(s) to $y_1, \ldots$ by $(y_1, \ldots) \leftarrow_{\$} \mathcal{A}(x_1, \ldots; r)$, and for a finite set X, we denote the action of sampling a uniformly random element x from X by $x \leftarrow_{\$} X$. Vectors are written in boldface $\mathbf{x}$ and by slight abuse of notation, running algorithms on vectors of elements indicates component-wise operation. A real-valued function $\mu(\lambda)$ is negligible if $\mu(\lambda) \in \mathcal{O}(\lambda^{-\omega(1)})$. The set of all negligible functions is denoted by NEGL.

2.1   Homomorphic Public-Key Encryption

Scheme $\Pi := (\mathbf{Gen}, \mathbf{Enc}, \mathbf{Dec}, \mathbf{Eval})$ denotes a homomorphic public-key encryption (HPKE) with message space $\{0, 1\}^\lambda$, where $\mathbf{Eval}$ is a *deterministic* algorithm. We require Π to be IND-CPA, perfectly correct, and compact, and also assume that the secret keys are the random coins used in key generation; this will allow to check key pairs for validity.

2.2   Obfuscators

An algorithm $\mathbf{Obf}$ is an *obfuscator* for circuit class $\mathcal{C} = \{\mathcal{C}_\lambda\}_{\lambda \in \mathbb{N}}$ if for any $m \in \{0, 1\}^\lambda$, $C \in \mathcal{C}_\lambda$, and $\overline{C} \leftarrow_{\$} \mathbf{Obf}(C)$ we have that $C(m) = \overline{C}(m)$. The security of $\mathbf{Obf}$ with respect a class $\mathcal{C}$ requires that no PPT adversary $\mathcal{A} := (\mathcal{A}_1, \mathcal{A}_2)$ can distinguish the obfuscation of two circuits in $\mathcal{C}$ with noticeable probability. We will consider two notions of obfuscation depending on the class of permissible adversaries. The first notion is *functional equivalence*, whereby the two circuits any sampled circuits C_1, C_2 must satisfy $C(m) = C(m)$ for all m. We will write **IO** for obfuscator whenever this level of security is assumed. The second notion is *X-ind sampling* [CLTV15], which, roughly speaking, requires the existence of a domain subset $\mathcal{X}$ of size at most X such that the two circuits are functionally equivalent outside $\mathcal{X}$ and furthermore within $\mathcal{X}$ the outputs are indistinguishable. We will write **PIO** for this case.

2.3   Dual-Mode NIZK Proof Systems

In our constructions we will be relying on special types of non-interactive zero-knowledge proof systems [GS08]. These systems have "dual-mode" common reference string (CRS) generation algorithms that produce indistinguishable CRSs in the "binding" and "hiding" modes. The standard prototype for such schemes are pairing-based Groth–Sahai proofs [GS08], and using a generic NP reduction to the satisfiability of quadratic equations we can obtain a suitable proof system for any NP language. We formalize the syntax and security of such proof systems next.

SYNTAX. A relation with setup is a pair of PPT algorithms $(\mathbf{S}, \mathbf{R})$ such that $\mathbf{S}(1^\lambda)$ outputs (gpk, gsk) and $\mathbf{R}(gpk, x, w)$ is a ternary relation and outputs a bit $b \in \{0, 1\}$. A dual-mode non-interactive zero-knowledge (NIZK) proof system Σ for $(\mathbf{S}, \mathbf{R})$ consists of five algorithms as follows. (1) Algorithm $\mathbf{BCRS}(gpk, gsk)$ outputs a (binding) common reference string crs and an extraction trapdoor td_{ext}; (2) $\mathbf{HCRS}(gpk, gsk)$ outputs a (hiding) common reference string crs and a simulation trapdoor td_{zk}; (3) $\mathbf{Prove}(gpk, crs, x, w)$, on input crs, an instance x, and a witness w for x, outputs a proof π; (4) $\mathbf{Verify}(gpk, crs, x, \pi)$ on input a bit string crs, an instance x, and a proof π, outputs accept or reject; (5) $\mathbf{WExt}(td_{ext}, x, \pi)$ on input an extraction trapdoor, an instance x, and a proof π, outputs a witness w^3; and (6) $\mathbf{Sim}(td_{zk}, crs, x)$ on input the simulation trapdoor td_{zk}, the CRS crs, and an instance x, outputs a simulated proof π.

SECURITY. We require a dual-mode NIZK to meet the following requirements. (1) binding and hiding CRS indistinguishability; (2) perfect completeness under the hiding and binding modes; (3) perfect soundness under the binding mode; (4) perfect extractability under the binding mode; (5) perfect witness-indistinguishability under the hiding mode; and (6) perfect zero-knowledge under the binding mode.

2.4 Hard Membership Problems

Finally, we will use languages with hard membership problems. More specifically, we say that a family $\mathcal{L} = \{\mathcal{L}_\lambda\}$ of families $\mathcal{L}_\lambda = \{L\}$ of languages $L \subseteq U$ in a universe $U = U_\lambda$ has a hard subset membership problem if the following holds. Namely, we require that no PPT algorithm can efficiently distinguish between $x \leftarrow_\$ L$ for $L \leftarrow_\$ \mathcal{L}_\lambda$, and $x \leftarrow_\$ U = U_\lambda$.

3 Multilinear Groups with Non-unique Encodings

Before presenting our constructions, we formally introduce what we mean by a multilinear group (MLG) scheme. Our abstraction is a direct adaptation of the "cryptographic" MLG setting of [BS03] to a setting where group elements have *non-unique* encodings. In our abstraction, on top of the procedures needed for

---

[3] We note that extraction in Groth–Sahai proofs does not for all types of statements recover a witness. (Instead, for some types of statements, only g^{w_i} for a witness variable $w_i \in \mathbb{Z}_p$ can be recovered.) Here, however, we will only be interested in witnesses $w = (w_1, \ldots, w_n) \in \{0, 1\}^n$ that are bit strings, in which case extraction always recovers w. (Specifically, extraction will recover g^{w_i} for all i, and thus all w_i.).

generating, manipulating and checking group elements, we introduce an *equality-checking* procedure which generalizes that for groups with unique encodings.

SYNTAX. A multilinear group (MLG) scheme Γ consists of six PPT algorithms as follows.

Setup$(1^\lambda, 1^\kappa)$: This is the setup algorithm. On input the security parameter 1^λ and the multilinearity 1^κ, it outputs the group parameters pp. These parameters include *generators* $g_1, \ldots, g_{\kappa+1}$, *identity elements* $1_1, \ldots, 1_{\kappa+1}$, and integers $N_1, \ldots, N_{\kappa+1}$ (which will represent group orders). We assume pp is provided to the various algorithms below.

Val$_i(h)$: This is the validity testing algorithm. On input (the group parameters and) a group index $1 \leq i \leq \kappa + 1$ and a string $h \in \{0,1\}^*$, it returns $b \in \{\top, \bot\}$. We define $\mathbb{G}_i$, which is also parameterized by pp, as the set of all h for which **Val**$_i(h)$ holds. We write $h \in \mathbb{G}_i$ when **Val**$_i(h)$ holds and refer to such strings as *group elements* (since we will soon impose a group structure on $\mathbb{G}_i$). We require that the bit-strings in $\mathbb{G}_i$ have lengths that are polynomial in 1^κ and 1^λ, a property that we refer to as *compactness*.

Eq$_i(h_1, h_2)$: This is the equality testing algorithm. On input two valid group elements $h_1, h_2 \in \mathbb{G}_i$, it outputs a Boolean value $b \in \{\top, \bot\}$.[4] We require **Eq**$_i$ to define an equivalence relation. We say that the group has unique encodings if **Eq**$_i$ simply checks the equality of bit strings. We write $\mathbb{G}_i(h)$ for the set of all $h' \in \mathbb{G}_i$ such that **Eq**$_i(h, h') = \top$; for any such h, h' in $\mathbb{G}_i$ we write $h = h'$; sometimes we write $h = h'$ in $\mathbb{G}_i$ for clarity. Since "$=$" refers to equality of bit-strings as well as equivalence under **Eq**$_i$ we will henceforth will write "as bit-strings" when we mean equality in that sense. We require $|\mathbb{G}_i/\mathbf{Eq}_i|$, the number of equivalence classes into which **Eq**$_i$ partitions $\mathbb{G}_i$, to be finite and equal to N_i (where N_i comes from pp). Note that equality testing algorithms **Eq**$_i$ for $1 \leq i \leq \kappa$ can be derived from one for **Eq**$_{\kappa+1}$ using the multilinear map **e** defined below, provided $N_{\kappa+1}$ is prime.

Op$_i(h_1, h_2)$: This algorithm will define our group operation. On input two valid group elements $h_1, h_2 \in \mathbb{G}_i$ it outputs $h \in \mathbb{G}_i$. We write $h_1 h_2$ in place of **Op**$_i(h_1, h_2)$ for simplicity. We require that **Op**$_i$ respect the equivalence relations **Eq**$_i$, meaning that if $h_1 = h_2$ in $\mathbb{G}_i$ and $h \in \mathbb{G}_i$, then $h_1 h = h_2 h$ in $\mathbb{G}_i$. We also demand that $h_1 h_2 = h_2 h_1$ in $\mathbb{G}_i$ (commutativity), for any third $h_3 \in \mathbb{G}_i$ we require $h_1(h_2 h_3) = (h_1 h_2)h_3$ in $\mathbb{G}_i$ (associativity) and $h_1 1_i = h_1$ in $\mathbb{G}_i$. These requirements ensure that $\mathbb{G}_i/\mathbf{Eq}_i$ acts as an Abelian group of order N_i with respect to the operation induced by **Op**$_i$ and identity element 1_i.

The algorithm **Op** gives rise to an exponentiation algorithm **Exp**$_i(h, z)$ that on input $h \in \mathbb{G}_i$ and $z \in \mathbb{N}$ outputs an $h' \in \mathbb{G}_i$ such that $h' = h \cdots h$ in $\mathbb{G}_i$ with z occurrences of h. When no h is specified, we assume $h = g_i$. This algorithm runs in polynomial time in the length of z. We denote **Exp**$_i(h, z)$ by h^z and define $h^0 := 1_i$. Note that under the definition of N_i for any $h \in \mathbb{G}_i$

---

[4] We assume, without loss of generality, that all algorithms return $\bot$ when run on invalid group elements.

we have that $\mathbf{Exp}_i(h, N_i) = 1_i$.[5] This in turn leads to an inversion algorithm $\mathbf{Inv}_i(h)$ that on input $h \in \mathbb{G}_i$ outputs h^{N_i-1}. We insist that g_i in fact has order N_i, so that (the equivalence class containing) g_i generates $\mathbb{G}_i/\mathbf{Eq}_i$. We do not treat the case where the N_i are unknown but the formalism is easily extended to include it by adding an explicit inversion algorithm and by replacing N_i in pp with an approximation (which may be needed for sampling purposes).

We use the *bracket* notion [EHK+13] to denote an element $h = g_i^x$ in $\mathbb{G}_i$ with $[x]_i$. When using this notation, we will write the group law additively. This notation will be convenient in the construction and analysis of our MLG schemes. For example $[z]_i + [z']_i$ succinctly denotes $\mathbf{Op}_i(\mathbf{Exp}(g_i, z), \mathbf{Exp}(g_i, z'))$. Note that when writing $[z]_i$ it is *not* necessarily the case that z is explicitly known.

$\mathbf{e}(h_1, \ldots, h_\kappa)$: This is the multilinear map algorithm. For κ group elements $h_i \in \mathbb{G}_i$ as input, it outputs $h_{\kappa+1} \in \mathbb{G}_{\kappa+1}$. We demand that for any $1 \leq j \leq \kappa$ and any $h'_j \in \mathbb{G}_j$

$$\mathbf{e}(h_1, \ldots, h_j h'_j, \ldots, h_\kappa) = \mathbf{e}(h_1, \ldots, h_j, \ldots, h_\kappa)\mathbf{e}(h_1, \ldots, h'_j, \ldots, h_\kappa).$$

We also require the map to be *non-degenerate* in the sense that for some tuple of elements as input the multilinear map outputs an element of $\mathbb{G}_{\kappa+1}$ not in the equivalence class of $1_{\kappa+1}$. (This implies that $\mathbf{e}$ is surjective onto $\mathbb{G}_{\kappa+1}/\mathbf{Eq}_{\kappa+1}$ when N_i is prime, but need not imply surjectivity when $N_{\kappa+1}$ is composite.) We call an MLG scheme *symmetric* if the group algorithms are independent of the group index for $1 \leq i \leq \kappa$ and the $\mathbf{e}$ algorithm is invariant under permutations of its inputs. That is for any permutation $\pi : [\kappa] \longrightarrow [\kappa]$ we have

$$\mathbf{e}(h_1, \ldots, h_\kappa) = \mathbf{e}(h_{\pi(1)}, \ldots, h_{\pi(\kappa)}).$$

We refer to all the other cases as being *asymmetric*. To distinguish the target group we frequently write $\mathbb{G}_T$ instead of $\mathbb{G}_{\kappa+1}$ (and similarly for 1_T and g_T in place of $1_{\kappa+1}$ and $g_{\kappa+1}$) as its structure in our construction will be different from that of the source groups $\mathbb{G}_1, \ldots, \mathbb{G}_\kappa$.

$\mathbf{Sam}_i(z)$: This is the sampling algorithm. On input $z \in \mathbb{N}$ it outputs $h \in \mathbb{G}_i$ whose distribution is "close" to that of uniform over the equivalence class $\mathbb{G}_i(g_i^z)$. Here "close" is formalized via computational, statistical or perfect indistinguishability. We also allow a special input ε to this algorithm, in which case the sampler is required to output a uniformly distributed $h \in \mathbb{G}_i$ together with a z such that $h \in \mathbb{G}_i(g_i^z)$. When outputting z is not required, we say that $\mathbf{Sam}_i(\varepsilon)$ is *discrete-logarithm oblivious*. Note that for groups with unique encodings these algorithms trivially exist. For notational convenience, for a known a we define $[a]_i$ to be an element sampled via $\mathbf{Sam}_i(a)$.

In some applications, we also rely on the following algorithm, which provides a canonical string for all group elements within an equivalence class.

---

[5] However, note that N_i need not be the least integer with this property.

Ext$_i(h)$: This is the extraction algorithm. On input $h \in \mathbb{G}_i$ it outputs a string $s \in \{0,1\}^{p(\lambda)}$ where $p(\cdot)$ denotes a polynomial function. We demand that for any $h_1, h_2 \in \mathbb{G}_i$ with $h_1 = h_2$ in $\mathbb{G}_i$ we have that **Ext**$_i(h_1) = $ **Ext**$_i(h_2)$ (as bit-strings). We also require that the distribution of **Ext**$_i([z]_i)$ is uniform over $\{0,1\}^{p(\lambda)}$, for $[z]_i \leftarrow_\$ $ **Sam**$_i(\varepsilon)$. For groups with unique encodings this algorithm trivially exists.

In the full version of the paper we provide possible extensions to this syntax.

COMPARISON WITH GGH. Our formalization differs from that of [GGH13a] which defines a *graded encoding scheme*. The main difference is that a graded encoding scheme defines a $\mathbf{e}_{i,j}$ algorithm that takes inputs from $\mathbb{G}_i$ and $\mathbb{G}_j$ and returns an element in $\mathbb{G}_{i+j}$ such that the result is linear in each input. Moreover, the abstraction and construction of graded encodings schemes in [GGH13a] do not provide any validity algorithms; these are useful in certain adversarial situations such as CCA security and signature verification. Further, all known candidate constructions of graded encoding schemes are noisy and only permit a limited number of operations.

4   The Construction

We now present our construction of an MLG scheme Γ according to the syntax introduced in Sect. 3. In the later sections we will consider special cases of the construction and prove the hardness of analogues of the multilinear DDH problem under various assumptions.

We rely on the following building blocks in our MLG scheme. (1) A cyclic group $\mathbb{G}_0$ of some order N_0 with generator g_0 and identity 1_0; formally we think of this as a 1-linear MLG scheme Γ_0 with unique encodings in which $\mathbf{e}$ is trivial; the algorithm **Val**$_0$ implies that elements of $\mathbb{G}_0$ are efficiently recognizable. (2) A general-purpose obfuscator **Obf**. (3) An additively homomorphic public-key encryption scheme $\Pi := (\mathbf{Gen}, \mathbf{Enc}, \mathbf{Dec}, \mathbf{Eval})$ with plaintext space $\mathbb{Z}_N$ (alternatively, a perfectly correct HPKE scheme). (4) A dual-mode NIZK proof system. (5) A family $\mathcal{TD}$ of (families of) languages TD which has a hard subset membership problem, and such that all TD have efficiently computable witness relations with unique witnesses.[6] (See Sect. 2 for more formal definitions.)

We reserve variables and algorithms with index 0 for the base scheme Γ_0; we also write $N = N_0$. We require that the algorithms of Γ_0 except for **Setup**$_0$ and **Sam**$_0$ are deterministic. We will also use the bracket notation to denote the group elements in $\mathbb{G}_0$. For example, we write $[z]_0, [z']_0 \in \mathbb{G}_0$ for two valid elements of the base group and $[z]_0 + [z']_0 \in \mathbb{G}_0$ for **Op**$_0([z]_0, [z']_0)$. Variables with nonzero indices correspond to various source and target groups. Given all of the above components, our MLG scheme Γ consists of algorithms as detailed in the sections that follow.

---

[6] An example of such a language is the Diffie–Hellman language TD $= \{(g_1^r, g_2^r) \mid r \in \mathbb{N}\}$ in a DDH group.

4.1   Setup

The setup algorithm for Γ samples parameters $pp_0 \leftarrow_\$ \mathbf{Setup}_0(1^\lambda)$ for the base MLG scheme, generates two encryption key pairs $(pk_j, sk_j) \leftarrow_\$ \mathbf{Gen}(1^\lambda)$ $(j = 1, 2)$, and a matrix $\mathbf{W} = (\boldsymbol{\omega}_1, \ldots, \boldsymbol{\omega}_k)^t \in \mathbb{Z}_N^{\kappa \times \ell}$ where κ is the linearity and $\ell \in \{2, 3\}$ is a parameter of our construction. It sets

$$gpk := (pp_0, pk_1, pk_2, [\mathbf{W}]_0, \mathsf{TD}, y),$$

where $[\mathbf{W}]_0$ denotes a matrix of $\mathbb{G}_0$ elements that entry-wise is written in the bracket notation, $\mathsf{TD} \leftarrow_\$ \mathcal{TD}$, and y is *not* in TD. In our MLG scheme we set $N_1 = \cdots = N_{\kappa+1} := N$, where N is the group order implicit in pp_0. The setup algorithm then generates a common reference string $crs = (crs', y)$ where $crs' \leftarrow_\$ \mathbf{BCRS}(gpk, gsk)$ for a relation $(\mathbf{S}, \mathbf{R})$ that will be defined in Sect. 4.2. It also constructs two obfuscated circuits $\overline{C}_{\mathrm{Map}}$ and $\overline{C}_{\mathrm{Add}}$ which we will describe in Sects. 4.3 and 4.4. For $1 \leq i \leq \kappa$, the identity elements 1_i and group generators g_i are sampled using $\mathbf{Sam}_i(0)$ and $\mathbf{Sam}_i(x_i)$ respectively for algorithm $\mathbf{Sam}_i$ described in Sect. 4.5 with $x_i \in [N]$ that is co-prime to N. We emphasize that this approach is well defined since the operation of $\mathbf{Sam}_i$ is defined independently of the generators and the identity elements and depends only on gpk and crs. We set $1_{\kappa+1} = 1_0$ and $g_{\kappa+1} = g_0$. The scheme parameters are

$$pp := (gpk, crs, \overline{C}_{\mathrm{Map}}, \overline{C}_{\mathrm{Add}}, g_1, \ldots, g_{\kappa+1}, 1_1, \ldots, 1_{\kappa+1}).$$

We note that this algorithm runs in polynomial time in λ as long as κ is polynomial in λ.

4.2   Validity and Equality

The elements of $\mathbb{G}_i$ for $1 \leq i \leq \kappa$ are tuples of the form $h = ([z]_0, \mathbf{c}_1, \mathbf{c}_2, \pi)$ where $\mathbf{c}_1, \mathbf{c}_2$ are encryptions of vectors from $\mathbb{Z}_N^\ell$ under , pk_1, pk_2, respectively (encryption algorithm $\mathbf{Enc}$ extends from plaintext space $\mathbb{Z}_N$ to $\mathbb{Z}_N^\ell$ in the obvious way) and where π is a NIZK to be defined below. We refer to $(\mathbf{c}_1, \mathbf{c}_2, \pi)$ as the *auxiliary information* for $[z]_0$. The elements of $\mathbb{G}_{\kappa+1}$ are just those of $\mathbb{G}_0$.

The NIZK proof system that we use corresponds to the following inclusive disjunctive relation $(\mathbf{S}, \mathbf{R} := \mathbf{R}_1 \vee \mathbf{R}_2)$. Algorithm $\mathbf{S}(1^\lambda)$ outputs $gpk = (pp_0, pk_1, pk_2, [\mathbf{W}]_0, \mathsf{TD})$ as defined above and sets $gsk = (sk_1, sk_2)$. Relation $\mathbf{R}_1$ on input gpk, tuple $([z]_0, \mathbf{c}_1, \mathbf{c}_2)$, and witness $(\mathbf{x}, \mathbf{y}, \mathbf{r}_1, \mathbf{r}_2, sk_1, sk_2)$ accepts iff $[z]_0 \in \mathbb{G}_0$, the *representations* of $[z]_0$ as $\mathbf{x}, \mathbf{y} \in \mathbb{Z}_N^\ell$ are valid with respect to $[\mathbf{W}]_0$ in the sense that

$$[z]_0 = [\langle \mathbf{x}, \boldsymbol{\omega}_i \rangle]_0 \wedge [z]_0 = [\langle \mathbf{y}, \boldsymbol{\omega}_i \rangle]_0,$$

(where $\langle \cdot, \cdot \rangle$ denotes inner product) and the following ciphertext validity condition (with respect to the inputs to the relation) is met:

$$(\mathbf{c}_1 = \mathbf{Enc}(\mathbf{x}, pk_1; \mathbf{r}_1) \wedge \mathbf{c}_2 = \mathbf{Enc}(\mathbf{x}, pk_2; \mathbf{r}_2))$$

$$\vee$$

$$(pk_1, sk_1) = \mathbf{Gen}(sk_1) \wedge (pk_2, sk_2) = \mathbf{Gen}(sk_2)$$

$$\wedge \mathbf{x} = \mathbf{Dec}(\mathbf{c}_1, sk_1) \wedge \mathbf{y} = \mathbf{Dec}(\mathbf{c}_2, sk_2))$$

Recall that we have assumed the secret key of the encryption scheme to be the random coins used in **Gen**. Note that the representation validity check can be efficiently performed "in the exponent" using $[\mathbf{W}]_0$ and the explicit knowledge of $\mathbf{x}$ and $\mathbf{y}$. Note also that for honestly generated keys and ciphertexts the two checks in the expression above are equivalent (although this not generally the case when ciphertexts are malformed).

Relation $\mathbf{R}_2$ depends on the language TD, and on input gpk, tuple $([z]_0, \mathbf{c}_1, \mathbf{c}_2)$, and witness w_y accepts iff $y \in$ TD.

For $1 \leq i \leq \kappa$, the $\mathbf{Val}_i$ algorithm for Γ, on input $([z]_0, \mathbf{c}_1, \mathbf{c}_2, \pi)$, first checks that the first component is in $\mathbb{G}_0$ using $\mathbf{Val}_0$ and then checks the proof π; if both tests pass, it then returns $\top$, else $\bot$. Observe that for an honest choice of $crs = (crs', y)$, the perfect completeness and the perfect soundness of the proof system ensure that only those elements which pass relation $\mathbf{R}_1$ are accepted. Algorithm $\mathbf{Val}_{\kappa+1}$ just uses $\mathbf{Val}_0$.

The equality algorithm $\mathbf{Eq}_i$ of Γ for $1 \leq i \leq \kappa$ first checks the validity of the two group elements passed to it and then returns true iff their first components match, according to $\mathbf{Eq}_0$, the equality algorithm from the base scheme Γ_0. Algorithm $\mathbf{Eq}_{\kappa+1}$ just uses $\mathbf{Eq}_0$. The correctness of this algorithm follows from the perfect completeness of Σ.

4.3   Group Operations

We provide a procedure that, given as inputs $h = ([z]_0, \mathbf{c}_1, \mathbf{c}_2, \pi)$ and $h' = ([z']_0, \mathbf{c}_1', \mathbf{c}_2', \pi') \in \mathbb{G}_i$, generates a tuple representing the product $h \cdot h'$. This, in particular, will enable our multilinear map to be run on the additions of group elements whose explicit representations are not necessarily known. We exploit the structure of the base group as well as the homomorphic properties of the encryption scheme to "add together" the first three components. We then use (sk_1, sk_2) as a witness to generate a proof π'' that the new tuple is well formed. (For technical reasons we check the validity of h and h' in two different ways: using proofs π, π', and also explicitly using (sk_1, sk_2). Note that, although useful in the analysis, the explicit check is redundant by the perfect soundness of the proof system under a binding crs'.)

In pp we include an obfuscation of the C_{Add} circuit shown in Fig. 1 (top), and again we emphasize that steps $5a$ or $5b$ are never reached with a binding crs' (but they may be reached with a hiding crs' later in the analysis). Either an **IO** or a **PIO** will be used to obfuscate this circuit. Note that although we have assumed the evaluation algorithm to be deterministic, algorithm **Prove** is randomized and we need to address how we deal with its coins. When using PIO to obfuscate $\overline{C}_{\mathsf{Add}}$, the obfuscator directly deals with the needed randomness.[7] When using IO, a random (but fixed) set of coins will be hardwired into the circuit and hence the same set of coins will be used for all inputs. (As we shall see, when using IO the proof system has to satisfy extra structural requirements; these

---

[7] Typically, the obfuscated circuit will have a PRF key hardwired in and derives the required randomness by applying the PRF to the circuit inputs.

CIRCUIT $C_{Add}[gpk, crs, sk_1, sk_2, td_{ext}; r](i, h, h')$:

1. if $\neg \mathbf{Val}_i(h) \vee \neg \mathbf{Val}_i(h')$ return $\bot$
2. parse $([z]_0, \mathbf{c}_1, \mathbf{c}_2, \pi) \leftarrow h$ and $([z']_0, \mathbf{c}'_1, \mathbf{c}'_2, \pi') \leftarrow h'$
3. $[z'']_0 \leftarrow [z]_0 + [z']_0$; $\mathbf{c}''_1 \leftarrow \mathbf{c}_1 + \mathbf{c}'_1$; $\mathbf{c}''_2 \leftarrow \mathbf{c}_2 + \mathbf{c}'_2$
4. (explicit validity check of h, h')
    4.1 $\mathbf{x} \leftarrow \mathbf{Dec}(\mathbf{c}_1, sk_1)$, $\mathbf{y} \leftarrow \mathbf{Dec}(\mathbf{c}_2, sk_2)$
        $\mathbf{x}' \leftarrow \mathbf{Dec}(\mathbf{c}'_1, sk_1)$, $\mathbf{y}' \leftarrow \mathbf{Dec}(\mathbf{c}'_2, sk_2)$
    4.2a if $([z]_0 \neq [\langle \mathbf{x}, \boldsymbol{\omega}_i \rangle]_0) \vee ([z]_0 \neq [\langle \mathbf{y}, \boldsymbol{\omega}_i \rangle]_0)$ goto 5a
    4.2b else if $([z']_0 \neq [\langle \mathbf{x}', \boldsymbol{\omega}_i \rangle]_0) \vee ([z']_0 \neq [\langle \mathbf{y}', \boldsymbol{\omega}_i \rangle]_0)$
        goto 5b
    4.2c else goto 5c (h, h' are valid)
5a. (h is invalid)
    5a.1 $w'_y \leftarrow_{\$} \mathbf{WExt}(td_{ext}, ([z]_0, \mathbf{c}_1, \mathbf{c}_2), \pi)$
    5a.2 if $\neg \mathbf{R}_2(gpk, (([z]_0, \mathbf{c}_1, \mathbf{c}_2)), w'_y)$ return $\bot$
    5a.3 $\pi'' \leftarrow \mathbf{Prove}(gpk, crs, ([z'']_0, \mathbf{c}''_1, \mathbf{c}''_2), w'_y; r)$
5b. (only h' is invalid) repeat 5a with h'
5c. $\pi'' \leftarrow \mathbf{Prove}(gpk, crs, ([z'']_0, \mathbf{c}''_1, \mathbf{c}''_2), (sk_1, sk_2); r)$
6. return $([z''], \mathbf{c}''_1, \mathbf{c}''_2, \pi'')$

CIRCUIT $C_{Map}[gpk, crs, \mathbf{W}, sk_1](h_1, \ldots, h_\kappa)$:

1. for $i = 1 \ldots \kappa$
    1.1 if $\neg \mathbf{Val}_i(h_i)$ return $\bot$
    1.2 $([z_i]_0, \mathbf{c}_{i,1}, \mathbf{c}_{i,2}, \pi_i) \leftarrow h_i$
    1.3 $\mathbf{x}_i \leftarrow \mathbf{Dec}(\mathbf{c}_{i,1}, sk_1)$
2. $z_{\kappa+1} \leftarrow \prod_{i=1}^{k} \langle \mathbf{x}_i, \boldsymbol{\omega}_i \rangle \pmod{N}$
3. return $[z_{\kappa+1}]_{\kappa+1}$

Fig. 1. Top: Circuit for addition of group elements. Explicit randomness r is used with an **IO** and is internally generated when using a **PIO**. **Bottom:** Circuit implementing the multilinear map. Recall that here $gpk = (pp_0, pk_1, pk_2, [\mathbf{W}]_0, \mathsf{TD}, y)$.

ensure that using the same coins throughout does not compromise security.) The $\mathbf{Op}_i$ algorithm for $1 \leq i \leq \kappa$ runs the obfuscated circuit on i, the input group elements. Algorithm $\mathbf{Op}_{\kappa+1}$ just uses $\mathbf{Op}_0$ as usual. The correctness of this algorithm follows from those of Γ_0 and Π, the completeness of Σ and the correctness, in our sense of, (the possibly probabilistic) obfuscator **Obf**; see Sect. 2 for the definitions.

4.4 The Multilinear Map

The multilinear map for Γ, on input κ group elements $h_i = [z_i]_i = ([z_i]_0, \mathbf{c}_{i,1}, \mathbf{c}_{i,2}, \pi_i)$, uses sk_1 to recover the representation $\mathbf{x}_i$. It then uses the explicit knowledge of the matrix $\mathbf{W}$ to compute the output of the map as

$$\mathbf{e}([z_1]_1, \ldots, [z_\kappa]_\kappa) := \left[\prod_{i=1}^{k} \langle \mathbf{x}_i, \boldsymbol{\omega}_i \rangle \right]_{\kappa+1}.$$

Recalling that $\mathbb{G}_{\kappa+1}$ is nothing other than $\mathbb{G}_0$, and $g_{\kappa+1} = g_0$, the output of the map is just the $\mathbb{G}_0$-element $(g_0)^{\prod_{i=1}^{k} \langle \mathbf{x}_i, \boldsymbol{\omega}_i \rangle}$. The product in the exponent can be efficiently computed over $\mathbb{Z}_N$ for *any* polynomial level of linearity κ and any ℓ as it uses $\mathbf{x}_i$ and $\boldsymbol{\omega}_i$ explicitly. The multilinearity of the map follows from the linearity of each of the multiplicands in the above product (and the completeness of Σ, the correctness of Π, and the correctness of the (possibly probabilistic) obfuscator **Obf**). An obfuscation $\overline{C}_{\mathrm{Map}}$ of the circuit implementing this operation (see Fig. 1, bottom) will be made available through the public parameters and **e** is defined to run this circuit on its inputs.

4.5   Sampling and Extraction

Given vectors $\mathbf{x}$ and $\mathbf{y}$ in $\mathbb{Z}_N^\ell$ satisfying $\langle \mathbf{x}, \boldsymbol{\omega}_i \rangle = \langle \mathbf{y}, \boldsymbol{\omega}_i \rangle$, we set $[z]_0 := [\langle \mathbf{y}, \boldsymbol{\omega}_i \rangle]_0$ (which can be computed using $[\mathbf{W}]_0$ and explicit knowledge of $\mathbf{x}$) and

$$[z]_i \leftarrow ([z]_0, \mathbf{c}_1 = \mathbf{Enc}(\mathbf{x}, pk_1; \mathbf{r}_1), \mathbf{c}_2 = \mathbf{Enc}(\mathbf{y}, pk_2; \mathbf{r}_2),$$
$$\pi = \mathbf{Prove}(gpk, crs, ([z]_i, \mathbf{c}_1, \mathbf{c}_2), (\mathbf{x}, \mathbf{y}, \mathbf{r}_1, \mathbf{r}_2)).$$

If $\mathbf{W}$ is explicitly known the vectors $\mathbf{x}$ and $\mathbf{y}$ can take arbitrary forms subject to validity. This matrix, however, is only implicitly known, and in our sampling procedure we set $\mathbf{x} = \mathbf{y} = (z, 0)$ when $\ell = 2$ and $\mathbf{x} = \mathbf{y} = (z, 0, 0)$ when $\ell = 3$. (We call these the canonical representations.) Note that the outputs of the sampler are *not* statistically uniform within $\mathbb{G}_i([z]_i)$. Despite this, under the IND-CPA security of the encryption scheme it can be shown that the outputs are computationally close to uniform.

   Since the target group has unique encodings, as noted in Sect. 3, an extraction algorithm for all groups can be derived from one for the target group. The latter can be implemented by applying a universal hash function to the group elements in $\mathbb{G}_T$, for example.

5   Indistinguishability of Encodings

In this section we will state two theorems that are essential tools in establishing the intractability of the κ-MDDH for our MLG scheme Γ constructed in Sect. 4. These theorems, roughly speaking, state that valid encodings of elements within a single equivalence class are computationally indistinguishable. We formalize this property via the κ-Switch game shown in Fig. 2. This game lets an adversary $\mathcal{A}$ choose an element $[z]_i \in \mathbb{G}_i$ by producing two valid representations $(\mathbf{x}_0, \mathbf{y}_0)$ and $(\mathbf{x}_1, \mathbf{y}_1)$ for it. The adversary is given an encoding of $[z]_i$ generated using $(\mathbf{x}_b, \mathbf{y}_b)$ for a random b, and has to guess the bit b. In this game, besides access to pp, which contains the obfuscated circuits for the group operation and the multilinear map, we also provide the matrix $\mathbf{W}$ in the clear to the adversary. This strengthens the κ-Switch game and is needed for our later analysis.

   To prove that the advantage of $\mathcal{A}$ in the κ-Switch game is negligible we rely on the security of the obfuscator, the IND-CPA security of the encryption

$\kappa\text{-Switch}_\Gamma^A(\lambda):$
$pp \leftarrow_\$ \mathbf{Setup}(1^\lambda, 1^\kappa)$
$((\mathbf{x}_0, \mathbf{y}_0), (\mathbf{x}_1, \mathbf{y}_1), i, st) \leftarrow_\$ \mathcal{A}_1(pp, \mathbf{W})$
$b \leftarrow_\$ \{0, 1\};\ \mathbf{r}_1, \mathbf{r}_2 \leftarrow_\$ (\{0, 1\}^{r(\lambda)})^{|\mathbf{x}_0|}$
$\mathbf{c}_1 \leftarrow \mathbf{Enc}(\mathbf{x}_b, pk_1; \mathbf{r}_1);\ \mathbf{c}_2 \leftarrow \mathbf{Enc}(\mathbf{y}_b, pk_2; \mathbf{r}_2)$
$\pi \leftarrow_\$ \mathbf{Prove}(gpk, crs, ([z]_0, \mathbf{c}_1, \mathbf{c}_2), (\mathbf{x}, \mathbf{y}, \mathbf{r}_1, \mathbf{r}_2, \bot, \bot))$
$b' \leftarrow_\$ \mathcal{A}_2 (([\langle \mathbf{x}_b, \boldsymbol{\omega}_i \rangle]_0, \mathbf{c}_1, \mathbf{c}_2, \pi), st)$
Return $(b = b')$

Fig. 2. Game formalizing the indistinguishability of encodings with an equivalence class. This game is specific to our construction Γ. An adversary is legitimate if $z = \langle \mathbf{x}_b, \boldsymbol{\omega}_i \rangle = \langle \mathbf{y}_b, \boldsymbol{\omega}_i \rangle$ for $b \in \{0, 1\}$. We note that $\mathcal{A}$ gets explicit access to matrix $\mathbf{W}$ generated during setup.

scheme, and the security of the NIZK proof system. Depending on the type of the obfuscator and proof system used, we show indistinguishability of encodings in two incomparable ways: (1) using a *probabilistic* obfuscator that is secure against X-IND adversaries and a dual-mode NIZK as defined in Sect. 2; and (2) using a (standard) indistinguishability obfuscator for deterministic circuits and a dual-mode NIZK that is required to satisfy a "witness-translation" property that we formalize in Sect. 5.2.

5.1 Using Probabilistic Indistinguishability Obfuscation

The indistinguishability of encodings using the first set of assumptions above is conceptually simpler to prove and we start with this case. Intuitively, the IND-CPA security of the encryption scheme will ensure that the encryptions of the two representations are indistinguishable. This argument, however, does not immediately work as the parameters pp contain component $\overline{C}_{\mathrm{Add}}$ that depends on *both* decryption keys. We deal with this by finding an alternative implementation of this circuit without the knowledge of the secret keys, in the presence of a slightly different public parameters (which are computationally indistinguishable to those described in Sect. 4). The next lemma, roughly speaking, says that *provided* parameters pp include an instance $y \in \mathsf{TD}$, then there exists an alternative implementation $\widehat{C}_{\mathrm{Add}}$ that does not use the secret keys, and whose obfuscation is indistinguishable to that of C_{Add} of Fig. 1 (top) for an adversary that *knows* the secret keys. It relies on the security of the obfuscator and the security of the NIZK proof system. A formal proof is in the full version, we give an overview of the proof below.

Lemma 1. *Let* **PIO** *be a secure obfuscator for X-IND samplers, and Σ be a dual-mode NIZK proof system. Additionally, let parameters $\widetilde{pp}$ sampled as in Sect. 4 but with $\widetilde{y} \in \mathsf{TD}$, and let $\widehat{pp}$ sampled as $\widetilde{pp}$ but with a hiding CRS $\widehat{crs}'$, and an obfuscation of circuit $\widehat{C}_{\mathrm{Add}}$ of Fig. 3. Then, for any PPT adversary $\mathcal{A}$,*

$$Pr[\mathcal{A}(\widetilde{pp}, sk_1, sk_2) = 1 : (sk_1, sk_2) \leftarrow_\$ \mathbf{Gen}(1^\lambda)]$$
$$- Pr[\mathcal{A}(\widehat{pp}, sk_1, sk_2) = 1 : (sk_1, sk_2) \leftarrow_\$ \mathbf{Gen}(1^\lambda)] \in \mathrm{NEGL}.$$

CIRCUIT $\widehat{C}_{\text{Add}}[gpk, crs, w_y; r](i, h, h')$:

1. if $\neg\mathbf{Val}_i(h) \vee \neg\mathbf{Val}_i(h')$ return $\bot$
2. parse $([z]_0, \mathbf{c}_1, \mathbf{c}_2, \pi) \leftarrow h$, and $([z']_0, \mathbf{c}'_1, \mathbf{c}'_2, \pi') \leftarrow h'$
3. $[z'']_0 \leftarrow [z]_0 + [z']_0$; $\mathbf{c}''_1 \leftarrow \mathbf{c}_1 + \mathbf{c}'_1$; $\mathbf{c}''_2 \leftarrow \mathbf{c}_2 + \mathbf{c}'_2$
4. $\pi'' \leftarrow \mathbf{Prove}(gpk, crs, ([z'']_0, \mathbf{c}''_1, \mathbf{c}''_2), w_y; r)$
6. return $([z''], \mathbf{c}''_1, \mathbf{c}''_2, \pi'')$

Fig. 3. Alternative circuit for addition of group elements. Recall that here $\widehat{pp}$ includes $gpk = (pp_0, pk_1, pk_2, [\mathbf{W}]_0, \text{TD}, \widetilde{y})$ where $\widetilde{y} \in \text{TD}$ (also includes a hiding CRS $\widehat{crs}'$). The circuit uses (the) witness w_y to $\widetilde{y} \in \text{TD}$ to produce π''.

Proof (Sketch). The crucial observation is that a witness w_y to $\widetilde{y} \in \text{TD}$ is also a witness to $x \in \mathbf{R}$, and therefore $\widehat{C}_{\text{Add}}$ can use w_y instead of sk_1, sk_2 to produce the output proof π''. Below we provide brief descriptions of the transformation from C_{Add} to $\widehat{C}_{\text{Add}}$, as well as some intuition for the justifications of each step.

Game$_0$: We start with (a PIO obfuscation of) circuit C_{Add} of Fig. 1 and with $\widehat{pp}$ including $\widetilde{y} \in \text{TD}$ and a binding crs'.

Game$_1$: The circuit has witness w_y to $\widetilde{y} \in \text{TD}$ hardcoded. If some input reaches the "invalid" branches (steps 5a or 5b of C_{Add}; see Fig. 1 (top)), C_{Add} does not extract a witness from the corresponding proof, but instead uses w_y to generate proof π''. Since the witness w_y is unique, and the CRS crs' guarantees perfect soundness, this leads to exactly the same behavior of C_{Add} in Game 0. Hence, this hop is justified by PIO. Note that Game 1 requires no extraction trapdoor td_{ext} anymore.

Game$_2$: The CRS $\widehat{crs}'$ included in the public parameters is now hiding (such that the generated proofs are perfectly witness-indistinguishable).

Game$_3$: Here, output proofs π'' for those inputs entering the "valid" branch (step 5c; see Fig. 1) use w_y (and not sk_1, sk_2) as witness. In particular, this game does not need to perform a explicit validity check (using sk_1, sk_2) anymore. This hop is justified by PIO, where the perfect witness indistinguishability of $\widehat{crs}'$ (when constructed as a hiding CRS) guarantees that the C_{Add} circuits in Games 2 and 3 have identically distributed outputs.

With the above lemma we can invoke IND-CPA security, and via a sequence of games obtain the result stated below. The proof can be found in the full version; here we give a high-level overview of the proof (see also Fig. 4).

Theorem 1 (Switching encodings using PIO). *Let Γ be the MLG scheme constructed in Sect. 4, where **PIO** is secure for X-IND samplers, Π is an IND-CPA-secure encryption scheme, and Σ is a dual-mode NIZK proof system. Then, encodings of equivalent group elements are indistinguishable. More precisely, for any PPT adversary $\mathcal{A}$ and all $\lambda \in \mathbb{N}$,*

$$\mathbf{Adv}_{\Gamma, \mathcal{A}}^{\kappa\text{-switch}}(\lambda) \in \text{NEGL}.$$

Proof (Sketch). The strategy of the proof is as follows. We start replacing parameters pp as described in Sect. 4 with parameters $\widetilde{pp}$ of Lemma 1, the latter include an instance $\tilde{y} \in \mathsf{TD}$, this hop is justified by the hardness of deciding membership in TD; then we apply Lemma 1 to replace parameters $\widetilde{pp}$ with $\widehat{pp}$, including an obfuscation of circuit $\widehat{C}_{\mathrm{Add}}$ of Fig. 3; at this point we invoke the IND-CPA security of the encryption scheme to change the representation vector encrypted under pk_2 of the challenge encoding (the challenge proof π^* is generated using simulator trapdoor td_{zk}, and hence is identically distributed to a real proof); next, we revert back to parameters pp, including a no-instance $y \notin \mathsf{TD}$ and an obfuscation of circuit C_{Add} of Fig. 1, which is justified again by the hardness of TD and Lemma 1; note that now it is possible to use sk_2 in C_{Map}, instead of sk_1, invoking the security of **PIO** (functional equivalence follows from the perfect soundness of the NIZK with a binding CRS); last, we repeat the same steps to change the representation vector encrypted under pk_1. This completes the proof. (See Fig. 4 for a sketch of the hybrids.)

5.2   Doing Without Probabilistic Obfuscation

In contrast to the PIO-based approach from Sect. 5.1, we can also only use (deterministic) indistinguishability obfuscation, but a stronger notion of NIZK proof system. Concretely, our proof works for any dual-mode NIZK proof system that enjoys perfect completeness, perfect soundness (when the CRS is generated using **BCRS**), perfect WI (when the CRS is generated by **HCRS**), and meets a structural requirement we explain below. This requirement is fulfilled by Groth–Sahai proofs [GS08] based on the DDH or k-Linear assumption.

A STRUCTURAL PROPERTY. To explain the required structural property, recall first that perfect WI guarantees that proofs that are honestly generated (under a hiding CRS) have a distribution that is independent of the used witness. For our purposes, we require a slightly more specialized property: we require that a change of the used witness (in **Prove**) can be compensated with a change of random coins. In other words, we require that for every hiding CRS crs, and for every statement x and pair of witnesses w, w' for x, there is a value Δ such that

$$\forall r : \qquad \mathbf{Prove}(gpk, crs, x, w; r) = \mathbf{Prove}(gpk, crs, x, w'; r + \Delta), \qquad (\star)$$

where "$+$" is a suitable homomorphic operation on random coins. Note that Δ may depend on w and w', but not on r. Furthermore, we require that Δ can be efficiently computed from x, w, w', and the zero-knowledge CRS trapdoor td_{zk} output by **HCRS**.

Again, we stress that Groth–Sahai proofs have the desired property (when restricting to statements with witnesses $w \in \{0,1\}^*$ that are bit strings). We give more details in the full version of this paper.

THE DETERMINISTIC CIRCUIT C_{Add}. We now comment on a necessary slight tweak to the multilinear map construction itself. Namely, we have to view both C_{Add} and C_{Map} as deterministic circuits (so they can be obfuscated using an

G.	public parameters	C_{Add} knows	C_{Map} knows	c_1 $(b=0)$ contains	c_2 $(b=0)$ contains	remark
0	pp	sk_1, sk_2, td_{ext}	sk_1	$(\mathbf{x}_0, \mathbf{y}_0)$	$(\mathbf{x}_0, \mathbf{y}_0)$	
1	$\widetilde{pp}$	sk_1, sk_2, td_{ext}	sk_1	$(\mathbf{x}_0, \mathbf{y}_0)$	$(\mathbf{x}_0, \mathbf{y}_0)$	TD indist.
2	$\widehat{pp}$	$\boxed{w_y}$	sk_1	$(\mathbf{x}_0, \mathbf{y}_0)$	$(\mathbf{x}_0, \mathbf{y}_0)$	Lemma 1
3	$\widehat{pp}$	w_y	sk_1	$(\mathbf{x}_0, \mathbf{y}_0)$	$\boxed{(\mathbf{x}_1, \mathbf{y}_1)}$	IND-CPA
4	$\widetilde{pp}$	$\boxed{sk_1, sk_2, td_{ext}}$	sk_1	$(\mathbf{x}_0, \mathbf{y}_0)$	$(\mathbf{x}_1, \mathbf{y}_1)$	Lemma 1
5	pp	sk_1, sk_2, td_{ext}	sk_1	$(\mathbf{x}_0, \mathbf{y}_0)$	$(\mathbf{x}_1, \mathbf{y}_1)$	TD indist.
6	pp	sk_1, sk_2, td_{ext}	$\boxed{sk_2}$	$(\mathbf{x}_0, \mathbf{y}_0)$	$(\mathbf{x}_1, \mathbf{y}_1)$	PIO
7	$\widetilde{pp}$	sk_1, sk_2, td_{ext}	sk_2	$(\mathbf{x}_0, \mathbf{y}_0)$	$(\mathbf{x}_1, \mathbf{y}_1)$	TD indist.
8	$\widehat{pp}$	$\boxed{w_y}$	sk_2	$(\mathbf{x}_0, \mathbf{y}_0)$	$(\mathbf{x}_1, \mathbf{y}_1)$	Lemma 1
9	$\widehat{pp}$	w_y	sk_2	$\boxed{(\mathbf{x}_1, \mathbf{y}_1)}$	$(\mathbf{x}_1, \mathbf{y}_1)$	IND-CPA
10	$\widetilde{pp}$	$\boxed{sk_1, sk_2, td_{ext}}$	sk_2	$(\mathbf{x}_1, \mathbf{y}_1)$	$(\mathbf{x}_1, \mathbf{y}_1)$	Lemma 1
11	$\widetilde{pp}$	sk_1, sk_2, td_{ext}	sk_2	$(\mathbf{x}_1, \mathbf{y}_1)$	$(\mathbf{x}_1, \mathbf{y}_1)$	TD indist.
12	pp	sk_1, sk_2, td_{ext}	$\boxed{sk_1}$	$(\mathbf{x}_1, \mathbf{y}_1)$	$(\mathbf{x}_1, \mathbf{y}_1)$	PIO

Fig. 4. Outline of the proof steps of Theorem 1. b is the random bit of the κ-Switch game (see Fig. 2). Changing between pp and $\widetilde{pp}$ is justified by the hardness of deciding membership of TD, and changing between $\widetilde{pp}$ and $\widehat{pp}$ by Lemma 1. The hops relying on PIO use the perfect soundness under binding crs' to argue function equivalence.

indistinguishability obfuscator **IO**). For C_{Map}, this is trivial, since it already is deterministic. Furthermore, we can view C_{Add} as a deterministic circuit that takes as input (among other things) random coins r, and outputs (among other things) a NIZK proof $\pi = \mathbf{Prove}(gpk, crs, x, w; r)$ for a fixed witness w hardwired into C_{Add}. For our purposes, we use a slight variation of C_{Add} that instead generates π as $\mathbf{Prove}(gpk, crs, x, w; R)$, where R is a *uniformly random* value that is *hardwired* (upon creation time) into C_{Add}. When we want to make the choice of R explicit, we also write C_{Add}^R.

For this slight variation of our construction, we claim:

Theorem 2 (Switching encodings using IO). *Let IO be an indistinguishability obfuscator, Π an IND-CPA encryption scheme, and Σ the specific dual-mode NIZK proof system of Groth and Sahai (see [GS08]). Let Γ be the MLG scheme of Sect. 4 obtained using these primitives. Then, for any PPT adversary $\mathcal{A}$,*

$$\mathbf{Adv}_{\Gamma, \mathcal{A}}^{\kappa\text{-switch}}(\lambda) \in \mathrm{NEGL}.$$

Here, we only give a brief intuition for the proof. A more detailed proof is given in the full version.

In a nutshell, the proof of Theorem 2 proceeds like that of Theorem 1, except of course in those steps that use the security of the probabilistic indistinguishability obfuscator **PIO**. There are two types of such steps (resp. changes of C_{Map} or C_{Add}): in the first type, functional equivalence is fully preserved (even when viewing C_{Add} as a deterministic circuit. This type of change occurs in the hop from

Game$_0$ to Game$_1$ in the proof of Lemma 1, and in the hops from Game$_5$ to Game$_6$ and from Game$_{11}$ to Game $_{12}$ in the proof of Theorem 1. Since the corresponding deterministic circuits are functionally equivalent (in case of $C_{\mathrm{Add}} = C_{\mathrm{Add}}^R$: when the same value of R is used), the security of **IO** can be directly utilized.

The second type of steps lets C_{Add} use a different witness (e.g., w_y instead of (sk_1, sk_2), or vice versa) to generate consistency proofs π''. This type of proof step occurs in the hop from Game$_2$ to Game$_3$ in the proof of Lemma 1. Note that at this point, the generated CRS is hiding, and $C_{\mathrm{Add}} = C_{\mathrm{Add}}^R$ uses a single hardcoded random string R as random coins to generate such proofs. By property $(\star)$ above, we have that

$$C_{\mathrm{Add},1}^R \equiv C_{\mathrm{Add},2}^{R+\Delta},$$

where $C_{\mathrm{Add},1}$ and $C_{\mathrm{Add},2}$ denote the C_{Add} variants before and after the step, and Δ denotes the randomness shift value from $(\star)$.

Hence, this change can be justified with a reduction to the (deterministic) indistinguishability property of **IO**. Specifically, a suitable circuit sampler would sample circuits $C_1 := C_{\mathrm{Add},1}^R$ and $C_2 := C_{\mathrm{Add},2}^{R+\Delta}$ for a uniform R, and a Δ generated from the corresponding witnesses. (We note that *during* this reduction, we can of course assume both relevant witnesses (sk_1, sk_2) and w_y to be known.)

The remaining parts of the proof of Theorem 2 (including the proof of Lemma 1) apply unchanged.

6   The Multilinear DDH Problem

In the full version we show that natural multilinear analogues of the decisional Diffie–Hellman (DDH) problem are hard for our MLG scheme Γ from Sect. 4. We will establish this for two specific **Setup** algorithms which give rise to symmetric and asymmetric multilinear maps in groups of prime order N. (See Sect. 3 for the formal definition.) In the symmetric case, we will base hardness on the q-strong DDH problem [BBS04] and in the asymmetric case on the standard DDH problem.

6.1   Intractable Problems

We start by formalizing the hard problems that we will be relying on and those whose hardness we will be proving. We do this in a uniform way using the language of group schemes of Sect. 3. Informally, the DDH problem requires the indistinguishability of g^{xy} from a random element given (g^x, g^y) for random x and y, the q-SDDH problem requires this for $g^{x^{q+1}}$ given $(g^x, g^{x^2}, \ldots, g^{x^q})$ and the κ-MDDH problem, whose hardness we will be establishing, generalizes the standard bilinear DDH problem (and its variants) and requires this for $g_T^{a_1 \cdots a_{\kappa+1}}$ in the presence of $(g^{a_1}, \ldots, g^{a_{\kappa+1}})$.

THE DDH PROBLEM. We say that a group scheme Γ_0 is DDH intractable if

$$\mathbf{Adv}_{\Gamma_0,\mathcal{A}}^{\mathrm{ddh}}(\lambda) := 2 \cdot \Pr\left[\mathrm{DDH}_{\Gamma_0}^{\mathcal{A}}(\lambda)\right] - 1 \in \mathrm{NEGL},$$

where game $\mathrm{DDH}_{\Gamma_0}^{\mathcal{A}}(\lambda)$ is shown in Fig. 5 (left).

$\text{DDH}_{\Gamma_0}^{\mathcal{A}}(\lambda):$	$q\text{-SDDH}_{\Gamma_0}^{\mathcal{A}}(\lambda):$	$(\kappa, I)\text{-MDDH}_{\Gamma}^{\mathcal{A}}(\lambda):$
$pp \leftarrow_{\$} \textbf{Setup}_0(1^\lambda, 1^0)$	$pp \leftarrow_{\$} \textbf{Setup}_0(1^\lambda, 1^0)$	$pp \leftarrow_{\$} \textbf{Setup}(1^\lambda, 1^\kappa)$
$b \leftarrow_{\$} \{0,1\}$	$q \leftarrow q(\lambda); \; b \leftarrow_{\$} \{0,1\}$	$b \leftarrow_{\$} \{0,1\}$
$x, y, z \leftarrow_{\$} \mathbb{Z}_N$	$x, z \leftarrow_{\$} \mathbb{Z}_N$	$a_1, \dots, a_T, z \leftarrow_{\$} \mathbb{Z}_N$
if $b = 1$ then	if $b = 1$ then	if $b = 1$ then
$\quad z \leftarrow x \cdot y$	$\quad z \leftarrow x^{q+1}$	$\quad [z]_T \leftarrow \mathbf{e}([a_1]_1, \dots, [a_i]_i)^{a_T}$
$b' \leftarrow_{\$} \mathcal{A}(pp, [x]_0, [y]_0,$	$b' \leftarrow_{\$} \mathcal{A}(pp,$	$b' \leftarrow_{\$} \mathcal{A}(pp, \{[a_i]_j\}_{(i,j) \in I},$
$\quad\quad\quad\quad\quad\quad [z]_0)$	$\quad\quad [x]_0, \dots, [x^q]_0,$	$\quad\quad\quad\quad\quad [z]_T)$
Return $(b = b')$	$\quad\quad [z]_0)$	Return $(b = b')$
	Return $(b = b')$	

Fig. 5. Left: The DDH problem. **Middle:** The strong DDH problem. **Right:** The multilinear DDH problem, where I specifies the available group elements. By slight abuse of notation, repeated use of $[a_i]_i$ denotes the same sample.

THE q-SDDH PROBLEM. For $q \in \mathbb{N}$ we say that a group scheme Γ_0 is q-SDDH intractable if

$$\textbf{Adv}_{\Gamma_0, \mathcal{A}}^{q\text{-sddh}}(\lambda) := 2 \cdot \Pr\left[q\text{-SDDH}_{\Gamma_0}^{\mathcal{A}}(\lambda)\right] - 1 \in \textsc{Negl},$$

where game q-SDDH$_{\Gamma_0}^{\mathcal{A}}(\lambda)$ is shown in Fig. 5 (middle).

THE (κ, I)-MDDH PROBLEM. For $\kappa \in \mathbb{N}$ we say that an MLG scheme Γ is κ-MDDH intractable with respect to the index set I if

$$\textbf{Adv}_{\Gamma, \mathcal{A}}^{(\kappa, I)\text{-mddh}}(\lambda) := 2 \cdot \Pr\left[(\kappa, I)\text{-MDDH}_{\Gamma}^{\mathcal{A}}(\lambda)\right] - 1 \in \textsc{Negl},$$

where game (κ, I)-MDDH$_{\Gamma}^{\mathcal{A}}(\lambda)$ is shown in Fig. 5 (right). Here I is a set of ordered pairs of integers (i, j) with $1 \le i \le \kappa + 1$, $1 \le j \le \kappa$. The adversary is provided with challenge group elements $[a_i]_j$ for $(i, j) \in I$, so that its challenge elements may lie in any combination of the groups. The standard MDDH problem corresponds to the case where

$$I = I^* := \{(1,1), \dots, (\kappa, \kappa), (\kappa + 1, \kappa)\}.$$

6.2 The Symmetric Setting

We describe a special variant of our general construction in Sect. 4 which gives rise to a *symmetric* MLG scheme as defined in Sect. 3. Recall that in the construction a matrix $\mathbf{W}$ was chosen uniformly at random in $\mathbb{Z}_N^{\kappa \times \ell}$. We set $\ell := 2$ and sample $\mathbf{W} = (\boldsymbol{\omega}_1, \dots, \boldsymbol{\omega}_\kappa)^t$ by setting $\boldsymbol{\omega}_i = (1, \omega)$ for a random $\omega \in \mathbb{Z}_N$. The generators and identity elements for all groups are set to be a single value generated for the first group. These modifications ensure that the scheme algorithms are independent of the index for $1 \le i \le \kappa$ and that $\mathbf{e}$ is invariant under all permutations of its inputs.

The following lemma, which provides a mechanism to compute polynomial values "in the exponent," will be helpful in the security analysis of our constructions.

Lemma 2 (Horner in the exponent). *Let $\boldsymbol{\omega} = (\omega_0, \omega_1, \omega_2) \in \mathbb{Z}_N$, and $\mathbf{x}_i = (x_{i,0}, x_{i,1}, x_{i,2}) \in \mathbb{Z}_N^3$ for $i = 1 \ldots \kappa$. Define $z_i := \langle \mathbf{x}_i, \boldsymbol{\omega} \rangle$. Then given only the implicit values $[\omega_0^i \omega_1^j \omega_2^k]_T$, for all i, j, k such that $i + j + k = \kappa$ and the explicit values $\mathbf{x}_i$ the element $[z_1 \cdots z_n]_T$ can be efficiently computed.*

Proof. Let

$$P(\omega_0, \omega_1, \omega_2) := \prod_{i=1}^{\kappa} (x_{i,0} \cdot \omega_0 + x_{i,1} \cdot \omega_1 + x_{i,2} \cdot \omega_2) = \sum_{i+j+k=\kappa} p_{ijk} \cdot \omega_0^i \omega_1^j \omega_2^k,$$

Clearly, if all p_{ijk} are known then $[P(\omega)]_T$ can be computed using $[\omega_0^i \omega_1^j \omega_2^k]_T$ with polynomially many operations. (There are $\mathcal{O}(\kappa^2)$ summands above.) To obtain these values we apply Horner's rule. Define

$$P_i(\omega_0, \omega_1, \omega_2) := \begin{cases} 1 & \text{if } i = 0 \text{ ;} \\ (x_{i,0} \cdot \omega_0 + x_{i,1} \cdot \omega_1 + x_{i,2} \cdot \omega_2) \cdot P_{i-1}(\omega_0, \omega_1, \omega_2) & \text{otherwise.} \end{cases}$$

The coefficients of P_κ are the required p_{ijk} values. Let t_i denote the number of terms in P_i. It takes at most $3t_i$ multiplications and $t_i - 1$ additions in $\mathbb{Z}_N$ to compute the coefficients of P_i from P_{i-1} and $\mathbf{x}_i$. Since $t_i \in \mathcal{O}(\kappa^2)$, at most $\mathcal{O}(\kappa^3)$ many operations in total are performed. We note that the lemma generalizes to any (constant) ℓ with computational complexity $\mathcal{O}(\kappa^\ell)$. $\qquad\square$

A formal statement and proof of the following result is in the full version of the paper, here we give a high level overview. Below $I = I^*$ denotes the index set with all the second components being 1.

Theorem 3 ($(\kappa - 1)$-SDDH hard $\implies$ symmetric (κ, I^*)-MDDH hard). *Let Γ^* denote scheme Γ of Sect. 4 constructed using base group Γ_0 and an indistinguishability obfuscator **IO** with modifications as described above, and let $\kappa \in \mathbb{N}$. Then for any PPT adversary $\mathcal{A}$ there are ppt adversaries $\mathcal{B}_1, \mathcal{B}_2$ of essentially the same complexity as $\mathcal{A}$ such that*

$$\mathbf{Adv}_{\Gamma^*, \mathcal{A}}^{(\kappa, I^*)\text{-mddh}}(\lambda) \leq 2 \cdot \mathbf{Adv}_{\Gamma_0, \mathcal{B}_1}^{(\kappa-1)\text{-sddh}}(\lambda) + (\kappa + 1) \cdot \mathbf{Adv}_{\Gamma^*, \mathcal{B}_2}^{\kappa\text{-switch}}(\lambda) + \mu(\lambda),$$

for all $\lambda \in \mathbb{N}$ and a suitable negligible function μ.

Proof (Sketch). In our reduction, the value ω used to generate $\mathbf{W}$ will play the role of the implicit value in the SDDH problem instance. We therefore change the implementation of C_{Map} to one that *does not know* ω in the clear and only uses the implicit values $[\omega^i]_0$ (recall that in our construction $\mathbb{G}_T$ is just $\mathbb{G}_0$, so these elements come from the SDDH instance). Such a circuit C_{Map}^* can be efficiently implemented using Horner's rule above. In more detail, C_{Map}^* has $[\omega^i]_T$ hardcoded in, recovers $\mathbf{x}_i$ from its inputs using sk_1, and then applies Lemma 2 with $(\omega_0, \omega_1, \omega_2) := (1, \omega, 0)$ to evaluate the multilinear map.

The proof proceeds along a sequence of $\kappa + 6$ games as follows.

$Game_0$: This is the κ-MDDH problem (Fig. 5, right). We use $\mathbf{x}_i$ and $\mathbf{y}_i$ to denote the representation vectors of a_i generated within the sampler $\mathbf{Sam}_{I(i)}(a_i)$, where $(i, I(i)) \in I$.

$Game_1$–$Game_\kappa$: In these games we gradually switch the representations of $[a_i]_1$ for $i \in [\kappa]$ so that they are of the form $(a_i - \omega, 1)$. Each hop can be bounded via the Switch game. (We have not (yet) changed the representation of $[a_{\kappa+1}]_1$.)

$Game_{\kappa+1}$: This game introduces a conceptual change: the a_i for $i \in [\kappa]$ are generated as $a_i + \omega$. Note that the distributions of these values are still uniform and that the exponent of the MDDH challenge when $b = 1$ is

$$a_{\kappa+1} \cdot \prod_{i=1}^{\kappa}(a_i + \omega).$$

This game prepares us for embedding a $(\kappa - 1)$-SDDH challenge and then to stepwise randomize the exponent above.

$Game_{\kappa+2}$: This game switches C_{Map} to C^*_{Map} as defined above. We use indistinguishability obfuscation and the fact that these circuits are functionally equivalent to bound this hop. We are now in a setting where ω is only implicitly known.

$Game_{\kappa+3}$: This game replaces $[\omega^\kappa]_0$ with a random value $[\tau]_0$ in C^*_{Map} and the computation of the challenge exponent. This hop can be bounded via the $(\kappa-1)$-SDDH game. Note that at this point the exponent is not information-theoretically randomized as τ is used within C^*_{Map}.

$Game_{\kappa+4}$: This game sets the representation of $[a_{\kappa+1}]_1$ to $(a_{\kappa+1} - \omega, 1)$. Once again, this hop can be bounded by the Switch game.

$Game_{\kappa+5}$: This game introduces a conceptual change analogous to that in $Game_{\kappa+1}$ for $a_{\kappa+1}$. Note that a linear factor $(a_{\kappa+1} + \omega)$ is introduced in this game. This will help to fully randomize the exponent next.

$Game_{\kappa+6}$: Analogously to $Game_{\kappa+3}$, this game replaces $[\omega^\kappa]_0$ with a random value $[\sigma]_0$. We bound this hop using the $(\kappa - 1)$-SDDH game.

In $Game_{\kappa+6}$, irrespective of the value of $b \in \{0, 1\}$, the challenge is uniformly and independently distributed as σ remains outside the view of the adversary. Hence the advantage of any (unbounded) adversary in this game is 0. This concludes the sketch proof.

6.3   The Asymmetric Setting

We describe a second variant of the construction in Sect. 4 that results in an asymmetric MLG scheme. We set $\ell := 2$ and choose the matrix $\mathbf{W} = (\boldsymbol{\omega}_1, \ldots, \boldsymbol{\omega}_\kappa)^t$ by setting $\boldsymbol{\omega}_i := (1, \omega_i)$ for random $\omega_i \in \mathbb{Z}_N$.

The following theorem shows that for index set $I = \{(i, I(i)) : 1 \le i \le \kappa+1\}$ given by an arbitrary function $I : [\kappa + 1] \longrightarrow [\kappa]$ of range at least 3, this construction is (κ, I)-MDDH intractable under the standard DDH assumption in the base group, the security of the obfuscator, and the κ-Switch game in Sect. 5. We present the proof intuition here and leave the details to the full version.

Theorem 4 (DDH hard $\implies$ asymmetric (κ, I^*)-MDDH hard). *Let Γ^* denote scheme Γ of Sect. 4 constructed using base group Γ_0 and an indistinguishability obfuscator **IO** with modifications as described above. Let $\kappa \geq 3$ be a polynomial and I^* as above. Then for any PPT adversary $\mathcal{A}$ there are ppt adversaries $\mathcal{B}_1$ and $\mathcal{B}_2$ such that*

$$\mathbf{Adv}_{\Gamma^*, \mathcal{A}}^{(\kappa, I^*)\text{-mddh}}(\lambda) \leq 2 \cdot \mathbf{Adv}_{\Gamma_0, \mathcal{B}_1}^{\mathrm{ddh}}(\lambda) + 3 \cdot \mathbf{Adv}_{\Gamma^*, \mathcal{B}_2}^{\kappa\text{-switch}}(\lambda) + \mu(\lambda),$$

for a all $\lambda \in \mathbb{N}$ and suitable negligible function μ.

Proof (Sketch). The general proof strategy is similar to that of the symmetric case, and proceeds along a sequence of 8 games as follows.

Game$_0$: This is the (κ, I)-MDDH problem. Without loss of generality we assume that $I(i) = i$ for $i \in [3]$.

Game$_1$–Game$_3$: In these games we gradually switch the representation vectors of $[a_i]_i$ for $i = 1, 2, 3$ to those of the form $(a_i - \omega_i, 1)$. Each of these hops can be bounded via the Switch game.

Game$_4$: This game introduces a conceptual change and generates a_i as $a_i + \omega_i$. The exponent of the MDDH challenge when $b = 1$ is

$$(a_1 + \omega_1)(a_2 + \omega_2)(a_3 + \omega_3) \cdot \prod_{j \geq 4}^{\kappa + 1} a_j.$$

Game$_5$: In this game we change the implementation of C_{Map} to one which uses all but two of the ω_i explicitly, the remaining two implicitly, and additionally $[\omega_1 \omega_2]_0$, i.e., $\omega_1 \omega_2$ given implicitly in the exponent. The new circuit C_{Map}^* will be implemented using Horner's rule and is functionally equivalent to the original circuit used in the scheme. We invoke the IO security of the obfuscator to conclude the hop. This game prepares us to embed a DDH challenge next.

Game$_6$: In this game we replace all the occurrences of $[\omega_1 \omega_2]_0$ with a random $[\tau]_0$ and the corresponding implicit values. We bound the distinguishing advantage in this hop down to the DDH game.

Game$_7$: Similarly to Game$_5$, we change the implementation of C_{Map}^* using $[\tau \omega_3]_0$ and argue via indistinguishability of obfuscations for functionally equivalent circuits.

Game$_8$: Finally, using the hardness of DDH, we replace all the occurrences of $[\tau \omega_3]_0$ with a random $[\sigma]_0$.

In Game$_8$, irrespective of the value of $b \in \{0, 1\}$, the challenge is uniformly and independently distributed as σ remains outside the view of the adversary. Hence the advantage of any (possibly unbounded) adversary in this game is 0.

7 The Rank Problem

The RANK problem is a generalization of DDH-like problems to matrices and has proven to be very useful in cryptographic constructions [BHHO08, NS09, GHV12, BLMR13, EHK+13]. Here we consider the problem in groups with non-unique

$$\boxed{\begin{aligned}
&(\kappa, m, n, r_0, r_1)\text{-RANK}_{\Gamma}^{\mathcal{A}}(\lambda): \\
&\underline{} \\
&pp \leftarrow_{\$} \mathbf{Setup}(1^{\lambda}, 1^{\kappa}) \\
&b \leftarrow_{\$} \{0, 1\} \\
&\mathbf{M}_0 \leftarrow_{\$} \mathrm{Rk}_{r_0}(\mathbb{Z}_N^{m \times n}); \ \mathbf{M}_1 \leftarrow_{\$} \mathrm{Rk}_{r_1}(\mathbb{Z}_N^{m \times n}) \\
&b' \leftarrow_{\$} \mathcal{A}(pp, [\mathbf{M}_b]) \\
&\text{Return } (b = b')
\end{aligned}}$$

Fig. 6. The RANK problem parameterized by integers κ, m, n, r_0 and r_1.

encodings equipped with a multilinear map. Our main result is to show that, subject to certain restrictions, the intractability of the rank problem for our construction of an MLG scheme Γ from Sect. 4 follows from that of the q-SDDH problem for Γ_0.

7.1 Formalization of the Problem

THE (κ, m, n, r_0, r_1)-RANK PROBLEM. For $\kappa, m, n, r_0, r_0 \in \mathbb{N}$ we say that an MLG scheme Γ is (κ, m, n, r_0, r_1)-RANK intractable if

$$\mathbf{Adv}_{\Gamma,\mathcal{A}}^{(\kappa,m,n,r_0,r_1)\text{-rank}}(\lambda) := 2 \cdot \Pr\left[(\kappa, m, n, r_0, r_1)\text{-RANK}_{\Gamma}^{\mathcal{A}}(\lambda)\right] - 1 \in \mathrm{NEGL},$$

where game (κ, m, n, r_0, r_1)-RANK$_{\Gamma}^{\mathcal{A}}(\lambda)$ is shown in Fig. 6.

In the presence of a κ-linear map the (κ, m, n, r_0, r_1)-RANK$_{\Gamma}^{\mathcal{A}}(\lambda)$ problem is easy for any $r_0 < r_1 < \kappa$, since the determinants of all the r_b-minors can be expressed as forms of degree at most κ, and the multilinear map can be used to distinguish their images in the target group. However, this does not invalidate the plausibility of the rank problem for $\kappa \leq r_0 < r_1$; indeed there are known reductions to the DDH, the decision linear problems [BHHO08, NS09].

7.2 The RANK Problem with Our MLG Scheme

Let pp denote the public parameters of such an MLG scheme, obtained by running **Setup** with input $(1^{\lambda}, 1^{\kappa})$. For simplicity, we focus on the case where N is prime. Let $\mathrm{Rk}_r(\mathbb{Z}_N^{m \times n})$ denote the set of $m \times n$ matrices over $\mathbb{Z}_N$ of rank r, where necessarily $r \leq \min(m, n)$. We use a variant of our construction in Sect. 4, setting $\ell := 3$ and sampling $\mathbf{W} = (\boldsymbol{\omega}_1, \ldots, \boldsymbol{\omega}_{\kappa})^t \in \mathbb{Z}_N^{\kappa \times 3}$ where $\boldsymbol{\omega}_i = (1, \omega, \omega^2)$ for $\omega \leftarrow_{\$} \mathbb{Z}_N$. Note that this results in a symmetric pairing and henceforth we omit subscripts from source group elements. Let $[\mathbf{M}]$ denote a matrix whose (i, j)th entry contains an encoding of the form $[m_{i,j}] = ([m_{i,j}]_0, \mathbf{c}_{i,j,1}, \mathbf{c}_{i,j,2}, \pi_{i,j})$, with $m_{i,j} \in \mathbb{Z}_N$.

We show that for our construction in Sect. 4, with the modification introduced above, the rank problem is indeed hard provided $\kappa \leq r_0 < r_1$. A standard hybrid argument shows that it is sufficient to establish this for $r_1 := r_0 + 1$, with a polynomial loss in the security. Our main result is stated below. The proof is in the full version of the paper, here we give only give some intuition.

Theorem 5 (SDDH $\implies$ RANK). *Let Γ denote scheme Γ of Sect. 3 with $\ell := 3$ and with respect to the base group Γ_0 and an indistinguishability obfuscator* **IO**. *Let κ, m, n, r be integers with $r \geq \kappa$. Then, for any* PPT *adversary $\mathcal{A}$ there are ppt adversaries $\mathcal{B}_1$ and $\mathcal{B}_2$ of essentially the same complexity as $\mathcal{A}$ such that for all $\lambda \in \mathbb{N}$ and a suitable negligible function μ*

$$\mathbf{Adv}_{\Gamma,\mathcal{A}}^{(\kappa,m,n,r,r+1)\text{-RANK}}(\lambda) \leq \sum_{q=1}^{2\kappa-1} \mathbf{Adv}_{\Gamma_0,\mathcal{B}_1}^{q\text{-sddh}}(\lambda) + (mn) \cdot \mathbf{Adv}_{\Gamma,\mathcal{B}_2}^{\kappa\text{-switch}}(\lambda) + \mu(\lambda).$$

7.3   Proof Intuition

The main difficulty comes in generating consistent encodings of a rank r challenge matrix $[\mathbf{M}]$ throughout its gradual transformation into a rank $r + 1$ challenge matrix. Contrast this with the MDDH reduction of Sect. 6, where the challenge that is transformed lives in the target group —a group with *unique* encodings. As we will see below, having encodings that are represented *also* with respect to ω^2 will help to overcome this problem and embed a 1-SDDH tuple.

EMBEDDING THE SDDH CHALLENGE. To reduce the rank problem to 1-SDDH, consider the following matrix

$$[\overline{\mathbf{W}}]_0 = \begin{bmatrix} [1]_0 & [\omega]_0 \\ [\omega]_0 & [\tau]_0 \end{bmatrix},$$

which is formed from an 1-SDDH challenge. We will exploit the fact that if $\tau = \omega^2$ then $\overline{\mathbf{W}}$ has rank 2, and if τ is uniform then it has rank 2 with overwhelming probability in λ.

LIFTING. To obtain an $m \times n$ matrix $\mathbf{M}$ of rank $r \geq \kappa$ or $r + 1$ we can use the standard trick of embedding the identity matrix $\mathbf{I}_{r-1}$ in the diagonal:

$$\mathbf{M} = \begin{bmatrix} \mathbf{S} & \\ & \mathbf{I}_{r-1} \\ & & \mathbf{0} \end{bmatrix},$$

where $\mathbf{0}$ denotes padding with zeroes from $\mathbb{Z}_N$ to bring the matrix up to the required size. Moreover, via the random self-reducibility of the rank problem the structure in $\mathbf{M}$ can be removed. An important point worth mentioning is that after the randomization we are still able to generate an encoded matrix $[\mathbf{M}]$ even when ω and τ are only known in the exponent.

BREAKING CORRELATION WITH C_{Map}. We follow a similar strategy to break the dependent between C_{Map} and ω. Using the powers $[\mathbf{h}]_0 = ([1]_0, [\omega]_0, \dots, [\omega^{2\kappa}]_0)$ we build circuit functionally equivalent to C_{Map}, indeed a circuit that outputs

$$\left[\prod_i^{\kappa} (x_{i,0} + x_{i,1}\omega + x_{i,2}\omega^2) \right]_T$$

via Lemma 2 (recall that $\mathbb{G}_T = \mathbb{G}_0$), and invoke the security of the obfuscator. We then use the q-SDDH assumptions for $2 \leq q \leq 2\kappa - 1$ in $\mathbb{G}_0$ to gradually transform $[\mathbf{h}]_0$ into $[\mathbf{q}]_0 = ([1]_0, [\omega]_0, [\omega^2]_0, [\tau_3]_0, \ldots, [\tau_{2\kappa}]_0)$ and embed a 1-SDDH tuple in the challenge matrix $[\mathbf{M}]$ as explained above.

Acknowledgements. Albrecht, Larraia and Paterson were supported by EPSRC grant EP/L018543/1. Hofheinz was supported by DFG grants HO 4534/2-2 and HO 4534/4-1.

References

[AB15] Applebaum, B., Brakerski, Z.: Obfuscating circuits via composite-order graded encoding. In: Dodis and Nielsen [DN15], pp. 528–556

[BBS04] Boneh, D., Boyen, X., Shacham, H.: Short group signatures. In: Franklin, M. (ed.) CRYPTO 2004. LNCS, vol. 3152, pp. 41–55. Springer, Heidelberg (2004)

[BHHO08] Boneh, D., Halevi, S., Hamburg, M., Ostrovsky, R.: Circular-secure encryption from decision Diffie-Hellman. In: Wagner, D. (ed.) CRYPTO 2008. LNCS, vol. 5157, pp. 108–125. Springer, Heidelberg (2008)

[BLMR13] Boneh, D., Lewi, K., Montgomery, H.W., Raghunathan, A.: Key homomorphic PRFs and their applications. In: Canetti and Garay [CG13a], pp. 410–428

[BLR+15] Boneh, D., Lewi, K., Raykova, M., Sahai, A., Zhandry, M., Zimmerman, J.: Semantically secure order-revealing encryption: multi-input functional encryption without obfuscation. In: Oswald and Fischlin [OF15], pp. 563–594

[BS03] Boneh, D., Silverberg, A.: Applications of multilinear forms to cryptography. Contemp. Math. **324**, 71–90 (2003)

[BWZ14] Boneh, D., Waters, B., Zhandry, M.: Low overhead broadcast encryption from multilinear maps. In: Garay, J.A., Gennaro, R. (eds.) CRYPTO 2014, Part I. LNCS, vol. 8616, pp. 206–223. Springer, Heidelberg (2014)

[CG13a] Canetti, R., Garay, J.A. (eds.): CRYPTO 2013, Part I. LNCS, vol. 8042. Springer, Heidelberg (2013)

[CG13b] Canetti, R., Garay, J.A. (eds.): CRYPTO 2013, Part II. LNCS, vol. 8043. Springer, Heidelberg (2013)

[CGH+15] Coron, J.-S., Gentry, C., Halevi, S., Lepoint, T., Maji, H.K., Miles, E., Raykova, M., Sahai, A., Tibouchi, M.: Zeroizing without low-level zeroes: new MMAP attacks and their limitations. In: Gennaro and Robshaw [GR15], pp. 247–266

[CHL+15] Cheon, J.H., Han, K., Lee, C., Ryu, H., Stehlé, D.: Cryptanalysis of the multilinear map over the integers. In: Oswald, E., Fischlin, M. (eds.) EUROCRYPT 2015. LNCS, vol. 9056, pp. 3–12. Springer, Heidelberg (2015)

[CLR15] Cheon, J.H., Lee, C., Ryu, H.: Cryptanalysis of the new CLT multilinear maps. Cryptology ePrint Archive, Report 2015/934 (2015). http://eprint.iacr.org/

[CLT13] Coron, J.-S., Lepoint, T., Tibouchi, M.: Practical multilinear maps over the integers. In: Canetti and Garay [CG13a], pp. 476–493

[CLT15] Coron, J.-S., Lepoint, T., Tibouchi, M.: New multilinear maps over the integers. In: Gennaro and Robshaw [GR15], pp. 267–286

[CLTV15] Canetti, R., Lin, H., Tessaro, S., Vaikuntanathan, V.: Obfuscation of probabilistic circuits and applications. In: Dodis and Nielsen [DN15], pp. 468–497

[Cor15] Coron, J.-S.: Cryptanalysis of GGH15 multilinear maps. Cryptology ePrint Archive, Report 2015/1037 (2015). http://eprint.iacr.org/

[DN15]   Dodis, Y., Nielsen, J.B. (eds.): TCC 2015, Part II. LNCS, vol. 9015. Springer, Heidelberg (2015)

[EHK+13] Escala, A., Herold, G., Kiltz, E., Ràfols, C., Villar, J.: An algebraic framework for Diffie-Hellman assumptions. In: Canetti and Garay [CG13b], pp. 129–147

[FHPS13] Freire, E.S.V., Hofheinz, D., Paterson, K.G., Striecks, C.: Programmable hash functions in the multilinear setting. In: Canetti and Garay [CG13a], pp. 513–530

[GGH13a] Garg, S., Gentry, C., Halevi, S.: Candidate multilinear maps from ideal lattices. In: Johansson, T., Nguyen, P.Q. (eds.) EUROCRYPT 2013. LNCS, vol. 7881, pp. 1–17. Springer, Heidelberg (2013)

[GGH+13b] Garg, S., Gentry, C., Halevi, S., Raykova, M., Sahai, A., Waters, B.: Candidate indistinguishability obfuscation and functional encryption for all circuits. In: 54th FOCS, pp. 40–49. IEEE Computer Society Press, October 2013

[GGH+13c] Garg, S., Gentry, C., Halevi, S., Sahai, A., Waters, B.: Attribute-based encryption for circuits from multilinear maps. In: Canetti and Garay [CG13b], pp. 479–499

[GGH15] Gentry, C., Gorbunov, S., Halevi, S.: Graph-induced multilinear maps from lattices. In: Dodis and Nielsen [DN15], pp. 498–527

[GGSW13] Garg, S., Gentry, C., Sahai, A., Waters, B.: Witness encryption and its applications. In: Boneh, D., Roughgarden, T., Feigenbaum, J. (eds.) 45th ACM STOC, pp. 467–476. ACM Press, June 2013

[GHV12] Galindo, D., Herranz, J., Villar, J.: Identity-based encryption with master key-dependent message security and leakage-resilience. In: Foresti, S., Yung, M., Martinelli, F. (eds.) ESORICS 2012. LNCS, vol. 7459, pp. 627–642. Springer, Heidelberg (2012)

[GR15]   Gennaro, R., Robshaw, M. (eds.): CRYPTO 2015, Part I. LNCS, vol. 9215. Springer, Heidelberg (2015)

[GS08]   Groth, J., Sahai, A.: Efficient non-interactive proof systems for bilinear groups. In: Smart, N.P. (ed.) EUROCRYPT 2008. LNCS, vol. 4965, pp. 415–432. Springer, Heidelberg (2008)

[HJ15]   Hu, Y., Jia, H.: Cryptanalysis of GGH map. Cryptology ePrint Archive, Report 2015/301 (2015). http://eprint.iacr.org/2015/301

[HSW13] Hohenberger, S., Sahai, A., Waters, B.: Full domain hash from (leveled) multilinear maps and identity-based aggregate signatures. In: Canetti and Garay [CG13a], pp. 494–512

[MF15]   Minaud, B., Fouque, P.-A.: Cryptanalysis of the new multilinear map over the integers. Cryptology ePrint Archive, Report 2015/941 (2015). http://eprint.iacr.org/

[NS09]   Naor, M., Segev, G.: Public-key cryptosystems resilient to key leakage. In: Halevi, S. (ed.) CRYPTO 2009. LNCS, vol. 5677, pp. 18–35. Springer, Heidelberg (2009)

[NY90]   Naor, M., Yung, M.: Public-key cryptosystems provably secure against chosen ciphertext attacks. In: 22nd ACM STOC, pp. 427–437. ACM Press, May 1990

[OF15]   Oswald, E., Fischlin, M. (eds.): EUROCRYPT 2015, Part II. LNCS, vol. 9057. Springer, Heidelberg (2015)

[PTT10]   Papamanthou, C., Tamassia, R., Triandopoulos, N.: Optimal authenticated data structures with multilinear forms. In: Joye, M., Miyaji, A., Otsuka, A. (eds.) Pairing 2010. LNCS, vol. 6487, pp. 246–264. Springer, Heidelberg (2010)

[TLL14]   Tang, F., Li, H., Liang, B.: Attribute-based signatures for circuits from multilinear maps. In: Chow, S.S.M., Camenisch, J., Hui, L.C.K., Yiu, S.M. (eds.) ISC 2014. LNCS, vol. 8783, pp. 54–71. Springer, Heidelberg (2014)

[YYHK14] Yamakawa, T., Yamada, S., Hanaoka, G., Kunihiro, N.: Self-bilinear map on unknown order groups from indistinguishability obfuscation and its applications. In: Garay, J.A., Gennaro, R. (eds.) CRYPTO 2014, Part II. LNCS, vol. 8617, pp. 90–107. Springer, Heidelberg (2014)

[YYHK15] Yamakawa, T., Yamada, S., Hanaoka, G., Kunihiro, N.: Self-bilinear map on unknown order groups from indistinguishability obfuscation and its applications. Cryptology ePrint Archive, Report 2015/128 (2015). http://eprint.iacr.org/2015/128

[Zim15]   Zimmerman, J.: How to obfuscate programs directly. In: Oswald and Fischlin [OF15], pp. 439–467

Perfect Structure on the Edge of Chaos

Trapdoor Permutations from Indistinguishability Obfuscation

Nir Bitansky[1]($\boxtimes$), Omer Paneth[2], and Daniel Wichs[3]

[1] MIT, Cambridge, USA
nbitansky@gmail.com
[2] Boston University, Boston, USA
[3] Northeastern University, Boston, USA

Abstract. We construct trapdoor permutations based on (sub-exponential) indistinguishability obfuscation and one-way functions, thereby providing the first candidate that is not based on the hardness of factoring.

Our construction shows that even highly structured primitives, such as trapdoor permutations, can be potentially based on hardness assumptions with *noisy structures* such as those used in candidate constructions of indistinguishability obfuscation. It also suggest a possible way to construct trapdoor permutations that resist quantum attacks, and that their hardness may be based on problems outside the complexity class SZK — indeed, while factoring-based candidates do not possess such security, future constructions of indistinguishability obfuscation might.

As a corollary, we eliminate the need to assume trapdoor permutations and injective one-way function in many recent constructions based on indistinguishability obfuscation.

1 Introduction

In the mid '70s and early '80s, powerful number-theoretic constructions related to factoring and discrete-logs kick-started modern cryptography. As these constructions gradually evolved into a comprehensive theory, generic primitives, such as one-way functions, collision-resistant hash functions, and trapdoor permutations, were defined with the aim of abstracting the properties needed in different applications. The aforementioned number-theoretic problems provided instantiations for each of these primitives, but at the same time appeared to offer a much richer algebraic structure. Whereas this structure is highly fruitful, it also limits the hardness of the corresponding problems to low complexity classes such as statistical zero-knowledge (SZK) [GK88] and makes them susceptible to quantum attacks [Sho97]. Therefore, a fundamental goal is to base cryptographic primitives on other less structured assumptions.

O. Paneth—Supported by the Simons award for graduate students in theoretical computer science and an NSF Algorithmic foundations grant 1218461.

D. Wichs—Supported by NSF grants CNS-1347350 and CNS-1314722.

E. Kushilevitz and T. Malkin (Eds.): TCC 2016-A, Part I, LNCS 9562, pp. 474–502, 2016.
DOI: 10.1007/978-3-662-49096-9_20

In some cases, such as one-way functions, it seems that we can avoid structured assumptions altogether. Indeed, one-way functions can be constructed generically from essentially any cryptographic primitive and have candidates from purely combinatorial assumptions [BFKL93, Gol11, JP00, AC08]. However, as we consider primitives that intrinsically require some structure, candidates become more scarce. For example, *injective* one-way functions are only known based on assumptions with some *algebraic homomorphism*, albeit these may feature noisy structures, such as the ones arising from lattices [PW08]. In particular, such assumptions can be placed in SZK, but are not known to be susceptible to quantum attacks. If we also require the one-way function to be a *permutation*, candidates become even more scarce, and only known based on the hardness of discrete-logs and factoring (or RSA) [RSA83, Rab79]. Further, *trapdoor permutations* (TDPs) are known exclusively based on factoring (or RSA).

Obfuscation. A promising source for new constructions, replacing ones that so far depended exclusively on specific algebraic assumptions, is *program obfuscation* — a method for shielding programs such that their implementation becomes hidden. Indeed, an ideal notion of obfuscation would allow us to securely express any required structure in the obfuscated programs. Understanding to what extent this intuition can be fulfilled requires looking into concrete notions of secure obfuscation. The question is what is the "right" notion to consider and under what kind of assumptions it can be achieved.

Of particular interest is the notion of *indistinguishability obfuscation* (iO), which have recently found candidate constructions [GGH+13b]. The notion of iO requires that the obfuscations of any two programs of the same size and functionality are indistinguishable. While this notion may not capture *ideal obfuscation*, it turns out to be sufficient for many known cryptographic primitives, suggesting an alternative for previous number theoretic constructions [SW14, BP14, CLTV14].

From an assumption perspective, the existing constructions of iO [GGH+13b, BR14, BGK+13, AB15, Zim15, AJ15, BV15, GLSW14] are instantiated based on multi-linear graded encodings [GGH13a, CLT13, CLT15] thus falling into Gentry's [Gen14] world of *computing on the edge of chaos* — they all rely on *noisy structures* in an essential way. Beyond the existing candidates, understanding on which assumptions iO can be based (and how structured they should be) is an open question; in particular, as far as we know, future constructions of iO may be based on problems outside $AM \cap coAM$ and/or outside BQP.

1.1 This Work

Our main result is a construction of trapdoor permutations based on subexponential indistinguishability obfuscation and one-way functions. As far as we know, this is the first construction of trapdoor permutations since the introduction of the RSA and Rabin trapdoor permutations [RSA83, Rab79] (and their variants). We also construct injective one-way functions based on standard iO and one-way functions. As a tool used in our constructions and a result of potentially independent interest, we show how to convert any one-way function into a *sometimes-injective one-way function* that is simultaneously injective and hard-to-invert on some sub-domain of noticeable density.

Properties. Our permutations have the following additional features. First, they are doubly-enhanced. Additionally, they can be generated so to have any *prescribed cycle structure* with the necessary property that small cycles are rare enough. So far this property has only been achieved for pseudo-random permutations [NR02]. Another feature is that inverting the permutation consists of simple symmetric-key operations (unlike in existing candidates). Finally, like in the RSA permutation, given the trapdoor it is possible to iterate the permutation (or its inverse) any number of times at the same cost as computing the function once.

One difference between the trapdoor permutations we construct and those typically defined in the literature [GR13] is that we only support sampling of *pseudo-uniform* elements in the domain rather than elements that are statistically close to uniform. The sampled elements are pseudo-uniform in a strong sense, namely, even given the trapdoor or more generally the coins used to sample the function. The double-enhancement requirement is relaxed in a somewhat similar manner (see details below). As far as we know, these relaxation are sufficient in known applications. Additionally we note that our permutations are not *certifiable*, meaning that we do not know of an efficient way to certify that a key is well-formed and describes a valid permutation.

iO as a hub. Based on our results, several constructions previously based on iO and additional structured assumptions, can now be based only on iO and one-way functions (or rather the assumption that NP $\neq$ coRP [KMN+14]). Examples include: non-interactive commitments [Blu81],[1] actively secure two-message oblivious transfer [SW14], non-interactive witness-indistinguishable proofs [BP14], obfuscation for Turing machines [KLW14],[2] hardness of the complexity class PPAD [BPR15], and more.

1.2   Technical Overview

Trapdoor permutations from obfuscation? Sounds easy. Thinking of obfuscation in ideal terms gives rise to a natural attempt at constructing TDPs: *simply obfuscate a pseudo-random permutation.* Clearly, with only black-box access to such a permutation, inversion is as impossible as inverting a random function. However, ideal virtual black-box obfuscation [BGI+01] of pseudo-random permutations is unknown, and is in fact subject to strong limitations [BCC+14]. Nevertheless, we show that some of this intuition can be recovered also when relying on the (not so ideal) notion of iO.

---

[1] See Appendix A for more details on the construction of non-interactive commitments.
[2] Formally, [KLW14] rely on injective PRGs, but these can be replaced with injective one-way functions using an observation of [BCP14], as explained in Sect. 4.2.

Outline of the construction. Our starting point is a recent construction suggested by Bitansky et al. [BPR15] to demonstrate the hardness of the complexity class PPAD. We observe that their construction can, in fact, be viewed as a trapdoor permutation family that lacks a crucial property: *it does not allow to sample elements from the permutation's domain.*

Our construction then follows the three steps below:

1. We construct a sampler for domain elements and prove the one-wayness of the permutation family even given this sampler. This involves extending the techniques developed in [BPR15].
2. We further augment the permutation family and sampler so that they will admit the requirements of enhanced and doubly-enhanced TDPs.
3. The construction of [BPR15] relies on injective one-way functions in addition to iO. We construct such injective one-way functions based on iO and one-way functions. We find this construction to be of independent interest.

We now elaborate on the construction and analysis in [BPR15], describe where it falls short of achieving an actual TDP, and then turn to describe our solutions.

A Closer Look into [BPR15]. Bitansky et al. construct a hard instance of the END-OF-THE-LINE problem based on iO. In the END-OF-THE-LINE problem we consider a sequence of nodes

$$x_1 \rightarrow x_2 \rightarrow \cdots \rightarrow x_T,$$

and a program F that maps x_i to x_{i+1} for $1 \le i < T$. The problem is given the source node x_1 and the program F find the sink node x_T. In the construction of [BPR15], each node x_i is a pair $(i, \mathsf{PRF}_S(i))$ where $\mathsf{PRF}_S : \mathbb{Z}_T \rightarrow \{0,1\}^\lambda$ is sampled from a family of pseudo-random functions, and $T \in \mathbb{N}$ is super-polynomial in the security parameter λ. The instance also contains an obfuscated program $\widetilde{\mathsf{F}}$ that maps x_i to x_{i+1} and outputs $\perp$ on any other input.

Bitansky et al. show that given strong enough iO and injective one-way functions (used only in the analysis) it is hard to find x_T given x_1 and the obfuscated program $\widetilde{\mathsf{F}}$. Intuitively, the path from x_1 to x_T can be thought of as an authenticated chain where a signature σ corresponding to some pair (i, σ) cannot be obtained without first obtaining all previous signatures on the path. It is not difficult to show that any efficient algorithm that only invokes $\widetilde{\mathsf{F}}$ as a black box cannot find the signature $\mathsf{PRF}_S(T)$. Their proof shows that the same hardness holds even given full access to the obfuscated program $\widetilde{\mathsf{F}}$.

Constructing trapdoor permutations. Indeed, the construction of [BPR15] described above gives rise to a natural candidate for a trapdoor permutation. A given permutation is over the set of nodes $\{x_i\}_{i \in \mathbb{Z}_T}$ and is defined by the cycle

$$x_1 \rightarrow x_2 \rightarrow \cdots \rightarrow x_T \rightarrow x_1.$$

The public key describing the permutation consists of the obfuscated program $\widetilde{\mathsf{F}}$ that maps x_i to x_{i+1} (where $i + 1$ is computed modulo T) and outputs $\perp$

on any other input. The trapdoor is simply the seed S of the pseudo-random function that allows us to efficiently invert the permutation. However, without the trapdoor, inverting the permutation on x_i is as hard as finding the end of the chain starting at x_i and ending at x_{i-1}.

To obtain a complete construction of a TDP, we need to specify how to sample random domain elements. The challenge here is that the domain of our permutation is very *sparse* and it is not clear how to sample from it without the trapdoor S. A naive suggestion is to include, as part of the public key, an obfuscated sampler program that given i outputs the node x_i. However, publishing such a program (obfuscated or not) clearly makes the permutation easy to invert. To explain, how this is solved, we now look more closely into the security proof of [BPR15].

The proof of [BPR15]. We sketch the argument from [BPR15] showing that the basic TDPs construction above (without any domain sampler) is one-way. That is, given $\widetilde{\mathsf{F}}$ and x_i for a random $i \in \mathbb{Z}_T$, it is hard to obtain x_{i-1} (in fact we prove this for every $i \in \mathbb{Z}_T$). To prove that finding the node x_{i-1} is hard it is sufficient to prove that the obfuscated circuit $\widetilde{\mathsf{F}}$ is computationally indistinguishable from a circuit that on input x_{i-1} returns $\bot$, rather than x_i as $\widetilde{\mathsf{F}}$ would. Indeed, any algorithm that can find x_{i-1} can also distinguish the two circuits. We next explain how indistinguishability of these two circuits is shown.

For every $\alpha, \beta \in \mathbb{Z}_T$ we consider the circuit $\widetilde{\mathsf{F}}_{\alpha,\beta}$ that is identical to $\widetilde{\mathsf{F}}$, except that for every j in the range from α to β (wrapping around T in case that $\alpha > \beta$) $\widetilde{\mathsf{F}}_{\alpha,\beta}$ on the input x_j outputs $\bot$. The argument proceeds in two steps.

Step 1: Split the chain into two parts. We show that for a *random* $u \in \mathbb{Z}_T$, the obfuscation $\widetilde{\mathsf{F}}_{u,u}$ is computationally indistinguishable from $\widetilde{\mathsf{F}}$. Intuitively this "splits" the authenticated chain into two parts: from x_i to x_u and from x_{u+1} to x_{i-1}. The proof of this step relies on the fact that the chain is of super-polynomial length T and therefore the index u of a random node in the chain is hard to guess.

Step 2: Erase the second part of the chain. We show that after the chain is split, is it hard to find any node x_j in the second part of the chain. Formally, we prove that the obfuscated circuits $\widetilde{\mathsf{F}}_{u,u}$ and $\widetilde{\mathsf{F}}_{u,i-1}$ are computationally indistinguishable. The proof is by a sequence of hybrids: for every j in the range between u and $i-2$, we rely on injective one-way functions and iO with super-polynomial hardness to show that the obfuscated circuits $\widetilde{\mathsf{F}}_{u,j}$ and $\widetilde{\mathsf{F}}_{u,j+1}$ are $T^{-\Theta(1)}$-indistinguishable. To prove that we can indistinguishably change the output of $\widetilde{\mathsf{F}}_{u,j}$ on the node x_{j+1} to $\bot$, we rely on the fact that in the circuit $\widetilde{\mathsf{F}}_{u,j}$ the successor of x_j is already erased and therefore, the circuit $\widetilde{\mathsf{F}}_{u,j}$ never explicitly outputs the node x_{j+1}.

Sampling from the Domain. As mentioned before, to allow sampling of elements in the domain we cannot simply provide a circuit that outputs x_i given

$i \in \mathbb{Z}_T$, as this would result in an obvious attack — given $x_i = (i, \mathsf{PRF}_S(i))$, one can directly obtain the preimage x_{i-1}. The idea is to provide instead an obfuscation $\widetilde{\mathsf{X}}$ of a sampler X_S that is supported on a very sparse, but still pseudo-random, subset of the domain. Concretely, X_S takes as input a seed s for a length doubling pseudo-random generator $\mathsf{PRG} : \mathbb{Z}_{\sqrt{T}} \to \mathbb{Z}_T$, and outputs x_i for $i = \mathsf{PRG}(s)$.

First, note that by pseudo-randomness, inverting x_{i+1} where $i = \mathsf{PRG}(s)$ is pseudorandom is as hard as inverting x_{i+1} when i is chosen truly at random. Thus, we fucus on showing that inverting x_i is hard for a truly random i even in the presence of the obfuscated sampler $\widetilde{\mathsf{X}}$.

The one-wayness proof described above, however, fails when the adversary is given the sampler $\widetilde{\mathsf{X}}$. The problem is that in the second step, when arguing that the obfuscated circuits $\widetilde{\mathsf{F}}_{u,j}$ and $\widetilde{\mathsf{F}}_{u,j+1}$ are indistinguishable, we used the fact that $\widetilde{\mathsf{F}}_{u,j}$ never explicitly outputs the node x_{j+1}. However, if $j+1$ is in the image of PRG, the sampler $\widetilde{\mathsf{X}}$ explicitly outputs x_{j+1} and we can no longer prove that $\widetilde{\mathsf{F}}_{u,j}$ and $\widetilde{\mathsf{F}}_{u,j+1}$ are indistinguishable.

Our solution is to consider, instead of the entire chain starting from x_i and ending at x_{i-1}, only a suffix of this chain of length $\sqrt[4]{T}$ starting from $x_{i-\sqrt[4]{T}}$ and ending at x_{i-1}. On the one hand, this chain segment is still of super-polynomial length, and therefore, we can still split the segment following Step 1 above. On the other hand, the segment is also not too large (of density $T^{-3/4}$ in $\mathbb{Z}_T$). Since that segment starts at a random index $i - \sqrt[4]{T}$, and since the image of PRG is of size only $\sqrt{T}$, we have that with overwhelming probability $1 - T^{-1/4}$ the segment interval will not contain any nodes in the support of the sampler $\widetilde{\mathsf{X}}$. When the segment and the support of $\widetilde{\mathsf{X}}$ are disjoint, we can again erase the entire chain segment following Step 2 above.

Enhancements. In applications of TDPs, it is often required that the TDPs are *enhanced* or even *doubly enhanced* [GR13]. We briefly recall these properties and explain how they are obtained. In enhanced TDPs, we essentially ask that it is possible to *obliviously* sample domain elements, without knowing their preimages. Translating to our setting, we require that $x_{\mathsf{PRG}(s)} \leftarrow \mathsf{X}_S$ is hard to invert, even given the coins s used to sample it. In the construction above, this may not be true. Indeed, given the seed s for the pseudo-random generator, we can no longer argue that inversion is as hard as for a truly uniform element. In fact, the PRG may be such that given s, it is easy to find s' such that $\mathsf{PRG}(s') = \mathsf{PRG}(s) - 1$ and thus easily invert. We observe that this can be circumvented if we make sure that PRG has a *discrete image* where a random image $\mathsf{PRG}(s)$ is likely to be isolated away from any other image. We show how to construct such PRGs from plain PRGs and pairwise-independent permutations.

In doubly enhanced TDPs, it is typically required that it is possible to sample an image-preimage pair (x, y) together with random coins used to sample the preimage y by the usual sampler. In our setting, we would like to sample an image $y = x_{\mathsf{PRG}(s)} \leftarrow \mathsf{X}_S$ together with randomness s and preimage $x_{\mathsf{PRG}(s)-1}$. We only achieve a relaxed form of this requirement, where s is pseudo-random rather than truly random, even given the trapdoor, or the coins used to sample the function. The idea is to slightly change the pseudo-random generator PRG in the previous constructions in a way that exploits the specific structure of our TDP. We only change PRG on a sparse set of seeds that has negligible density, and thus previous properties are preserved (see more details in Sect. 4.3).

Injective One-Way Functions from iO. We now describe the main ideas behind constructing injective one-way function from iO and plain one-way functions. We rely on two-message statistically-binding commitment schemes [Nao91] and puncturable PRFs (both known from any one-way function). In the constructed family, every function $\mathsf{OWF}_{M_1,S}$ is associated with a first message M_1 for the commitment scheme and a pseudo-random function PRF_S. The public description of the function contains an obfuscated circuit $\widetilde{C}$ that on input x outputs a commitment $\mathsf{COM}_{M_1}(x; \mathsf{PRF}_S(x))$ with respect to the first commitment message M_1, plaintext x and randomness $\mathsf{PRF}_S(x)$. The fact that the function is injective (with overwhelming probability over M_1) follows directly from the statistical binding of the commitment. We focus on arguing one-wayness.

Our goal is to show that it is hard to recover a random x given $\widetilde{C}$ and $\widetilde{C}(x)$. We start by considering a hybrid circuit defined similarly to $\widetilde{C}$ except that it contains the punctured key $S\{x\}$ and given input x, it outputs a hardcoded commitment; since we did not change the functionality of the circuit indistinguishability follows by iO. Using pseudo-randomness at the punctured point x and the hiding of the commitment we can now argue that the hardcoded commitment hides x, replacing it with a commitment to some arbitrary plaintext, using true randomness. The problem is that now, even if we unpuncture $S\{x\}$, x itself still needs to appear in the clear as part of the code of the circuit in order to trigger the output of the hardcoded commitment.

Nevertheless, we may try to apply a similar strategy to the one previously used for our TDPs. Concretely, we note that x is only used to test if an input x' satisfies $x' = x$, and this comparison can be performed in an "encrypted form" — instead of hardcoding x in the clear we can hardcode $g(x)$ for some one-way function g and compare images instead of preimages. Unfortunately, to argue that this does not change functionality *the function g must itself be injective* which seems to bring us back to square one.

The key observation is that we may gain by using a function g that is only *sometimes injective*; namely, it is enough that g is simultaneously injective and hard to invert only on some noticeable subset of its domain. We show that such functions can be constructed from any one-way function. Now, leveraging the iO requirement only on the corresponding injective sub-domain, we can show that the above construction results in a *weak* one-way function that is *fully injective*; indeed, we only invoke sometimes-injective of g in the proof of one-wayness. Then, to obtain a (strong) injective one-way function, we can apply standard direct-product amplification [Yao82].

Constructing Sometimes Injective One-Way Functions. We outline the main idea behind constructing a sometimes-injective one-way function g, as above, based on any one way function f. First, consider for simplicity the case that the function f is r-regular. Roughly, the idea is extract the $\log(r)$ bits of randomness that remain in x conditioned on $f(x)$ and append them to the function output as in [HILL99]. However, due to their inherent entropy loss, standard randomness extractors cannot extract enough random bits to guarantee any meaningful injectiveness. Nevertheless, for our purpose, the extracted bits need not be statistically-close to uniform, they only need to preserve one-wayness. Accordingly, we use the *unpredictability extractors* of [DPW14], which allow extracting more bits so to guarantee injectiveness, while still preserving meaningful one-wayness.

To deal with f that is not regular, we may set r to be the most frequent regularity of f. This only shrinks the portion of the domain where f is both injective and hard to invert by some polynomial factor. A uniform construction is obtained by choosing r at random.

2 Preliminaries

The cryptographic definitions in the paper follow the convention of modeling security against non-uniform adversaries. An efficient adversary $\mathcal{A}$ is modeled as a sequence of circuits $\mathcal{A} = \{\mathcal{A}_\lambda\}_{\lambda \in \mathbb{N}}$, such that each circuit $\mathcal{A}_\lambda$ is of polynomial size $\lambda^{O(1)}$ with $\lambda^{O(1)}$ input and output bits; we shall also consider adversaries of some super polynomial size $t(\lambda) = \lambda^{\omega(1)}$. We often omit the subscript λ when it is clear from the context. The resulting hardness will accordingly be against non-uniform algorithms. The result can be cast into the uniform setting, with some adjustments to the analysis.

2.1 Indistinguishability Obfuscation

We define indistinguishability obfuscation (iO) with respect to a give class of circuits. The definition is formulated as in [BGI+01].

Definition 2.1 (Indistinguishability obfuscation *[BGI+01]*). *A PPT algorithm iO is said to be an* indistinguishability obfuscator *for a class of circuits $\mathcal{C}$, if it satisfies:*

1. **Functionality:** *for any $C \in \mathcal{C}$,*

$$\Pr_{i\mathcal{O}}\left[\forall x : i\mathcal{O}(C)(x) = C(x)\right] = 1.$$

2. **Indistinguishability:** *for any polysize distinguisher $\mathcal{D}$ there exists a negligible function $\mu(\cdot)$, such that for any two circuits $C_0, C_1 \in \mathcal{C}$ that compute the same function and are of the same size λ:*

$$\left|\Pr[\mathcal{D}(i\mathcal{O}(C_0)) = 1] - \Pr[\mathcal{D}(i\mathcal{O}(C_1)) = 1]\right| \leq \mu(\lambda),$$

where the probability is over the coins of $\mathcal{D}$ and $i\mathcal{O}$.

We further say that $i\mathcal{O}$ is (t, δ)-secure, for some function $t(\cdot)$ and concrete negligible function $\delta(\cdot)$, if for all $t(\lambda)^{O(1)}$ distinguishers the above indistinguishability gap $\mu(\lambda)$ is smaller than $\delta(\lambda)^{\Omega(1)}$.

2.2 Puncturable Pseudorandom Functions

We consider a simple case of the puncturable pseudo-random functions (PRFs) where any PRF may be punctured at a single point. The definition is formulated as in [SW14], and is satisfied by the GGM [GGM86] PRF [BW13,KPTZ13, BGI14],

Definition 2.2 (Puncturable PRFs). *Let* n, k *be polynomially bounded length functions. An efficiently computable family of functions*

$$\mathcal{PRF} = \left\{ \mathsf{PRF}_S : \{0,1\}^{n(\lambda)} \to \{0,1\}^{\lambda} : S \in \{0,1\}^{k(\lambda)}, \lambda \in \mathbb{N} \right\},$$

associated with an efficient (probabilistic) key sampler $\mathcal{K}_{\mathcal{PRF}}$, *is a puncturable PRF if there exists a poly-time puncturing algorithm* Punc *that takes as input a key* S, *and a point* x^*, *and outputs a punctured key* $S\{x^*\}$, *so that the following conditions are satisfied:*

1. **Functionality is preserved under puncturing:** *For every* $x^* \in \{0,1\}^{n(\lambda)}$,

$$\Pr_{S \leftarrow \mathcal{K}_{\mathcal{PRF}}(1^\lambda)} \left[\forall x \neq x^* : \mathsf{PRF}_S(x) = \mathsf{PRF}_{S\{x^*\}}(x) : S\{x^*\} = \mathsf{Punc}(S, x^*) \right] = 1.$$

2. **Indistinguishability at punctured points:** *for any polysize distinguisher* $\mathcal{D}$ *there exists a negligible function* $\mu(\cdot)$, *such that for all* $\lambda \in \mathbb{N}$, *and any* $x^* \in \{0,1\}^{n(\lambda)}$,

$$\left| \Pr[\mathcal{D}(x^*, S\{x^*\}, \mathsf{PRF}_S(x^*)) = 1] - \Pr[\mathcal{D}(x^*, S\{x^*\}, u) = 1] \right| \leq \mu(\lambda),$$

where $S \leftarrow \mathcal{K}_{\mathcal{PRF}}(1^\lambda), S\{x^*\} = \mathsf{Punc}(S, x^*)$, *and* $u \leftarrow \{0,1\}^\lambda$.

   We further say that $\mathcal{PRF}$ *is* (t, δ)-*secure, for some function* $t(\cdot)$ *and concrete negligible function* $\delta(\cdot)$, *if for all* $t(\lambda)^{O(1)}$ *distinguishers the above indistinguishability gap* $\mu(\lambda)$ *is smaller than* $\delta(\lambda)^{\Omega(1)}$.

2.3 Injective One-Way Functions

We shall also rely on (possibly keyed) injective one-way functions.

Definition 2.3 (Injective OWF). *Let* k *be polynomially bounded length function. An efficiently computable family of functions*

$$\mathcal{OWF} = \left\{ \mathsf{OWF}_K : \{0,1\}^\lambda \to \{0,1\}^* : K \in \{0,1\}^{k(\lambda)}, \lambda \in \mathbb{N} \right\},$$

associated with an efficient (probabilistic) key sampler $\mathcal{K}_{\mathcal{OWF}}$, *is an injective OWF if it satisfies*

1. **Injectiveness:** *With overwhelming probability over the choice of* $K \leftarrow \mathcal{K}_{\mathcal{OWF}}(1^\lambda)$, *the function* OWF_K *is injective.*

2. **One-wayness:** *For any polysize inverter $\mathcal{A}$ there exists a negligible function $\mu(\cdot)$, such that for all $\lambda \in \mathbb{N}$,*

$$\Pr\left[\mathcal{A}(K, \mathsf{OWF}_K(x)) = x : \begin{array}{c} K \leftarrow \mathcal{K}_{\mathcal{OWF}}(1^\lambda) \\ x \leftarrow \{0,1\}^\lambda \end{array}\right] \leq \mu(\lambda).$$

We further say that $\mathcal{OWF}$ is (t, δ)-secure, for some function $t(\cdot)$ and concrete negligible function $\delta(\cdot)$, if for all $t(\lambda)^{O(1)}$ inverters the above inversion probability $\mu(\lambda)$ is smaller than $\delta(\lambda)^{\Omega(1)}$.

3   Injective One-Way Functions from iO

In this section, we construct injective one-way functions from iO and plain injective one-way functions. We start by defining and constructing *sometimes injective one-way functions*.

3.1   Sometimes Injective One-Way Functions

For a function $f : \{0,1\}^\lambda \to \{0,1\}^*$ and any input $x \in \{0,1\}^\lambda$, we denote by

$$\mathbf{H}_f(x) := \log |\{x' : f(x') = f(x)\}| = \mathbf{H}_\infty\left(x' \leftarrow f^{-1}(f(x))\right),$$
$$\mathbf{Inj}(f) := \{x : \mathbf{H}_f(x) = 0\}$$

the min-entropy of a random preimage of $f(x)$, and the subset of inputs on which f is injective, respectively.

We next define sometimes-injective OWFs (SIOWFs). Roughly speaking, such functions are injective and hard to invert over a noticeable fraction of their domain.

Definition 3.1 (Sometimes-Injective OWF). *Let k be polynomially bounded length function. An efficiently computable family of functions*

$$\mathcal{SIOWF} = \left\{\mathsf{SIOWF}_K : \{0,1\}^\lambda \to \{0,1\}^* : K \in \{0,1\}^{k(\lambda)}, \lambda \in \mathbb{N}\right\},$$

associated with an efficient (probabilistic) key sampler $\mathcal{K}_{\mathcal{SIOWF}}$, is a sometimes injective OWF if for every key $K \in \{0,1\}^{k(\lambda)}$ there exists an injective subset $\mathbf{I}_K \subseteq \mathbf{Inj}(\mathsf{SIOWF}_K)$, satisfying the following conditions:

1. **Sometimes injectiveness:** *There exists a polynomial $p(\cdot)$ such that for any $\lambda \in \mathbb{N}$:*

$$\Pr\left[x \in \mathbf{I}_K : \begin{array}{c} K \leftarrow \mathcal{K}_{\mathcal{SIOWF}}(1^\lambda) \\ x \leftarrow \{0,1\}^\lambda \end{array}\right] \geq 1/p(\lambda).$$

2. **One-wayness over injective subdomain:** *for any polysize inverter $\mathcal{A}$ there is a negligible function $\mu(\cdot)$ such that for any $\lambda \in \mathbb{N}$:*

$$\Pr\left[\mathcal{A}(K, \mathsf{SIOWF}_K(x)) = x : \begin{array}{c} K \leftarrow \mathcal{K}_{\mathcal{SIOWF}}(1^\lambda) \\ x \leftarrow \mathbf{I}_K \end{array}\right] \leq \mu(\lambda).$$

We further say that $\mathcal{SIOWF}$ is t-secure, for some super-polynomial function $t(\cdot)$, if the one-wayness requirement holds for all $t(\lambda)^{O(1)}$ inverters.

The Construction. Let $f : \{0,1\}^* \rightarrow \{0,1\}^*$ be any one-way function. We construct an SIOWF

$$\mathcal{SIOWF} = \left\{ \mathsf{SIOWF}_K : \{0,1\}^\lambda \rightarrow \{0,1\}^* : K \in \{0,1\}^{k(\lambda)}, \lambda \in \mathbb{N} \right\},$$

with a corresponding key sampler $\mathcal{K}_{\mathcal{OWF}}$ as follows:

- A random key $K := (S, e) \leftarrow \mathcal{K}_{\mathcal{SIOWF}}(1^\lambda)$ consists of a random $e \leftarrow [\lambda]$ and a random seed S for a hash function $\mathsf{h}_S : \{0,1\}^\lambda \rightarrow \{0,1\}^{e+1}$ drawn from a q-wise independent family, where we set $q = \lambda$ to be the security parameter.
- For $x \in \{0,1\}^\lambda$, the function is defined by $\mathsf{SIOWF}_K(x) := (f(x), \mathsf{h}_S(x))$.

Proposition 3.1. *$\mathcal{SIOWF}$ is a sometimes injective one-way function.*

Proof. Throughout, we denote by $E_e \subseteq \{0,1\}^\lambda$ the subset of values x such that $\mathbf{H}_f(x) \in [e - 1, e)$. For $K = (S, e)$, we define $\mathbf{I}_{S,e} = E_e \cap \mathbf{Inj}(\mathsf{SIOWF}_{S,e}) \subseteq \mathbf{Inj}(\mathsf{SIOWF}_{S,e})$. We start by proving the following preliminary claim saying that the function is injective with high-probability over the set E_e.

Claim 3.1. *For any $\lambda \in \mathbb{N}, e \in [\lambda], x \in E_e$*

$$\Pr_S \left[x \in \mathbf{Inj}(\mathsf{SIOWF}_{S,e}) \right] \geq \frac{1}{2}.$$

Proof (Proof of Claim 3.1). Fix any $\lambda \in \mathbb{N}, e \in [\lambda], x \in E_e$, and let $y = f(x)$. Since the output of $\mathsf{SIOWF}_{S,e}(x)$ includes y, it suffices to show that x does not collide with any other $x' \in f^{-1}(y)$. By q-wise independence (in fact, pairwise is sufficient here), for any such x',

$$\Pr_S \left[\mathsf{h}_S(x) = \mathsf{h}_S(x') \right] \leq 2^{-e-1}.$$

Thus, the expected number of such x' that collide with x is:

$$2^{-e-1}(\left| f^{-1}(y) \right| - 1) \leq 2^{-e-1} \cdot 2^{\mathbf{H}_f(x)} \leq 1/2,$$

and the claim now follows by Markov's inequality.

Sometimes injectiveness follows directly:

$$\Pr \left[x \in \mathbf{I}_{S,e} : \begin{array}{c} (S, e) \leftarrow \mathcal{K}_{\mathcal{SIOWF}}(1^\lambda) \\ x \leftarrow \{0,1\}^\lambda \end{array} \right] \geq$$

$$\Pr_{e,x} \left[x \in E_e \right] \cdot \min_{e,x \in E_e} \Pr_S \left[x \in \mathbf{Inj}(\mathsf{SIOWF}_{S,e}) \right] \geq \frac{1}{2\lambda}.$$

where $\Pr_{e,x} \left[x \in E_e \right] = 1/\lambda$ since for any fixed x there is a unique $e \in [\lambda]$ such that $x \in E_e$, and $\min_{e,x \in E_e} \Pr_S \left[x \in \mathbf{Inj}(\mathsf{SIOWF}_{S,e}) \right] \geq 1/2$ by the previous claim.

Next, we prove one-wayness over the injective subdomain. Fix any polysize inverter $\mathcal{A}$ and security parameter λ. Firstly, we notice that

$$
\begin{aligned}
&\Pr_{\substack{S,e \\ x \leftarrow \mathbf{I}_{S,e}}} [\mathcal{A}(S, e, f(x), \mathsf{h}_S(x)) = x] \\
&\leq \Pr_{\substack{S,e \\ x \leftarrow E_e}} [\mathcal{A}(S, e, f(x), \mathsf{h}_S(x)) = x] \Big/ \Pr_{\substack{S,e \\ x \leftarrow E_e}} [x \in \mathbf{I}_{S,e}] \qquad (1) \\
&\leq 2 \Pr_{\substack{S,e \\ x \leftarrow E_e}} [\mathcal{A}(S, e, f(x), \mathsf{h}_S(x)) = x],
\end{aligned}
$$

where we can bound the denominator in Eq. (1) by at least $\frac{1}{2}$ by Claim 3.1. Therefore, it remains to show that $\Pr_{\substack{S,e \\ x \leftarrow E_e}} [\mathcal{A}(S, e, f(x), \mathsf{h}_S(x)) = x]$ is negligible. To prove this, we rely on a theorem from [DPW14] showing that any q-wise independent family essentially preserves uninvertability.

Theorem 3.2 (*[DPW14, Theorem 4.1]* **(restated)**). *Let* $\{\mathsf{h}_S : \{0,1\}^n \to \{0,1\}^m : S \in \{0,1\}^d\}$ *be a q-wise independent hashing family. For any* $D : \{0,1\}^m \times \{0,1\}^d \to \{0,1\}$ *and any random variable* $X \in \{0,1\}^n$ *with min-entropy* $\mathbf{H}_\infty(X) \geq k$, *if* $\Pr[D(U, S) = 1] = \delta$, *then* $\Pr[D(\mathsf{h}_S(X), S) = 1] \leq O(q 2^{m-k}) \max\{\delta, 2^{-q}\}$.

Using the above theorem, we have:

$$
\begin{aligned}
&\Pr_{\substack{S,e \\ x \leftarrow E_e}} [\mathcal{A}(S, e, f(x), \mathsf{h}_S(x)) = x] \\
&\leq \Pr_{\substack{S,e \\ x \leftarrow E_e}} [\mathcal{A}(S, e, f(x), \mathsf{h}_S(x)) \in f^{-1}(f(x))] \\
&= \Pr_{\substack{S,e,x \leftarrow E_e \\ x' \leftarrow f^{-1}(f(x))}} [\mathcal{A}(S, e, f(x), \mathsf{h}_S(x')) \in f^{-1}(f(x))] \qquad (2) \\
&= \mathbb{E}_{\substack{e,x \leftarrow E_e}} \Pr_{\substack{S \\ x' \leftarrow f^{-1}(f(x))}} [\mathcal{A}(S, e, f(x), \mathsf{h}_S(x')) \in f^{-1}(f(x))] \\
&\leq \mathbb{E}_{\substack{e,x \leftarrow E_e}} O(\lambda) \max \left\{ \Pr_S [\mathcal{A}(S, e, f(x), U) \in f^{-1}(f(x))], 2^{-\lambda} \right\} \qquad (3) \\
&\leq \mathbb{E}_{\substack{e,x \leftarrow E_e}} O(\lambda) \left(\Pr_S [\mathcal{A}(S, e, f(x), U) \in f^{-1}(f(x))] + 2^{-\lambda} \right) \\
&\leq O(\lambda) 2^{-\lambda} + O(\lambda) \Pr_{\substack{S,e \\ x \leftarrow E_e}} [\mathcal{A}(S, e, f(x), U) \in f^{-1}(f(x))] \\
&\leq O(\lambda) 2^{-\lambda} + O(\lambda) \frac{\Pr_{S,e,x} [\mathcal{A}(S, e, f(x), U) \in f^{-1}(f(x))]}{\Pr_{e,x}[x \in E_e]} \leq \mu(\lambda). \qquad (4)
\end{aligned}
$$

Equation (2) follows since we can think of sampling the pair $x, f(x)$ as equivalent to sampling $x', f(x)$ where $x \leftarrow E_e, x' \leftarrow f^{-1}(f(x))$. Equation (2) follows by applying Theorem 3.2 with the variable $x' \leftarrow f^{-1}(f(x))$ having entropy

$k = \mathbf{H}_f(x) \geq e - 1$ and with hash output-length $m = e + 1$ and independence $q = \lambda$. To apply the theorem, we think of a distinguisher $D_{\mathcal{A},e,f(x)}$ that given (z, S) tests whether $\mathcal{A}(S, e, f(x), z)$ inverts $f(x)$. In the equation, the random variable U is uniformly random $e + 1$ bit string. In Eq. (4) we can bound the numerator $\Pr_{S,e,x} \left[\mathcal{A}(S, e, f(x), U) \in f^{-1}(f(x)) \right]$ by some negligible by the one-wayness of f and we can bound the denominator $\Pr_{e,x}[x \in E_e] \geq 1/\lambda$ by the same argument we used previously. Therefore $\mu(\lambda)$ is negligible as we wanted to show.

Remark 3.1 (super-polynomial security). In the above construction, starting from a one-way function f that is t-secure directly yields t-security of $\mathcal{SIOWF}$.

3.2 Injective OWFs from iO and SIOWFs

We now construct a family of injective one-way functions based on iO and one-way functions. We first construct a weak but (fully) injective one-way function, and then use standard direct product amplification.

Ingredients. Let $i\mathcal{O}$ be an indistinguishability obfuscator for P/poly, and let $\mathcal{PRF}$ be a family of puncturable pseudo-random functions, where for $S \leftarrow \mathcal{K}_{\mathcal{PRF}}(1^\lambda)$, PRF_S maps $\{0,1\}^\lambda \to \{0,1\}^\lambda$. Let $(\mathrm{COM}_1, \mathrm{COM}_2)$ be a two message statistically-binding commitment scheme, where $\mathrm{COM}_1(1^\lambda)$ samples a first message M_1, and $\mathrm{COM}_2(x, M_1; r)$ computes a commitment M_2 to plaintext $x \in \{0,1\}^\lambda$, with respect to the first message M_1 and random coins $r \in \{0,1\}^\lambda$.

The Function Family. For $M_1 \leftarrow \mathrm{COM}_1(1^\lambda)$, $S \leftarrow \mathcal{K}_{\mathcal{PRF}}(1^\lambda)$, consider the circuit $C_{M_1,S} : \{0,1\}^\lambda \to \{0,1\}^*$ defined by

$$C_{M_1,S}(x) := \mathrm{COM}_2(x, M_1; \mathrm{PRF}_S(x)),$$

padded to some polynomial size $\ell(\lambda)$ to be determined later in the analysis.

The constructed family of one-way functions $\mathcal{OWF}$ consists of all obfuscations of such circuits:

1. A random key $\mathrm{OWF}_K \leftarrow \mathcal{K}_{\mathcal{OWF}}(1^\lambda)$ consists of an obfuscation $\widetilde{C} \leftarrow i\mathcal{O}(C_{M_1,S})$, for a first commitment message $M_1 \leftarrow \mathrm{COM}_1(1^\lambda)$ and PRF seed $S \leftarrow \mathcal{K}_{\mathcal{PRF}}(1^\lambda)$.
2. The function is given by $\mathrm{OWF}_K(x) = \widetilde{C}(x)$.

The fact that the construction gives an injective family follows directly from the statistical binding of the commitment. We next show that it is also weakly one-way.

Proposition 3.2. *Assume there exists a family $\mathcal{SIOWF}$ of sometimes-injective one-way functions. Then the above construction is a weak one-way function.*

Proof. (Proof sketch.). Let $\mathbf{I}_K$ and $p(\cdot)$ be as in Definition 3.1 such that $\mathcal{SIOWF}$ has an injective sub-domain $\mathbf{I}_K$ of density $1/p(\lambda)$. We show that any poly-size adversary $\mathcal{A}$ fails to invert the constructed $\mathcal{OWF}$ with probability at least $\frac{1}{p(\lambda)} - \mu(\lambda)$ for some negligible $\mu(\cdot)$. For this purpose we consider a sequence of hybrids.

Hyb_1: The real experiment. Here $\mathcal{A}$ is given as input $\widetilde{C}, \widetilde{C}(x)$ for a random input $x \leftarrow \{0,1\}^\lambda$ and random key $\widetilde{C} \leftarrow \mathcal{K}_{\mathcal{OWF}}(1^\lambda)$ and tries to obtain x.

Hyb_2: Here $\widetilde{C}$ is an obfuscation of an augmented circuit. In the new circuit, the PRF seed S is replaced with $S\{x\}$, which is punctured at x. In addition, $M_2 = \mathsf{COM}_2(x, M_1; \mathsf{PRF}_S(x))$ is hardwired as the output on input x (the input x itself is also hardwired). This circuit computes the same function as the previous $C_{M_1, S}$, thus by the iO guarantee, $\mathcal{A}$ inverts the function with the same probability up to a negligible difference.

Hyb_3: Here $M_2 = \mathsf{COM}_2(x, M_1; r)$ is generated with truly uniform randomness r, rather than $\mathsf{PRF}_S(x)$. (This includes both the hardwired M_2 as well as the output of the function $\widetilde{C}(x) = M_2$ given to $\mathcal{A}$.) By pseudorandomness at punctured points, the inversion probability is again maintained up to a negligible difference.

Hyb_4: Here $M_2 = \mathsf{COM}_2(0^\lambda, M_1; r)$ is a commitment to 0^λ, rather than to x. By the computational hiding of the commitment, the inversion probability is again maintained up to a negligible difference.

Hyb_5: Here we unpuncture S. The point x itself is still hardwired into the circuit in the clear. This does not change functionality, and thus the inversion probability is maintained, up to a negligible difference, by iO.

Hyb_6: In this hybrid, we also sample a random key $K \leftarrow \mathcal{K}_{\mathcal{SIOWF}}(1^\lambda)$ for a sometimes-injective OWF, and instead of sampling $x \leftarrow \{0,1\}^\lambda$ uniformly at random, we sample it from the injective sub-domain $x \leftarrow \mathbf{I}_K$. Since the density of $\mathbf{I}_K$ is at least $1/p(\lambda)$,

$$\Pr\left[\mathcal{A} \text{ fails to obtain } x \text{ in } \mathsf{Hyb}_5\right] \geq \frac{1}{p} \cdot \Pr\left[\mathcal{A} \text{ fails to obtain } x \text{ in } \mathsf{Hyb}_6\right]$$

Hyb_7: In this hybrid, instead of storing x in the clear and comparing it to the input (in order to decide whether to return M_2), we store its image $\mathsf{SIOWF}_K(x)$. Comparison of x with an input x' is now done by first computing $\mathsf{SIOWF}(x')$ and then comparing the images. Since $x \in \mathbf{I}_K \subseteq \mathbf{Inj}(\mathsf{SIOWF}_K)$ this does not change functionality and the inversion probability is preserved by iO.

Finally, we note that in Hyb_7 the view of $\mathcal{A}$ can be efficiently simulated from $K, \mathsf{SIOWF}_K(x)$. Thus, from one-wayness $\mathcal{SIOWF}$ over $\mathbf{I}_K$, it follows that $\mathcal{A}$ fails to obtain x in this hybrid with except with negligible probability. Therefore, overall, we deduce that $\mathcal{A}$ fails to obtain x in the original experiment with probability at least $\frac{1}{p(\lambda)} - \mu(\lambda)$, for some negligible $\mu(\lambda)$, as required.

The Padding Parameter. $\ell(\lambda)$ is chosen to be the maximum size among all circuits we went through in the analysis, so that iO can always be applied.

4   Trapdoor Permutations from iO

In this section we define Trapdoor Permutations (TDPs) and their enhancements, and construct them from sub-exponentially-secure iO. At large the definitions follow [GR13], with some exceptions discussed below.

4.1   Standard TDPs

We start by defining standard (non-enhanced) TDPs.

Definition 4.1 (TDP). *Let k be polynomially bounded length function. An efficiently computable family of functions*

$$\mathcal{TDP} = \left\{ \mathsf{TDP}_{PK} : D_{PK} \to D_{PK} : PK \in \{0,1\}^{k(\lambda)}, \lambda \in \mathbb{N} \right\},$$

associated with efficient (probabilistic) key and domain samplers $(\mathcal{K}, \mathcal{S})$, is a (standard) TDP if it satisfies

1. **Trapdoor invertibility:** *For any (PK, SK) in the support of $\mathcal{K}(1^\lambda)$, the function TDP_{PK} is a permutation of a corresponding domain D_{PK}. The inverse $\mathsf{TDP}_{PK}^{-1}(y)$ can be efficiently computed for any $y \in D_{PK}$, using the trapdoor SK.*

2. **Domain sampling:** *$\mathcal{S}(PK)$ samples a pseudo-uniform element in the domain D_{PK}; that is, for any polysize distinguisher $\mathcal{D}$, there exists a negligible $\mu(\cdot)$ such that for all $\lambda \in \mathbb{N}$,*

$$\left| \Pr\left[\mathcal{D}(r_{\mathcal{K}}, x) = 1 : \begin{array}{c} r_{\mathcal{K}} \leftarrow \{0,1\}^{\mathrm{poly}(\lambda)} \\ (PK, SK) \leftarrow \mathcal{K}(1^\lambda; r_{\mathcal{K}}) \\ x \leftarrow \mathcal{S}(PK) \end{array} \right] - \Pr\left[\mathcal{D}(r_{\mathcal{K}}, x) = 1 : \begin{array}{c} r_{\mathcal{K}} \leftarrow \{0,1\}^{\mathrm{poly}(\lambda)} \\ (PK, SK) \leftarrow \mathcal{K}(1^\lambda; r_{\mathcal{K}}) \\ x \leftarrow D_{PK} \end{array} \right] \right| \leq \mu(\lambda).$$

3. **One-wayness:** *For any polysize inverter $\mathcal{A}$ there exists a negligible function $\mu(\cdot)$, such that for all $\lambda \in \mathbb{N}$,*

$$\Pr\left[\mathcal{A}(PK, \mathsf{TDP}_{PK}(x)) = x : \begin{array}{c} (PK, SK) \leftarrow \mathcal{K}(1^\lambda) \\ x \leftarrow \mathcal{S}(PK) \end{array} \right] \leq \mu(\lambda).$$

The above definition is similar to the one in [GR13] with the exception that $\mathcal{S}(PK)$ in [GR13] is required to sample a domain element that is statistically close to a uniform domain element, whereas we only require computational indistinguishability. Importantly, we require that computational-indistinguishability holds even given the random coins used to generate (PK, SK). This property is required in applications (e.g., the EGL oblivious transfer protocol) and follows automatically (and thus not required explicitly) in the case of statistical-indistinguishability.

Also, we note that like in trapdoor permutations with statistical (rather than computational) domain sampling, the one-wayness requirement can be restated in any of the following equivalent forms:

3.a. The pre-image x is sampled uniformly from the domain:

$$\Pr\left[\mathcal{A}(PK, \mathsf{TDP}_{PK}(x)) = x : \begin{array}{c}(PK, SK) \leftarrow \mathcal{K}(1^\lambda) \\ x \leftarrow D_{PK}\end{array}\right] \leq \mu(\lambda).$$

3.b. The adversary inverts a random domain element x:

$$\Pr\left[\mathcal{A}(PK, x) = \mathsf{TDP}_{PK}^{-1}(x) : \begin{array}{c}(PK, SK) \leftarrow \mathcal{K}(1^\lambda) \\ x \leftarrow D_{PK}\end{array}\right] \leq \mu(\lambda).$$

3.c. The adversary inverts a domain element sampled by $\mathcal{S}(PK)$:

$$\Pr\left[\mathcal{A}(PK, x) = \mathsf{TDP}_{PK}^{-1}(x) : \begin{array}{c}(PK, SK) \leftarrow \mathcal{K}(1^\lambda) \\ x \leftarrow \mathcal{S}(PK)\end{array}\right] \leq \mu(\lambda).$$

The Construction. We now proceed to describe the construction of a TDP. The construction relies on super-polynomial hardness assumptions; for a convenient setting of parameters we assume that the underlying cryptographic primitives are sub-exponentially hard. In Sect. 4.4, we discuss relaxations to more mild (but still super-polynomial) hardness.

Ingredients. Fix any constant $\varepsilon < 1$, and let $T = T(\lambda) = 2^{\lambda^{\varepsilon/2}}$. We require the following primitives:

- $i\mathcal{O}$ is a $(\lambda, 2^{-\lambda^\varepsilon})$-secure indistinguishability obfuscator for P/poly.
- $\mathcal{PRF}$ is a $(\lambda, 2^{-\lambda^\varepsilon})$-secure family of puncturable pseudo-random functions, which for $\lambda \in \mathbb{N}$ maps $\mathbb{Z}_T$ to $\{0,1\}^\lambda$.
- $\mathcal{OWF}$ is a $(2^{\lambda^\varepsilon}, 2^{-\lambda^\varepsilon})$-secure family of injective one-way functions, which for $\lambda \in \mathbb{N}$ maps $\{0,1\}^\lambda$ to $\{0,1\}^*$. (Will only come up in the analysis, and not in the construction itself.)
- PRG is a (polynomially-secure) length-doubling pseudo-random generator.

The Function Family. The core of the construction will be obfuscations of circuits $(\mathsf{F}_S, \mathsf{X}_S)$ for computing the function forward and sampling domain elements, respectively. These obfuscations will be embedded in the function key PK and their corresponding secret S will be the trapdoor. The circuits are defined next. For $S \leftarrow \mathcal{K}_{\mathcal{PRF}}(1^\lambda)$:

1. $\mathsf{F}_S(i, \sigma)$: takes as input $i \in \mathbb{Z}_T$ and $\sigma \in \{0,1\}^\lambda$ and checks whether $\sigma = \mathsf{PRF}_S(i)$. If so it returns $i+1, \mathsf{PRF}_S(i+1)$, where $i+1$ is computed modulo T. Otherwise it returns $\bot$.
2. $\mathsf{X}_S(s)$: takes as input a seed $s \in \{0,1\}^{\log \sqrt{T}}$ and outputs $(i, \sigma) = (\mathsf{PRG}(s), \mathsf{PRF}_S(\mathsf{PRG}(s)))$, where i is interpreted as a residue in $\mathbb{Z}_T$.

Both circuits are padded so that their total size is $\ell(\lambda)$, for a fixed polynomial $\ell(\cdot)$ specified later.

The constructed family $\mathcal{TDP}$ is now defined as follows.

1. A random key PK consists of obfuscations $\widetilde{\mathsf{F}} \leftarrow i\mathcal{O}(\mathsf{F}_S)$ and $\widetilde{\mathsf{X}} \leftarrow i\mathcal{O}(\mathsf{X}_S)$, for $S \leftarrow \mathcal{K}_{\mathcal{PRF}}(1^\lambda)$. The corresponding trapdoor SK is S.
2. The domain D_{PK} is $\{(i, \sigma) : i \in \mathbb{Z}_T, \sigma = \mathsf{PRF}_S(i)\}$.
3. To compute $\mathsf{TDP}_{PK}(i, \sigma)$, return $\widetilde{\mathsf{F}}(i, \sigma)$.
4. To compute $\mathsf{TDP}_{PK}^{-1}(i, \sigma)$ given SK, return $(i - 1, \mathsf{PRF}_S(i - 1))$, where $i - 1$ is computed modulo T.
5. The domain sampler $\mathcal{S}(PK; s)$ takes as input PK and randomness $s \in \{0, 1\}^{\log \sqrt{T}}$ and outputs $\widetilde{\mathsf{X}}(s)$.

Proposition 4.1. *The above construction of $\mathcal{TDP}$ is a trapdoor permutation.*

Proof. The fact that TDP is trapdoor-invertible follows readily from the construction. The fact that the domain sampler $\mathcal{S}(PK)$ samples domain elements that are computationally-indistinguishable from uniform domain elements, even given the coins of $\mathcal{K}$ used to generate (PK, SK), follows directly from the pseudo-randomness guarantee of PRG.

From hereon, we focus on showing one-wayness. It would be simplest to work with the formulation (3.b) of the one-wayness requirement. Concretely, fix any polysize $\mathcal{A}$, we show that there exists a negligible $\mu(\cdot)$ such that for every $\lambda \in \mathbb{N}$,

$$\Pr\left[\mathsf{PRF}_S(i - 1) \leftarrow \mathcal{A}(\widetilde{\mathsf{F}}, \widetilde{\mathsf{X}}, i, \mathsf{PRF}_S(i)) : \begin{array}{l} S \leftarrow \mathcal{K}_{\mathcal{PRF}}(1^\lambda) \\ \widetilde{\mathsf{F}} \leftarrow i\mathcal{O}(\mathsf{F}_S) \\ \widetilde{\mathsf{X}} \leftarrow i\mathcal{O}(\mathsf{X}_S) \\ i \leftarrow \mathbb{Z}_T \end{array}\right] \leq \mu(\lambda).$$

We show that except with sub-exponentially-small probability $\mathcal{A}(\widetilde{\mathsf{F}}, \widetilde{\mathsf{X}}, i, \mathsf{PRF}_S(i))$ cannot output σ^* such that $\widetilde{\mathsf{F}}(i - 1, \sigma^*) \neq \bot$, which is equivalent to showing that $\sigma^* \neq \mathsf{PRF}_S(i - 1)$. We prove this via a sequence of indistinguishable hybrid experiments where the obfuscated $\widetilde{\mathsf{F}}$ is gradually augmented to return $\bot$ on an increasing interval, until it eventually returns $\bot$ on some interval $[i - u, i - 1]$ (for every possible signature), meaning in particular that $\mathcal{A}(\widetilde{\mathsf{F}}, \widetilde{\mathsf{X}}, i, \mathsf{PRF}_S(i))$ cannot find an accepting signature σ^* for $i - 1$. Throughout the hybrids we change the obfuscated circuits and assume that they are always padded so that their total size is $\ell(\lambda)$, for a fixed polynomial $\ell(\cdot)$ specified later.

Hyb_1: The original experiment.

Hyb_2: Here $\widetilde{\mathsf{F}}$ is an obfuscation of a circuit $\mathsf{F}_{i,v,S,K'}^{(2)}$. The circuit has hardwired a key $K' \leftarrow \mathcal{K}_{\mathcal{OWF}}(1^{\lambda'})$ for and injective OWF defined on inputs of length $\lambda' = \log \sqrt[4]{T}$, and a random image $v = \mathsf{OWF}_{K'}(u)$, for $u \leftarrow \{0, 1\}^{\lambda'} \cong \mathbb{Z}_{\sqrt[4]{T}}$. The circuit behaves like F, with the exception that given any input (k, σ) such that $k \in [i - \sqrt[4]{T}, i - 1]$ and $\mathsf{OWF}_{K'}(i - k) = v$, the circuit returns $\bot$.

$\mathsf{Hyb}_{3,j}, j \in [0, \sqrt[4]{T} - 1]$: Here $\widetilde{\mathsf{F}}$ is an obfuscation of a circuit $\mathsf{F}_{i,u,S}^{(3,j)}$. The circuit has a random index $u \leftarrow \mathbb{Z}_{\sqrt[4]{T}}$. On any input (k, σ), it returns $\bot$ if $k \in [i - u, i - u + j]$, where we truncate j so that $j = \min\{j, u - 1\}$. On any other input it behaves just like F_S.

$\mathsf{Hyb}_{4,j}, j \in [0, \sqrt[4]{T} - 1]$: Here $\widetilde{\mathsf{F}}$ is an obfuscation of a circuit $\mathsf{F}^{(4,j)}_{i,u,S\{i-u+j\},\sigma_{i-u+j}}$. The circuit is the same as $\mathsf{F}^{(3,j)}_{i,u,S}$, only that it has a punctured PRF key $S\{i - u + j\}$, and the value $\sigma_{i-u+j} = \mathsf{PRF}_S(i - u + j)$ is hardwired. In addition, $\widetilde{\mathsf{X}}$ is an obfuscation of a circuit $\mathsf{X}^{(4,j)}_{S\{i-u+j\}}$. The circuit is the same as X_S, only that it has the punctured $S\{i - u + j\}$, and whenever $\mathsf{PRF}_S(i - u + j)$ is required the circuit returns $\bot$ (no value is hardwired instead).

$\mathsf{Hyb}_{5,j}, j \in [0, \sqrt[4]{T} - 1]$: Here $\widetilde{\mathsf{F}}$ is an obfuscation of a circuit $\mathsf{F}^{(5,j)}_{i,u,S\{i-u+j\},\sigma_{i-u+j}}$. The circuit is the same as $\mathsf{F}^{(4,j)}_{i,u,S\{i-u+j\},\sigma_{i-u+j}}$, only that the hardwired σ_{i-u+j} is not set to $\mathsf{PRF}_S(i - u + j)$, but sampled uniformly at random from $\{0,1\}^\lambda$.

$\mathsf{Hyb}_{6,j}, j \in [0, \sqrt[4]{T} - 1]$: Here $\widetilde{\mathsf{F}}$ is an obfuscation of a circuit $\mathsf{F}^{(6,j)}_{i,u,S,v,K}$. The circuit is the same as $\mathsf{F}^{(5,j)}_{i,u,S\{i-u+j\},\sigma_{i-u+j}}$, only that instead of storing σ_{i-u+j} in the clear $v = \mathsf{OWF}_K(\sigma_{i-u+j})$ is stored, and comparison of σ and σ_{i-u+j} is done by comparing $\mathsf{OWF}_K(\sigma)$ and $\mathsf{OWF}_K(\sigma_{i-u+j})$. Here $K \leftarrow \mathcal{K}_{\mathcal{OWF}}(1^\lambda)$ is a key for an injective OWF from the family $\mathcal{OWF}$. Also, the PRF seed S is no longer punctured. In addition, $\widetilde{\mathsf{X}}$ is again an obfuscation of X_S (where S is no longer punctured).

We prove the following:

Claim 4.1. *For any polysize distinguisher* $\mathcal{D}$, *all* $\lambda \in \mathbb{N}$, *and all* $j \in [0, \sqrt[4]{T(\lambda)} - 1]$:

1. $\left|\Pr[\mathcal{D}(\mathsf{Hyb}_1) = 1] - \Pr[\mathcal{D}(\mathsf{Hyb}_2) = 1]\right| \le 2^{-\Omega(\lambda^{\varepsilon^2})}$,
2. $\left|\Pr[\mathcal{D}(\mathsf{Hyb}_2) = 1] - \Pr[\mathcal{D}(\mathsf{Hyb}_{3,0}) = 1]\right| \le 2^{-\Omega(\lambda^\varepsilon)}$,
3. $\left|\Pr[\mathcal{D}(\mathsf{Hyb}_{3,j}) = 1] - \Pr[\mathcal{D}(\mathsf{Hyb}_{4,j}) = 1]\right| \le T^{-1/2} + 2^{-\Omega(\lambda^\varepsilon)}$,
4. $\left|\Pr[\mathcal{D}(\mathsf{Hyb}_{4,j}) = 1] - \Pr[\mathcal{D}(\mathsf{Hyb}_{5,j}) = 1]\right| \le 2^{-\Omega(\lambda^\varepsilon)}$,
5. $\left|\Pr[\mathcal{D}(\mathsf{Hyb}_{5,j}) = 1] - \Pr[\mathcal{D}(\mathsf{Hyb}_{6,j}) = 1]\right| \le T^{-1/2} + 2^{-\Omega(\lambda^\varepsilon)}$,
6. $\left|\Pr[\mathcal{D}(\mathsf{Hyb}_{6,j}) = 1] - \Pr[\mathcal{D}(\mathsf{Hyb}_{3,j+1}) = 1]\right| \le 2^{-\Omega(\lambda^\varepsilon)}$,

where the view of $\mathcal{D}$ *in each hybrid consists of the corresponding obfuscated* $\widetilde{\mathsf{F}}, \widetilde{\mathsf{X}}$ *and* $(i, \mathsf{PRF}_S(i))$.

Proving the above claim will conclude the proof of Proposition 4.1 since it implies that

$$\Pr\left[\begin{array}{l} \sigma \leftarrow \mathcal{A}(\widetilde{\mathsf{F}}, \widetilde{\mathsf{X}}, i, \mathsf{PRF}_S(i)) \\ \mathsf{F}(i-1, \sigma) \ne \bot \end{array} : \begin{array}{l} S \leftarrow \mathcal{K}_{\mathcal{PRF}}(1^\lambda) \\ \widetilde{\mathsf{F}} \leftarrow i\mathcal{O}(\mathsf{F}_S) \\ \widetilde{\mathsf{X}} \leftarrow i\mathcal{O}(\mathsf{X}_S) \\ i \leftarrow \mathbb{Z}_T \end{array}\right] \le$$

$$\Pr\left[\begin{array}{l} \sigma \leftarrow \mathcal{A}(\widetilde{\mathsf{F}}, \widetilde{\mathsf{X}}, i, \mathsf{PRF}_S(i)) \\ \mathsf{F}(i-1, \sigma) \neq \bot \end{array} : \begin{array}{l} S \leftarrow \mathcal{K}_{\mathcal{PRF}}(1^\lambda) \\ \widetilde{\mathsf{F}} \leftarrow \boxed{i\mathcal{O}(\mathsf{F}_{i,S,u}^{(3, \sqrt[4]{T})})} \\ \widetilde{\mathsf{X}} \leftarrow i\mathcal{O}(\mathsf{X}_S) \\ i \leftarrow \mathbb{Z}_T \end{array}\right]$$

$$+ \lambda^{-\omega(1)} + 2^{-\Omega(\lambda^{\varepsilon^2})} + \sqrt[4]{T} \cdot (T^{-1/2} + 2^{-\Omega(\lambda^{\varepsilon})}) =$$

$$0 + \lambda^{-\omega(1)} + 2^{-\Omega(\lambda^{\varepsilon^2})} + 2^{\lambda^{\frac{\varepsilon}{2}}/4} \cdot (2^{-\lambda^{\frac{\varepsilon}{2}}/2} + 2^{-\Omega(\lambda^{\varepsilon})}) =$$

$$\lambda^{-\omega(1)},$$

where the first to last equality follows from the fact that $\mathsf{F}_{S,u}^{(3, \sqrt[4]{T})}(i-1, \sigma) = \bot$ for any σ.

Proof (Proof of Claim 4.1.). We prove each of the items in the claim. The proof is at most part similar to the one in [BPR15], with several exceptions.

Proof of 1 and 6. Recall that here we need to show that

1. $|\Pr[\mathcal{D}(\mathsf{Hyb}_1) = 1] - \Pr[\mathcal{D}(\mathsf{Hyb}_2) = 1]| \leq 2^{-\Omega(\lambda^{\varepsilon^2})}$,
6. $|\Pr[\mathcal{D}(\mathsf{Hyb}_{6,j}) = 1] - \Pr[\mathcal{D}(\mathsf{Hyb}_{3,j+1}) = 1]| \leq 2^{-\Omega(\lambda^{\varepsilon})}$.

In both cases, one obfuscated program differs from the other on exactly a single point, which is the unique (random) preimage of the corresponding image v (in the first case $v = \mathsf{OWF}_{K'}(u)$, and in the second $v = \mathsf{OWF}_K(\sigma_{i-u+j})$).

To prove the claim, we rely on a lemma proven in [BCP14] that roughly shows that, for circuits that only differ on a single input, iO implies what is known as *differing input obfuscation* [BGI+01], where it is possible to efficiently extract from any iO distinguisher an input on which the underlying circuits differ.

Lemma 4.1 (special case of [BCP14]). *Let $i\mathcal{O}$ be a (t, δ)-secure indistinguishability obfuscator for P/poly. There exists a PPT oracle-aided extractor $\mathcal{E}$, such that for any $t^{O(1)}$-size distinguisher $\mathcal{D}$, and two equal size circuits C_0, C_1 differing on exactly one input x^*, the following holds. Let C_0', C_1' be padded versions of C_0, C_1 of size $s \geq 3 \cdot |C_0|$.*

If $\quad |\Pr[\mathcal{D}(i\mathcal{O}(C_0')) = 1] - \Pr[\mathcal{D}(i\mathcal{O}(C_1')) = 1]| = \eta \geq \delta(s)^{o(1)}$,

then $\quad \Pr\left[x^* \leftarrow \mathcal{E}^{\mathcal{D}(\cdot)}(1^{1/\eta}, C_0, C_1)\right] \geq 1 - 2^{-\Omega(s)}$.

Using the lemma, we show that if either item 2 or 7 do not hold, we can invoke the distinguisher $\mathcal{D}$ to invert the underlying one-way function. The argument is similar in both cases up to different parameters; for concreteness, we focus on the first.

Assume that for infinitely many $\lambda \in \mathbb{N}$, $\mathcal{D}$ distinguishes Hyb_0 from Hyb_1 with gap $\eta(\lambda) = 2^{-o(\lambda^{\varepsilon^2})}$. Then, by averaging, with probability $\eta(\lambda)/2$ over the choice of (u, K'), $\mathcal{D}$ distinguishes the two distributions conditioned on these choices with gap $\eta(\lambda)/2$. Thus, we can invoke the extractor $\mathcal{E}$ given by Lemma 4.1 to invert the one-way function family $\mathcal{OWF}$ with probability $\frac{\eta(\lambda)}{2} \cdot (1 - 2^{-\Omega(\lambda)}) \geq 2^{-o(\lambda^{\varepsilon^2})}$ in time $t_{\mathcal{E}}(\lambda) \cdot t_{\mathcal{D}}(\lambda) \leq \eta(\lambda)^{-O(1)} \cdot \lambda^{O(1)} = 2^{O(\lambda^{\varepsilon^2})}$. Note that, indeed, given the image and the one-way function key, the inverter can construct the corresponding circuits efficiently. Recall that OWF'_K is defined on inputs of size $\lambda' = \log \sqrt[4]{T} = \lambda^{\varepsilon/2}/4$, and is $(2^{-\lambda'^{\varepsilon}}, 2^{\lambda'^{\varepsilon}})$-secure. Thus we get a contradiction to its one-wayness.

Proof of 2. Recall that here we need to show that

2. $\left| \Pr[\mathcal{D}(\mathsf{Hyb}_2) = 1] - \Pr[\mathcal{D}(\mathsf{Hyb}_{3,0}) = 1] \right| \leq 2^{-\Omega(\lambda^{\varepsilon})}.$

Here the obfuscated $\widetilde{\mathsf{F}}$ compute the exact same function in both hybrids. Specifically, for any input (k, σ), a comparison in the clear of $i - k$ and u is replaced by comparison of their corresponding values $\mathsf{OWF}_{K'}(i-k)$ and $\mathsf{OWF}_{K'}(u)$ under an injective one-way function. Thus, the required indistinguishability follows from the $(\lambda, 2^{-\lambda^{\varepsilon}})$-security of $i\mathcal{O}$.

Proof of 3. Recall that here we need to show that

3. $\left| \Pr[\mathcal{D}(\mathsf{Hyb}_{3,j}) = 1] - \Pr[\mathcal{D}(\mathsf{Hyb}_{4,j}) = 1] \right| \leq T^{-1/2} + 2^{-\Omega(\lambda^{\varepsilon})}.$

Here also, the obfuscated $\widetilde{\mathsf{F}}$ computes the exact same function in both hybrids. Specifically, rather than computing $\sigma_{i-u+j} = \mathsf{PRF}_S(i - u + j)$ using the PRF key S, the value σ_{i-u+j} is hardwired and directly compared to σ. For any other index, the punctured key $S\{i - u + j\}$ is used.

We now claim that the obfuscated $\widetilde{\mathsf{X}}$ also computes the same function in both hybrids with overwhelming probability $1 - T^{-1/2}$. Indeed, since X_S only computes PRF_S on values in the image of PRG, the probability that X_S and $\mathsf{X}^{(4,j)}_{S\{i-u+j\}}$ do not compute the same function can be bounded by the probability that $i - u + j$ is not in the image of PRG. Recall that i is sampled uniformly from $\mathbb{Z}_T$; thus, $i - u + j$ is also uniformly random in $\mathbb{Z}_T$, and we can bound the probability that it is in the image of $\mathsf{PRG} : \mathbb{Z}_{\sqrt{T}} \to \mathbb{Z}_T$ by $\sqrt{T} \cdot T^{-1} = T^{-1/2}$.

The required indistinguishability now follows from iO security.

Proof of 4. Recall that here we need to show that

4. $\left| \Pr[\mathcal{D}(\mathsf{Hyb}_{4,j}) = 1] - \Pr[\mathcal{D}(\mathsf{Hyb}_{5,j}) = 1] \right| \leq 2^{-\Omega(\lambda^{\varepsilon})}.$

The only difference between the two obfuscated circuit distributions is that in the first the hardwired value σ_{i-u+j} in $\widetilde{\mathsf{F}}$ is $\mathsf{PRF}_S(i - u + j)$, whereas in the second it is sampled independently uniformly at random. Indistinguishability follows from the $(2^{\lambda^{\varepsilon}}, 2^{-\lambda^{\varepsilon}})$-pseudo-randomness at the punctured point guarantee. Note that, indeed, given punctured key $S\{i - u + j\}$ and σ_{i-u+j}, a distinguisher can construct the corresponding circuits $\widetilde{\mathsf{F}}, \widetilde{\mathsf{X}}$ efficiently.

Proof of 5. Recall that here we need to show that

5. $\left|\Pr[\mathcal{D}(\mathsf{Hyb}_{5,j}) = 1] - \Pr[\mathcal{D}(\mathsf{Hyb}_{6,j}) = 1]\right| \leq T^{-1/2} + 2^{-\Omega(\lambda^\varepsilon)}.$

Here also, the two obfuscated $\widetilde{\mathsf{F}}$ in both hybrids compute the exact same function. First, the comparison of σ and σ_{i-u+j} is replaced by comparison of their corresponding values under an injective one-way function. In addition, the punctured key $S\{i - u + j\}$ is replaced with a non-punctured key S. This does not affect functionality as the two keys compute the same function on all points except $i - u + j$, and the circuit in the two hybrids treats any input $i - u + j, \sigma$, independently of the PRF key.

Also, $\widetilde{\mathsf{X}}$ now obfuscates the unpunctured version X_S instead of $\mathsf{X}^{(4,j)}_{S\{i-u+j\}}$. As before this does not change functionality with overwhelming probability $1 - T^{-1/2}$.

Overall, the required indistinguishability follows from iO.

This concludes the proof of the Claim 4.1 and Proposition 4.1.

The Padding Parameter $\ell(\lambda)$. We choose $\ell(\lambda)$ so that each of the circuits $\widetilde{\mathsf{F}}_{\cdots}$ considered above can be implemented by a circuit of size at most $\ell(\lambda)/3$. (The extra $1/3$ slack is taken to satisfy Lemma 4.1 in the analysis below.)

4.2 Enhanced TDPs

We next define enhanced TDPs. These are basically (standard) TDPs where it is possible to obliviously sample hard-to-invert images; concretely, given $x \leftarrow \mathcal{S}(PK; r_\mathcal{S})$, it is hard to find $\mathsf{TDP}^{-1}_{PK}(x)$, even given the coins $r_\mathcal{S}$ of $\mathcal{S}$.

Definition 4.2 (Enhanced TDP). *A TDP family $\mathcal{TDP}$ is said to be enhanced if for any polysize inverter $\mathcal{A}$ there exists a negligible function $\mu(\cdot)$, such that for all $\lambda \in \mathbb{N}$,*

$$\Pr\left[\mathcal{A}(PK, r_\mathcal{S}) = \mathsf{TDP}^{-1}_{PK}(x) : \begin{array}{c} (PK, SK) \leftarrow \mathcal{K}(1^\lambda) \\ r_\mathcal{S} \leftarrow \{0,1\}^{\mathrm{poly}(\lambda)} \\ x \leftarrow \mathcal{S}(PK; r_\mathcal{S}) \end{array}\right] \leq \mu(\lambda).$$

Enhancing the Previous Construction. We now describe how to enhance the construction presented in the previous section.

Is the previous TDP already enhanced? We start by noting that the family $\mathcal{TDP}$ constructed in the previous section may not be enhanced. Specifically, recall that the randomness $r_\mathcal{S}$ used by $\mathcal{S}$ in the construction is a seed s for PRG which is extended to an index i, and the corresponding domain element is $(i, \mathsf{PRF}_S(i))$. Note that it may very well be that given the seed s such that $i = \mathsf{PRG}(s)$, it is not hard to find another seed s' such that $\mathsf{PRG}(s') = i - 1$. In this case, the inverter may invoke the sampler $\mathcal{S}$ with this randomness and invert $(i, \mathsf{PRF}_S(i))$.

Looking more closely into the analysis, in the previous section, we could replace i with a truly random index, which with high-probability had no images of PRG in its close surrounding, due to the sparseness of PRG's image. This no longer works, as given the seed s used to generate i, we can no longer replace it with a truly random index.

Discrete-Image PRGs. To circumvent the above, we rely on a pseudo-random generator *with discrete image*, meaning that with overwhelming probability over the choice of the seed s, the corresponding image PRG(s) has no other image PRG(s') in its close surrounding. We show how to construct such pseudo-random generators from plain pseudo-random generators. More accurately, we construct a family of pseudo-random generators indexed by some public seed h, where the discrete image requirement holds with overwhelming probability for a random seed h.

Definition 4.3 (Discrete-image PRG). *Let k and ℓ be polynomially bounded length functions. An efficiently computable family of functions*

$$\mathcal{PRG} = \left\{ \mathsf{PRG}_h : \{0,1\}^\lambda \to \{0,1\}^{\ell(\lambda)} : h \in \{0,1\}^{k(\lambda)}, \lambda \in \mathbb{N} \right\},$$

associated with an efficient (probabilistic) key sampler $\mathcal{K}_{\mathcal{PRG}}$, is a discrete-image PRG if it satisfies:

1. **Pseudo-randomness:** *For any polysize distinguisher $\mathcal{D}$ there is a negligible μ such that for any $\lambda \in \mathbb{N}$:*

$$\left| \Pr\left[\mathcal{D}(h, \mathsf{PRG}_h(s)) = 1 : \begin{array}{c} h \leftarrow \mathcal{K}_{\mathcal{PRG}}(1^\lambda) \\ s \leftarrow \{0,1\}^\lambda \end{array} \right] - \Pr\left[\mathcal{D}(h, u) = 1 : \begin{array}{c} h \leftarrow \mathcal{K}_{\mathcal{PRG}}(1^\lambda) \\ u \leftarrow \{0,1\}^{\ell(\lambda)} \end{array} \right] \right| \leq \mu(\lambda).$$

2. **Discrete image:** *for any $\lambda \in \mathbb{N}$ and any $t \in \mathbb{Z}_{2^{\ell(\lambda)}} \setminus \{0\}$:*

$$\Pr\left[\exists s' \neq s : \mathsf{PRG}_h(s) - \mathsf{PRG}_h(s') = t \bmod 2^{\ell(\lambda)} : \begin{array}{c} h \leftarrow \mathcal{K}_{\mathcal{PRG}}(1^\lambda) \\ s \leftarrow \{0,1\}^\lambda \end{array} \right] \leq 2^{-\ell(\lambda)+\lambda}.$$

A construction of discrete-image PRGs. Let $\mathsf{PRG} : \{0,1\}^\lambda \to \{0,1\}^{\ell(\lambda)}$ be a (plain) pseudo-random generator, and let

$$\mathcal{H}_\lambda = \left\{ h : \{0,1\}^{\ell(\lambda)} \to \{0,1\}^{\ell(\lambda)} : \lambda \in \mathbb{N} \right\},$$

be a family of pair-wise independent permutations. We construct a discrete-image family

$$\mathcal{PRG} = \left\{ \mathsf{PRG}_h : \{0,1\}^\lambda \to \{0,1\}^{\ell(\lambda)} \right\},$$

as follows.

- The public seed h is a random hash in the family $\mathcal{H}_\lambda$.
- The generator is given by

$$\mathsf{PRG}_h(s) := h(\mathsf{PRG}(s)).$$

Claim 4.2. $\mathcal{PRG}$ *is a discrete-image pseudo-random generator.*

Proof. The pseudo-randomness property follows directly from the fact that PRG is a pseudo-random generator and h is an efficiently computable permutation.

To prove discrete-image, it suffices to show that for any fixed $s \in \{0,1\}^\lambda$ and any $t \in \mathbb{Z}_{2^{\ell(\lambda)}} \setminus \{0\}$,

$$\Pr\left[\exists s' \neq s : \mathsf{PRG}_h(s) - \mathsf{PRG}_h(s') = t \bmod 2^{\ell(\lambda)} : h \leftarrow \mathcal{K}_{\mathcal{PRG}}(1^\lambda)\right] \leq 2^{-\ell(\lambda)+\lambda}.$$

Indeed, by pairwise-independence, conditioning on the value of $\mathsf{PRG}_h(s) = h(\mathsf{PRG}(s))$, for every $s' \in \{0,1\}^\lambda$ such that $\mathsf{PRG}(s') \neq \mathsf{PRG}(s)$, the value $h(\mathsf{PRG}(s'))$ is uniformly random in $\mathbb{Z}_{2^{\ell(\lambda)}}$ and thus $h(\mathsf{PRG}(s')) = h(\mathsf{PRG}(s)) + t \bmod 2^{\ell(\lambda)}$ with probability at most $2^{-\ell(\lambda)}$. Taking union-bound over all $s' \in \{0,1\}^\lambda$, the claim follows.

The Augmented Construction. The construction of enhanced TDPs is now identical to the one in Sect. 4.1, except that we augment the obfuscated domain sampling circuit X_S to a circuit $\mathsf{X}_{S,h}$ that also has hardwired a random public seed h for a discrete-image PRG. The new sampling circuit is now defined as the previous ones, except that instead of using a plain $\mathsf{PRG} : \mathbb{Z}_{\sqrt{T}} \to \mathbb{Z}_T$ we use the discrete image $\mathsf{PRG}_h : \mathbb{Z}_{\sqrt{T}} \to \mathbb{Z}_T$.

Proposition 4.2. *The augmented construction is an enhanced trapdoor permutation.*

Proof (Proof sketch). The proof is identical to that of Proposition 4.1 with two exceptions to the proof of one-wayness. Whereas in Proposition 4.1, we consider, in Hyb_1 an adversary that tries to invert $(i, \mathsf{PRF}_S(i))$ for a truly uniform $i \leftarrow \mathbb{Z}_T$. Now, $i \leftarrow \mathsf{PRG}_h(s) \in \mathbb{Z}_T$ is a pseudo-random element, and the adversary also obtains the seed s, which are the coins of the sampler $\mathcal{S}(PK)$.

The second difference is when switching between $\mathsf{X}_{S,h}$ and $\mathsf{X}_{S\{i-u+j\},h}$ (in the proofs of items 3 and 5). In Proposition 4.1, we relied on the fact that i is uniformly random and thus $i - u + j \bmod T$ is not in the image of PRG with probability $T^{-1/2}$, implying that puncturing does not affect functionality and letting us invoke the iO guarantee. Now i is no longer random, but the same holds based on the discrete image property of PRG_h (when choosing $t = u - j \bmod T$).

4.3 Doubly Enhanced TDPs

We now define doubly-enhanced TDPs. These are enhanced TDPs where given the key PK, it is possible to sample coins r_S together with a preimage x of $y = \mathcal{S}(PK, r_S)$. In [GR13], it is required that r_S is distributed as uniformly random coins for $\mathcal{S}$. We relax this requiring that $r_{\mathcal{S}}$ is only pseudo-random even given the randomness used to sample (PK, SK). Indeed, this relaxation suffices for applications of doubly-enhanced TDPs such as non-interactive zero-knowledge.

Definition 4.4 (Doubly-enhanced TDP). *An enhanced TDP family $\mathcal{TDP}$ is said to be doubly-enhanced there exists a sampler $\mathcal{R}$ satisfying the following two requirements.*

1. **Correlated preimage sampling.** *For any PK in the support of $\mathcal{K}(1^\lambda)$:*

$$(x, r_{\mathcal{S}}) \leftarrow \mathcal{R}(PK) \ such \ that \ \mathsf{TDP}_{PK}(x) = \mathcal{S}(PK, r_{\mathcal{S}}).$$

2. **Pseudorandomness.** *For any polysize distinguisher $\mathcal{D}$ there is a negligible μ such that for any $\lambda \in \mathbb{N}$:*

$$\left| \Pr\left[\mathcal{D}(x, r_{\mathcal{S}}, r_{\mathcal{K}}) = 1 : \begin{matrix} PK \leftarrow \mathcal{K}(1^\lambda, r_{\mathcal{K}}) \\ (x, r_{\mathcal{S}}) \leftarrow \mathcal{R}(PK) \end{matrix} \right] - \Pr\left[\mathcal{D}(x, r_{\mathcal{S}}, r_{\mathcal{K}}) = 1 : \begin{matrix} PK \leftarrow \mathcal{K}(1^\lambda; r_{\mathcal{K}}) \\ r_{\mathcal{S}} \leftarrow \{0,1\}^{\mathrm{poly}(\lambda)} \\ y \leftarrow \mathcal{S}(PK, r_{\mathcal{S}}) \\ x \leftarrow \mathsf{TDP}_{PK}^{-1}(y) \end{matrix} \right] \right| \leq \mu(\lambda).$$

Doubly Enhancing the Previous Construction. To make the previous construction doubly enhanced we show how to slightly augment the discrete-image PRG used in the construction on some sparse subset of seeds (thus not hurting previous properties), while taking advantage of the particular structure of our TDP.

Concretely, we augment the code of PRG_h to compute a new PRG_h^* as follows. Let $\mathsf{PRG}' : \mathbb{Z}_{\sqrt[4]{T}} \rightarrow \mathbb{Z}_{\sqrt{T}}$ be a length doubling pseudorandom generator that expands small seeds $s' \in \mathbb{Z}_{\sqrt[4]{T}}$ to longer seeds $s \in \mathbb{Z}_{\sqrt{T}}$ for PRG_h. PRG_h^* acts as follows:

Given a (private) seed $s \in \mathbb{Z}_{\sqrt{T}}$ as input, parse it as $(s', r') \in \mathbb{Z}_{\sqrt[4]{T}} \times \mathbb{Z}_{\sqrt[4]{T}}$.

1. If $r' = 0$, compute $\mathsf{PRG}'(s')$, and output $\mathsf{PRG}_h^*(s', r') := \mathsf{PRG}_h(\mathsf{PRG}'(s')) - 1 \bmod T$.
2. Otherwise, output as before $\mathsf{PRG}_h^*(s', r') = \mathsf{PRG}_h(s', r')$.

The Augmented Construction. The construction of doubly-enhanced TDPs is now identical to the one of enhanced TDPs, except that we instantiate the pseudo-random generator with the new $\mathcal{PRG}^* = \{\mathsf{PRG}_h^*\}$.

Proposition 4.3. *The augmented construction is a doubly-enhanced trapdoor permutation.*

Proof (Proof sketch). First notice that we did not harm the pseudo-randomness and discrete-image properties of the original family $\mathcal{PRG}$. Indeed, the augmented PRG_h^* only behaves differently from PRG_h on the set $\{s = (s', r') : r' = 0\}$, which has negligible density $T^{-1/4}$. The pseudo-randomness and discrete-image properties, however, are defined for a uniformly random $(s', r') \in \mathbb{Z}_{\sqrt[4]{T}} \times \mathbb{Z}_{\sqrt[4]{T}}$, and thus remain unaffected.

We can now define the sampler $\mathcal{R}(PK)$:

1. Pick a random (short) seed $s' \leftarrow \mathbb{Z}_{\sqrt[4]{T}}$.
2. Compute $r_\mathcal{S} = \mathsf{PRG}'(s') \in \mathbb{Z}_{\sqrt[4]{T}} \times \mathbb{Z}_{\sqrt[4]{T}}$ and $r_\mathcal{S}^x = (s', 0) \in \mathbb{Z}_{\sqrt[4]{T}} \times \mathbb{Z}_{\sqrt[4]{T}}$.
3. Return $(x, r_\mathcal{S})$ where $x = \mathcal{S}(PK; r_\mathcal{S}^x)$.

The pseudorandomness of $r_\mathcal{S}$, conditioned on $(r_\mathcal{K}, x)$, follows directly from the pseudo-randomness guarantee of PRG'. We now note that x is the preimage of $y = \mathcal{S}(PK, r_\mathcal{S})$. We shall assume for simplicity that $\mathsf{PRG}'(s')$ never outputs $s = (s''; r'')$ such that $r'' = 0$ (PRG' can always be augmented to satisfy this property). Then, by construction $x = (i - 1, \mathsf{PRF}_S(i-1))$ where $i = \mathsf{PRG}_h(\mathsf{PRG}'(s'))$ and $y = (i, \mathsf{PRF}_S(i))$.

This completes the proof.

4.4   Relaxing Subexponential Security

In all constructions above, we assumed all cryptographic primitives are subexponentially hard. We now explain how this can be relaxed, and what are the tradeoffs between the hardness of the different primitives. Let $f(\cdot), g(\cdot), h(\cdot)$ be sub-linear functions and assume that $\mathcal{OWF}$ is $(2^{f(\lambda)}, 2^{-f(\lambda)})$-secure, $\mathcal{PRF}$ is $(\lambda, 2^{-g(\lambda)})$-secure, and $i\mathcal{O}$ is $(\lambda, 2^{-h(\lambda)})$-secure. We can restate Claim 4.1 as follows.

Claim 4.3 (Claim *4.1* generalized). *For any polysize distinguisher $\mathcal{D}$, all $\lambda \in \mathbb{N}$, and all $j \in [0, \sqrt[4]{T}]$:*

1. $\left|\Pr[\mathcal{D}(\mathsf{Hyb}_1) = 1] - \Pr[\mathcal{D}(\mathsf{Hyb}_2) = 1]\right| \leq 2^{-\Omega(f(\log \sqrt[4]{T}))} + 2^{-\Omega(h(\lambda))}$,
2. $\left|\Pr[\mathcal{D}(\mathsf{Hyb}_2) = 1] - \Pr[\mathcal{D}(\mathsf{Hyb}_{3,0}) = 1]\right| \leq 2^{-\Omega(h(\lambda))}$,
3. $\left|\Pr[\mathcal{D}(\mathsf{Hyb}_{3,j}) = 1] - \Pr[\mathcal{D}(\mathsf{Hyb}_{4,j}) = 1]\right| \leq T^{-1/2} + 2^{-\Omega(h(\lambda))}$,
4. $\left|\Pr[\mathcal{D}(\mathsf{Hyb}_{4,j}) = 1] - \Pr[\mathcal{D}(\mathsf{Hyb}_{5,j}) = 1]\right| \leq 2^{-\Omega(g(\lambda))}$,
5. $\left|\Pr[\mathcal{D}(\mathsf{Hyb}_{5,j}) = 1] - \Pr[\mathcal{D}(\mathsf{Hyb}_{6,j}) = 1]\right| \leq T^{-1/2} + 2^{-\Omega(h(\lambda))}$,
6. $\left|\Pr[\mathcal{D}(\mathsf{Hyb}_{6,j}) = 1] - \Pr[\mathcal{D}(\mathsf{Hyb}_{3,j+1}) = 1]\right| \leq 2^{-\Omega(f(\lambda))} + 2^{-\Omega(h(\lambda))}$.

The overall inversion probability can be bounded by

$$2^{-\Omega(f(\log \sqrt[4]{T}))} + \sqrt[4]{T} \cdot (T^{-1/2} + 2^{-\Omega(f(\lambda))} + 2^{-\Omega(g(\lambda))} + 2^{-\Omega(h(\lambda))}).$$

In particular, letting $m(\lambda) = \min\{f(\lambda), g(\lambda), h(\lambda)\}$, we can guarantee hardness of the resulting TDP as long as

1. $T(\lambda) = \lambda^{-\omega(1)}$.
2. $m(\lambda) = \omega(\log(T))$.
3. $f(\log \sqrt[4]{T}) = \omega(\log \lambda)$.

For instance, for any constant $\varepsilon < 1$, we can set

- $T = 2^{(\log \lambda)^{2/\varepsilon}}$,
- $f(\lambda) = \lambda^\varepsilon$ ($\mathcal{OWF}$ is still sub-exponential),
- $g(\lambda) = h(\lambda) = (\log \lambda)^{2+2/\varepsilon}$ ($\mathcal{PRF}$ and $i\mathcal{O}$ are quasi-polynomial).

Alternatively, we can set

- $T = 2^{2^{(\log \lambda)^\varepsilon}}$,
- $f(\lambda) = g(\lambda) = h(\lambda) = 2^{(\log \lambda)^{\frac{1+\varepsilon}{2}}}$ (all primitives are only $2^{\lambda^{o(1)}}$-secure).

Acknowledgements. We thank Mark Zhandry for bringing to our attention the question of injective OWFs from indistinguishability obfuscation.

A On Non-interactive Commitments from iO

As mentioned above, our result allows to remove additional assumptions in iO based construction of several fundamental primitives including non-interactive commitments. Here we elaborate on this implication.

In the construction of non-interactive commitments based on injective one-way functions [Blu81] it is crucial that a malicious sender cannot send a malformed function key defining a function that is not injective. We observe however, that it is sufficient to require that every key in the support of the key-generation algorithm defines an injective function. If the function family has this property, we can modify the decommitment algorithm and have sender present its random coins, proving that the key is indeed in the support of the key-generation algorithm. While our injective one-way function do not satisfy the above property, our trapdoor permutations do.

References

[AB15] Applebaum, B., Brakerski, Z.: Obfuscating circuits via composite-order graded encoding. In: Dodis, Y., Nielsen, J.B. (eds.) TCC 2015, Part II. LNCS, vol. 9015, pp. 528–556. Springer, Heidelberg (2015)

[AC08] Achlioptas, D., Coja-Oghlan, A.: Algorithmic barriers from phase transitions. In: 49th Annual IEEE Symposium on Foundations of Computer Science, FOCS 2008, October 25–28, 2008, Philadelphia, PA, USA, pp. 793–802 (2008)

[AJ15] Ananth, P., Jain, A.: Indistinguishability obfuscation from compact functional encryption. In: Crypto (2015)

[BCC+14] Bitansky, N., Canetti, R., Cohn, H., Goldwasser, S., Kalai, Y.T., Paneth, O., Rosen, A.: The impossibility of obfuscation with auxiliary input or a universal simulator. In: Garay, J.A., Gennaro, R. (eds.) CRYPTO 2014, Part II. LNCS, vol. 8617, pp. 71–89. Springer, Heidelberg (2014)

[BCP14] Boyle, E., Chung, K.-M., Pass, R.: On extractability obfuscation. In: Lindell, Y. (ed.) TCC 2014. LNCS, vol. 8349, pp. 52–73. Springer, Heidelberg (2014)

[BFKL93] Blum, A., Furst, M.L., Kearns, M., Lipton, R.J.: Cryptographic primitives bon hard learning problems. In: Stinson, D.R. (ed.) CRYPTO 1993. LNCS, vol. 773, pp. 278–291. Springer, Heidelberg (1993)

[BGI+01] Barak, B., Goldreich, O., Impagliazzo, R., Rudich, S., Sahai, A., Vadhan, S.P., Yang, K.: On the (im)possibility of obfuscating programs. In: Kilian, J. (ed.) CRYPTO 2001. LNCS, vol. 2139, pp. 1–18. Springer, Heidelberg (2001)

[BGI14] Boyle, E., Goldwasser, S., Ivan, I.: Functional signatures and pseudorandom functions. In: Krawczyk, H. (ed.) PKC 2014. LNCS, vol. 8383, pp. 501–519. Springer, Heidelberg (2014)

[BGK+13] Barak, B., Garg, S., Kalai, Y.T., Paneth, O., Sahai, A.: Protecting obfuscation against algebraic attacks. In: Cryptology ePrint Archive, Report 2013/631 (2013). http://eprint.iacr.org/

[Blu81] Blum, M.: Coin flipping by telephone. In: Proceedings of the 18th Annual International Cryptology Conference, pp. 11–15 (1981)

[BP14] Bitansky, N., Paneth, O.: Zaps and non-interactive witness indistinguishability from indistinguishability obfuscation. In: Cryptology ePrint Archive, Report 2014/295 (2014). http://eprint.iacr.org/

[BPR15] Bitansky, N., Paneth, O., Rosen, A.: On the cryptographic hardness of finding a nash equilibrium. In: FOCS (2015)

[BR14] Brakerski, Z., Rothblum, G.N.: Virtual black-box obfuscation for all circuits via generic graded encoding. In: Lindell, Y. (ed.) TCC 2014. LNCS, vol. 8349, pp. 1–25. Springer, Heidelberg (2014)

[BV15] Bitansky, N., Vaikuntanathan, V.: Indistinguishability obfuscation from functional encryption. In: FOCS (2015)

[BW13] Boneh, D., Waters, B.: Constrained pseudorandom functions and their applications. In: Sako, K., Sarkar, P. (eds.) ASIACRYPT 2013, Part II. LNCS, vol. 8270, pp. 280–300. Springer, Heidelberg (2013)

[CLT13] Coron, J.-S., Lepoint, T., Tibouchi, M.: Practical multilinear maps over the integers. In: Canetti, R., Garay, J.A. (eds.) CRYPTO 2013, Part I. LNCS, vol. 8042, pp. 476–493. Springer, Heidelberg (2013)

[CLT15] Coron, J.-S., Lepoint, T., Tibouchi, M.: New multilinear maps over the integers. In: Crypto (2015)

[CLTV14] Canetti, R., Lin, H., Tessaro, S., Vaikuntanathan, V.: Obfuscation of probabilistic circuits and applications. In: Cryptology ePrint Archive, Report 2014/882 (2014). http://eprint.iacr.org/

[DPW14] Dodis, Y., Pietrzak, K., Wichs, D.: Key derivation without entropy waste. In: Nguyen, P.Q., Oswald, E. (eds.) EUROCRYPT 2014. LNCS, vol. 8441, pp. 93–110. Springer, Heidelberg (2014)

[Gen14] Craig Gentry. Computing on the edge of chaos: Structure and randomness in encrypted computation. Electronic Colloquium on Computational Complexity (ECCC), 21:106, 2014

[GGH13a] Garg, S., Gentry, C., Halevi, S.: Candidate multilinear maps from ideal lattices. In: Johansson, T., Nguyen, P.Q. (eds.) EUROCRYPT 2013. LNCS, vol. 7881, pp. 1–17. Springer, Heidelberg (2013)

[GGH+13b] Garg, S., Gentry, C., Halevi, S., Raykova, M., Sahai, A., Waters, B.: Candidate indistinguishability obfuscation and functional encryption for all circuits. In: 54th Annual IEEE Symposium on Foundations of Computer Science, FOCS 2013, 26–29 October, 2013, Berkeley, CA, USA, pp. 40–49 (2013)

[GGM86] Goldreich, O., Goldwasser, S., Micali, S.: How to construct random functions. J. ACM **33**(4), 792–807 (1986)

[GK88] Goldreich, O., Kushilevitz, E.: A perfect zero-knowledge proof for a problem equivalent to discrete logarithm. In: Proceedings of Advances in Cryptology - CRYPTO 1988, 8th Annual International Cryptology Conference, Santa Barbara, California, USA, August 21–27, 1988, pp. 57–70 (1988)

[GLSW14] Gentry, C., Lewko, A.B., Sahai, A., Waters, B.: Indistinguishability obfuscation from the multilinear subgroup elimination assumption. In: IACR Cryptology ePrint Archive 2014, p. 309 (2014)

[Gol11] Goldreich, O.: Candidate one-way functions based on expander graphs. In: Goldreich, O., et al. (eds.) Studies in Complexity and Cryptography. LNCS, vol. 6650, pp. 76–87. Springer, Heidelberg (2011)

[GR13] Goldreich, O., Rothblum, R.D.: Enhancements of trapdoor permutations. J. Cryptol. **26**(3), 484–512 (2013)

[HILL99] Håstad, J., Impagliazzo, R., Levin, L.A., Luby, M.: A pseudorandom generator from any one-way function. SIAM J. Comput. **28**(4), 1364–1396 (1999)

[JP00] Juels, A., Peinado, M.: Hiding cliques for cryptographic security. Des. Codes Crypt. **20**(3), 269–280 (2000)

[KLW14] Koppula, V., Lewko, A.B., Waters, B.: Indistinguishability obfuscation for turing machines with unbounded memory. In: Cryptology ePrint Archive, Report 2014/925 (2014). http://eprint.iacr.org/

[KMN+14] Komargodski, I., Moran, T., Naor, M., Pass, R., Rosen, A., Yogev, E.: One-way functions and (im)perfect obfuscation. In: IACR Cryptology ePrint Archive 2014, p. 347 (2014)

[KPTZ13] Kiayias, A., Papadopoulos, S., Triandopoulos, N., Zacharias, T.: Delegatable pseudorandom functions and applications. In: ACM Conference on Computer and Communications Security, pp. 669–684 (2013)

[Nao91] Naor, M.: Bit commitment using pseudorandomness. J. Cryptol. **4**(2), 151–158 (1991)

[NR02] Naor, M., Reingold, O.: Constructing pseudo-random permutations with a prescribed structure. J. Cryptol. **15**(2), 97–102 (2002)

[PW08] Peikert, C., Waters, B.: Lossy trapdoor functions and their applications. In: Proceedings of the 40th Annual ACM Symposium on Theory of Computing, Victoria, British Columbia, Canada, May 17–20, 2008, pp. 187–196 (2008)

[Rab79] Rabin, M.O.: Digitalized signatures and public-key functions as intractable as factorization. Technical report LCR/TR-212, MIT Laboratory of Computer Science (1979)

[RSA83] Rivest, R.L., Shamir, A., Adleman, L.M.: A method for obtaining digital signatures and public-key cryptosystems (reprint). Commun. ACM **26**(1), 96–99 (1983)

[Sho97] Shor, P.W.: Polynomial-time algorithms for prime factorization and discrete logarithms on a quantum computer. SIAM J. Comput. **26**(5), 1484–1509 (1997)

[SW14] Sahai, A., Waters, B.: How to use indistinguishability obfuscation: deniable encryption, and more. In: Symposium on Theory of Computing, STOC 2014, May 31 – June 03, 2014, New York, NY, USA, pp. 475–484 (2014)

[Yao82] Yao, A.C.-C.: Theory and applications of trapdoor functions (extended abstract). In: 23rd Annual Symposium on Foundations of Computer Science, 3–5 November 1982, Chicago, Illinois, USA, pp. 80–91 (1982)

[Zim15] Zimmerman, J.: How to obfuscate programs directly. In: Oswald, E., Fischlin, M. (eds.) EUROCRYPT 2015. LNCS, vol. 9057, pp. 439–467. Springer, Heidelberg (2015)

Cryptographic Assumptions
(Invited Talk followed by Panel)

Cryptographic Assumptions: A Position Paper

Shafi Goldwasser[1,2]([✉]) and Yael Tauman Kalai[3]

[1] MIT, Cambridge, USA
shafi@theory.csail.mit.edu
[2] Weizmann Institute, Rehovot, Israel
[3] Microsoft Research, Cambridge, USA
yael@microsoft.com

Abstract. The mission of theoretical cryptography is to define and construct *provably secure* cryptographic protocols and schemes. Without proofs of security, cryptographic constructs offer no guarantees whatsoever and no basis for evaluation and comparison. As most security proofs necessarily come in the form of a reduction between the security claim and an intractability assumption, such proofs are ultimately only as good as the assumptions they are based on. Thus, the complexity implications of every assumption we utilize should be of significant substance, and serve as the yard stick for the value of our proposals.

Lately, the field of cryptography has seen a sharp increase in the number of new assumptions that are often complex to define and difficult to interpret. At times, these assumptions are hard to untangle from the constructions which utilize them.

We believe that the lack of standards of what is accepted as a reasonable cryptographic assumption can be harmful to the credibility of our field. Therefore, there is a great need for *measures* according to which we classify and compare assumptions, as to which are *safe* and which are not. In this paper, we propose such a classification and review recently suggested assumptions in this light. This follows the footsteps of Naor (Crypto 2003).

Our governing principle is relying on hardness assumptions that are independent of the cryptographic constructions.

1   Introduction

Conjectures and assumptions are instrumental for the advancement of science. This is true in physics, mathematics, computer science, and almost any other discipline. In mathematics, for example, the Riemann hypothesis (and its extensions) have far reaching applications to the distribution of prime numbers. In computer science, the assumption that $P \neq NP$ lies in the foundations of complexity theory. The more recent Unique Games Conjecture [40] has been instrumental to our ability to obtain tighter bounds on the hardness of approximation of several problems. Often, such assumptions contribute tremendously to our understanding of certain topics and are the force moving research forward.

Assumptions are paramount to cryptography. A typical result constructs schemes for which breaking the scheme is an NP computation. As we do not

© International Association for Cryptologic Research 2016
E. Kushilevitz and T. Malkin (Eds.): TCC 2016-A, Part I, LNCS 9562, pp. 505–522, 2016.
DOI: 10.1007/978-3-662-49096-9_21

know that P $\neq$ NP, an assumption to that effect (and often much more) must be made. Thus, essentially any cryptographic security proof is a reduction from the existence of an adversary that violates the security definition to dispelling an underlying conjecture about the intractability of some computation. Such reductions present a "win-win" situation which gives provable cryptography its beauty and its power: either we have designed a scheme which resists all polynomial time adversaries or an adversary exists which contradicts an existing mathematical conjecture. Put most eloquently, "Science wins either way[1]".

Naturally, this is the case *only if* we rely on mathematical conjectures whose statement is scientifically interesting independently of the cryptographic application itself. Most definitely, the quality of the assumption determines the value of the proof.

Traditionally, there were a few well-studied computational assumptions under which cryptographic schemes were proven secure. These assumptions can be partitioned into two groups: *generic* and *concrete*. Generic assumptions include the existence of one-way functions, the existence of one-way permutations, the existence of a trapdoor functions, and so on. We view generic assumptions as postulating the existence of a cryptographic primitive. Concrete assumptions include the universal one-way function assumption [31],[2] the assumption that Goldreich's expander-based function is one-way [32], the Factoring and RSA assumptions [47,49], the Discrete Log assumption over various groups [24], the Quadratic Residuosity assumption [37], the DDH assumption [24], the parity with Noise (LPN) assumption [2,10], the Learning with Error (LWE) assumption [48], and a few others.

A construction which depends on a generic assumption is generally viewed as superior to that of a construction from a concrete assumption, since the former can be viewed as an unconditional result showing how abstract cryptographic primitives are reducible to one another, setting aside the question of whether a concrete implementation of the generic assumption exists. And yet, a generic assumption which is not accompanied by at least one proposed instantiation by a concrete assumption is often regarded as useless. Thus, most of the discussion in this paper is restricted to concrete assumptions, with the exception of Sect. 2.5, which discusses generic assumptions.

Recently, the field of cryptography has been overrun by numerous assumptions of radically different nature than the ones preceding. These assumptions are often nearly impossible to untangle from the constructions which utilize them. The differences are striking. Severe restrictions are now assumed on the class of algorithms at the disposal of any adversary, from assuming that the adversary is only allowed a restricted class of operations (such as the Random Oracle Model restriction, or generic group restrictions), to assuming that any adversary who breaks the cryptosystem must do so in a particular way (this includes various

---

[1] Silvio Micali, private communication.

[2] A universal one-way function· is a candidate one-way function f such that if one-way functions exist then f itself is one-way [31]. The universal one-way function assumption asserts that this universal f is indeed one-way.

knowledge assumptions). The assumptions often make mention of the cryptographic application itself and thus are not of independent interest. Often the assumptions come in the form of an exponential number of assumptions, one assumption for every input, or one assumption for every size parameter. Overall, whereas the constructions underlied by the new assumptions are ingenious, their existence distinctly lacks a "win-win" consequence.

Obviously, in order to make progress and move a field forward, we should occasionally embrace papers whose constructions rely on newly formed assumptions and conjectures. This approach marks the birth of modern cryptography itself, in the landmark papers of [24,49]. However, any conjecture and any new assumption must be an open invitation to refute or simplify, which necessitates a clear understanding of what is being assumed in the first place. The latter has been distinctly lacking in recent years.

Our Thesis. We believe that the lack of standards in what is accepted as a reasonable cryptographic assumption is harmful to our field. Whereas in the past, a break to a provably secure scheme would lead to a mathematical breakthrough, there is a danger that in the future the proclaimed guarantee of provable security will lose its meaning. We may reach an absurdum, where the underlying assumption is that the scheme itself is secure, which will eventually endanger the mere existence of our field.

We are in great need of *measures* which will capture which assumptions are "safe", and which assumptions are "dangerous". Obviously, safe does not mean correct, but rather captures that regardless of whether a safe assumption is true or false, it is of interest. Dangerous assumptions may be false and yet of no independent interest, thus using such assumptions in abundance poses the danger that provable security will lose its meaning.

One such measure was previously given by Naor [43], who classified assumptions based on the complexity of falsifying them. Loosely speaking,[3] an assumption is said to be *falsifiable*, if one can efficiently check whether an adversary is successful in breaking it.

We argue that the classification based on falsifiability alone has proved to be too inclusive. In particular, assumptions whose mere statement refers to the cryptographic scheme they support can be (and have been) made falsifiable. Thus, falsifiability is an important feature but not sufficient as a basis for evaluating current assumptions,[4] and in particular, it does not exclude assumptions that are construction dependent.

In this position paper, we propose a stricter classification. Our governing principle is the goal of relying on hardness assumptions that are independent of the constructions.

---

[3] We refer here to the notion of falsifiability as formalized by Gentry and Wichs [30], which is slightly different from the original notions proposed by Naor. We elaborate on these notions, and on the difference between them, in Sect. 2.6 and in Appendix A.

[4] We note that this was also explicitly pointed out by Naor who advocated falsifiability as an important feature, not as a sufficient one.

2   Our Classification

We formalize the notion of a *complexity assumption*, and argue that such assumptions is what we should aim for.

Intuitively, complexity assumptions are non-interactive assumptions that postulate that given an input, distributed according to an efficiently sampleable distribution $\mathcal{D}$, it is hard to compute a valid "answer" (with non-negligible advantage), where checking the validity of the answers can be done in polynomial time.

More specifically, we distinguish between two types of complexity assumptions:

1. *Search* complexity assumptions, and
2. *Decision* complexity assumptions.

Convention: Throughout this manuscript, for the sake of brevity, we refer to a family of poly-size circuits $\mathcal{M} = \{\mathcal{M}_n\}$ as a polynomial time non-uniform algorithm $\mathcal{M}$.

2.1   Search Complexity Assumptions

Each assumption in the class of search complexity assumptions consists of a pair of probabilistic polynomial-time algorithms $(\mathcal{D}, \mathcal{R})$, and asserts that there does not exist an efficient algorithm $\mathcal{M}$ that on input a random challenge x, distributed according $\mathcal{D}$, computes any value y such that $\mathcal{R}(x, y) = 1$, with non-negligible probability. Formally:

Definition 1. *An assumption is a* search complexity assumption *if it consists of a pair of probabilistic polynomial-time algorithms $(\mathcal{D}, \mathcal{R})$, and it asserts that for any efficient[5] algorithm $\mathcal{M}$ there exists a negligible function μ such that for every $n \in \mathbb{N}$,*

$$\Pr_{x \leftarrow \mathcal{D}(1^n)} [\mathcal{M}(x) = y \text{ s.t. } \mathcal{R}(x, y) = 1] \leq \mu(n). \tag{1}$$

Note that in Definition 1 above, we require that there is an efficient algorithm $\mathcal{R}$ that takes as input a pair (x, y) and outputs 0 or 1. One could consider a more liberal definition, of a *privately-verifiable* search complexity assumption, which is similar to the definition above, except that algorithm $\mathcal{R}$ is given not only the pair (x, y) but also the randomness r used by $\mathcal{D}$ to generate x.

Definition 2. *An assumption is a* privately-verifiable search complexity assumption *if it consists of a pair of probabilistic polynomial-time algorithms $(\mathcal{D}, \mathcal{R})$, and it asserts that for any efficient algorithm $\mathcal{M}$ there exists a negligible function μ such that for every $n \in \mathbb{N}$,*

$$\Pr_{r \leftarrow \{0,1\}^n} [\mathcal{M}(x) = y \text{ s.t. } \mathcal{R}(x, y, r) = 1 \mid x = \mathcal{D}(r)] \leq \mu(n). \tag{2}$$

---

[5] "Efficient" can be interpreted in several ways. We elaborate on the various interpretations below.

The class of privately-verifiable search complexity assumptions is clearly more inclusive.

What is an Efficient Algorithm? Note that in Definitions 1 and 2 above, we restricted the adversary $\mathcal{M}$ to be an *efficient algorithm*. One can interpret the class of efficient algorithms in various ways. The most common interpretation is that it consists of all non-uniform polynomial time algorithms. However, one can interpret this class as the class of all *uniform* probabilistic polynomial time algorithms, or parallel NC algorithms, leading to the notions of search complexity assumption with *uniform* security or with *parallel* security, respectively. One can also strengthen the power of the adversary $\mathcal{M}$ and allow it to be a *quantum* algorithm.

More generally, one can define a (t, ϵ) search complexity assumption exactly as above, except that we allow $\mathcal{M}$ to run in time $t(n)$ (non-uniform or uniform, unbounded depth or bounded depth, with quantum power or without) and require that it cannot succeed with probability $\epsilon(n)$ on a random challenge $x \leftarrow \mathcal{D}(1^n)$. For example, $t(n)$ may be sub-exponentially large, and $\epsilon(n)$ may be sub-exponentially small. Clearly the smaller t is, and the larger ϵ is, the weaker (and thus more reasonable) the assumption is.

Uniformity of $(\mathcal{D}, \mathcal{R})$. In Definition 1 above, we require that the algorithms $\mathcal{D}$ and $\mathcal{R}$ are *uniform* probabilistic polynomial-time algorithms. We could have considered the more general class of *non-uniform* search complexity assumptions, where we allow $\mathcal{D}$ and $\mathcal{R}$ to be *non-uniform* probabilistic polynomial-time algorithms. We chose to restrict to uniform assumptions for two reasons. First, we are not aware of any complexity assumption in the cryptographic literature that consists of *non-uniform* $\mathcal{D}$ or $\mathcal{R}$. Second, allowing these algorithms to be non-uniform makes room for assumptions whose description size grows with the size of the security parameter, which enables them to be construction specific and not of independent interest. We would like to avoid such dependence. We note that one could also consider search complexity assumptions where $\mathcal{D}$ and $\mathcal{R}$ are allowed to be quantum algorithms, or algorithms resulting from any biological process.

Examples. The class of (publicly-verifiable) search complexity assumptions includes almost all traditional search-based cryptographic assumptions, including the Factoring and RSA assumptions [47,49], the strong RSA assumption [6,26], the Discrete Log assumption (in various groups) [24], the Learning Parity with Noise (LPN) assumption [10], and the Learning with Error (LWE) assumption [48]. An exception is the computational Diffie-Hellman assumption (in various groups) [24], which is a *privately-verifiable* search complexity assumption, since given (g^x, g^y, z) it is hard to test whether $z = g^{xy}$, unless we are given x and y, which constitutes the randomness used to generate (g^x, g^y).

We note that the LPN assumption and the LWE assumption each consists of a family of complexity assumptions,[6] one assumption for each m, where m is the number of examples of noisy equations given to the adversary. However, as was observed by [29], there is a reduction between the LPN (repectively LWE) assumption with a fixed m to the LPN (repectively LWE) assumption with an arbitrary m, that incurs essentially no loss in security.

t-Search Complexity Assumptions. The efficient algorithm $\mathcal{R}$ associated with a search complexity assumption can be thought of as an NP relation algorithm. We believe that it is worth distinguishing between search complexity assumptions for which with overwhelming probability, $x \leftarrow \mathcal{D}(1^n)$ has at most polynomially many witnesses, and assumptions for which with non-negligible probability, $x \leftarrow \mathcal{D}(1^n)$ has exponentially many witnesses. We caution that the latter may be too inclusive, and lead to an absurdum where the assumption assumes the security of the cryptographic scheme itself, as exemplified below.

Definition 3. *For any function $t = t(n)$, a search complexity assumption $(\mathcal{D}, \mathcal{R})$ is said to be a t-search complexity assumption if there exists a negligible function μ such that*

$$\Pr_{x \leftarrow \mathcal{D}(1^n)} [|\{y : (x, y) \in \mathcal{R}\}| > t] \leq \mu(n) \tag{3}$$

Most traditional search-based cryptographic assumptions are 1-search complexity assumptions; i.e., they are associated with a relation $\mathcal{R}$ for which every x has a *unique* witness. Examples include the Factoring assumption, the RSA assumption, the Discrete Log assumption (in various groups), the LPN assumption, and the LWE assumption. The square-root assumption in composite order group is an example of a 4-search complexity assumption, since each element has at most 4 square roots modulo $N = pq$.

An example of a traditional search complexity assumption that is a t-search assumption only for an exponentially large t, is the strong RSA assumption. Recall that this assumption assumes that given an RSA modulus N and a random element $y \leftarrow \mathbb{Z}_N^*$, it is hard to find *any* exponent $e \in \mathbb{Z}_N^*$ together with the e'th root $y^{e^{-1}} \bmod N$. Indeed, in some sense, the strong RSA assumption is "exponentially" stronger, since the standard RSA assumption assumes that it is

---

[6] Loosely speaking, the LPN assumption with error parameter $p \in (0, 1)$ (where p is a constant), asserts that for any poly-size adversary that observes polynomially many noisy linear equations of the form $\{(a_i, a_i \cdot x + e_i)\}_{i=1}^{\text{poly}(n)}$, outputs x with at most negligible probability, where $x \in_R \{0, 1\}^n$ is random, all the linear equations $a_i \in_R \{0, 1\}^n$ are independent and random, and each e_i is an independent Bernoulli random variable, where $e_i = 1$ with probability p and $e_i = 0$ otherwise. The LWE assumption is similar to the LPN assumption, except that it is associated with a (possibly large) field $\mathbb{F}$. It assumes that as above, given noisy linear equations $\{(a_i, a_i \cdot x + e_i)\}_{i=1}^{\text{poly}(n)}$ it is hard to find x, where now the equations are over the field $\mathbb{F}$, and $x \in_R \mathbb{F}$, each $a_i \in_R \mathbb{F}^n$, and each error e_i is independently distributed according to a discrete Gaussian distribution. We refer the reader to [48] for the precise definition.

hard to find the e'th root, for a single e, whereas the strong RSA assumption assumes that this is hard for exponentially many e's.

Whereas the strong RSA assumption is considered quite reasonable in our community, the existence of exponentially many witnesses allows for assumptions that are overly tailored to cryptographic primitives, as exemplified below.

Consider for example the assumption that a given concrete candidate two-message delegation scheme for a polynomial-time computable language L is *adaptively* sound. This asserts that there does not exist an efficient non-uniform algorithm $\mathcal{M}$ that given a random challenge from the verifier, produces an instance $x \notin L$ together with an accepting answer to the challenge. By our definition, this is a t-complexity assumption for an exponential t, which is publicly verifiable if the underlying delegation scheme is publicly verifiable, and is privately verifiable if the underlying delegation scheme is privately verifiable. Yet, this complexity assumption is an example of an absurdum where the assumption assumes the security of the scheme itself. This absurdum stems from the fact that t is exponential. If we restricted t to be polynomial this would be avoided.

We emphasize that we are not claiming that 1-search assumptions are necessarily superior to t-search assumptions for exponential t. This is illustrated in the following example pointed out to us by Micciancio and Ducas. Contrast the Shortest Integer Solution (SIS) assumption [41], which is a t-search assumption for an exponential t, with the Learning with Error (LWE) assumption, which is 1-complexity assumption. It is well known that the LWE assumption is reducible to the SIS assumption [48]. Loosely speaking, given an LWE instance one can use an SIS breaker to find short vectors in the dual lattice, and then use these vectors to solve the LWE instance. We note that a reduction in the other direction is only known via a quantum reduction [53].

More generally, clearly if Assumption A possesses properties that we consider desirable, such as being 1-search, falsifiable, robust against quantum adversaries, etc., and Assumption A is reducible to Assumption B, then the latter should be considered at least as reasonable as the former.

2.2   Decisional Complexity Assumptions

Each assumption in the class of decisional complexity assumptions consists of two probabilistic polynomial-time algorithms $\mathcal{D}_0$ and $\mathcal{D}_1$, and asserts that there does not exist an efficient algorithm $\mathcal{M}$ that on input a random challenge $x \leftarrow \mathcal{D}_b$ for a random $b \leftarrow \{0,1\}$, outputs b with non-negligible advantage.

Definition 4. *An assumption is a* decisional complexity assumption *if it is associated with two probabilistic polynomial-time distributions* $(\mathcal{D}_0, \mathcal{D}_1)$, *such that for any efficient[7] algorithm* $\mathcal{M}$ *there exists a negligible function* μ *such that for any* $n \in \mathbb{N}$,

$$\Pr_{b \leftarrow \{0,1\}, x \leftarrow \mathcal{D}_b(1^n)} [\mathcal{M}(x) = b] \leq \frac{1}{2} + \mu(n). \tag{4}$$

---

[7] "Efficient algorithms" can be interpreted in several ways, as we elaborated on in Sect. 2.1.

Example 1. This class includes all traditional decisional assumptions, such as the DDH assumption [24], the Quadratic Residuosity (QR) assumption [37], the N'th Residuosity assumption [44], the decisional LPN assumption [2], the decisional LWE assumption [48], the decisional linear assumption over bilinear groups [11], and the Φ-Hiding assumption [15]. Thus, this class is quite expressive. The Multilinear Subgroup Elimination assumption, which was recently proposed and used to construct IO obfuscation in [28], is another member of this class. To date, however, this assumption has been refuted in all proposed candidate (multilinear) groups [18,19,42].

An example of a decisional assumption that *does not* belong to this class is the strong DDH assumption over a prime order group G [16]. This assumption asserts that *for every* distribution $\mathcal{D}$ with min-entropy $k = \omega(\log n)$, it holds that

$$(g^r, g^x, g^{rx}) \approx (g^r, g^x, g^u),$$

where $x \leftarrow \mathcal{D}$ and $r, u \leftarrow \mathbb{Z}_p$, where p is the cardinality of G, and g is a generator of G.

This assumption was introduced by Canetti [16], who used it to prove the security of his point function obfuscation construction. Since for point function obfuscation the requirement is to get security for *every* point x, it is impossible to base security under a polynomial complexity assumption. This was shown by Wee [54], who constructed a point function obfuscation scheme under a complexity assumption with an extremely small ϵ. We note that if instead of requiring security to hold for *every* point x, we require security to hold for every distribution on inputs with min-entropy n^ϵ, for some constant $\epsilon > 0$, then we can rely on standard (polynomial) complexity assumptions, such as the LWE assumption [36], and a distributional assumption as above is not necessary.

Many versus two distributions. One can consider an "extended" decision complexity assumption which is associated with polynomially many distributions, as opposed to only two distributions. Specifically, one can consider the decision complexity assumption that is associated with a probabilistic polynomial-time distribution $\mathcal{D}$ that encodes $t = \mathsf{poly}(n)$ distributions, and the assumption is that for any efficient algorithm $\mathcal{M}$ there exists a negligible function μ such that for any $n \in \mathbb{N}$,

$$\Pr_{i \leftarrow [t], x \leftarrow \mathcal{D}(1^n, i)}[\mathcal{M}(x) = i] \leq \frac{1}{t} + \mu(n). \tag{5}$$

We note however that such an assumption can be converted into an equivalent decision assumption with two distributions $\mathcal{D}_0$ and $\mathcal{D}_1$, using the Goldreich-Levin hard-core predicate theorem [34], as follows: The distribution $\mathcal{D}_0$ will sample at random $i \leftarrow [t]$, sample at random $x \leftarrow \mathcal{D}(1^n, i)$, sample at random $r \leftarrow [t]$, and output $(x, r, r \cdot i)$. The algorithm $\mathcal{D}_1$ will similarly sample i, x, r but will output (x, r, b) for a random bit $b \leftarrow \{0, 1\}$.

2.3 Worst-Case vs. Average-Case Hardness

Note that both Definitions 1 and 4 capture *average-case* hardness assumptions, as opposed to *worst-case* hardness assumptions. Indeed, at first sight, relying on average-case hardness in order to prove the security of cryptographic schemes seems to be necessary, since the security requirements for cryptographic schemes require adversary attacks to fail with *high probability*, rather than in the worst case.

One could have considered the stricter class of *worse-case* (search or decision) complexity assumptions. A worst-case search assumption, is associated with a polynomial time computable relation $\mathcal{R}$, and requires that no polynomial-time non-uniform algorithm $\mathcal{M}$ satisfies that *for every* $x \in \{0,1\}^*$, $\mathcal{R}(x, \mathcal{M}(x)) = 1$. A worst-case decisional assumption is a promise assumption which is associated with two sets of inputs S_0 and S_1, and requires there is no polynomial-time non-uniform algorithm $\mathcal{M}$, that *for every* $x \in \{0,1\}^*$, given the promise that it is in $S_0 \cup S_1$, guesses correctly whether $x \in S_0$ or $x \in S_1$.

There are several cryptographic assumptions for which there are random self-reductions from worst-case to average-case for *fixed-parameter problems*[8]. Examples include the Quadratic-Residuosity assumption, the Discrete Logarithm assumption, and the RSA assumption [37]. In fact, the Discrete Log assumption over fields of size 2^n has a (full) worst-case to average case reduction [7].[9] Yet, we note that the Discrete Log assumption over fields of small characteristic (such as fields of size 2^n) have been recently shown to be solvable in quasi-polynomial time [5], and as such are highly vulnerable.

There are several lattice based assumptions that have a worst-case to average-case reduction [1,13,46,48]. Such worst-case assumptions are usable for cryptography, and include the GapSVP assumption [33] and the assumption that it is hard to approximate the Shortest Independent Vector Problem (SIVP) within polynomial approximation factors [41].

Whereas being a worst-case complexity assumption is a desirable property and average to worst case reductions are a goal in itself, we believe that at this point in the life-time of our field establishing the security of novel cryptographic schemes (e.g., IO obfuscation) based on an average case complexity assumption would be a triumph. We note that traditionally cryptographic hardness assumptions were average-case assumptions (as exemplified above).

2.4 Search versus Decision Complexity Assumptions

An interesting question is whether search complexity assumptions can always be converted to decision complexity assumptions and vice versa.

---

[8] By a "worst-case to average-case reduction for a fixed-parameter problem", we think of a problem instance as a pair (n, x) and a reduction which holds per fixed n.

[9] More generally, such a worst-case to average case reduction exists if the security parameter determines the field, its representation, and a generator of the field. As was shown by Shoup in [50,51], finding a representation (i.e., an irreducible polynomial) and a generator for fields of small characteristic can be done in polynomial time.

We note that any decision complexity assumption can be converted into a *privately-verifiable* search complexity assumption that is sound assuming the decision assumption is sound, but not necessarily into a *publicly verifiable* search complexity assumption. Consider, for example, the DDH assumption. Let f_{DDH} be the function that takes as input n tuples (where n is the security parameter), each tuple is either a DDH tuple or a random tuple, and outputs n bits, predicting for each tuple whether it is a DDH tuple or a random tuple. The direct product theorem [39] implies that if the DDH assumption is sound then it is hard to predict f_{DDH} except with negligible probability. The resulting search complexity assumption is privately-verifiable, since in order to verify whether a pair $((x_1, \ldots, x_n), (b_1, \ldots, b_n))$ satisfies that $(b_1, \ldots, b_n) = f_{DDH}(x_1, \ldots, x_n)$, one needs the private randomness used to generate $(x_1, \ldots, x_n)$.

In the other direction, it would seem at first that one can map any (privately-verifiable or publicly verifiable) search complexity assumption into an equivalent decision assumption, using the hard-core predicate theorem of Goldreich and Levin [34]. Specifically, given any (privately-verifiable) search complexity assumption $(\mathcal{D}, \mathcal{R})$, consider the following decision assumption: The assumption is associated with two distributions $\mathcal{D}_0$ and $\mathcal{D}_1$. The distribution $\mathcal{D}_b$ generates (x, y), where $x \leftarrow \mathcal{D}(1^n)$ and where $\mathcal{R}(x, y) = 1$, and outputs a triplet (x, r, u) where r is a random string, and if $b = 0$ then $u = r \cdot y (\mathrm{mod}\ 2)$ and if $b = 1$ then $u \leftarrow \{0, 1\}$. The Goldreich-Levin hard-core predicate theorem states that the underlying search assumption is sound if and only if $x \leftarrow \mathcal{D}_0$ is computationally indistinguishable from $x \leftarrow \mathcal{D}_1$. However, $\mathcal{D}_0$ and $\mathcal{D}_1$ are efficiently sampleable only if generating a pair (x, y), such that $x \leftarrow \mathcal{D}(1^n)$ and $\mathcal{R}(x, y) = 1$, can be done efficiently. Since the definition of search complexity assumptions only assures that $\mathcal{D}$ is efficiently sampleable and does not mandate that the pair (x, y) is efficiently sampleable, the above transformation from search to decision complexity assumption does not always hold.

2.5   Concrete versus Generic Assumptions

The examples of assumptions we mentioned above are concrete assumptions. Another type of assumption made in cryptography is a *generic* assumption, such as the assumption that one-way functions exist, collision resistant hash families exist, or IO secure obfuscation schemes exist.

We view generic assumptions as *cryptographic primitives* in themselves, as opposed to cryptographic assumptions. We take this view for several reasons. First, in order to ever make use of a cryptographic protocol based on a generic assumption, we must first instantiate it with a concrete assumption. Thus, in a sense, a generic assumption is only as good as the concrete assumptions it can be based on. Second, generic assumptions are *not falsifiable*. The reason is that in order to falsify a generic assumption one needs to falsify *all* the candidates.

The one-way function primitive has the unique feature that it has a *universal* concrete instantiation, and hence is falsifiable. Namely, there exists a (universal) concrete one-way function candidate f such that if one-way functions exist then f itself is one-way [31]. This state of affairs would be the gold standard for any generic assumption; see discussion in Sect. 2.7. Moreover, one-way functions can be constructed based on any complexity assumption, search or decision.

In the other extreme, there are generic assumptions that have no instantiation under any (search or decisional) complexity assumption. Examples include the generic assumption that there exists a 2-message delegation scheme for NP, the assumption that P-certificates exist [20], the assumption that extractable collision resistant hash functions exist [8,21,23], and the generic assumption that IO obfuscation exists.[10]

2.6 Falsifiability of Complexity Assumptions

Naor [43] defined the class of falsifiable assumptions. Intuitively, this class includes all the assumptions for which there is a constructive way to demonstrate that it is false, if this is the case. Naor defined three notions of falsifiability: *efficiently falsifiable*, *falsifiable*, and *somewhat falsifiable*. We refer the reader to Appendix A for the precise definitions.

Gentry and Wichs [30] re-formalized the notion of a falsifiable assumption. They provide a single formulation, that arguably more closely resembles the intuitive notion of falsifiability. According to [30] an assumption is falsifiable if it can be modeled as an interactive game between an efficient challenger and an adversary, at the conclusion of which the challenger can efficiently decide whether the adversary won the game. Almost all followup work that use the term of falsifiable assumptions use the falsifiability notion of [30], which captures the intuition that one can efficiently check (using randomness and interaction) whether an attacker can indeed break the assumption. By now, when researchers say that an assumption is falsifiable they most often refer to the falsifiability notion of [30]. In this paper we follow this convention.

Definition 5. *[30] A falsifiable cryptographic assumption consists of a probabilistic polynomial-time interactive challenger C. On security parameter n, the challenger $C(1^n)$ interacts with a non-uniform machine $\mathcal{M}(1^n)$ and may output a special symbol* win. *If this occurs, we say that $\mathcal{M}(1^n)$ wins $C(1^n)$. The assumption states that for any efficient non-uniform $\mathcal{M}$,*

$$\Pr[\mathcal{M}(1^n) \text{ wins } C(1^n)] = \mathsf{negl}(n),$$

where the probability is over the random coins of C. For any $t = t(n)$ and $\epsilon = \epsilon(n)$, an (t, ϵ) assumption is falsifiable if it is associated with a probabilistic polynomial-time C as above, and for every $\mathcal{M}$ of size at most $t(n)$, and for every $n \in \mathbb{N}$,

$$\Pr[\mathcal{M}(1^n) \text{ wins } C(1^n)] \leq \epsilon(n).$$

The following claim is straightforward.

Claim 1. Any (search or decision) complexity assumption is also a falsifiable assumption (according to Definition 5), but not vice versa.

---

[10] We note that this assumption was recently reduced to the subgroup elimination assumption [28], which is a new decisional complexity assumptions. To date, however, this assumption has been refuted in all proposed candidate (multi-linear) groups.

2.7   Desirable Properties of Complexity Assumptions

We emphasize that our classification described above is minimal and does not take into account various measures of how "robust" the assumption is. We mention two such robustness measures below.

Robustness to auxiliary inputs. One notion of robustness that was considered for *search* assumptions is that of robustness to *auxiliary inputs*.

Let us consider which auxiliary inputs may be available to an adversary of a complexity assumption. Recall that search complexity assumptions are associated with a pair of probabilistic polynomial time algorithms $(\mathcal{D}, \mathcal{R})$ where the algorithm $\mathcal{D}$ generates instances $x \leftarrow \mathcal{D}$ and the assumption is that given $x \leftarrow \mathcal{D}$ it is computationally hard to find y such that $(x, y) \in \mathcal{R}$. As it turns out however, for all known search assumptions that are useful in cryptography, it is further the case that one can efficiently generate not only an instance $x \leftarrow \mathcal{D}$, but pairs (x, y) such that $(x, y) \in \mathcal{R}$. Indeed, it is what most often makes the assumption useful in a cryptographic context. Typically, in a classical adversarial model, y is part of the secret key, whereas x is known to the adversary. Yet due to extensive evidence a more realistic adversarial model allows the adversary access to partial knowledge about y which can be viewed generally as access to an auxiliary input.

Thus, one could have defined a search complexity assumption as a pair $(\mathcal{D}, \mathcal{R})$ as above, but where the algorithm $\mathcal{D}$ generates pairs (x, y) (as opposed to only x), such that $(x, y) \in \mathcal{R}$ and the requirement is that any polynomial-size adversary who is given only x, outputs some y' such that $(x, y') \in \mathcal{R}$, only with negligible probability. This definition is appropriate when considering robustness to auxiliary information. Informally, such a search assumption is said to be resilient to auxiliary inputs if given an instance x sampled according to $\mathcal{D}$, and given some auxiliary information about the randomness used by $\mathcal{D}$ (and in particular, given some auxiliary information about y), it remains computationally hard to find y' such that $(x, y') \in \mathcal{R}$.

Definition 6. *A search complexity assumption $(\mathcal{D}, \mathcal{R})$ as above is said to be resilient to $t(n)$-hard-to-invert auxiliary inputs if for any $t(n)$-hard-to-invert function $L : \{0, 1\}^n \to \{0, 1\}^*$,*

$$\Pr_{r \leftarrow \{0,1\}^n, (x,y) \leftarrow \mathcal{D}(r)} [\mathcal{M}(x, L(r)) = y' \text{ s.t. } \mathcal{R}(x, y') = 1] \leq \mu(n), \qquad (6)$$

where L is said to be $t(n)$-hard-to-invert if for every $t(n)$-time non-uniform algorithm $\mathcal{M}$ there exists a negligible μ such that for every $n \in \mathbb{N}$,

$$\Pr_{z \leftarrow L(U_n)} [\mathcal{M}(z) = r : L(r) = z] = \mu(n). \qquad (7)$$

It was shown in [36] that the decisional version of the LWE assumption is resilient to $t(n)$-hard-to-invert auxiliary inputs for $t(n) = 2^{n^{\delta}}$, for any constant $\delta > 0$. In particular, this implies that the LWE assumption is robust to leakage attacks. In contrast, the RSA assumptions is known to be completely broken even if only 0.27 fraction of random bits of the secret key are leaked [38].

Universal assumptions. We say that a (concrete) complexity assumption A is *universal* with respect to a generic assumption if the following holds: If A is false then the generic assumption is false. In other words, if the generic assumption has a concrete sound instantiation then A is it. Today, the only generic assumption for which we know a universal instantiation is one-way functions [31].

Open Problem: We pose the open problem of finding a universal instantiations for other generic assumptions, in particular for IO obfuscation, witness encryption, or 2-message delegation for NP.

3 Recently Proposed Cryptographic Assumptions

Recently, there has been a proliferation of cryptographic assumptions. We next argue that many of the recent assumptions proposed in the literature, *even the falsifiable* ones, are *not* complexity assumptions.

IO Obfuscation constructions. Recently, several constructions of IO obfuscation have been proposed. These were proved under ad-hoc assumptions [27], meta assumptions [45], and ideal-group assumptions [4,14]. These assumptions are not complexity assumptions, for several reasons: They are either overly tailored to the construction, or artificially restrict the adversaries.

The recent result of [28] constructed IO obfuscation under a new complexity assumption, called Subgroup Elimination assumption. This is a significant step towards constructing IO under a standard assumption. However, to date, this assumption is known to be false in all candidate (multi-linear) groups.

Assuming IO obfuscation exists. A large body of work which emerged since the construction of [27], constructs various cryptographic primitives assuming IO obfuscation exists. Some of these results require only the existence of IO obfuscation for circuits with only polynomially many inputs (eg., [9]). Note that any instantiation of this assumption is falsifiable. Namely, the assumption that a given obfuscation candidate $\mathcal{O}$ (for circuits with polynomially many inputs) is IO secure, is falsifiable. The reason is that to falsify it one needs to exhibit two circuits C_0 and C_1 in the family such that $C_0 \equiv C_1$, and show that it can distinguish between $\mathcal{O}(C_0)$ and $\mathcal{O}(C_1)$. Note that since the domain of C_0 and C_1 consists of polynomially many elements one can efficiently test whether indeed $C_0 \equiv C_1$, and of course the falsifier can efficiently prove that $\mathcal{O}(C_0) \not\approx \mathcal{O}(C_1)$ by showing that one can distinguish between these two distributions. On the other hand, this is not a complexity assumption. Rather, such an assumption consists of many (often exponentially many) decision complexity assumptions: For every $C_0 \equiv C_1$ in the family $\mathcal{C}_n$ (there are often exponentially many such pairs), the corresponding decision complexity assumption is that $\mathcal{O}(C_0) \approx \mathcal{O}(C_1)$. Thus, intuitively, such an assumption is exponentially weaker than a decisional complexity assumption.

Artificially restricted adversaries assumptions. We next consider the class of assumptions that make some "artificial" restriction on the adversary. Examples include the Random Oracle Model (ROM) [25] and various generic group models [12,52]. The ROM restricts the adversary to use a given hash function only in a black-box manner. Similarly, generic group assumptions assume the adversary uses the group structure only in an "ideal" way. Another family of assumptions that belongs to this class is the family knowledge assumptions. Knowledge assumptions artificially restrict the adversaries to compute things in a certain way. For example, the Knowledge-of-Exponent assumption [22] assumes that any adversary that given (g, h) computes (g^z, h^z), must do so by "first" computing z and then computing (g^z, h^z).

We note that such assumptions cannot be written even as exponentially many complexity assumptions. Moreover, for the ROM and the generic group assumptions, we know of several examples of *insecure* schemes that are proven secure under these assumptions [3,17,35].

We thus believe that results that are based on such assumption should be viewed as *intermediate* results, towards the goal of removing such artificial constraints and constructing schemes that are provably secure under complexity assumptions.

4   Summary

Theoretical cryptography is in great need for a methodology for classifying assumptions. In this paper, we define the class of search and decision *complexity assumptions*. An overall guiding principle in the choices we made was to rule out hardness assumptions which are construction dependent.

We believe that complexity assumptions as we defined them are general enough to capture all "desirable" assumptions, and we are hopeful that they will suffice in expressive power to enable proofs of security for sound constructions. In particular, all traditional cryptographic assumptions fall into this class.

We emphasize, that we do not claim that all complexity-based complexity assumptions are necessarily desirable or reasonable. For example, false complexity assumptions are clearly not reasonable. In addition, our classification does not incorporate various measures of how "robust" an assumption is, such as: how well studied the assumption is, whether it is known to be broken by quantum attacks, whether it has a worst-case to average-case reduction, or whether it is known to be robust to auxiliary information.

Acknowledgements. We would like to thank many colleagues for their comments on an earlier draft of this work, including Ducas, Goldreich Micianccio, Peikert, Regev, Sahai, and Vaikuntanathan. In particular, we are grateful to Micciancio for extensive illuminating discussions on many aspects of hardness assumptions on lattices, and to Sahai and Vaikuntanathan for clarifying discussions on the strength of the sub-group elimination assumption used for IO obfuscation. Thanks to Bernstein for pointing out a long overlooked worst-case to average-case reduction for discrete-log in fields of small characteristic.

A   Falsifiable Assumptions

Naor [43] defined three notions of falsifiability: *efficiently falsifiable, falsifiable,* and *somewhat falsifiable.*

Definition 7. *A (t, ϵ) assumption is* efficiently falsifiable *if there exists a family of distributions $\{\mathcal{D}_n\}_{n \in \mathbb{N}}$, a verifier $V : \{0,1\}^* \times \{0,1\}^* \to \{0,1\}$, such that the following holds for any parameter $\delta > 0$:*

1. If the assumption is false then there exists a falsifier $\mathcal{B}$ that satisfies

$$\Pr_{x \to \mathcal{D}_n} [\mathcal{B}(x) = y \ s.t. \ V(x, y) = 1] \geq 1 - \delta. \tag{A.1}$$

Moreover, the runtime of $\mathcal{B}$ is polynomial in the runtime of the adversary that breaks the assumption and polynomial in $n, \log 1/\epsilon, \log 1/\delta$.

2. The runtime of V and the time it takes to sample an element from $\mathcal{D}_n$ is $\mathsf{poly}(n, \log 1/\epsilon, \log 1/\delta)$.

3. If there exists a falsifier $\mathcal{B}$ that runs in time t and solves random challenges $x \leftarrow \mathcal{D}_n$ with probability γ, then there exists an adversary $\mathcal{A}$ that runs in time $\mathsf{poly}(t)$ and breaks the original assumption with probability $\mathsf{poly}(\gamma)$.

Definition 8. *A (t, ϵ) assumption is* falsifiable *if everything is as in Definition 7 except that the runtime of V and of sampling $\mathcal{D}_n$ may depend on $1/\epsilon$ (as opposed to $\log 1/\epsilon$).*

Definition 9. *A (t, ϵ) assumption is* somewhat falsifiable *if everything is as in Definition 7 except that the runtime of V and of sampling $\mathcal{D}_n$ may depend on $1/\epsilon$ (as opposed to $\log 1/\epsilon$), and on the runtime of $\mathcal{B}$. In particular, this means that V may simulate $\mathcal{B}$.*

Remark 1. We note that any efficiently falsifiable assumption is also a relation-based complexity assumption. However, we find the notion of efficiently falsifiable to be very restrictive, since intuitively it only includes assumptions that are random self reducible. The definition of falsifiable is less restrictive, however a falsifiable assumption is not necessarily a complexity assumption, since in order to verify a break of the assumption one needs to run in time $1/\epsilon$ which is super-polynomial. We view the notion of somewhat falsifiable to be too weak. Allowing the runtime of the verifier to depend on the runtime of the falsifier $\mathcal{B}$ makes this class very inclusive, and it includes many interactive assumptions (we refer the reader to [43] for details).

References

1. Ajtai, M.: Generating hard instances of lattice problems (extended abstract). In: Proceedings of the Twenty-Eighth Annual ACM Symposium on the Theory of Computing, 22–24 May 1996, Philadelphia, Pennsylvania, USA, pp. 99–108 (1996)

2. Alekhnovich, M.: More on average case vs approximation complexity. In: Proceedings of the 44th Symposium on Foundations of Computer Science (FOCS 2003), 11–14 October 2003, Cambridge, MA, USA, pp. 298–307 (2003)

3. Barak, B.: How to go beyond the black-box simulation barrier. In: 42nd Annual Symposium on Foundations of Computer Science, FOCS 2001, 14–17 October 2001, Las Vegas, Nevada, USA, pp. 106–115 (2001)

4. Barak, B., Garg, S., Kalai, Y.T., Paneth, O., Sahai, A.: Protecting obfuscation against algebraic attacks. In: Nguyen, P.Q., Oswald, E. (eds.) EUROCRYPT 2014. LNCS, vol. 8441, pp. 221–238. Springer, Heidelberg (2014)

5. Barbulescu, R., Gaudry, P., Joux, A., Thomé, E.: A heuristic quasi-polynomial algorithm for discrete logarithm in finite fields of small characteristic. In: Nguyen, P.Q., Oswald, E. (eds.) EUROCRYPT 2014. LNCS, vol. 8441, pp. 1–16. Springer, Heidelberg (2014)

6. Barić, N., Pfitzmann, B.: Collision-free accumulators and fail-stop signature schemes without trees. In: Fumy, W. (ed.) EUROCRYPT 1997. LNCS, vol. 1233, pp. 480–494. Springer, Heidelberg (1997)

7. Bernstein, D.J.: Private communication (2015)

8. Bitansky, N., Canetti, R., Chiesa, A., Goldwasser, S., Lin, H., Rubinstein, A., Tromer, E.: The hunting of the SNARK. IACR Cryptology ePrint Archive 2014, p. 580 (2014)

9. Bitansky, N., Garg, S., Lin, H., Pass, R., Telang, S.: Succinct randomized encodings and their applications. In: Proceedings of the Forty-Seventh Annual ACM on Symposium on Theory of Computing, STOC 2015, 14–17 June 2015, Portland, OR, USA, pp. 439–448 (2015)

10. Blum, A., Kalai, A., Wasserman, H.: Noise-tolerant learning, the parity problem, and the statistical query model. In: Proceedings of the Thirty-Second Annual ACM Symposium on Theory of Computing, May 21–23, 2000, Portland, OR, USA, pp. 435–440 (2000)

11. Boneh, D., Boyen, X., Shacham, H.: Short group signatures. In: Franklin, M. (ed.) CRYPTO 2004. LNCS, vol. 3152, pp. 41–55. Springer, Heidelberg (2004)

12. Boneh, D., Lipton, R.J.: Algorithms for black-box fields and their application to cryptography. In: Koblitz, N. (ed.) CRYPTO 1996. LNCS, vol. 1109, pp. 283–297. Springer, Heidelberg (1996)

13. Brakerski, Z., Langlois, A., Peikert, C., Regev, O., Stehlé, D.: Classical hardness of learning with errors. In: Symposium on Theory of Computing Conference, STOC 2013, 1–4 June 2013, Palo Alto, CA, USA, pp. 575–584 (2013)

14. Brakerski, Z., Rothblum, G.N.: Virtual black-box obfuscation for all circuits via generic graded encoding. In: Lindell, Y. (ed.) TCC 2014. LNCS, vol. 8349, pp. 1–25. Springer, Heidelberg (2014)

15. Cachin, C., Micali, S., Stadler, M.A.: Computationally private information retrieval with polylogarithmic communication. In: Stern, J. (ed.) EUROCRYPT 1999. LNCS, vol. 1592, pp. 402–414. Springer, Heidelberg (1999)

16. Canetti, R.: Towards realizing random oracles: hash functions that hide all partial information. In: Kaliski Jr., B.S. (ed.) CRYPTO 1997. LNCS, vol. 1294, pp. 455–469. Springer, Heidelberg (1997)

17. Canetti, R., Goldreich, O., Halevi, S.: The random oracle methodology, revisited. J. ACM **51**(4), 557–594 (2004). http://doi.acm.org/10.1145/1008731.1008734

18. Cheon, J.H., Han, K., Lee, C., Ryu, H., Stehlé, D.: Cryptanalysis of the multilinear map over the integers. In: Oswald, E., Fischlin, M. (eds.) EUROCRYPT 2015. LNCS, vol. 9056, pp. 3–12. Springer, Heidelberg (2015)

19. Cheon, J.H., Lee, C., Ryu, H.: Cryptanalysis of the new CLT multilinear maps. IACR Cryptology ePrint Archive 2015, p. 934 (2015)
20. Chung, K., Lin, H., Pass, R.: Constant-round concurrent zero knowledge from p-certificates. In: 54th Annual IEEE Symposium on Foundations of Computer Science, FOCS 2013, 26–29 October, 2013, Berkeley, CA, USA, pp. 50–59 (2013)
21. Di Crescenzo, G., Lipmaa, H.: Succinct NP proofs from an extractability assumption. In: Beckmann, A., Dimitracopoulos, C., Löwe, B. (eds.) CiE 2008. LNCS, vol. 5028, pp. 175–185. Springer, Heidelberg (2008)
22. Damgård, I.B.: Towards practical public key systems secure against chosen ciphertext attacks. In: Feigenbaum, J. (ed.) CRYPTO 1991. LNCS, vol. 576, pp. 445–456. Springer, Heidelberg (1992)
23. Damgård, I., Faust, S., Hazay, C.: Secure two-party computation with low communication. In: Cramer, R. (ed.) TCC 2012. LNCS, vol. 7194, pp. 54–74. Springer, Heidelberg (2012)
24. Diffie, W., Hellman, M.E.: New directions in cryptography. IEEE Trans. Inf. Theory $22(6)$, 644–654 (1976)
25. Fiat, A., Shamir, A.: How to prove yourself: practical solutions to identification and signature problems. In: Odlyzko, A.M. (ed.) CRYPTO 1986. LNCS, vol. 263, pp. 186–194. Springer, Heidelberg (1987)
26. Fujisaki, E., Okamoto, T.: Statistical zero knowledge protocols to prove modular polynomial relations. In: Kaliski Jr., B.S. (ed.) CRYPTO 1997. LNCS, vol. 1294, pp. 16–30. Springer, Heidelberg (1997)
27. Garg, S., Gentry, C., Halevi, S., Raykova, M., Sahai, A., Waters, B.: Candidate indistinguishability obfuscation and functional encryption for all circuits. In: 54th Annual IEEE Symposium on Foundations of Computer Science, FOCS 2013, 26–29 October, 2013, Berkeley, CA, USA, pp. 40–49 (2013)
28. Gentry, C., Lewko, A.B., Sahai, A., Waters, B.: Indistinguishability obfuscation from the multilinear subgroup elimination assumption. IACR Cryptology ePrint Archive 2014, p. 309 (2014)
29. Gentry, C., Peikert, C., Vaikuntanathan, V.: Trapdoors for hard lattices and new cryptographic constructions. In: Proceedings of the 40th Annual ACM Symposium on Theory of Computing, 17–20 May 2008, Victoria, British Columbia, Canada, pp. 197–206 (2008)
30. Gentry, C., Wichs, D.: Separating succinct non-interactive arguments from all falsifiable assumptions. In: STOC, pp. 99–108 (2011)
31. Goldreich, O.: The Foundations of Cryptography: Basic Techniques, vol. 1. Cambridge University Press, Cambridge (2001)
32. Goldreich, O.: Randomness and computation. In: Goldreich, O. (ed.) Studies in Complexity and Cryptography. LNCS, vol. 6650, pp. 507–539. Springer, Heidelberg (2011)
33. Goldreich, O., Goldwasser, S.: On the limits of nonapproximability of lattice problems. J. Comput. Syst. Sci. $60(3)$, 540–563 (2000)
34. Goldreich, O., Levin, L.A.: A hard-core predicate for all one-way functions. In: Proceedings of the 21st Annual ACM Symposium on Theory of Computing, 14–17 May 1989, Seattle, Washigton, USA, pp. 25–32 (1989)
35. Goldwasser, S., Kalai, Y.T.: On the (in)security of the Fiat-Shamir paradigm. In: Proceedings of the 44th Symposium on Foundations of Computer Science (FOCS 2003), 11–14 October 2003, Cambridge, MA, USA, pp. 102–113 (2003)

36. Goldwasser, S., Kalai, Y.T., Peikert, C., Vaikuntanathan, V.: Robustness of the-learning with errors assumption. In: Proceedings of the Innovations in Computer Science, ICS 2010, 5–7 January 2010, Tsinghua University, Beijing, China, pp. 230–240 (2010). http://conference.itcs.tsinghua.edu.cn/ICS2010/content/papers/19.html

37. Goldwasser, S., Micali, S.: Probabilistic encryption. J. Comput. Syst. Sci. **28**(2), 270–299 (1984)

38. Heninger, N., Shacham, H.: Reconstructing RSA private keys from random key bits. In: Halevi, S. (ed.) CRYPTO 2009. LNCS, vol. 5677, pp. 1–17. Springer, Heidelberg (2009)

39. Impagliazzo, R., Jaiswal, R., Kabanets, V., Wigderson, A.: Uniform direct product theorems: simplified, optimized, and derandomized. SIAM J. Comput. **39**(4), 1637–1665 (2010)

40. Khot, S.: On the unique games conjecture. In: Proceedings of the 46th Annual IEEE Symposium on Foundations of Computer Science (FOCS 2005), 23–25 October 2005, Pittsburgh, PA, USA, p. 3 (2005)

41. Micciancio, D., Goldwasser, S.: Complexity of Lattice Problems: A Cryptographic Perspective, vol. 671. Springer Science & Business Media, New York (2012)

42. Minaud, B., Fouque, P.: Cryptanalysis of the new multilinear map over the integers. IACR Cryptology ePrint Archive 2015, p. 941 (2015)

43. Naor, M.: On cryptographic assumptions and challenges. In: Boneh, D. (ed.) CRYPTO 2003. LNCS, vol. 2729, pp. 96–109. Springer, Heidelberg (2003)

44. Paillier, P.: Public-key cryptosystems based on composite degree residuosity classes. In: Stern, J. (ed.) EUROCRYPT 1999. LNCS, vol. 1592, pp. 223–238. Springer, Heidelberg (1999)

45. Pass, R., Seth, K., Telang, S.: Indistinguishability obfuscation from semantically-secure multilinear encodings. In: Garay, J.A., Gennaro, R. (eds.) CRYPTO 2014, Part I. LNCS, vol. 8616, pp. 500–517. Springer, Heidelberg (2014)

46. Peikert, C.: Public-key cryptosystems from the worst-case shortest vector problem: extended abstract. In: Proceedings of the 41st Annual ACM Symposium on Theory of Computing, STOC 2009, May 31 - June 2, 2009, Bethesda, MD, USA, pp. 333–342 (2009)

47. Rabin, M.O.: Digitalized signatures and public-key functions as intractable as factorization. Technical report, MIT Laboratory for Computer Science (1979)

48. Regev, O.: On lattices, learning with errors, random linear codes, and cryptography. In: Proceedings of the 37th Annual ACM Symposium on Theory of Computing, 22–24 May 2005, Baltimore, MD, USA, pp. 84–93 (2005)

49. Rivest, R.L., Shamir, A., Adleman, L.M.: A method for obtaining digital signatures and public-key cryptosystems. Commun. ACM **21**(2), 120–126 (1978)

50. Shoup, V.: New algorithms for finding irreducible polynomials over finite fields. In: 29th Annual Symposium on Foundations of Computer Science, 24–26 October 1988, White Plains, New York, USA, pp. 283–290 (1988)

51. Shoup, V.: Searching for primitive roots in finite fields. In: Proceedings of the 22nd Annual ACM Symposium on Theory of Computing, 13–17 May 1990, Baltimore, Maryland, USA, pp. 546–554 (1990)

52. Shoup, V.: Lower bounds for discrete logarithms and related problems. In: Fumy, W. (ed.) EUROCRYPT 1997. LNCS, vol. 1233, pp. 256–266. Springer, Heidelberg (1997)

53. Stehlé, D., Steinfeld, R., Tanaka, K., Xagawa, K.: Efficient public key encryption based on ideal lattices. In: Matsui, M. (ed.) ASIACRYPT 2009. LNCS, vol. 5912, pp. 617–635. Springer, Heidelberg (2009)

54. Wee, H.: On obfuscating point functions. In: Proceedings of the 37th Annual ACM Symposium on Theory of Computing, 22–24 May 2005, Baltimore, MD, USA, pp. 523–532 (2005). http://doi.acm.org/10.1145/1060590.1060669

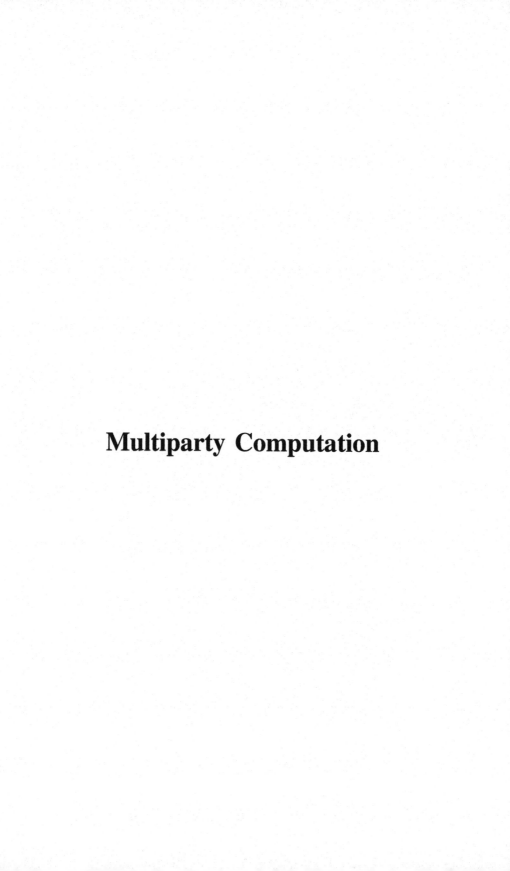

Multiparty Computation

Adaptive Security with Quasi-Optimal Rate

Brett Hemenway[1]([✉]), Rafail Ostrovsky[2], Silas Richelson[3], and Alon Rosen[4]

[1] University of Pennsylvania, Philadelphia, USA
fbrett@cis.upenn.edu
[2] UCLA, Los Angeles, USA
rafail@cs.ucla.edu
[3] MIT, Cambridge, USA
sirichel@csail.mit.edu
[4] Herzliya Interdisciplinary Center, Herzliya, Israel
alon.rosen@idc.ac.il

Abstract. A multiparty computation protocol is said to be adaptively secure if it retains its security in the presence of an adversary who can adaptively corrupt participants as the protocol proceeds. This is in contrast to a static corruption model where the adversary is forced to choose which participants to corrupt before the protocol begins. A central tool for constructing adaptively secure protocols is non-committing encryption (Canetti, Feige, Goldreich and Naor, STOC '96). The original protocol of Canetti et al. had ciphertext expansion $\mathcal{O}(k^2)$ where k is the security parameter, and prior to this work, the best known constructions had ciphertext expansion that was either $\mathcal{O}(k)$ under general assumptions, or alternatively $\mathcal{O}(\log(n))$, where n is the length of the message, based on a specific factoring-based hardness assumption.

In this work, we build a new non-committing encryption scheme from lattice problems, and specifically based on the hardness of (Ring) Learning With Errors (LWE). Our scheme achieves ciphertext expansion as small as $\mathrm{polylog}(k)$. Moreover when instantiated with Ring-LWE, the public-key is of size $\mathcal{O}(n\,\mathrm{polylog}(k))$. All previously proposed schemes had public-keys of size $\Omega(n^2\mathrm{polylog}(k))$.

R. Ostrovsky—Work done in part while visiting Simons Institute in Berkeley and supported in part by NSF grants 09165174, 1065276, 1118126 and 1136174, US-Israel BSF grant 2008411, OKAWA Foundation Research Award, IBM Faculty Research Award, Xerox Faculty Research Award, B. John Garrick Foundation Award, Teradata Research Award, and Lockheed-Martin Corporation Research Award. This material is based upon work supported in part by DARPA Safeware program. The views expressed are those of the author and do not reflect the official policy or position of the Department of Defense or the U.S. Government.

S. Richelson—Part of this work done while visiting IDC Herzliya, supported by the European Research Council under the European Unions Seventh Framework Programme (FP 2007–2013), ERC Grant Agreement n. 307952.

A. Rosen—Efi Arazi School of Computer Science, IDC Herzliya, Israel. Work supported by ISF grant no.1255/12 and by the ERC under the EU's Seventh Framework Programme (FP/2007–2013) ERC Grant Agreement n. 307952. Work in part done while the author was visiting the Simons Institute for the Theory of Computing, supported by the Simons Foundation and by the DIMACS/Simons Collaboration in Cryptography through NSF grant #CNS-1523467.

Keywords: Adaptive security · Non-committing encryption · LWE · Ring-LWE

1   Introduction

Secure multiparty computation (MPC) allows a group of players to compute any joint function of their private inputs, even when some of the players are adversarial [GMW87, BGW88, CCD88]. MPC protocols are often categorized based on the security properties they offer. One natural and well-studied distinction is between protocols which are secure against an *adaptive* adversary, and those which are only secure when the adversary is *static*. A static adversary is one who chooses which parties to corrupt before the protocol begins, while an adaptive adversary can choose which parties to corrupt on the fly, and thus the corruption pattern may depend on the messages exchanged during the protocol. Adaptive security is desirable as it models real-world adversarial behavior more honestly. Unfortunately, adaptively secure protocols are significantly harder to construct as several techniques from the literature for proving security in the static model do not seem to carry over to the adaptive model.

Adaptively Secure MPC. The information-theoretic MPC protocol of [BGW88] is adaptively secure when each pair of parties is connected by a secure channel (so communication between honest parties may not be observed by the adversary). Roughly, this is because for a security threshold of t, the views of any t players are statistically independent of the secret inputs of all other parties. Thus the information the adversary obtains from corrupting fewer than t parties can be simulated and so adaptively choosing new parties to corrupt does not provide an advantage. One might hope to obtain an adaptively secure protocol in the plain model (*i.e.* without ideal channels) by using semantically secure encryption. Specifically, each pair of parties might first exchange public keys and then communicate "privately" by publicly broadcasting encryptions of their messages. The ideal adversary might then be able to emulate a real interaction by broadcast encryptions of zero, and prove indistinguishability using semantic security. However, as pointed out in [CFGN96], this intuition does not exactly work. Essentially the problem is that the ciphertext in ordinary encryption "commits" the sender to one message. An ideal adversary, therefore, would be unable to open an encryption of zero to anything except zero. This is problematic for adaptive security because if a party is corrupted after it has already sent encryptions, then upon learning the secret keys and previously used randomness, the adversary will be able to tell if it is in the ideal or real world based on whether the entire communication is encryptions of zero or not. [CFGN96] goes on to define and construct a stronger type of encryption called *non-committing encryption* (NCE) which allows the above intuition to go through.

Non-Committing Encryption. An encryption scheme is non-committing if a simulator can geterate a public key/ciphertext pair that is indistinguishable from a real public key/ciphertext, but for which it can later produce a secret key/encryption randomness pair which "explains" the ciphertext as an encryption of *any adversarially chosen message*. This provides a natural method for creating adaptively secure MPC protocols: first design a statically secure protocol in the private channels model, then instantiate the private channels with NCE.

Prior Work on NCE. The original work of [CFGN96] gives an NCE protocol based on the existence of a special type of trapdoor permutations and is relatively inefficient: requiring a sender to send a ciphertext of size $\mathcal{O}(k^2)$ to encrypt a single bit (for security parameter k). In other words, its *ciphertext expansion factor* is $\mathcal{O}(k^2)$. Choi, Dachman-Soled, Malkin and Wee [CDSMW09] construct an NCE protocol with ciphertext expansion $\mathcal{O}(k)$ starting from any *obviliously-samplable cryptosystem*. The paradigm of using obliviously-samplable encryption to achieve adaptive security goes back to [DN00] who give a three-round protocol which adaptively, securely realizes the ideal message transmission functionality (thus, allowing for adaptively secure MPC). [DN00] show how to instantiate obvliviously-samplable encryption based on a variety of assumptions, building on an earlier work of Beaver [Bea97], which essentially constructs obvliviously-samplable encryption assuming DDH. Very recently, Hemenway, Ostrovsky and Rosen [HOR15] construct an NCE protocol with logarithmic expansion based on the Φ-hiding assumption, which is related to (though generally believed to be stronger than) RSA.

See Fig. 1 for a comparison of past and current work on NCE.

Reference	Rounds	CT Expansion	Public-Key Size	Assumption
[CFGN96]	2	$\mathcal{O}(k^2)$	$\mathcal{O}(n^3\mathrm{polylog}(k))$	Common-Domain TDPs
[Bea97]	3	$\mathcal{O}(k)$	$\mathcal{O}(n^2\mathrm{polylog}(k))$	DDH
[DN00]	3	$\mathcal{O}(k)$	$\mathcal{O}(n^2\mathrm{polylog}(k))$	Oblivious samplable PKE
[CDSMW09]	2	$\mathcal{O}(k)$	$\mathcal{O}(n^2\mathrm{polylog}(k))$	Oblivious samplable PKE
[HOR15]	2	$\mathcal{O}(\log n)$	$\mathcal{O}(n^2\mathrm{polylog}(k))$	Φ-hiding
This work	2	$\mathrm{polylog}(k)$	$\mathcal{O}(n^2\mathrm{polylog}(k))$	LWE
This work	2	$\mathrm{polylog}(k)$	$\mathcal{O}(n\ \mathrm{polylog}(k))$	Ring-LWE

Fig. 1. Comparison to prior work. The parameter k denotes the security parameter, and n denotes the message length.

1.1 Our Contribution

In this work, we construct an NCE scheme with polylogarithmic ciphertext expansion and which improves upon the recent work of [HOR15] in a number of ways.

Assumption: Learning with errors (LWE) [Reg09] is known to be as hard as worst-case lattice problems, and is widely accepted as a cryptographic hardness assumption. Its ring variant is as hard as worst-case problems on ideal lattices and is widely used in practice as it allows representing a vector consicely as a single ring element. For comparison, the Φ−hiding assumption is not as widely used or accepted, and certain choices of parameters must be carefully avoided as they are susceptible to polynomial time attacks.

Smaller Public Keys: When instantiated with Ring-LWE, the public-key of our scheme is of size $\mathcal{O}(n\mathrm{polylog}(k))$. All previously proposed schemes had public-keys of size $\Omega(n^2\mathrm{polylog}(k))$.

No Sampling Issues: One subtle shortcoming of the [HOR15] work is that the non-committing property of their encryption scheme necessitates the existence of a public modulus N whose factorization is not known. This means that in order to attain full simulatability the modulus N will have to be sampled jointly by the parties, using a secure protocol (which itself needs to be made adaptively secure). Our current work, in contrast, does not suffer from this shortcoming and does not necessitate joint simulation of the public parameters.

1.2   Our Construction

For this high-level description of our protocol we assume familiarity with the definition of non-committing encryption as well as Regev's LWE-based encryption scheme [Reg09], and Micciancio-Peikert trapdoors for LWE [MP12]. We refer the reader to Sect. 2 or the papers themselves for more information on these topics.

KeyGen. Let q, m, k be LWE parameters and n a parameter linear in the message length ℓ. The receiver chooses a random subset $I_R \subset \{1,\ldots,n\}$ of size $n/8$, a matrix $\mathbf{A} \in \mathbb{Z}_q^{m \times k}$ and vectors $\mathbf{v}_1,\ldots,\mathbf{v}_n \in \mathbb{Z}_q^m$, where $\mathbf{v}_i$ is an LWE instance if $i \in I_R$ and is random otherwise. The public key is $(\mathbf{A}, \mathbf{v}_1,\ldots,\mathbf{v}_n)$ and the secret key is $(I_R, \{\mathbf{s}_i\}_{i \in I_R})$ where $\mathbf{s}_i \in \mathbb{Z}_q^k$ is the LWE secret for $\mathbf{v}_i$. So receiver has generated n Regev public keys, but for which it only knows $n/8$ of the corresponding secret keys.

Enc. Let $\mathsf{msg} \in \{0,1\}^\ell$ be a plaintext, and let $y = (y_1,\ldots,y_n) = \mathbf{ECC}(\mathsf{msg}) \in \{0,1\}^n$ be the image of msg under a suitable error-correcting code. The sender chooses a random subset $I_S \subset \{1,\ldots,n\}$ of size $n/8$ and generates Regev encryptions, under public key $(\mathbf{A}, \mathbf{v}_i)$ of y_i if $i \in I_S$ or of a random bit if $i \notin I_S$. Important for the efficiency of our protocol is that these encryptions be generated using shared randomness. Specifically, the sender chooses a random short $\mathbf{r} \in \mathbb{Z}_q^m$ and constructs the ciphertext $(\mathbf{u}, w_1,\ldots,w_n)$ where $\mathbf{u} = \mathbf{r}^{\mathsf{t}}\mathbf{A} \in \mathbb{Z}_q^k$ and $w_i = \mathbf{r}^{\mathsf{t}}\mathbf{v}_i + e_i + (q/2)z_i$ where $z_i = y_i$ if $i \in I_S$ and is random otherwise (the e_i are short Gaussian errors). So sender has encrypted a string $z \in \{0,1\}^n$ which agrees with y in $9n/16$ of the positions on expectation. The encryption randomness consists of $\mathbf{r}, I_S$ as well as the Gaussian errors.

Dec. Given a ciphertext $(\mathbf{u}, w_1,\ldots,w_n)$ and secret key $(I_R, \{\mathbf{s}_i\}_{i \in I_R})$, receiver constructs $y' \in \{0,1\}^n$ by setting $y'_i = \lfloor (2/q)(w_i - \mathbf{u}^{\mathsf{t}}\mathbf{s}_i) \rceil$ if $i \in I_R$ and y'_i to be a random bit otherwise. Then receiver decodes y_i' and outputs $\mathsf{msg}' \in \{0,1\}^\ell$. So receiver is decrypting the ciphertexts for which he knows the secret key, and is completing this to a string in $\{0,1\}^n$ by filling in the remaining positions randomly.

Correctness. Let $y = \mathbf{ECC}(\mathsf{msg}) \in \{0,1\}^n$ be the coded message and let $y' \in \{0,1\}^n$ be the string obtained during decryption. Note that whenever $i \in I_R \cap I_S$, $y_i = y_i'$ with high probability, and whenever $i \notin I_R \cap I_S$ $y_i = y_i'$ with probability $1/2$. Therefore, $y_i = y_i'$ for $65n/128$ of the values $i \in \{1, \ldots, n\}$ on expectation. It can be shown using the tail bound (Lemma 1), that there exists a constant $\delta > 0$ such that $y_i = y_i'$ for at least $(1/2 + \delta)n$ values of i with high probability. By our choice of error correcting code, we can decode given such a tampered codeword.

Adversary's Real World View. The non-committing adversary receives the secret key and encryption randomness and tries to use these values to distinguish the real and ideal worlds. The most difficult aspects of the real world view to simulate are the subsets $I_R, I_S \subset \{1, \ldots, n\}$. In the real world, both sets are random of size $n/8$ and so the size of their intersection is a hypergeometric random variable. This must be replicated in the ideal world in order for indistinguishability to hold. To complicate matters, it is important that the right number of $i \in I_R \cap I_S$ are such that $y_i = 0$ and $y_i' = 0$. Likewise, we must make sure that the right number of $i \in I_R \cap \overline{I_S}$ are such that $y_i = 0$ and $y_i' = 1$, and so on. This involves carefully computing the multivariate hypergeometric distribution which arises from the real world execution so that we may emulate it in the ideal world. We will leave out most of the details for this overview; the specifics are given in Sect. 3.

Simulating the Public Key and Ciphertext. The simulator chooses a partition $I_{\mathsf{good}} \cup I_{\mathsf{bad}} = \{1, \ldots, n\}$ at random, and chooses vectors $\mathbf{v}_1, \ldots, \mathbf{v} \in \mathbb{Z}_q^m$ so it knows an LWE secret $\mathbf{s}_i$ for $\mathbf{v}_i$ whenever $i \in I_{\mathsf{good}}$, and so that it knows an MP trapdoor for the matrix $\hat{\mathbf{A}} = [\mathbf{A}|\mathbf{V}]$ where the columns of $\mathbf{V}$ are $\{\mathbf{v}_i\}_{i \in I_{\mathsf{bad}}}$. The sets I_{good} and I_{bad} correspond to the $i \in \{1, \ldots, n\}$ for which it knows/doesn't know a secret key. It is important for simulation that I_{good} is much larger than I_R. A good choice, for example is $|I_{\mathsf{good}}| = 3n/4$. The simulator then sets its public key to $(\mathbf{A}, \mathbf{v}_1, \ldots, \mathbf{v}_n)$. It further partitions I_{good} into $I_{\mathsf{good},0} \cup I_{\mathsf{good},1}$ and sets the ciphertext $(\mathbf{u}, w_1, \ldots, w_n)$ to be so that $\mathbf{u} = \mathbf{r}^t \mathbf{A}$ and w_i is a valid Regev encryption of 0 (resp. 1) if $i \in I_{\mathsf{good},0}$ (resp. $i \in I_{\mathsf{good},1}$), and w_i is random if $i \in I_{\mathsf{bad}}$.

Simulating the Secret Key and Decryption Randomness. Upon receiving msg, the simulator sets $(y_1, \ldots, y_n) = \mathbf{ECC}(\mathsf{msg}) \in \{0,1\}^n$ and must produce $I_R, I_S \subset \{1, \ldots, n\}$, string $y' \in \{0,1\}^n$ and short $\mathbf{r} \in \mathbb{Z}_q^m$ which look like the quantities which arise from a real world execution (we ignore the Gaussian errors in this discussion). The simulator must choose $I_R \subset I_{\mathsf{good}}$ since the $\mathbf{v}_i$ for $i \in I_{\mathsf{bad}}$ are "lossy" and so have no secret keys (this is why we chose I_{good} much larger than I_R). The simulator must also choose $y_i' = b$ for $i \in I_{\text{good},b}$, since these w_i are valid encryptions of b. More subtly, the simulator must make sure to choose y' so that the number of i for which $y_i' = 0$ and $i \in I_R$ is as in the real world. As mentioned above, this more delicate than one might think; see the paragraph below for an example. Finally, the simulator sets I_S to be a subset of $\{i : y_i = y_i'\}$ of appropriate size and so the sizes of its intersections with the sets chosen so far are distributed as in the real world. Finally, the simulator sets y_i' for $i \in I_{\text{bad}}$ to be as needed to complete the real world view. Note the simulator has a trapdoor for $\hat{\mathbf{A}}$ and so may choose short $\mathbf{r} \in \mathbb{Z}_q^m$ so that $\mathbf{r}^t \hat{\mathbf{A}}$ is as he chooses.

We conclude this discussion with an example which illustrates the care required to make the above proof go through. For ease of this discussion, we let $y' = y$, even though this is not true for our construction (which makes it even more complicated). Consider only the choice of I_R and the number of $i \in I_R$ such that $y_i = 0$. In the real world, I_R is chosen randomly from $\{1, \ldots, n\}$, and since $y_i = 0$ for exactly $n/2$ of the $i \in \{1, \ldots, n\}$ (**ECC** is balanced), $\#\{i \in I_R : y_i = 0\}$ is distributed according to the hypergeometric distribution $\mathsf{H}(\frac{n}{8}, \frac{n}{2}, n)$. In the ideal world, I_R is chosen randomly from I_{good} which is itself partitioned into $I_{\text{good},0}$ and $I_{\text{good},1}$ of equal size so that $y_i = b$ for all $i \in I_{\text{good},b}$. Therefore, in the ideal world $\#\{i \in I_R : y_i = 0\}$ is distributed according to the hypergeometric distribution $\mathsf{H}(\frac{n}{8}, \frac{3n}{8}, \frac{3n}{4})$. While the expectations are equal, the random variables themselves are not and so I_R must be chosen in the ideal world carefully in order to emulate the real world successfully. Details are in Sect. 3.

2   Preliminaries

2.1   Notation

If A is a Probabilistic Polynomial Time (PPT) machine, then we use $a \xleftarrow{\$} A$ to denote running the machine A and obtaining an output, where a is distributed according to the internal randomness of A. If R is a set, we use $r \xleftarrow{\$} R$ to denote sampling uniformly from R. If R and X are sets then we use the notation $\Pr_{r,x}[A(x,r) = c]$ to denote the probability that A outputs c when x is sampled uniformly from X and r is sampled uniformly from R. A function is said to be *negligible* if it vanishes faster than the inverse of any polynomial. For simplicity, we often suppress random inputs to functions. In such cases, we use a semicolon to separate optional random inputs. Thus $c \xleftarrow{\$} \mathsf{Enc}(pk, m)$ and $c = \mathsf{Enc}(pk, m; r)$ both indicate an encryption of m under the public key pk, but in the first case, we consider Enc as a randomized algorithm, and in the second we consider Enc as a deterministic algorithm depending on the randomness r.

2.2   Non-Committing Encryption

Non-committing encryption was introduced by Canetti, Feige, Goldreich and Naor in [CFGN96] as a primitive which allows one compile a protocol which is adaptively secure as long as all pairs of parties are connected with a secure channel, into an adaptively secure protocol in the plain model. The following definition is from [DN00] and is consistent with this viewpoint.

Definition 1 (Non-Committing Encryption). *We say that a two party protocol Π is a* non-committing encryption scheme *if it adaptively, securely realizes the message transmission functionality:*

$$f(m, \perp) = (\perp, m).$$

The following indistinguishability based definition is sufficient and easier to work with. In our proof of security we will this second definition in its game form.

Definition 2. *A cryptosystem $\mathcal{PKE} = (\mathsf{Gen}, \mathsf{Enc}, \mathsf{Dec})$ is called non-committing, if there exists a PPT simulator $\mathsf{Sim} = (\mathsf{Sim}_1, \mathsf{Sim}_2)$ with the following properties:*

1. **Efficiency:** *The algorithms $\mathsf{Gen}, \mathsf{Enc}, \mathsf{Dec}$ and Sim are all PPT.*
2. **Correctness:** *For any message $m \in \mathcal{M}(\mathsf{pp})$*

$$\Pr\left[\mathsf{Dec}(sk, c) = m : (pk, sk) \overset{\$}{\leftarrow} \mathsf{Gen}(1^k), c \overset{\$}{\leftarrow} \mathsf{Enc}(pk, m)\right] = 1 - \mathsf{negl}$$

3. **Simulatability:** *For any PPT adversary $\mathcal{A}$, the distributions Λ^{Ideal} and Λ^{Real} are computationally indistinguishable where*

$$\Lambda^{\mathrm{Ideal}} = \{(m, pk, c, r_1, r_2) : (pk, c, t) \overset{\$}{\leftarrow} \mathsf{Sim}_1(1^k), m \overset{\$}{\leftarrow} \mathcal{A}(pk), (sk, r_1, r_2) \overset{\$}{\leftarrow} \mathsf{Sim}_2(m, t)\}$$

and

$$\Lambda^{\mathrm{Real}} = \{(m, pk, c, r_1, r_2) : (pk, sk) \overset{\$}{\leftarrow} \mathsf{Gen}(\mathsf{pp}; r_1), m \overset{\$}{\leftarrow} \mathcal{A}(pk), c \overset{\$}{\leftarrow} \mathsf{Enc}(pk, m; r_2)\}$$

Note that semantic security follows from simulatability.

2.3 Learning with Errors

The learning with errors (LWE) problem [Reg09] is specified by the security parameter k, a modulus q and an error distribution χ over $\mathbb{Z}_q$. In this paper our errors will be drawn exclusively from discrete Gaussians. We specify the discrete Gaussian with standard deviation σ by χ_σ. In its decisional form, the problem asks one to distinguish, for a random $\mathbf{s} \in \mathbb{Z}_q^k$, between the distribution $\mathsf{A}_{\mathbf{s},\chi}$ from $\mathsf{Unif}(\mathbb{Z}_q^k \times \mathbb{Z}_q)$ where $\mathsf{A}_{\mathbf{s},\chi} = \{(\mathbf{a}, \langle \mathbf{a}, \mathbf{s} \rangle + e)\}_{\mathbf{a} \leftarrow \mathbb{Z}_q^k, e \leftarrow \chi}$. Several important results [Reg09, Pei09, BLP13] establish the hardness of decisional LWE based on worst-case lattice problems. The following fact is standard.

Fact 1. Let $q = k^{\omega(1)}$ be superpolynomial in the security parameter k. Let $B, \sigma < q$ be such that $\sigma/B = k^{\omega(1)}$, and let χ be a $B-$bounded distribution. Then with high probability over $e \leftarrow \chi$, we have $\chi_\sigma \approx_{\mathsf{s}} e + \chi_\sigma$.

Ring LWE. The ring variant of LWE [LPR13] is often used in practice as it allows representing vectors succinctly as ring elements. The vectors in the usual LWE problem formulated above are replaced by elements in the quotient ring $R = \mathbb{Z}[x]/\Phi(x)$ for an irreducible cyclotomic polynomial Φ (often $\Phi(x) = x^\ell + 1$ for ℓ a power of 2). Cryptographic schemes instantiated using ring LWE often are considerably more efficient than the corresponding constructions over ordinary LWE. If we instantiate our basic NCE scheme (which is based on LWE) on top of ring LWE instead, we can shrink the public key size from $\tilde{\mathcal{O}}(k^2)$ to $\tilde{\mathcal{O}}(k)$. Finally, we remark that the hardness of ring LWE can be based on the worst-case hardness of lattice problems on ideal lattices.

Trapdoors for LWE. Micciancio and Peikert [MP12] show how to embed a trapdoor into a matrix $\mathbf{A} \in \mathbb{Z}_q^{m \times k}$ which allows solving several tasks which are usually believed to be hard. In their construction, the trapdoor of $\mathbf{A}$ is a matrix $\mathbf{T} \in \{0,1\}^{(k \log q) \times m}$ such that $\mathbf{TA} = \mathbf{G}$, where $\mathbf{G} \in \mathbb{Z}_q^{(k \log q) \times k}$ is the so-called "gadget matrix". To be precise, [MP12] shows (among other things) how to sample the pair $(\mathbf{A}, \mathbf{T})$ in such a way so that (1) $\mathbf{A}$ is statistically close to uniform in $\mathbb{Z}_q^{m \times k}$ and (2) there is an efficient algorithm Sample_σ which takes as input the tuple $(\mathbf{u}, \mathbf{A}, \mathbf{T})$ where $\mathbf{u} \in \mathbb{Z}_q^k$ is arbitrary and outputs a vector $\mathbf{r} \in \mathbb{Z}_q^m$ from a distribution which is statistically close to $D_{\sigma, \mathbf{u}, \mathbf{A}}$, the discrete Gaussian of standard deviation σ on the lattice

$$\Lambda_{\mathbf{u}}(\mathbf{A}) = \{\mathbf{v} \in \mathbb{Z}_q^m : \mathbf{v}^t \mathbf{A} = \mathbf{u}^t\}.$$

Their construction carries over to the ring setting as well.

2.4   Error Correcting Codes

Our construction makes use of constant-rate binary codes which are uniquely and efficiently decodeable from a $(1/2 - \delta)$−fraction of computationally bounded errors. Such codes are constructed in [MPSW05] by using efficient list-decodeable codes along with computationally secure signatures. We will further assume that our codes are *balanced*, in the sense that exactly half of the bits of all codewords are 0 s and the other half are 1s. This can be arranged, for example, by concatenating a list decodeable code with a suitable binary error correcting code.

2.5   The Binomial and Hypergeometric Distributions

Binomial Distribution. A binomial random variable $X_{p,n}$ is equal to the number of successes when an experiment with success probability p is independently repeated n times. The density function is given by $\Pr(X_{p,n} = k) = \binom{n}{k} p^k (1 - p)^{n-k}$ with expectation $\mathbb{E}[X_{p,n}] = pn$. We denote the binomial distribution by $\mathsf{B}_p(n)$.

Hypergeometric Distribution. Our construction involves the randomized process: given $a, b < n$ independently choose random subsets $A, B \subset \{1, \ldots, n\}$ of sizes a

and b. We will be interested in the size of the intersection $A \cap B$, which defines a hypergeometric random variable $X_{a,b,n}$ with density function $\Pr(X_{a,b,n} = k) = \binom{b}{k}\binom{n-b}{a-k}/\binom{n}{a}$ and expectation $\mathbb{E}[X_{a,b,n}] = ab/n$. We denote the hypergeometric distribution of $X_{a,b,n}$ by $\mathsf{H}(a,b,n)$.

Multivariate Hypergeometric Distribution. We will also use a variant of the above process when $\{1, \ldots, n\}$ has been partitioned $\{1, \ldots, n\} = B_1 \cup \cdots \cup B_t$ where $|B_i| = b_i$. Then if $A \subset \{1, \ldots, n\}$ of size a is chosen randomly, independent of the partition, the tuple $(|A \cap B_1|, \ldots, |A \cap B_t|)$ is a multivariate hypergeometric random variable $X_{a,\{b_i\},n}$ with density function

$$\Pr(X_{a,\{b_i\},n} = (k_1, \ldots, k_t)) = \frac{\binom{b_1}{k_1} \cdots \binom{b_t}{k_t}}{\binom{n}{a}},$$

where $k_1 + \cdots + k_t = a$. The expectation is $(ab_1/n, \ldots, ab_t/n)$. We denote the multivariate hypergeometric distribution $\mathsf{H}^t(a, \{b_1, \ldots, b_t\}, n)$. Note that the single variable hypergeometric distribution $\mathsf{H}(a, b, n)$ is the same as $\mathsf{H}^2(a, \{b, n - b\}, n)$ corresponding to the partition $\{1, \ldots, n\} = B \cup \overline{B}$. We will make use of the following tail bounds on $\mathsf{B}_p(n)$, $\mathsf{H}(a, b, n)$ and $\mathsf{H}^t(a, \{b_i\}, n)$ proved by Hoeffding [Hoe63].

Lemma 1 (Tail Bounds). *Let $\alpha, \beta, \epsilon \in (0, 1)$ be constants. Also for a constant t, choose constants $\beta_1, \ldots, \beta_t \in (0, 1)$ such that $\beta_1 + \cdots + \beta_t = 1$. Set $a = \alpha n$, $b = \beta n$ and $b_i = \beta_i n$.*

1. *Let X be a random variable drawn either from $\mathsf{B}_{\alpha\beta}(n)$ or $\mathsf{H}(a, b, n)$. Then*

$$Pr(X \geq (\alpha\beta + \epsilon)n \ OR \ X \leq (\alpha\beta - \epsilon)n) = e^{-\Omega(n)}$$

2. *Let $(X_1, \ldots, X_t)$ be a random variable drawn from $\mathsf{H}^t(a, \{b_1, \ldots, b_t\}, n)$. Then*

$$Pr(\exists \ i \ st \ X_i \geq (\alpha\beta_i + \epsilon)n \ OR \ X_i \leq (\alpha\beta_i - \epsilon)n) = e^{-\Omega(n)}.$$

The constants hidden by Ω depend quadratically on ϵ.

3   Non-Committing Encryption from LWE

3.1   The Basic Scheme

Params. Our scheme involves the following parameters:

- integers k, n, q and $m > 2k \log q$;
- real numbers σ, σ' such that $2\sqrt{k} < \sigma < \sigma' < q/\sqrt{k}$ and such that $\sigma^2/\sigma' = \mathsf{negl}(k)$;
- integers $c_R, c_S \leq n$ and $\delta \in (0, 1)$ such that $\delta < c_R c_S/2n^2$. To be concrete, we set $c_R = c_S = n/8$.

KeyGen. Draw $\mathbf{A} \xleftarrow{\$} \mathbb{Z}_q^{m \times k}$ and let $I_R \subset \{1, \ldots, n\}$ be a random subset of size c_R. Define vectors $\mathbf{v}_i \in \mathbb{Z}_q^m$ for $i = 1, \ldots, n$:

$$\mathbf{v}_i = \begin{cases} \mathbf{As}_i + \mathbf{e}_i, & i \in I_R \\ \text{uniform in } \mathbb{Z}_q^m, & i \notin I_R \end{cases}$$

where $\mathbf{s}_i \xleftarrow{\$} \mathbb{Z}_q^k$ and $\mathbf{e}_i \xleftarrow{\$} \chi_\sigma^m$. Output $(\mathsf{pk}, \mathsf{sk}) = ((\mathbf{A}, \mathbf{v}_1, \ldots, \mathbf{v}_n); \{\mathbf{s}_i\}_{i \in I_R})$.

Encryption. Given $\mathsf{msg} \in \{0, 1\}^\ell$ and $\mathsf{pk} = (\mathbf{A}, \mathbf{v}_1, \ldots, \mathbf{v}_n)$, let $y = (y_1, \ldots, y_n) = \mathbf{ECC}(\mathsf{msg}) \in \{0, 1\}^n$ where $\mathbf{ECC}$ is a balanced binary error-correcting code with constant rate, which is uniquely decodeable from a $(1/2 - \delta)$−fraction of computationally bounded errors, as described in Sect. 2.3. Choose a random subset $I_S \subset \{1, \ldots, n\}$ of size c_S. Also, for each $i \notin I_S$, choose a random bit $z_i \leftarrow \{0, 1\}$. Finally, choose $\mathbf{r} \leftarrow \chi_\sigma^m$ and $e'_1, \ldots, e'_n \leftarrow \chi_{\sigma'}$. Output ciphertext $\mathsf{ct} = (\mathbf{u}, w_1, \ldots, w_n)$ where $\mathbf{u}^t = \mathbf{r}^t \mathbf{A} \in \mathbb{Z}_q^{1 \times k}$, and $w_i \in \mathbb{Z}_q$ is given by

$$w_i = \begin{cases} \mathbf{r}^t \mathbf{v}_i + e'_i + (q/2) y_i, & i \in I_S \\ \mathbf{r}^t \mathbf{v}_i + e'_i + (q/2) z_i, & i \notin I_S \end{cases}.$$

The encryption randomness is $\left(I_S, \{z_i\}_{i \notin I_S}, \mathbf{r}, e'_1, \ldots, e'_n\right)$.

Decryption. Given $\mathsf{ct} = (\mathbf{u}, w_1, \ldots, w_n)$ and $\mathsf{sk} = \{\mathbf{s}_i\}_{i \in I_R}$, set

$$y'_i = \left\lfloor \frac{2(w_i - \mathbf{u}^t \mathbf{s}_i)}{q} \right\rceil$$

for all $i \in I_R$, and extend to a string $y' \in \{0, 1\}^n$ via $y'_i \xleftarrow{\$} \{0, 1\}$ when $i \notin I_R$. Output $\mathsf{msg}' \in \{0, 1\}^\ell$ obtained by applying the decoding algorithm of $\mathbf{ECC}$ to y'.

3.2   Correctness and Real World Subsets

Correctness. Let $y' = (y'_1, \ldots, y'_n) \in \{0, 1\}^n$ be the faulty codeword obtained during decryption. We must show that the decoding algorithm correctly outputs msg with overwhelming probability. We have:

- $i \in I_R \cap I_S$: then $y'_i = y_i$. The number of such i is $k \leftarrow \mathsf{H}(c_R, c_S, n)$.
- $i \notin I_R \cap I_S$: then $y'_i = y_i$ with probability $1/2$ independently of all other i.

It follows that the codeword y' has k' errors and $n - k'$ correct symbols where $k' \leftarrow \mathsf{B}_{1/2}(n - k)$. Fix a constant $\epsilon > 0$ with $3\epsilon < c_R c_S / n^2 - 2\delta$. We have, by Lemma 1, that with all but negligible probability in n,

$$k' \leq (1/2 + \epsilon)(n - k) \leq n(1/2 + \epsilon)(1 + \epsilon - c_R c_S / n^2) \leq n(1/2 - \delta).$$

In this case, the fraction of errors in the faulty codeword y' is less than $1/2 - \delta$ and so y' decodes correctly and decryption succeeds.

Efficiency. The ciphertext size of our scheme is $(n + k) \log q = \mathcal{O}(n\text{polylog}(k))$, while the public key is of size $m(k+n) \log q = \mathcal{O}(k^2\text{polylog}(k))$. We remark that when this construction is instantiated using a ring-LWE based encryption scheme as described in [LPR13], the public key size can be reduced to $\mathcal{O}(k\text{polylog}(k))$. This requires using the ring based version of the trapdoors from [MP12].

Real World Subsets. The most technically delicate issue with our construction is that the faulty codeword y' produced in simulation must have the same distribution of errors as in the real world. In particular, the adversary learns four sets

- ECC_0 : The set of coordinates where the codeword is 0.
- $\overline{I_S}$: The set of coordinates "honestly" generated by the sender.
- $\overline{D_0}$: The set of coordinates outside of I_S that "randomly" encrypt a 0.
- $\overline{I_R}$: The set of coordinates where the receiver has the decryption key.

These sets also define their complements, $\overline{\mathsf{ECC}_0} = \mathsf{ECC}_1$ is the set of coordinates where the codeword is one, and D_1 is the set of coordinates that were random encryptions of a one, thus $D_0 \cap D_1 = \overline{I_S}$. Note that ECC_0 and ECC_1 are defined by the message, I_S, D_0 and D_1 are defined by the sender's randomness and I_R is defined by the receiver's randomness. Therefore, in an honest execution, the sets I_S, D_0, D_1 will be independent of ECC_0, and I_R will be independent of everything. We let $\mathsf{F}_{\mathsf{real}}$ be the resulting distribution on (I_R, I_S, D_0, D_1). So whenever $|I_R| = c_R$, $|I_S| = c_S$ and $\{1, \ldots, n\} = I_S \cup D_0 \cup D_1$ is a partition, the probability density function is given by

$$\Pr\big(\mathsf{F}_{\mathsf{real}} = (I_R, I_S, D_0, D_1)\big) = \frac{1}{2^n \binom{n}{c_R}\binom{n}{c_S}}.$$

Even though $\mathsf{F}_{\mathsf{real}}$ is independent of msg, we often write $(I_R, I_S, D_0, D_1) \longleftarrow \mathsf{F}_{\mathsf{real}}(\mathsf{msg})$ when we are interested in how the sets intersect ECC_0 and ECC_1, defined by msg. The three different partitions

$$\{1, \ldots, n\} = \mathsf{ECC}_0 \cup \mathsf{ECC}_1 = I_S \cup D_0 \cup D_1 = I_R \cup \overline{I_R},$$

let us further partition $\{1, \ldots, n\}$ into 12 subsets by choosing one set from each partition. We compute now the sizes of the various intersections as this information will be important in defining our simulator.

As the error correcting code is balanced, we have $|\mathsf{ECC}_0| = |\mathsf{ECC}_1| = \frac{n}{2}$. We have, therefore, that $|I_S \cap \mathsf{ECC}_0| = k \longleftarrow \mathsf{H}(c_S, \frac{n}{2}, n)$, and $|I_S \cap \mathsf{ECC}_1| = c_S - k$. Similarly, $|D_0 \cap \mathsf{ECC}_0| = k' \longleftarrow \mathsf{B}_{1/2}(\frac{n}{2} - k)$ and $|D_0 \cap \mathsf{ECC}_1| = k'' \longleftarrow \mathsf{B}_{1/2}(\frac{n}{2} - c_S + k)$ which fixes $|D_1 \cap \mathsf{ECC}_0| = \frac{n}{2} - k - k'$ and $|D_1 \cap \mathsf{ECC}_1| = \frac{n}{2} - c_S + k - k''$. As I_R is chosen independently to be a random subset of $\{1, \ldots, n\}$ of size c_R, if we set

$$\alpha_b = |I_R \cap I_S \cap \mathsf{ECC}_b|; \quad \beta_b = |I_R \cap D_0 \cap \mathsf{ECC}_b|; \quad \gamma_b = |I_R \cap D_1 \cap \mathsf{ECC}_b|,$$

(thus fixing $|\overline{I_R} \cap I_S \cap \mathsf{ECC}_0| = k - \alpha_0$ and so on), then

$$(\alpha_0, \beta_0, \gamma_0, \alpha_1, \beta_1, \gamma_1) \longleftarrow \mathsf{H}^6\left(c_R, \left\{k, k', \frac{n}{2} - k - k', c_S - k, k'', \frac{n}{2} - c_S + k - k''\right\}, n\right).$$

This calculation will be useful when building our simulator.

3.3   The Simulator

Simulated Public Key and Ciphertext. Fix $c_{\text{good}} = 3n/4$. The simulator chooses $\hat{\mathbf{A}} \in \mathbb{Z}_q^{m \times (k+n-c_{\text{good}})}$ along with a trapdoor $\mathbf{T} \in \{0,1\}^{n \log q \times m}$ such that $\mathbf{T}\hat{\mathbf{A}} = \mathbf{G}$ according to [MP12]. He picks a random subset $I_{\text{good}} \subset \{1, \dots, n\}$ of size c_{good}, and sets $I_{\text{bad}} = \{1, \dots, n\} - I_{\text{good}}$. Write $\hat{\mathbf{A}} = [\mathbf{A}|\mathbf{V}]$ where $\mathbf{A} \in \mathbb{Z}_q^{m \times k}$ and $\mathbf{V} \in \mathbb{Z}_q^{m \times (n-c_{\text{good}})}$. For $i \in I_{\text{good}}$, draw $\mathbf{s}_i \leftarrow \mathbb{Z}_q^k$ and $\mathbf{e}_i \leftarrow \chi_\sigma^m$ and define $\mathbf{v}_1, \dots, \mathbf{v}_n \in \mathbb{Z}_q^m$:

$$\mathbf{v}_i = \begin{cases} \mathbf{A}\mathbf{s}_i + \mathbf{e}_i, & i \in I_{\text{good}} \\ \text{column of } \mathbf{V}, & i \in I_{\text{bad}} \end{cases},$$

so that the vectors $\{\mathbf{v}_i\}_{i \in I_{\text{bad}}}$ are the columns of $\mathbf{V}$. The public key is $\mathsf{pk} = (\mathbf{A}, \mathbf{v}_1, \dots, \mathbf{v}_n)$ and the data $\{\mathbf{s}_i\}_{i \in I_{\text{good}}}$ is stored as it will be used when generating the secret key: I_R will be a proper subset of I_{good}.

The simulater then chooses $\mathbf{r} \overset{\$}{\leftarrow} \chi_\sigma^m$, $e_i^* \overset{\$}{\leftarrow} \chi_{\sigma'}$ for $i \in I_{\text{good}}$, and randomly partitions I_{good} into subsets of equal size $I_{\text{good},0}$ and $I_{\text{good},1}$. The ciphertext is $\mathsf{ct} = (\mathbf{u}, w_1, \dots, w_n)$ where $\mathbf{u}^t = \mathbf{r}^t \mathbf{A} \in \mathbb{Z}_q^{1 \times k}$ and

$$w_i = \begin{cases} \mathbf{r}^t \mathbf{v}_i + e_i^*, & i \in I_{\text{good},0} \\ \mathbf{r}^t \mathbf{v}_i + e_i^* + (q/2), & i \in I_{\text{good},1} \\ w' \overset{\$}{\leftarrow} \mathbb{Z}_q, & i \in I_{\text{bad}} \end{cases}$$

The subsets $I_{\text{good},0}, I_{\text{good},1}, I_{\text{bad}}$ are stored for use when generating the encryption randomness.

Simulated Secret Key and Randomness. $\mathcal{S}$ draws $(I_R, I_S, D_0, D_1) \leftarrow \mathsf{F}_{\text{ideal}}(\mathsf{msg}, I_{\text{good},0}, I_{\text{good},1})$, where the ideal world subset function $\mathsf{F}_{\text{ideal}}$ is defined below (so in particular, $|I_R| = c_R$, $|I_S| = c_S$ and I_S, D_0, D_1 is a partition of $\{1, \dots, n\}$). The simulator then sets the secret key to $\{\mathbf{s}_i\}_{i \in I_R}$ using the vectors $\{\mathbf{s}_i\}_{i \in I_{\text{good}}}$ computed during public key generation. To compute the randomness, $\mathcal{S}$ sets $z_i = b$ for all $i \in D_b$. Then for each $i \in I_{\text{bad}}$, it draws $e_i' \leftarrow \chi_{\sigma'}$ and uses the trapdoor $\mathbf{T}$ to sample a Gaussian $\bar{\mathbf{r}} \in \mathbb{Z}_q^m$ such that $\bar{\mathbf{r}}^t \mathbf{A} = \mathbf{u}$ and $\bar{\mathbf{r}}^t \mathbf{v}_i = w_i - e_i' - (q/2)z_i$ for all $i \in I_{\text{bad}}$. Finally, for $i \in I_{\text{good}}$, $\mathcal{S}$ sets $e_i' = e_i^* + (\mathbf{r} - \bar{\mathbf{r}})^t \mathbf{e}_i$ and defines the encryption randomness $\mathsf{rand} = (I_S, \{z_i\}_{i \notin I_S}, \bar{\mathbf{r}}, e_1', \dots, e_n')$.

Ideal World Subsets. We now describe the distribution $\mathsf{F}_{\text{ideal}}$ which the simulator chooses to define the sets I_R, I_S, D_0, D_1. For simplicity, we assume that the target message msg is given at the beginning before all random choices are made. This is not exactly what happens in the ideal world, where the non-committing adversary $\mathcal{A}$ gets to see pk before specifying msg. However, it follows directly from the hardness of LWE that $\mathcal{A}$ cannot gain advantage by specifying msg after seeing pk.

Upon receiving $\mathsf{msg} \in \{0,1\}^\ell$ as input, $\mathsf{F}_{\text{ideal}}$ sets $y = (y_1, \dots, y_n) = \mathbf{ECC}(\mathsf{msg}) \in \{0,1\}^n$ to be the target codeword, defining $\mathsf{ECC}_0 = \{i : y_i = 0\}$

and $\mathsf{ECC}_1 = \{i : y_i = 1\}$. It then chooses a random $I_{\mathsf{good}} \subset \{1, \ldots, n\}$ of size c_{good} and $I_{\mathsf{bad}} = \{1, \ldots, n\} - I_{\mathsf{good}}$ as in the real world. It further divides I_{good} randomly into two halves $I_{\mathsf{good},0}$ and $I_{\mathsf{good},1}$ defining two partitions

$$\{1, \ldots, n\} = \mathsf{ECC}_0 \cup \mathsf{ECC}_1 = I_{\mathsf{good},0} \cup I_{\mathsf{good},1} \cup I_{\mathsf{bad}}.$$

The resulting six intersections have sizes:

- $|I_{\mathsf{good},0} \cap \mathsf{ECC}_0| = t \leftarrow \mathsf{H}\left(\frac{n}{2}, \frac{c_{\mathsf{good}}}{2}, n\right)$; $|I_{\mathsf{good},0} \cap \mathsf{ECC}_1| = \frac{c_{\mathsf{good}}}{2} - t$;
- $|I_{\mathsf{good},1} \cap \mathsf{ECC}_0| = t' \leftarrow \mathsf{H}\left(\frac{n}{2} - t, \frac{c_{\mathsf{good}}}{2}, n - \frac{c_{\mathsf{good}}}{2}\right)$; $|I_{\mathsf{good},1} \cap \mathsf{ECC}_1| = \frac{c_{\mathsf{good}}}{2} - t'$;
- $|I_{\mathsf{bad}} \cap \mathsf{ECC}_0| = \frac{n}{2} - t - t'$; $|I_{\mathsf{bad}} \cap \mathsf{ECC}_1| = \frac{n}{2} - c_{\mathsf{good}} + t + t'$.

$\mathsf{F}_{\mathsf{ideal}}$ needs to output $I_R, I_S, D_0, D_1 \subset \{1, \ldots, n\}$ such that the various intersections have the same sizes as in the real world. It proceeds as follows:

1. $\mathsf{F}_{\mathsf{ideal}}$ draws random variables $k \leftarrow \mathsf{H}\left(c_S, \frac{n}{2}, n\right)$, $k' \leftarrow \mathsf{B}_{1/2}\left(\frac{n}{2} - k\right)$, $k'' \leftarrow \mathsf{B}_{1/2}\left(\frac{n}{2} - c_S + k\right)$ and $(\alpha_0, \beta_0, \gamma_0, \alpha_1, \beta_1, \gamma_1) \leftarrow \mathsf{H}^6\left(c_R, \{k, k', \frac{n}{2} - k - k', c_S - k, k'', \frac{n}{2} - c_S + k - k''\}, n\right)$.
2. $\mathsf{F}_{\mathsf{ideal}}$ defines:
   - $I_R \cap I_S \cap \mathsf{ECC}_b$: a random subset of $I_{\mathsf{good},b} \cap \mathsf{ECC}_b$ of size α_b;
   - $I_R \cap D_0 \cap \mathsf{ECC}_b$: a random subset of $I_{\mathsf{good},0} \cap \mathsf{ECC}_b$ of size β_b;
   - $I_R \cap D_1 \cap \mathsf{ECC}_b$: a random subset of $I_{\mathsf{good},1} \cap \mathsf{ECC}_b$ of size γ_b;

   in such a way so that all six sets are disjoint. We prove in Claim 3.3 that the subsets $I_{\mathsf{good},b} \cap \mathsf{ECC}_{b'}$ are large enough to allow the above definitions with high probability. This fully defines I_R, but not I_S, D_0, D_1 (we still need their intersections with $\overline{I_R}$). Let $\mathsf{Rem}_b \subset I_{\mathsf{good},b}$ be the $i \in I_{\mathsf{good},b}$ that remain unassigned after this process.

   Remark: As I_R is now fully defined, we will not need the secret keys for the $i \in \mathsf{Rem}_0 \cup \mathsf{Rem}_1$. The only difference moving forward between $\mathsf{Rem}_0, \mathsf{Rem}_1$ and I_{bad} is that the ciphertexts v_i for $i \in \mathsf{Rem}_b$ can only be decrypted to b, whereas the ciphertexts v_i for $i \in I_{\mathsf{bad}}$ can be decrypted to 0 or 1 as they were generated with lossy public keys.
3. The sizes computed so far, along with the requirements $|I_S| = c_S$, $|I_S \cap \mathsf{ECC}_0| = k$, $|D_0 \cap \mathsf{ECC}_0| = k'$, and $|D_0 \cap \mathsf{ECC}_1| = k''$ determine the sizes of the remaining six sets. For example,

$$|\overline{I_R} \cap I_S \cap \mathsf{ECC}_0| = |I_S \cap \mathsf{ECC}_0| - |I_R \cap I_S \cap \mathsf{ECC}_0| = k - \alpha_0.$$

$\mathsf{F}_{\mathsf{ideal}}$ sets
   - $\overline{I_R} \cap I_S \cap \mathsf{ECC}_b$: subset of $(\mathsf{Rem}_b \cup I_{\mathsf{bad}}) \cap \mathsf{ECC}_b$;
   - $\overline{I_R} \cap D_0 \cap \mathsf{ECC}_b$: subset of $(\mathsf{Rem}_0 \cup I_{\mathsf{bad}}) \cap \mathsf{ECC}_b$;
   - $\overline{I_R} \cap D_1 \cap \mathsf{ECC}_b$: subset of $(\mathsf{Rem}_1 \cup I_{\mathsf{bad}}) \cap \mathsf{ECC}_b$;
   randomly such that (1) all six sets are disjoint and of the required size, (2) $\mathsf{Rem}_b \cap \mathsf{ECC}_b$ is fully contained in $\overline{I_R} \cap (I_S \cup D_b) \cap \mathsf{ECC}_b$, $\mathsf{Rem}_b \cap \mathsf{ECC}_{1-b}$ is fully contained in $\overline{I_R} \cap D_b \cap \mathsf{ECC}_{1-b}$. We prove in Claim 3.3 below that this is possible whp.
4. $\mathsf{F}_{\mathsf{ideal}}$ outputs (I_R, I_S, D_0, D_1).

Claim. If we set $c_{good} = 3n/4$, $c_R = c_S = n/8$ then whp over the choice of $I_{good,0}, I_{good,1}$ and the random variables drawn in step 1, it is possible to define the subsets in steps 2 and 3 above.

Proof. Step 2 requires

$$(I_R \cap (I_S \cup D_b) \cap ECC_b) \subset I_{good,b} \cap ECC_b; \quad (I_R \cap D_{1-b} \cap ECC_b) \subset I_{good,1-b} \cap ECC_b$$

for $b = 0, 1$ which is possible if and only if the four inequalities are satisfied:

$$\alpha_0 + \beta_0 \le t; \; \alpha_1 + \gamma_1 \le \frac{c_{good}}{2} - t'; \; \gamma_0 \le t'; \; \beta_1 \le \frac{c_{good}}{2} - t.$$

To see that all four are satisfied with high probability, note that the expectation of each right side is $c_{good}/4$, while the largest expectation of a left side is $c_R + 3c_R c_S/2n$.

On the other hand, step 3 requires

$$Rem_b \cap ECC_b \subset \overline{I_R} \cap (I_S \cup D_b) \cap ECC_b; \quad Rem_{1-b} \cap ECC_b \subset \overline{I_R} \cap D_{1-b} \cap ECC_b,$$

for $b = 0, 1$, which is possible if and only if the four inequalities are satisfied:

$$t \le k + k'; \; \frac{c_{good}}{2} - t' \le \frac{n}{2} - k''; \; t' \le \frac{n}{2} - k - k'; \; \frac{c_{good}}{2} - t \le k''.$$

The expectations of all four left hand sides is $c_{good}/4$, while the smallest right hand side has expectation $(n - c_S)/4$. If we set $\epsilon = 1/64$ then $c_{good} = 3n/4$, $c_R = c_S = n/8$ satisfy

$$c_R + \frac{3c_R c_S}{2n} + \epsilon n < \frac{c_{good}}{4} < \frac{n - c_S}{4} - \epsilon n,$$

and so the tail bound in Lemma 1, implies that all of the inequalities are satisfied with high probability.

We note that while $\epsilon = 1/64$ might be unsatisfactory in practice since the confidence offered by Lemma 1 is $1 - \exp(-\epsilon^2 n/2)$ (recall n is the message length which is a large constant times the security parameter, so $\epsilon = 1/64$ might well be fine), different values of ϵ may be obtained by varying c_R, c_S, and c_{good}. $\square$

Claim. The subsets I_R, I_S, D_0, D_1 output by the above process are distributed within negligible statistical distance of the corresponding subsets which arise in the real world, with high probability.

Proof. We compute the probability that the tuple (I_R, I_S, D_0, D_1) is output in the ideal worlds and check that it equals

$$\frac{1}{2^n \binom{n}{c_R}\binom{n}{c_S}},$$

like in the real world. We make two observations. Note first that for any $msg \in \{0,1\}^\ell$ which defines ECC_0 and ECC_1, a process which outputs (I_R, I_S, D_0, D_1)

can be equivalently thought of as a process which outputs 12 pairwise disjoint subsets corresponding to the twelve intersections of the three partitions

$$\{1, \ldots, n\} = \mathsf{ECC}_0 \cup \mathsf{ECC}_1 = I_S \cup D_0 \cup D_1 = I_R \cup \overline{I_R}.$$

The second observation is that choosing a random subset $A \subset \{1, \ldots, n\}$ of size a and then outputting a random subset $B \subset A$ of size b is the same as just outputting a random subset of $\{1, \ldots, n\}$ of size b. With these observations in mind, it is not difficult to complete the computation that

$$\Pr_{\mathsf{ideal}}(I_R, I_S, D_0, D_1) = \frac{1}{2^n \binom{n}{c_R}\binom{n}{c_S}},$$

for any $\mathsf{msg} \in \{0, 1\}^\ell$. The details are left to the reader. $\square$

3.4   Proof of Security

H_0 − *The Ideal World.*

- $\mathcal{C}$ chooses a random $I_{\mathsf{good}} \subset \{1, \ldots, n\}$ of size c_{good} and $\hat{\mathbf{A}} = [\mathbf{A}|\mathbf{V}] \in \mathbb{Z}_q^{m \times (k+n-c_{\mathsf{good}})}$ along with a trapdoor $\mathbf{T} \in \{0, 1\}^{n \log q \times m}$ such that $\mathbf{T}\hat{\mathbf{A}} = \mathbf{G}$ according to [MP12]. Then for each $i \in I_{\mathsf{good}}$, $\mathcal{C}$ draws $\mathbf{s}_i \leftarrow \mathbb{Z}_q^n$ and $\mathbf{e}_i \leftarrow \chi_\sigma^m$ and sets $\mathbf{v}_i = \mathbf{A}\mathbf{s}_i + \mathbf{e}_i$. For $i \in I_{\mathsf{bad}}$ $\mathcal{C}$ lets $\mathbf{v}_i$ be a column of $\mathbf{V}$. $\mathcal{C}$ sets $\mathsf{pk} = (\mathbf{A}, \mathbf{v}_1, \ldots, \mathbf{v}_n)$ and saves $\{\mathbf{s}_i\}_{i \in I_{\mathsf{good}}}$.
- $\mathcal{C}$ randomly partitions I_{good} into two halves of equal sizes $I_{\mathsf{good},0}$ and $I_{\mathsf{good},1}$ and chooses $\mathbf{r} \leftarrow \chi_\sigma^m$, setting $\mathbf{u}^\mathsf{t} = \mathbf{r}^\mathsf{t}\mathbf{A} \in \mathbb{Z}_q^{1 \times k}$. For $i \in I_{\mathsf{good},b}$, $\mathcal{C}$ sets $w_i = \mathbf{r}^\mathsf{t}\mathbf{v}_i + e_i^* + (q/2)b$ where each $e_i^* \leftarrow \chi_{\sigma'}$. For $i \in I_{\mathsf{bad}}$, $\mathcal{C}$ lets $w_i \in \mathbb{Z}_q$ be random. $\mathcal{C}$ sets $\mathsf{ct} = (\mathbf{u}, w_1, \ldots, w_n)$.
- $\mathcal{C}$ sends pk to $\mathcal{A}$ and receives msg.
- $\mathcal{C}$ computes $(I_R, I_S, D_0, D_1) \leftarrow \mathsf{F}_{\mathsf{ideal}}(\mathsf{msg}, I_{\mathsf{good},0}, I_{\mathsf{good},1})$ and sets $\mathsf{sk} = (I_R, \{\mathbf{s}_i\}_{i \in I_R})$.
- Finally, for each $i \in I_{\mathsf{bad}} \cap D_b$, $\mathcal{C}$ draws $e_i' \leftarrow \chi_{\sigma'}$ and sets $w_i' = w_i - e_i' - (q/2)b$, then $\mathcal{C}$ draws $\bar{\mathbf{r}} \leftarrow \mathsf{Sample}_\sigma(\mathbf{u}', \hat{\mathbf{A}}, \mathbf{T})$ according to [MP12] where $\mathbf{u}' = (\mathbf{u}, \{w_i'\}_{i \in I_{\mathsf{bad}}}) \in \mathbb{Z}_q^{k+n-c_{\mathsf{good}}}$. For each $i \in I_{\mathsf{good}}$, $\mathcal{C}$ sets $e_i' = e_i^* + (\mathbf{r} - \bar{\mathbf{r}})^\mathsf{t}\mathbf{e}_i$. Lastly, for each $i \in D_b$, $\mathcal{C}$ sets $z_i = b$. He collects all of this information into $\mathsf{rand} = (I_S, \{z_i\}_{i \notin I_S}, \bar{\mathbf{r}}, e_1', \ldots, e_n')$.
- $\mathcal{C}$ sends $(\mathsf{ct}, \mathsf{sk}, \mathsf{rand})$.

H_1 − The main difference between this world and H_0 is that here $\mathcal{C}$ does not choose ct until after he sends pk to $\mathcal{A}$ and receives msg. This allows us to avoid selecting $\bar{\mathbf{r}}$ or the e_i^*.

- $\mathcal{C}$ chooses $I_{\mathsf{good}} \subset \{1, \ldots, n\}$, $\hat{\mathbf{A}} = [\mathbf{A}|\mathbf{V}] \in \mathbb{Z}_q^{m \times (k+n-c_{\mathsf{good}})}$, $\{\mathbf{s}_i\}_{i \in I_{\mathsf{good}}}$ and $\{\mathbf{e}_i\}_{i \in I_{\mathsf{good}}}$ and $\{\mathbf{v}_i\}_{i=1,\ldots,n}$ just as in H_0 and sets $\mathsf{pk} = (\mathbf{A}, \mathbf{v}_1, \ldots, \mathbf{v}_n)$, saving $\{\mathbf{s}_i\}_{i \in I_{\mathsf{good}}}$.
- $\mathcal{C}$ sends pk to $\mathcal{A}$ and receives msg.

- $\mathcal{C}$ randomly chooses $I_{\text{good},0}$ and $I_{\text{good},1}$ and computes $(I_R, I_S, D_0, D_1) \leftarrow$ $\mathsf{F}_{\text{ideal}}(\text{msg}, I_{\text{good},0}, I_{\text{good},1})$, and sets $\text{sk} = (I_R, \{\mathbf{s}_i\}_{i \in I_R})$.
- $\mathcal{C}$ chooses $\mathbf{r} \leftarrow \chi_\sigma^m$ and for $i \in (I_S \cap I_{\text{good},b}) \cup D_b$, sets $z_i = b$ and $w_i = \mathbf{r}^t \mathbf{v}_i + e_i' + (q/2)z_i$. $\mathcal{C}$ sets $\text{ct} = (\mathbf{u}, w_1, \ldots, w_n)$ and $\text{rand} = (I_S, \{z_i\}_{i \notin I_S}, \mathbf{r}, e_1', \ldots, e_n')$.
- $\mathcal{C}$ sends $(\text{ct}, \text{sk}, \text{rand})$.

Claim. $H_1 \approx_{\mathsf{s}} H_0$.

Proof. We must show that the pair $(\bar{\mathbf{r}}, \{e_i'\}_{i=1,\ldots,n}) \leftarrow H_0$ is statistically close to $(\mathbf{r}, \{e_i'\}_i) \leftarrow H_1$. Note that $\bar{\mathbf{r}}$ is chosen by first drawing $\mathbf{r} \leftarrow \chi_\sigma^m$ and then using the trapdoor preimage sampler to draw Gaussian $\bar{\mathbf{r}}$ such that $\bar{\mathbf{r}}^t \hat{\mathbf{A}} = \mathbf{r}^t \hat{\mathbf{A}}$. The induced distribution on $\bar{\mathbf{r}}$ is statistically close to simply drawing $\mathbf{r} \leftarrow \chi_\sigma^m$ as in H_1. Second note that $e_i' \leftarrow \chi_{\sigma'}$ for all i in H_1, while in H_0, this is only the case for $i \in I_{\text{bad}}$. For $i \in I_{\text{good}}$, $e_i' = e_i^* + (\mathbf{r} - \bar{\mathbf{r}})^t \mathbf{e}_i$ where $e_i^* \leftarrow \chi_{\sigma'}$ and $\mathbf{e}_i \leftarrow \chi_\sigma^m$. This is statistically close to $\chi_{\sigma'}$ as $\sigma^2/\sigma' = \mathsf{negl}(k)$, using Fact 1. □

H_2 – In this world we draw $\mathbf{A} \in \mathbb{Z}_q^{m \times k}$ and $\{\mathbf{v}_i\}_{i \in I_{\text{bad}}}$ randomly instead of along with a trapdoor.

- $\mathcal{C}$ chooses $I_{\text{good}} \subset \{1, \ldots, n\}$, $\mathbf{A} \in \mathbb{Z}_q^{m \times k}$, and sets $\mathbf{v}_i = \mathbf{A}\mathbf{s}_i + \mathbf{e}_i$ for $i \in I_{\text{good}}$ and $\mathbf{v} \leftarrow \mathbb{Z}_q^m$ for $i \in I_{\text{bad}}$, where $\{\mathbf{s}_i\}_{i \in I_{\text{good}}}$ and $\{\mathbf{e}_i\}_{i \in I_{\text{good}}}$ are as in H_1. $\mathcal{C}$ sets $\text{pk} = (\mathbf{A}, \mathbf{v}_1, \ldots, \mathbf{v}_n)$, and saves $\{\mathbf{s}_i\}_{i \in I_{\text{good}}}$.
- $\mathcal{C}$ sends pk to $\mathcal{A}$ and receives msg.
- $\mathcal{C}$ randomly chooses $I_{\text{good},0}$ and $I_{\text{good},1}$ and computes $(I_R, I_S, D_0, D_1) \leftarrow$ $\mathsf{F}_{\text{ideal}}(\text{msg}, I_{\text{good},0}, I_{\text{good},1})$, and sets $\text{sk} = (I_R, \{\mathbf{s}_i\}_{i \in I_R})$.
- $\mathcal{C}$ chooses $\mathbf{r} \leftarrow \chi_\sigma^m$ and for $i \in (I_S \cap I_{\text{good},b}) \cup D_b$, sets $z_i = b$ and $w_i = \mathbf{r}^t \mathbf{v}_i + e_i' + (q/2)z_i$. $\mathcal{C}$ sets $\text{ct} = (\mathbf{u}, w_1, \ldots, w_n)$ and $\text{rand} = (I_S, \{z_i\}_{i \notin I_S}, \mathbf{r}, e_1', \ldots, e_n')$.
- $\mathcal{C}$ sends $(\text{ct}, \text{sk}, \text{rand})$.

Claim. $H_2 \approx_{\mathsf{s}} H_1$.

Proof. This follows immediately from the fact that matrices drawn along with their trapdoors as in [MP12] are statistically close to uniform. As we weren't using the trapdoor in H_1 anyway, changing $\hat{\mathbf{A}}$ to a uniform matrix, this does not affect anything functionally. □

H_3 – *The Real World.* In this world we change the way the subsets (I_R, I_S, D_0, D_1) are drawn; we draw them from F_{real} instead of $\mathsf{F}_{\text{ideal}}$.

- $\mathcal{C}$ draws $(I_R, I_S, D_0, D_1) \leftarrow \mathsf{F}_{\text{real}}$ and a random $\mathbf{A} \in \mathbb{Z}_q^{m \times k}$ and sets $\mathbf{v}_i = \mathbf{A}\mathbf{s}_i + \mathbf{e}_i$ for $i \in I_R$ and $\mathbf{v} \leftarrow \mathbb{Z}_q^m$ for $i \notin I_R$, where $\{\mathbf{s}_i\}_{i \in I_R}$ and $\{\mathbf{e}_i\}_{i \in I_R}$ are as in H_2. $\mathcal{C}$ sets $\text{pk} = (\mathbf{A}, \mathbf{v}_1, \ldots, \mathbf{v}_n)$, and $\text{sk} = \{\mathbf{s}_i\}_{i \in I_R}$.
- $\mathcal{C}$ sends pk to $\mathcal{A}$ and receives msg and sets $y = \mathbf{ECC}(\text{msg})$.
- $\mathcal{C}$ draws $\mathbf{r} \leftarrow \chi_\sigma^m$ and sets $\mathbf{u}^t = \mathbf{r}^t \mathbf{A}$ and $w_i = \mathbf{r}^t \mathbf{v}_i + e_i' + (q/2)y_i$ for $i \in I_S$, where $e_i' \leftarrow \chi_{\sigma'}$. Then for each $i \in D_b$, $\mathcal{C}$ sets $z_i = b$ and $w_i = \mathbf{r}^t \mathbf{v}_i + e_i' + (q/2)z_i$. Finally $\mathcal{C}$ sets $\text{ct} = (\mathbf{u}, w_1, \ldots, w_n)$ and $\text{rand} = (I_S, \{z_i\}_{i \notin I_S}, \mathbf{r}, e_1', \ldots, e_n')$.
- $\mathcal{C}$ sends $(\text{ct}, \text{sk}, \text{rand})$.

Claim. $H_3 \approx_c H_2$.

Proof Sketch. This follows from Claim 3.3, which states that the (I_R, I_S, D_0, D_1) from $\mathsf{F_{real}}$ is identical to the tuple drawn from $\mathsf{F_{ideal}}$, combined with the fact that a PPT adversary cannot gain advantage by choosing msg after seeing pk rather than before or else it can be used to break LWE. $\square$

References

[Bea97] Beaver, D.: Plug and play encryption. In: Kaliski Jr., B.S. (ed.) CRYPTO 1997. LNCS, vol. 1294, pp. 75–89. Springer, Heidelberg (1997)

[BGW88] Ben-Or, M., Goldwasser, S., Wigderson, A.: Completeness theorems for non-cryptographic fault-tolerant distributed computation. In: STOC, pp. 1–10. ACM, New York (1988)

[BLP13] Brakerski, Z., Langlois, A., Peikert, C., Regev, O., Stehle, D.: Classical hardness of learning with errors. In: STOC 2013, pp. 575–584 (2013)

[CCD88] Chaum, D., Crépeau, C., Damgård, I.: Multiparty unconditionally secure protocols. In: STOC, pp. 11–19 (1988)

[CDSMW09] Choi, S.G., Dachman-Soled, D., Malkin, T., Wee, H.: Improved non-committing encryption with applications to adaptively secure protocols. In: Matsui, M. (ed.) ASIACRYPT 2009. LNCS, vol. 5912, pp. 287–302. Springer, Heidelberg (2009)

[CFGN96] Canetti, R., Feige, U., Goldreich, O., Naor, M.: Adaptively secure multi-party computation. In: Proceedings of the Twenty-Eighth Annual ACM Symposium on Theory of Computing, STOC 1996, pp. 639–648. ACM, New York (1996)

[DN00] Damgård, I.B., Nielsen, J.B.: Improved non-committing encryption schemes based on a general complexity assumption. In: Bellare, M. (ed.) CRYPTO 2000. LNCS, vol. 1880, pp. 432–450. Springer, Heidelberg (2000)

[GMW87] Goldreich, O., Micali, S., Wigderson, A.: How to play any mental game. In: STOC 1987, pp. 218–229 (1987)

[Hoe63] Hoeffding, W.: Probability inequalities for sums of bounded random variables. J. Am. Stat. Assoc. **58**, 13–30 (1963)

[HOR15] Hemenway, B., Ostrovsky, R., Rosen, A.: Non-committing encryption from Φ-hiding. In: Dodis, Y., Nielsen, J.B. (eds.) TCC 2015, Part I. LNCS, vol. 9014, pp. 591–608. Springer, Heidelberg (2015)

[LPR13] Lyubashevsky, V., Peikert, C., Regev, O.: A toolkit for ring-LWE cryptography. In: Johansson, T., Nguyen, P.Q. (eds.) EUROCRYPT 2013. LNCS, vol. 7881, pp. 35–54. Springer, Heidelberg (2013)

[MP12] Micciancio, D., Peikert, C.: Trapdoors for lattices: simpler, tighter, faster, smaller. In: Pointcheval, D., Johansson, T. (eds.) EUROCRYPT 2012. LNCS, vol. 7237, pp. 700–718. Springer, Heidelberg (2012)

[MPSW05] Micali, S., Peikert, C., Sudan, M., Wilson, D.A.: Optimal error correction against computationally bounded noise. In: Kilian, J. (ed.) TCC 2005. LNCS, vol. 3378, pp. 1–16. Springer, Heidelberg (2005)

[Pei09] Peikert, C.: Public-key cryptosystems from the worst-case shortest vector problem: extended abstract. In: Proceedings of the 41st Annual ACM Symposium on Theory of Computing, STOC 2009, pp. 333–342. ACM, New York (2009)

[Reg09] Regev, O.: On lattices, learning with errors, random linear codes, and cryptography. J. ACM **56**(6) (2009)

On the Complexity of Additively Homomorphic UC Commitments

Tore Kasper Frederiksen, Thomas P. Jakobsen, Jesper Buus Nielsen,
and Roberto Trifiletti[✉]

Department of Computer Science, Aarhus University, Aarhus, Denmark
{jot2re,tpj,jbn,roberto}@cs.au.dk

Abstract. We present a new constant round additively homomorphic commitment scheme with (amortized) computational and communication complexity linear in the size of the string committed to. Our scheme is based on the non-homomorphic commitment scheme of Cascudo *et al.* presented at PKC 2015. However, we manage to add the additive homomorphic property, while at the same time reducing the constants. In fact, when opening a large enough batch of commitments we achieve an amortized communication complexity converging to the length of the message committed to, *i.e.*, we achieve close to rate 1 as the commitment protocol by Garay *et al.* from Eurocrypt 2014. A main technical improvement over the scheme mentioned above, and other schemes based on using error correcting codes for UC commitment, we develop a new technique which allows to based the extraction property on erasure decoding as opposed to error correction. This allows to use a code with significantly smaller minimal distance and allows to use codes without efficient decoding.

Our scheme only relies on standard assumptions. Specifically we require a pseudorandom number generator, a linear error correcting code and an ideal oblivious transfer functionality. Based on this we prove our scheme secure in the Universal Composability (UC) framework against a static and malicious adversary corrupting any number of parties.

On a practical note, our scheme improves significantly on the non-homomorphic scheme of Cascudo *et al.* Based on their observations in regards to efficiency of using linear error correcting codes for commitments we conjecture that our commitment scheme might in practice be more efficient than all existing constructions of UC commitment, even non-homomorphic constructions and even constructions in the random oracle model. In particular, the amortized price of computing one of our commitments is less than that of evaluating a hash function once.

Keywords: Commitments · UC · Homomorphic · Minimal assumptions · Linear error correcting codes · Erasure codes

The authors acknowledge support from the Danish National Research Foundation and The National Science Foundation of China (under the grant 61361136003) for the Sino-Danish Center for the Theory of Interactive Computation and from the Center for Research in Foundations of Electronic Markets (CFEM), supported by the Danish Strategic Research Council.

R. Trifiletti—Partially supported by the European Research Commission Starting Grant 279447.

E. Kushilevitz and T. Malkin (Eds.): TCC 2016-A, Part I, LNCS 9562, pp. 542–565, 2016.
DOI: 10.1007/978-3-662-49096-9_23

1 Introduction

Commitment schemes are the digital equivalent of a securely locked box: it allows a sender P_s to hide a secret from a receiver P_r by putting the secret inside the box, sealing it, and sending the box to P_r. As the receiver cannot look inside we say that the commitment is *hiding*. As the sender is unable to change his mind as he has given the box away we say the commitment is also *binding*. These simple, yet powerful properties are needed in countless cryptographic protocols, especially when guaranteeing security against a *malicious* adversary who can arbitrarily deviate from the protocol at hand. In the stand-alone model, commitment schemes can be made very efficient, both in terms of communication and computation and can be based entirely on the existence of one-way functions. These can *e.g.* be constructed from cheap symmetric cryptography such as pseudorandom generators [Nao90].

In this work we give an additively homomorphic commitment scheme secure in the UC-framework of [Can01], a model considering protocols running in a concurrent and asynchronous setting. The first UC-secure commitment schemes were given in [CF01, CLOS02] as feasibility results, while in [CF01] it was also shown that UC-commitments cannot be instantiated in the standard model and therefore require some form of setup assumption, such as a CRS. Moreover a construction for UC-commitments in such a model implies public-key cryptography [DG03]. Also, in the UC setting the previously mentioned hiding and binding properties are augmented with the notions of *equivocality* and *extractability*, respectively. These properties are needed to realize the commitment functionality we introduce later on. Loosely speaking, a scheme is equivocal if a single commitment can be opened to any message using special trapdoor information. Likewise a scheme is extractable if from a commitment the underlying message can be extracted efficiently using again some special trapdoor information.

Based on the above it is not surprising that UC-commitments are significantly less efficient than constructions in the stand-alone model. Nevertheless a plethora of improvements have been proposed in the literature, *e.g.* [DN02, NFT09, Lin11, BCPV13, Fuj14, CJS14] considering different number theoretic hardness assumptions, types of setup assumption and adversarial models. Until recently, the most efficient schemes for the adversarial model considered in this work were that of [Lin11, BCPV13] in the CRS model and [HMQ04, CJS14] in different variations of the random oracle model [BR93].

Related Work. In [GIKW14] and independently in [DDGN14] it was considered to construct UC-commitments in the OT-hybrid model and at the same time confining the use of the OT primitive to a once-and-for-all setup phase. After the setup phase, the idea is to only use cheap symmetric primitives for each commitment thus amortizing away the cost of the initial OTs. Both approaches strongly resembles the "MPC-in-the-head" line of work of [IKOS07, HIKN08, IPS08] in that the receiver is watching a number of communication channels not disclosed to the sender. In order to cheat meaningfully in this paradigm the sender needs to cheat in many channels, but since he is unaware where the receiver is watching he will get caught with high probability. Concretely these schemes build on VSS and allow the receiver to learn an unqualified set of shares for a secret s. However the setup is such that the sender does not know which unqualified set is being "watched", so when opening he is forced to open to enough positions with consistent shares to avoid getting caught. The scheme of [GIKW14] focused primarily on the rate of the commitments in an asymptotic setting while [DDGN14] focused on the computational complexity. Furthermore the secret sharing scheme of the latter is based on Reed-Solomon codes and the scheme achieved both additive and multiplicative homomorphisms.

The idea of using OTs and error correction codes to realize commitments was also considered in [FJN+13] in the setting of two-party secure computation using garbled circuits. Their scheme also allowed for additively homomorphic operations on commitments, but requires a code with a specific privacy property. The authors pointed to [CC06] for an example of such a code, but it turns out this achieves quite low constant rate due to the privacy restriction. Care also has to be taken when using this scheme, as binding is not guaranteed for all committed messages. The authors capture this by allowing some message to be "wildcards". However, in their application this is acceptable and properly dealt with.

Finally in [CDD+15] a new approach to the above OT watch channel paradigm was proposed. Instead of basing the underlying secret sharing scheme on a threshold scheme the authors proposed a scheme for a particular access structure. This allowed realization of the scheme using additive secret sharing and any linear code, which achieved very good concrete efficiency. The only requirement of the code is that it is linear and the minimum distance is at least $2s + 1$ for statistical security s. To commit to a message m it is first encoded into a codeword c. Then each field element c_i of c is additively shared into two field elements c_i^0 and c_i^1 and the receiver learns one of these shares via an oblivious transfer. This in done in the watch-list paradigm where the same shares c_i^0 are learned for all the commitments, by using the OTs only to transfer short seeds and then masking the share c_i^0 and c_i^1 for all commitments from these pairs of seeds. This can be seen as reusing an idea ultimately going back to [Kil88, CvT95]. Even if the adversary commits to a string c' which is not a codeword, to open to another message, it would have to guess at least s of the random choice bits of the receiver. Furthermore the authors propose an additively homomorphic version of their scheme, however at the cost of using VSS which imposes higher constants than their basic non-homomorphic construction.

Motivation. As already mentioned, commitment schemes are extremely useful when security against a malicious adversary is required. With the added support for additively homomorphic operations on committed values even more applications become possible. One is that of maliciously secure two-party computation using the LEGO protocols of [NO09, FJN+13, FJNT15]. These protocols are based on cut-and-choose of garbled circuits and require a large amount of homomorphic commitments, in particular one commitment for each wire of all garbled gates. In a similar fashion the scheme of [AHMR15] for secure evaluation of RAM programs also make use of homomorphic commitments to transform

---

$\mathcal{F}_{ROT}$ interacts with a sender P_s, a receiver P_r and an adversary $\mathcal{S}$ and it proceeds as follows:

Transfer: Upon receiving $(\texttt{transfer}, \texttt{sid}, \texttt{otid}, k)$ from both P_s and P_r, forward this message to $\mathcal{S}$ and wait for a reply. If $\mathcal{S}$ sends back $(\texttt{no-corrupt}, \texttt{sid}, \texttt{otid})$, sample $l^0, l^1 \in_R \{0,1\}^k$ and $b \in_R \{0,1\}$ and output $\left(\texttt{deliver}, \texttt{sid}, \texttt{otid}, \left(l^0, l^1\right)\right)$ to P_s and $\left(\texttt{deliver}, \texttt{sid}, \texttt{otid}, \left(l^b, b\right)\right)$ to P_r. If $\mathcal{S}$ instead sends back $\left(\texttt{corrupt-sender}, \texttt{sid}, \texttt{otid}, \left(\tilde{l}^0, \tilde{l}^1\right)\right)$ or $\left(\texttt{corrupt-receiver}, \texttt{sid}, \texttt{otid}, \left(\tilde{l}^{\tilde{b}}, \tilde{b}\right)\right)$ and the sender, respectively the receiver is corrupted, proceed as above, but instead of sampling all values at random, use the values provided by $\mathcal{S}$.

Fig. 1. Ideal functionality $\mathcal{F}_{ROT}$.

garbled wire labels of one garbled circuit to another. Thus any improvement in the efficiency of homomorphic commitments is directly transferred to the above settings as well.

Our Contribution. We introduce a new, very efficient, additively homomorphic UC-secure commitment scheme in the $\mathcal{F}_{ROT}$-hybrid model. The $\mathcal{F}_{ROT}$-functionality is fully described in Fig. 1. Our scheme shows that:

1. The asymptotic complexity of additively homomorphic UC commitment is the same as the asymptotic complexity of non-homomorphic UC commitment, *i.e.*, the achievable rate is $1 - o(1)$. In particular, the homomorphic property comes for free.
2. In addition to being asymptotically optimal, our scheme is also more practical (smaller hidden constants) than any other existing UC commitment scheme, even non-homomorphic schemes and even schemes in the random oracle model.

In more detail our main contributions are as follows:

- We improve on the basic non-homomorphic commitment scheme of [CDD+15] by reducing the requirement of the minimum distance of the underlying linear code from $2s + 1$ to s for statistical security s. At the same time our scheme becomes additively homomorphic, a property not shared with the above scheme. This is achieved by introducing an efficient consistency check at the end of the commit phase, as described now. Assume that the corrupted sender commits to a string c' which has Hamming distance 1 to some codeword c_0 encoding message m_0 and has Hamming distance $s-1$ to some other codeword c_1 encoding message m_1. For both the scheme in [CDD+15] and our scheme this means the adversary can later open to m_0 with probability $\frac{1}{2}$ and to m_1 with probability 2^{-s+1}. Both of these probabilities are considered too high as we want statistical security 2^{-s}. So, even if we could decode c' to for instance m_0, this might not be the message that the adversary will open to later. It is, however, the case that the adversary cannot later open to both m_0 and m_1, except with probability 2^{-s} as this would require guessing s of the random choice bits. The UC simulator, however, needs to extract which of m_0 and m_1 will be opened to already at commitment time. We introduce a new consistency check where we after the commitment phase ask the adversary to open a random linear combination of the committed purported codewords. This linear combination will with overwhelming probability in a well defined manner "contain" information about every dirty codeword c' and will force the adversary to guess some of the choice bits to successfully open it to some close codeword c. The trick is then that the simulator can extract which of the choice bits the adversary had to guess and that if we puncture the code and the committed strings at the positions at which the adversary guessed the choice bits, then the remaining strings can be proven to be codewords in the punctured code. Since the adversary guesses at most $s - 1$ choice bits, except with negligible probability 2^{-s} we only need to puncture $s - 1$ positions, so the punctured code still has distance 1. We can therefore erasure decode and thus extract the committed message. If the adversary later open to another message he will have to guess additional choice bits, bringing him up to having guessed at least s choice bits. With the minimal distance lowered the required code length is also reduced and therefore also the amount of required initial OTs. As an example, for committing to messages of size $k = 256$ with statistical security $s = 40$ this amounts to roughly 33 % less initial OTs than required by [CDD+15].

- We furthermore propose a number of optimizations that reduce the communication complexity by a factor of 2 for each commitment compared to [CDD+15] (without taking into account the smaller code length required). We give a detailed comparison to the schemes of [Lin11, BCPV13, CJS14] and [CDD+15] in Sect. 4 and show that for the above setting with $k = 256$ and $s = 40$ our new construction outperforms all existing schemes in terms of communication if committing to 304 messages or more while retaining the computational efficiency of [CDD+15]. This comparison includes the cost of the initial OTs. If committing to 10,000 messages or more we see the total communication is around 1/3 of [BCPV13], around 1/2 of the basic scheme of [CDD+15] and around 1/21 of the homomorphic version.

- Finally we give an extension of any additively homomorphic commitment scheme that achieves an amortized rate close to 1 in the opening phase. Put together with our proposed scheme and breaking a long message into many smaller blocks we achieve rate close to 1 in both the commitment and open phase of our protocol. This extension is interactive and is very similar in nature to the introduced consistency check for decreasing the required minimum distance. Although based on folklore techniques this extension allows for very efficiently homomorphic commitment to long messages without requiring correspondingly many OTs.

2   The Protocol

We use κ and s to denote the computational and statistical security parameter respectively. This means that for any fixed s and any polynomial time bounded adversary, the advantage of the adversary is $2^{-s} + \text{negl}(\kappa)$ for a negligible function negl. *i.e.*, the advantage of any adversary goes to 2^{-s} faster than any inverse polynomial in the computational security parameter. If $s = \Omega(\kappa)$ then the advantage is negligible. We will be working over an arbitrary finite field $\mathbb{F}$. Based on this, along with s, we define $\hat{s} = \lceil s/\log_2(|\mathbb{F}|)\rceil$.

We will use as shorthand $[n] = \{1, 2, \ldots, n\}$, and $e \in_R S$ to mean: sample an element e uniformly at random from the set S. When $\boldsymbol{r}$ and $\boldsymbol{m}$ are vectors we write $\boldsymbol{r}\|\boldsymbol{m}$ to mean the vector that is the concatenation of $\boldsymbol{r}$ and $\boldsymbol{m}$. We write $y \leftarrow P(x)$ to mean: perform the (potentially randomized) procedure P on input x and store the output in variable y. We will use $x := y$ to denote an assignment of x to y. We will interchangeably use subscript and bracket notation to denote an index of a vector, *i.e.* x_i and $\boldsymbol{x}[i]$ denotes the i'th entry of a vector $\boldsymbol{x}$ which we will always write in bold. Furthermore we will use $\pi_{i,j}$ to denote a projection of a vector that extracts the entries from index i to index j, *i.e.* $\pi_{i,j}(\boldsymbol{x}) = (x_i, x_{i+1}, \ldots, x_j)$. We will also use $\pi_l(\boldsymbol{x}) = \pi_{1,l}(\boldsymbol{x})$ as shorthand to denote the first l entries of $\boldsymbol{x}$.

In Fig. 2 we present the ideal functionality $\mathcal{F}_{\text{HCOM}}$ that we UC-realize in this work. The functionality differs from other commitment functionalities in the literature by only allowing the sender P_s to decide the number of values he wants to commit to. The functionality then commits him to *random* values towards a receiver P_r and reveals the values to P_s. The reason for having the functionality commit to several values at a time is to reflect the batched nature of our protocol. That the values committed to are random is a design choice to offer flexibility for possible applications. In Appendix A we show an efficient black-box extension of $\mathcal{F}_{\text{HCOM}}$ to chosen-message commitments.

2.1 Protocol Π_{HCOM}

Our protocol Π_{HCOM} is cast in the $\mathcal{F}_{\text{ROT}}$-hybrid model, meaning the parties are assumed access to the ideal functionality $\mathcal{F}_{\text{ROT}}$ in Fig. 1. The protocol UC-realizes the functionality $\mathcal{F}_{\text{HCOM}}$ and is presented in full in Figs. 4 and 5. At the start of the protocol a once-and-for-all **Init** step is performed where P_s and P_r only need to know the size of the committed values k and the security parameters. We furthermore assume that the parties agree on a $[n, k, d]$ linear code $\mathcal{C}$ in systematic form over the finite field $\mathbb{F}$ and require that the minimum distance $d \geq s$ for statistical security parameter s. The parties then invoke n copies of the ideal functionality $\mathcal{F}_{\text{ROT}}$ with the computational security parameter κ as input, such that P_s learns n pairs of κ-bit strings l_i^0, l_i^1 for $i \in [n]$, while P_r only learns one string of each pair. In addition to the above the parties also introduce a commitment counter T which simply stores the number of values committed to. Our protocol is phrased such that multiple commitment phases are possible after the initial ROTs have been performed, and the counter is simply incremented accordingly.

Next a **Commit** phase is described where at the end, P_s is committed to γ pseudorandom values. The protocol instructs the parties to expand the previously learned κ-bit strings, using a pseudorandom generator PRG, into

$\mathcal{F}_{\text{HCOM}}$ interacts with a sender P_s, a receiver P_r and an adversary $\mathcal{S}$, working over a finite field $\mathbb{F}$.

Init: Upon receiving a message $(\texttt{init}, \texttt{sid}, k)$ from both parties P_s and P_r, store the message length k.

Commit: Upon receiving a message $(\texttt{commit}, \texttt{sid}, \gamma)$ from P_s, forward this message to $\mathcal{S}$ and wait for a reply. If $\mathcal{S}$ sends back $(\texttt{no-corrupt}, \texttt{sid})$ proceed as follows: Sample γ uniformly random values $r_j \in \mathbb{F}^k$ and associate to each of these a unique unused identifier j and store the tuple $(\texttt{random}, \texttt{sid}, j, r_j)$. We let $\mathcal{J}$ denote the set of these identifiers. Finally send $\big(\texttt{committed}, \texttt{sid}, \mathcal{J}, \{r_j\}_{j \in \mathcal{J}}\big)$ to P_s and $(\texttt{receipt}, \texttt{sid}, \mathcal{J})$ to P_r and $\mathcal{S}$.
If P_s is corrupted and $\mathcal{S}$ instead sends back $\big(\texttt{corrupt-commit}, \texttt{sid}, \{\tilde{r}_j\}_{j \in \mathcal{J}}\big)$, proceed as above, but instead of sampling the values at random, use the values provided by $\mathcal{S}$.

Open: Upon receiving a message $\big(\texttt{open}, \texttt{sid}, \{(c, \alpha_c)\}_{c \in C}\big)$ from P_s, if for all $c \in C$, a tuple $(\texttt{random}, \texttt{sid}, c, r_c)$ was previously recorded and $\alpha_c \in \mathbb{F}$, send $\big(\texttt{opened}, \texttt{sid}, \{(c, \alpha_c)\}_{c \in C}, \sum_{c \in C} \alpha_c \cdot r_c\big)$ to P_r and $\mathcal{S}$. Otherwise, ignore.

Fig. 2. Ideal functionality $\mathcal{F}_{\text{HCOM}}$.

row-vectors $\bar{s}_i^b \in \mathbb{F}^{\mathcal{T}+\gamma+1}$ for $b \in \{0,1\}$ and $i \in [n]$. The reason for the extra length will be apparent later. We denote by $\mathcal{J} = \{\mathcal{T}+1, \ldots, \mathcal{T}+\gamma+1\}$ the set of indices of the $\gamma+1$ commitments being setup in this invocation of **Commit**. After the expansion P_s knows all of the above $2n$ row-vectors, while P_r only knows half. The parties then view these row-vectors as matrices S^0 and S^1 where row i of S^b consists of the vector $\bar{s}_i^b$. We let $s_j^b \in \mathbb{F}^n$ denote the j'th column vector of the matrix S^b for $j \in \mathcal{J}$. These column vectors now determine the committed pseudorandom values, which we define as $r_j = r_j^0 + r_j^1$ where $r_j^b = \pi_k(s_j^b)$ for $j \in \mathcal{J}$. The above steps are also pictorially described in Fig. 3.

The goal of the commit phase is for P_r to hold one out of two shares of each entry of a codeword of $\mathcal{C}$ that encodes the vector r_j for all $j \in \mathcal{J}$. At this point of the protocol, what P_r holds is however not of the above form. Though, because the code is in systematic form we have by definition that P_r holds such a sharing for the first k entries of each of these codewords. To ensure the same for the rest of the entries, for all $j \in \mathcal{J}$, P_s computes $t_j \leftarrow \mathcal{C}(r_j)$ and lets $c_j^0 = \pi_{k+1,n}(s_j^0)$. It then computes the correction value $\bar{c}_j = \pi_{k+1,n}(t_j) - c_j^0 - \pi_{k+1,n}(s_j^1)$ and sends this to P_r. Figure 3 also gives a quick overview of how these vectors are related.

When receiving the correction value $\bar{c}_j$, we notice that for the columns s_j^0 and s_j^1, P_r knows only the entries $w_j^i = s_j^{b_i}[i]$ where b_i is the choice-bit it received from $\mathcal{F}_{\text{ROT}}$ in the i'th invocation. For all $l \in [n-k]$, if $b_{k+l} = 1$ it is instructed to update its entry as follows:

$$w_j^{k+l} := \bar{c}_j[l] + w_j^{k+l} = t_j[k+l] - c_j^0[l] - s_j^1[k+l] + w_j^{k+l} = t_j[k+l] - c_j^0[l] .$$

Due to the above corrections, it is now the case that for all $l \in [n-k]$ if $b_{k+l} = 0$, then $w_j^{k+l} = c_j^0[l]$ and if $b_{k+l} = 1$, $w_j^{k+l} = t_j[k+l] - c_j^0[l]$. This means that at

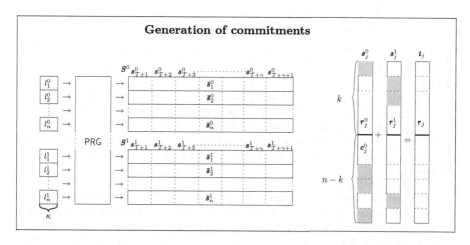

Fig. 3. On the left hand side we see how the initial part of the **Commit** phase of Π_{HCOM} is performed by P_s when committing to γ messages. On the right hand side we look at a single column of the two matrices S^0, S^1 and how they define the codeword t_j for column $j \in \mathcal{J}$, where $\mathcal{J} = \{\mathcal{T}+1, \ldots, \mathcal{T}+\gamma+1\}$.

this point, for all $j \in \mathcal{J}$ and all $i \in [n]$, P_r holds exactly one out of two shares for each entry of the codeword t_j that encodes the vector r_j.

The **Open** procedure describes how P_s can open to linear combinations of previously committed values. We let C be the indices to be opened and α_c for $c \in C$ be the corresponding coefficients. The sender then computes $r^0 = \sum_{c \in C} \alpha_c \cdot r_c^0$, $r^1 = \sum_{c \in C} \alpha_c \cdot r_c^1$, and $c^0 = \sum_{c \in C} \alpha_c \cdot c_c^0$ and sends these to P_r. When receiving the three values, the receiver computes the codeword $t \leftarrow \mathcal{C}(r^0 + r^1)$ and from c^0 and t it computes c^1. It also computes $w = \sum_{c \in C} \alpha_c \cdot w_c$ and verifies that r^0, r^1, c^0, and c^1 are consistent with these. If everything matches it accepts $r^0 + r^1$ as the value opened to.

If the sender P_s behaves honestly in **Commit** of Π_{HCOM}, then the scheme is UC-secure as it is presented until now. In fact it is also additively homomorphic due to the linearity of the code $\mathcal{C}$ and the linearity of additive secret sharing. However, this only holds because P_r holds shares of valid codewords. If we consider a malicious corruption of P_s, then the shares held by P_r might not be of valid codewords, and then it is undefined at commitment time what the value committed to is.

2.2   Optimizations over [CDD+15]

The work of [CDD+15] describes two commitment schemes, a basic and a homomorphic version. For both schemes therein the above issue of sending correct shares is handled by requiring the underlying code $\overline{\mathcal{C}}$ with parameters $[\overline{n}, k, \overline{d}]$ to have minimum distance $\overline{d} \geq 2s + 1$, as then the committed values are always defined to be the closest valid codewords of the receivers shares. This is however not enough to guarantee binding when allowing homomorphic operations.

Π_{HCOM} describes a protocol between a sender P_s and a receiver P_r. We let PRG : $\{0,1\}^{\kappa} \to \mathbb{F}^{\mathrm{poly}(\kappa)}$ be a pseudorandom generator with arbitrary polynomial stretch.

Init:

1. On common input $(\mathtt{init}, \mathtt{sid}, k)$ we assume the parties agree on a linear code $\mathcal{C}$ in systematic form over $\mathbb{F}$ with parameters $[n, k, d]$. The parties also initialize an internal commitment counter $\mathcal{T} = 0$.
2. For $i \in [n]$, both parties send $(\mathtt{transfer}, \mathtt{sid}, i, \kappa)$ to $\mathcal{F}_{\mathsf{ROT}}$. It replies with $\left(\mathtt{deliver}, \mathtt{sid}, i, \left(l_i^0, l_i^1\right)\right)$ to P_s and $\left(\mathtt{deliver}, \mathtt{sid}, i, \left(l_i^{b_i}, b_i\right)\right)$ to P_r.

Commit:

1. On common input $(\mathtt{commit}, \mathtt{sid}, \gamma)$, for $i \in [n]$, both parties use PRG to extend their received seeds into vectors of length $\mathcal{T} + \gamma + 1$. These are denoted $\bar{\boldsymbol{s}}_i^0, \bar{\boldsymbol{s}}_i^1 \in \mathbb{F}^{\mathcal{T}+\gamma+1}$ where P_s knows both and P_r knows $\bar{\boldsymbol{s}}_i^{b_i}$. Next define the matrices $\boldsymbol{S}^0, \boldsymbol{S}^1 \in \mathbb{F}^{n \times (\mathcal{T}+\gamma+1)}$ such that for $i \in [n]$ the i'th row of $\boldsymbol{S}^b$ is $\bar{\boldsymbol{s}}_i^b$ for $b \in \{0,1\}$.
2. Let $\mathcal{J} = \{\mathcal{T}+1, \dots, \mathcal{T}+\gamma+1\}$. For $j \in \mathcal{J}$ let the column vector of these matrices be $\boldsymbol{s}_j^0$, respectively $\boldsymbol{s}_j^1$. For $b \in \{0,1\}$, P_s lets $\boldsymbol{r}_j^b = \pi_k\left(\boldsymbol{s}_j^b\right)$ and lets $\boldsymbol{r}_j = \boldsymbol{r}_j^0 + \boldsymbol{r}_j^1$. Also P_r lets $\boldsymbol{w}_j = \left(w_j^1, \dots, w_j^n\right)$ and $(b_1, \dots, b_n) \leftarrow \boldsymbol{b}$ where $w_j^i = \boldsymbol{s}_j^{b_i}[i]$ for $i \in [n]$.
3. For $j \in \mathcal{J}$, P_s computes $\boldsymbol{t}_j \leftarrow \mathcal{C}\left(\boldsymbol{r}_j\right)$ and lets $\boldsymbol{c}_j^0 = \pi_{k+1,n}\left(\boldsymbol{s}_j^0\right)$. It then computes the correction value $\bar{\boldsymbol{c}}_j = \pi_{k+1,n}\left(\boldsymbol{t}_j\right) - \boldsymbol{c}_j^0 - \pi_{k+1,n}\left(\boldsymbol{s}_j^1\right)$.
4. Finally P_s sends the set $\left\{\bar{\boldsymbol{c}}_j\right\}_{j \in \mathcal{J}}$ to P_r. For $l \in [n-k]$ if $b_{k+l} = 1$, P_r updates $w_j^{k+l} := \bar{\boldsymbol{c}}_j[l] + w_j^{k+l}$.

Consistency Check [a]

5. P_r samples $(x_1, \dots, x_\gamma) \in_R \mathbb{F}^\gamma$ and sends these to P_s.
6. P_s then computes

$$\tilde{r}^0 = r_{\mathcal{T}+\gamma+1}^0 + \sum_{j=1}^{\gamma} x_j r_{\mathcal{T}+j}^0,$$

$$\tilde{r}^1 = r_{\mathcal{T}+\gamma+1}^1 + \sum_{j=1}^{\gamma} x_j r_{\mathcal{T}+j}^1, \quad \tilde{c}^0 = c_{\mathcal{T}+\gamma+1}^0 + \sum_{j=1}^{\gamma} x_j c_{\mathcal{T}+j}^0$$

and sends $\left(\tilde{r}^0, \tilde{r}^1, \tilde{c}^0\right)$ to P_r.

7. P_r computes $\tilde{w} = w_{\mathcal{T}+\gamma+1} + \sum_{j=1}^{\gamma} x_j w_{\mathcal{T}+j}$ and $\tilde{t} \leftarrow \mathcal{C}\left(\tilde{r}^0 + \tilde{r}^1\right)$. It lets $\tilde{c} \leftarrow \pi_{k+1,n}\left(\tilde{t}\right)$ and lets $\tilde{c}^1 = \tilde{c} - \tilde{c}^0$. Finally for $u \in [k]$ and $v \in [n-k]$, P_r verifies that $\tilde{r}^{b_u}[u] = \tilde{w}[u]$ and $\tilde{c}^{b_{k+v}}[v] = \tilde{w}[k+v]$. If not, P_r outputs abort and halts.

Output

8. Both parties increment their local counter $\mathcal{T} := \mathcal{T} + \gamma$. P_s now holds opening information $\left\{\left(r_j^0, r_j^1, c_j^0\right)\right\}_{j \in [\mathcal{T}]}$ and P_r holds the verifying information $\left\{w_j\right\}_{j \in [\mathcal{T}]}$. Let $\overline{\mathcal{J}} = \mathcal{J} \setminus \{\mathcal{T}+\gamma+1\}$. P_s outputs $\left(\mathtt{committed}, \mathtt{sid}, \overline{\mathcal{J}}, \{r_j\}_{j \in \overline{\mathcal{J}}}\right)$ and P_r outputs $\left(\mathtt{receipt}, \mathtt{sid}, \overline{\mathcal{J}}\right)$.

---

[a] The check is repeated $\hat{s}$ times, where $\hat{s}$ depends on $|\mathbb{F}|$.

Fig. 4. Protocol Π_{HCOM} UC-realizing $\mathcal{F}_{\mathsf{HCOM}}$ in the $\mathcal{F}_{\mathsf{ROT}}$-hybrid model – part 1.

Open:

1. On input $\left(\textbf{open}, \textbf{sid}, \{(c, \alpha_c)\}_{c \in C}\right)$ where each $\alpha_c \in \mathbb{F}$, if for all $c \in C$, P_s holds $\left(r_c^0, r_c^1, c_c^0\right)$ it computes

$$r^0 = \sum_{c \in C} \alpha_c \cdot r_c^0, \quad r^1 = \sum_{c \in C} \alpha_c \cdot r_c^1, \quad c^0 = \sum_{c \in C} \alpha_c \cdot c_c^0$$

   and sends $\left(\textbf{opening}, \{c, \alpha_c\}_{c \in C}, \left(r^0, r^1, c^0\right)\right)$ to P_r. Else it ignores the input message.

2. Upon receiving the message $\left(\textbf{opening}, \{c, \alpha_c\}_{c \in C}, \left(r^0, r^1, c^0\right)\right)$ from P_s, if for all $c \in C$, P_r holds w_c it lets $r = r^0 + r^1$ and computes $w = \sum_{c \in C} \alpha_c \cdot w_c$ and $t \leftarrow C(r)$. It lets $c = \pi_{k+1,n}(t)$ and computes $c^1 = c - c^0$. Finally for $i \in [k]$ and $l \in [n-k]$, P_r verifies that $r^{b_i}[i] = w[i]$ and $c^{b_{k+l}}[l] = w[k+l]$. If all checks are valid P_r outputs $\left(\textbf{opened}, \textbf{sid}, \{(c, \alpha_c)\}_{c \in C}, r\right)$. Else it aborts and halts.

Fig. 5. Protocol $\Pi_{\textsf{HCOM}}$ UC-realizing $\mathcal{F}_{\textsf{HCOM}}$ in the $\mathcal{F}_{\textsf{ROT}}$-hybrid model – part 2.

To support this, the authors propose a version of the scheme that involves the sender P_s running a "MPC-in-the-head" protocol based on a verifiable secret sharing scheme of which the views of the simulated parties must be sent to P_r.

Up until now the scheme we have described is very similar to the basic scheme of [CDD+15]. The main difference is the use of $\mathcal{F}_{\textsf{ROT}}$ as a starting assumption instead of $\mathcal{F}_{\textsf{OT}}$ and the way we define and send the committed value corrections. In [CDD+15] the corrections sent are for both the 0 and the 1 share. This means they send $2\bar{n}$ field elements for each commitment in total. Having the code in systematic form implies that for all $j \in \mathcal{J}$ and $i \in [k]$ the entries w_j^i are already defined for P_r as part of the output of the PRG, thus saving $2k$ field elements of communication per commitment. Together with only sending corrections to the 1-share, we only need to send $n - k$ field elements as corrections. Meanwhile this only commits the sender to a pseudorandom value, so to commit to a chosen value another correction of k elements needs to be sent. In total we therefore save a factor 2 of communication from these optimizations.

However the main advantage of our approach comes from ensuring that the shares held by P_r binds the sender P_s to his committed value, while only requiring a minimum distance of s. On top of that our approach is also additively homomorphic. The idea is that P_r will challenge P_s to open a random linear combination of all the committed values and check that these are valid according to $\mathcal{C}$. Recall that $\gamma + 1$ commitments are produced in total. The reason for this is to guarantee hiding for the commitments, even when P_r learns a random linear combination of them. Therefore, the linear combination is "blinded" by a pseudorandom value only used once and thus it appears pseudorandom to P_r as well. This consistency check is sufficient if $|\mathbb{F}|^{-1} \leq 2^{-s}$, however if the field is too small then the check is simply repeated $\hat{s} = \lceil s/\log_2(|\mathbb{F}|) \rceil$ times such that $|\mathbb{F}|^{-\hat{s}} \leq 2^{-s}$. In total this approach requires setting up commitments to $\hat{s}$ additional values for each invocation of **Commit**.

The intuition why the above approach works is that if the sender P_s sends inconsistent corrections, it will get challenged on these positions with high probability. In order to pass the check, P_s must therefore guess which choice-bit P_r holds for each position for which it sent inconsistent values. The challenge therefore forces P_s to make a decision at commitment time which underlying value to send consistent openings to, and after that it can only open to that value successfully. In fact, the above approach also guarantees that the scheme is homomorphic. This is because all the freedom P_s might have had by sending maliciously constructed corrections is removed already at commitment time *for all values*, so after this phase commitments and shares can be added together without issue.

To extract *all* committed values when receiving the opening to the linear combination the simulator identifies which rows of S^0 and S^1 P_s is sending inconsistent shares for. For these positions it inserts erasures in all positions of t_j (as defined by $S^0, S^1, \tilde{c}_j$ and C). As there are at most $s - 1$ positions where P_s could have cheated and the distance of the linear code is $d \geq s$ the simulator can erasure decode all columns to a unique value, and this is the only value P_s can successfully open to.[1]

2.3   Protocol Extension

The protocol Π_{HCOM} implements a commitment scheme where the sender commits to pseudorandom values. In many applications however it is needed to commit to chosen values instead. It is know that for any UC-secure commitment scheme one can easily turn a commitment from a random value into a commitment of a chosen one using the random value as a one-time pad encryption of the chosen value. For completeness, in Appendix A, we show this extension for any protocol implementing $\mathcal{F}_{\mathsf{HCOM}}$.

In addition we also highlight that all additively homomorphic commitment schemes support the notion of batch-opening. For applications where a large amount of messages need to be opened at the same time this has great implications on efficiency. The technique is roughly that P_s sends the values he wants to open directly to P_r. To verify correctness the receiver then challenges the sender to open to $\hat{s}$ random linear combinations of the received messages. For the same reason as for the consistency check of **Commit** this optimization retains binding. Using this method the overhead of opening the commitments is independent of the number of messages opened to and therefore amortizes away in the same manner as the consistency check and the initial OTs. However this way of opening messages has the downside of making the opening phase interactive, which is not optimal for all applications. See Appendix A for details.

The abovementioned batch-opening technique also has applicability when committing to large messages. Say we want to commit to a message m of length M. The naive approach would be to instantiate our scheme using a $[n_M, M, s]$ code. However this would require $n_M \geq M$ initial OTs and in addition only

---

[1] All linear codes can be efficiently erasure decoded if the number of erasures is $\leq d-1$.

achieve rate $M/(M+n_M) \geq 1/2$ in the opening phase. Instead of the above, the idea is to break the large message of length M into blocks of length l for $l \ll M$. There will now be $N = \lceil M/l \rceil$ of these blocks in total. We then instantiate our scheme with a $[n_s, l, s]$ code and commit to m in blocks of size l. When required to open we use the above-mentioned batch-opening to open all N blocks of m. It is clear that the above technique remains additively homomorphic for commitments to the large messages. In [GIKW14] they show an example for messages of size 2^{30} where they achieve rate $1.046^{-1} \approx 0.95$ in both the commit and open phase. In Appendix A we apply our above approach to the same setting and conclude that in the commit phase we achieve rate ≈ 0.974 and even higher in the opening phase. This is including the cost of the initial OTs.

3 Security

In this section we prove the following theorem.

Theorem 1. *The protocol* Π_{HCOM} *in Figs. 4 and 5 UC-realizes the* $\mathcal{F}_{HCOM}$ *functionality of Fig. 2 in the* $\mathcal{F}_{ROT}$-*hybrid model against any number of static corruptions.*

Proof. We prove security for the case with a dummy adversary, so that the simulator is outputting simulated values directly to the environment and is receiving inputs directly from the environment. We focus on the case with one call to **Commit**. The proof trivially lifts to the case with multiple invocations. The case with two static corruptions is trivial. The case with no corruptions follows from the case with a corrupted receiver, as in the ideal functionality $\mathcal{F}_{HCOM}$ the adversary is given all values which are given to the receiver, so one can just simulate the corrupted receiver and then output only the public transcript of the communication to the environment. We now first prove the case with a corrupted receiver and then the case with a corrupted sender.

Assume that P_r is corrupted. We use $\breve{P}_r$ to denote the corrupted receiver. This is just a mnemonic pseudonym for the environment $\mathcal{Z}$. The main idea behind the simulation is to simply run honestly until the opening phase. In the opening phase we then equivocate the commitment to the value received from the ideal functionality $\mathcal{F}_{HCOM}$ by adjusting the bits $\bar{s}_j^{1-b_i}$ not being watched by the receiver. This will be indistinguishable from the real world as the vectors $\bar{s}_i^{1-b_i}$ are indistinguishable from uniform in the view of $\breve{P}_r$ and if all the vectors $\bar{s}_i^{1-b_i}$ were uniform, then adjusting the bits not watched by $\breve{P}_r$ would be perfectly indistinguishable.

We first describe how to simulate the protocol without the step *Consistency Check*. We then discuss how to extend the simulation to this case.

The simulator $\mathcal{S}$ will run **Init** honestly, simulating $\mathcal{F}_{ROT}$ to $\breve{P}_r$. It then runs **Commit** honestly. On input $\big(\texttt{opened}, \texttt{sid}, \{(c, \alpha_c)\}_{c \in C}, r\big)$ it must simulate an opening.

In the simulation we use the fact that in the real protocol P_r can recompute all the values received from P_s given just the value r and the values w_c, which it already knows, and assuming that the checks $r^{b_i}[i] = w[i]$ and $c^{b_k+l}[l] = w[k+l]$ at the end of Fig. 5 are true. This goes as follows: First compute $w = \sum_{c \in C} \alpha_c \cdot w_c$, $t = C(r)$ and $c = \pi_{k+1,n}(t)$, as in the protocol. Then for $i \in [k]$ and $l \in [n-k]$ define

$$r^{b_i}[i] = w[i] , \quad c^{b_k+l}[l] = w[k+l] . \tag{1}$$

$$r^{1-b_i}[i] = r[i] - r^{b_i}[i] , \quad c^{1-b_k+l}[l] = c[l] - c^{b_k+l}[l] . \tag{2}$$

In (1) we use that the checks are true. In (2) we use that $r = r^0 + r^1$ and $c^1 = c - c^0$ by construction of P_r. This clearly correctly recomputes (r^0, r^1, c^0).

On input $(\text{opened}, \text{sid}, \{(c, \alpha_c)\}_{c \in C}, r)$ from $\mathcal{F}_{\text{HCOM}}$, the simulator will compute (r^0, r^1, c^0) from r and the values w_c known by $\check{P}_r$ as above and send $(\text{opening}, \{c, \alpha_c\}_{c \in C}, (r^0, r^1, c^0))$ to $\check{P}_r$.

We now argue that the simulation is computationally indistinguishable from the real protocol. We go via two hybrids.

We define *Hybrid I* as follows. Instead of computing the rows $\bar{s}_i^{1-b_i}$ from the seeds $l_i^{1-b_i}$ the simulator samples $\bar{s}_i^{1-b_i}$ uniformly at random of the same length. Since $\check{P}_r$ never sees the seeds $l_i^{1-b_i}$ and P_s only uses them as input to PRG, we can show that the view of $\check{P}_r$ in the simulation and Hybrid I are computationally indistinguishable by a black box reduction to the security of PRG.

We define *Hybrid II* as follows. We start from the real protocol, but instead of computing the rows $\bar{s}_i^{1-b_i}$ from the seeds $l_i^{1-b_i}$ we again sample $\bar{s}_i^{1-b_i}$ uniformly at random of the same length. As above, we can show that the view of $\check{P}_r$ in the protocol and Hybrid II are computationally indistinguishable.

The proof then concludes by transitivity of computational indistinguishability and by observing that the views of $\check{P}_r$ in Hybrid I and Hybrid II are perfectly indistinguishable. The main observation needed for seeing this is that in Hybrid I all the bits $r_j[i]$ are chosen uniformly at random and independently by $\mathcal{F}_{\text{HCOM}}$, whereas in Hybrid II they are defined by $r_j[i] = r_j^0[i] + r_j^1[i] = r_j^{b_i}[i] + r_j^{1-b_i}[i]$, where all the bits $r_j^{1-b_i}[i]$ are chosen uniformly at random and independently by $\mathcal{S}$. This yields the same distributions of the values r_j. All other value clearly have the same distribution.

We now address the step *Consistency Check*. The simulation of this step follows the same pattern as above. Define $\tilde{r} = \tilde{r}^0 + \tilde{r}^1$. This is the value from which $\tilde{t}$ is computed in Step 7 in Fig. 4. In the simulation and Hybrid I, instead pick $\tilde{r}$ uniformly at random and then recompute the values sent to $\check{P}_r$ as above. In Hybrid II compute $\tilde{r}$ as in the protocol (but still starting from the uniformly random $\bar{s}_i^{1-b_i}$). Then simply observe that $\tilde{r}$ has the same distribution in Hybrid I and Hybrid II. In Hybrid I it is uniformly random. In Hybrid II it is computed as $\tilde{r}^0 + \tilde{r}^1 = (r_{T+\gamma+1}^0 + r_{T+\gamma+1}^1) + \sum_{j=1}^{\gamma} x_j r_{T+j}$, and it is easy to see that $r_{T+\gamma+1}^0 + r_{T+\gamma+1}^1$ is uniformly random and independent of all other values in the view of $\check{P}_r$.

We now consider the case where the sender is corrupted who we denote $\check{P}_s$. The simulator will run the code of P_s honestly, simulating also $\mathcal{F}_{\text{ROT}}$ honestly. It will record the values (b_i, l_i^0, l_i^1) from **Init**. The remaining job of the simulator is then to extract the values $\tilde{r}_j$ to send to $\mathcal{F}_{\mathsf{HCOM}}$ in the command $\left(\texttt{corrupt-commit}, \mathsf{sid}, \{\tilde{r}_j\}_{j \in \mathcal{J}}\right)$. This should be done such that the probability that the receiver later outputs $(\texttt{opened}, \mathsf{sid}, \{(c, \alpha_c)\}_{c \in C}, r)$ for $r \neq \sum_{c \in C} \alpha_c \tilde{r}_c$ is at most 2^{-s}. We first describe how to extract the values $\tilde{r}_j$ and then show that the commitments are binding to these values.

We use the *Consistency Check* performed in the second half of Fig. 4 to define a set $E \subseteq \{1, \ldots, n\}$. We call this the erasure set. This name will make sense later, but for now think of E as the set of indices for which the corrupted sender $\check{P}_s$ after the consistency checks knows the choice bits b_i for $i \in E$ and for which the bits b_i for $i \notin E$ are still uniform in the view of $\check{P}_s$.

Define the column vectors s_j^0 and s_j^1 as in the protocol. This is possible as the seeds from $\mathcal{F}_{\mathsf{ROT}}$ are well defined. Following the protocol, and adding a few more definitions, define

$$r_j^0 = \pi_k(s_j^0) \ , \ r_j^1 = \pi_k(s_j^1) \ , \ r_j = r_j^0 + r_j^1 \ , \ u_j^0 = \pi_{k+1,n}(s_j^0) \ , \ u_j^1 = \pi_{k+1,n}(s_j^1) \ ,$$
$$u_j = u_j^0 + u_j^1 \ , \ t_j = \mathcal{C}(r_j) \ , \ c_j = \pi_{k+1,n}(t_j) \ , \ c_j^0 = u_j^0 \ , \ c_j^1 = c_j - c_j^0 \ ,$$
$$d_j^0 = u_j^0 \ , \ d_j^1 = u_j^1 + \bar{c}_j \ , \ d_j = d_j^0 + d_j^1 = u_j + \bar{c}_j \ , \ w_j^0 = r_j^0 \| d_j^0 \ , \ w_j^1 = r_j^1 \| d_j^1 .$$

Notice that if P_s is honest, then $\bar{c}_j = c_j - u_j$ and therefore $d_j = d_j^0 + d_j^1 = u_j^0 + u_j^1 + \bar{c}_j = c_j$. Hence d_j^0 and d_j^1 are the two shares of the non-systematic part c_j the same way that r_j^0 and r_j^1 are the two shares of the systematic part r_j. If the sender was honest we would in particular have that $w_j^0 + w_j^1 = r_j \| d_j = r_j \| c_j = \mathcal{C}(r_j)$, i.e., w_j^0 and w_j^1 would be the two shares of the whole codeword.

We can define the values that an honest P_s *should* send as

$$\tilde{r}^0 = r_{T+\gamma+1}^0 + \sum_j x_j r_j^0 \ , \ \tilde{r}^1 = r_{T+\gamma+1}^1 + \sum_j x_j r_j^1 \ , \ \tilde{c}^0 = c_{T+\gamma+1}^0 + \sum_j x_j c_j^0 \ .$$

These values can be used to define values

$$\tilde{r} = \tilde{r}^0 + \tilde{r}^1 \ , \ \tilde{t} = \mathcal{C}(\tilde{r}) \ , \ \tilde{c} = \pi_{k+1,n}(\tilde{t}) \ ,$$
$$\tilde{c}^1 = \tilde{c} - \tilde{c}^0 \ , \ \tilde{w}^0 = \tilde{r}^0 \| \tilde{c}^0 \ , \ \tilde{w}^1 = \tilde{r}^1 \| \tilde{c}^1 \ .$$

We use $(\check{r}^0, \check{r}^1, \check{c}^0)$ to denote the values actually sent by $\check{P}_s$ and we let the following denote the values computed by P_r (plus some extra definitions).

$$\check{r} = \check{r}^0 + \check{r}^1 \ , \ \check{t} = \mathcal{C}(\check{r}) \ , \ \check{c} = \pi_{k+1,n}(\check{t}) \ ,$$
$$\check{c}^1 = \check{c} - \check{c}^0 \ , \ \check{w}^0 = \check{r}^0 \| \check{c}^0 \ , \ \check{w}^1 = \check{r}^1 \| \check{c}^1 \ , \ \check{w} = \check{w}^0 + \check{w}^1 \ .$$

The simulator computes

$$\tilde{w} = w_{T+\gamma+1} + \sum_j x_j w_{T+j} \tag{3}$$

as P_r in the protocol. For later use, define $\tilde{w}^0 = w^0_{\mathcal{T}+\gamma+1} + \sum_j x_j w^0_{\mathcal{T}+j}$ and $\tilde{w}^1 = w^1_{\mathcal{T}+\gamma+1} + \sum_j x_j w^1_{\mathcal{T}+j}$.

The check performed by P_r is then simply to check for $u = 1, \ldots, n$ that

$$\check{w}^{b_u}[u] = \tilde{w}[u] . \tag{4}$$

Notice that in the protocol we have that $w_j = b * (w^1_j - w^0_j) + w^0_j$, where $*$ denotes the Schur product also known as the positionwise product of vectors. To see this notice that $(b*(w^1_j - w^0_j) + w^0_j)[i] = b_i(w^1_j[i] - w^0_j[i]) + w^0_j[i] = w^{b_i}_j[i]$. In other words, $w_j[i] = w^{b_i}_j[i]$. It then follows from (3) that $\tilde{w} = b*(\tilde{w}^1 - \tilde{w}^0) + \tilde{w}^0$, from which it follows that $\tilde{w}[u] = \tilde{w}^{b_u}[u]$. From (4) it then follows that $\check{P}_s$ passes the consistency check if and only if for $u = 1, \ldots, n$ it holds that $\check{w}^{b_u}[u] = \tilde{w}^{b_u}[u]$. We make some definitions related to this check. We say that a position $u \in [n]$ is *silly* if $\check{w}^0[u] \neq \tilde{w}^0[u]$ and $\check{w}^1[u] \neq \tilde{w}^1[u]$. We say that a position $u \in [n]$ is *clean* if $\check{w}^0[u] = \tilde{w}^0[u]$ and $\check{w}^1[u] = \tilde{w}^1[u]$. We say that a position $u \in [n]$ is *probing* if it is not silly or clean. Let E denote the set of probing positions u. Notice that if there is a silly position u, then $\check{w}^{b_u}[u] \neq \tilde{w}^{b_u}[u]$ so $\check{P}_s$ gets caught. We can therefore assume without loss of generality that there are no silly positions. For the probing positions $u \in E$, there is by definition a bit c_u such that $\check{w}^{1-c_u}[u] \neq \tilde{w}^{1-c_u}[u]$ and such that $\check{w}^{c_u}[u] = \tilde{w}^{c_u}[u]$. This means that $\check{P}_s$ passes the test only if $c_u = b_u$ for all $u \in E$. Since $\check{P}_s$ knows c_u it follows that if $\check{P}_s$ does not get caught, then it can guess b_u for $u \in E$ with probability 1.

Before we proceed to describe the extractor, we are now going to show two facts about E. First we will show that $|E| < s$, except with probability 2^{-s}. This follows from the simple observation that each b_u for $u \in E$ is uniformly random and $\check{P}_s$ passes the consistency test if and only if $c_u = b_u$ for $u \in E$ and the only information that $\check{P}_s$ has on the bits b_u is via the probing positions. Hence $\check{P}_s$ passes the consistency test with probability at most $2^{-|E|}$.

Second, let $\mathcal{C}_{-E}$ be the code obtained from $\mathcal{C}$ by puncturing at the positions $u \in E$, i.e., a codeword of $\mathcal{C}_{-E}$ can be computed as $t = \mathcal{C}(r)$ and then outputting t_{-E}, i.e., the vector t where we remove the positions $u \in E$. We show that for all $j = \mathcal{T} + 1, \ldots, \mathcal{T} + \gamma$ it holds that $(w^0_j + w^1_j)_{-E} \in \mathcal{C}_{-E}(\mathbb{F}^k)$, except with probability 2^{-s}. To see this, assume for the sake of contradiction that there exists such j where $(w^0_j + w^1_j)_{-E} \notin \mathcal{C}_{-E}(\mathbb{F}^k)$. Then the probability that it does not happen that $(\tilde{w}^0 + \tilde{w}^1)_{-E} \notin \mathcal{C}_{-E}(\mathbb{F}^k)$ is at most $|\mathbb{F}|^{-1}$, as a random linear combination of non-codewords become a codeword with probability at most $|\mathbb{F}|^{-1}$.[2] We repeat this test a number $\hat{s}$ of times such that $|\mathbb{F}|^{-\hat{s}} \leq 2^{-s}$. Since the tests succeed independently with probability at most $|\mathbb{F}|^{-1}$ it follows that $(\tilde{w}^0 + \tilde{w}^1)_{-E} \notin \mathcal{C}_{-E}(\mathbb{F}^k)$ except with probability 2^{-s}. Since by construction $\check{w}^0 + \check{w}^1 \in \mathcal{C}(\mathbb{F}^k)$, we have that $(\check{w}^0 + \check{w}^1)_{-E} \in \mathcal{C}_{-E}(\mathbb{F}^k)$, so when $(\tilde{w}^0 + \tilde{w}^1)_{-E} \notin \mathcal{C}_{-E}(\mathbb{F}^k)$ we either have that $(\tilde{w}^0)_{-E} \neq (\check{w}^0)_{-E}$ or $(\tilde{w}^1)_{-E} \neq (\check{w}^1)_{-E}$. Since there are no silly positions, this implies that we have a new probing position $u \notin E$, a contradiction to the definition of E.

---

[2] In it easy to see that in general, a random linear combination of vectors from outside some linear subspace will end up in the subspace with probability at most $|\mathbb{F}|^{-1}$.

We can now assume without loss of generality that $|E| < s$ and that $(\boldsymbol{w}_j^0 + \boldsymbol{w}_j^1)_{-E} \in \mathcal{C}_{-E}(\mathbb{F}^k)$. From $|E| < s$ and $\mathcal{C}$ having minimal distance $d \geq s$ we have that $\mathcal{C}_{-E}$ has minimal distance ≥ 1. Hence we can from each j and each $(\boldsymbol{w}_j^0 + \boldsymbol{w}_j^1)_{-E} \in \mathcal{C}_{-E}(\mathbb{F}^k)$ compute $\tilde{r}_j \in \mathbb{F}^k$ such that $(\boldsymbol{w}_j^0 + \boldsymbol{w}_j^1)_{-E} = \mathcal{C}_{-E}(\tilde{r}_j)$. These are the values that $\mathcal{S}$ will send to $\mathcal{F}_{\mathsf{HCOM}}$.

We then proceed to show that for all $\{(c, \alpha_c)\}_{c \in C}$ the environment can open to $(\mathsf{opened}, \mathsf{sid}, \{(c, \alpha_c)\}_{c \in C}, \tilde{r})$ for $\tilde{r} = \sum_{c \in C} \alpha_c \tilde{r}_c$ with probability 1. The reason for this is that if $\breve{P}_s$ computes the values in the opening correctly, then clearly $(\breve{w}^0)_{-E} = (\tilde{w}^0)_{-E}$ and $(\breve{w}^1)_{-E} = (\tilde{w}^1)_{-E}$. Furthermore, for the positions $u \in E$ it can open to any value as it knows b_u. It therefore follows that if $\breve{P}_s$ can open to $(\mathsf{opened}, \mathsf{sid}, \{(c, \alpha_c)\}_{c \in C}, r)$ for $r \neq \sum_{c \in C} \alpha_c \tilde{r}_c$, then it can open $\{(c, \alpha_c)\}_{c \in C}$ to two different values. Since the code has distance $d \geq s$, it is easy to see that after opening some $\{(c, \alpha_c)\}_{c \in C}$ to two different values, the environment can compute with probability 1 at least s of the choice bits b_u, which it can do with probability at most 2^{-s}, which is negligible.   □

4   Comparison with Recent Schemes

In this section we compare the efficiency of our scheme to the most efficient schemes in the literature realizing UC-secure commitments with security against a static and malicious adversary. In particular, we compare our construction to the schemes of [Lin11, BCPV13, CJS14, CDD+15].

The scheme of [BCPV13] (Fig. 6) is a slightly optimized version of [Lin11] (Protocol 2) which implement a multi-commitment ideal functionality. Along with [CJS14] these schemes support commitments between multiple parties natively, a property not shared with the rest of the protocols in this comparison. We therefore only consider the two party case where a sender commits to a receiver. The schemes of [Lin11, BCPV13] are in the CRS-model and their security relies on the DDH assumption. As the messages to be committed to are encoded as group elements the message size and the level of security are coupled in these schemes. For large messages this is not a big issue as the group size would just increase as well, or one can break the message into smaller blocks and commit to each block. However, for shorter messages, it is not possible to decrease the group size, as this would weaken security. The authors propose instantiating their scheme over an elliptic curve group over a field size of 256-bits so later in our comparison we also consider committing to values of this length. This is optimal for these schemes as the overhead of working with group elements of 256-bits would become more apparent if committing to smaller values.

The scheme of [CJS14] in the global random oracle model can be based on any stand-alone secure trapdoor commitment scheme, but for concreteness we compare the scheme instantiated with the commitment scheme of [Ped92] as also proposed by the authors. As [Ped92] is also based on the DDH assumption we use the same setting and parameters for [CJS14] as for the former two schemes.

We present our detailed comparison in Table 1. The table shows the costs of all the previously mentioned schemes in terms of OTs required, communication, number of rounds and computation. For the schemes of [CDD+15] we have fixed the sharing parameter t to 2 and 3 for the basic and homomorphic version, respectively. To the best of our knowledge this is also the optimal choice in all settings. Also for the scheme of [CJS14] we do not list the queries to the random oracle in the table, but remark that their scheme requires 6 queries per commitment. For our scheme, instead of counting the cost of sending the challenges $(x_1, x_2, \ldots, x_\gamma) \in \mathbb{F}$, we assume the receiver sends a random seed of size κ instead. This is then used as input to a PRG whose output is used to determine the challenges.

Table 1. Comparison of the most efficient UC-secure schemes for committing to γ messages of k components. Sizes are in bits. Legend: g is size of a group element, l is size of a scalar in the exponent, h is the output length of the random oracle, f is the size of a finite field element, $\hat{s}$ is the number of consistency checks performed, Exp. denotes the number of modular exponentiations, Enc. denotes the number of encoding procedures of the corresponding codes which have length $\bar{n}$ and n. The schemes of [CDD+15] are presented with the sharing parameter t set to 2 for the basic and 3 for the homomorphic.

Scheme	Homo	OTs $\binom{2}{1}$ $\binom{3}{2}$		Communication Commit	Open	Rounds Commit	Open	Computation Commit Exp.	Enc.	Open Exp.	Enc.
[Lin11]	✗	0	0	$4g$	$6g + 4l + k$	1	5	5	0	$18\frac{1}{3}$	0
[BCPV13]	✗	0	0	$4g$	$5g + 3l + k$	1	3	10	0	12	0
[CJS14]	✗	0	0	$4g + 2l + h$	$3l + 2h + 3\kappa + k$	2	3	5	0	5	0
[CDD+15], basic	✗	$\bar{n}$	0	$2\bar{n}f$	$(k + \bar{n} + 1)f$	1	1	0	1	0	1
[CDD+15], homo	✓	0	$\bar{n}$	$6(k+2\bar{n})\bar{n}f/k$	$(k + 2\bar{n} + 1)f$	1	1	0	$8\bar{n}/k + 2$	0	1
This Work	✓	n	0	$(2\hat{s}nf+\kappa)/\gamma + nf$	$(k + n + 1)f$	3	1	0	$2\hat{s}/\gamma + 1$	0	1

To give a flavor of the actual numbers we compute Table 1 for specific parameters in Table 2. We fix the field to $\mathbb{F}_2$ and look at computational security $\kappa = 128$, statistical security $s = 40$ and instantiate the random oracle required by [CJS14] with SHA-256. As the schemes of [Lin11, BCPV13, CJS14] rely on the hardness of the DDH assumption, a 256-bit EC group is assumed sufficient for 128-bit security [SRG+14]. As already mentioned we look at message length $k = 256$ as this is well suited for these schemes.[3] The best code we could find for the schemes of [CDD+15] in this setting has parameters $[631, 256, 81]$ and is a shortened BCH code. For our scheme, the best code we have identified for the above parameters is a $[419, 256, 40]$ expurgated BCH code [SS06]. Also, we recall the experiments performed in [CDD+15] showing that exponentiations in a EC-DDH group of the above size require roughly 500 times more computation time compared to encoding using a BCH code for parameters of the above type.[4] In their brief comparison with [HMQ04], another commitment scheme in

---

[3] We here assume a perfect efficient encoding of 256-bit values to group elements of a 256-bit EC group.

[4] They run the experiments with a shortened BCH code with parameters $[796, 256, 121]$, which therefore suggests their observations are also valid for our choice of parameters.

Table 2. Concrete efficiency comparison of the most efficient UC-secure schemes for committing to messages of size $k = 256$, $\kappa = 128$, $h = 256$ and $s = 40$ where the field is $\mathbb{F}_2$ and hence $\hat{s} = 40$. In the table γ represents the number of commitments the parties perform. These numbers include the cost of performing the initial OTs, both in terms of communication and computation.

Scheme	Homo	OTs $\binom{2}{1}$	$\binom{3}{2}$	Communication Commit	Open	Rounds Commit	Open	Computation Commit Exp.	Enc.	Open Exp.	Enc.
[Lin11]	✗	0	0	1,024	2,816	1	5	5	0	$18\frac{1}{3}$	0
[BCPV13]	✗	0	0	1,024	2,304	1	3	10	0	12	0
[CJS14]	✗	0	0	1,792	1,920	2	3	5	0	5	0
[CDD$^+$15], basic, $\gamma = 304$	✗	631	0	4,451	888	1	1	24	1	0	1
[CDD$^+$15], homo, $\gamma = 304$	✓	0	631	36,265	1,519	1	1	96	22	0	1
This Work, $\gamma = 304$	✓	419	0	2,647	676	3	1	16	1.3	0	1
[CDD$^+$15], basic, $\gamma = 1,000$	✗	631	0	2,232	888	1	1	7	1	0	1
[CDD$^+$15], homo, $\gamma = 1,000$	✓	0	631	26,649	1,519	1	1	28	22	0	1
This Work, $\gamma = 1,000$	✓	419	0	1,097	676	3	1	5	1	0	1
[CDD$^+$15], basic, $\gamma = 10,000$	✗	631	0	1,359	888	1	1	0	1	0	1
[CDD$^+$15], homo, $\gamma = 10,000$	✓	0	631	22,869	1,519	1	1	3	22	0	1
This Work, $\gamma = 10,000$	✓	419	0	487	676	3	1	0	1	0	1
[CDD$^+$15], basic, $\gamma = 100,000$	✗	631	0	1,272	888	1	1	0	1	0	1
[CDD$^+$15], homo, $\gamma = 100,000$	✓	0	631	22,491	1,519	1	1	0	22	0	1
This Work, $\gamma = 100,000$	✓	419	0	426	676	3	1	0	1	0	1

the random oracle model, the experiments showed that one of the above BCH encodings is roughly 1.6 times faster than 4 SHA-256 invocations, which is the number of random oracle queries required by [HMQ04]. This therefore suggests that one BCH encoding is also faster than the 6 random oracle queries required by [CJS14] if indeed instantiated with SHA-256.

To give as meaningful comparisons as possible we also instantiate the initial OTs and include the cost of these in Table 2. As the homomorphic version of [CDD+15] require 2-out-of-3 OTs in the setup phase, using techniques described in [LOP11, LP11], we have calculated that these require communicating 26 group elements and 44 exponentiations per invocation. The standard 1-out-of-2 OTs we instantiate with [PVW08] which require communicating 6 group elements and computing 11 exponentiations per invocation.

In Table 2 we do not take into consideration OT extension techniques [Bea96, IKNP03, Nie07, NNOB12, Lar15, ALSZ15, KOS15], as we do so few OTs that even the most efficient of these schemes *might* not improve the efficiency in practice. We note however that if in a setting where OT extension is already used, this would have a very positive impact on our scheme as the OTs in the setup phase would be much less costly. On a technical note some of the ideas used in this work are very related to the OT extension techniques introduced in [IKNP03] (and used in all follow-up work that make black-box use of a PRG). However an important and interesting difference is that in our work we do not "swap" the roles of the sender and receiver for the initial OTs as otherwise the case for current OT extension protocols. This observation means that the related work of [GIKW14], which makes use of OT extension, would look inherently different from our protocol, if instantiated with one of the OT extension protocols that follow the [IKNP03] blueprint.

As can be seen in Table 2, our scheme improves as the number of committed values γ grows. In particular we see that at around 304 commitments, for the above message sizes and security parameters, our scheme outperforms all previous schemes in total communication, while at the same time offering additive homomorphism.

A   Protocol Extension

As the scheme presented in Sect. 2 only implements commitments to random values we here describe an efficient extension to chosen message commitments. Our extension Π_{EHCOM} is phrased in the $\mathcal{F}_{\mathsf{HCOM}}$-hybrid model and it is presented in Fig. 6. The techniques presented therein are folklore and are known to work for any UC-secure commitment scheme, but we include them as a protocol extension for completeness. The **Chosen-Commit** step shows how one can turn a commitment of a random value into a commitment of a chosen value. This is done by simply using the committed random value as a one-time pad on the chosen value and sending this to P_r. The **Extended-Open** step describes how to open to linear combinations of either random commitments, chosen commitments or both. It works by using $\mathcal{F}_{\mathsf{HCOM}}$ to open to the random commitments and the commitments used to one-time pad the chosen commitments. Together with the previously sent one-time pad the receiver can then learn the designated linear combination.

Finally we present a **Batch-Open** step that achieves very close to optimal amortized communication complexity for opening to a set of messages. The technique is similar to the consistency check of Π_{HCOM}. When required to open to a set of messages, the sender P_s will start by sending the messages directly to the receiver P_r. Next, the receiver challenges the sender to open to a random linear combination of all the received messages. When receiving the opening from $\mathcal{F}_{\mathsf{HCOM}}$, P_r verifies that it is consistent with the previously received messages and if this is the case it accepts these. For the exact same reasons as covered in the proof of Theorem 1 it follows that this approach of opening values is secure. For clarity and ease of presentation the description of batch-opening does not take into account opening to linear combinations of random and chosen commitments. However the procedure can easily be extended to this setting using the same approach as in **Extended-Open**.

In terms of efficiency, to open N commitments with message-size l, the sender needs to send lN field elements along with the verification overhead $\hat{s}\hat{O} + \kappa$ where $\hat{O}$ is the cost of opening to a commitment using $\mathcal{F}_{\mathsf{HCOM}}$. Therefore if the functionality is instantiated with the scheme Π_{HCOM}, the total communication for batch-opening is $\hat{s}(k + n)f + \kappa + kNf$ bits where k is the length of the message, n is the length of the code used, f is the size of a field element and $\hat{s}$ is the number of consistency checks needed.

We now elaborate on the applicability of batch-opening for committing to large messages as mentioned in Sect. 2.3. Recall that there we split the large

Π_{EHCOM} describes a protocol between a sender P_s and a receiver P_r.

Chosen-Commit:

1. On input $(\text{chosen-commit}, \text{sid}, \text{cid}, m)$, P_s picks an already committed to value r_j and computes $\widetilde{m} = m - r_j$. It then sends $(\text{chosen}, \text{sid}, \text{cid}, j, \widetilde{m})$ to P_r. Else it ignores the message.
2. P_r stores $(\text{chosen}, \text{sid}, \text{cid}, j, \widetilde{m})$ and outputs $(\text{chosen-receipt}, \text{sid}, \text{cid})$.

Extended-Open:

1. On input $\left(\text{extended-open}, \text{sid}, \{(j, \alpha_j)\}_{j \in C_r}, \{(l, \beta_l)\}_{l \in C_c}\right)$ with $\beta_l \in \mathbb{F}$ for $l \in C_c$ and $\alpha_j \in \mathbb{F}$, for $j \in C_r$, P_s verifies that and it has previously committed to a value r_j using $\mathcal{F}_{\text{HCOM}}$ for $j \in C_r$. Else it ignores the message. For all $l \in C_c$ P_s verifies that it previously sent the message $\left(\text{chosen}, \text{sid}, l, \overline{j}, \widetilde{m}_l\right)$ to P_r. Let $\overline{\mathcal{J}}$ be the set of the corresponding indices $\overline{j}$, similarly let $\overline{\beta_{\overline{j}}} = \beta_l$ for the corresponding ID l. P_s then sends $\left(\text{open}, \text{sid}, \{(j, \alpha_j)\}_{j \in C_r} \cup \left\{\left(\overline{j}, \overline{\beta_{\overline{j}}}\right)\right\}_{\overline{j} \in \overline{\mathcal{J}}}\right)$ to $\mathcal{F}_{\text{HCOM}}$.
2. Upon receiving $\left(\text{open}, \text{sid}, \{(j, \alpha_j)\}_{j \in C_r} \cup \left\{\left(\overline{j}, \overline{\beta_{\overline{j}}}\right)\right\}_{\overline{j} \in \overline{\mathcal{J}}}, r\right)$ from $\mathcal{F}_{\text{HCOM}}$, P_r identifies the previously received messages $\left(\text{chosen}, \text{sid}, l, \overline{j}, \widetilde{m}_l\right)$ and outputs $\left(\text{extended-opened}, \text{sid}, \{(j, \alpha_j)\}_{j \in C_r}, \{(l, \beta_l)\}_{l \in C_c}, r + \sum_{l \in C_c} \beta_l \cdot \widetilde{m}_l\right)$.

Batch-Open:[a]

1. On input $(\text{batch-open}, \text{sid}, C_r, C_c)$. For all $j \in C_r$ P_s verifies that it has previously committed to a value r_j using $\mathcal{F}_{\text{HCOM}}$. Else it ignores the message. For all $l \in C_c$ P_s verifies that it previously sent the message $\left(\text{chosen}, \text{sid}, l, \overline{j}, \widetilde{m}_l\right)$ to P_r. P_s then sends $\left(\text{batch-open}, \text{sid}, \{(j, r_j)\}_{j \in C_r}, \{(l, m_l)\}_{l \in C_c}\right)$ to P_r, where r_j and m_l are random and chosen messages, respectively, previously committed to.
2. Let $t_r = |C_r|$ and $t_c = |C_c|$. P_r then samples uniformly random values $x_1, \ldots, x_{t_r}, y_1, \ldots, y_{t_c} \in \mathbb{F}$ and sends these to P_s.
3. P_s and P_r run **Extended-Open** with input

$$\left(\text{extended-open}, \text{sid}, \{(j_u, x_u)\}_{u \in [t_r]}, \{(l_v, y_v)\}_{v \in [t_c]}\right)$$

where j_u and l_v are the u'th and v'th element of C_r and C_c respectively, under an arbitrary ordering.
4. P_r lets $\left(\text{extended-opened}, \text{sid}, \{(j_u, x_u)\}_{u \in [t_r]}, \{(l_v, y_v)\}_{v \in [t_c]}, n\right)$ be the output of running **Extended-Open**. P_r now verifies that

$$n = \sum_{u \in [t_r]} x_u \cdot r_{j_u} + \sum_{v \in [t_c]} y_v \cdot m_{l_v} .$$

If true then P_r outputs $\left(\text{batch-opened}, \text{sid}, \{(j, r_j)\}_{j \in C_r} \cup \{(l, m_l)\}_{l \in C_c}\right)$. Else it aborts and halts.

---

[a] The check is repeated $\hat{s}$ times, where $\hat{s}$ depends on $|\mathbb{F}|$.

Fig. 6. Protocol Π_{EHCOM} in the $\mathcal{F}_{\text{HCOM}}$-hybrid model.

message m of size M into N blocks of size l and the idea is to instantiate Π_{HCOM} with a $[n_s, l, s]$ code and commit to m in blocks of size l. This requires n_s initial OTs to setup and requires sending $(2\hat{s}n_s)f + \kappa + lNf$ bits to commit to all blocks. For a fixed s this has rate close to 1 for large enough l. In the opening phase we can then use the above batch-opening technique to open to all the blocks of the original message, and thus achieve a rate of $Mf/\hat{s}(l+n_s)f+\kappa+lNf \approx 1$ in the opening phase as well.

In [GIKW14] the authors present an example of committing to strings of length 2^{30} with statistical security $s = 30$ achieving rate $1.046^{-1} \approx 0.95$ in both the commit and open phase. To achieve these number the field size is required to be very large as well. The authors propose techniques to reduce the field size, however at the cost of reducing the rate. We will instantiate the approach described above using a binary BCH code over the field $\mathbb{F}_2$ and recall that these have parameters $[n - 1, n - \lceil \frac{d-1}{2} \rceil \log(n + 1), \geq d]$. Using a block length of 2^{13} and $s = 30$ therefore gives us a code with parameters $[8191, 7996, 30]$. Thus we split the message into $134,285 = \lceil 2^{30}/7996 \rceil$ blocks. In the commitment phase we therefore achieve rate $2^{30}/2 \cdot 30 \cdot 8191+128+8191 \cdot 134,285 \approx 0.976$. Using the batch-opening technique the rate in the opening phase is even higher than in the commit phase, as this does not require any "blinding" values. In the above calculations we do not take into account the 8191 initial OTs required to setup our scheme. However using the OT-extension techniques of [KOS15], each OT for κ-bit strings can be run using only κ initial "seed" OTs and each extended OT then requires only κ bits of communication. Instantiating the seed OTs with the protocol of [PVW08] for $\kappa = 128$ results in $6 \cdot 256 \cdot 128 + 8191 \cdot 128 = 1,245,056$ extra bits of communication which lowers the rate to 0.974.

Finally, based on local experiments with BCH codes with the above parameters, we observe that the running time of an encoding operation using the above larger parameters is roughly 2.5 times slower than an encoding using a BCH code with parameters $[796, 256, 121]$. This suggests that the above approach remains practical for implementations as well.

References

[AHMR15] Afshar, A., Hu, Z., Mohassel, P., Rosulek, M.: How to efficiently evaluate RAM programs with malicious security. In: Oswald, E., Fischlin, M. (eds.) EUROCRYPT 2015. LNCS, vol. 9056, pp. 702–729. Springer, Heidelberg (2015)

[ALSZ15] Asharov, G., Lindell, Y., Schneider, T., Zohner, M.: More efficient oblivious transfer extensions with security for malicious adversaries. In: Oswald, E., Fischlin, M. (eds.) EUROCRYPT 2015. LNCS, vol. 9056, pp. 673–701. Springer, Heidelberg (2015)

[BCPV13] Blazy, O., Chevalier, C., Pointcheval, D., Vergnaud, D.: Analysis and improvement of Lindell's UC-secure commitment schemes. In: Jacobson, M., Locasto, M., Mohassel, P., Safavi-Naini, R. (eds.) ACNS 2013. LNCS, vol. 7954, pp. 534–551. Springer, Heidelberg (2013)

[Bea96] Beaver, D.: Correlated pseudorandomness and the complexity of private computations. In: 28th ACM STOC, pp. 479–488. ACM Press (1996)

[BR93]   Bellare, M., Rogaway, P.: Random oracles are practical: a paradigm for designing efficient protocols. In: Ashby, V. (ed.) ACM CCS 1993, pp. 62–73. ACM Press (1993)

[Can01]   Canetti, R.: Universally composable security: a new paradigm for cryptographic protocols. In: 42nd FOCS, pp. 136–145. IEEE Computer Society Press (2001)

[CC06]   Chen, H., Cramer, R.: Algebraic geometric secret sharing schemes and secure multi-party computations over small fields. In: Dwork, C. (ed.) CRYPTO 2006. LNCS, vol. 4117, pp. 521–536. Springer, Heidelberg (2006)

[CDD+15] Cascudo, I., Damgård, I., David, B., Giacomelli, I., Nielsen, J.B., Trifiletti, R.: Additively homomorphic UC commitments with optimal amortized overhead. In: Katz, J. (ed.) PKC 2015. LNCS, vol. 9020, pp. 495–515. Springer, Heidelberg (2015)

[CF01]   Canetti, R., Fischlin, M.: Universally composable commitments. In: Kilian, J. (ed.) CRYPTO 2001. LNCS, vol. 2139, pp. 19–40. Springer, Heidelberg (2001)

[CJS14]   Canetti, R., Jain, A., Scafuro, A.: Practical UC security with a global random oracle. In: Ahn, G.-J., Yung, M., Li, N. (ed.) ACM CCS 2014, pp. 597–608. ACM Press (2014)

[CLOS02] Canetti, R., Lindell, Y., Ostrovsky, R., Sahai, A.: Universally composable two-party and multi-party secure computation. In: 34th ACM STOC, pp. 494–503. ACM Press (2002)

[CvT95]   Crépeau, C., van de Graaf, J., Tapp, A.: Committed oblivious transfer and private multi-party computation. In: Coppersmith, D. (ed.) CRYPTO 1995. LNCS, vol. 963, pp. 110–123. Springer, Heidelberg (1995)

[DDGN14] Damgård, I., David, B., Giacomelli, I., Nielsen, J.B.: Compact VSS and efficient homomorphic UC commitments. In: Sarkar, P., Iwata, T. (eds.) ASIACRYPT 2014, Part II. LNCS, vol. 8874, pp. 213–232. Springer, Heidelberg (2014)

[DG03]   Damgård, I., Groth, J.: Non-interactive and reusable non-malleable commitment schemes. In: 35th ACM STOC, pp. 426–437. ACM Press (2003)

[DN02]   Damgård, I.B., Nielsen, J.B.: Perfect hiding and perfect binding universally composable commitment schemes with constant expansion factor. In: Yung, M. (ed.) CRYPTO 2002. LNCS, vol. 2442, pp. 581–596. Springer, Heidelberg (2002)

[FJN+13] Frederiksen, T.K., Jakobsen, T.P., Nielsen, J.B., Nordholt, P.S., Orlandi, C.: MiniLEGO: efficient secure two-party computation from general assumptions. In: Johansson, T., Nguyen, P.Q. (eds.) EUROCRYPT 2013. LNCS, vol. 7881, pp. 537–556. Springer, Heidelberg (2013)

[FJNT15] Frederiksen, T.K., Jakobsen, T.P., Nielsen, J.B., Trifiletti, R.: TinyLEGO: an interactive garbling scheme for maliciously secure two-party computation. Cryptology ePrint Archive, Report 2015/309 (2015). http://eprint.iacr.org/2015/309

[Fuj14]   Fujisaki, E.: All-but-many encryption. In: Sarkar, P., Iwata, T. (eds.) ASIACRYPT 2014, Part II. LNCS, vol. 8874, pp. 426–447. Springer, Heidelberg (2014)

[GIKW14] Garay, J.A., Ishai, Y., Kumaresan, R., Wee, H.: On the complexity of UC commitments. In: Nguyen, P.Q., Oswald, E. (eds.) EUROCRYPT 2014. LNCS, vol. 8441, pp. 677–694. Springer, Heidelberg (2014)

[HIKN08] Harnik, D., Ishai, Y., Kushilevitz, E., Nielsen, J.B.: OT-combiners via secure computation. In: Canetti, R. (ed.) TCC 2008. LNCS, vol. 4948, pp. 393–411. Springer, Heidelberg (2008)

[HMQ04] Hofheinz, D., Müller-Quade, J.: Universally composable commitments using random oracles. In: Naor, M. (ed.) TCC 2004. LNCS, vol. 2951, pp. 58–76. Springer, Heidelberg (2004)

[IKNP03] Ishai, Y., Kilian, J., Nissim, K., Petrank, E.: Extending oblivious transfers efficiently. In: Boneh, D. (ed.) CRYPTO 2003. LNCS, vol. 2729, pp. 145–161. Springer, Heidelberg (2003)

[IKOS07] Ishai, Y., Kushilevitz, E., Ostrovsky, R., Sahai, A.: Zero-knowledge from secure multiparty computation. In: Johnson, D.S., Feige,U. (eds.) 39th ACM STOC, pp. 21–30. ACM Press (2007)

[IPS08] Ishai, Y., Prabhakaran, M., Sahai, A.: Founding cryptography on oblivious transfer – efficiently. In: Wagner, D. (ed.) CRYPTO 2008. LNCS, vol. 5157, pp. 572–591. Springer, Heidelberg (2008)

[Kil88] Kilian, J.: Founding cryptography on oblivious transfer. In: 20th ACM STOC, pp. 20–31. ACM Press (1998)

[KOS15] Keller, M., Orsini, E., Scholl, P.: Actively secure OT extension with optimal overhead. In: Gennaro, R., Robshaw, M.J.B. (eds.) CRYPTO 2015, Part I. LNCS, vol. 9215, pp. 724–741. Springer, Heidelberg (2015)

[Lar15] Larraia, E.: Extending oblivious transfer efficiently. In: Aranha, D.F., Menezes, A. (eds.) LATINCRYPT 2014. LNCS, vol. 8895, pp. 368–386. Springer, Heidelberg (2015)

[Lin11] Lindell, Y.: Highly-efficient universally-composable commitments based on the DDH assumption. In: Paterson, K.G. (ed.) EUROCRYPT 2011. LNCS, vol. 6632, pp. 446–466. Springer, Heidelberg (2011)

[LOP11] Lindell, Y., Oxman, E., Pinkas, B.: The IPS compiler: optimizations, variants and concrete efficiency. In: Rogaway, P. (ed.) CRYPTO 2011. LNCS, vol. 6841, pp. 259–276. Springer, Heidelberg (2011)

[LP11] Lindell, Y., Pinkas, B.: Secure two-party computation via cut-and-choose oblivious transfer. In: Ishai, Y. (ed.) TCC 2011. LNCS, vol. 6597, pp. 329–346. Springer, Heidelberg (2011)

[Nao90] Naor, M.: Bit commitment using pseudo-randomness. In: Brassard, G. (ed.) CRYPTO 1989. LNCS, vol. 435, pp. 128–136. Springer, Heidelberg (1990)

[NFT09] Nishimaki, R., Fujisaki, E., Tanaka, K.: Efficient non-interactive universally composable string-commitment schemes. In: Pieprzyk, J., Zhang, F. (eds.) ProvSec 2009. LNCS, vol. 5848, pp. 3–18. Springer, Heidelberg (2009)

[Nie07] Nielsen, J.B.: Extending oblivious transfers efficiently - how to get robustness almost for free. Cryptology ePrint Archive, Report 2007/215 (2007). http://eprint.iacr.org/2007/215

[NNOB12] Nielsen, J.B., Nordholt, P.S., Orlandi, C., Burra, S.S.: A new approach to practical active-secure two-party computation. In: Safavi-Naini, R., Canetti, R. (eds.) CRYPTO 2012. LNCS, vol. 7417, pp. 681–700. Springer, Heidelberg (2012)

[NO09] Nielsen, J.B., Orlandi, C.: LEGO for two-party secure computation. In: Reingold, O. (ed.) TCC 2009. LNCS, vol. 5444, pp. 368–386. Springer, Heidelberg (2009)

[Ped92] Pedersen, T.P.: Non-interactive and information-theoretic secure verifiable secret sharing. In: Feigenbaum, J. (ed.) CRYPTO 1991. LNCS, vol. 576, pp. 129–140. Springer, Heidelberg (1992)

[PVW08] Peikert, C., Vaikuntanathan, V., Waters, B.: A framework for efficient and composable oblivious transfer. In: Wagner, D. (ed.) CRYPTO 2008. LNCS, vol. 5157, pp. 554–571. Springer, Heidelberg (2008)

[SRG+14] Smart, N.P., Rijmen, V., Gierlichs, B., Paterson, K.G., Stam, M., Warinschi, B., Gaven, W.: Algorithms, key size and parameters report 2014 (2014)

[SS06] Schürer, R., Schmid, W.C.: Mint: a database for optimal net parameters. In: Niederreiter, H., Talay, D. (eds.) Monte Carlo and Quasi-Monte Carlo Methods 2004, pp. 457–469. Springer, Heidelberg (2006)

Simplified Universal Composability Framework

Douglas Wikström[(✉)]

KTH Royal Institute of Technology, Stockholm, Sweden
dog@kth.se

Abstract. We introduce a simplified universally composable (UC) security framework in our thesis (2005). In this paper we present an updated more comprehensive and illustrated version. The introduction of our simplified model is motivated by the difficulty to describe and analyze concrete protocols in the full UC framework due to its generality and complexity.

The main differences between our formalization and the general UC security framework are that we consider: a fixed number of parties, static corruption, and simple ways to bound the running times of the adversary and environment. However, the model is easy to extend to adaptive adversaries. Authenticated channels become a trivial ideal functionality.

We generalize the framework to allow protocols to securely realize other protocols. This allows a natural and modular description and analysis of protocols.

We introduce invertible transforms of models that allow us to reduce the proof of the composition theorem to a simple special case and transform any hybrid protocol into a hybrid protocol with at most one ideal functionality. This factors out almost all of the technical details of our framework to be considered when relating our framework to any other security framework, e.g., the UC framework, and makes this easy.

1 Introduction

Canetti [3], and independently Pfitzmann and Waidner [11] propose security frameworks for reactive processes. Both frameworks have composition theorems, and are based on older definitional work. The initial ideal-model based definitional approach for secure function evaluation is informally proposed by Goldreich, Micali, and Wigderson in [6]. The first formalizations appear in Goldwasser and Levin [7], Micali and Rogaway [10], and Beaver [1]. Canetti [2] presents the first definition of security that is preserved under composition. See [2,3] for an excellent background.

The basic approach of all these models is the same. An ideal functionality is defined that implicitly captures the functionality and security properties we expect from a real protocol. The real protocol is then said to be secure if it is indistinguishable from the ideal functionality by any efficient distinguisher. However, in an execution of the real protocol the adversary may influence the execution or extract information that it passes on to the distinguisher. Thus, we introduce an simulation adversary (simulator) that is given the same task,

© International Association for Cryptologic Research 2016
E. Kushilevitz and T. Malkin (Eds.): TCC 2016-A, Part I, LNCS 9562, pp. 566–595, 2016.
DOI: 10.1007/978-3-662-49096-9_24

but when interacting with the ideal functionality. The ideal functionality is secure by inspection, so the simulation adversary can by definition not attack the ideal functionality in any meaningful way. Instead it must simulate a real attack to the distinguisher. The definition of security then says that if for every real adversary there exists an simulation adversary such that no efficient distinguisher can distinguish: (1) an interaction with the real protocol and the real adversary from (2) an interaction with the ideal functionality and the simulation adversary, then the real protocol is said to securely realize the ideal functionality.

The UC framework is an ambitious attempt to capture the security of a wide range of settings in a uniform way, but the original UC framework was flawed in several ways. The most recent version of the online paper [3] contains a discussion about the issues and pointers to relevant literature. However, the core ideas of the UC framework are correct, and there are no flaws in the basic instantiations needed to prove the security of practical protocols. In this paper we detail one possible instantiation, but before we do so, we point out the main areas where our particular instantiation is more restricted, and hence less complex, than the general framework.

Canetti assumes the existence of an "operating system" that takes care of the instantiation of subprotocols when needed. This is necessary to handle dynamically instantiated subprotocols, but in our application we may assume that all subprotocols are instantiated at the start of the execution. This means that we can view each instance of a subprotocol as a separate Turing machine that exists from scratch that interacts with the invoking protocol with a predefined session identifier.

Canetti models an asynchronous communication network, where the adversary has the power to delete, modify, and insert any messages of his choice. To do this he is forced to give details for exactly what the adversary is allowed to do to messages passed in different ways between interactive Turing machines, which quickly becomes quite complex. We instead factor out all aspects of the communication network into a separate concrete "communication model"-machine. The real, ideal, and hybrid models are then defined solely by how certain machines are linked. The adversary is defined as any interactive Turing machine, and how the adversary can interact with other machines also follows implicitly from the definitions of the real and ideal communication models. With our approach there is also no need for session identifiers.

The above means that the real, ideal, and hybrid models can not only be illustrated by a graph of connected parties, they *are* graphs of Turing machines in a very tangible way, which makes the composition theorem almost trivial.

There are several ways to model corruption in cryptographic protocols. In this paper, we only consider *static* corruption, i.e., the adversary must decide which parties to corrupt before the execution starts. However, it is straightforward to extend the model to adaptive corruption as explained in Remark 2. Even dynamic adversaries could be handled in a similar way, so there is no inherent restriction to static adversaries.

1.1   Contribution

We present a precise and workable security framework using modularized definitions that are easily verified to be sound. Abstractions emerge in a natural way that are firmly grounded in the underlying definitions. Although our treatment may initially seem more complex than the description of the UC framework, the actual content is captured faithfully in simple drawings that are enough to understand the framework, and the composition theorem becomes almost trivial.

Explicit invertible transforms are introduced that can turn any hybrid model into a hybrid model with a single ideal functionality (or a real model). Thus, it suffices to consider how the security of such a protocol in our simplified UC framework relates to its security in any other security framework, in particular the UC framework. This also immediately generalizes the single composition theorem to allow multiple compositions.

We introduce a novel generalization the UC framework and other frameworks we are aware of in that the definition of security captures the case where a hybrid protocol securely realizes another hybrid protocol, and not only ideal functionalities. This allows a novel type of proof that is not only based on securely realizing ideal functionalities and applying the composition theorem. We give natural examples where this technique is applicable.

The essential restriction in our framework compared to general UC is that the set of parties and the protocol, including all subprotocols and ideal functionalities used, are determined at the start of the execution.

1.2   Related Work

Several frameworks have been proposed today, but we only mention two frameworks that perhaps are closest to our framework at a philosphical level.

Constructive cryptography was developed and proposed by Maurer and Renner [8,9] independently of our work. The design of cryptographic primitives and protocols in this framework is viewed as the construction of an ideal resource from assumed or real resources. It shares with our framework the aims of achieving simplicity and eliminating irrelevant artefacts. We have not carried out a detailed analysis of the relations between their model and ours, but we are currently corresponding with the authors.

In subsequent, but independent work, Canetti et al. [4] propose an alternative formalization of a simplified UC framework motivated by the same problems as we do, and to some extent they use also the same approach as we do. Their motivation and the restrictions they introduce compared to the full UC framework are the same. Several features of the formalization that distinguishes it from the UC framework are also similar, e.g., their explicit "router" corresponds to our "communication model".

We consider the main difference between our framework and theirs to be that they use top-down approach, whereas we gradually build the model from the bottom up. They explicitly relate their model to the general UC model. We instead provide transforms that allow us to relate our framework to any other framework with ease, since today there are many proposals of security framework and it is nearly impossible to understand each framework sufficiently well to perform a valid comparison.

That said, we hope that the reader takes the time to read both papers, since they both attempt to capture the core ideas of the UC framework in a way that is easier to understand and use.

2   Interactive Turing Machines

Parties and algorithms are modeled as probabilistic Turing machines, but to be able to talk about multiple parties that interact with each other we need to augment this model with a notion of communication. We follow the approach of Goldreich [5] and Canetti [3] and define *interactive Turing machines*, but we replace the activation bit used by Goldreich by a slightly more complicated gadget to allow seamless treatment of multiparty protocols.

Definition 1 (Interactive Turing Machine). *An* interactive Turing machine *(ITM) is a Turing machine with the following tapes and tape heads in addition to its work tapes: a read-only identity tape, a read-only security parameter tape, a read-once input tape, a write-once output tape, a read-once random tape, a write-once send head s, a read-once receive head r, and two single-bit read/write activity heads a_s and a_r. The following restrictions apply to an ITM, where we use brackets to indicate the value stored in the cell pointed at by a tape head.*

1. *If $([a_s], [a_r]) \in \{(0,0), (1,0)\}$, then it is* inactive *and can not change its state in a state transition, or read, write, or move on any tape.*
2. *If $([a_s], [a_r]) = (0,1)$, then it is* active *and can change its state in a state transition.*
3. *A special instruction allows it to atomically: set $([a_s], [a_r]) = (1,0)$ and become inactive.*

Note that a single ITM is not a complete computational model, since some tape heads do not have matching tapes. Two ITM's are connected by adding

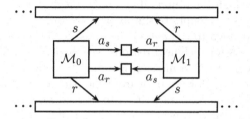

Fig. 1. The ITM's $\mathcal{M}_0$ and $\mathcal{M}_1$ share activation and send/receive tapes. The send head of $\mathcal{M}_0$ points to same tape as the receive head of $\mathcal{M}_1$ and vice versa. A corresponding configuration is used for the activation tapes. The figure does not contain the other tapes of the ITM's.

the missing tapes and pairing the write-once send head of one party with the read-once receive head of the other and the activity head a_s of one party with the activity head a_r of the other. Intuitively, the activation tapes implement an "activation token" that is passed back and forth between the parties. This is illustrated in Fig. 1. We denote the set of all ITM's by ITM.

3   Graph of Interactive Turing Machines

To connect multiple ITM's with each other without introducing extra tapes for each machine and thereby change the computational model, we introduce a gadget that plays the role of a router. A router is a Turing machine with several sets of tape heads that can share tapes with interactive Turing machines (ITM) or other routers.

Definition 2 (Router). *An l-router is a Turing machine with write-once send heads denoted $s_0, \ldots, s_l$, read-once receive heads, denoted $r_0, \ldots, r_l$, and single-bit read/write activity heads $a_{s,i}$ and $a_{r,i}$ for $i = [0, l]$ such that $\sum_{i=0}^{k}([a_{s,i}] + [a_{r,i}]) \in \{0, 1\}$.*

Active. *If $[a_{r,i}] = 1$ for some $i \in [0, k]$, then it is active and proceeds as follows.*

1. *To form a string w it reads and stores symbols from its ith receive tape using r_i until it encounters $\bot$.*
2. *If $i = 0$ then*
   - *if $|w| \geq n$ and the last n bits of w is an integer $j \in [k]$, then it writes w except the last n bits to its jth send tape using s_j, and*
   - *otherwise it writes $\Diamond \| w$ to its 0th send tape using s_0.*

   If $i \neq 0$, then it sets $j = 0$ and writes w and a n-bit representation of i to its 0th send tape using s_0.
3. *It sets $([a_{s,j}], [a_{r,i}]) = (1, 0)$ (as an atomic operation) to pass the activity token to the jth party.*

Inactive. *If $[a_{r,i}] = 0$ for all $i \in [0, k]$, then it is inactive and keeps its state and does not read, write, or move on any tape.*

The use of routers inbetween ITM's makes sure that an ITM activates another ITM (indirectly through the router) if and only if it first sends it a message. The message may of course be empty to simply pass activation. Note that the address of a message is appended to the *end* of the message. This may seem odd, but it turns out to be useful for technical reasons (see Appendix A.4 for details).

Due to the test in step 2, a message can only be copied from the 0th receive tape to the ith send tape for $i > 0$, or from the ith receive tape for $i > 0$ to the 0th send tape. Furthermore, data written to or read from the 0th tape contains the index of another pair of tapes as an n-bit appendix, whereas it does not for other tapes. Thus, data written to the 0th write-once tape may be badly formed in which case the data is simply written back to the 0th write-once tape with the prefix $\Diamond$. This prefix is a special symbol used only for this purpose that indicates badly formed inputs.

Remark 1 (Concatenation). Concatenations such as that in Step 2 are common in this chapter and the chapters that follows. Care has to be taken to avoid that such concatenation, directly or indirectly, give rise to strings that can not be decoded uniquely into the original components. We can not solve this by simply stating that concatenation is a short hand for an invertible encoding algorithm, since we need the associative property of concatenation to prove that routers and communication models "commute". Fortunately, it is easy to see that there is no risk of ambiguous representations for most uses of concatenation.

To connect routers and ITM's with each other we let them share tapes pairwise. We formalize this as follows.

Definition 3 (Slot of Interactive Turing Machine or Router). *A tuple of heads of an ITM (s, r, a_s, a_r) or a tuple of heads of a router $(s_i, r_i, a_{s,i}, a_{r,i})$ is a slot. (Using notation from Definitions 1 and 2.)*

Definition 4 (Linked). *Two slots (s, r, a_s, a_r) and (s', r', a'_s, a'_r) are linked if there are four tapes such that the heads of each pair (s, r'), (s', r), (a_s, a'_r), and (a'_s, a_r) point to the same tape and no other heads point to any of these tapes.*

An ITM graph is simply a number of ITM's that are linked to each other indirectly using routers. Note that a router of which the 0th slot is linked to an ITM effectively increases the number of slots of the ITM. From now on we take this view. A basic requirement of an ITM graph to be executable, is that no ITM has any "dangling" tape heads.

Definition 5 (ITM Graph). *An ITM graph is a set V of ITM's, a set R of routers, and a set of additional tapes such that the slot of each ITM is linked to the 0th slot of a router, the 0th slot of each router is linked to the slot of an ITM, and every other slot of every router in R is linked to a slot of a different router in R. The set of all ITM graphs is denoted $\mathsf{G_{ITM}}$.*

In other words, we use the routers to increase the number of slots of ITM's and then link the slots of routers to each other to allow the ITM's to communicate. Figure 2 illustrates this. The idea behind this approach is to restrict the notion of an ITM to Turing machines that have a fixed number of tapes. This avoids the need to change the computational model by adding tapes for parties in a protocol depending on how many parties there are.

Definition 6 (Initializing an ITM Graph). *To initialize an ITM graph with ITM's $\mathcal{M}_1, \ldots, \mathcal{M}_k$, the identity tape of $\mathcal{M}_j$ is assigned the integer j in binary, every cell of every activity tape is set to zero, every cell of every random tape is set to a randomly chosen bit, every cell of every other tape is set to $\bot$, and tape heads pointing to the same tape are set to point to the same cell.*

We say that a tape of an initialized ITM graph is assigned a string x when we fill the consecutive cells starting at the cell pointed to by the tape heads with x. This is done in the reachable direction for directed tape heads and in some canonical direction for other tape heads.

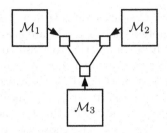

Fig. 2. An ITM graph consisting of parties $\mathcal{M}_1$, $\mathcal{M}_2$, and $\mathcal{M}_3$ linked by three unnamed routers providing three slots each. The 0th slot of each router is marked by an arrow. We use this convention throughout this paper.

To simplify the analysis of running times, we ignore the state transitions occuring in routers when stating running times. This does not change any results about concrete protocols in any essential way, since only a small constant number of routers are used and they all run in linear time in the messages forwarded.

Definition 7 (Executing an ITM Graph). *An ITM graph with ITM's $\mathcal{M}_1, \ldots, \mathcal{M}_k$, that has been initialized, is executed starting at $\mathcal{M}_1$ on security parameter n and input z to $\mathcal{M}_1$ as follows.*

1. Assign 1^n to the security parameter tape of $\mathcal{M}_j$ for $j \in [k]$.
2. Set the input tape of $\mathcal{M}_1$ to z.
3. Set $[a_r] = 1$, where a_r is the receiving activity head of $\mathcal{M}_1$.
4. Repeatedly execute the transition functions of all ITM's in unison.

Note that due to the demand that an ITM or a router is active to change its state, or read, write, or move on a tape, this effectively means that a single machine is executing at any time.

Definition 8 (Bounding the Running Time). *Let G be an ITM graph and let X be a subset of the ITM's in G. We say that the running time of G is bounded at X by T_X if the number of active state transitions taking place in ITM's in X is bounded by T_X.*

The above gives a solid foundation for defining a simple and explicit version of the UC framework, but the notation is cumbersome. From now on we say that two ITM's are *linked* if two or more slots of their routers are linked. This allows us to take an abstract view of an ITM graph as a set of ITM's V and a set of links E describing how the ITM's are connected. If two machines are linked, then they can exchange messages and activate each other.

However, an ITM with a set of slots not only expects to be linked to some other ITM's, it expects that particular slots are used to form links to particular slots of other ITM's. Thus, we must label the slots of each ITM and introduce notation for forming a link using two such slots. Suppose that the ITM's $\mathcal{M}_1$ and $\mathcal{M}_2$ have slots [a] and [b] respectively. Then $\langle \mathcal{M}_1[a], \mathcal{M}_2[b] \rangle$ denotes a link formed between slot [a] of $\mathcal{M}_1$ and slot [b] of $\mathcal{M}_2$. Due to the restrictions on ITM's, the definition of a router, and the starting state of an initialized ITM graph, this guarantees that exactly one ITM is active at any given time. In figures, we now draw the machines as circles instead of squares to indicate that we have abstracted from the details of communication.

Throughout we use the convention that a small letter in a slot, e.g., a in $[a]$, is a variable over the set of all labels of slots, and a capital letter is the label given verbatim, e.g., $\mathcal{M}$ in $[\mathcal{M}]$.

4   Entities of Models

Before we introduce the real, ideal, and hybrid models, we introduce the ITM's used to form these models. To be able to talk about different types of ITM's below without ambiguity we *mark* them. This can be formalized by adding an additional read-only tape on which the marking is written when the ITM is initialized, but we avoid formalizing this to avoid cluttering. Furthermore, each ITM of a given type has dedicated named slots.

An implementation of a function in software typically checks that the input is of a given form and returns an error code or throws an exception otherwise. It is then the responsibility of the caller of the function to deal with the error or exception. We mirror this in that if an ITM receives a message w on a slot $[a]$ that does not match the explicitly stated format of valid messages, then $\Diamond \| w$ is written to $[a]$. We have already used this convention in Definition 2.

A communication model captures how the parties of a protocol can communicate in the presence of an adversary.

Definition 9 (Communication Model). *A k-communication model $\mathcal{C}$ is an ITM marked as a "communication model" with one ideal functionality slot $[\mathcal{F}]$, party slots $[\mathcal{P}_1], \ldots, [\mathcal{P}_k]$, and an adversary slot $[\mathcal{A}]$. If $\Diamond \| w$ is read from $[\mathcal{P}_i]$ or $[\mathcal{F}]$, then $\Diamond \| w$ is written to $[\mathcal{A}]$.*

The adversary slot is used by an adversary to influence the behaviour of the communication model, e.g., if the communication model represents the Internet, then the adversary can insert, delay, or remove messages. The party slots are used by parties to communicate through the communication model. The ideal functionality slot is used to communicate with an ideal functionality. Note that the above definition implies that whenever a party or an ideal functionality refuses to accept an input, then the adversary is informed about this incident and activated. When no ideal functionality is needed we tacitly assume that an ideal functionality that refuses any input is used.

Definition 10 (Ideal Functionality). *An ideal functionality $\mathcal{F}$ is an ITM marked as an "ideal functionality" with a single communication slot $[\mathcal{C}]$.*

The communication slot is used by the ideal functionality both to accept inputs and to return outputs.

Definition 11 (Party). *An f-party $\mathcal{P}$ is an ITM marked "party" with an environment slot $[\mathcal{Z}]$, a communication slot $[\mathcal{C}]$, f subparty slots $[\mathcal{U}_1], \ldots, [\mathcal{U}_f]$, and an adversary slot $[\mathcal{A}]$. When $f = 0$ we simply say that $\mathcal{P}$ is a party.*

The subprotocol slots are used in the hybrid model to formalize access to subprotocols and ideal functionalities. The adversary slot is only used by corrupted parties. If it is not used in the formation of a model, then we assume that it is simply linked to an ITM that does not accept any input.

Definition 12 (Protocol). *A (k, f)-protocol π is a list $(\mathcal{P}_1, \ldots, \mathcal{P}_k)$ of f-parties. When $f = 0$ we simply say that π is a k-protocol (or protocol when k is clear from the context).*

Definition 13 (Adversary). *A (k, f)-adversary $\mathcal{A}$ is an ITM marked as an "adversary" with a communication slot $[\mathcal{C}]$, an environment slot $[\mathcal{Z}]$, f subadversary slots $[\mathcal{A}_1], \ldots, [\mathcal{A}_f]$, and k corrupted party slots $[\mathcal{P}_1^*], \ldots, [\mathcal{P}_k^*]$. When $f = 0$ we simply say that $\mathcal{A}$ is a k-adversary.*

The corrupted party slots are used to communicate with corrupted parties in protocols. Depending on which parties, and how many parties, are corrupted some of these slots may remain unused. To meet the requirement that a model is an ITM graph we assume that each such slot is linked to an ITM that does not accept any input. Figures 3 and 4 illustrate a communication model, an ideal functionality, a party, an adversary, and a corrupt party.

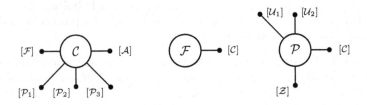

Fig. 3. To the left a 3-communication model $\mathcal{C}$ with ideal functionality slot $[\mathcal{F}]$, adversary slot $[\mathcal{A}]$, and party slots $[\mathcal{P}_1]$, $[\mathcal{P}_2]$, and $[\mathcal{P}_3]$. In the middle an ideal functionality $\mathcal{F}$ with a single communication slot $[\mathcal{C}]$. To the right a 2-party with subparty slots $[\mathcal{U}_1]$ and $[\mathcal{U}_2]$, communication slot $[\mathcal{C}]$, and environment slot $[\mathcal{Z}]$.

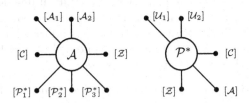

Fig. 4. To the left a $(3, 2)$-adversary with a communication slot $[\mathcal{C}]$, subadversary slots $[\mathcal{A}_1]$ and $[\mathcal{A}_2]$, an environment slot $[\mathcal{Z}]$, and corrupted party slots $[\mathcal{P}_1^*]$, $[\mathcal{P}_2^*]$, and $[\mathcal{P}_3^*]$. To the right a corrupted 2-party $\mathcal{P}^*$ with a communication slot $[\mathcal{C}]$, subparty slots $[\mathcal{U}_1]$ and $[\mathcal{U}_2]$, an adversary slot $[\mathcal{A}]$, and an environment slot $[\mathcal{Z}]$.

5   Real Free Models

The real communication model formalizes a network in which the adversary can read, delete, modify, and insert any message of its choice. The Internet is an example of such a network.

Definition 14 (Real Communication Model). *The* real k-communication model $\mathcal{N}_k$ *is defined as follows.*

- *If w is read from* $[\mathcal{P}_i]$, *where $i \in [k]$, then $\mathcal{P}_i\|w$ is written to* $[\mathcal{A}]$.
- *If $\mathcal{P}_i\|w$ is read from* $[\mathcal{A}]$, *where $i \in [k]$, then w is written to* $[\mathcal{P}_i]$.

A real free model describes a protocol that executes over a real communication model. We define a map that combines a communication model, parties, and an adversary into a graph of linked ITM's. Recall that $\langle \mathcal{M}_0[a], \mathcal{M}_1[b] \rangle$ denotes a link between slot $[a]$ of $\mathcal{M}_0$ and slot $[b]$ of $\mathcal{M}_1$.

Definition 15 (Real Model Map). *The* real (k, I, f)-model map *is the map* $\mathscr{R}_{k,I,f} : (\pi, \mathcal{A}, \pi^*) \mapsto (V, E)$, *where $\pi = (\mathcal{P}_1, \dots, \mathcal{P}_k)$ is a (k, f)-protocol, $\mathcal{A}$ is an f-adversary, and $\pi^* = \{\mathcal{P}_i^*\}_{i \in I}$ is a set of corrupted f-parties, defined by*

$$V = \{\mathcal{N}_k, \mathcal{A}\} \cup \bigcup\nolimits_{i \notin I} \{\mathcal{P}_i\} \cup \bigcup\nolimits_{i \in I} \{\mathcal{P}_i^*\} \quad and$$

$$E = \{\langle \mathcal{A}[\mathcal{C}], \mathcal{N}_k[\mathcal{A}] \rangle\} \cup \bigcup\nolimits_{i \notin I} \{\langle \mathcal{P}_i[\mathcal{C}], \mathcal{N}_k[\mathcal{P}_i] \rangle\}$$

$$\cup \bigcup\nolimits_{i \in I} \{\langle \mathcal{P}_i^*[\mathcal{C}], \mathcal{N}_k[\mathcal{P}_i] \rangle, \langle \mathcal{P}_i^*[\mathcal{A}], \mathcal{A}[\mathcal{P}_i] \rangle\}.$$

Definition 16 (Real Free Model). *A* real free (k, I, f)-model M *is an output of the real free (k, I, f)-model map. If $f = 0$, then we simply say that M is a real free (k, I)-model.*

We say that the real model is *free*, since the parties and the adversary in it have free environment slots (and possibly free subparty or subadversary slots), i.e., a real free model is not an ITM graph and can not be executed. Figures 5 and 6 illustrate real free models without and with corruption.

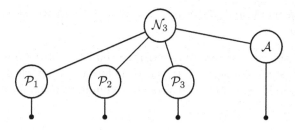

Fig. 5. A real free $(3, \emptyset)$-model $\mathscr{R}_{3,\emptyset,0}(\pi, \mathcal{A}, \emptyset)$ with a real 3-communication model $\mathcal{N}_3$, 3-protocol $\pi = (\mathcal{P}_1, \mathcal{P}_2, \mathcal{P}_3)$, and real 3-adversary $\mathcal{A}$.

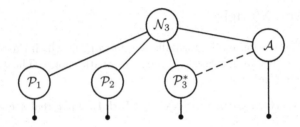

Fig. 6. The real $(3, I)$-model $\mathscr{R}_{3,I,0}(\pi, \mathcal{A}, \pi^*)$ with indices of corrupted parties $I = \{3\}$, real 3-communication model $\mathcal{N}_3$, 3-protocol $\pi = (\mathcal{P}_1, \mathcal{P}_2, \mathcal{P}_3)$, real 3-adversary $\mathcal{A}$, and set of corrupted parties $\pi^* = \{\mathcal{P}_3^*\}$. Note the link between $\mathcal{A}$ and the corrupted party $\mathcal{P}_3^*$.

6   Ideal Free Models

The ideal model formalizes a protocol execution in an ideal world where there is an ideal functionality, i.e., a trusted party that performs some service. The trusted party is simply an ITM executing a program, and it communicates with the parties through the ideal communication model.

The ideal communication model below captures the fact that the adversary may decide if and when it would like to deliver a message from the ideal functionality to a party, but it cannot read the contents of the communication between parties and the ideal functionality.

Definition 17 (Ideal Communication Model). *The* ideal k-communication model $\mathcal{I}_k$ *is defined as follows.*

- *If $\mathcal{F}\|m$ is read from $[\mathcal{A}]$, then $\mathcal{S}\|m$ is written to $[\mathcal{F}]$.*
- *If $\mathcal{S}\|m$ is read from $[\mathcal{F}]$, then $\mathcal{F}\|m$ is written to $[\mathcal{A}]$.*
- *If w is read from $[\mathcal{P}_i]$, then $\mathcal{P}_i\|w$ is written to $[\mathcal{F}]$.*
- *If $w\|(\mathcal{P}_j, w_j)_{j \in J}\|e$ is read from $[\mathcal{F}]$, where $J \subset [k]$, then for $j \in J$:*
  1. τ_j is chosen randomly, and
  2. $(\mathcal{P}_j, w_j\|e)$ is stored in a database under τ_j.
  Then $w\|(\mathcal{P}_j, \tau_j)_{j \in J}\|e$ is written to $[\mathcal{A}]$.
- *If τ is read from $[\mathcal{A}]$ and $(\mathcal{P}_j, w\|e)$ is stored under τ in the database, then $w\|e$ is written to $[\mathcal{P}_j]$.*

In our thesis we use an authenticated bulletin board for communication. Authenticated channels are trivial to define using an ideal functionality. Although we could absorb this into a separate communication model, this makes little sense.

Definition 18 (Authenticated Channels Functionality). *The* authenticated channels functionality $\mathcal{F}_{auth}$ *repeatedly reads an input of the form $\mathcal{P}_i\|(\mathcal{P}_j, m)$ from $[\mathcal{C}]$ and writes $(\mathcal{P}_j, \mathcal{P}_i\|m)\|(\mathcal{P}_j, \mathcal{P}_i\|m)$ to $[\mathcal{C}]$.*

In most formalizations the lengths of messages are provided to the simulation adversary by the communication model. This is needed to prove the security of most protocols, since without it the ideal functionality could hide the lengths of messages from the simulation adversary (something that would be impossible to achieve in a real protocol). Our formalization requires the definition of each ideal functionality to provide the lengths explicitly. However, for concrete protocols this is rarely needed, since the lengths of messages can be derived by the simulation adversary from the security parameter.

Definition 19 (Dummy Party). *A dummy party is a party that writes any input on* [$\mathcal{Z}$] *to* [$\mathcal{C}$], *and writes any input on* [$\mathcal{C}$] *to* [$\mathcal{Z}$].

Dummy parties are introduced to provide identical interfaces to the parties in real models and to ideal functionalities. There may be many copies of the dummy party. Dummy parties are denoted by $\mathcal{Q}_i$ to distinguish them from real parties and may be thought of as labels for links. We denote a dummy k-protocol by $(\mathcal{Q}_1, \ldots, \mathcal{Q}_k)$.

The ideal free model below captures the setup one wishes to realize, i.e., the environment may interact with the ideal functionality $\mathcal{F}$, except that the adversary $\mathcal{S}$ has some control over how the communication model behaves.

Definition 20 (Ideal Free Model Map). *The* ideal free (k, I)-model map *is the map* $\mathscr{I}_{k,I} : (\mathcal{F}, \mathcal{S}, \sigma^*) \mapsto (V, E)$, *where* $I \subset [k]$ *is a set of indices of corrupted parties,* $\mathcal{F}$ *is an ideal functionality,* $\mathcal{S}$ *is a simulation k-adversary, and* $\sigma^* = \{\mathcal{Q}_i^*\}_{i \in I}$ *is a set of corrupted parties, defined by*

$$V = \{\mathcal{I}_k, \mathcal{F}, \mathcal{S}\} \cup \bigcup_{i \notin I} \{\mathcal{Q}_i\} \cup \bigcup_{i \in I} \{\mathcal{Q}_i^*\}, \quad and$$

$$E = \left\{ \langle \mathcal{I}_k[\mathcal{F}], \mathcal{F}[\mathcal{C}] \rangle, \langle \mathcal{S}[\mathcal{C}], \mathcal{I}_k[\mathcal{A}] \rangle \right\} \cup \bigcup_{i \notin I} \left\{ \langle \mathcal{Q}_i[\mathcal{C}], \mathcal{I}_k[\mathcal{P}_i] \rangle \right\}$$

$$\cup \bigcup_{i \in I} \left\{ \langle \mathcal{Q}_i^*[\mathcal{C}], \mathcal{I}_k[\mathcal{P}_i] \rangle, \langle \mathcal{Q}_i^*[\mathcal{A}], \mathcal{S}[\mathcal{P}_i] \rangle \right\}.$$

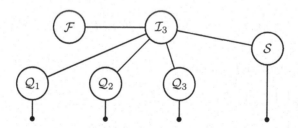

Fig. 7. An ideal free $(3, \emptyset)$-model $\mathscr{I}_{3,\emptyset}(\mathcal{F}, \mathcal{S}, \emptyset)$ with ideal 3-communication model $\mathcal{I}_3$, dummy 3-protocol $(\mathcal{Q}_1, \mathcal{Q}_2, \mathcal{Q}_3)$, ideal functionality $\mathcal{F}$, and simulation 3-adversary $\mathcal{S}$.

Definition 21 (Ideal Free Model). *An* ideal free (k, I)-model *is an output of the ideal free (k, I)-model map.*

Figure 7 illustrate an ideal free model without corruption.

7 Hybrid Free Models

A hybrid free model formalizes the execution of a real protocol that has access to other real subprotocols, ideal functionalities, or hybrid protocols. It can both be used to describe protocols that need setup assumptions (or trusted parties) for specific tasks and as a tool to construct protocols in a modular way.

Note that the following definitions give a joint inductive definition of the hybrid free model map and hybrid free models.

Definition 22 (Hybrid Free Model). *A hybrid free (k, I, f)-model is an output of the hybrid free k-model map $\mathscr{H}_{k,I,f}$ of Definition 26 below. We drop f from our notation if it is zero.*

Definition 23 (Free Model). *A free (k, I, f)-model is a real free (k, I, f)-model, a hybrid free (k, I, f)-model, or provided $f = 0$, an ideal free (k, I)-model.*

A free model is complete if it does not have any dangling subparty slots. Thus, every free ideal model and every real/hybrid (k, I)-model is complete.

Definition 24 (Complete Free Model). *A free $(k, I, 0)$-model is complete.*

Definition 25 (Root of Free Model). *The root of a free (k, I, f)-model (V, E) is the unique pair of a protocol and adversary $((\mathcal{X}_1, \ldots, \mathcal{X}_k), \mathcal{A})$ such that $\mathcal{X}_i \in V$ is a party with a free slot $[\mathcal{Z}]$ for $i \in [k]$ and $\mathcal{A} \in V$ is an adversary with a free slot $[\mathcal{Z}]$.*

We stress that if $i \in I$, then $\mathcal{X}_i$ is a corrupted party usually denoted $\mathcal{P}_i^*$ (or $\mathcal{Q}_i^*$), and otherwise it is an uncorrupted party $\mathcal{P}_i$ (or $\mathcal{Q}_i$) defined by the original protocol or dummy protocol of the ideal functionality.

Definition 26 (Hybrid Free Model Map). *The hybrid free (k, I, f)-model map is the map $\mathscr{H}_{k,I,f}$ with $f > 0$ that takes as input:*

- *A real free (k, I, f)-model (V, E) with root $((\mathcal{X}_1, \ldots, \mathcal{X}_k), \mathcal{A})$.*
- *A complete free (k, I)-model (V_j, E_j) with root $((\mathcal{X}_{j,1}, \ldots, \mathcal{X}_{j,k}), \mathcal{A}_j)$ for $j \in [f]$.*

and outputs a complete free model (V', E') where

$$V' = V \cup \bigcup_{j \in [f]} V_j \quad \text{and}$$

$$E' = E \cup \bigcup_{j \in [f]} \left(E_j \cup \{\langle \mathcal{A}[\mathcal{A}_j], \mathcal{A}_j[\mathcal{Z}] \rangle\} \cup \bigcup_{i \in [k]} \{\langle \mathcal{X}_i[\mathcal{U}_j], \mathcal{X}_{j,i}[\mathcal{Z}] \rangle\} \right).$$

8 Environments and Models

To be able to execute a free model we need an *environment* that connects to the free slots of the root protocol and root adversary. We formalize the environment in which a protocol is executed as an ITM (Fig. 8).

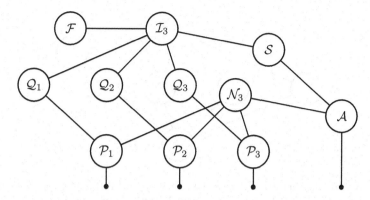

Fig. 8. A hybrid free model $\mathscr{H}_{3,I,1}\big(\mathscr{R}_{3,I,1}(\pi,\mathcal{A},\emptyset),\mathscr{I}_{3,I}(\mathcal{F},\mathcal{S},\emptyset)\big)$ with indices of corrupted parties $I = \emptyset$, real 3-communication model $\mathcal{N}_3$, root $(3,1)$-protocol $\pi = (\mathcal{P}_1,\mathcal{P}_2,\mathcal{P}_3)$, root $(3,1)$-adversary $\mathcal{A}$, ideal 3-communication model $\mathcal{I}_3$, dummy 3-protocol $(\mathcal{Q}_1,\mathcal{Q}_2,\mathcal{Q}_3)$, ideal functionality $\mathcal{F}$, and simulation 3-subadversary $\mathcal{S}$.

Definition 27 (Environment). *A k-environment is an ITM marked as an "environment" with party slots* $[\mathcal{P}_1],\ldots,[\mathcal{P}_k]$ *and an adversary slot* $[\mathcal{A}]$.

Figure 9 illustrates an environment. The environment provides the data used by the parties in the protocol and is always the first ITM to be activated during the execution of the model.

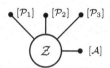

Fig. 9. A k-environment with party slots $[\mathcal{P}_1]$, $[\mathcal{P}_2]$, and $[\mathcal{P}_3]$, and an adversary slot $[\mathcal{A}]$.

Definition 28 (Environment Map). *The (k,I)-environment map $\mathscr{Z}_k :$ $(M,\mathcal{Z}) \mapsto (V',E')$ takes a complete free (k,I)-model $M = (V,E)$ with root $((\mathcal{X}_1,\ldots,\mathcal{X}_k),\mathcal{A})$ and a k-environment $\mathcal{Z}$ as input and outputs (V',E') where*

$$V' = V \cup \{\mathcal{Z}\} \quad and$$

$$E' = E \cup \big\{\langle \mathcal{Z}[\mathcal{A}],\mathcal{A}[\mathcal{Z}]\rangle\big\} \cup \bigcup\nolimits_{i\in[k]} \big\{\langle \mathcal{Z}[\mathcal{P}_i],\mathcal{X}_i[\mathcal{Z}]\rangle\big\}.$$

Definition 29 (Model). *A (k,I)-model is an output of the (k,I)-environment map.*

Note that a model is an ITM graph, which means that it can be executed. In an execution of a model the environment is always activated first with some auxiliary input. Figures 10, 11, and 12 illustrate a real model, an ideal model, and a hybrid model respectively. We abuse notation and write $\mathscr{R}_{k,I,f}(\pi,\mathcal{A},\pi^*,\mathcal{Z})$ instead of $\mathscr{Z}_k(\mathscr{R}_{k,I,f}(\pi,\mathcal{A},\pi^*),\mathcal{Z})$ and correspondingly for ideal and hybrid free model maps.

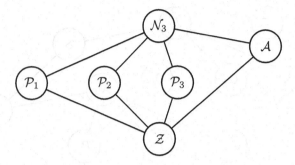

Fig. 10. A real $(3, \emptyset)$-model $\mathscr{R}_{3,\emptyset,0}(\pi, \mathcal{A}, \emptyset, \mathcal{Z})$ with real 3-communication model $\mathcal{N}_3$, 3-protocol $\pi = (\mathcal{P}_1, \mathcal{P}_2, \mathcal{P}_3)$, real 3-adversary $\mathcal{A}$, and 3-environment $\mathcal{Z}$.

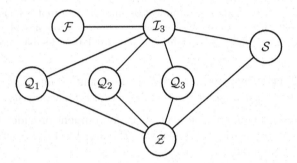

Fig. 11. An ideal $(3, \emptyset)$-model $\mathscr{I}_{3,\emptyset}(\mathcal{F}, \mathcal{S}, \emptyset, \mathcal{Z})$ with ideal 3-communication model $\mathcal{I}_3$, dummy 3-protocol $(\mathcal{Q}_1, \mathcal{Q}_2, \mathcal{Q}_3)$, ideal functionality $\mathcal{F}$, simulation 3-adversary $\mathcal{S}$, and 3-environment $\mathcal{Z}$.

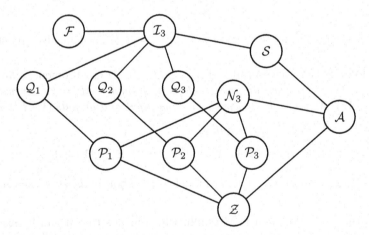

Fig. 12. A hybrid model $\mathscr{H}_{3,I,1}(\mathscr{R}_{3,I,1}(\pi, \mathcal{A}, \emptyset), \mathscr{I}_{3,I}(\mathcal{F}, \mathcal{S}, \emptyset), \mathcal{Z})$ with real 3-communication model $\mathcal{N}_3$, root $(3, 1)$-protocol $\pi = (\mathcal{P}_1, \mathcal{P}_2, \mathcal{P}_3)$, root $(3, 1)$-adversary $\mathcal{A}$, ideal 3-communication model $\mathcal{I}_3$, dummy 3-protocol $(\mathcal{Q}_1, \mathcal{Q}_2, \mathcal{Q}_3)$, ideal functionality $\mathcal{F}$, simulation 3-subadversary $\mathcal{S}$, and 3-environment $\mathcal{Z}$.

9   Classes of Adversaries

We need to bound the running times of the adversary, the simulation adversary, and the environment to give a definition of security. Several ways to do this have been proposed in the literature. We choose a simple solution that gives concrete bounds on the security reductions. Given a model $M = (V, E)$ with an adversary $\mathcal{H}$ (real, ideal, or hybrid) and environment $\mathcal{Z}$ we say that:

1. $\mathcal{H}$ has running time $T_{\mathcal{H}}$ if the running time of M is bounded by $T_{\mathcal{H}}$ at $V \setminus \{\mathcal{Z}\}$.
2. $\mathcal{Z}$ has running time $T_{\mathcal{Z}}$ if the running time of M is bounded by $T_{\mathcal{Z}}$ at $\{\mathcal{Z}\}$.

We remark that this approach differs from the simpler approach used in our thesis [12] and in [4], where the running time of each ITM was simply bounded by a polynomial in the security parameter. The advantage with the current approach is that ideal functionalities and protocols never halt until they are explicitly asked to by the adversary or the environment. However, both approaches are possible in our formalization.

10   Simplified Notation

At this point we have defined the models of the simplified UC framework rigorously, but it is convenient to introduce some alternative notation more in line with the literature to emphasize protocols, ideal functionalities, and adversaries instead of the technical details of how these are linked. We stress that we do not abandon the original notation; the freedom to change notation when convenient greatly simplifies describing and analyzing protocols.

It is easy to see that we may assume that all corrupted parties and all adversaries except the one linked to the environment are simulations of the router of Definition 2 with a suitable number of heads. This is illustrated in Fig. 13.

The subprotocols and ideal functionalities of a hybrid model are arranged in a tree of subprotocols where every ideal functionality is a leaf. Thus, given the set of indices of corrupted parties and the tree of subprotocols and ideal functionalities, an adversary, and an environment we can introduce an indexing scheme and recover the hybrid model. We denote a tree of subprotocols and ideal functionalities by inductively applying the rules that:

1. An ideal free model based on an ideal functionality $\mathcal{F}$ is denoted by $\mathcal{F}$.
2. A real free model based on a protocol π is denoted by π.
3. A hybrid free model based on a real protocol π, and complete free models based on hybrid protocols $\rho_1, \ldots, \rho_t$ is denoted $\pi(\rho_1, \ldots, \rho_t)$.

We may consider the set of indices of corrupted parties to be embedded in the description of the adversary and simply say that we consider an adversary that corrupts a certain set of parties. This convention gives less concrete notation than the original, but it is more in line with the literature.

Suppose that ρ is such a description of a protocol, $\mathcal{Z}$ is an environment, and $\mathcal{A}$ is an adversary (where the indices of corrupted parties have been encoded).

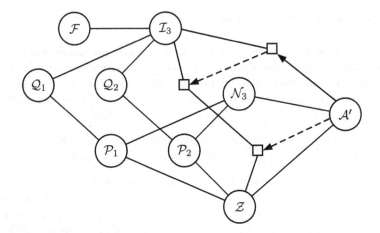

Fig. 13. A modification of a hybrid free model with corruption, where $\mathcal{Q}_3^*$, $\mathcal{P}_3^*$, and $\mathcal{S}$ are replaced by routers and $\mathcal{A}'$ is a corresponding modification of $\mathcal{A}$, but with an environment $\mathcal{Z}$ turning it into a model. The 0th slot of each router is marked by an arrow. We stress that strictly speaking each router is simulated by an ITM to adhere to our definitions. The routers needed for this ITM to have multiple links are hidden by our abstractions.

Then we denote by $\mathcal{Z}_z(\rho, \mathcal{A})$ the output of the environment $\mathcal{Z}$ running on auxiliary input z when executing the model recovered from ρ, $\mathcal{Z}$, and $\mathcal{A}$. Sometimes we structure the adversary to match the topology of the protocols and ideal functionalities, i.e., we denote each simulation subadversary by $\mathcal{S}$ and each hybrid or real subadversary by $\mathcal{A}$ with suitable subscripts.

We remark that in hybrid models the number of dummy parties linked to any ideal functionalities that are used is easily derived. Thus, there is no need to state this explicitly. This is not the case for ideal models, but the number of parties is always clear from the context.

Example 1. Suppose that π is a protocol that uses real subprotocols π_0 and π_1, and an ideal functionality $\mathcal{F}$, where π_1 in turn uses an ideal functionality $\mathcal{F}_1$. Suppose further that $\mathcal{A}$ is the overall adversary that attacks π, and orchestrates: (1) subadversaries $\mathcal{S}$ and $\mathcal{S}_1$ of $\mathcal{F}$ and $\mathcal{F}_1$ respectively, and (2) real subadversaries $\mathcal{A}_1$ and $\mathcal{A}_2$ of π_1 and π_2 respectively. Then the output of the corresponding model executed with auxiliary input z is denoted by $\mathcal{Z}_z(\pi(\pi_0, \pi_1(\mathcal{F}_1), \mathcal{F}), \mathcal{A}(\mathcal{A}_0, \mathcal{A}_1(\mathcal{S}_1), \mathcal{S}))$. If we are not interested in the internal structure of $\mathcal{A}$, then we simply write $\mathcal{A}$ instead of $\mathcal{A}(\mathcal{A}_0, \mathcal{A}_1(\mathcal{S}_1), \mathcal{S})$.

11   Definition of Security

Following the approach outlined at the beginning of the paper we now formalize the security of protocols. In this paper we only consider static corruption, i.e., an adversary may only choose a set of parties to corrupt before execution starts.

Remark 2. Adaptive corruption is easy to add to our framework as follows. (1) Add a link between each party and the adversary. There are already slots prepared for this. (2) Wrap each party in an ITM that simulates the party until it receives "corrupt" from the adversary, at which point it writes the state of the party to the adversary, and waits for a new ITM with a given state in return that it executes instead. The adversary may now use the link to the wrapped replacement freely. (3) Stipulate to which sets of parties the adversary may send "corrupt". Another wrapper of the adversary can be used to enforce this to avoid restrictions when quantifying over adversaries.

One would typically assume a uniform adversarial structure for subprotocols as for static corruption, but the approach works even when this is not the case.

Most proofs of security only hold as long as the adversary does not corrupt certain parties or some subsets of parties. An adversarial structure is a collection of sets, where each set is a set of indices of parties that the adversary can corrupt. We use J to denote an adversarial structure.

Example 2. If we have five parties $\mathcal{P}_1, \ldots, \mathcal{P}_5$ in a protocol and we are able to prove that the protocol is secure provided that at most one out of $\mathcal{P}_1$ and $\mathcal{P}_2$ is corrupted and two out of $\mathcal{P}_3$, $\mathcal{P}_4$, and $\mathcal{P}_5$ are corrupted. Then the adversarial structure we consider is $J = \{\{1, 3, 4\}, \{1, 4, 5\}, \{1, 3, 5\}, \{2, 3, 4\}, \{2, 4, 5\}, \{2, 3, 5\}\}$.

Here we only consider the case where corruption takes place in a uniform way in all free models within a model, i.e., if a party is corrupted, then so are all its subparties recursively. However, it is quite natural to generalize this in certain situations.

We use A to denote a class of adversaries with running time bounded by T_A, where the number of parties k and the topology of hybrid adversaries are implicit. Furthermore, the subset of such adversaries that corrupt the parties with indices in a set J are denoted by A_J. We use the same conventions for a class of simulation adversaries S and the corresponding class S_J of adversaries that corrupt dummy parties with indices in J. Finally, we use Z to denote a class of environments with running time bounded by T_Z. Given two classes A and A' of adversaries with the same topology, we simply write $A + A'$ to denote the class of adversaries with the same topology and running time $T_A + T_{A'}$.

For standard asymptotic security we can simply require that T_A, T_S, and T_Z are polynomially bounded, but for concrete security claims we can give explicit upper bounds.

Definition 30 (Secure Realization). *A protocol ρ is a* (J, A, S, Z, μ)*-secure realization of a target protocol τ if for every $J \in J$ and every adversary $\mathcal{A} \in A_J$, there exists a simulation adversary $\mathcal{S} \in S_J$ such that for every environment $\mathcal{Z} \in Z$ and every auxiliary input $z \in \{0, 1\}^*$:*

$$|\Pr[\mathcal{Z}_z(\rho, \mathcal{A}) = 1] - \Pr[\mathcal{Z}_z(\tau, \mathcal{S}) = 1]| \le \mu.$$

The above definition is considerably more general than other flavours of the UC framework in that *a protocol can securely realize another protocol* and not only an ideal functionality. This may seem contrived at first glance, but is in fact an important generalization that simplifies the description and analysis of concrete protocols.

Consider for example an ideal functionality for distributed key generation and decryption. It outputs a public key and can then be used to decrypt ciphertexts if asked to do so by the parties using its service. This works well with a CCA2-secure cryptosystem, but for IND-CPA secure cryptosystems the functionality can not be securely realized, since a simulator has no way of limiting access to the plaintexts needed to simulate decryption. Thus, any application of such a functionality must ensure that this information is otherwise available, but there are several ways to do this, e.g., a trusted party, secret sharing, and proofs of knowledge, and these are actually used in various electronic voting systems (see [13] for a discussion).

We can formalize an intuitive ideal functionality $\mathcal{F}$ for distributed key generation and decryption, and several different ideal functionalities $\mathcal{F}_1, \ldots, \mathcal{F}_l$ for submitting a ciphertext as an input to the ideal functionality. The individual functionalities may be impossible to securely realize in isolation, but we can consider a hybrid protocol $\pi(\mathcal{F}, \mathcal{F}_i)$, where π forces any inputs to $\mathcal{F}$ to first be processed by $\mathcal{F}_i$ (possibly along with other information or through interaction) in such a way that $\mathcal{F}_i$, and hence the simulator, knows the plaintext of any ciphertexts decrypted by $\mathcal{F}$. This hybrid protocol can then be securely realized by a protocol of the form $\pi(\sigma, \sigma_i)$, where σ and σ_i are the natural and often classic implementations in practice. The hybrid protocol $\pi(\mathcal{F}, \mathcal{F}_i)$ may either be viewed as a type of ideal functionality that is secure by inspection, in which π should be a "thin" middle layer that is trivial to understand, or there could be another ideal functionality $\mathcal{F}'$ that it securely realizes. Thus, this approach avoids some of the artificial complexity of the UC framework and allows a more modular approach.

12   Universal Composition Theorem

Canetti [3] proves a powerful composition theorem. Loosely speaking it says that if a protocol π securely realizes some functionality $\mathcal{F}$, then the protocol π can be used instead of the ideal functionality regardless of how the functionality $\mathcal{F}$ is employed. The general composition theorem can handle polynomially many instances of a constant number of ideal functionalities for many different adversarial models, but we only need the following weaker special case due to the results in Appendix A.

Theorem 1 (Special Universal Composition Theorem). *If ρ_0 is a $(\mathsf{J}, \mathsf{A}, \mathsf{S}, \mathsf{Z}, \mu)$-secure realization of τ_0 and $\pi(\tau_0, \mathcal{F}_1)$ is a $(\mathsf{J}, \mathsf{A} + \mathsf{S}, \mathsf{S}', \mathsf{Z}, \mu)$-secure realization of τ, then $\pi(\rho_0, \mathcal{F}_1)$ is a $(\mathsf{J}, \mathsf{A}, \mathsf{S}', \mathsf{Z}, \mu + \mu')$-secure realization of τ.*

Proof. The triangle inequality implies that for every simulation adversary $\mathcal{S}_0$, every hybrid adversary $\mathcal{A}(\mathcal{A}_0, \mathcal{S}_1)$, every simulation adversary $\mathcal{S}$, every environment $\mathcal{Z}$ and every auxiliary input $z \in \{0, 1\}^*$

$$\left| \Pr\left[\mathcal{Z}_z\big(\pi(\rho_0,\mathcal{F}_1),\mathcal{A}(\mathcal{A}_0,\mathcal{S}_1)\big) = 1 \right] - \Pr\left[\mathcal{Z}_z(\tau,\mathcal{S}) = 1 \right] \right|$$

$$\leq \left| \Pr\left[\mathcal{Z}_z\big(\pi(\rho_0,\mathcal{F}_1),\mathcal{A}(\mathcal{A}_0,\mathcal{S}_1)\big) = 1 \right] - \Pr\left[\mathcal{Z}_z\big(\pi(\tau_0,\mathcal{F}_1),\mathcal{A}(\mathcal{S}_0,\mathcal{S}_1)\big) = 1 \right] \right|$$

$$+ \left| \Pr\left[\mathcal{Z}_z\big(\pi(\tau_0,\mathcal{F}_1),\mathcal{A}(\mathcal{S}_0,\mathcal{S}_1)\big) = 1 \right] - \Pr\left[\mathcal{Z}_z(\tau,\mathcal{S}) = 1 \right] \right| \tag{1}$$

We now denote by $\mathcal{Z}_z(\mathcal{A},\mathcal{S}_1)$ the environment that simulates the environment $\mathcal{Z}$ on auxiliary input z, the real free model $\mathscr{R}_{k,J,2}(\pi,\mathcal{A},\pi^*)$, and the ideal free model $\mathscr{I}_{k,J}(\mathcal{F}_1,\mathcal{S}_1,\sigma_1^*)$. Here π^* and σ_1^* are the sets of corrupted subparties, but without loss of generality we may assume that they are routers. This allows us to rewrite the right side of Inequality (1) as

$$\left| \Pr\left[\mathcal{Z}_z(\mathcal{A},\mathcal{S}_1)(\rho_0,\mathcal{A}_0) = 1 \right] - \Pr\left[\mathcal{Z}_z(\mathcal{A},\mathcal{S}_1)(\tau_0,\mathcal{S}_0) = 1 \right] \right|$$

$$+ \left| \Pr\left[\mathcal{Z}_z\big(\pi(\tau_0,\mathcal{F}_1),\mathcal{A}(\mathcal{S}_0,\mathcal{S}_1)\big) = 1 \right] - \Pr\left[\mathcal{Z}_z(\tau,\mathcal{S}) = 1 \right] \right|,$$

without restricting the quantification.

Note that if $\mathcal{A}(\mathcal{A}_0,\mathcal{S}_1) \in \mathsf{A}_J$ and $\mathcal{S}_0 \in \mathsf{S}_J$, then $\mathcal{A}_0 \in \mathsf{A}_J$ and $\mathcal{A}(\mathcal{S}_0,\mathcal{S}_1) \in \mathsf{A}_J + \mathsf{S}_J$. Morover, if $\mathcal{Z}(\mathcal{A},\mathcal{S}_1) \in \mathsf{Z}$, then $\mathcal{Z} \in \mathsf{Z}$. From the hypothesis of the theorem we know that for every hybrid adversary $\mathcal{A}(\mathcal{A}_0,\mathcal{S}_1) \in \mathsf{A}_J$ there exists a simulation adversary $\mathcal{S}_0 \in \mathsf{S}_J$ such that for the hybrid adversary $\mathcal{A}(\mathcal{S}_0,\mathcal{S}_1) \in (\mathsf{A}_J + \mathsf{S}_J)$ there exists a simulation adversary $\mathcal{S} \in \mathsf{S}'_J$ such that for every environment $\mathcal{Z}_z(\mathcal{A},\mathcal{S}_1) \in \mathsf{Z}$ and every auxiliary input $z \in \{0,1\}^*$

$$\left| \Pr\left[\mathcal{Z}_z(\mathcal{A},\mathcal{S}_1)(\rho_0,\mathcal{A}_0) = 1 \right] - \Pr\left[\mathcal{Z}_z(\mathcal{A},\mathcal{S}_1)(\tau_0,\mathcal{S}_0) = 1 \right] \right| \leq \mu \quad \text{and}$$

$$\left| \Pr\left[\mathcal{Z}_z\big(\pi(\tau_0,\mathcal{F}_1),\mathcal{A}(\mathcal{S}_0,\mathcal{S}_1)\big) = 1 \right] - \Pr\left[\mathcal{Z}_z(\tau,\mathcal{S}) = 1 \right] \right| \leq \mu'.$$

We conclude that for every $\mathcal{A}(\mathcal{A}_0,\mathcal{S}_1) \in \mathsf{A}_J$ there exists a simulation adversary $\mathcal{S} \in \mathsf{S}'$ such that for every $\mathcal{Z} \in \mathsf{Z}$ and every auxiliary input $z \in \{0,1\}^*$

$$\left| \Pr\left[\mathcal{Z}_z\big(\pi(\rho_0,\mathcal{F}_1),\mathcal{A}(\mathcal{A}_0,\mathcal{S}_1)\big) = 1 \right] - \Pr\left[\mathcal{Z}_z(\tau,\mathcal{S}) = 1 \right] \right| \leq \mu + \mu'.$$

13   Transforms of Models

It is intuitively clear that we can absorb any real subprotocols into the main protocol by simply combining each real party and its subparties into single new real party, but this does not give a valid model according to our definitions, since each such party is linked to *multiple* real communication models. A similar problem appears when bundling multiple ideal communication models.

In Appendix A we describe and analyze three explicit faithful transforms that allow us to: (1) simulate multiple ITM's in a single ITM, (2) simulate multiple links between two ITM's using a single link, and (3) simulate multiple identical communication models using a single communication model. The first two are straightforward, but the third depends on the details of the definitions of the communication models. A transform is faithful if it is invertible and preserves functionality.

These transforms give us the freedom to view protocols with subprotocols and ideal functionalities in the most convenient way for each situation without sacrificing rigor. In particular, it means that we can apply Theorem 1 to protocols with more than two ideal functionalities. More precisely, we can transform any protocol and adversary into a protocol of the form $\pi(\mathcal{F}_0, \mathcal{F}_1)$, as required by the composition theorem and a corresponding adversary $\mathcal{A}$. Suppose that π_0 securely realizes $\mathcal{F}_0$. Then, due to the composition theorem we know that there is a simulation adversary $\mathcal{S}$ which shows that $\pi(\pi_0, \mathcal{F}_1)$ securely realizes $\pi(\mathcal{F}_0, \mathcal{F}_1)$. Due to faithfulnesss, we may then recover the original protocol along with a modified simulation adversary $\mathcal{S}'$, which implies that the composition is secure for the original protocol. We provide details in Appendix A.

14   Relation to Other Security Frameworks

It is natural to ask if the simplified UC framework captures the same notion of security as other security frameworks. Instead of providing relations and proofs for particular other frameworks we exploit our transforms to make this easy for any security framework.

The faithful transforms allow us to turn any protocol into a protocol with at most one ideal functionality. If a protocol securely realizes an ideal functionality, then its transform does as well. Thus, proving that it securely realizes the functionality in another security framework is reduced to the special case where the protocol has at most one ideal functionality. More precisely, to relate the simplified UC framework to an alternative framework it suffices that: (1) protocols with at most one ideal functionality can be expressed in the alternative framework (with suitable restrictions), and (2) if there is an adversary that contradicts the security of such a protocol in the alternative framework, then there is an adversary that violates the security in the simplified UC framework.

In particular, relating the simplified UC framework to *any* reasonable presentation of the UC framework is straightforward. This should be contrasted with the analysis of Canetti et al. [4] which relates their presentation of the simplified UC framework with a *particular* presentation of the UC framework. Determining if their proof still holds after further modifications of the UC framework or for other alternative presentations is cumbersome.

A   Transforms of Models

This section is dedicated to define and analyze the transforms informally described in the body of the paper. Although the definitions are somewhat technical in nature, the ideas and concepts are simple and illustrated in Figs. 14, 15, 16, 17, and 18. For all practical purposes, i.e., when analyzing concrete protocols, browsing these illustrations should be enough.

Throughout, we assume without loss of generality that if a Turing machine M_i simulates some other Turing machines for $i = 1, \ldots, m$ and M is said to simulate the M_i's, then during execution M instead simulates the machines simulated by each M_i directly. Thus, we may freely argue in terms of nested simulations without any computational penalty. To avoid cluttering we also assume that simulation of multiple Turing machines can be done without any overhead.

A.1 Faithful Transforms of ITM Graphs

We are interested in transforms of ITM graphs that preserve the functionality of the original, but we must also be able to invert each transform and recover the original ITM graph. Below we give rigorous definitions that captures these properties, but for all our transforms it is straightforward to see that this is the case.

Intuitively, the first component of an input to a transform is the ITM graph to be transformed and the second component parametrizes the transform, e.g., it may pinpoint particular ITMs to remove, move, or link in a specific way.

Definition 31 (Transform). *An ITM graph transform is a map $\Phi : S \rightarrow$* $\mathsf{G_{ITM}}$, *where $S \subset \mathsf{G_{ITM}} \times \mathsf{ITM}^*$.*

Given an ITM graph $G = (V, E)$ we may assume that we can list the ITMs in V in a canonical order. Thus, it is meaningful to view any inputs and random tapes of these ITMs as lists $m = (m_1, \ldots, m_{|V|})$ and $r = (r_1, \ldots, r_{|V|})$, respectively, and denote by $\mathcal{Z}_G(n, m, r)$ the output of $\mathcal{Z} \in V$ in an execution of G starting at $\mathcal{Z}$, using security parameter n, on inputs m and random tapes r.

We need to argue about the behaviour of both an ITM graph and its transform on the "same" random tapes, but the latter may have more or less ITMs. Thus, for every integers $a, b > 0$, we need a bijection $\epsilon : (\{0,1\}^*)^a \rightarrow (\{0,1\}^*)^b$ such that both ϵ and its inverse are efficiently computable. Such bijections are readily constructed, e.g., we can use interleaving of bits.

Definition 32 (Faithful Transforms). *An ITM graph transform $\Phi : S \rightarrow$* $\mathsf{G_{ITM}}$ *is* faithful *if*

1. **Preservation of functionality.** *For every $(G, C) \in S$, where $G = (V, E)$ with communication models U, every $\mathcal{Z} \in V$, every security parameter n, every random tapes $r \in (\{0,1\}^*)^{|V| - |U|}$ to non-communication models, and every inputs $m \in (\{0,1\}^*)^{|V|}$ the transformed ITM graph $\Phi(G, C)$ computes the same function at $\mathcal{Z}$ with overwhelming probability, i.e.,*

$$\Pr\left[\mathcal{Z}_G(n, m, r) = \mathcal{Z}_{\Phi(G,C)}(n, m', r')\right] < 2^{-\mathsf{poly}(n)},$$

where $m' = \epsilon(m)$, $r' = \epsilon(r)$, and the probability is taken over the random tapes of the communication models.

2. **Invertibility.** *There exists a transform $\Theta : S \rightarrow \mathsf{G_{ITM}}$ that computes the original ITM graph, i.e., for every $(G, C) \in S$ we have $\Theta(\Phi(G, C), C) = G$.*

In the following we compose transforms and it is not possible in general to invert each step without access to the parameter C used in the transform, e.g., we may modify different parts of an ITM graph and it is impossible to know afterwords which part was modified first.

However, for composed transforms we may view the sequence of parameters used as a transcript and recover the previous ITM graph of each step due to the invertibility property. Thus, if $G' = \Phi(G, C)$, then without loss of generality we abuse notation and simply write $G = \Phi^{-1}(G')$ instead of $G = \Theta(G', C)$ and assume that the parameter C is available.

A.2   Simulating Multiple Interactive Turing Machines

The most obvious simplification of the description of an ITM graph is to let a single ITM simulate several other ITMs as well as their links. This is illustrated in Fig. 14 and defined below. Given subsets A and B of a set V of ITMs, we denote by $\mathsf{E}(A, B)$ the set of links between slots of ITMs in A and B respectively and set $\mathsf{E}(A) = \mathsf{E}(A, A)$.

Definition 33 (Simulation of ITMs). *Let $G = (V, E)$ be an ITM graph and let $A \subset V$. Denote by $S_{\mathcal{X}}$ the set of slots of $\mathcal{X} \in A$ that are not part of a link in $\mathsf{E}(A)$. Then $\Omega_{ITM}(A)$ denotes the ITM that simulates all ITMs in A with slots $\bigcup_{\mathcal{X} \in A} \bigcup_{[a] \in S_{\mathcal{X}}} \{[\mathcal{X}|a]\}$, where $[\mathcal{X}|a]$ is identified with the slot $[a]$ of $\mathcal{X}$ in the simulation for every $\mathcal{X} \in A$.*

Definition 34 (Simulation Transform). *Define the* simulation transform *$\Phi_{ITM}(G, A) = (V', E')$, where $G = (V, E)$ is an ITM graph and $A \subset V$, by*

$$B = V \setminus A$$
$$\mathcal{X}_A = \Omega_{ITM}(A)$$
$$V' = B \cup \{\mathcal{X}_A\}$$
$$E' = \mathsf{E}(B) \cup \bigcup\nolimits_{\langle \mathcal{X}[a], \mathcal{Y}[b] \rangle \in \mathsf{E}(A, B)} \{\langle \mathcal{X}_A[\mathcal{X}|a], \mathcal{Y}[b] \rangle\}.$$

We abuse notation and write $G' = \Phi_{ITM}(G, A_1, A_2)$ instead of the more cumbersome $G' = \Phi_{ITM}(\Phi_{ITM}(G, A_1), A_2)$ and correspondingly for multiple sets $A_1, \dots, A_l$.

Theorem 2. *The simulation transform is faithful.*

Proof. It is clear that the transform is faithful, since the parties in A are simply simulated and we merely replace the links to parties outside A with corresponding links with differently labeled slots.

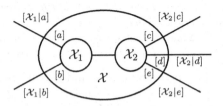

Fig. 14. Two ITMs $\mathcal{X}_1$ and $\mathcal{X}_2$ with links to other ITMs (not shown in the figure) are simulated by a single ITM $\mathcal{X}$ that inherits the links of all the original parties. More precisely, the slots [a] and [b] of $\mathcal{X}_1$ are exposed as the slots [$\mathcal{X}_1$|a] and [$\mathcal{X}_1$|b], and the slots [c] and [d], and [e] of $\mathcal{X}_2$ are exposed as [$\mathcal{X}_2$|c], [$\mathcal{X}_2$|d], and [$\mathcal{X}_2$|e].

A.3   Simulate Multiple Links

Suppose that two ITMs have multiple direct links between them. Then we simply plug in two routers and absorb these routers into the respective ITMs using wrappers. This is illustrated in Fig. 15.

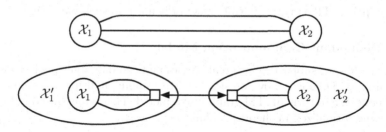

Fig. 15. The upper part shows two ITMs $\mathcal{X}_1$ and $\mathcal{X}_2$ that are linked by three links. The lower part shows how two routing wrappers can be used to form slightly modified ITMs $\mathcal{X}_1'$ and $\mathcal{X}_2'$ that are connected by a single link.

Definition 35 (Wrapper). *Let $\mathcal{X}$ be an ITM with slots $[a_1], \ldots, [a_l]$, let $\mathcal{R}$ be an l-router, define links $E = \{\langle \mathcal{X}[a_j], \mathcal{R}[j] \rangle\}_{j \in [l]}$, and define $\mathcal{X}' = \Omega_{wrap}(\mathcal{X}, ([a_1], \ldots, [a_l]), [a])$ to be the* wrapper *ITM that simulates $\mathcal{X}$ and $\mathcal{R}$ including the links in E, and identifies $\mathcal{R}[0]$ with a new slot $[a]$ of $\mathcal{X}'$. All other slots of $\mathcal{X}$ are exposed by $\mathcal{X}'$.*

Definition 36 (Swap). *Let $G = (V, E)$ be an ITM graph, let $\mathcal{X} \in V$, and define $\Omega_{swap}(E, \mathcal{X}, \mathcal{Y})$, where L is the set of common labels of slots of $\mathcal{X}$ and $\mathcal{Y}$, by*

$$\Omega_{swap}(E, \mathcal{X}, \mathcal{Y}) = \bigcup_{a \in L} \bigcup_{\langle \mathcal{X}[a], \mathcal{Z}[a] \rangle \in E} \{\langle \mathcal{Y}[a], \mathcal{Z}[a] \rangle\}.$$

We generalize Ω_{swap} to lists of ITMs in the natural way, i.e., we simply write $\Omega_{swap}(E, (\mathcal{X}_1, \mathcal{X}_2), (\mathcal{Y}_1, \mathcal{Y}_2))$ instead of $\Omega_{swap}(\Omega_{swap}(E, \mathcal{X}_1, \mathcal{Y}_1), \mathcal{X}_2, \mathcal{Y}_2)$ and similarly for longer lists.

Definition 37 (Link Simulation Transform). *Define the* link simulation transform *$\Phi_{links}(G, A) = (V', E')$, where $G = (V, E)$ is an ITM graph and $A = \{\mathcal{X}_1, \mathcal{X}_2\}$ with $A \subset V$ and $E_A = \mathsf{E}(\mathcal{X}_1, \mathcal{X}_2) = \{\langle \mathcal{X}_1[a_i], \mathcal{X}_2[b_i] \rangle\}_{i \in [l]}$, by*

$$\mathcal{X}_1' = \Omega_{wrap}(\mathcal{X}_1, ([a_i])_{i \in [l]}, [a]) \quad \text{where } [a] \text{ is not a slot of } \mathcal{X}_1$$
$$\mathcal{X}_2' = \Omega_{wrap}(\mathcal{X}_2, ([b_i])_{i \in [l]}, [b]) \quad \text{where } [b] \text{ is not a slot of } \mathcal{X}_2$$
$$V' = (V \setminus A) \cup \{\mathcal{X}_1', \mathcal{X}_2'\}$$
$$E' = \mathsf{E}(V \setminus A) \cup \Omega_{swap}(E \setminus E_A, (\mathcal{X}_1, \mathcal{X}_2), (\mathcal{X}_1', \mathcal{X}_2')) \cup \{\langle \mathcal{X}_1'[a], \mathcal{X}_2'[b] \rangle\}.$$

Theorem 3. *The link simulation transform is faithful.*

Proof. The flow of information between slots $[a_i]$ and $[b_i]$ is identical in the original ITM graph and its transform, since routers are deterministic and take no input, and we can recover the original ITM graph from its transform given A.

We abuse notation and simply write $\Phi_{links}(G)$ for the repeated application of the link simulation transform to, starting from G, a sequence of ITM graphs and any pair of ITMs with multiple links in it until no such pair exists.

A.4   Redundant Communication Models

Even if we absorb subparties into real parties and simulate multiple links with a single link as explained above we still need to combine multiple communication models into one to turn an ITM graph into a model. This is illustrated in Figs. 16 and 17 and formalized in the next definition.

Definition 38 (Redundant Communication Models). *Let $G = (V, E)$ be an ITM graph and let $B = \{C_{k,1}, \ldots, C_{k,l}\}$ be a set of ideal/real communication models in V. Then B is a set of l-redundant ideal/real k-communication models of G if $l > 1$ and there is a subset $A = \{X_1, \ldots, X_k, \mathcal{H}, \mathcal{Y}\}$ of V, such that $\mathsf{E}(A \cup B)$ is of the form*

$$\bigcup_{j \in [l]} \left\{ \langle \mathcal{H}[c_j], C_{k,j}[A] \rangle, \langle \mathcal{Y}[c_j], C_{k,j}[\mathcal{F}] \rangle \right\} \cup \bigcup_{i \in [k]} \left\{ \langle C_{k,j}[P_i], X_i[c_j] \rangle \right\}.$$

Note that X_j plays the role of a party, except that it is linked to multiple communication models. Similarly, $\mathcal{H}$ and $\mathcal{Y}$ represent an adversary and an ideal functionality, respectively, except that they are linked to multiple communication models.

We stress that the definition should be interpreted to say that all communication models of a set of redundant communication models must either be ideal or real and never a mix of both.

Definition 39 (Redundant Communication Model Transform). *Define the redundant communication model transform $\Phi_{red}(G, B) = (V', E')$, where $G = (V, E)$ is an ITM graph with a set of l-redundant k-communication models B (with notation from Definition 38), $c = ([c_1], \ldots, [c_l])$, $[c_i] \neq [C]$, and C_k is a k-communication model, by*

$$X_i' = \Omega_{wrap}(X_i, c, [C]), \mathcal{H}' = \Omega_{wrap}(\mathcal{H}, c, [C]), \text{ and }\quad \mathcal{Y}' = \Omega_{wrap}(\mathcal{Y}, c, [C]),$$
$$V' = (V \setminus A) \cup \{C_k, X_1', \ldots, X_k', \mathcal{H}', \mathcal{Y}'\}$$
$$E' = \left\{ \langle \mathcal{H}'[C], C_k[A] \rangle, \langle \mathcal{Y}'[C], C_k[\mathcal{F}] \rangle \right\} \cup \bigcup_{i \in [k]} \left\{ \langle X_i'[C], C_k[P_i] \rangle \right\}$$
$$\cup \Omega_{swap}\left(E, (\mathcal{H}, \mathcal{Y}, X_1, \ldots, X_k), (\mathcal{H}', \mathcal{Y}', X_1', \ldots, X_k')\right).$$

Note that if the redundant communication models are ideal, then each X_i is the result of a combining multiple dummy parties with the simulation transform, which means that X_i' is equivalent to a single dummy party. We abuse notation and simply write $\Phi_{red}(G)$ for the repeated application of the redundant communication model transform until there is no longer any set of redundant communication models.

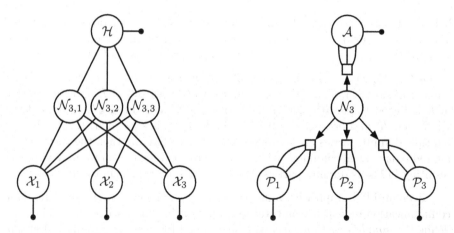

Fig. 16. The left side shows a real free model except that each party and the adversary is linked by a set of 3-redundant real 3-communication models. The right side shows how a single equivalent real communication model can be formed. Here it is understood in the figure that the routers with multiple links to a party or adversary would be absorbed into the party to reduce the number of links.

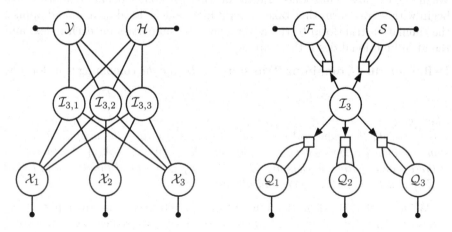

Fig. 17. The left side shows an ideal free model except that each party and the simulation adversary is linked by a set of 3-redundant ideal 3-communication models. The right side shows how a single equivalent ideal communication model can be formed. Here it is understood that the routers with multiple links to a party or adversary would be absorbed into the party.

Theorem 4. *The redundant communication model transform is faithful.*

Proof. Note that a communication model is routing messages based on the *prefixes* of messages. The routers used in the wrappers on the other hand route messages based on the *postfixes* of messages. This means that adding/removing a prefix commutes with adding/removing a postfix. Ideal communication models behave in the same way. The probability that the same randomly chosen tag appears in two ideal communication models is exponentially small. (This is where we need the extra leg room in the definition of preservation of functionality.)

Remark 3. Consider a dummy party Q and two routers R and R' with the same number of slots l. If a slot [a] of Q is linked to the 0th slot of R and the ith slot of R is linked to the ith slot of R' for $i \in [l]$, then an ITM Q' that simulates Q, R, and R' and exposes the 0th slot of R' as [a] is equivalent to Q. Thus, if multiple dummy parties are simulated by a single ITM and then the corresponding redundant ideal communication models are combined, then we may view the resulting ITMs as dummy parties. We tacitly ignore this technicality below.

Given an ITM graph with multiple links between some parties or redundant communication models it is natural to simplify it by eliminating them. Thus, we define the *simplifying transform* as a short hand for cleaning up an ITM graph

$$\Phi_{sim}(G) = \Phi_{red}(\Phi_{links}(G)).$$

A.5   Transforms of Models

We are now ready to introduce transforms that turn one model into another. We begin with a transform that takes several free models, ideal or real, and applies the simulation transform to the adversaries, the subparties for each index, and the ideal functionalities if there are any.

Definition 40 (Combining Transform). *Define the* combining transform *by*

$$\Phi_{com}(G, H) = \Phi_{sim}(\Phi_{ITM}(G, H, X_1, \ldots, X_k, F))$$

where G is a (k, I)-model, $H = \{\mathcal{H}_1, \ldots, \mathcal{H}_l\}$ is a set of real/simulation subadversaries with the same real parent adversary, $\mathcal{X}_{j,i}$ is the ith party linked to the same communication model $\mathcal{C}_{k,j}$ as $\mathcal{H}_j$, and $X_i = \{\mathcal{X}_{1,i}, \ldots, \mathcal{X}_{l,i}\}$. Furthermore, if the communication models are ideal, then $\mathcal{F}_j$ is the ideal functionality linked to $\mathcal{C}_{k,j}$ and $F = \{\mathcal{F}_1, \ldots, \mathcal{F}_l\}$, and otherwise $F = \emptyset$.

We stress that the definition must be interpreted to say that the input adversaries are either all real or all ideal and never a mix. We sometimes abuse notation and write $\Phi_{com}(G, F)$, where F is a set of ideal functionalities, to denote $\Phi_{com}(G, H)$ where H is the set of simulation adversaries linked to the ideal communication models linked to the ideal functionalities in F. We also write $\Phi_{com}(G)$ to denote the transform that repeatedly applies the combining transform to a sequence of models starting with G until there are no real/simulation subadversaries with the same parent in a model.

Definition 41 (Absorbing Transform). *Define the* absorbing transform *by*

$$\Phi_{abs}(G, A) = \Phi_{sim}(\Phi_{ITM}(G, A, P_1, \ldots, P_k))$$

where G is a (k, I)-model, $A = \{\mathcal{A}_1, \ldots, \mathcal{A}_l\}$ is a set of real subadversaries such that $\mathcal{A}_{j+1}$ is a real subadversary of $\mathcal{A}_j$ for $j = 1, \ldots, l - 1$, $\mathcal{P}_{j,i}$ is the ith party linked to the same real communication model as $\mathcal{A}_j$, and $P_i = \{\mathcal{P}_{1,i}, \ldots, \mathcal{P}_{l,i}\}$.

Note that the absorbing transform not only absorbs the subprotocol. It also absorbs real subadversaries into the root adversary. We abuse notation and write $\Phi_{abs}(G)$ for the transform that repeatedly applies the absorbing transform to a sequence of models starting with G until no real subadversary exists in a resulting model. We say that a model without subprotocols is normalized, i.e., a model for which the absorbing transformation can not be applied.

Definition 42 (Normalized Model). *A (k, I)-model is* normalized *if it has no subprotocols, or equivalently no real subadversaries.*

Another natural transform is to collapse parts of models. This is useful to focus on particular parts of a model and allows generalizing the composition theorem.

Definition 43 (Collapsing Transform). *Define the* collapsing transform $\Phi_{col}(G, \rho)$, *where G is a (k, I)-model and ρ is a hybrid protocol embedded in G, as the transform that repeatedly applies the combining and/or the absorbing transforms to G except the free model uniquely identified by ρ until no longer possible.*

Note that if G is a model, then $\Phi_{col}(G, \rho)$ has at most one ideal functionality outside of the free model in G based on ρ, i.e., the corresponding protocol is of the form $\pi(\rho, \mathcal{F})$ for some root protocol π and ideal functionality $\mathcal{F}$. In particular, we can use the collapsing transform without restriction to put a model into a minimal form where all ideal functionalities have been combined into a single ideal functionality and all subprotocols have been absorbed.

Definition 44 (Minimal Model). *A (k, I)-model is* minimal *if it is normalized and has at most one ideal functionality.*

A.6   Adversary Converter

We need an explicit way to map an arbitrary adversary (not constructed through our transforms) for a transformed protocol back into an equivalent adversary of the protocol.

Fortunately, the transforms leave a blue print for what to do. Note that a template adversary resulting from our transforms of models consists of an original adversary and routers forming trees where the leaves of the trees are linked to the original adversary and the roots of the trees are exposed as slots of the adversary. We may think of the trees as mapped onto an annulus where the inner circle represents the original adversary and the outer circle represents the template adversary. This is illustrated in Fig. 18 and made precise below.

To convert an adversary with identical slots to the template adversary, we simply fold the annulus inside out, link the roots of the trees of routers to the adversary and relabel the slots of the leaves of the trees of routers to the labels of the original adversary. If we plug this converted adversary into the original model and transform it, then we get a transformed converted adversary that is

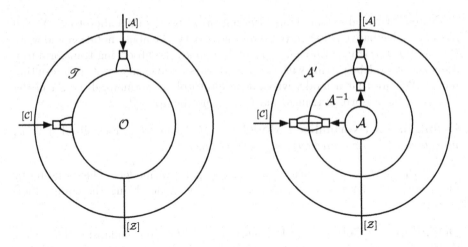

Fig. 18. The left side illustrates a template adversary $\mathcal{T}$ resulting from applying some of the above transforms on an original adversary $\mathcal{O}$ that does not corrupt any parties. The original model could for example have had two subprotocols that were absorbed and two ideal functionalities that were combined.

equivalent to the adversary, since each tree of routers is effectively canceled by its mirror embedded in the converted adversary.

We say that a set of routers form a tree if exactly one router has a free 0th slot and the 0th slot of every other router is linked to the ith slot of another router for some $i > 0$. We say that the free 0th slot is the root of the tree and all other free slots are leaves of the tree.

Definition 45 (Adversary Converter). *The adversary converter Φ_{adv} is defined as follows. Let $\mathcal{T}$ be an adversary that simulates an original adversary $\mathcal{O}$ and trees $t_1, \ldots, t_r$ of routers with roots exposed as slots $[a_1], \ldots, [a_r]$ of $\mathcal{T}$ and leaves $[b_{j,1}], \ldots, [b_{j,s_j}]$ of t_j linked to slots $[c_{j,1}], \ldots, [c_{j,s_j}]$ of $\mathcal{O}$. Let $\mathcal{A}$ be an adversary with slots $[a_1], \ldots, [a_r]$. Then $\Phi_{adv}(\mathcal{T}, \mathcal{A})$ is the adversary that simulates $\mathcal{A}$ and $t_1, \ldots, t_r$ with the set of links $\left\{\langle \mathcal{A}[a_j], t_j[0]\rangle\right\}_{j \in [r]}$ and exposes $[b_{j,i}]$ as $[c_{j,i}]$ for $j = 1, \ldots, r$ and $i \in 1, \ldots, s_j$.*

The importance of the adversary converter can be illustrated as follows. Suppose we are given a model M and wish to prove that its embedded hybrid protocol securely realizes some ideal functionality $\mathcal{F}$. To do this we need to show that for every adversary $\mathcal{A}$, there is a suitable simulator $\mathcal{S}$. Given an adversary we can of course apply our transforms and get a new model M' along with a transformed adversary $\mathcal{A}'$ for which the simulator is still suitable.

More interesting is to consider the transformed protocol of M' directly. If this securely realizes $\mathcal{F}$, then for every adversary $\mathcal{A}'$ in M', there exists a suitable simulator $\mathcal{S}'$. We may plug in a place-holder adversary $\mathcal{O}$ and apply the transforms to get a template adversary $\mathcal{T}$ as in the definition. Then we can use this to construct an adversary $\mathcal{A}$ such that if we transform M with $\mathcal{A}$ we will recover M' and an adversary that is functionally identical to $\mathcal{A}'$.

Thus, we can safely prove the security for any transformed protocol and conclude that any other transformation of it is secure as well.

References

1. Beaver, D.: Foundations of secure interactive computing. In: Feigenbaum, J. (ed.) CRYPTO 1991. LNCS, vol. 576, pp. 377–391. Springer, Heidelberg (1992)
2. Canetti, R.: Security and composition of multi-party cryptographic protocols. J. Cryptol. **13**(1), 143–202 (2000)
3. Canetti, R.: Universally composable security: a new paradigm for cryptographic protocols. In: 42nd IEEE Symposium on Foundations of Computer Science (FOCS), pp. 136–145. IEEE Computer Society Press (2001). (Full version at Cryptology ePrint Archive, Report 2000/067. http://eprint.iacr.org, October 2001)
4. Canetti, R., Cohen, A., Lindell, Y.: A simpler variant of universally composable security for standard multiparty computation. Cryptology ePrint Archive, Report 2014/553 (2014). http://eprint.iacr.org/
5. Goldreich, O.: Foundations of Cryptography: Basic Tools. Cambridge University Press, New York (2000)
6. Goldreich, O., Micali, S., Wigderson, A.: How to play any mental game. In: 19th ACM Symposium on the Theory of Computing (STOC), pp. 218–229. ACM Press (1987)
7. Goldwasser, S., Levin, L.A.: Fair computation of general functions in presence of immoral majority. In: Menezes, A., Vanstone, S.A. (eds.) CRYPTO 1990. LNCS, vol. 537, pp. 77–93. Springer, Heidelberg (1991)
8. Maurer, U.: Constructive cryptography – a new paradigm for security definitions and proofs. In: Mödersheim, S., Palamidessi, C. (eds.) TOSCA 2011. LNCS, vol. 6993, pp. 33–56. Springer, Heidelberg (2012)
9. Maurer, U., Renner, R.: Abstract cryptography. In: The Second Symposium on Innovations in Computer Science, ICS 2011, pp. 1–21, January 2011
10. Micali, S., Rogaway, P.: Secure computation. In: Feigenbaum, J. (ed.) CRYPTO 1991. LNCS, vol. 576, pp. 392–404. Springer, Heidelberg (1992)
11. Pfitzmann, B., Waidner, M.: Composition and integrity preservation of secure reactive systems. In: 7th ACM Conference on Computer and Communications Security (CCS), pp. 245–254. ACM Press (2000)
12. Wikström, D.: On the security of mix-nets and hierarchical group signatures. Doctoral thesis, Department of Numerical Analysis and Computer Science, Royal Institute of Technology, TRITA NA 05–38, ISSN 0348–2952, ISRN KTH/NA/R–05/38–SE, ISBN 91-7283-717-9, December 2005. http://www.kth.se
13. Wikström, D.: Simplified submission of inputs to protocols. In: Ostrovsky, R., De Prisco, R., Visconti, I. (eds.) SCN 2008. LNCS, vol. 5229, pp. 293–308. Springer, Heidelberg (2008)

Characterization of Secure Multiparty Computation Without Broadcast

Ran Cohen[1]([⊠]), Iftach Haitner[2], Eran Omri[3], and Lior Rotem[4]

[1] Department of Computer Science, Bar-Ilan University, Ramat Gan, Israel
cohenrb@cs.biu.ac.il
[2] School of Computer Science, Tel Aviv University, Tel Aviv, Israel
iftachh@cs.tau.ac.il
[3] Department of Computer Science and Mathematics, Ariel University, Ariel, Israel
omrier@ariel.ac.il
[4] School of Computer Science, Tel Aviv University, Tel Aviv, Israel
lior.rotem@gmail.com

Abstract. A major challenge in the study of cryptography is characterizing the necessary and sufficient assumptions required to carry out a given cryptographic task. The focus of this work is the necessity of a broadcast channel for securely computing symmetric functionalities (where all the parties receive the same output) when one third of the parties, or more, might be corrupted. Assuming all parties are connected via a peer-to-peer network, but no broadcast channel (nor a secure setup phase) is available, we prove the following characterization:

- A symmetric n-party functionality can be securely computed facing $n/3 \leq t < n/2$ corruptions (i.e., honest majority), if and only if it is $(n-2t)$-*dominated*; a functionality is k-dominated, if *any* k-size subset of its input variables can be set to *determine* its output.
- Assuming the existence of one-way functions, a symmetric n-party functionality can be securely computed facing $t \geq n/2$ corruptions (i.e., no honest majority), if and only if it is 1-dominated and can be securely computed with broadcast.

It follows that, in case a third of the parties might be corrupted, broadcast is necessary for securely computing non-dominated functionalities (in which "small" subsets of the inputs cannot determine the output), including, as interesting special cases, the Boolean XOR and coin-flipping functionalities.

Keywords: Broadcast · Point-to-point communication · Multiparty computation · Coin flipping · Fairness · Impossibility result

R. Cohen—Work supported by THE ISRAEL SCIENCE FOUNDATION (grant No. 189/11), the Ministry of Science, Technology and Space and by the National Cyber Bureau of Israel.
I. Haitner—Research supported by ERC starting grant 638121, ISF grant 1076/11, I-CORE grant 4/11, BSF grant 2010196, and Check Point Institute for Information Security.
E. Omri—Research supported by ISF grant 544/13.

E. Kushilevitz and T. Malkin (Eds.): TCC 2016-A, Part I, LNCS 9562, pp. 596–616, 2016.
DOI: 10.1007/978-3-662-49096-9_25

1   Introduction

Broadcast (introduced by Lamport et al. [20] as the Byzantine Generals problem) allows any party to deliver a message of its choice to all parties, such that all honest parties will receive the same message even if the broadcasting party is corrupted. Broadcast is an important resource for implementing secure multiparty computation. Indeed, much can be achieved when broadcast is available (hereafter, the broadcast model); in the computational setting, assuming the existence of oblivious transfer, every efficient functionality can be securely computed *with abort*,[1] facing an arbitrary number of corruptions [15,25]. Some functionalities can be computed with *full security*,[2] e.g., Boolean OR and three-party majority [17], or $1/p$-security,[3] e.g., coin-flipping protocols [18,21]. In the information-theoretic setting, considering ideally-secure communication lines between the parties, every efficient functionality can be computed with full security against unbounded adversaries,[4] facing any minority of corrupted parties [24].

The above drastically changes when broadcast or a secure setup phase are not available.[5] Specifically, when considering multiparty protocols (involving more than two parties), in which the parties are connected only via a peer-to-peer network (hereafter, the point-to-point model) and one third of the parties, or more, might be corrupted.[6] Considering authenticated channels and assuming the existence of oblivious transfer, every efficient functionality can be securely computed with abort, facing an arbitrary number of corruptions [12]. In the full-security model, some important functionalities *cannot* be securely computed (e.g., Byzantine agreement [22] and three-party majority [9]), whereas other functionalities can (e.g., *weak* Byzantine agreement [12] and Boolean OR [9]). The characterization of many other functionalities, however, was unknown. For instance, it was unknown whether the coin-flipping functionality or the Boolean XOR functionality can be computed with full securely, even when assuming an honest majority.

---

[1] An efficient attack in the real world is computationally indistinguishable, via a simulator, from an attack on an "ideal computation", in which malicious parties are allowed to prematurely abort.

[2] The malicious parties in the "ideal computation" are *not* allowed to prematurely abort.

[3] The real model is $1/p$-indistinguishable from an "ideal computation" without abort.

[4] The real and ideal models are statistically close: indistinguishable even in the eyes of an all-powerful distinguisher.

[5] In case a secure setup phase is available, *authenticated broadcast* can be computed facing $t < n$ corrupted parties; Authenticated broadcast exists in the computational setting over authenticated channels assuming one-way functions exist [10] and in the information-theoretic setting over secure channels assuming a limited access to a broadcast channel in the offline phase [23].

[6] For two-party protocols, the broadcast model is equivalent to the point-to-point model (and thus all the results mentioned in the broadcast model hold also in the point-to-point model). If less than a third of the parties are corrupted, broadcast can be implemented using a protocol, and every functionality can be computed with information-theoretic security [2,5].

1.1   Our Result

A protocol is *t-consistent*, if in any execution of the protocol, in which at most t parties are corrupted, *all* honest parties output the same value. Our main technical result is the following attack on consistent protocols.

Lemma 1 (main lemma, informal). *Let $n \geq 3$, $t \geq \frac{n}{3}$ and let $s = n - 2t$ if $t < \frac{n}{2}$ and $s = 1$ otherwise. Let π be an efficient n-party, t-consistent protocol in the point-to-point model with secure channels. Then, there exists an efficient adversary that by corrupting any s-size subset $\mathcal{I}$ of the parties can do the following: first, before the execution of π, output a value $y^* = y^*(\mathcal{I})$. Second, during the execution of π, force the remaining honest parties to output y^*.*

The lemma extends to expected polynomial-time protocols, and to protocols that only guarantee consistency to hold with high probability. We prove the lemma by extending the well-known hexagon argument of Fischer et al. [11], originally used for proving the impossibility of reaching (strong and weak) Byzantine agreement in the point-to-point model.

A corollary of Lemma 1 is the following lower bound on symmetric functionalities (i.e., all parties receive the same output value). A functionality is *k-dominated*, if there exists an efficiently computable value y^* such that *any* k-size subset of the functionality input variables, can be manipulated to make the output of the functionality be y^* (e.g., the Boolean OR functionality is 1-dominated with value $y^* = 1$).

Corollary 1 (Informal). *Let $n \geq 3$, $t \geq \frac{n}{3}$, and let $s = n - 2t$ if $t < \frac{n}{2}$ and $s = 1$ otherwise. A symmetric n-party functionality that can be computed with full security in the point-to-point model with secure channels, facing up to t corruptions, is s-dominated.*[7]

Interestingly, the above lower bound is tight. Cohen and Lindell [9] (following Fitzi et al. [12]) showed that assuming one-way functions exist, any 1-dominated functionality (e.g., Boolean OR) that can be securely computed in the broadcast model with authenticated channels, can be securely computed in the point-to-point model with authenticated channels. This shows tightness when an honest majority is not assumed. We generalize the approach of [9], using the two-threshold detectable precomputation of Fitzi et al. [13], to get the following upper bound.

Proposition 1 (Informal). *Let $n \geq 3$ and $\frac{n}{3} \leq t < \frac{n}{2}$. Assuming up to t corruptions, any efficient symmetric n-party functionality that is $(n-2t)$-dominated can be computed in the secure-channels point-to-point model with information-theoretic security.*

---

[7] Stating the lower bound in the secure-channels model is stronger than stating it in the authenticated-channels model, since if a functionality can be computed with authenticated channels then it can be computed with secure channels.

Combining Corollary 1, Proposition 1 and [9, Theorem 7], yields the following characterization of symmetric functionalities.

Theorem 1 (main theorem, informal). *Let $n \geq 3$, $t \geq \frac{n}{3}$ and let f be an efficient symmetric n-party functionality.*

1. *For $t < \frac{n}{2}$, f can be t-securely computed (with information-theoretic security) in the secure-channels point-to-point model, if and only if f is $(n - 2t)$-dominated.*
2. *For $t \geq \frac{n}{2}$, assuming one-way functions exist, f can be t-securely computed (with computational security) in the authenticated-channels point-to-point model, if and only if f is 1-dominated and can be t-securely computed (with computational security) in the authenticated-channels broadcast model.*

Another application of Lemma 1 regards coin-flipping protocols. A coin-flipping protocol [3] allows the honest parties to jointly flip an unbiased coin, where even a coalition of (efficient) cheating parties cannot bias the outcome of the protocol by too much. We focus on protocols in which honest parties must output the same bit. Although Theorem 1 shows that fully-secure coin flipping cannot be achieved facing one-third corruptions, we provide a stronger impossibility result under a weaker security requirement that only assumes $\frac{n}{3}$-consistency and a non-trivial bias. In particular, we show that $1/p$-secure coin flipping cannot be achieved using consistent protocols in case a third of the parties might be corrupted.

Corollary 2 (impossibility of many-party coin flipping in the point-to-point model, informal). *In the secure-channels point-to-point model, there exists no $(n \geq 3)$-party coin-flipping protocol that guarantees a non-trivial bias (i.e., smaller than $\frac{1}{2}$) against an efficient adversary controlling one third of the parties.*

The above is in contrast to the broadcast model, in which coin flipping can be computed with full security if an honest majority exists [4,6], and $1/p$-security when no honest majority is assumed [1,7,18].

1.2 Our Technique

We present the ideas underlying our main technical result, showing that the following holds in the point-to-point model. For any efficient consistent protocol involving more than two parties, if one third of the parties (or more) might be corrupted, then there exists an efficient adversary that can make the honest parties output a predetermined value. In the following discussion we focus on three-party protocols with a single corrupted party.

Let $\pi = (A, B, C)$ be an efficient 1-consistent three-party protocol, and let q be its round complexity on inputs of fixed length κ. Consider the following ring network $R = (A^1, B^1, C^1, \ldots, A^q, B^q, C^q)$, where each two consecutive parties, as well as the first and last, are connected via a secure channel, and party P^j, for $P \in \{A, B, C\}$, has the code of P (see Fig. 1).

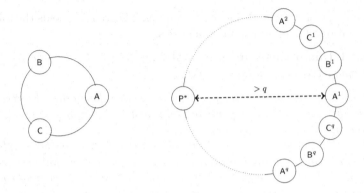

Fig. 1. The original 3-party protocol $\pi = (\mathsf{A}, \mathsf{B}, \mathsf{C})$ is on the left. On the right is the $3q$-Ring — q copies of π concatenated. Communication time between parties of opposite sides is larger than $3q/2 > q$.

Consider an execution of R on input $\boldsymbol{w} = (w_{\mathsf{A}}^1, w_{\mathsf{B}}^1, w_{\mathsf{C}}^1, \ldots, w_{\mathsf{A}}^q, w_{\mathsf{B}}^q, w_{\mathsf{C}}^q) \in (\{0,1\}^\kappa)^{3q}$ (i.e., party P^i has input w_{P}^i, containing its actual input and random coins). A key observation is that the view of party A^j, for instance, in this execution, is a *valid* view of the party A on input w_{A}^j in an interaction of π in which B acts honestly on input w_{B}^j. It is also a valid view of A, on input w_{A}^j, in an interaction of π in which C acts honestly on input $w_{\mathsf{C}}^{j-1 \pmod q}$. Hence, the consistency of π yields that any two consecutive parties in R output the same value, and thus *all* parties of R output the *same* value.

Consider for concreteness an attack on the parties $\{\mathsf{A}, \mathsf{B}\}$. The efficient adversary D first selects a value $\boldsymbol{w} \in (\{0,1\}^\kappa)^{3q}$, emulates (in its head) an execution of R on $\boldsymbol{w}$, and sets y^* to be the output of the party $\mathsf{P}^* = \mathsf{A}^{q/2}$ in this execution. To interact with the parties $\{\mathsf{A}, \mathsf{B}\}$ in π, the adversary D corrupts party C and emulates an execution of R, in which all but $\{\mathsf{A}^1, \mathsf{B}^1\}$ have their inputs according to $\boldsymbol{w}$ (the roles of all parties but $\{\mathsf{A}^1, \mathsf{B}^1\}$ are played by the corrupted C), and $\{\mathsf{A}, \mathsf{B}\}$ take (without knowing it) the roles of $\{\mathsf{A}^1, \mathsf{B}^1\}$.

We claim that the output of $\{\mathsf{A}, \mathsf{B}\}$ under the above attack is y^*. Observe that the emulation of R, induced by the interaction of D with $\{\mathsf{A}, \mathsf{B}\}$, is just a valid execution of R on some input $\boldsymbol{w}'$ (not completely known to the adversary). Hence, by the above observation, all parties in R (including $\{\mathsf{A}, \mathsf{B}\}$) output the same value at the end of this emulation. Since the execution of R ends after at most q rounds, and since the number of communication links between $\{\mathsf{A}^1, \mathsf{B}^1\}$ and P^* is $\approx 3q/2 > q$, the actions of $\{\mathsf{A}^1, \mathsf{B}^1\}$ have *no effect* on the view of P^*. In particular, the output of P^* in the attack is also y^*, and by the above this is also the output of $\{\mathsf{A}, \mathsf{B}\}$.

Extension to Expected Polynomial-Time Protocols. The above attack works perfectly if π runs in (strict) polynomial time. For expected polynomial-time protocols, one has to work slightly harder to come up with an attack that is (almost) as good.

Let q be the expected round complexity of π. That is, an honest party of π halts after q rounds in expectation, regardless of what the other parties do, where the expectation is over its random coins. Consider the ring $R = (A^1, B^1, C^1, \ldots, A^m, B^m, C^m)$, for $m = 2q$. By Markov bound, in a random execution of R, a party halts after m rounds with probability at least $\frac{1}{2}$.

The adversary D attacking $\{A, B\}$ is defined as follows. For choosing a value for y^*, it emulates an execution of R on arbitrary inputs and uniformly-distributed random coins. If the party $P^* = A^{m/2}$ halts in at most m rounds, D sets y^* to be P^*'s output, and continues to the second stage of the attack. Otherwise, it emulates R on new inputs and random coins. Note that in k attempts, D finds a good execution with probably (at least) $1 - 2^{-k}$. After finding y^*, the adversary D continues as in the strict polynomial case discussed above.

The key observation here is that in the emulated execution of R, induced by the interaction of D with $\{A, B\}$, the party P^* *never* interacts in more than m communication rounds. Therefore, again, being far from $\{A, B\}$, their actions do not affect P^* in the first m rounds, and so do not affect it at all. Hence, P^* outputs y^* also in the induced execution, and so do the parties $\{A, B\}$.

1.3   Additional Related Work

Negative Results. In their seminal work, Lamport et al. [20] defined the problem of simulating a broadcast channel in the point-to-point model in terms of the Byzantine agreement problem. They showed that a broadcast protocol exists if and only if more than two-thirds of the parties are honest. Lamport [19] defined the weak Byzantine agreement problem, and showed that even this weak variant of agreement cannot be computed, using deterministic protocols, facing one-third corruptions. Fischer et al. [11] presented simpler proofs to the above impossibility results using the so-called hexagon argument, which is also the basis of our lower bound (see Sect. 1.2). They assumed a protocol exists for the three-party case, and composed multiple copies of this protocol into a ring system that contains an internal conflict. Since the ring system cannot exist, it follows that the three-party protocol does not exist. We remark that the result of [11] extends to public-coins protocols, where parties have access to a common random string. It follows that coin flipping is not sufficient for solving Byzantine agreement, and thus the impossibility result for coin flipping stated in Corollary 2 is not implied by the aforementioned impossibility of Byzantine agreement.

Cohen and Lindell [9] analyzed the relation between security in the broadcast model and security in the point-to-point model, and showed that some (non 1-dominated) functionalities, e.g., three-party majority, that can be computed in the broadcast model cannot be securely computed in the point-to-point model, since they imply the existence of broadcast.

Positive Results. If the model is augmented with a trusted setup phase, e.g., a public-key infrastructure (PKI), then Byzantine agreement can be computed facing any number of corrupted parties [20]. Pfitzmann and Waidner [23] presented an information-theoretic broadcast protocol assuming a temporary broadcast channel is available during the setup phase. Fitzi et al. [12] presented a probabilistic protocol that securely computes weak Byzantine agreement facing an arbitrary number of corrupted parties. Cohen and Lindell [9] showed (using the protocol from [12]) that assuming the existence of one-way functions, any 1-dominated functionality that can be securely computed in the broadcast model, can also be securely computed in the point-to-point model.

Goldwasser and Lindell [16] presented a weaker definition for MPC without agreement, in which non-unanimous abort is permitted, i.e., some of the honest parties may receive output while other honest parties might abort. Using this weaker definition, they utilized non-consistent protocols and constructed secure protocols in the point-to-point model, assuming an arbitrary number of corrupted parties.

1.4 Open Questions

Our result for the non honest-majority case (second item of Theorem 1), requires the existence of one-way functions. In particular, given a protocol π for computing a 1-dominated functionality f with full security in the broadcast model, one-way functions are used for compiling π into a protocol for computing f with full security in the point-to-point model.[8] It might be, however, that the existence of such a broadcast-model protocol (for non-trivial functionalities) implies the existence of one-way functions, and thus adding this extra assumption is not needed.

A different interesting challenge is characterizing which *non*-symmetric functionalities can be computed in the point-to-point model, in the spirit of what we do here for symmetric functionalities. For example, can a three-party coin flipping in which only two parties learn the outcome coin, be computed with full security facing a single corruption?

Paper Organization. Basic definitions can be found in Sect. 2. Our attack is described in Sect. 3, and its applications are given in Sect. 4. The characterization is presented in Sect. 5.

2   Preliminaries

2.1   Notations

We use calligraphic letters to denote sets, uppercase for random variables, lowercase for values, boldface for vectors, and sans-serif (e.g., A) for algorithms (i.e., Turing Machines). For $n \in \mathbb{N}$, let $[n] = \{1, \cdots, n\}$. Let poly denote the set all positive polynomials and let PPT denote a probabilistic algorithm that runs in *strictly* polynomial time. A function $\nu \colon \mathbb{N} \mapsto [0, 1]$ is *negligible*, denoted $\nu(\kappa) = \mathrm{neg}(\kappa)$, if $\nu(\kappa) < 1/p(\kappa)$ for every $p \in$ poly and large enough κ.

---

[8] For some trivial functionalities, e.g., constant functions, there exist information-theoretically secure protocols in the point-to-point model that are not based on such a compilation, and this extra assumption is not needed.

The statistical distance between two random variables X and Y over a finite set $\mathcal{U}$, denoted $\mathrm{SD}(X,Y)$, is defined as $\frac{1}{2} \cdot \sum_{u \in \mathcal{U}} |\Pr[X = u] - \Pr[Y = u]|$. We say that X and Y are δ-close if $\mathrm{SD}(X,Y) \leq \delta$ and statistically close (denoted $X \overset{\mathrm{s}}{\equiv} Y$) is they are δ-close and δ is negligible.

Two distribution ensembles $X = \{X(a,\kappa)\}_{a \in \{0,1\}^*, \kappa \in \mathbb{N}}$ and $Y = \{Y(a,\kappa)\}_{a \in \{0,1\}^*, \kappa \in \mathbb{N}}$ are computationally indistinguishable (denoted $X \overset{\mathrm{c}}{\equiv} Y$) if for every non-uniform polynomial-time distinguisher D there exists a function $\nu(\kappa) = \mathrm{neg}(\kappa)$, such that for every $a \in \{0,1\}^*$ and all sufficiently large κ's

$$|\Pr[\mathsf{D}(X(a,\kappa), 1^\kappa) = 1] - \Pr[\mathsf{D}(Y(a,\kappa), 1^\kappa) = 1]| \leq \nu(\kappa).$$

2.2 Protocols

An n-party protocol $\pi = (\mathsf{P}_1, \ldots, \mathsf{P}_n)$ is an n-tuple of probabilistic interactive TMs. The term *party* P_i refers to the i'th interactive TM. Each party P_i starts with input $x_i \in \{0,1\}^*$ and random coins $r_i \in \{0,1\}^*$. Without loss of generality, the input length of each party is assumed to be the security parameter κ. An *adversary* D is another interactive TM describing the behavior of the corrupted parties. It starts the execution with input that contains the identities of the corrupted parties and their private inputs, and possibly an additional auxiliary input. The parties execute the protocol in a synchronous network. That is, the execution proceeds in rounds: each round consists of a *send phase* (where parties send their message from this round) followed by a *receive phase* (where they receive messages from other parties).

In the *point-to-point (communication) model*, which is the one we assume by default, all parties are connected via a *fully-connected point-to-point network*. We consider two models for the communication lines between the parties: In the *authenticated-channels* model, the communication lines are assumed to be ideally authenticated but not private (and thus the adversary cannot modify messages sent between two honest parties but can read them). In the *secure-channels* model, the communication lines are assumed to be ideally private (and thus the adversary cannot read or modify messages sent between two honest parties). In the *broadcast model*, all parties are given access to a physical broadcast channel in addition to the point-to-point network. In both models, no preprocessing phase is available.

Throughout the execution of the protocol, all the honest parties follow the instructions of the prescribed protocol, whereas the corrupted parties receive their instructions from the adversary. The adversary is considered to be *malicious*, meaning that it can instruct the corrupted parties to deviate from the protocol in any arbitrary way. At the conclusion of the execution, the honest parties output their prescribed output from the protocol, the corrupted parties output nothing and the adversary outputs an (arbitrary) function of its view of the computation (containing the views of the corrupted parties). The view of a party in a given execution of the protocol consists of its input, its random coins, and the messages it sees throughout this execution.

2.2.1   Time and Round Complexity

We consider both *strict* and *expected* bounds on time and round complexity.

Definition 1 (time complexity). *Protocol* $\pi = (\mathsf{P}_1, \ldots, \mathsf{P}_n)$ *is a T-time protocol, if for every $i \in [n]$ and every input $x_i \in \{0,1\}^*$, random coins $r_i \in \{0,1\}^*$, and sequence of messages P_i receives during the course of the protocol, the running time of an honest party P_i is at most $T(|x_i|)$. If $T \in$ poly, then π is of* (strict) polynomial time.

Protocol π has an expected running time T, *if for every $i \in [n]$, every input $x_i \in \{0,1\}^*$ and sequence of messages P_i receives during the course of the protocol, the expected running time of an honest party P_i, over its random coins r_i, is at most $T(|x_i|)$. If $T \in$ poly, then π has* expected polynomial running time.

Definition 2 (round complexity). *Protocol $\pi = (\mathsf{P}_1, \ldots, \mathsf{P}_n)$ is a q-round protocol, if for every $i \in [n]$ and every input $x_i \in \{0,1\}^*$, random coins $r_i \in \{0,1\}^*$, and sequence of messages P_i receives during the course of the protocol, the round number in which an honest party P_i stops being active (i.e., stops sending and receiving messages) is at most $q(|x_i|)$. If $q \in$ poly, then π has* (strict) polynomial round complexity.

Protocol π has an expected round complexity q, *if for every $i \in [n]$, every input $x_i \in \{0,1\}^*$ and sequence of messages P_i receives during the course of the protocol, the expected round number in which an honest party P_i stops being active, over its random coins r_i, is at most $q(|x_i|)$. If $q \in$ poly, then π has* expected polynomial round complexity.

3   Attacking Consistent Protocols

In this section, we present a lower bound for secure protocols in the secure-channels point-to-point model. Protocols in consideration are only assumed to have a very mild security property (discussing the more standard notion of security is deferred to Sect. 4). Specifically, we only require the protocol to be consistent – all honest parties output the same value. We emphasize that in a consistent protocol, a party may output the special error symbol $\bot$ (i.e., abort), but it can only do so if all honest parties output $\bot$ as well.

Definition 3 (consistent protocols). *A protocol π is (δ, t)-consistent against C-class (e.g., polynomial-time, expected polynomial-time) adversaries, if the following holds. Consider an execution of π on security parameter κ, and any vector of inputs of length κ for the parties, in which a C-class adversary controls at most t parties. Then with probability at least $\delta(\kappa)$, all honest parties output the same value, where the probability is taken over the random coins of the adversary and of the honest parties.*

We now present an attack on consistent protocols whose round complexity is strictly bounded. An extension of the attack to consistent protocols with a bound on their *expected* number of rounds appears in the full version of this paper [8].

Lemma 2. *Let* $n \geq 3$, *let* $t \geq \frac{n}{3}$, *and let* $s = n - 2t$ *if* $t < \frac{n}{2}$ *and* $s = 1$ *otherwise.* *Let* π *be an* n-*party,* T-*time,* q-*round protocol in the secure-channels point-to-point model that is* $(1 - \delta, t)$-*consistent against* $(T_{\mathsf{D}} = 2nqT)$-*time adversaries.* *Then, there exists a* T_{D}-*time adversary* D *such that given the control over any* s-*size subset* $\mathcal{I}$ *of parties, the following holds: on security parameter* κ, D *first outputs a value* $y^* = y^*(\mathcal{I})$. *Next,* D *interacts with the remaining honest parties of* π *on inputs of length* κ, *and except for probability at most* $\left(\frac{3}{2} \cdot q(\kappa) + 1\right) \cdot \delta(\kappa)$, *the output of every honest party in this execution is* y^*.[9]

For a polynomial-time protocol that is $(1 - \mathrm{neg}, t)$-consistent against PPT adversaries and assuming an honest majority, Lemma 2 yields a PPT adversary that by controlling $n - 2t$ of the parties can manipulate the outputs of the honest parties (i.e., forcing them all to be y^*) with all but a negligible probability. If an honest majority is not assumed, the adversary can manipulate the outputs of the honest parties, by controlling any single party, except for a negligible probability.

We start by proving the lemma for three-party protocols, and later prove the multiparty case using a reduction to the three-party case. We actually prove a stronger statement for the three-party case, where the value y^* is independent of the set of corrupted parties.

Lemma 3. *Let* π *be a* 3-*party,* q-*round protocol in the secure-channels point-to-point model, let* T *be the combined running-time of all three parties.*[10] *If* π *is* $(1 - \delta, 1)$-*consistent against* $(T_{\mathsf{D}} = 2qT)$-*time adversaries, then there exists a* T_{D}-*time adversary* D *such that the following holds. On security parameter* κ, D *first outputs a value* y^*. *Next, given the control over any non-empty set of parties,* D *interacts with the remaining honest parties of* π *on inputs of length* κ, *and except for probability at most* $\frac{3}{2} \cdot q(\kappa) \cdot \delta(\kappa)$, *the output of every honest party in this execution is* y^*.

Proof. We fix the input-length parameter κ and omit it from the notation when its value is clear from the context. Let $\pi = (\mathsf{A}, \mathsf{B}, \mathsf{C})$ and let $m = q$ (assume for ease of notation that m is even). Consider, without loss of generality, that a single party is corrupted (the case of two corrupted parties follows from the proof) and assume for concreteness that the corrupted party is C. Consider the following ring network $\mathsf{R} = (\mathsf{A}^1, \mathsf{B}^1, \mathsf{C}^1, \ldots, \mathsf{A}^m, \mathsf{B}^m, \mathsf{C}^m)$, in which each two consecutive parties, as well as the first and last, are connected via a secure channel, and party P^j, for $\mathsf{P} \in \{\mathsf{A}, \mathsf{B}, \mathsf{C}\}$, has the code of P. Let $v = \kappa + T(\kappa)$, and consider an execution of R with arbitrary inputs and uniformly-distributed random coins for the parties being $\boldsymbol{w} = (w_{\mathsf{A}}^1, w_{\mathsf{B}}^1, w_{\mathsf{C}}^1, \ldots, w_{\mathsf{A}}^m, w_{\mathsf{B}}^m, w_{\mathsf{C}}^m) \in (\{0,1\}^v)^{3m}$ (i.e., party P^i has input w_{P}^i, containing its actual input and random coins).

A key observation is that the point of view of the party A^j, for instance, in such an execution, is a *valid* view of the party A on input w_{A}^j in an execution of

---

[9] We would get slightly better parameters using an attack in which at least one honest party (but not necessarily all) outputs y^*.

[10] This is more general than T-time 3-party protocols, as it captures asymmetry between the running time of the parties; this measure will turn out to be useful for proving Lemma 2.

π in which B acts honestly on input w_B^j. It is also a valid view of A, on input w_A^j, in an execution of π in which C acts honestly on input $w_C^{j-1 \pmod{m}}$. This observation yields the following consistency property of R.

Claim 1. *Consider an execution of R on joint input $w \in (\{0,1\}^v)^{3m}$, where the parties' coins in w are chosen uniformly at random, and the parties' (actual) inputs are chosen arbitrarily. Then parties of distance d in R, measured by the (minimal) number of communication links between them, as well as all $d-1$ parties between them, output the same value with probability at least $1 - d\delta$.*

Proof. Consider the pair of neighboring parties $\{A^j, B^j\}$ in the ring R (an analogous argument holds for any two neighboring parties). Let D be an adversary, controlling the party C of π that interacts with $\{A, B\}$ by emulating an execution of R on arbitrary inputs and uniform random coins (apart from the roles of $\{A^j, B^j\}$), and let $\{A, B\}$ take (without knowing that) the roles of $\{A^j, B^j\}$ in this execution. The joint view of $\{A, B\}$ in this emulation has the same distribution as the joint view of $\{A^j, B^j\}$ in an execution of R with uniform random coins. Hence, the $(1 - \delta)$-consistency of π yields that A^j and B^j output the same value in an execution of R on $w \in (\{0,1\}^v)^{3m}$ (where the random coins within w of each party are chosen uniformly at random) with probability at least $1 - \delta$. The proof follows by a union bound.

The adversary D first selects a value for $w \in (\{0,1\}^v)^{3m}$, consisting of arbitrary input values (e.g., zeros) and uniformly-distributed random coins, and sets y^* to be the output of $P^* = A^{m/2}$ in the execution of R on w. To interact with $\{A, B\}$ in π, D emulates an execution of R in which all but $\{A^1, B^1\}$ have their inputs according to w, and $\{A, B\}$ take the roles of $\{A^1, B^1\}$. The key observation is that the view of party P^* in the emulation induced by the above attack, is the *same* as its view in the execution of R on w (regardless of the inputs of $\{A, B\}$). This is true since the execution of R ends after at most m communication rounds. Thus, the actions of $\{A, B\}$ have no effect on the view of P^*, and therefore the output of P^* is y^* also in the emulated execution of R. Finally, since all the parties in the emulated execution of R have uniformly-distributed random coins, and since the distance between P^* and $\{A, B\}$ is (less than) $\frac{3m}{2}$, Claim 1 yields that with probability at least $1 - \frac{3m}{2} \cdot \delta$, the output of $\{A, B\}$ under the above attack is y^*.

Note that the value y^* does not depend on the identity of the corrupted party, since in the first step y^* is set independently of C, and in the second step the attack follows without any change when the honest parties play the roles of $\{B^1, C^1\}$ if A is corrupted or $\{A^2, C^1\}$ if B is corrupted.

We now proceed to prove Lemma 2 in the many-party case.

Proof. Let $\pi = (P_1, \ldots, P_n)$ be a T-time, q-round, n-party protocol that is $(1 - \delta, t)$-consistent against $2nqT$-time adversaries. We will show an adversary that by controlling any s corrupted parties, manipulates all honest parties to output a predetermine value. We separately handle the case that $\frac{n}{3} \leq t < \frac{n}{2}$ and the case $\frac{n}{2} \leq t < n$.

Case $\frac{n}{3} \leq t < \frac{n}{2}$. Let $\mathcal{I} \subseteq [n]$ be a subset of size $s = n - 2t$, representing the indices of the corrupted parties in π. Consider the three-party protocol $\pi' = (A', B', C')$, defined by partitioning the set $[n]$ into three subsets $\{\mathcal{I}_{A'}, \mathcal{I}_{B'}, \mathcal{I}\}$, where $\mathcal{I}_{A'}$ and $\mathcal{I}_{B'}$ are each of size t, and letting A' run the parties $\{P_i\}_{i \in \mathcal{I}_{A'}}$, B' run the parties $\{P_i\}_{i \in \mathcal{I}_{B'}}$ and C' run the parties $\{P_i\}_{i \in \mathcal{I}}$. Each of the parties in π' waits until all the virtual parties it is running halt, arbitrarily selects one of them and outputs the virtual party's output value.

Since the subsets $\mathcal{I}_{A'}, \mathcal{I}_{B'}, \mathcal{I}$ are of size at most t, the q-round, 3-party protocol π' is $(1 - \delta, 1)$-consistent against $2nqT$-adversaries (otherwise there exists a $2nqT$-time adversary against the consistency of π, corrupting at most t parties). In addition, since the combined time complexity of all three parties is nT, by Lemma 3 there exists a $2nqT$-time adversary D' that first determines a value y^*, and later, given control over any party in π' (in particular C'), can force the two honest parties to output y^* with probability at least $1 - \frac{3q\delta}{2}$.

The attacker D for π, controlling the parties indexed by $\mathcal{I}$, is defined as follows: In the first step, D runs D' and outputs the value y^* that D' outputs. In the second step, D interacts with the honest parties in π by simulating the parties $\{A', B'\}$ to D', i.e., D runs D' and sends every message it receives from D' to the corresponding honest party in π, and similarly, whenever D receives a message from an honest party in π it forwards it to D'. It is immediate that there exists $i \in \mathcal{I}_{A'}$ such that P_i outputs y^* in the execution of π with the same probability that A' outputs y^* in the execution of π', i.e., with probability at least $1 - \frac{3q\delta}{2}$. From the consistency property of π, all honest parties output the same value with probability at least $1 - \delta$, and using the union bound we conclude that the output of all honest parties in π under the above attack is y^* with probability at least $1 - (\frac{3q\delta}{2} + \delta)$.

Case $\frac{n}{2} \leq t < n$. Let $i^* \in [n]$ be the index of the corrupted party in π and consider the three-party protocol $\pi' = (A', B', C')$ defined by partitioning the set $[n]$ into three subsets $\{\mathcal{I}_{A'}, \mathcal{I}_{B'}, \{i^*\}\}$, for $|\mathcal{I}_{A'}| = \lceil \frac{n-1}{2} \rceil$ and $|\mathcal{I}_{B'}| = \lfloor \frac{n-1}{2} \rfloor$. As in the previous case, the size of each subset $\mathcal{I}_{A'}, \mathcal{I}_{B'}, \{i^*\}$ is at most t, and the proof proceeds as above.

4   Impossibility Results for Secure Computation

In this section, we present applications of the attack of Sect. 3 to secure multiparty computations in the secure-channels point-to-point model.[11] In Sect. 4.1, we show that the only symmetric functionalities that can be securely realized, according to the real/ideal paradigm, in the presence of $n/3 \leq t < n/2$ corrupted parties (i.e., honest majority), are $(n - 2t)$-dominated functionalities. The only symmetric functionalities that can be securely realized in the presence of $n/2 \leq t < n$ corrupted parties (i.e., no honest majority), are 1-dominated

---

[11] Note that a lower bound in the secure-channels model is stronger than in the authenticated-channels model.

functionalities. In Sect. 4.2, we show that non-trivial $(n > 3)$-party coin-flipping protocols, in which the honest parties must output a bit, are impossible when facing $t \geq n/3$ corrupted parties.

For concreteness, we focus on strict polynomial-time protocols secure against strict polynomial-time adversaries, but all the results readily extend to the expected polynomial-time regime.

4.1 Symmetric Functionalities Secure According to the Real/Ideal Paradigm

The model of secure computation we consider is defined in Sect. 4.1.1, dominated functionalities are defined in Sect. 4.1.2 and the impossibility results are stated and proved in Sect. 4.1.3.

4.1.1 Model Definition

We provide the basic definitions for secure multiparty computation according to the real/ideal paradigm, for further details see [14]. Informally, a protocol is secure according to the real/ideal paradigm, if whatever an adversary can do in the real execution of protocol, can be done also in an ideal computation, in which an uncorrupted trusted party assists the computation. We consider *full security*, meaning that the ideal-model adversary cannot prematurely abort the ideal computation.

Functionalities.

Definition 4 (functionalities). *An n-party* functionality *is a random process that maps vectors of n inputs to vectors of n outputs.*[12] *Given an n-party functionality $f \colon (\{0,1\}^*)^n \mapsto (\{0,1\}^*)^n$, let $f_i(\boldsymbol{x})$ denote its i'th output coordinate, i.e., $f_i(\boldsymbol{x}) = f(\boldsymbol{x})_i$. A functionality f is* symmetric, *if the output values of all parties are the same, i.e., for every $\boldsymbol{x} \in (\{0,1\}^*)^n$, $f_1(\boldsymbol{x}) = f_2(\boldsymbol{x}) = \ldots = f_n(\boldsymbol{x})$.*

Real-Model Execution. A real-model execution of an n-party protocol proceeds as described in Sect. 2.2.

Definition 5 (real-model execution). *Let $\pi = (\mathsf{P}_1, \ldots, \mathsf{P}_n)$ be an n-party protocol and let $\mathcal{I} \subseteq [n]$ denote the set of indices of the parties corrupted by D. The* joint execution of π under $(\mathsf{D}, \mathcal{I})$ in the real model, *on input vector $\boldsymbol{x} = (x_1, \ldots, x_n)$, auxiliary input z and security parameter κ, denoted $\mathrm{REAL}_{\pi, \mathcal{I}, \mathsf{D}(z)}(\boldsymbol{x}, \kappa)$, is defined as the output vector of $\mathsf{P}_1, \ldots, \mathsf{P}_n$ and $\mathsf{D}(z)$ resulting from the protocol interaction, where for every $i \in \mathcal{I}$, party P_i computes its messages according to D, and for every $j \notin \mathcal{I}$, party P_j computes its messages according to π.*

---

[12] We assume that a functionality can be computed in polynomial time.

Ideal-Model Execution. An ideal computation of an n-party functionality f on input $\boldsymbol{x} = (x_1, \ldots, x_n)$ for parties $(\mathsf{P}_1, \ldots, \mathsf{P}_n)$ in the presence of an ideal-model adversary D controlling the parties indexed by $\mathcal{I} \subseteq [n]$, proceeds via the following steps.

Sending inputs to trusted party: An honest party P_i sends its input x_i to the trusted party. The adversary may send to the trusted party arbitrary inputs for the corrupted parties. Let x_i' be the value actually sent as the input of party P_i.

Trusted party answers the parties: If x_i' is outside of the domain for P_i, for some index i, or if no input was sent for P_i, then the trusted party sets x_i' to be some predetermined default value. Next, the trusted party computes $f(x_1', \ldots, x_n') = (y_1, \ldots, y_n)$ and sends y_i to party P_i for every i.

Outputs: Honest parties always output the message received from the trusted party and the corrupted parties output nothing. The adversary D outputs an arbitrary function of the initial inputs $\{x_i\}_{i \in \mathcal{I}}$, the messages received by the corrupted parties from the trusted party $\{y_i\}_{i \in \mathcal{I}}$ and its auxiliary input.

Definition 6 (ideal-model computation). *Let $f \colon (\{0,1\}^*)^n \mapsto (\{0,1\}^*)^n$ be an n-party functionality and let $\mathcal{I} \subseteq [n]$. The* joint execution of f under (D, I) *in the ideal model, on input vector $\boldsymbol{x} = (x_1, \ldots, x_n)$, auxiliary input z to D and security parameter κ, denoted $\mathrm{IDEAL}_{f,\mathcal{I},\mathsf{D}(z)}(\boldsymbol{x}, \kappa)$, is defined as the output vector of $\mathsf{P}_1, \ldots, \mathsf{P}_n$ and $\mathsf{D}(z)$ resulting from the above described ideal process.*

Security Definition. Having defined the real and ideal models, we can now define security of protocols according to the real/ideal paradigm.

Definition 7. *Let $f \colon (\{0,1\}^*)^n \mapsto (\{0,1\}^*)^n$ be an n-party functionality, and let π be a probabilistic polynomial-time protocol computing f. The protocol π t-securely computes f (with computational security), if for every non-uniform polynomial-time real-model adversary D, there exists a non-uniform (expected) polynomial-time adversary S for the ideal model, such that for every $\mathcal{I} \subseteq [n]$ of size at most t, it holds that*

$$\Big\{\mathrm{REAL}_{\pi,\mathcal{I},\mathsf{D}(z)}(\boldsymbol{x}, \kappa)\Big\}_{(\boldsymbol{x},z) \in (\{0,1\}^*)^{n+1}, \kappa \in \mathbb{N}} \overset{c}{\equiv} \Big\{\mathrm{IDEAL}_{f,\mathcal{I},\mathsf{S}(z)}(\boldsymbol{x}, \kappa)\Big\}_{(\boldsymbol{x},z) \in (\{0,1\}^*)^{n+1}, \kappa \in \mathbb{N}}.$$

The protocol π t-securely computes f (with information-theoretic security), if for every real-model adversary D, there exists an adversary S for the ideal model, whose running time is polynomial in the running time of D, such that for every $\mathcal{I} \subseteq [n]$ of size at most t,

$$\Big\{\mathrm{REAL}_{\pi,\mathcal{I},\mathsf{D}(z)}(\boldsymbol{x}, \kappa)\Big\}_{(\boldsymbol{x},z) \in (\{0,1\}^*)^{n+1}, \kappa \in \mathbb{N}} \overset{s}{\equiv} \Big\{\mathrm{IDEAL}_{f,\mathcal{I},\mathsf{S}(z)}(\boldsymbol{x}, \kappa)\Big\}_{(\boldsymbol{x},z) \in (\{0,1\}^*)^{n+1}, \kappa \in \mathbb{N}}.$$

4.1.2 Dominated Functionalities

A special class of symmetric functionalities are those with the property that every subset of a certain size can fully determine the output. For example, the multiparty Boolean AND and OR functionalities both have the property that every individual party can determine the output (for the AND functionality any party can always force the output to be 0, and for the OR functionality any party can always force the output to be 1). We distinguish between the case where there exists a single value for which every large enough subset can force the output and the case where different subsets can force the output to be different values.

Definition 8 (dominated functionalities). *A symmetric n-party functionality f is* weakly k-dominated, *if for every k-size subset $\mathcal{I} \subseteq [n]$ there exists a polynomial-time computable value $y^*_{\mathcal{I}}$, for which there exist inputs $\{x_i\}_{i \in \mathcal{I}}$, such that $f(x_1, \ldots, x_n) = y^*_{\mathcal{I}}$ for* any *complementing subset of inputs $\{x_j\}_{j \notin \mathcal{I}}$. The functionality f is k-dominated, if there exists a polynomial-time computable value y^* such that for every k-size subset $\mathcal{I} \subseteq [n]$ there exist inputs $\{x_i\}_{i \in \mathcal{I}}$, for which $f(x_1, \ldots, x_n) = y^*$ for* any *subset of inputs $\{x_j\}_{j \notin \mathcal{I}}$.*

Example 1. The function $f(x_1, x_2, x_3, x_4) = (x_1 \wedge x_2) \vee (x_3 \wedge x_4)$ is an example of a 4-party function that is weakly 2-dominated but not 2-dominated. Every pair of input variables can be set to determine the output value. However, there is no single output value that can be determined by all pairs, for example, $\{x_1, x_2\}$ can force the output to be 1 (by setting $x_1 = x_2 = 1$) whereas $\{x_1, x_3\}$ can force the output to be 0 (by setting $x_1 = x_3 = 0$). The function

$$f_{2\text{-of-}4}(x_1, x_2, x_3, x_4) = (x_1 \wedge x_2) \vee (x_1 \wedge x_3) \vee (x_1 \wedge x_4) \vee (x_2 \wedge x_3) \vee (x_2 \wedge x_4) \vee (x_3 \wedge x_4)$$

is 2-dominated with value $y^* = 1$.

Claim 2. *Let f be an n-party functionality and let $m \leq \frac{n}{3}$. If f is weakly m-dominated, then it is m-dominated.*

Proof. Let $\mathcal{I}_1, \mathcal{I}_2 \subseteq [n]$ be two subsets of size m. In case $\mathcal{I}_1$ and $\mathcal{I}_2$ are disjoint, consider the corresponding sets of input variables $\{x_i\}_{i \in \mathcal{I}_1}$ and $\{x_i\}_{i \in \mathcal{I}_2}$, and fix an arbitrary complementing subset of inputs $\{x_j\}_{j \notin \mathcal{I}_1 \cup \mathcal{I}_2}$. On the one hand it holds that $f(x_1, \ldots, x_n) = y^*_{\mathcal{I}_1}$ and on the other hand it holds that $f(x_1, \ldots, x_n) = y^*_{\mathcal{I}_2}$, hence $y^*_{\mathcal{I}_1} = y^*_{\mathcal{I}_2}$.

In case $\mathcal{I}_1$ and $\mathcal{I}_2$ are not disjoint, it holds that $|\mathcal{I}_1 \cup \mathcal{I}_2| < 2m \leq \frac{2n}{3}$ and since $m \leq \frac{n}{3}$, there exists a subset $\mathcal{I}_3 \subseteq [n] \setminus (\mathcal{I}_1 \cup \mathcal{I}_2)$ of size m. Denote by $y^*_{\mathcal{I}_3}$ the output value that can be determined by the input variables $\{x_i\}_{i \in \mathcal{I}_3}$ ($y^*_{\mathcal{I}_3}$ is guaranteed to exist since f is weakly m-dominated). $\mathcal{I}_3$ is disjoint from $\mathcal{I}_1$ and from $\mathcal{I}_2$, so it follows that $y^*_{\mathcal{I}_1} = y^*_{\mathcal{I}_3}$ and $y^*_{\mathcal{I}_2} = y^*_{\mathcal{I}_3}$, therefore $y^*_{\mathcal{I}_1} = y^*_{\mathcal{I}_2}$.

4.1.3   The Lower Bound

Lemma 4. *Let $n \geq 3$, let $t \geq \frac{n}{3}$ and let f be a symmetric n-party functionality that can be t-securely computed in the secure-channels point-to-point model.*

1. *If $\frac{n}{3} \leq t < \frac{n}{2}$, then f is $(n - 2t)$-dominated.*
2. *If $\frac{n}{2} \leq t < n$, then f is 1-dominated.*

Proof. Assume that $\frac{n}{3} \leq t < \frac{n}{2}$ (the proof for $\frac{n}{2} \leq t < n$ is similar). Let π be a protocol that t-securely computes f in the point-to-point model with secure channels. Since f is symmetric, all honest parties output the same value (except for a negligible probability), hence π is $(1 - \text{neg}, t)$-consistent; let D be the PPT adversary guaranteed from Lemma 2 and let $\mathcal{I} \subseteq [n]$ be any subset of size $n - 2t$. It follows that given control over $\{P_i\}_{i \in \mathcal{I}}$, D can first fix a value $y_{\mathcal{I}}^*$, and later force the output of the honest parties to be $y_{\mathcal{I}}^*$ (except for a negligible probability). Since π t-securely computes f and $n - 2t \leq t$, there exists an ideal-model adversary S that upon corrupting $\{P_i\}_{i \in \mathcal{I}}$, can force the output of the honest parties in the ideal-model computation to be $y_{\mathcal{I}}^*$. All S can do is to select the input values of the corrupted parties, hence, there must exist input values $\{x_i\}_{i \in \mathcal{I}}$ that determine the output of the honest parties to be $y_{\mathcal{I}}^*$, i.e., f is weakly $(n - 2t)$-dominated. Since $n - 2t \leq \frac{n}{3}$ and following Claim 2 we conclude that f is $(n - 2t)$-dominated.

4.2 Coin-Flipping Protocols

A coin-flipping protocol [3] allows the honest parties to jointly flip an unbiased coin, where even a coalition of cheating (efficient) parties cannot bias the outcome of the protocol by much. Our focus is on coin flipping, where the honest parties *must* output a bit. Although Lemma 4 immediately shows that coin flipping cannot be securely computed according to the real/ideal paradigm, we present a stronger impossibility result by considering weaker security requirements.

Definition 9. *A polynomial-time n-party protocol π is a (γ, t)-bias coin-flipping protocol, if the following holds.*

1. *π is $(1, t)$-consistent against PPT adversaries.*[13]
2. *When interacting on security parameter κ (for sufficiently large κ's) with a PPT adversary controlling at most t corrupted parties, the common output of the honest parties is $\gamma(\kappa)$-close to the being a uniform bit.*[14]

The following is a straightforward application of Lemma 2.

Lemma 5. *In the secure-channels point-to-point model, for $n \geq 3$ and $\gamma(\kappa) < \frac{1}{2} - 2^{-\kappa}$, there exists no n-party, $(\gamma, \lceil \frac{n}{3} \rceil)$-bias coin-flipping protocol.*

Proof. Let π be a point-to-point n-party $(\gamma, \lceil \frac{n}{3} \rceil)$-bias coin-flipping protocol. Let D be the PPT adversary that is guaranteed by Lemma 2 (since π is $(1, \lceil \frac{n}{3} \rceil)$-consistent against PPT adversaries). Consider some fixed set of $\lceil \frac{n}{3} \rceil$ corrupted parties of π and let $Y(\kappa)$ denote the random variable of $D(\kappa)$'s output in the first step of the attack. Without loss of generality, for infinitely many values of κ it holds that $\Pr[Y(\kappa) = 0] \leq \frac{1}{2}$. Consider the adversary D' that on security

---

[13] Our negative result readily extends to protocols where consistency is only guaranteed to hold with high probability.

[14] In particular, the honest parties are allowed to output $\bot$, or values other than $\{0, 1\}$, with probability at most γ.

parameter κ, repeats the first step of $\mathsf{D}(\kappa)$ until the resulting value of y^* is non-zero or κ failed attempts have been reached, where if the latter happens D' aborts. Next, D' continues the non-zero execution of D to make the honest parties of π output y^*. It is immediate that for infinitely many values of κ, the common output of the honest parties under the above attack is 0 with probability at most $2^{-\kappa}$, and hence the common output of the honest parties is $\frac{1}{2} - 2^{-\kappa}$ far from uniform. Thus, π is not a $(\gamma, \lceil \frac{n}{3} \rceil)$-bias coin-flipping protocol.

5 Characterizing Secure Computation Without Broadcast

In this section we show that the lower bounds presented in Lemma 4 is tight. We treat separately the case where an honest majority is assumed and the case where no honest majority is assumed.

5.1 No Honest Majority

Cohen and Lindell [9, Theorem 7] showed that, assuming the existence of one-way functions, any 1-dominated functionality that can be t-securely computed in the broadcast model with authenticated channels, can also be t-securely computed in the point-to-point model with authenticated channels.[15] Combining with Lemma 4, we establish the following result.

Theorem 2 (restating second part of Theorem 1). *Let $n \geq 3$, let $\frac{n}{2} \leq t < n$ and assume that one-way functions exist. An n-party functionality can be t-securely computed in the authenticated-channels point-to-point model, if and only if it is 1-dominated and can be t-securely computed in the authenticated-channels broadcast model.*

Proof. Immediately by Lemma 4 and Cohen and Lindell [9, Theorem 7].

5.2 Honest Majority

Proposition 2. *Let $n \geq 3$, let $\frac{n}{3} \leq t < \frac{n}{2}$, and let f be a symmetric n-party functionality. If f is $(n - 2t)$-dominated, then it can be t-securely computed in the secure-channels point-to-point model with information-theoretic security.*

To prove Proposition 2 we use the *two-threshold multiparty protocol* of Fitzi et al. [13, Theorem 6]. This protocol with parameters t_1, t_2 runs in the point-to-point model with secure channels, and whenever $t_1 \leq t_2$ and $t_1 + 2t_2 < n$, the following holds. Let $\mathcal{I}$ be the set of parties that the (computationally unbounded)

---

[15] The result in [9] is based on the computationally-secure protocol in [12, Theorem 2]. In the authenticated-channels point-to-point model, this protocol requires one-way functions for constructing a consistent public-key infrastructure between the parties, to be used for authenticated broadcast.

adversary corrupts. If $|\mathcal{I}| \leq t_1$, then the protocol computes f with full security. If $t_1 < |\mathcal{I}| \leq t_2$, then the protocol securely computes f with fairness (i.e., the adversary may force *all* honest parties to output $\perp$, provided that it learns no new information). In Sect. 5.2.1, we formally define the notion of two-threshold security. This notion captures the security achieved by the protocol of Fitzi et al. [13, Theorem 6].

Theorem 3 [13, Theorem 6]. *Let $n \geq 3$, let t_1, t_2 be parameters such that $t_1 \leq t_2$ and $t_1 + 2t_2 < n$, and let f be an n-party functionality. Then, f can be (t_1, t_2)-securely computed in the secure-channels point-to-point model with information-theoretic security.*

We now proceed to the proof of Proposition 2.

Proof (Proof of Proposition 2*).* Let f be an $(n - 2t)$-dominated functionality with default output value y^*. If $n - 2t = 1$, then f is 1-dominated, and since $t < \frac{n}{2}$, f can be t-securely computed with information-theoretic security in the secure-channels broadcast model (e.g., using Rabin and Ben-Or [24]). Hence, the proposition follows from [9, Theorem 7].[16]

For $n - 2t \geq 2$, set $t_1 = n - 2t - 1$ and $t_2 = t$, and let π' be the n-party protocol, guaranteed to exist by Theorem 3, that (t_1, t_2)-securely computes f. We define π to be the following n-party protocol for computing f in the point-to-point model with secure channels.

Protocal 1 *(π)*

1. *The parties run the protocol π'. Let y_i be the output of P_i at the end of the execution.*
2. *If $y_i \neq \perp$, party P_i outputs y_i, otherwise it outputs y^*.*

Let D be an adversary attacking the execution of π and let $\mathcal{I} \subseteq [n]$ be a subset of size at most t. It follows from Theorem 3 that there exists a (possibly aborting) adversary S' for D in the t_1-threshold ideal model such that

$$\left\{\mathrm{REAL}_{\pi', \mathcal{I}, \mathsf{D}(z)}(\boldsymbol{x}, \kappa)\right\}_{(\boldsymbol{x}, z) \in (\{0,1\}^*)^{n+1}, \kappa \in \mathbb{N}} \stackrel{\mathrm{s}}{\equiv} \left\{\mathrm{IDEAL}^{t_1}_{f, \mathcal{I}, \mathsf{S}'(z)}(\boldsymbol{x}, \kappa)\right\}_{(\boldsymbol{x}, z) \in (\{0,1\}^*)^{n+1}, \kappa \in \mathbb{N}}.$$

Using S', we construct the following non-aborting adversary S for the full-security ideal model. On inputs $\{x_i\}_{i \in \mathcal{I}}$ and auxiliary input z, S starts by emulating S' on these inputs, playing the role of the trusted party (in the t_1-threshold ideal model). If S' sends an **abort** command, it is guaranteed that $|\mathcal{I}| \geq n - 2t$ and since f is $(n - 2t)$-dominated, there exist input values $\{x'_i\}_{i \in \mathcal{I}}$ that determine the output of f to be y^*. So in this case, S sends these $\{x'_i\}_{i \in \mathcal{I}}$ to the trusted party (in the full-security ideal model) and returns $\perp$ to S'. Otherwise, S' does not abort and S forwards the message from S' to the trusted party and

---

[16] When an honest majority is assumed, the result in [9] can be adjusted to use the information-theoretically secure protocol in [12, Theorem 3]. In the secure-channels point-to-point model, this protocol uses information-theoretically pseudo-signatures [23] for computing a setup, to be used for authenticated broadcast.

the answer from the trusted party back to S'. In both cases S outputs whatever S' outputs and halts.

A main observation is that the views of the adversary D in an execution of π and in an execution of π' (with the same inputs and random coins) are identical. This holds since the only difference between π and π' is in the second step of π that does not involve any interaction. It follows that in case the output of the parties in Step 1 of π is not $\perp$, the joint distribution of the honest parties' output and the output of D in π is statistically close to the output of the honest parties and of S in the full-security ideal model (since the later is exactly the output of the honest parties and of S' in the t_1-threshold ideal model). If the output in Step 1 of π is $\perp$, then all honest parties in π output y^*. In this case S' sends abort (except for a negligible probability) and since S sends to the trusted party the input values $\{x_i'\}_{i \in \mathcal{I}}$ that determine the output of f to be y^*, the honest parties' output is y^* also in the ideal computation. We conclude that

$$\left\{ \mathrm{REAL}_{\pi,\mathcal{I},\mathsf{D}(z)}(\boldsymbol{x},\kappa) \right\}_{(\boldsymbol{x},z) \in (\{0,1\}^*)^{n+1}, \kappa \in \mathbb{N}} \overset{\mathrm{s}}{\equiv} \left\{ \mathrm{IDEAL}_{f,\mathcal{I},\mathsf{S}(z)}(\boldsymbol{x},\kappa) \right\}_{(\boldsymbol{x},z) \in (\{0,1\}^*)^{n+1}, \kappa \in \mathbb{N}}.$$

Theorem 4 (restating the first part of Theorem 1). *Let $n \geq 3$ and $\frac{n}{3} \leq t < \frac{n}{2}$. A symmetric n-party functionality can be t-securely computed in the secure-channels point-to-point model, if and only if it is $(n - 2t)$-dominated.*

Proof. Immediately follows by Lemma 4 and Proposition 2.

5.2.1 Defining Two-Threshold Security

We present a weaker variant of the ideal model that allows for a premature (and fair) abort, in case sufficiently many parties are corrupted. Next, we define two-threshold security of protocols.

Threshold Ideal-Model Execution. A t-threshold ideal computation of an n-party functionality f on input $\boldsymbol{x} = (x_1, \dots, x_n)$ for parties $(\mathsf{P}_1, \dots, \mathsf{P}_n)$, in the presence of an ideal-model adversary D controlling the parties indexed by $\mathcal{I} \subseteq [n]$, proceeds via the following steps.

Sending inputs to trusted party: An honest party P_i sends its input x_i to the trusted party. The adversary may send to the trusted party arbitrary inputs for the corrupted parties. If $|\mathcal{I}| > t$, then the adversary may send a special abort command to the trusted party. Let x_i' be the value actually sent as the input of party P_i.

Trusted party answers the parties: If the adversary sends the special abort command (specifically, $|\mathcal{I}| > t$), then the trusted party sends $\perp$ to all the parties. Otherwise, if x_i' is outside of the domain for P_i, for some index i, or if no input is sent for P_i, then the trusted party sets x_i' to be some predetermined default value. Next, the trusted party computes $f(x_1', \dots, x_n') = (y_1, \dots, y_n)$ and sends y_i to party P_i for every i.

Outputs: Honest parties always output the message received from the trusted party and the corrupted parties output nothing. The adversary D outputs an arbitrary function of the initial inputs $\{x_i\}_{i \in \mathcal{I}}$, the messages received by the corrupted parties from the trusted party $\{y_i\}_{i \in \mathcal{I}}$ and its auxiliary input.

Definition 10 (Threshold ideal-model computation). *Let* $f\colon (\{0,1\}^*)^n \mapsto$ $(\{0,1\}^*)^n$ *be an n-party functionality and let* $\mathcal{I} \subseteq [n]$. *The* joint execution of f under (D, I) in the t-threshold ideal model, *on input vector* $\boldsymbol{x} = (x_1, \ldots, x_n)$, *auxiliary input* z *to* D *and security parameter* κ, *denoted* $\mathrm{IDEAL}^t_{f,\mathcal{I},\mathsf{D}(z)}(\boldsymbol{x}, \kappa)$, *is defined as the output vector of* $\mathsf{P}_1, \ldots, \mathsf{P}_n$ *and* $\mathsf{D}(z)$ *resulting from the above described ideal process.*

Definition 11. *Let* $f\colon (\{0,1\}^*)^n \mapsto (\{0,1\}^*)^n$ *be an n-party functionality, and let* π *be a probabilistic polynomial-time protocol computing* f. *The* protocol π (t_1, t_2)-securely computes f (with information-theoretic security), *if for every real-model adversary* D, *there exists an adversary* S *for the* t_1-*threshold ideal model, whose running time is polynomial in the running time of* D, *such that for every* $\mathcal{I} \subseteq [n]$ *of size at most* t_2

$$\left\{ \mathrm{REAL}_{\pi,\mathcal{I},\mathsf{D}(z)}(\boldsymbol{x}, \kappa) \right\}_{(\boldsymbol{x},z) \in (\{0,1\}^*)^{n+1}, \kappa \in \mathbb{N}} \overset{\mathrm{s}}{\equiv} \left\{ \mathrm{IDEAL}^{t_1}_{f,\mathcal{I},\mathsf{S}(z)}(\boldsymbol{x}, \kappa) \right\}_{(\boldsymbol{x},z) \in (\{0,1\}^*)^{n+1}, \kappa \in \mathbb{N}}.$$

References

1. Beimel, A., Omri, E., Orlov, I.: Protocols for multiparty coin toss with dishonest majority. In: Rabin, T. (ed.) CRYPTO 2010. LNCS, vol. 6223, pp. 538–557. Springer, Heidelberg (2010)
2. Ben-Or, M., Goldwasser, S., Wigderson, A.: Completeness theorems for non-cryptographic fault-tolerant distributed computation (extended abstract). In: Proceedings of the 29th Annual Symposium on Foundations of Computer Science (FOCS), pp. 1–10 (1988)
3. Blum, M.: Coin flipping by telephone. In: Advances in Cryptology - CRYPTO 1981, pp. 11–15 (1981)
4. Broder, A.Z., Dolev, D.: Flipping coins in many pockets (Byzantine agreement on uniformly random values). In: Proceedings of the 25th Annual Symposium on Foundations of Computer Science (FOCS), pp. 157–170 (1984)
5. Chaum, D., Crépeau, C., Damgård, I.: Multiparty unconditionally secure protocols (extended abstract). In: Proceedings of the 10th Annual ACM Symposium on Theory of Computing (STOC), pp. 11–19 (1988)
6. Chor, B., Goldwasser, S., Micali, S., Awerbuch, B.: Verifiable secret sharing and achieving simultaneity in the presence of faults (extended abstract). In: Proceedings of the 26th Annual Symposium on Foundations of Computer Science (FOCS), pp. 383–395 (1985)
7. Cleve, R.: Limits on the security of coin flips when half the processors are faulty. In: Proceedings of the 18th Annual ACM Symposium on Theory of Computing (STOC), pp. 364–369 (1986)
8. Cohen, R., Haitner, I., Omri, E., Rotem, L.: Characterization of secure multiparty computation without broadcast. Cryptology ePrint Archive, Report 2015/846 (2015). http://eprint.iacr.org/
9. Cohen, R., Lindell, Y.: Fairness versus guaranteed output delivery in secure multiparty computation. In: Sarkar, P., Iwata, T. (eds.) ASIACRYPT 2014, Part II. LNCS, vol. 8874, pp. 466–485. Springer, Heidelberg (2014)
10. Dolev, D., Strong, R.: Authenticated algorithms for Byzantine agreement. SIAM J. Comput. **12**(4), 656–666 (1983)

616   R. Cohen et al.

11. Fischer, M.J., Lynch, N.A., Merritt, M.: Easy impossibility proofs for distributed consensus problems. In: Proceedings of the Fourth Annual ACM Symposium on Principles of Distributed Computing (PODC), pp. 59–70 (1985)
12. Fitzi, M., Gottesman, D., Hirt, M., Holenstein, T., Smith, A.: Detectable Byzantine agreement secure against faulty majorities. In: Proceedings of the 21st Annual ACM Symposium on Principles of Distributed Computing (PODC), pp. 118–126 (2002)
13. Fitzi, M., Hirt, M., Holenstein, T., Wullschleger, J.: Two-threshold broadcast and detectable multi-party computation. In: Biham, E. (ed.) EUROCRYPT 2003. LNCS, vol. 2656, pp. 51–67. Springer, Heidelberg (2003)
14. Goldreich, O.: Foundations of Cryptography Basic Applications, vol. 2. Cambridge University Press, New York (2004)
15. Goldreich, O., Micali, S., Wigderson, A.: How to play any mental game or a completeness theorem for protocols with honest majority. In: Proceedings of the 19th Annual ACM Symposium on Theory of Computing (STOC), pp. 218–229 (1987)
16. Goldwasser, S., Lindell, Y.: Secure computation without agreement. In: Malkhi, D. (ed.) DISC 2002. LNCS, vol. 2508, pp. 17–32. Springer, Heidelberg (2002)
17. Gordon, S.D., Katz, J.: Complete fairness in multi-party computation without an honest majority. In: Reingold, O. (ed.) TCC 2009. LNCS, vol. 5444, pp. 19–35. Springer, Heidelberg (2009)
18. Haitner, I., Tsfadia, E.: An almost-optimally fair three-party coin-flipping protocol. In: Proceedings of the 46th Annual ACM Symposium on Theory of Computing (STOC), pp. 817–836 (2014)
19. Lamport, L.: The weak Byzantine generals problem. J. ACM **30**(3), 668–676 (1983)
20. Lamport, L., Shostak, R.E., Pease, M.C.: The Byzantine generals problem. ACM Trans. Program. Lang. Syst. **4**(3), 382–401 (1982)
21. Moran, T., Naor, M., Segev, G.: An optimally fair coin toss. In: Reingold, O. (ed.) TCC 2009. LNCS, vol. 5444, pp. 1–18. Springer, Heidelberg (2009)
22. Pease, M.C., Shostak, R.E., Lamport, L.: Reaching agreement in the presence of faults. J. ACM **27**(2), 228–234 (1980)
23. Pfitzmann, B., Waidner, M.: Unconditional Byzantine agreement for any number of faulty processors. In: Proceedings of the 9th Annual Symposium on Theoretical Aspects of Computer Science (STACS), pp. 339–350 (1992)
24. Rabin, T., Ben-Or, M.: Verifiable secret sharing and multiparty protocols with honest majority (extended abstract). In: Proceedings of the 30th Annual Symposium on Foundations of Computer Science (FOCS), pp. 73–85 (1989)
25. Yao, A.C.: Protocols for secure computations. In: Proceedings of the 23th Annual Symposium on Foundations of Computer Science (FOCS), pp. 160–164 (1982)

Author Index

Printed in the United States
By Bookmasters